Primary Productivity and Biogeochemical Cycles in the Sea

ENVIRONMENTAL SCIENCE RESEARCH

Series Editor:

Herbert S. Rosenkranz
Department of Environmental and Occupational Health
Graduate School of Public Health
University of Pittsburgh
130 DeSoto Street
Pittsburgh, Pennsylvania

Founding Editor:
Alexander Hollaender

Recent Volumes in this Series

A Continuation Order Plan is available for this series. A continuation order will bring delivery of each new volume immediately upon publication. Volumes are billed only upon actual shipment. For further information please contact the publisher.

PREFACE

Biological processes in the oceans play a crucial role in regulating the fluxes of many important elements such as carbon, nitrogen, sulfur, oxygen, phosphorus, and silicon. As we come to the end of the 20th century, oceanographers have increasingly focussed on how these elements are cycled within the ocean, the interdependencies of these cycles, and the effect of the cycle on the composition of the earth's atmosphere and climate. Many techniques and tools have been developed or adapted over the past decade to help in this effort. These include satellite sensors of upper ocean phytoplankton distributions, flow cytometry, molecular biological probes, sophisticated moored and shipboard instrumentation, and vastly increased numerical modeling capabilities.

This volume is the result of the 37th Brookhaven Symposium in Biology, in which a wide spectrum of oceanographers, chemists, biologists, and modelers discussed the progress in understanding the role of primary producers in biogeochemical cycles. The symposium is dedicated to Dr. Richard W. Eppley, an intellectual giant in biological oceanography, who inspired a generation of scientists to delve into problems of understanding biogeochemical cycles in the sea.

We gratefully acknowledge support from the U.S. Department of Energy, the National Aeronautics and Space Administration, the National Science Foundation, the National Oceanic and Atmospheric Administration, the Electric Power Research Institute, and the Environmental Protection Agency. Special thanks to Claire Lamberti for her help in producing this volume.

Symposium Committee

Paul Falkowski, Chairman
Robert Bidigare
Sally Chisholm
Dale Kiefer
Sharon Smith

v

CONTENTS

ESTIMATION OF GLOBAL OCEAN PRODUCTION

THE ROLE OF MARINE ORGANISMS IN PRIMARY PRODUCTION

NEW PRODUCTION AND BIOGEOCHEMICAL CYCLES

LOSS PROCESSES AND MATERIAL RECYCLING

PHYTOPLANKTON IN THE GLOBAL CONTENT

TOWARD UNDERSTANDING THE ROLES OF PHYTOPLANKTON IN BIOGEOCHEMICAL CYCLES: PERSONAL NOTES

Richard W. Eppley

1969 Loring Street
San Diego, CA 92109

INTRODUCTION

This is a collection of remembrances of those who whetted my interest in biogeochemical cycles in my formative years of the 1960s and 1970s. I used to ask myself "How can anyone justify earning a living by studying phytoplankton?" Colleagues kept pointing out that phytoplankton production for fish food was not a suitable justification. The links between phytoplankton production and fish harvest, at least for local fisheries, were not obvious. In search of more compelling reasons I began to lean on the role of phytoplankton in C and N cycling as justification and to learn, especially from geochemists and micropaleontologists, of the great importance of phytoplankton for all of us for all time.

The people and events described are more or less in chronological order rather than by subject matter. This order will lead to some awkwardness for the reader as will the personal nature of this account. This is not intended to be a scientific report but rather portions of a personal history. The pathway described starts in the fall of 1963 with my arrival at the Scripps Institution of Oceanography (Scripps) and ends with the global estimate of new production that Bruce Peterson and I reported in 1979 (Eppley and Peterson, 1979).

EARLY DAYS IN THE FOOD CHAIN RESEARCH GROUP

Oceanography grew quickly in the early 1960s, resulting in the importation of many people from elsewhere into the field; for example, I had studied seaweed physiology with L. R. Blinks at Stanford's Hopkins Marine Station. The rapid growth of research funds stopped in 1968 when Congress passed the Mansfield Amendment, an act that restricted the funding of academic oceanography by the so-called "mission agencies" of the U. S. government. Meanwhile, in 1962, Luigi Provasoli invited me to one of the AIBS workshops on marine biology when the Soviet invitee was unable to attend. The workshop included the leading phytoplankton researchers of that time: E. Steemann Nielsen, T. Braarud, G. E. Hutchinson, G. A. Riley, G. E. Fogg, M. R. Droop, R. A. Vollenweider, J. H. Ryther, and John Strickland. As the junior researcher in the group, I listened a

Primary Productivity and Biogeochemical Cycles in the Sea
Edited by P.G. Falkowski and A.D. Woodhead, Plenum Press, New York, 1992

1

great deal and said very little in the discussion that was recorded and edited by C. H. Oppenheimer (1966).

John D. H. Strickland

John Strickland came to Scripps from Nanaimo, British Columbia, in summer 1963 to organize the Food Chain Research Group (FCRG). The FCRG was funded largely by the Atomic Energy Commission and was modeled after the Ketchum-Ryther-Menzel-Yentsch-Guillard group at the Woods Hole Oceanographic Institution (WHOI), Massachusetts. I was delighted that Strickland hired me as a specialist in phytoplankton physiology, with emphasis on one of his many interests, growth kinetics. My work started with first measuring nitrate with cadmium reduction columns (after Grasshof), and then assessing the extracellular release of photoassimilated carbon during photosynthesis. Strickland organized cruises to the Peru upwelling region, initiated the Scripps "deep tank" experiments, and stimulated a rebirth of plankton dynamics studies at Scripps. The deep tank enclosure experiments emphasized budgets for the added nutrients and the fate of photosynthetic carbon, as Strickland and Tim Parsons had done earlier in the giant plastic bag experiments at Nanaimo. The work was a marvelous introduction to a career in phytoplankton-nutrient relationships. Moreover, Strickland, by providing both the intellectual and executive leadership of the group, was an inspiring role model.

Besides nutrient budgets for the deep tank experiments, Strickland's interest in nitrogen also led to papers on nitrification and the subsurface nitrite maxima off Peru, with colleague A. F. Carlucci and M. Fiadero, a student at Scripps, and on hydroxylamine in seawater with Lucia Solorzano (Strickland, 1972).

Strickland's early use of the autoanalyzer for continuous nutrient measurements, both in depth-profiling and underway mapping modes, introduced me to mesoscale oceanography. Earlier, while at Nanaimo, he had considered aerial color photography as a means to assess the spatial distribution of phytoplankton blooms on scales of kilometers. Strickland would have been delighted with the satellite ocean color images that became available in the 1980s and with the trace metal 'clean' techniques that have allowed John Martin and others to look into the role of iron in phytoplankton ecology.

John H. Ryther

John Ryther led the group at WHOI that provided the model for the FCRG. In the late 1950s and early 1960s, he was the intellectual leader of phytoplankton physiology in the United States. Nearly all his papers of that period were seminal for me, especially his broad-brush syntheses on global ocean photosynthesis and the importance of photosynthesis-irradiance (P-I) curves. One of his calculations, on the time required for ocean photosynthesis to produce as much oxygen as was present in the earth's atmosphere, was the first in my experience to imply that phytoplankton physiologists might contribute to understanding global-scale biogeochemistry.

Robert W. Holmes

Bob Holmes and I shared an office and a laboratory in the early years of the FCRG. An American student of Trygve Braarud, University of Oslo (like Ted Smayda), Holmes was my only colleague in phytoplankton at Scripps with formal training in biological oceanography and a knowledge of its historical literature. Eystein Paasche visited the FCRG for a year and shared an office with us. Holmes, Paasche, and Strickland introduced me to the papers of Gran and Braarud, G. A. Riley, and Harvey,

Atkins and Cooper of the Plymouth school. I knew of Steemann Nielsen's work on ocean photosynthesis earlier, but only Strickland had a complete file of his reprints. (I did not learn much about the earlier Kiel School under Hensen, then Brandt, or of Nathanson's contributions to the nitrogen cycle until years later. Eric Mill's recent book (Mills, 1989) on the history of biological oceanography before 1960 gives a full account.)

Holmes also introduced me to the "austausch coefficient" in the continuity equation. We discussed vertical mixing and stratification, especially with respect to Talling's (lakes) and Steemann Nielsen's (oceans) P-I curves, which provided a biological time scale for the duration of stratification through the changes in Talling's I_k parameter (a measure of light saturation for photosynthesis) and its depth variability with changes with seasonal stratification. Strickland also did studies of photosynthesis and growth as functions of irradiance that were part of the discussions. As a result, I measured P-I relationships (crudely by today's standards) as well as nitrate on the first FCRG cruise to Peru, where we saw marked contrasts in the variation of I_k with depth in mixed vs. stratified waters, as Nielsen had reported. The recognition of the contrast in photosynthetic light curves between mixed and stratified waters was essential for my later understanding of the seasonal, regional, and episodic differences in the proportions of new and regenerated production.

Theodore J. Smayda

Ted Smayda's papers on the measurement of sinking rates of phytoplankton cells from cultures (summarized in Smayda, 1970) stimulated Holmes and me to try using Carl Lorenzen's (1966) *in vivo* chlorophyll fluorescence method to measure sinking rates. Holmes established an extensive culture collection of phytoplankton that provided our experimental material. The collection relied heavily on Robert Guillard (WHOI) and Luigi Provasoli (Haskins Laboratory, New York) for many species, but there were a few local isolates in the collection as well. Besides the phytoplankton in the culture collection, we also measured sinking rates of copepod fecal pellets, provided by Gus Paffenhofer who was visiting in the group. Learning seems to be easier from first-hand observation than from books. Thus, it was brought home to me what many others already knew: individual phytoplankton cells do not sink fast enough to be major carriers of material to depth. Our aggregates of diatoms sank a bit faster than individual cells, but the fecal pellets sank fastest. However, about that time, both Tim Parsons and John Steele independently published reports of coastal spring blooms of diatoms sinking en mass to the sea floor. It seemed unlikely to me then that this finding was important for the open sea - How times change (Peinert et al., 1989).

NITROGEN CYCLING, NEW AND REGENERATED PRODUCTION

Richard C. Dugdale

My interest in nitrogen began under Strickland, but it was greatly stimulated by Dick Dugdale's hypothesis that nutrient uptake kinetics might play a major role in regulating phytoplankton production (Dugdale, 1967). I first met Dugdale when he visited the FCRG in the late 1960s to see John Strickland about using autoanalyzer methods of nutrient analyses, including underway mapping. F. A. J. Armstrong and Strickland, with Solorzano, had already published on the use of such systems with interesting results. Dugdale's classic papers on the significance of nutrient uptake kinetics for phytoplankton ecology (Dugdale, 1967), and, with John Goering, on new and regenerated production and the ^{15}N methodology (Dugdale and Goering, 1967) became major inspirations for the rest of my career.

The first paper (Dugdale, 1967) was largely theoretical, so with the analytical methods available in the FCRG and our culture collection, it was possible to see if Michaelis-Menten enzyme kinetics actually held for nitrate and ammonium uptake of phytoplankton cultures (they did).

James J. McCarthy

Meanwhile, in the summer of 1968, Jim McCarthy, an FCRG student, took Dugdale's phytoplankton ecology course at Friday Harbor. McCarthy returned to Scripps and the FCRG full of enthusiasm and insisted upon setting up the technology for ^{15}N work in my laboratory. Theodore Enns, a laboratory neighbor, offered his homemade mass spectrometer for the work, and McCarthy was off and running. Before he left with his new Ph.D., he taught the technology to Ed Renger, my partner since 1970. Thanks to Renger, and later to Glen Harrison, the methodology became an integral part of the work in our laboratory. McCarthy's thesis was on urea, one of the recycled forms of N used by many phytoplankters. He found urea only in surface waters, providing us with a clue as to where regeneration was likely to be most intense. In 1969, Dugdale invited McCarthy and me to join him on one of his early cruises to Peru, where McCarthy measured urea and I assayed nitrate reductase activity. Dick Barber, John Goering, Ted Packard, Jane MacIsaac, Lou Hobson, and Dolores Blasco, representing several U.S. oceanographic institutions, were our shipmates. Later Paul Falkowski, then a graduate student at the University of British Columbia, read the nitrate reductase papers and came to Scripps to talk about assaying phytoplankton enzymes.

W. Glen Harrison

Glen Harrison initially came to the FCRG because of our shared interest in the use of nitrate by migrating dinoflagellates and to work in the CEPEX program (originally the Controlled Ecosystem Pollution Experiment) led by Dave Menzel of Skidaway Institute of Oceanography. The CEPEX bag (mesocosm enclosure) experiments provided opportunity to assess new and regenerated production in a confined, but rather large volume, and to compare sedimentation of particulate nitrogen with added inorganic N (Parsons et al., 1977), i.e., new production vs. export.

OCEAN PHOTOSYNTHESIS AND THE CARBON CYCLE

Peter M. Williams

Pete Williams came to the FCRG in 1963 to be the group chemist. His continuing interest in dissolved organic carbon, the carbon cycle, and in isotope geochemistry made him an important resource from the beginning (for example, when in 1963 Strickland assigned me the job of evaluating the release of DOC in phytoplankton cultures). One of the seminal events in my career was reading, at Williams' suggestion, W. S. Broecker's *Chemical Oceanography*, where in an early chapter, he provided global ocean estimates of vertical fluxes, including that of biogenic carbon from the surface to the ocean interior (Broecker, 1974). At about the same time, I read G. A. Riley's famous flux paper on oxygen, phosphate, and nitrate in the Atlantic (Riley, 1951) and began to think of the FCRG's work in the central North Pacific and the Southern California Bight in a larger context.

ENTER SEDIMENT TRAPPING

John H. Martin and George A. Knauer

Buoyed by the budding recognition that nitrate, its assimilation, nitrate reductase activity and photosynthetic new production were probably important to the sinking of organic particles out of the euphotic zone, and that this was a topic of wide interest, I began to read more widely (not widely enough, of course, but widely relative to the limited scope of the previous work). A close friend, Carl Lorenzen at the University of Washington, used sediment traps to study the fate of photosynthetic pigments and found that much was carried as chlorophyll degradation products in fecal material. S. Honjo at WHOI published a seminal paper on fecal pellets as agents of transport (Honjo, 1978). A paper that I read earlier, during the phytoplankton sinking rate experiments, reporting the rapid transport of particle active radionuclides to the ocean floor, apparently *via* fecal pellets (Osterburg et al., 1963) was brought to my attention. However, the first paper I read on sediment trap fluxes of particulate organic C and N near the base of the euphotic zone and in the open ocean was by Martin et al. (1979). I was so enthusiastic about it that I sent a letter to Martin in July 1978 that included the following:

> Some free time finally appeared and I got a chance to read your lovely flux paper. Two of the numbers generated here serve as an independent check of your flux measurements for N. The first is that the vertical eddy diffusion of nitrate provides 35-40% of the nitrogen assimilated by phytoplankton in the local coastal waters. This agrees with your N flux of 39% at 50m for coastal upwelling...viz., what comes up must go down.

> Your open ocean sampling in the gyre can also be checked against our data. Your 75m sample might be in the euphotic zone, and hence contaminated a bit by wandering phytoplankton, so 10-15% of N production might be a good estimate of sinking at the bottom of the euphotic zone...

> Your flux measurements are precisely the information needed here to draw up budgets of N & P for the euphotic zone in our work areas...

("Our data" refers to data in Eppley et al., 1979. "The gyre" refers to the central gyre of the North Pacific. Our work areas were coastal southern California and the North Pacific gyre). A couple of years later, George Knauer took Renger and me to sea with him so that we could directly compare the sediment trap C and N fluxes with new production measured at the same time.

Bruce J. Peterson

In the 1970s the data on nitrate-based new production were extremely limited in time and space; they still are limited. The ocean remains undersampled with respect to measurements of photosynthesis and the situation is worse for nitrate uptake measurements. However, there are more data for the former than the latter, and global assessments have been made. In the collaboration with Bruce Peterson, we sought to use the limited nitrate-based new production data, along with the global syntheses of ocean photosynthesis, to yield an ocean estimate of new production. It came about this way, and I quote from a letter that I wrote to Peterson on December 5, 1978:

Some months ago I reviewed a proposal for DOE that included an appendix "Perspectives on the importance of oceanic particle flux in the global carbon cycle" written by you. This piece has greatly clarified my thinking as to the significance of the ^{15}N phytoplankton work going on here and in Goering, McCarthy and Dugdale's laboratories. I would like to cite it as a manuscript in preparation. Have you any plans to publish it? If not may I cite it as personal communication?

You will be pleased, along with Wally Broecker, that the ^{15}N results support your estimates of the flux of sinking particulate matter and that my estimate of this flux is tentatively $2\text{-}3 \times 10^9$ tons C y^{-1}. I will send the manuscript when it is done, and that will hopefully be in January.

The manuscript became our joint paper (Eppley and Peterson, 1979). It included a graph showing a non-linear relationship between new production and total production (the sum of new and regenerated production). Using that relationship and the existing summaries of ocean photosynthetic production, we were able to estimate global new production. That such a relationship was non-linear has been supported by more recent work (cf. Berger et al., 1989). New production in rich coastal regions is often a high fraction of the total production, while that in oligotrophic oceanic gyres constitutes only about 5% of the total production (Chiswell et al., 1990). However, a major exception to the rule has been discovered in the equatorial Pacific, where new (and export) production are unexpectedly low compared to photosynthetic production (Murray et al., 1990; Wilkerson and Dugdale, personal communication).

As always, departures from the expected are what drive new observations and theory. A goal of this symposium is to better understand and estimate the export fluxes of C and N in relation to photosynthetic production. I hope also, perhaps in the not distant future, to learn how ocean climate and its temporal change will influence such relationships in the coming decades.

REFERENCES

Berger, W. H., Smetacek, V. S., and Wefer, G., 1989, Ocean productivity and paleoproductivity-an overview, in: "Productivity of the Ocean: Past and Present," W. H. Berger, V. S. Smetacek and G. Wefer, eds., Wiley, Chichester.

Broecker, W. S., 1974, "Chemical Oceanography," Harcourt, Brace, Johanovich, New York.

Chiswell, S., Firing, E., Karl, D., Lukas, R., and Winn, C., 1990, Hawaii Ocean Time-series Data Report 1, 1988-1989, Univ. Hawaii, Honolulu.

Dugdale, R. C., 1967, Nutrient limitation in the sea: dynamics, identification, and significance, Limnol. Oceanogr., 12:685.

Dugdale, R. C., and Goering, J. J., 1967, Uptake of new and regenerated forms of nitrogen in primary productivity, Limnol. Oceanogr., 23:196.

Eppley, R. W., and Peterson, B. J., 1979, Particulate organic matter flux and planktonic new production in the deep ocean, Nature, 282:677.

Eppley, R. W., Renger, E. H., and Harrison, W. G., 1979, Nitrate and phytoplankton production in southern California coastal waters, Limnol. Oceanogr., 24:483.

Honjo, S. J., 1978, Sedimentation of materials in the Sargasso Sea at a 5,637 m deep station, *J. Mar. Res.*, 36:469.

Lorenzen, C. J., 1966, A method for the continuous measurement of *in vivo* chlorophyll concentration, *Deep-Sea Res.*, 13:223-227.

Martin, J. H, Knauer, G. A., and Bruland, K., 1979, Fluxes of particulate carbon, nitrogen, and phosphorus in the upper water column of the northeast Pacific, *Deep-Sea Res.*, 26:97.

Mills, E. L., 1989, "Biological Oceanography, An Early History, 1870-1960," Cornell Univ. Press, Ithaca.

Murray, J. W., Downs, J. N., Strom, S., Wei, C.-L., and Jannasch, H. W, 1990, Nutrient assimilation, export production and ^{234}Th scavenging in the eastern equatorial Pacific, *Deep-Sea Res.*, 36:1471.

Oppenheimer, C. H., ed., 1966, "Marine Biology II," New York Acad. Sci., New York.

Osterburg, C. A., Carey, A. G., and Curl, H., 1963, Acceleration of sinking rates of radionuclides in the ocean, *Nature*, 200:1276.

Parsons, T. R., von Brockel, K., Koeller, P., Takahashi, M., Reeve, M. R., and Holm-Hansen, O., 1977, The distribution of organic carbon in a planktonic food web following nutrient enrichment, *J. exp. mar. Biol. Ecol.*, 26:235.

Peinert, R., von Bodungen, B., and Smetacek, V. S., 1989, Food web structure and loss rate, *in*: "Productivity of the Ocean: Past and Present," W. H. Berger, V. S. Smetacek, and G. Wefer, eds., Wiley, Chichester.

Riley, G. A., 1951, Oxygen, phosphate, and nitrate in the Atlantic Ocean, *Bull. Bingham Oceanogr. Coll.*, 13:1.

Smayda, T. J., 1970, The suspension and sinking of phytoplankton in the sea, *Mar. Biol. Ann. Rev.*, 8:353.

Strickland, J. D. H., 1972, Research on the marine planktonic food web at the Institute of Marine Resources: A review of the past seven years work, *Oceanogr. Mar. Biol. Ann. Rev.*, 10:349.

THE NATURE AND MEASUREMENT OF THE LIGHT
ENVIRONMENT IN THE OCEAN

John T.O. Kirk

CSIRO Division of Plant Industry
Canberra, Australia

INTRODUCTION

The nature - that is, the characteristics, and the properties - of the light environment in the ocean is determined by two things: first, by the nature of the light flux incident on the surface of the ocean from above, and second, by the optical properties of the oceanic water itself. The underwater light environment is what results from the operation of the latter on the former. In this paper we shall consider (a) the nature of the incident solar flux; (b) the inherent optical properties of the ocean, how they are measured and what components of the aquatic medium they are due to; and (c) how the characteristics of the underwater light field are measured, and what these measurements reveal about the light environment in the ocean.

THE SOLAR RADIATION INCIDENT ON THE OCEAN

In the context of primary production, what are the relevant features of the solar radiation stream incident on the surface of the ocean? They are the way its energy is distributed across the electromagnetic spectrum, and the ways in which this energy supply varies with time, on a daily, and a seasonal basis.

Under clear-sky conditions, most of the energy is in the direct solar beam. Fig. 1 shows the spectral distribution of direct solar radiation at a solar altitude of 42°. About half of the radiation is in the visible/photosynthetic waveband (400-700 nm), and because this volume is about photosynthetic primary production, we might think that this band is the only one of concern and that all the energy in the infrared can be disregarded. In fact, this is not the case. Although it is not used directly by phytoplankton for photosynthesis, the infrared component of the solar flux, nevertheless, exerts a very important indirect controlling influence upon it, through its effects on the hydrodynamic behavior of the water.

The spectral distribution of the directly utilizable fraction of the solar flux - the photosynthetically available radiation, or PAR - is presented in more detail in Fig. 2. These spectral distributions were measured by Tyler and Smith in the 1960s, at three different locations - Crater Lake, Oregon; the Gulf Stream, off the Bahamas; and

Primary Productivity and Biogeochemical Cycles in the Sea
Edited by P.G. Falkowski and A.D. Woodhead, Plenum Press, New York, 1992

San Diego, all under clear skies. One important difference between these spectral distributions and the previous one (Fig. 1), is that here it is the quantum irradiance which is plotted, whereas the other curve is of irradiance expressed in energy units. In the context of photosynthesis, the quantum irradiance is more appropriate since all absorbed quanta are equally effective in photosynthesis, whatever their wavelength. Plotting spectral irradiance in the quantum form has the effect of dragging the curve down at the short-wavelength end of the spectrum, since short-wavelength photons are more energetic, and so there are fewer for a given amount of energy.

If we ignore the minor wiggles on the curves, the significant feature is that the spectral distribution of the solar radiation incident upon the ocean surface, in the photosynthetic waveband, is rather flat from about 450 nm upwards, but falls away quite sharply with decreasing wavelength below 450 nm. A useful figure, which can be derived from these curves is that in bright summer sunlight, photosynthetically available photons are falling on the ocean surface at a rate of about 10^{21} m^{-2} s^{-1}.

The incident solar irradiance is a function both of the time of day and the time of year. On a cloudless day, the irradiance varies in an approximately sinusoidal manner, from dawn to dusk (Fig. 3). The mathematical simplicity of this behavior is very convenient when it comes to modelling primary production, or solar heating. Much of the time the behavior is not so simple, as on the days shown by the other two curves in the figure, one corresponding to intermittent cloud, the other to overcast conditions.

The integrated area under any of these curves gives the total solar radiant energy received per square metre during the day; this is known as the daily insolation. Daily insolation varies during the year in accordance with the change in solar altitude, and seasonal changes in cloud cover. Fig. 4 shows the variation observed at latitude 35° S, averaged over 3 years, the insolation in mid-summer being nearly four times the mid-winter value.

Having arrived at the surface of the ocean, the solar photons then have to pass through the air-water interface. While some are reflected upwards again at the surface, losses are only 2 to 3% for solar altitudes between 45° and 90°. Surface reflection increases at lower solar altitudes, but only becomes really serious from about 20° downwards, rising from about 13 to 100% as solar angle diminishes from 20 to zero degrees. Roughening of the water surface by wind substantially decreases the surface reflection losses that would otherwise occur at low solar altitude.

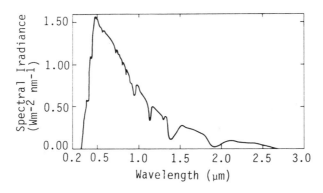

Fig. 1.   Direct solar spectral irradiance. Solar altitude 41.8°. Air mass 1.5. (Plotted from data of M.P. Thekaekara, reported by Mecherikunnel and Duncan, 1982).

THE INHERENT OPTICAL PROPERTIES OF THE OCEAN

Now the solar photons have got into the ocean, what kind of light field do they create within the water column? Just under the surface, the light field is much the same as that just above the surface, except that, because of refraction it is oriented somewhat more vertically downward. Further down through the water column, the light changes markedly in intensity, in spectral composition, and in angular distribution. This is because oceanic water has certain optical properties that operate upon the solar radiation stream to change it as it travels downwards.

The optical properties of most interest are the scattering coefficient and the spectral absorption coefficients (spectral, because the absorption coefficient varies markedly across the spectrum). Following the usage proposed by Preisendorfer 30 years ago (1961), I refer to these as inherent optical properties. They are so described because a sample of a given sea water has a certain value of absorption coefficient or scattering coefficient regardless of whether it is measured *in situ* in the sea, or back in the laboratory. These are inherent properties of the medium itself, and depend only on its composition, not on whether the day is sunny or cloudy, the sun is high or low, nor any other factor of the environment.

These inherent optical properties are so called to distinguish them from what Preisendorfer (1961) referred to as the apparent optical properties of the ocean, such as the vertical attenuation coefficient for downward irradiance, or the irradiance reflectance, which are, in reality, local properties of the underwater light field existing at a certain point, at a certain time, but which are, nevertheless, often used, and talked about, as though they were indeed properties of the oceanic water itself. A more extensive discussion of inherent and apparent optical properties may be found in Kirk (1983).

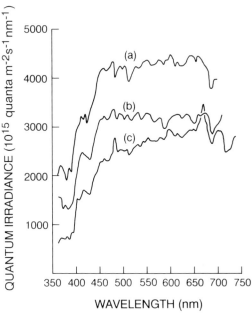

Fig. 2.   Spectral distribution of total solar quantum irradiance under clear sky conditions measured at (a) Crater Lake, Oregon, (b) Gulf Stream, Bahamas, and (c) San Diego, California. Plotted from data of Tyler and Smith (1970).

Since it is through the inherent optical properties that the aquatic medium acts upon the incoming solar radiation, then to understand how the underwater light environment gets to be the way it is, these optical properties must be quantified. Other important reasons for knowing the inherent optical properties are to calculate the rate of energy absorption by a given component - such as the phytoplankton, if we want quantum yields - and also, to carry out computer modelling of the underwater light field, to determine those aspects that are hard to measure directly.

The actual values that the absorption and scattering coefficients have in any particular region of the ocean depends on the composition of the water in that region: on how much dissolved yellow color, how much phytoplankton and of what specific pigment composition, how much particulate organic detritus, and how many inorganic particles, are present. We may expect all these to vary markedly from one part of the sea to another. Since we cannot assume constancy of composition, methods are needed for measuring the inherent optical properties, that can be used in any part of the ocean. I shall come in a moment to current developments in the measurement of some of these properties, but before doing so I note that there is one constant optical factor, throughout the world's oceans, and that is water itself. So what does water - pure water - contribute to the inherent optical properties of the ocean? At this juncture I consider only its contribution to light absorption: I shall return to scattering later.

Light Absorption by Water

Figure 5 depicts the absorption spectrum of pure water, and shows that water absorbs quite strongly at the red end of the visible spectrum. This is, in fact, just the tail of a very much stronger absorption at longer wavelengths - water has several intense absorption bands in the infrared. These shoulders in the red region are believed to be higher harmonics of the hydrogen-oxygen bond vibrational absorption of the water molecule, the fundamental of which is well out in the infrared at wavelength 3 microns. Oceanic water owes its blue color to this absorption in the red. Down in the blue region, water has very little absorption, but there is an indication of a slow rise into the ultraviolet.

Dissolved Color, Phytoplankton, and Detritus

Absorption by water itself is, thus, not only a constant, but also a significant contributor to light absorption by the ocean. There are three other components of

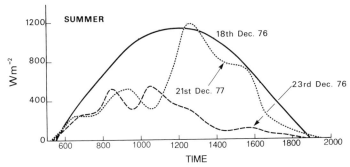

Fig. 3. Diurnal variation of total solar irradiance on long summer days, under different atmospheric conditions (———) clear sky; ..... intermittent cloud; ----- overcast). Measured by F.X. Dunin at 35°49'S, NSW, Australia.

significance. First, there is dissolved yellow color ('gelbstoff', 'gilvin') consisting of soluble humic material derived, in part, from soil humic substances carried by rivers into the sea, and in part, by decomposition of phytoplankton. The absorption spectrum of marine yellow color is much the same shape as that of the freshwater material, with absorption rising exponentially towards the short-wavelength end of the spectrum, but the concentration is generally much lower: the coastal seawater shown in Fig. 6 (bottom curve) has much less dissolved color than any of the freshwater bodies, and oceanic water has even less - perhaps a fifth or a tenth of this level, with an absorption coefficient at 440 nm of only in the region of 0.05 m^{-1}. But considering that pure water has an absorption coefficient of only about 0.015 m^{-1} at 440 nm, then it is apparent that dissolved yellow color contributes importantly to absorption of blue light in the ocean.

The second absorbing component is the phytoplankton. Figure 7 shows the absorption spectrum of mixed oceanic phytoplankton, taken from data of Morel and Prieur (1977). What is plotted is the specific absorption coefficient, which can be regarded as equivalent to the actual absorption coefficient for a concentration of 1 mg chlorophyll per cubic meter. That value is high for most oceanic waters. For an unproductive water with, for example, 0.2 mg chlorophyll *a* per cubic meter, the absorption coefficient at 440 nm due to phytoplankton pigments is only about 0.005 m^{-1}, which is only about one-third the absorption due to water itself. Thus, in productive waters with 1 mg or more of chlorophyll per cubic meter, the phytoplankton is a major contributor to light absorption, but in the most oligotrophic ocean waters, it is a minor, but not trivial, component.

The third light-absorbing component of the ocean includes all the non-living particulate matter. Most of the color in this fraction is due to yellow-brown organic detritus, derived mainly from breakdown of phytoplankton. Its absorption spectrum is quite similar to that of the dissolved yellow color, but with some shoulders due to the presence of breakdown products of photosynthetic pigments. Figure 8 shows absorption spectra, measured by Iturriaga and Siegel (1989), of the detrital and phytoplankton

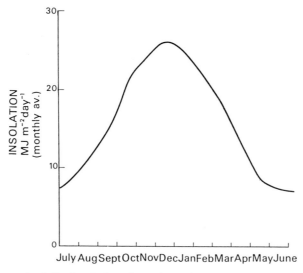

Fig. 4.   Change in daily insolation throughout the year. Average for period 1978-80, measured at 35°S, ACT, Australia.

fractions from a mesotrophic station in the Sargasso Sea. The actual amount of absorption due to detritus seems to be about the same as absorption due to the phytoplankton from which it is derived. Figure 9 shows, for an idealized, rather productive, ocean water, the individual absorption spectra of dissolved color, phytoplankton, detritus, at plausible levels, and of water itself, together with the total absorption spectrum of the water due to all four components together.

Measurement of the Absorption Spectrum of Seawater

Supposing that in order to understand or predict the light environment in some part of the ocean, we wish to measure the absorption coefficients of the water across the photosynthetic waveband, to determine just such a curve as the uppermost curve in Fig. 9. How can this be done? Accurate spectra cannot be obtained by putting some of our ocean water in a spectrophotometer cell, and running a spectrum. Unfortunately, even with all the components combined, ocean waters absorb very feebly in the blue and green parts of the spectrum, and in the sort of pathlengths - typically not more than 10 cm - that can be accommodated in laboratory spectrophotometers, very little absorption occurs, and accurate spectra are hard to obtain. However, over the pathlengths of many meters that exist in the ocean, the absorption is very significant.

One good approach would be to have specially designed absorption meters with long pathlengths. But even if the pathlength were long enough for measurable absorption to occur, there remains a further technical problem that must be solved, namely, how to distinguish the attenuation of the signal which is genuinely due to absorption, from that which is due to photons in the measuring beam simply being scattered away from the detector? Or, to put it another way, how can we be sure that we are measuring the absorption coefficient, and not the absorption coefficient plus some unknown part of the scattering coefficient?

This is a long-standing problem in marine optics, and various solutions have been tried. I will briefly describe two very recent ones. Zaneveld and his colleagues in Oregon have resurrected the idea of shining the measuring beam down an internally reflecting tube

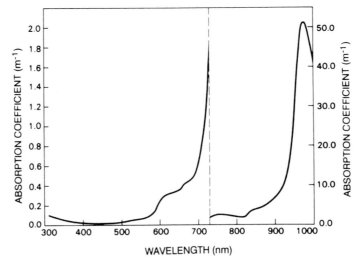

Fig. 5.   Absorption spectrum of pure water. Data of Smith and Baker (1981), and Palmer and Williams (1974).

(Zaneveld et al., 1990). The principle is that photons, which are scattered to one side, will be reflected back again, and so can still be detected at the other end. I am currently carrying out Monte Carlo modelling of the behavior of photons in such a reflective tube absorption meter, to assess possible errors, and their probable causes.

Fry and coworkers developed an instrument based on a different principle, an integrating cavity. The water sample is enclosed within a cavity made of a translucent, and diffusely reflecting, material (Fry and Kattawar, 1988; Pope et al., 1990). Photons are introduced into the cavity uniformly from all round, and a completely diffuse light field is set up within it, the intensity of which can be measured. The anticipated advantages are, first, that since the light field is already highly diffuse, additional diffuseness caused by scattering will have little effect; second, because the photons undergo many multiple reflections from one part of the inner wall to another, the effective pathlength within the instrument is very long, thus solving the pathlength problem. The prototype instrument gives promising results.

For photosynthetic primary production we may want to measure, not only the total absorption coefficients, but the absorption which is specifically due to the phytoplankton. Such information is needed, for example, to accurately calculate quantum yield. Two major technical problems must be solved in measuring phytoplankton absorption spectra. The first, is because the material to be measured is particulate, and therefore scatters and absorbs light. The second is associated with the very low concentration of phytoplankton.

Good absorption spectra of particles can be obtained using a spectrophotometer with an integrating sphere or a diffusing plate, with the help of which it is possible to measure true absorption, and not confound it with attenuation due to scattering; such methods have been applied to marine and freshwater phytoplankton (Shibata, 1959; Kirk,

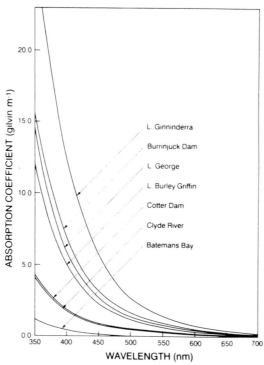

Fig. 6.   Absorption spectra of dissolved yellow color in various Australian natural waters (Kirk, 1976). The ordinate scale corresponds to the true *in situ* absorption coefficient due to soluble humic color.

1980; Bricaud et al., 1983; Davies-Colley et al., 1986; Sathyendranath et al., 1987; Haardt and Maske, 1987). Because of the second technical problem, however, the usually low concentration of phytoplankton, it will generally be necessary to concentrate the phytoplankton first, and if the true *in situ* absorption coefficient is required, this must be done quantitatively. It is possible to filter a substantial volume of seawater, and then resuspend the harvested phytoplankton, but the percent recovery is likely to be rather poor with present methods and materials. Transverse membrane filtration, or perhaps continuous flow centrifugation, might be worth exploring.

A very popular alternative approach - pioneered by Yentsch, and used extensively by Kiefer and coworkers - which on the face of it solves both problems at once, is to pass a large volume of seawater through a filter, and then measure the spectrum of the particulate material on the filter itself, without resuspending it. The problem with this method is that, as a consequence of multiple internal reflection within the layer of algae, there is a very marked amplification of absorption (up to 6-fold) and to calculate the true absorption coefficients of phytoplankton freely suspended in the ocean, then this amplification factor must be determined reasonably accurately; this determination is not easy, especially because it is far from constant, varying from about 2.5 to 6. So this is a potentially powerful technique but requires considerable care and expertise in determining the pathlength amplification factor to give accurate results.

A different experimental approach was developed by Iturriaga and Siegel (1989), who used a monochromator in conjunction with a microscope to measure the absorption spectra of individual phytoplankton cells. If sufficient cells are measured, and their numbers in the water are known, then the true *in situ* absorption coefficients due to the phytoplankton can be calculated.

In measuring the absorption spectra of oceanic phytoplankton, there is always the problem of distinguishing the true algal absorption from that due to particulate organic detritus. Iturriaga and Siegel solved the problem by carrying out separate measurements on detrital particles and phytoplankton cells. Morrow et al. (1989) and Roesler et al. (1989), using the filter method, developed statistical techniques, based on the shape of the spectra, for separating out phytoplankton absorption from that by the detritus and other components. Quite a different route to the determination of *in situ* phytoplankton absorption was taken by Bidigare et al. (1987) and Smith et al. (1989). Making the plausible assumption that essentially all the undegraded photosynthetic pigments in the

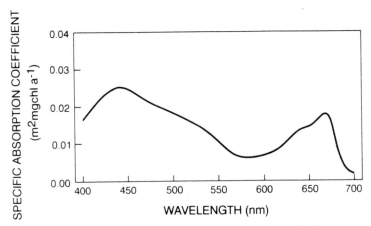

Fig. 7.   Specific *in situ* absorption coefficient for oceanic phytoplankton, corresponding to 1 mg chlorophyll *a* m^{-3} (after Morel and Prieur, 1977).

water column originate in living cells, they carried out a complete pigment analysis of the total particulate fraction, using HPLC; then, using literature data on the spectral properties of pigment-protein complexes, they calculated the absorption coefficients due to phytoplankton.

Light Scattering in the Ocean

Water itself is a very significant contributor to light absorption in the ocean. How much does it contribute to scattering? The first point to make is that water contributes very little to the total scattering coefficient of the aquatic medium which, even in the clearest oceanic waters, is almost entirely accounted for by both living particles (mainly phytoplankton), and non-living particles in the water. However, this is not the whole story.

Scattering by particles is very much concentrated in the forward direction, only about 2% occurring at angles greater than 90°. However, pure water scatters equal amounts of light backwards and forwards (Morel, 1974), and although water itself makes a small contribution in the ocean to the total scattering coefficient - which is the sum of scattering in all directions - it can account for a third or more of the backscattering. The particular significance of this is that backscattering is responsible for most of the upwelling light flux that exists in the ocean, some of which emerges up through the surface to be seen by a human observer looking down, or a remote sensing radiometer on a satellite.

Although scattering by oceanic particles does not vary much with wavelength, scattering by pure water does. Like Rayleigh scattering in the atmosphere (although liquid water does not carry out Rayleigh scattering) scattering by pure water increases with the reciprocal of the fourth power of the wavelength, and so is much more intense at the blue, than at the red end of the spectrum. Thus, the predominantly blue color of the upwelling light flux seen is a consequence both of absorption by water at the red end of the spectrum, and the more intense backscattering at the blue end of the spectrum.

To recapitulate, in the ocean nearly all the total scattering coefficient is accounted for by suspended particles, but water itself makes an important contribution to backscattering, and consequently, to the upwelling part of the light field.

How can the scattering coefficient be measured for a particular part of the ocean? Ideally, some small volume of the water could be illuminated with a parallel light beam,

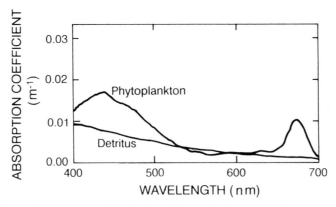

Fig. 8.   Absorption spectra of detrital and phytoplankton particles from a mesotrophic station in the Sargasso Sea (after Iturriaga and Siegel, 1989).

the light scattered from that volume measured at all angles, and all the scattering added together to arrive at the total scattering per increment of distance in the water. In reality, very few such complete sets of measurements have been carried out. More commonly, the amount of light scattered at just one angle to the incident beam is measured, and it is assumed that this is proportional to the total scattering over all angles. Measurement at an angle of 10° to the incident beam is thought to be particularly suitable because at this angle there is relatively little error due to variation in the shape of the angular distribution of scattering in different waters (Oishi, 1990). Thus, the direct measurement of the total scattering coefficient is something that can be carried out reasonably well.

THE UNDERWATER LIGHT FIELD

Much more could be said about the nature and measurement of the inherent optical properties of the ocean; I have not, for example, touched upon the volume scattering function, which describes the angular distribution of scattering, or the measurement of the beam attenuation coefficient, c, which is equal to the sum of the absorption and scattering coefficients (a and b). I want now to move on to the actual light field that is created within the ocean, by the operation of these inherent optical properties upon the incoming stream of solar photons.

Measurement of the Underwater Light Field

First, how can this underwater light environment be measured? If we want to know simply the total radiant energy available for photosynthesis, then an irradiance meter can be used whose response has, by judicious selection of photodetector and filters, been confined to the photosynthetic band, 400-700 nm, and made proportional to the flux of quanta rather than of energy, at any wavelength in this waveband.

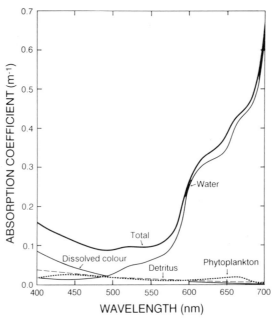

Fig. 9.   Total absorption spectrum of an idealized, productive (1 mg chlorophyll a m^{-3}) ocean water, together with spectra of the individual absorbing components.

PAR meters of this type can be built with a flat diffusing collector to measure irradiance, which is the radiant flux per unit area of a surface: and by having them facing up, or down, they can be used to measure downward (E_d) or upward (E_u) irradiance, respectively. Instead of a flat collector, they can be made with a spherical collector, so that the total light coming from all directions is measured. This is the scalar irradiance. The next step beyond a PAR meter is a spectral irradiance meter, or spectroradiometer, which will scan through the spectrum, giving a plot of irradiance as a function of wavelength. Like PAR meters, spectroradiometers can be made with flat collectors so that they measure normal irradiance, or with spherical collectors so that they measure scalar irradiance.

This raises the interesting and very important question for someone about to purchase light-measuring equipment: which is it best to measure - normal irradiance or scalar irradiance? It is sometimes argued that, since to the floating phytoplankton cell all photons are equally useful no matter from what direction they come, then scalar irradiance - which is a measure of the radiant intensity from all directions, is more appropriate than normal irradiance. This seems reasonable, especially considering that the exact equation for the rate of absorption of radiant energy by any one component of the medium, such as the phytoplankton, uses the scalar irradiance (Eo), not the normal irradiance:

$$\frac{d\Phi_i(z)}{dv} = E_o(z)a_i \tag{1}$$

The rate of absorption of energy per unit volume by any individual component of the medium is the product of the scalar irradiance and the absorption coefficient for that component. Therefore, the greatest possible accuracy in calculating, for example, the rate of absorption of photosynthetic quanta by the phytoplankton population, requires the spectral scalar irradiance, together with a set of values for the phytoplankton absorption coefficients; this information is not easy to obtain. In most situations, the detailed extra information will not be available to make use of the scalar irradiance data. It might be as

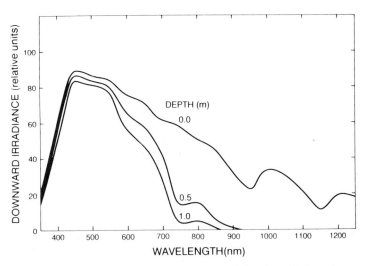

Fig. 10.   Changes in the spectral distribution of the visible/near infrared waveband with depth in the oceanic surface layer. Calculated using Monte Carlo program WATER 1 (Kirk, 1988) for water with b = 0.1 m^{-1}, g_{440} = 0.05 m^{-1}, and a solar altitude of 45°.

satisfactory, or better, to settle for the normal irradiance data, obtained with a flat collector.

The reason I suggest it might be better is that if, with the normal irradiance meter, measurements are made of the upward and the downward irradiance, then the data obtained on the underwater light field can be used to arrive at good estimates of the absorption and scattering coefficients - i.e., the inherent optical properties of the water - without having to measure them directly. Thus, if a choice must be made, I suggest the normal rather then the scalar irradiance meter is selected (but there is much to be said for having both).

So far, the properties of the oceanic light environment whose measurement I have considered, all concern various measures of light intensity - total photosynthetic irradiance, spectral distribution of irradiance, upwelling, downwelling, and scalar. However, as I indicated earlier, the angular distribution of the light changes with depth: this is not of mere academic interest because the angular distribution at any depth markedly influences the rate of attenuation of the light with depth, as well as being intimately related to the amount of light which flows upward, to emerge through the surface and thus be available for remote sensing of the ocean.

How do we measure the angular distribution? The most complete body of information would be what is called the radiance distribution. Radiance is the radiant flux per unit solid angle per unit area in a specified direction, and so the radiance distribution at any point in the water is the set of values of radiance measured at suitable angular intervals, over the full 180° of vertical angle and 360° of azimuth angle. This measurement represents a lot of data, and complete radiance distributions are mainly of interest to specialists in underwater light. Radiance distributions are most likely to be measured with special cameras with fisheye lenses, a technique pioneered by Smith et al. (1970), and further developed by Voss (1989).

There are two aspects of the angular distribution that are less than a complete radiance distribution, which can more readily be determined, but which are interesting and useful in their own right. The first is the irradiance reflectance (R), which is the ratio at any depth, of the upward irradiance to the downward irradiance: this is measured with the normal irradiance meter, pointing down, then up:

$$R(z) = \frac{E_u(z)}{E_d(z)} \tag{2}$$

The irradiance reflectance, combined with certain other irradiance data, can provide an independent way of estimating the absorption and scattering coefficients.

The second parameter of the angular distribution which can fairly easily be determined is the average cosine ($\bar{\mu}$). This is the average value at a given point in the water, of the cosine of the zenith angle of all the photons in a volume element at that point. This measure will provide an average value for the whole angular distribution. The reason why this can fairly easily be determined is that it is equal to the net downward irradiance divided by the scalar irradiance at that depth in the water, net downward irradiance simply being (E_d-E_u), the downward minus the upward irradiance:

$$\bar{\mu}(z) = \frac{E_d(z)-E_u(z)}{E_o(z)} \tag{3}$$

20

So, if a scalar irradiance meter is used as well as a normal irradiance meter, the average cosine can be obtained. But that is not all. If, as well as the average cosine, the vertical attenuation coefficient for net downward irradiance is determined:

$$K_E(z) = -\frac{d \ln(E_d-E_u)}{dz} \tag{4}$$

then the absorption coefficient can be calculated, making use of the fact:

$$a = K_E(z)\bar{\mu}(z) \tag{5}$$

that if this attenuation coefficient is multiplied by the average cosine the absorption coefficient, a, is obtained, a relationship first demonstrated in the 1930s by the Russian physicist, Gershun. While I am discussing net downward irradiance, I draw to your attention something very useful that can be calculated from it, and its vertical attenuation coefficient, namely, the rate of water heating as a function of depth:

$$\frac{d\Phi(z)}{dv} = [E_d(z)-E_u(z)]K_E(z) \tag{6}$$

The total rate of energy absorption per unit volume at depth z, is simply the product of these two quantities.

Before I leave this topic of how to characterize the underwater light field, I mention that as well as doing it by direct measurement, it can be done by calculation. The data required for the calculation are the inherent optical properties - the absorption coefficients over the spectral range of interest, the scattering coefficient, and the volume scattering function. Computer modelling of the underwater light field using, for example, Monte Carlo methods, can be carried out quite accurately; if realistic values are supplied for the inherent optical properties, and appropriate data for the solar flux incident on the surface, a complete picture of the underwater light environment can be generated. In a given real-

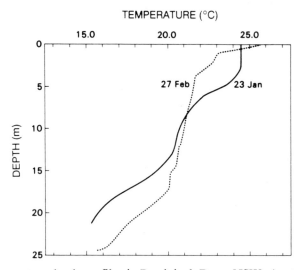

Fig. 11.   Temperature-depth profiles in Burrinjuck Dam, NSW, Australia, on two dates in the summer of 1974.

world situation, direct *in situ* measurement is preferable. Where the modelling option is particularly valuable is in answering what if questions. For example, what will happen if the concentration or pigment makeup of the phytoplankton population changes? What will happen if there is an increase in dissolved yellow color, or scattering particles associated with river outflow? In addition, modelling can provide information on parameters that are hard to measure, such as the complete radiance distribution.

Characteristics of the Underwater Light Field

Having outlined the significant properties of the underwater light environment and how they can be measured or calculated, I now look at the actual light field that exists in the ocean, and in particular, how it changes with depth. The light field immediately below the surface is not very different from that in the air just above the surface. As noted earlier, it is directed somewhat more vertically downwards because of refraction at the surface, and perhaps a few percent of the photons are lost by reflection at the air-water interface, but nearly all the energy is still there, with essentially the same spectral distribution.

There is a rather important feature of this spectral distribution just under the surface that must not be ignored, namely, that about half of it is in the infrared. With this in mind, let us look at what happens to the solar radiation just within the top meter or so of the ocean. Figure 10 shows calculated curves for the spectral distribution of downward irradiance in a typical oceanic water, just below the surface, and at 0.5 and 1.0 meters depth. I present the data only as far out into the infrared as 1.25 microns; this is sufficient to make the essential point, which is that virtually all the infrared radiation is removed by absorption, by the water molecules themselves, in about the first half-meter. This means that at depths beyond a meter or so, attention can legitimately be confined to that waveband extending from the near UV to what is sometimes called the far-red, just beyond 700 nm.

In terms of ocean functioning, it means that about half of the solar heating, with all its implications for the hydrodynamic behavior of the water mass, is initially delivered to, or captured by, the 1-meter layer at the surface. Under very still conditions, this heat can remain trapped in the surface layer for some hours. One of the temperature profiles

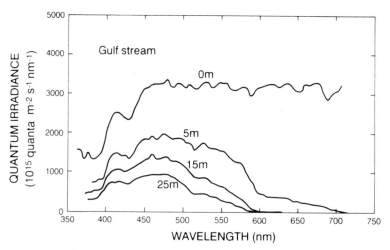

Fig. 12.   Spectral distribution of downward quantum irradiance in the very pure oceanic water of the Gulf Stream. Plotted from data of Tyler and Smith (1970).

(27 February) in Fig. 11 shows such a situation for an inland water body, but in most circumstances, wind and wave-induced turbulence mix the heat in with the water beneath.

Of the radiant energy left, something like one third is in the orange-red band, 600-700 nm, and this too is absorbed quite strongly by water itself; not as strongly as the infrared, but sufficient to ensure its complete removal by about 10 meters. Fig. 12 shows spectral distributions at a series of depths in the very pure waters of the Gulf Stream off the Bahamas, measured by Tyler and Smith (1970). It is apparent that by 15 meters, there is no radiation at wavelengths greater than 600 nm left.

Thus, about two-thirds of the total solar radiant energy penetrating the ocean surface is absorbed within the upper 10 meters, and it is mainly this radiant energy absorption, combined with wind-induced turbulence, which gives rise to the upper, warm, mixed layer, the existence of which has enormous implications for the functioning of the whole oceanic ecosystem, especially for phytoplankton primary production.

Most of the light below about 15 meters in very pure oceanic waters, such as these, is confined to the blue-green waveband, 400-550 nm, with a broad peak occurring in the blue region at 440-490 nm. Since water absorbs very weakly in this blue spectral region, the local heating rate at any depth below about 15 meters is low. However, this radiant energy is nevertheless ultimately all absorbed, and in the context of solar heating of the ocean, this blue-green waveband is important since it is photons at these wavelengths which carry thermal energy down to the lower levels of the photic zone.

In coastal waters, levels of dissolved yellow color are higher than in most of the ocean. Consequently, there is strong absorption in the blue region. For example, although the water is virtually oceanic in Jervis Bay, southeastern Australia, there is a noticeable depletion in the blue region (Fig. 13), compared to the Gulf Stream spectral distributions. In Batemans Bay (SE Australia), which has a substantial river inflow, there is more yellow color, absorption in the blue is more intense, and the spectral distribution

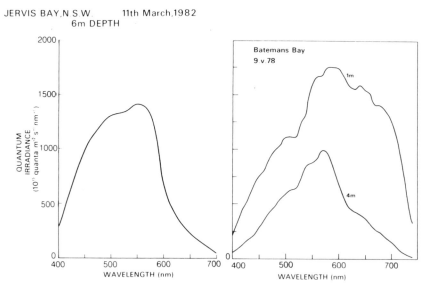

Fig. 13. Spectral distribution of downward quantum irradiance in coastal waters (southeast Australia). Contribution of yellow color from river inflow is significant in Batemans Bay, but slight in Jervis Bay.

23

at a depth of several meters, although it extends into the blue, becomes quite sharply peaked in the yellow region at about 570 nm, a situation which is common in coastal waters influenced by river discharge.

In the context of primary production, the significance of these changes in spectral distribution with depth is that phytoplankton, and benthic algae in coastal waters, must have pigment systems which can efficiently harvest light from radiation fields that are spectrally very biassed in different ways, according to the light absorption characteristics of the particular water mass in which they find themselves. This is presumably why dinoflagellates, diatoms, and other phytoplankters invest so much in specialized carotenoid-protein complexes which absorb in the blue-green region, and the same is likely to be true of the biliproteins of the cyanophytes.

The phytoplankton also use the upwelling light stream for photosynthesis. In the ocean, the upwelling light at any given depth is even more concentrated into the blue or blue-green wavebands than is the downwelling light at the same depth: within just a few meters from the surface, orange-red light is absent from the upwelling stream even when it is still a significant component of the downwelling stream (Fig. 14).

Let us now examine how the total energy available for photosynthesis varies with depth. The spectral distributions for the Gulf Stream (Fig. 12), show that there is a particularly intense removal of energy between 0 and 5 m depth. This is because of the strong absorption by water in the 600-700 nm, orange to red, waveband. At greater depths, certainly beyond about 15 m, only the weakly absorbed wavelengths penetrate, and so the rate of removal of energy per unit depth decreases. This shows up clearly in plots of downward irradiance of total PAR with depth. For example, the curve for the Tasman Sea, in Fig. 15, is approximately biphasic, showing a higher rate of attenuation of PAR in the upper 8 m, but settling down to a lower, and virtually constant, rate of attenuation,

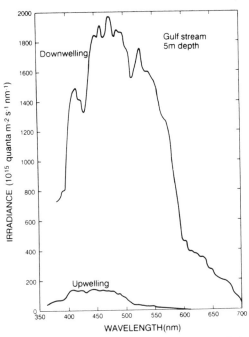

Fig. 14.   Spectral distributions of upwelling and downwelling light streams at 5 m depth in the Gulf Stream. Plotted from data of Tyler and Smith (1970).

with depth further down. From about 10 m downwards the spectral distribution does not seem to change much with depth, all the strongly absorbed wavelengths having already been removed.

The total amount of upwelling light available for photosynthesis is not large. In the oceanic waters studied by Tyler and Smith (1970), the upward irradiance for total PAR at 5 meters depth was typically 2 to 5% of the downward irradiance. Down in the blue waveband, however, the upwelling light can be about 10% of the downwelling light, as shown in the data from the Gulf Stream referred to above (Fig. 14).

Mechanism of Attenuation of the Light Stream

The attenuation of the downwelling light stream with depth is primarily due to absorption of the solar photons, by water and by pigments, as discussed. But it is also influenced by scattering, in some situations quite strongly. Scattering intensifies attenuation in two ways. First, by making the light field more diffuse, it increases the average pathlength that the photons must traverse per unit depth, and, in this way, increases the probability of their being absorbed. Second, it removes some photons from the downwelling stream by scattering them upwards. In this way the upwelling light stream within the ocean is created, and upward, as well as downward irradiance can be measured.

Some of this upwelling light stream passes up through the water surface, taking with it, in the form of its characteristic spectral distribution, information about the composition of the water mass. This information is what is used in remote sensing.

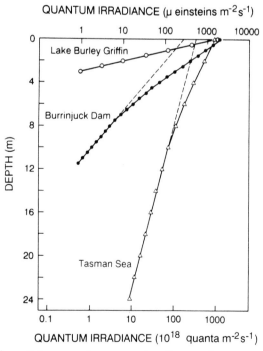

Fig. 15. Attenuation of downward quantum irradiance of PAR with depth in a coastal seawater (Tasman Sea, off Batemans Bay, NSW, Australia) and two inland waters (Lake Burley Griffin, ACT; Burrinjuck Dam, NSW).

By how much does scattering intensify vertical attenuation beyond the attenuation that would take place if the water only absorbed, and did not scatter, the light? The answer is given by a relatively simple equation, derived empirically by computer modelling (Kirk, 1981a; 1984):

$$K_d = \frac{1}{\mu_o}[a^2 + G(\mu_o)ab]^{1/2} \qquad (7)$$

K_d is the vertical attenuation coefficient for downward irradiance, μ_o is the cosine of the refracted solar photons just beneath the surface, a is the absorption coefficient and b is the scattering coefficient. $G(\mu_o)$ is a coefficient whose value is determined by the shape of the volume scattering function. You can see that if b is set equal to zero, i.e. there is no scattering, then Eq 7 reduces to $K_d = a/\mu_o$. With this equation the vertical attenuation coefficient can be calculated for downward irradiance for any combination of absorption and scattering.

The effect of scattering is certainly significant. It is a function of the actual ratio of the scattering to the absorption coefficient. A b/a value, that is scattering to absorption ratio, of 5, for example, which would not be unusual, would increase the vertical attenuation coefficient by about 50% over what it would be if there were only absorption and no scattering.

From the Field Back to the Inherent Optical Properties

The underwater light field at any depth, especially its angular distribution, is determined by the inherent optical properties of the medium, and by the properties of the solar radiation flux incident on the surface. With increasing depth in the sea, the more the properties of the field become a function just of the inherent optical properties of the

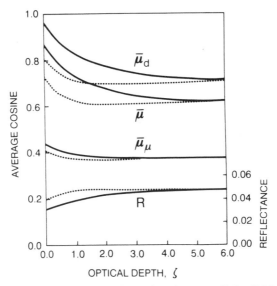

Fig. 16.   Variation of angular distribution of underwater light field with optical depth ($\zeta = K_d z$) in water with b/a = 5. The angular parameters are the average cosines for downwelling ($\bar{\mu}_d$), upwelling ($\bar{\mu}_u$) and total ($\bar{\mu}$) flux, and irradiance reflectance (R). Data obtained by Monte Carlo calculation (Kirk, 1981a) for light incident vertically (———) and at 45° (----).

water - the absorption and scattering coefficients, and volume-scattering function - and less a function of the surface-incident flux (other than just its total intensity) (Preisendorfer, 1961; Jerlov, 1976). For example, in Fig. 16, various angular parameters of the underwater light field are plotted as a function of optical depth for two different light fluxes at the surface: one vertical, the other at 45°. It is apparent that with increasing depth, each parameter - whether reflectance, or the average cosines for the downwelling, upwelling, or total, light - settles down to a constant value regardless of the angle of the incident light: the constant values actually arrived at are determined by, and are a function of, the values of the inherent optical properties of this particular water.

If the characteristics of the field are mainly determined by the inherent optical properties of the medium, can observations on the field perhaps provide information about these optical properties? Yes, they can. I have briefly touched on one example of this already. Thus, from measurements of the net downward irradiance we can, using Gershun's equation, arrive at an estimate of the absorption coefficient.

Further information can be derived from measurements of reflectance - the ratio of upward to downward irradiance - at an appropriate optical depth. Reflectance is determined largely by b over a, the ratio of scattering to absorption, and from measurements of reflectance, and other functions of the irradiance in the underwater environment, it is possible to arrive at estimates of the scattering, backscattering, and absorption coefficients of the water (Kirk, 1981b; 1989).

What this means is that measurements of the underwater light environment - which oceanographers carry out because of their direct relevance to photosynthesis, or visibility - can also sometimes be used to provide valuable additional information about some of the fundamental properties of the ocean (Gordon et al., 1975; Kirk, 1981b; 1989; Preisendorfer and Mobley, 1984; McCormick and Rinaldi, 1989; Zaneveld, 1989). Thus, the circle is complete. Not only can we proceed from known, or assumed, inherent optical properties to the consequent nature of the light field, but, we also can, given information about that light field, make the return journey to deduce the optical properties of the ocean.

ACKNOWLEDGEMENTS

This work received support from the Ocean Optics Program of the Office of Naval Research, under grant N00014-91-J-1366.

REFERENCES

Bidigare, R.R., Smith, R.C., Baker, K.S., and Marra, J., 1987, Oceanic primary production estimates from measurements of spectral irradiance and pigment con-centrations, *Global Biogeochem. Cycles,* 1:171.

Bricaud, A., Morel, A., and Prieur, L., 1983, Optical efficiency factors of some phytoplankters, *Limnol. Oceanogr.,* 28:816.

Davies-Colley, R.J., Pridmore, R.D., and Hewitt, J.E., 1986, Optical properties of some freshwater phytoplanktonic algae, *Hydrobiologia,* 133:165.

Fry, E.S., and Kattawar, G.W., 1988, Measurement of the absorption coefficient of ocean water using isotropic illumination, *Ocean Optics 9, Proc. SPIE,* 925:142.

Gordon, H.R., Brown, O.B., and Jacobs, M.M., 1975, Computed relationships between the inherent and apparent optical properties of a flat homogeneous ocean, *Appl. Optics,* 14:417.

Haardt, H., and Maske, H., 1987, Specific *in vivo* absorption coefficient of chlorophyll *a* at 675 nm, *Limnol. Oceanogr.,* 32:608.

Iturriaga, R., and Siegel, D.A., 1989, Microphotometric characterization of phytoplankton and detrital absorption properties in the Sargasso Sea, *Limnol. Oceanogr.,* 34:1706.

Jerlov, N.G., 1976, "Marine Optics," Elsevier, Amsterdam.

Kirk, J.T.O., 1976, Yellow substance (gelbstoff) and its contribution to the attenuation of photosynthetically active radiation in some inland and coastal southeastern Australian waters, *Aust. J. Mar. Freshwater Res.,* 27:61.

Kirk, J.T.O., 1980, Spectral absorption properties of natural waters: contribution of the soluble and particulate fractions to light absorption in some inland waters of southeastern Australia, *Aust. J. Mar. Freshwater Res.,* 31:287.

Kirk, J.T.O., 1981a, A Monte Carlo study of the nature of the underwater light field in, and the relationship between optical properties of, turbid yellow waters, *Aust. J. Mar. Freshwater Res.,* 32:517.

Kirk, J.T.O., 1981b, Estimation of the scattering coefficient of natural waters using underwater irradiance measurements, *Aust. J. Mar. Freshwater Res.,* 32:533.

Kirk, J.T.O., 1983, "Light and Photosynthesis in Aquatic Ecosystems," Cambridge University Press, Cambridge.

Kirk, J.T.O., 1984, Dependence of relationship between inherent and apparent optical properties of water on solar altitude, *Limnol. Oceanogr.,* 29:350.

Kirk, J.T.O., 1988, Effect of scattering and absorption on solar pond efficiency, *Solar Energy,* 40:107.

Kirk, J.T.O., 1989, The upwelling light stream in natural waters, *Limnol. Oceanogr.,* 34:1410.

McCormick, N.J., and Rinaldi, G.E., 1989, Seawater optical property estimation from *in situ* irradiance measurements, *Appl. Optics,* 28, 2605.

Mecherikunnel, A., and Duncan, C.H., 1982, Total and spectral solar irradiance measured at ground surface, *Appl. Optics,* 21:554.

Morel, A., 1974, Optical properties of pure water and pure seawater, *in* "Optical Aspects of Oceanography," N.G. Jerlov, and E. Steemann Nielsen, eds., Academic Press, London.

Morel, A., and Prieur, L., 1977, Analysis of variations in ocean colour, *Limnol. Oceanogr.,* 22: 709.

Morrow, J.H., Chamberlin, W.S., and Kiefer, D.A., 1989, A two-component description of spectral absorption by marine particles, *Limnol. Oceanogr.,* 34:1500.

Oishi, T., 1990, Significant relationship between the backward scattering coefficient of sea water and the scatterance at 120°, *Appl. Optics,* 29:4658.

Palmer, K.F., and Williams, D., 1974, Optical properties of water in the near infrared, *J. Opt. Soc. Amer.,* 64:1107.

Pope, R.M., Fry, E.S., Montgomery, R.L., and Sogandares, F., 1990, Integrating cavity absorption meter; measurement results, *Ocean Optics 10, Proc. SPIE,* 1302:165.

Preisendorfer, R.W., 1961, Application of radiative transfer theory to light measurements in the sea, *Union Geod. Geophys. Inst. Monogr.,* 10:11.

Preisendorfer, R.W., and Mobley, C.D., 1984, Direct and inverse irradiance models in hydrologic optics, *Limnol. Oceanogr.,* 29:903.

Roesler, C.S., Perry,M.J., and Carder, K.L., 1989, Modeling *in situ* phytoplankton absorption from total absorption spectra in productive inland marine waters, *Limnol. Oceanogr.,* 34:1510.

Sathyendranath, S., Lazzara, L., and Prieur, L., 1987, Variations in the spectral values of specific absorption of phytoplankton, *Limnol. Oceanogr.*, 32:403.

Shibata, K., 1959, Spectrophotometry of translucent biological materials - opal glass transmission method, *Methods Biochem. Anal.*, 7:77.

Smith, R.C., Austin, R.W., and Tyler, J.E., 1970, An oceanographic radiance distribution camera system, *Appl. Optics*, 9:2015.

Smith, R.C., and Baker, K.S., 1981, Optical properties of the clearest natural waters (200-800 nm), *Appl. Optics*, 20:177.

Smith, R.C., Prezelin, B.B., Bidigare, R.R., and Baker, K.S., 1989, Bio-optical modeling of photosynthetic production in coastal waters, *Limnol. Oceanogr.*, 34:1524.

Tyler, J.E., and Smith, R.C., 1970, "Measurements of Spectral Irradiance Underwater," Gordon and Breach, New York.

Voss, K.J., 1989, Use of the radiance distribution to measure the optical absorption coefficient in the ocean, *Limnol. Oceanogr.*, 34:1614.

Zaneveld, J.R.V., 1989, An asymptotic closure theory for irradiance in the sea and its inversion to obtain the inherent optical properties, *Limnol. Oceanogr.*, 34:1442.

Zaneveld, J.R.V., Bartz, R., and Kitchen, J.C., 1990, A reflective-tube absorption meter, *Ocean Optics 10, Proc. SPIE*, 1302.

THE FUNCTIONAL AND OPTICAL ABSORPTION CROSS-SECTIONS OF PHYTOPLANKTON PHOTOSYNTHESIS

Zvy Dubinsky

Department of Life Sciences
Bar Ilan University
Ramat Gan 52900, Israel

INTRODUCTION

Phytoplankton, like all photosynthetic organisms, have highly complex light-harvesting systems, consisting of pigment beds arranged on asymmetrical membranes. These pigment arrays, or "antennae", as well as any other cellular, light-absorbing structures and compounds have a definite probability "cloud" for absorbing impinging photons. This wavelength-dependent probability may be quantified as a cross-section, with dimensions of area per unit of compound. These *in vivo* cross-sections are invariably smaller than those of the same substance, for example, chlorophyll *a*, when extracted by any suitable solvent, brought into solution, and purified. The optical *in-vivo* cross-section of these pigments varies considerably between species, as well as within the same species, in response to such environmental factors as ambient light and nutrient status.

In addition to these purely "physical" optical, or total, cross-sections, system-specific, functional cross-sections of photosynthetic entities can be measured, such as PSII (photosystem II) and PSI (photosystem I). Such cross-sections may be computed from the light saturation curve for the production of any quantifiable product of the functional entity being studied, such as oxygen-evolution, or fluorescence. These functional cross-sections and their ratios are affected by various aspects of the physiological status of the phytoplankton cells, determined by their history in respect to key environmental parameters.

Under low-photon fluxes, the ratio of the functional to optical cross-section, being the ratio of the light energy channeled to a particular process to the total light harvested by the cell, is related to the maximal quantum yield of this process.

THE EFFECTIVE IN-VIVO OPTICAL CROSS-SECTION OF PHYTOPLANKTON

The importance of the pigmented phytoplankton cells in determining the behavior of underwater light was examined in both laboratory cultures (Tyler and Smith, 1970; Kirk, 1975a; b; 1976) and in the field (Kiefer and Austin, 1974; Smith and Baker, 1978a; b). From these studies, it was suggested that both freshwater (Kirk, 1980) and marine water bodies (Smith and Baker, 1978b) may be classified according to their optical

Primary Productivity and Biogeochemical Cycles in the Sea
Edited by P.G. Falkowski and A.D. Woodhead, Plenum Press, New York, 1992

characteristics, and the relation of these to phytoplankton abundance. This classification was done by partitioning the total attenuation of underwater light among the major-light absorbing fractions: water, gilvin, and particulates, which include both tripton and phytoplankton (Prieur and Sathyendranath, 1981; Kirk, 1983).

The effect of phytoplankton on the apparent properties of underwater light was quantified by regressing the attenuation coefficient for diffuse downwelling irradiance, K_d, against the concentration of chlorophyll a. This treatment allows one to correlate a major apparent property of the light field to phytoplankton distribution, as expressed in the concentration of chlorophyll a. This quantification also allows us to differentiate between the attenuation of light due to phytoplankton, $k_c C$, and that due to all other non-phytoplankton, light-absorbing fractions, k_w,

$$K_d = k_w + k_c C \tag{1}$$

as discussed by Bannister (1974). Here k_c is the specific extinction coefficient of chlorophyll a, and C is chlorophyll a concentration. This equation may be further elaborated by partitioning the average attenuation coefficient, $K_d(PAR)$ for the visible light domain, PAR (as defined by Morel, 1978), into additional partial attenuation coefficients due to water alone, K_W, gilvin, K_G, tripton, K_{TR}, and phytoplankton, K_{PH}. For most cases, the following convenient approximation is valid:

$$K_d(PAR) = K_W + K_G + K_{TR} + K_{PH} \tag{2}$$

In an homogeneously mixed water column, the partial attenuation coefficient of phytoplankton, K_{PH}, is approximately proportional to the product of the concentration of phytoplankton, B_c ($=C$), in mg chlorophyll a m^{-3}, and k_c, the specific vertical attenuation coefficient per unit phytoplankton, in m^2 mg^{-1} chlorophyll a. This attenuation coefficient has the dimensions of a specific cross-section of chlorophyll a. Dubinsky and Berman (1979) and Dubinsky (1980) compared some values and symbols used in the early field determinations of this and related parameters.

Because of their complex relationship, the discussion and definitions of optical cross-sections of phytoplankton in the early biological literature does not differentiate between their apparent and inherent aspects. However, k_c is actually more closely related to the inherent optical properties of the water-body in question, than to the apparent ones. For a rigorous discussion of the theoretical aspects of aquatic optics and their relation to phytoplankton, specialized publications such as Preisendorfer (1961; 1976), Morel and Bricaud (1981), Kirk (1983; 1984), and Gordon (1989) should be consulted.

Both k_c, the specific vertical attenuation coefficient per unit chlorophyll a, which is an apparent optical property of the underwater light field, and $\circ a^*$, as defined by Kiefer et al. (1979), and by Wilson and Kiefer (1979), which is an inherent property of the phytoplankton cells in the water, have the dimensions of a cross-section of chlorophyll a, m^2 mg^{-1} chl a.

The optical cross-sections of phytoplankton are normalized to chlorophyll a, which is the most intuitively correct choice because this ubiquitous pigment is found in all photosynthetic organisms, except bacteria, other than cyanobacteria. However, this choice creates problems. Chlorophyll a absorbs light, presenting a distinct absorption cross-section to light, but it is not the only light-absorbing compound in phytoplankton cells. Other pigments, such as chlorophylls b and c; phycobiliproteins like phycocyanin, allophycocyanin, phycoerythrin; and carotenoids such as fucoxanthin, peridinin, and lutein are an integral part of the light harvesting antennae. Others including carotenoids such as β-carotene (not all of it) and astaxanthin, and some of the xanthophylls, such as violaxanthin seem to serve as photo-protective pigments, preventing excess light energy from reaching the photosynthetic apparatus. Whatever their function may be, these pigments contribute to the cross-section of the single cell, and to the attenuation of underwater light due to absorption of light quanta. The conceptual problem this contribution creates is that by normalizing all light attenuation, or cross-sections of phytoplankton cells and populations to chlorophyll a, we actually assign the light absorbed by any phytoplankton pigment to chlorophyll a, thereby virtually increasing effective cross-sections of this pigment. Falkowski et al., (1985) attempted to avoid this difficulty when they also normalized the cross-sections to cellular carbon, thereby defining an *in-vivo* carbon-specific cross-section, σ_c, and analyzing how this differed from the commonly used chlorophyll a-based k_c.

Additional optical cross-sections relevant to the study of phytoplankton photosynthesis may be defined. These are the cell cross-section, σ_B, which is obtained by multiplying cellular chlorophyll a by k_c and the photosynthetic unit (PSU) cross-section, σ_{PSU}. These cross-sections were calculated by dividing $\circ a^*$, or k_c, by the number of molecules of chlorophyll a in one mg of this pigment, which resulted in a cross-section per single molecule of this compound. This value was then multiplied by either the number of molecules of chlorophyll a in the cell, or by the PSU "size".

Every light-absorbing compound, or pigment within the phytoplankton cell, has characteristic probabilities of absorbing photons at different wavelengths, which determine the absorption spectrum of this compound. This is also true for any mixture of such pigments, or in the case at hand, for live phytoplankton. Accordingly, we can partition the spectral attenuation coefficients for diffuse downwelling irradiance $K_{d\lambda}$ into the same absorbing components into which $K_d(PAR)$ was partitioned, either $k_{..}$ and k_c, or K_W, K_G, K_{TR}, and K_{PH}. Such treatment produces a spectral distribution of K_{PH}, and k_c, as was done in the field for Lake Kinneret by Dubinsky and Berman (1979), and more elaborately in the laboratory by Dubinsky et al. (1986) and by Berner et al. (1989).

Since k_c is a spectral average coefficient, its value depends on the absorption of the cell at each particular wavelength, and on the amount of quanta corresponding to this wavelength, in the total radiant flux. A predominantly green cell will have a higher probability of absorbing photons in the red region of the spectrum, than in the green region. In other words, for this cell, $k_{c650} > k_{c560}$. Atlas and Bannister (1980) computed the spectral-averaged values of k_c for a few phytoplankton species with different pigment compositions, at different depths in the main types of ocean waters classified by color. They showed the importance of the spectral distribution of underwater light in determining phytoplankton cross-sections. Dubinsky et al. (1986) using the same approach in laboratory studies, determined the spectral-averaged values of k_c in several phytoplankton species under a specific light source (quartz-halogen), according to the following equation:

$$k_c = \frac{\Sigma(k_{c\lambda} I_\lambda \Delta_{\lambda n})}{\Sigma(I_\lambda \Delta_{\lambda n})} \qquad (3)$$

where $k_{c\lambda}$ is the chlorophyll a specific cross section measured at wavelength λ, I_λ is irradiance at λ, and $\Delta_{\lambda n}$ is the discrete interval between n wavelengths.

On the other hand, $_{o}a^*$, and its spectral components, $_{o}a^*\lambda$, which have to be measured under monochromatic light, affect the underwater light field, but are not affected by its spectral distribution.

FACTORS AFFECTING THE OPTICAL CROSS-SECTIONS OF PHYTOPLANKTON

Since anything affecting $_{o}a^*$ will invariably also affect k_c in the same direction, in the following discussion we use the term cross-section, meaning $_{o}a^*$, and therefore, k_c as well. Unlike when a pure compound is extracted into a solvent, the cross-sections of intact phytoplankton cells are strongly affected by many biochemical, ultrastructural and geometrical factors and properties, which, in turn, depend on the taxonomic affiliation of the species, or the taxonomic distribution of the phytoplankton population and its physiological status. There are a few main properties of the phytoplankton cells which affect the values of cross-sections.

Photoadaptation and Chlorophyll _a_ Content of Cells and Phytoplankton Populations

Because of the importance of photoadaptation for phytoplankton, this process received considerable attention in the study of aquatic primary production, as reviewed by Falkowski (1980; 1981; 1984) and Richardson et al. (1983). In the numerous studies on the mechanisms of photoadaptation in phytoplankton, a common trend was reported of increase in chlorophyll _a_ and in other light-harvesting pigments as growth irradiance decreasese. Depending on algal species, this change may be moderate, as in the "_Cyclotella_" type, or very large, up to tenfold, in the "_Chlorella_" type, as defined in the

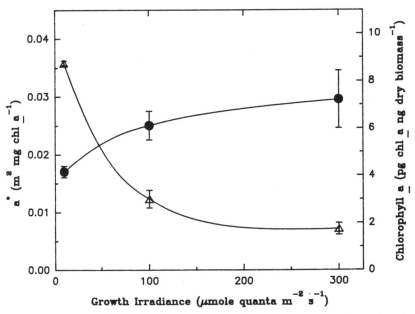

Fig. 1.   Effect of growth irradiance on the chlorophyll _a_ content (●--●) and on the optical absorption cross section (Δ--Δ) in _Tetraedron minimum_.

pioneering studies of photoadaptation by Steeman-Nielsen and Jorgensen (1968), and their co-workers. Whatever the photoadaptive type may be, the cross-section of phytoplankton invariably decreases, as cellular pigmentation increases during photoadaptation to low light (Fig. 1). This change in cross-section reduces the increase in light-harvesting resulting from increases in cellular pigment level. This counter-intuitive response results from the inevitable increase in mutual shading between cellular light-harvesting entities as their density increases. This shading happens on all scales, among individual pigment molecules and among thylakoids (Berner et al., 1989), among chloroplasts within cells, and among cells. The latter phenomenon has been reported from the super-dense populations of the mutualistic, endosymbiotic zooxanthellae that are ubiquitous in reef-building corals (Dubinsky et al., 1984; 1990). In these cases, the observed reduction in the effective optical cross-sections may be viewed as increased "overlap" among cross-sections. This overlap sets optical upper boundaries on cellular chlorophyll levels, and on integrated concentrations of chlorophyll a in a water column, as well as on the areal concentration of chlorophyll a, in the specialized case of zooxanthellae. As the product of the cross-section, as determined under lowest pigment, or cell concentration, and the areal concentration of the pigment exceeds unity, we always see a decrease in cross-sections, resulting from their mutual self-shading, or overlap.

Nutrient Status and Pigment Ratios

Nutrient status affects the optical cross-sections of phytoplankton, because it controls the ability of the cells to synthesize various components of the light-harvesting apparatus. If for example, a phytoplankton population grows, or survives under nutrient limitation, its pigment composition will differ from that of a population that is similar, but replete with nutrients. This difference stems from the fact that while chlorophyll a and the associated light-harvesting proteins are all nitrogen-containing compounds, the various carotenoids are not. The resulting increase in the carotenoid/chlorophyll ratio was described and proposed as a diagnostic indicator of nutrient depletion (Ketchum et al., 1958; Yentsch and Vaccaro, 1958; Dubinsky and Polna, 1976; Dubinsky and Berman, 1979; Heath et al., 1990). These changes in the ratio of chlorophyll a to other pigments results in an increase in cross-sections, because these are normalized to the concentration of chlorophyll a, as discussed above. The increase in light absorption by the higher content of carotenoid of the cells is "assigned" to the reduced content of chlorophyll a, resulting in an increase in $\circ a^*$ and k_c. In a chemostat study with *Isochrysis galbana*, Herzig and Falkowski (1989) describe the systematic increase in $\circ a^*$, as μ decreases, under reduced nitrogen levels.

A similar phenomenon was observed during the transition of the chlorophyte *Haematococcus pluvialis* from the green to the red form, upon transfer to a nitrogen-free medium (Zlotnik, personal communication). This increase in cross-section in nutrient-stressed cells may precipitate photodynamic damage in the photosynthetic apparatus, unless some of the added carotenoids actually dissipate the harvested light energy rather than transfer it to reaction centers. This latter is the case with most of the β-carotene accumulating in globules outside the chloroplasts, in *Dunaliella salina*, or with astaxanthin in *Haematococcus pluvialis*.

Since the major single feature on which phytoplankton taxonomy is based is the pigment composition of the cells, phytoplankton of predominantly different systematic composition, or unialgal cultures belonging to various divisions, will show different specific cross-sections of chlorophyll a, even under similar environmental conditions, affecting nutrient status and photoadaptation.

The Ultrastructural Organization of the Cell

The ultrastructural organization of phytoplankton has a marked effect on cross-sections; however, this aspect has hardly been investigated. From the known effects of changes in refractive index on the optical properties of particles, we expect diatoms with siliceous walls to have different optical properties than naked flagellates, even if they did contain the same amount of cellular chlorophyll a. The same applies to cell materials like lipids, glycerol, and many other common cell constituents, such as the gas vacuoles of some cyanobacteria, all of which are likely to influence the absorption of light by cells.

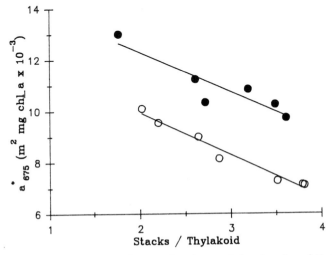

Fig. 2.   The relationship between changes in a^*_{675} and the density of thylakoid stacks during the first 24h after a low to high ($\circ$) and high to low ($\bullet$) light intensity transition in *Dunaliella tertiolecta*. (After Berner et al., 1989).

Furthermore, the internal architecture of the cell determines whether and how the light-absorbing pigments are packaged within it. Differences in the ultrastructure of the photosynthetic apparatus and of the cell, in general, act on the overall optical properties of the cells via microscale packaging effects, not unlike those described for whole phytoplankton populations (Duysens, 1956; Kirk, 1975a,b;1976; Morel and Bricaud, 1981). Are there chloroplasts, and if there are, as in most phytoplankton except cyanobacteria, is there a single large one, or are there numerous small ones? These aspects are bound to affect cross-sections, although they have hardly been investigated. Berner et al. (1989), showed that in the chlorophyte *Dunaliella tertiolecta* during photoadaptation, the number of thylakoids per stack in the grana changed, which considerably affected $\circ a^*$ (Fig. 2). In addition, Berner et al. (1989) distinguished the contribution of the ultrastructural changes described above from those caused by the changes in pigment ratios, also taking place during photoadaptation.

If the changes in cross-sections found occurring in one species due to photoadaptive changes in the ultrastructural organization within the thylakoids, are as marked as those found by Berner et al. (1989), much larger differences in cross-sections should be expected, if we were to compare cells of taxa with different organization of the cell and of the photosynthetic apparatus within it.

The Size and Geometry of the Cells

Kirk (1975a;b; 1976; 1983) showed that changes in cell size affect the cross-sections of phytoplankton populations. For the same chlorophyll a concentration per unit volume of the cell suspension, the larger the size of the cells, the smaller their absorbance, and hence, their cross-sections will be smaller. Conversely, as cells become smaller, their absorbance approaches that of a solution, of the same concentration of chlorophyll a, although, we always should expect to find the relationship:

$$D_{sus} < D_{sol} \tag{4}$$

where D_{sus} is the absorption of a cell suspension, and D_{sol}, that of a solution.

Kirk (1976; 1983) also described an opposite trend when cells of increasing deviation from a spherical shape are compared, again assuming the same concentration of chlorophyll a. Kirk showed that the more elongated the cells are, the ratio of D_{sus}/D_{sol} increases. These conclusions underscore the importance of changes in the size of pytoplankton cells occurring during their life cycle, or during changes in nutrient status and photoadaptation, in determining their cross-sections, and therefore their light-harvesting properties. The same is true for the changes in shape sometimes occurring as part of a life-cycle of a single algal species, but much more frequently associated with the species succession so typical of phytoplankton communities.

There are a few additional methods for determining optical cross-sections, besides those derived from measurements in the field and based on the attenuation of the underwater light field by phytoplankton. In the field, the major problem is the partitioning of attenuation by the total particulate fraction between phytoplankton and tripton. This partitioning was estimated by measuring the absorption of all the particulates on a filter, and repeating the reading after extracting the filters. The measurement on filters requires correction due to amplification of pathlength by multiple scattering in the layer as described by Mitchell and Kiefer (1984).

In the laboratory, absorption measurements were made on dilute suspensions by Ley and Mauzerall (1982), in a normal spectrophotometer, using opal glass spectrophotometry (e.g., Dubinsky et al., 1986) and with integrating sphere attachments (Kiefer et al., 1979), and from beam attenuation within a cuvette for measuring oxygen evolution (Dubinsky et al., 1987). Cleveland et al.(1989) distinguished the absorption of phytoplankton from that of non-living particulates based on their mathematical treatment of absorption spectra of single-species cultures grown in the laboratory.

Perry and Porter (1989) applied flow-cytometry to determine the cross-sections of individual cells in the field and compared their results with those obtained from numerous laboratory cultures. They based their analysis on the relationship between the fluorescence of the phytoplankton cells and their cross-sections, and found very good correspondence.

One of the most interesting recent applications of the optical cross-sections of phytoplankton was the attempt by Smith et al. (1989) to construct a photon budget for the upper ocean. They used their values for cross-sections of phytoplankton to estimate the fate of all light absorbed in the ocean. This approach will no doubt be attempted in the future in studies of water bodies.

FUNCTIONAL CROSS-SECTIONS OF PHYTOPLANKTON

In addition to the optical cross-sections of phytoplankton, functional cross-sections have been defined and measured. These are associated with photosynthesis, or with its partial reactions and related phenomena; in principle, they may be determined for any photochemical reaction in which the formation of a product or response, R, may be quantified. These cross-sections are best understood in the context of the photon flux needed to saturate the reaction. From the saturation curve of any such reaction, in respect of the photon flux delivered as a variable intensity, brief flash, within the duration of a single occurrence of the reaction, the so-called, single-turnover flash, a target size may be calculated, representing the probability of eliciting a photochemical response, according to the equation (Ley and Mauzerall, 1982):

$$Y = R/R_{max} = 1-e^{-\sigma E} \tag{5}$$

which describes a cumulative single-hit Poisson probability distribution. If Y is the yield of the reaction, in terms of the ratio of the measured response to its maximal ratio, and E is the number of photons per unit area delivered during the flash, then σ is the functional cross-section being estimated.

The Cross-section of PSII

In the case of the functional cross-sections for photosynthetic oxygen evolution by PSII, Ley and Mauzerall (1982) used a laser at 596nm to deliver submicrosecond flashes, which were varied over five orders of magnitude to obtain a saturation curve. They measured oxygen evolution with a bare platinum, Pickett-type polarograph.

Using the equation

$$P = P_{max}(1-e^{-\sigma_{O_2} E}) \tag{6}$$

P and P_{max} are the measured and maximal measured oxygen evolution, and E, the measured number of photons per cm^2 delivered during the flash. The cross-section of PSII for oxygen production, σ_{O_2}, remains the only unknown. Using this method, Ley and Mauzerall determined σ_{O_2} in *Chlorella vulgaris* with great precision ($\pm 2\%$). They found it to be 70 A^2 (7000 nm^2) at 596nm, which was the wavelength of the laser they used. Ley (1984) used the same method to determine the effective cross-sections for oxygen production in the phytoplanktonic rhodophyte *Porphyridium cruentum*. Ley measured σ_{O_2} values at 546 and 596 nm, before and following state II to state I transitions, and found a 50% increase. From these data he analyzed the changes in energy transfer between phycobilisomes and the reaction centers of PSII (RCII) occurring in this organism upon state transition. This approach was used in a few subsequent studies with other phytoplankton and higher-plant chloroplast suspensions, and is reviewed by Mauzerall and Greenbaum (1988).

Recent studies used photoacoustic techniques to measure oxygen evolution from phytoplankton monolayers, and other plant material. This technique uses a microphone to convert the pressure wave generated by the oxygen evolved after a light pulse into an electronic signal. This method requires the separation of the signal due to oxygen evolution from pressure resulting from dissipation of thermal energy (Canaani et al., 1988). Work is in progress with photoacoustic techniques to measure σ_{O_2} in *Chlorella vulgaris* (Cha and Mauzerall, 1990).

Less direct methods were also used to determine σ_{O_2} or the related cross-section σ_{PSII}. There is an inverse relationship between chlorophyll fluorescence and the ongoing rate of photosynthesis. This relation is based on the increasing ratio of closed/total number of RCII as photosynthetic rate approaches P_{max}, its maximal, light- saturated value. When a RCII is excited by energy of a photon absorbed in its antenna an electron is transferred to a primary acceptor, Q_A. This "trap" remains "closed" until it is reduced by a primary donor and the electron moves on from the primary acceptor. Any light reaching the closed trap between the time of reduction of the reaction center and the reoxidation of the acceptor will elicit fluorescence. Therefore, fluorescence will increase from a minimum, F_o, when all traps are open, such as in a phytoplankton sample preincubated in the dark, to a maximal value, F_{max}, when all centers are closed as photosynthesis proceeds at its maximal, light-saturated rate. Falkowski et al. (1986) used a pump-and-probe technique first proposed by Mauzerall (1972) to obtain a saturation curve of photosynthesis. In this method the change between fluorescence yield of a weak "probe" flash preceding (F_o) and following (F_I) an actinic "pump" flash, is measured. Kolber et al. (1990) developed this approach, and designed laboratory and sea-going versions of a double flash (actually triple flash) fluorometer, based on it. With this apparatus, in addition to the ongoing rate of photosynthesis, the functional cross-section of PSII is computed from the variable fluorescence curve. The variable fluorescence, $\Delta\Phi$, is defined as:

$$\Delta\Phi = (F_I - F_o)/F_o \qquad (7)$$

At a saturating actinic flash:

$$\Delta\Phi_{sat} = (F_{max} - F_o)/F_o \qquad (8)$$

By varying the intensity of the "pump" flash, a functional cross-section for PSII, σ_{PSII}, may be calculated from a non-linear regression, again assuming the same cumulative, single-hit Poisson distribution as with oxygen evolution,

$$\Delta\Phi/\Delta\Phi_{sat} = 1 - e^{-\sigma_{PSII}E} \qquad (9)$$

Dubinsky et al. (1986) calculated the changes occurring in cross-sections of PSII during phoptoadaptation of three phytoplankton species. They fitted data relating photosynthesis under continuous, white irradiance to a single-hit, cumulative Poisson distribution. They estimated the turnover time τ from P_{max} values and PSU "sizes," and calculated the number of photons delivered at each irradiance level per unit area, during the interval τ. The transition from discrete flashes to a continuous light raises difficult theoretical and experimental problems, since the continuous illumination may cause additional complicating changes such as photoinhibition and increased respiration rates, all of which are avoided when using single turnover flashes. Nevertheless, they showed that while in *Isochrysis galbana*, adaptation to low light was accompanied by a consistent increase in σ_{PSII}, no such change occurred either in *Thalassiosira weisflogii* or in

Prorocentrum micans. Because they determined photosynthetic rates under a "white" beam, their cross-sections are weighed spectral-averaged cross-sections, for the specific light source with which they worked.

The Functional Cross-section of PSI

The measurement of this cross-section proved rather elusive since, as was pointed out by Mauzerall and Greenbaum (1989), unlike in the case of PSII, there is no readily quantifiable product indicative of its activity. These authors (Greenbaum and Mauzerall, 1987) suggested that the amplitude of the oscillation in oxygen uptake following a single turnover flash may be used to estimate the cross-section of PSI. They attributed the respiratory transients in O_2 uptake occurring following the flash to PSI activity, and constructed a saturation curve for the amplitude of the oscillation as a function of flash intensity. By fitting the relative yield, in this case the relative amplitude of the respiratory transient, to the cumulative single-hit Poisson distribution, they estimated the cross-section of PSI. In another study they fit their data to a sum of two Poissonians, indicating the presence of a mixture of two different cross-sections (Greenbaum et al., 1987). Subsequent work and analysis proved that these cross-sections are of both photosystems. Thus the respiratory transient measures the cross-section of both PSI and PSII (Mauzerall and Greenbaum, 1989; Greenbaum and Mauzerall, 1991).

A different approach to estimate the cross-section of PSI is based on the observation that under anaerobic conditions in many green alga an hydrogenase is induced. Hydrogen production by this enzyme involves only PSI. From the saturation curve of hydrogen production in *Chlamydomonas*, Greenbaum (1977) was able to determine the cross-section of PSI. A similar technique was used by Boichenko and Litvin (1986) to determine the cross-sections of PSI and PSII. They used pulsed H_2 and O_2 production measured with a platinum polarographic electrode polarized for O_2 detection or, under anaerobic conditions, for H_2 oxidation.

QUANTUM YIELDS OF PHOTOSYNTHESIS

Quantum yields of photosynthesis are of great interest in the study of the photosynthetic process (Mauzerall and Greenbaum, 1989) and in ecological studies (Dubinsky, 1980), where their relation to various ecosystem properties and environmental factors has been explored. Early determinations were based on the ratio between the energy stored in a given water volume through the photosynthetic process, and the quanta absorbed by the same volume during the same time (e.g., Dubinsky and Berman, 1976; 1981a,b; Dubinsky et al., 1984). For the specialized case of zooxanthellate corals, an analogous areal computation was applied to determine quantum yields (Dubinsky et al., 1984; 1990; Wyman et al., 1987).

In recent oceanographic work, attempts were made to incorporate cross-sections and quantum yields in algorithms for determining photosynthesis from distributions of chlorophyll *a* as determined by satellite imagery, and from the properties of the underwater light field (Kiefer and Mitchell, 1983; Falkowski; Lewis; Kiefer, this volume). This need rekindled the interest in the determination of quantum yields, using current experimental and conceptual approaches.

If the value for the optical cross-section $\circ a^*$, or k_c, which is determined per mg of chlorophyll *a*, is divided by the number of chlorophyll *a* molecules in one mg of chlorophyll, we obtain the optical cross-section of a single chlorophyll *a* molecule, σ_{chl}.

When this value is multiplied by the Emerson-Arnold photosynthetic unit (PSU) size (Emerson and Arnold, 1932), we obtain an optical absorption cross-section for the PSU, σ_{PSU}. This cross-section relates to the light absorbed by the entire PSU, which, by the Emerson and Arnold (1932) definition, is the oxygen-producing functional entity. The light absorbed by any area exposed to a flash equals the product of the area by the number of photons delivered by the flash per unit area. For a single PSU this is:

$$\sigma_{PSU} \, E \tag{10}$$

By virtue of the definition of the PSU, the maximum oxygen produced by a sample containing n chlorophyll a molecules, illuminated by a single turnover flash, is:

$$n/PSU. \tag{11}$$

Under non-saturating light this equals:

$$\frac{n}{PSU} \, (1-e^{-\sigma_{O_2} E}) \tag{12}$$

Under low light:

$$1-e^{-\sigma_{O_2} E} \; -----> \; \sigma_{O_2} \, E$$

Therefore, at low E, the maximum quantum yield Φ_{max}, equals the oxygen produced by a sample, divided by the light absorbed by all the chlorophyll in it:

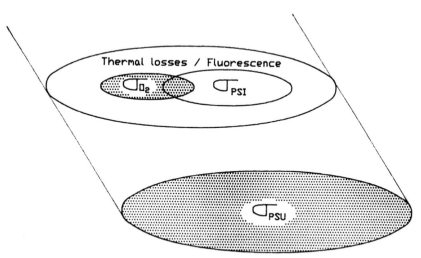

Fig. 3. Schematic representation of Φ as the ratio of $\sigma_{O_2}/\sigma_{PSU}$. The large elliptical area is the total energy harvested by the Emerson and Arnold (1932) PSU. The σ_{O_2} ellipse is the fraction of that energy resulting in PSII mediated oxygen evolution. The difference between these areas ($\sigma_{PSU} - \sigma_{O_2}$) consists of energy utilized in PSI activity, fluorescence, thermal decay, and other losses.

$$\Phi_{max} = \frac{\frac{n}{PSU} \sigma_{O_2} E}{\frac{n}{PSU} \sigma_{PSU} E} = \frac{\sigma_{O_2}}{\sigma_{PSU}} = \frac{\sigma_{O_2}}{PSU \ \sigma_{chl}} \qquad (13)$$

Since the photon number cancels out, Φ_{max} equals the ratio between the optical and functional cross-sections (Ley and Mauzerall, 1982; Dubinsky et al., 1986; Mauzerall and Greenbaum, 1989) (Fig. 3). The advent of the sea-going pump and probe fluorometer allows us to determine the maximal quantum yield, based on the ratio of cross-sections, on an unprecedented bathymetric and temporal resolution (Kolber et al., 1990; Falkowski, this volume).

CONCLUSIONS

The sensitivity of the optical cross-section to cell size, photoadaptation, and nutrient limitation, determines the importance of this parameter in coupling the oceanic photon fluxes to the material fluxes. This coupling is accomplished on the single-cell level, or on the phytoplankton community structure level (Chisholm, this volume) by optimizing the ratios of light-harvesting to nutrient assimilation rates. These adjustments result in the spatial and temporal variation in the maximal quantum yield of global oceanic photosynthesis (Falkowski; Lewis, this volume).

REFERENCES

Atlas, D., and Bannister, T. T., 1980, Dependence of mean spectral extinction coefficient of phytoplankton on depth, water colour and species, *Limnol. Oceanogr.*, 25:157.

Bannister, T. T., 1974, Production equations in terms of chlorophyll concentration, quantum yield and upper limits on production, *Limnol. Oceanogr.*, 19:1.

Berner, T., Wyman, K., Dubinsky, Z., and Falkowski, P.G., 1989, Photoadaptation and the "package effect" in *Dunaliella tertiolecta* (Chlorophyceae), *J. Phycol.*, 25:70.

Boichenko, V.A., Ladygin, V.G., and Litvin, F.F., 1989, Structural and functional organization of photosynthetic units in the cells of *Chlamydomonas reinhardii*, *Mol. Biol. (Moscow)*, 23:107.

Cha, J. and Mauzerall, D., 1990, The photosystem II inhibitor DCMU decreases the observed turnover time of photosystem I in *Chlorella* yet retains the high thermodynamic efficiency determined by pulsed photoacoustics, *Proc. 10th International Biophysics Congress, Vancouver, BC*, Abs. No. P7.3.26.

Canaani, O., Malkin, S., and Mauzerall, D., 1988, Pulsed photoacoustic detection of flash induced oxygen evolution from intact leaves and its oscillations, *Proc. Natl. Acad. Sci. USA.*, 85:4725.

Chisholm, S., this volume.

Cleveland, J. S., Perry, M. J., Kiefer, D. A., and Talbot, M. C., 1989, Maximal quantum yield of photosynthesis in the northwestern Sargasso Sea, *J. Mar. Res.*, 47:869.

Dubinsky, Z., 1980, Light utilization efficiency in natural phytoplankton communities, *in*: "Primary Productivity in the Sea," P.G. Falkowski, ed., Plenum, New York.

Dubinsky, Z., and Polna, M., 1976, Pigment composition during a Peridinium bloom in Lake Kinneret (Israel), *Hydrobiologia*, 51:234.

Dubinsky, Z., and Berman, T., 1976, Light utilization efficiencies of phytoplankton in Lake Kinneret (Sea of Galilee), *Limnol. Oceanogr.*, 21:226.

Dubinsky, Z., and Berman, T., 1979, Seasonal changes in the spectral composition of downwelling irradiance in Lake Kinneret (Israel), *Limnol. Oceanogr.*, 24:652.

Dubinsky, Z., and Berman, T., 1981a, Light utilization by phytoplankton in Lake Kinneret (Israel), *Limnol. Oceanogr.*, 26:660.

Dubinsky, Z., and Berman, T., 1981b, Photosynthetic efficiencies in aquatic ecosystems, *Verh. Internat. Verein. Limnol.*, 21:205.

Dubinsky, Z., Berman, T., and Schanz, F., 1984, Field experiments for *in situ* measurement of photosynthetic efficiency and quantum yield, *J. Plankton Res.*, 6:339.

Dubinsky, Z., Falkowski, P. G., Porter, J. W., and Muscatine, L., 1984, The absorption and utilization of radiant energy by light and shade adapted colonies of the hermatypic coral, *Stylophora pistillata*, *Proc. Roy. Soc. Lond.*, 222B:203.

Dubinsky, Z., Falkowski, P. G., and Wyman, K., 1986, Light harvesting and utilization in phytoplankton, *Plant Cell Physiol.*, 27:1335.

Dubinsky, Z., Falkowski, P. G., Post, A. F., and van Nes, U. M., 1987, A system for measuring phytoplankton photosynthesis in a defined light field with an oxygen electrode, *J. Plankton Res.*, 9:607.

Dubinsky, Z., Stambler, N., Ben-Zion, M., McCloskey, L. R., Falkowski, P. G., and Muscatine, L., 1990, The effects of external nutrient resources on the optical properties and photosynthetic efficiency of *Stylophora pistillata*, *Proc. Roy. Soc. Lond.*, B 239:231.

Duysens, L. N. M., 1956, The flattening of the absorption spectrum of suspensions as compared to that of solutions, *Biochim. Biophys. Acta.*, 19:1.

Emerson, R., and Arnold, W., 1932, A separation of the reactions in photosynthesis by means of intermittent light, *Jour. Gen. Physiol.*, 15:391.

Falkowski, P. G., 1980, Light-shade adaptation in marine phytoplankton, *in*: "Primary Productivity in the Sea", P. G. Falkowski, ed., Plenum, New York.

Falkowski, P. G., 1981, Light-shade adaptation and assimilation numbers, *J. Plankton Res.*, 3:203.

Falkowski, P. G., 1984, Physiological responses of phytoplankton to natural light regimes, *J. Plankton Res.*, 6:295.

Falkowski, P.G., this volume.

Falkowski, P. G., Dubinsky, Z., and Wyman, K., 1985, Growth-irradiance relationships in phytoplankton, *Limnol. Oceanogr.*, 30: 311.

Falkowski, P. G., Wyman, K., Ley, A., and Mauzerall, D., 1986, Relationship of steady-state photosynthesis to fluorescence in eucaryotic algae, *Biochim. Biophys. Acta.*, 849:183.

Gordon, H. R., 1989, Can the Lambert-Beer law be applied to the diffuse attenuation coefficient of ocean water?, *Limnol. Oceanogr.*, 34:1389.

Greenbaum, E., 1977, The photsynthetic unit of hydrogen evolution, *Science,* 196:879.

Greenbaum, N.L. and Mauzerall, D., 1987, Measurement of the optical cross section of photosystem I in *Chlorella*, *in*: "Progress in Photosynthesis Research," Vol. II, J. Biggens, ed., Martinus Nijhoff Publishers, Dordrecht, The Netherlands.

Greenbaum, N.L. and Mauzerall, D., 1991, Effects of irradiance level on distribution of chlorophylls between PSII and PSI as determined from optical cross-sections, *Biochim. Biophys. Acta*, 1057:195.

Greenbaum, N. L., Ley, A. C., and Mauzerall, D. C., 1987, Use of a light-induced respiratory transient to measure the optical cross section of photosystem I in *Chlorella*, *Plant Physiol.*, 84:879.

Heath, M. R., Richardson, K., and Kiorboe, T., 1990, Optical assessment of phytoplankton nutrient depletion, *J. Plankton Res.*, 12:381.

Herzig, R. and Falkowski, P. G., 1989, Nitrogen limitation in *Isochrysis galbana* (Haptophyceae). I. photosynthetic energy conversion and growth efficiencies, *J. Phycol.*, 25:462.

Ketchum, B. H., Ryther, J. H., Yentch, C. S., and Corwin, N, 1958, Productivity in relation to nutrients, *Rapp. P.- V. Reun. Cons. Perm. Int. Expl. Mer.*, 144:132.

Kiefer, D.A., this volume.

Kiefer, D. A., and Mitchell, B. G., 1983, A simple steady state description of phytoplankton growth based on absorption cross section and quantum efficiency, *Limnol. Oceanogr.*, 24:770.

Kiefer, D. A., and Austin, R.W., 1974, The effect of varying phytoplankton concentration on submarine light transmission in the Gulf of California, *Limnol. Oceanogr.*, 19:55.

Kiefer, D. A., Olson, R. J., and Wilson, W. H., 1979, Reflectance spectroscopy of marine phytoplankton, Part 1. Optical properties as related to age and growth rate, *Limnol. Oceanogr.*, 24:664.

Kirk, J. T. O., 1975a, A theoretical analysis of the contribution of algal cells to the attenuation of light within natural waters. I. General treatment of suspensions of pigmented cells, *New Phytol.*, 75:11.

Kirk, J. T. O., 1975b, A theoretical analysis of the contribution of algal cells to the attenuation of light within natural waters. II. Spherical cells, *New Phytol.*, 75:21.

Kirk, J. T. O., 1976, A theoretical analysis of the contribution of algal cells to the attenuation of light within natural waters. III. Cylindrical and spheroidal cells, *New Phytol.*, 77:341.

Kirk, J. T. O., 1980, Spectral absorption properties of natural waters: contribution of the soluble and particulate fractions to light absorption in some inland waters of southeastern Australia, *Aust. J. Mar. Freshwater Res.*, 32:287.

Kirk, J. T. O., 1983, "Light and Photosynthesis in Aquatic Ecosystems," Cambridge University Press, Cambridge.

Kirk, J. T. O., 1984, Dependence of the relationship between inherent and apparent optical properties of water on solar altitude, *Limnol. Oceanogr.*, 29:350.

Kolber, Z., Wyman, K. D., and Falkowski, P. G., 1990, Natural variability in photosynthetic energy conversion efficiency: A study in the Gulf of Maine, *Limnol. Oceanogr.*, 35:72.

Lewis, M., this volume.

Ley, A. C., 1984, Effective absorption cross-sections in *Porphyridium cruentum* implications for energy transfer between phycobilisomes and photosystem II reaction centers, *Plant Physiol.*, 74:451.

Ley, A. C., and Mauzerall, D. C., 1982, Absolute absorption cross-sections for photosystem II and the minimum quantum requirement for photosynthesis in *Chlorella vulgaris, Biochim. Biophys. Acta.*, 680:95.

Mauzerall, D., 1972, Light-induced fluorescence changes in *Chlorella,* and the primary photoreactions for the production of oxygen, *Proc. Nat. Acad. Sci. USA*, 69:1358.

Mauzerall, D., and Greenbaum, N. L., 1989, The absolute size of a photosynthetic unit, *Biochim. Biophys. Acta.*, 974:119.

Mitchell, B. G., and Kiefer, D. A., 1984, Determination of absorption and fluorescence absorption spectra for phytoplankton, *in*: "Marine Phytoplankton and Productivity," O. Holm-Hansen, L. Bolis, and R. Gilles, eds., Springer Verlag, Berlin.

Morel, A., 1978, Available, usable and stored radiant energy in relation to marine photosynthesis, *Deep-Sea Res.*, 25:673.

Morel, A., and Bricaud, A., 1981, Theoretical results concerning light absorption in a discrete medium, and application to specific absorption of phytoplankton, *Deep-Sea Res.*, 28:1375.

Perry, M. J., and Porter, S. M., 1989, Determination of the cross-section absorption co-efficient of individual phytolankton cells by analytical flow cytometry, *Limnol. Oceanogr.*, 34:1727.

Preisendorfer, R. W., 1961, Application of radiative transfer theory to light measurements in the sea, *Int. Union Geod. Geophys. Monogr.*, 10:11.

Preisendorfer, R. W., 1976, "Hydrologic Optics," V. I. NTIS PB 259793/8ST. Natl. Tech. Inform. Serv., Springfield, Va.

Prieur, L., and Sathyendranath, S., 1981, An optical classification of coastal and oceanic waters based on the specific absorption curves of phytoplankton pigments, dissolved organic matter, and other particulate materials, *Limnol. Oceanogr.*, 26:671.

Richardson, K., Beardall, J., and Raven, J. A., 1983, Adaptation of unicellular algae to irradiance: an analysis of strategies, *New Phytol.*, 93:157.

Smith, R. C., and Baker, K. S., 1978a, The bio-optical state of ocean waters and remote sensing, *Limnol. Oceanogr.*, 23:247.

Smith, R. C., and Baker, K. S., 1978b, Optical classification of natural waters, *Limnol. Oceanogr.*, 23:260.

Smith, R. C., Marra, J., Perry, M. J., Baker, K. S., Swift, E., Buskey, E., and Kiefer, D. A., 1989, Estimation of a photon budget for the upper ocean in the Sargasso Sea, *Limnol. Oceanogr.*, 34:1673.

Steemen-Nielsen, E., and Jorgensen, E.G., 1968, The adaptation of algae, I, General part, *Physiol. Pl.*, 21:401.

Tyler, J. E., and Smith, R. C., 1970, "Measurements of Spectral Irradiance Underwater," Gordon and Breach.

Wilson, W. H., and Kiefer, D. A., 1979, Reflectance spectroscopy of marine phytoplankton, Part 2, A simple model of ocean color, *Limnol. Oceanogr.*, 24:673.

Wyman, K.D., Dubinsky, Z., Porter, J. W., and Falkowski, P. G., 1987, Light absorption and utilization among hermatypic corals: a study in Jamaica, West Indies, *Marine Biology*, 283-292.

Yentsch, C. S., and Vaccaro, R. Y., 1958, Phytoplankton and nitrogen in the oceans, *Limnol. Oceanogr.*, 3:443.

MOLECULAR ECOLOGY OF PHYTOPLANKTON PHOTOSYNTHESIS

Paul G. Falkowski

Department of Applied Science
Brookhaven National Laboratory
Upton, NY 11973

INTRODUCTION

The ability to derive basin-scale maps of phytoplankton chlorophyll in the upper ocean from satellite color sensors (see Lewis, this volume) has led increasingly to the development of models relating biomass to primary production (Eppley et al., 1985; Falkowski, 1981; Platt, 1986; Platt and Sathyendranath, 1988; Morel, 1991). Chlorophyll, however, represents a pool size, while primary production is a flux. To derive a flux from a pool, a time-dependent variable must be incorporated. The simplest models relating carbon fixation to chlorophyll incorporate irradiance (Bidigare et al., this volume); the transfer function is a quantum yield. These so-called light-chlorophyll models (Ryther and Yentsch, 1957; Cullen, 1990) are virtually impossible to verify in the ocean, hence their credulity presumably lies in understanding the underlying biological processes and how those processes are regulated. Here, I examine how some of the key parameters which are implicitly or explicitly incorporated in rational light-chlorophyll models are regulated at a fundamental, molecular level.

While light-chlorophyll models may appear to differ in notation and some detail (e.g. Bannister and Laws, 1980; Kirk, 1983; Kiefer and Mitchell, 1983; Falkowski et al., 1985; Geider et al., 1986; Sakshaug et al., 1989), they basically are similar. All of these models calculate the absorption of radiation as the product of the spectral optical absorption cross-section normalized to chlorophyll, a*, and incident spectral irradiance, I_o. The rate of photosynthesis is simply calculated from the product of the quantum yield, ϕ, and the rate of light absorption. The allocation of carbon to cell biomass can then be predicted from the photosynthetic quotient and respiratory losses (Falkowski et al., 1985). However, at light saturation, the rate of light absorption becomes irrelevant to photosynthesis or growth, and the minimum turnover time, τ, for electrons to be transferred from water to carbon dioxide becomes rate determining. At present, no model predicts τ with accuracy, and thus the phytoplankton models do not predict, from first principles, the maximum photosynthetic or growth rate of phytoplankton.

Paralleling the progress in modeling the photophysiological responses of phytoplankton, molecular biologists and biophysicists have made considerable progress over the past decade in understanding the basic structure of photosynthetic systems. A major watershed was the elucidation of the amino-acid sequences of the L, M, and H

Primary Productivity and Biogeochemical Cycles in the Sea
Edited by P.G. Falkowski and A.D. Woodhead, Plenum Press, New York, 1992

subunits, and subsequently, the X-ray structure of the functional apparatus in *Rhodopseudomonas viridis*. That structural investigation provided a model for the reaction center of photosystem II in oxygenic photosynthesis (Michel and Deisenhofer, 1988). There is no comparable model of PS I, but general features of the reaction center structure may be presumed. An understanding of the structure of light-harvesting chlorophyll proteins is largely based on X-ray analysis of the chlorophyll a/b binding protein from peas (Kuhlbrandt and Wang, 1991); its relationship or relevance to the LHCs in phytoplankton is unclear. However, basic schematic models for both photosystems have been developed, and the compartmental location of the genes encoding for these elements for some cyanobacteria and eucaryotic microalgae is known (Table 1).

Application of physiological models to estimate phytoplankton production and growth in the sea depends on parameters derived from either laboratory or field measurements. Both such data have revealed that variations in absorption cross sections, quantum yields, and turnover times are related to variations in growth irradiance and nutrient limitation (Falkowski et al., 1985, 1991; Dubinsky et al., 1986; Kolber et al., 1988, 1990). The ability to describe the climatological variability of these parameters in the ocean would conceptually lead to much more accurate descriptions of primary production from knowledge of incident irradiance and the distribution of chlorophyll.

Using abstract mathematical models in conjunction with structural biological models, it is possible to gain some insight into how, on a molecular level, phytoplankton perceive the outside world, transduce that information, and respond by altering components in the photosynthetic apparatus. I will focus on how molecular alterations of the photosynthetic apparatus which affect absorption cross-sections, quantum yields, and the maximum rate of electron transport, play an important role in determining the photosynthetic flux of carbon in the ocean.

ABSORPTION CROSS-SECTIONS

Most studies of variations in absorption cross-sections in cultured phytoplankton have been related to changes associated with irradiance. The functional absorption cross-sections of PS I and PS II can change on time scales of a few minutes (state transitions) to a few hours (photoadaptation) in response to photon flux densities (Ley, 1980; Falkowski, 1984; Falkowski and LaRoche, 1991; Mauzerall and Greenbaum, 1989). PS II cross sections can be measured by following the change in the flash intensity saturation curve for O_2 evolution (Ley and Mauzerall, 1982; Dubinsky, this volume) or for the change in variable fluorescence (Falkowski et al., 1986). The two curves are virtually identical (Falkowski et al., 1988), indicating they have the same functional antenna. PS I cross-sections can be measured by following the light-intensity-dependent changes in P_{700} oxidation/reduction kinetics (Zipfel and Owens, 1991; Telfer et al., 1984), the changes in EPR signal strength (Weaver and Weaver, 1969), the changes in respiratory transients (Greenbaum et al., 1987; Greenbaum and Mauzerall, 1991), or the changes in hydrogen evolution (Greenbaum, 1977).

At a molecular level, the absorption cross-section is related to the average "size" of the antenna serving the reaction center. The antenna system of phytoplankton is composed of either phycobilisomes (in some cyanobacteria and cryptomonads), or light-harvesting chlorophyll proteins. One way to increase the cross section is to add more antenna molecules to each of the reaction center complexes. Algae accomplish this in two ways. On short time-scales of a few minutes, some fraction of the light-harvesting chlorophyll proteins may become reversibly phosphorylated (Bennett, 1991). Allen

Table 1. Some Major Photosynthetic Genes in Cyanobacteria and Eucaryotic Microalgae and Their Products

Complex	Gene Designation	Gene Product	Location	Codons
Photosystem II	psbA	32 kDa Q$_B$protein, D1	chloroplast	353
	psbB	47 kDa Chl$a$ protein	chloroplast	508
	psbC	44 kDa Chl$a$ protein	chloroplast	461
	psbD	34 kDa protein, D2	chloroplast	353
	psbE	9 kDa cytochrome b-559	chloroplast	83
	psbF	4 kDa cytochrome b-559	chloroplast	39
	psbG	24 kDa polypeptide	chloroplast	248
	psbH	10 kDa phosphoprotein	chloroplast	73
	psbI		chloroplast	36
	psbJ		chloroplast	53
	psbK		chloroplast	55
	psbL	3.2 kDa polypeptide	chloroplast	38
	OEE1	33 kDa polypeptide	nucleus	~1000
	OEE2	23 kDa polypeptide	nucleus	~900
Cytochrome b$_6$-f complex	petA	cytochrome f	chloroplast	320
	petB	cytochrome b-563	chloroplast	215
	petC	Reiske Fe-S protein	nucleus	~900
	petD	17 kDa polypeptide	chloroplast	160
Photosystem I	psaA	P700-Chl$a$ protein	chloroplast	751
	psaB	P700-Chl$a$ protein	chloroplast	735
	psaC	8 kDa 2(4Fe-4S) protein	chloroplast	81
	psaD	Ferredoxin docking protein	nucleus	
	psaE		nucleus	
	psaF	plastocyanin docking protein	nucleus	
	psaH-J	unknown functions	nucleus	
ATP Synthase	atpA	CF$_1$ subunit	chloroplast	504
	atpB	CF$_1$ subunit	chloroplast	498
	atpC	CF$_1$ γ subunit	nucleus	~1000
	atpD	CF$_1$ δ subunit	nucleus	~750
	atpE	CF$_1$ subunit	chloroplast	137
	atpF	CF$_0$ subunit I	chloroplast	183
	atpH	CF$_0$ subunit III	chloroplast	81
	atpL	CF$_0$ subunit IV	chloroplast	247
RuBisCO	rbcL	large subunit	chloroplast	550
	rbcS	small subunit	nucleus in chlorophytes chloroplast in chromophytes	120
Light Harvesting	cab	major light-harvesting chlorophyll-binding proteins associated with PS I and PS II	nucleus	various

et al., (1981) suggested that phosphorylation is mediated by the redox potential of the intersystem electron carrier, plastoquinone. When PQ is oxidized, phosphatase activity is greater than kinase activity, and the LHCP becomes dephosphorylated. The dephosphorylated pigment-protein complex associates with PS II, leading to an increase in the effective absorption cross-section of that photosystem. Conversely, upon reduction of PQ, which is caused by a high rate of photon absorption by PS II, a kinase phosphorylates a fraction of LHC II, and the pigment protein complex detaches from the photosystem, so decreasing the cross-section of PS II. It was thought that the LHC II migrated to PS I, leading to a simultaneous increase in the absorption cross-section of that photosystem. However, biophysical evidence suggests that in cyanobacteria and eucaryotic algae the changes in PS II cross-sections are not accompanied by simultaneous changes in PS I cross-sections (Owens, 1986).

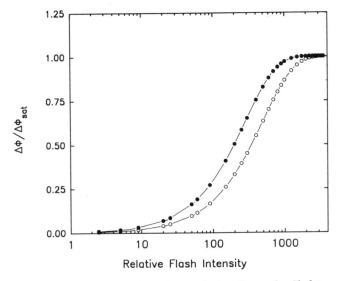

Fig. 1. Changes in σ_{PSII} as a function of growth irradiance in *Skeletonema costatum* grown in nutrient-replete conditions. Cells grown at lower irradiance (●) have larger cross-sections. The data were obtained with a pump and probe fluorometer, as described by Kolber et al., 1988.

The role of phosphorylation-related changes in cross-sections, or state transitions in natural phytoplankton communities has never been understood. They are usually experimentally promoted by changes in light; however, in the chromophyte, *Olisthodiscus danica*, phosphorylation of the chlorophyll *a/c* fucoxanthin proteins is independent of illumination and insensitive to the photosynthetic electron-transport inhibitor, DCMU (Gibbs and Biggins, 1991). The paradigm that red light promotes an increase in the functional absorption cross-section of PS II, while blue light leads to a decrease in that cross-section is based on experiments with chlorophytes (Bonaventura and Myers, 1969). In the ocean, higher intensities of light are associated with a red-enriched region of the spectrum. Thus, increased irradiance would tend to promote a state I transition on the basis of the spectral composition, while simultaneously promoting a state II transition on the basis of photon fluence.

State transitions may lead to alterations in the ratio of linear to cyclic electron flow, where state II essentially promotes more cyclic flow around PS I, and hence, more ATP production relative to NADPH formation. Turpin and Bruce (1990) found that in *Selenastrum minutum*, NH_4^+ assimilation may also lead to a state transition, apparently by altering the ratio of demand for ATP/NADPH. Specifically, addition of NH_4^+ simulated a state II transition by forcing an increased demand for ATP. NO_3^- and NO_2^- have similar ATP/NADPH requirements to CO_2 assimilation (Turpin, 1991) and do not induce state transitions. It may be that in phytoplankton, nutrient limitation and pulsed supply may influence short-term scale changes in the functional absorption cross-sections of PS I and PSII more than the irradiance regime.

On longer time scales, of hours to days, phytoplankton may synthesize more or less antenna pigment-protein complexes relative to reaction center proteins. This process, which may be mediated by changes in the average irradiance to which cells are exposed, is usually phenomenologically considered a component of photoadaptation. Generally, the absorption cross-sections of PS II and PS I increase with decreasing irradiance (Fig. 1). In some species, however, such as *Thalassiosira weisflogii* (Kolber et al., 1988), photoadaptation does not result in any change in the absorption cross-section of PSII. Changes in the absorption cross-section are often due to irradiance-dependent changes in the accumulation of light-harvesting chlorophyll protein complexes relative to reaction centers (Falkowski et al., 1981). Sukenik et al. (1988) showed that in the chlorophyte, *Dunaliella tertiolecta*, a decrease in growth irradiance leads to an increase in both LHC II and LHC I apoproteins. How a change in irradiance level leads to a change in the synthesis of light-harvesting chlorophyll proteins is unknown.

MOLECULAR BASES OF PHOTOADAPTATION

The synthesis and assembly of functional light-harvesting chlorophyll complexes requires the coordination of two biosynthetic pathways, namely that for pigments and that for protein. The pigments are synthesized in the chloroplast. The LHC apoproteins are encoded in the nucleus, translated in the cytoplasm, and transported across the chloroplast membrane as larger precursors, where specific proteases cleave the transit peptide. The mature apoprotein then becomes inserted in the thylakoid membranes in specific orientations, and subsequently binds pigments.

One possible mechanism for regulating the synthesis and assembly of light-harvesting chlorophyll protein complexes during photoadaptation is via post-translational stabilization by pigments of the nascent apoprotein molecules. This hypothesis of post-translational control is based on the observation that in mutants of (e.g.) *Chlamydomonas*, in which chlorophyll *b* synthesis is genetically blocked, LHC apoproteins are synthesized but degraded before they are inserted into thylakoid membranes (Michel et al., 1983). Thus, if chlorophyll synthesis was regulated by light intensity, the apoproteins could be synthesized at a constant rate. If the pigment production rate was equal to or exceeded that of the apoprotein, the resulting complex would be stabilized and become incorporated in the membrane as a functional antenna. However, if, pigment synthesis lagged behind that of the apoprotein, the uncomplexed protein would be rapidly degraded.

Mortain-Bertrand et al. (1990) tested this hypothesis in *Dunaliella tertiolecta* by blocking the synthesis of chlorophyll with the inhibitor gabaculine. The synthesis and stability of LHC II was followed in pulse-case studies with $^{35}SO_4$. Following a shift from high to low irradiance, LHC II apoproteins were radiolabeled and remained labeled, even after a 96 h case, in both control and gabaculine-treated cells. However, in the latter LHC II did not increase relative to other membrane proteins, as it did in control cells. These results suggest that chlorophyll synthesis is required for the accumulation, but not directly for the stabilization, of LHC II.

LaRoche et al. (1991) examined how light intensity affects the abundance of messenger RNA encoding for LHC II in *Dunaliella*. They found that within 1.5 hours following a shift to either low light, darkness, or low light in the presence of gabaculine, the LHC II mRNA levels increased markedly. Nine hours following a shift to low light, the mRNA for LHC II had increased four-fold, while transcripts for photosynthetic proteins (e.g., D1, Rubisco large subunit) did not change (Fig. 2). In cells transferred to low light, however, the mRNA levels remained elevated after 36 h, while, in cells kept in darkness or in the presence of gabaculine, the transcripts declined. Additionally, LHC II apoproteins only accumulated in cells kept in low light. Thus, chlorophyll synthesis is not required for an increase in LHC II transcript abundance, but is required for the accumulation of LHC II apoproteins.

Thus, the evidence suggests that the messenger RNA levels for LHC II apoproteins are affected by irradiance, while translation of the message appears to require chlorophyll

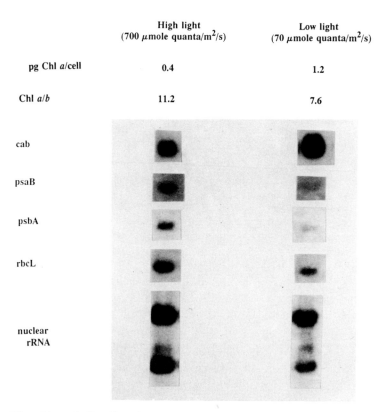

	High light (700 μmole quanta/m^2/s)	Low light (70 μmole quanta/m^2/s)
pg Chl *a*/cell	0.4	1.2
Chl *a*/*b*	11.2	7.6
cab		
psaB		
psbA		
rbcL		
nuclear rRNA		

Fig. 2.   The chlorophyll *a*/*b* ratio and relative abundance of mRNA for several of photosynthetic genes in *Dunaliella tertiolecta*. Note the large increase in cab mRNA relative to psaB and psbA (data courtesy of Julie LaRoche).

biosynthesis. How the messenger RNA levels are promoted by low irradiance is unclear; however, as submicromolar concentrations of DCMU also promote an increase in chlorophyll (Beale and Appleman, 1971), the transcluetion mechanism would appear to be related to linear photosynthetic electron flow.

Both the optical (a*) and functional absorption cross-sections (σ_{PSII}) change in response to nutrient limitation (Kolber et al., 1988; Herzig and Falkowski, 1989). Both nitrogen and iron limitation lead to increases in a* (Greene et al, 1991). The increase in a* occurs because nutrient limitation often leads to a reduction in intracellular pigment content, and a concomitant lessening of self-shading within the thylakoid membranes (Berner et al., 1989). However, nutrient limitation can also lead to an increase in σ_{PSII} (Kolber et al., 1988; Greene et al., 1991), a result that is counter-intuitive. In the case of iron or nitrogen limitation, the nuclear-encoded light-harvesting chlorophyll protein complexes appear to be preferentially synthesized relative to the chloroplast-encoded reaction center proteins (Falkowski et al., 1989). While nutrient limitation leads to an overall reduction in pigment per cell, there is more LHC II present relative to reaction centers. Thus, under nutrient-replete conditions, increased irradiance tends to lead to a reduction in σPSII, while nitrogen and iron limitation tend to lead to an increase in σ_{PSII}. How do these two processes interact in the ocean?

There are few measurements of PS II absorption cross-sections in the ocean. One data set from the Gulf of Maine (Kolber et al., 1990) suggested that irradiance was not the only factor determining the spatial distribution of PS II cross-sections, and that nutrient availability also played an important role. Off the coast of Hawaii the absorption cross-sections decrease with depth, reaching a minimum at the nitricline (Fig. 3). Clearly, such a phenomenon suggests that light is not the only factor controlling PS II cross-sections *in situ*.

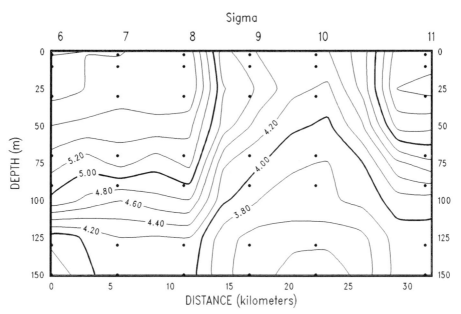

Fig. 3.   Section showing the spatial distribution of σ_{PSII} off the coast of Hawaii. Note the decrease in σ_{PSII} with depth, which is opposite to that expected if the variability in σ_{PSII} was due to photoadaptation. At station 10, a cyclonic eddy pumped nitrate into the euphotic zone; this phenomenon was associated with decreased σ_{PSII}.

MOLECULAR BASES OF CHANGES IN QUANTUM YIELDS OF PHOTOSYNTHESIS

The quantum yield is a mathematical parameter in rational models relating irradiance to photosynthesis or growth. The determination of the maximum or limit quantum yield in photosynthesis has a long and tumultuous history. It is now generally accepted that the limit yield of the reaction for oxygen evolution is 0.125 O_2/quanta; however, that yield is seldom obtained except under ideal circumstances (Myers, 1980). Some physiological models, such as that of Kiefer and Mitchell (1983), assume a constant limit yield of 0.10, and subsequently model the relationship between the quantum yield and irradiance by a hyperbolic function. Similarly, so-called ψ functions, relating surface incident irradiance and integrated euphotic-zone chlorophyll concentrations to integrated carbon fixation (Falkowski, 1981; Platt, 1986; Morel, 1991), inherently assume or explicitly derive a quantum yield. Both of these general types of models are attractive because of their potential ability to derive primary production from satellite chlorophyll data sets or from moored instrumentation. However, if the limit yields of photosynthesis vary significantly in the ocean, then the rates of carbon fixation estimated by these models could have significant error. For the moment, I will not discuss further whether or not a hyperbolic function best fits the relationship (Cullen et al., this volume), but instead focus on the variability in the maximum quantum yield, ϕ_{max}.

The quantum yield is not directly measured; it is a derived parameter. Usually it is derived from two independent measurements, the rate of light absorption, which is, in turn, calculated from measurement of incident spectral irradiance and a* (Dubinsky, this volume), and the initial slope of the photosynthesis irradiance curve, which is derived from some curve-fitting procedure (e.g., Jassby and Platt, 1976). It is difficult to measure the absorption of light by algae in the laboratory, let alone in the ocean. In most of the early determinations of quantum yield, algal cultures were so dense that all the visible light was absorbed by the cells, thereby negating the problem of having to measure scattering. In the ocean, the measurement of a* is complicated by the relatively dilute concentration of phytoplankton, and consequently scattering and the absorption of photosynthetically active radiation by non-photosynthetic particles poses a formidable technical problem (see Kirk, this volume). Hence, a* itself must be derived indirectly (Bidigare et al., this volume). Almost all have reported limit yields in excess of the theoretical maximum. Interestingly, however, Cleveland et al. (1989) suggested that variations in the yield were not randomly distributed, and therefore, presumably not due to measurement error. These workers correlated the variations in yields to distance from the nitricline, and inferred that nutrient limitation affected the limit yield. That inference is supported by laboratory data showing that the maximum quantum yield of oxygen evolution is affected by nitrogen limitation (Welschmeyer and Lorenzen, 1991; Herzig and Falkowski, 1989; Kolber et al., 1988; Osborne and Geider, 1986; Greene et al., 1991).

As described by Ley and Mauzerall (1982), Dubinsky et al. (1986), and Dubinsky (1991), the quantum yield of photosynthetic oxygen evolution is the ratio of the optical absorption cross-section to the functional absorption cross-section of PS II. The functional absorption cross-section of PS II can be measured in the ocean with a pump- and-probe fluorometer, without interference from non-phytoplankton absorption (Kolber et al., 1990). Absolute calibration of the actinic flash intensity allows for calculation of absolute quantum yield, if a* can be measured (Greene et al., 1991). Thus, even though this technique circumvents the problem of measuring P vs I curves and deriving an initial slope, the accurate measurement of a* remains an obstacle to accurate and rapid assessment of the variability of ϕ_{max}.

Maintenance of the maximum quantum yield requires the integration of two processes, the efficient transfer of excitation energy from the antenna pigments to the reaction centers, and the existence of functional and open reaction centers. In PS II, if either the transfer process or reaction center function is impaired, the maximum change in the quantum yield of fluorescence will decrease. In the former case, excitation energy absorbed by the pigment bed will not be trapped, and subsequently, a larger fraction will be re-emitted as fluorescence, raising the so-called F_o level. In the latter case, both F_o and F_m will be reduced (Butler and Katajima, 1975; Kiefer and Reynolds, this volume). Thus, measurement of the maximum change in the quantum yield of fluorescence is a surrogate index of the maximum quantum yield of photosynthesis. The maximum change in the quantum yield of fluorescence can be readily measured *in situ* (Kolber et al., 1990).

Empirically, the numerical value of the change in the quantum yield of fluorescence has an upper bound. If calculated by the expression, $\Delta\phi_{sat} = (F_m - F_o)/F_o$, the maximum value is *ca.* 1.6; the value is independent of species and growth irradiance, as long as the cells are nutrient replete (Falkowski et al., 1986; Kolber et al., 1988). It is unclear why the maximum values of $\Delta\phi_{sat}$ converge on 1.6; however, it presumably reflects the ratio of fluorescence lifetimes between open and closed PSII reaction centers (Schatz et al., 1988; Kiefer and Reynolds, this volume).

In the summer in the Gulf of Maine, Kolber et al. (1990) found that $\Delta\phi_{sat}$ varied by a factor of three in the euphotic zone, but, moreover, was correlated with distance from the nutricline. Thus, their work independently supported that of Cleveland et al. (1989), and implied that even in "nutrient rich" coastal waters, photosynthetic rates were not simply limited by the availability of light, but could also be limited by the availability of

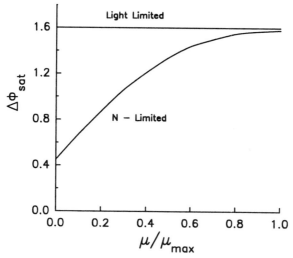

Fig. 4.   Effects of growth irradiance and nutrient limitation on the change in the maximum quantum yield of fluorescence in phytoplankton. If the relative growth rate is regulated by irradiance, and nutrients remain saturating, $\Delta\phi_{sat}$ is constant, averaging about 1.6. If the relative growth rate is regulated by nutrients, but light is saturating, $\Delta\phi_{sat}$ is a hyperbolic function of μ/μ_{max}.

nutrients. Subsequently, Falkowski et al. (1991) found that in the subtropical Pacific, $\Delta\phi_{sat}$ was elevated by nutrient pumping in an eddy, approaching the maximum value, but outside of the eddy, was much lower than the maximum value. Why should $\Delta\phi_{sat}$ vary?

$\Delta\phi_{sat}$ varies with nutrient status. Specifically, Kolber et al. (1988) found that in three species of phytoplankton grown in nitrogen-limited chemostat cultures, $\Delta\phi_{sat}$ was a hyperbolic function of growth rate (Fig. 4). Moreover, when growth rates for the three species were normalized to the respective maxima, one curve described the relationship between $\Delta\phi_{sat}$ and the relative specific growth rate, independent of species. Clearly, more work needs to be done to determine how robust that relationship is; however, it implies that under nitrogen limiting conditions, $\Delta\phi_{sat}$ can be related to the relative specific growth rate of phytoplankton (Falkowski et al., 1991). Greene et al. (1991) also found that $\Delta\phi_{sat}$ was affected by iron availability in batch culture, and they used $\Delta\phi_{sat}$ as one diagnostic of iron limitation. We have done preliminary experiments with phosphorous limitation, but our few results suggest that phosphate limitation may lead to a reduction in $\Delta\phi_{sat}$, although not nearly to the extent that nitrogen or iron limitation do.

MOLECULAR BASES OF VARIATIONS IN $\Delta\phi_{sat}$

The transfer of excitation energy from the light-harvesting chlorophyll proteins (or from phycobilisomes for that matter) to reaction centers is mediated by a few highly conserved chlorophyll-protein complexes, which are encoded in the chloroplast (Table 1). Two in particular, CP43 and CP47, are thought to be essential for energy transfer. Western blots of nitrogen-limited cells suggest that there is a reduction in CP43, CP47 as well as in the reaction center proteins D1 and D2 relative to the light-harvesting chlorophyll proteins. That observation suggests nutrient limitation may reduce synthesis of chloroplast-encoded proteins more than nuclear-encoded proteins (Kolber et al., 1988). Picosecond measurements of fluorescence lifetime measurements (Geider, Greene, and Mauzerall, unpublished) suggest that excitation energy is not transferred effectively to PS II reaction centers in iron-limited cells. This finding is further supported by a decrease in F_0 upon addition of iron. Thus, it would appear that the major cause of reduction in $\Delta\phi_{sat}$ under iron, and possibly, nitrogen limitation is due to the reduction in the efficiency of transfer of excitation energy to reaction centers. Much more work needs to be done to clarify the molecular mechanisms by which nutrient limitation alters the quantum yield of fluorescence.

On an ecological level, however, the results of Cleveland et al. (1989), Kolber et al. (1990), and Falkowski et al. (1991) have profound implications. For the past decade, biological oceanographers have vigorously debated and explored whether phytoplankton growth rates are nutrient-limited. One school of thought, championed by such workers as Caperon (1968), Thomas (1970), and Dugdale (1967), suggested that phytoplankton growth could be related to external nutrient concentration by a hyperbolic function, and because the concentration of fixed inorganic nitrogen was vanishingly low throughout most of the upper oligotrophic ocean, chlorophyll-specific primary production (and, by inference, specific growth rate) was nutrient-limited. Another school, led by Goldman (1980), and Laws et al. (1987), suggested that although the pool size of nitrogen was low, the regenerative flux of nitrogen through a microbial food web was so rapid as to sustain phytoplankton at or near their maximum relative specific growth rates. Both schools have argued from indirect evidence, as maximum specific growth rates have not been directly and unambiguously measured *in situ*. The variations in quantum yield found for photosynthesis (Cleveland et al., 1989) and fluorescence (Kolber et al., 1990; Falkowski et al., 1991) in both coastal and oligotrophic ocean waters suggest that, in fact, average specific growth rates and primary productivity are nutrient-limited. While this evidence is also indirect, it (a) is directly measured with minimum experimental manipulation, and (b) cannot be explained by any other environmental or biological process that we are aware of. Clearly, however, many more measurements of the spatial and temporal variations in the functional absorption cross-sections of PS II and quantum yields are needed, and undoubtedly the controversy over whether phytoplankton are at or near their maximum relative specific growth rates will continue.

Another implication of the variability in quantum yields is the application of simple irradiance - chlorophyll models to predict carbon fixation. If nitrogen limitation leads to a decrease in quantum yields, but it is assumed (as is presently the case) that the limit is constant, then the incorporation of a fixed limit yield into production models will potentially overestimate production. Most of the light-chlorophyll model validation or parameterization has been from coastal regions, where nutrient limitation may not be severe. I examined the relationship between surface irradiance, integrated euphotic zone chlorophyll, and integrated euphotic zone carbon fixation for historical data sets from Northwest Africa, Baja, and off the island of Hawaii, and cannot find a fixed value for ψ. Much of the variability in ψ probably arises from the effects of nutrients on the quantum yield of photosynthesis. One possible means developing more accurate estimates of primary production from satellite measurements is to produce "climatological" maps of quantum yields. Incidently, passive, solar-induced measurements of fluorescence cannot be used to derive the yield variability. In the present models, variability in the yields is ascribed to variations in photosynthetic rate, assuming a fixed limit yield (Kiefer et al., 1989). The use of passive fluorescence to estimate variability in the maximum quantum yields of photosynthesis would be an exercise in circular reasoning within the context of the present models.

VARIATIONS IN P_{MAX}

While there is a theoretical limit yield for photosynthetic oxygen evolution, which is based on the so-called "Z" scheme, theoretical calculations of maximum photosynthetic rate (e.g., Falkowski, 1981) are based on empirical parameterizations. It would seem rather straightforward to measure P^B_{max}; however, the literature on the interpretation of the parameter is confusing, as are the causes of its variability (Cullen et al., this volume).

P_{max} can be defined as a ratio of the concentration of photosynthetic units to their turnover time. The concentration of photosynthetic units can be defined as the number of reaction centers per cell; usually, but not always, concentration is defined as the number of PS II reaction centers. If the ratio of PS II to PS I reaction centers is not unity (Falkowski et al., 1981; Dubinsky et al., 1986), then the calculated turnover times will differ, and have different meanings. However, for the sake of simplicity, let us define the concentration, n, of reaction centers based on PS II.

In practice, the turnover time of interest in determining P_{max} is the minimum time required for an electron to be extracted from water on the donor side of PS II, to pass through the intersystem electron transport chain to PS I, and subsequently to reduce either CO_2 or some other terminal electron acceptor (e.g., NO_3^- or SO_4^{2-}). This turnover time is usually calculated from P_{max} and oxygen flash yields (e.g., Falkowski et al., 1981; Herzig and Falkowski, 1989), and not directly measured. It should be noted that the time constant for the decay of variable fluorescence (Falkowski et al., 1986; Kolber et al., 1988) differs from the turnover time for whole chain electron transfer. The former corresponds to electron transfer reactions between the primary and secondary electron acceptors in PS II (specifically between Q_a and Q_b), and not to the overall rate-limiting reactions in carbon fixation. The time constants associated with the transfer of electrons between Q_a and Q_b are an order of magnitude faster than the whole chain electron transfer reaction. Henceforth, to avoid confusion, I shall refer to the whole chain turnover time as the "slow" turnover time.

Myers and Graham (1971) first noted that the slow turnover time increased in *Chlorella pyrenoidosa* grown in trubidostat cultures as cells acclimated to lower irradiance levels. Simultaneously, the number of PS II reaction centers increased. Thus, as cells became shade adapted, the maximum rate at which the average reaction center could process electrons decreased.

Sukenik et al. (1987) examined the effect of growth irradiance on the slow turnover time in *Dunaliella tertiolecta* and found that the turnover time increased almost five-fold between cells grown in turbidostats at five separate irradiance levels between 80 and 1900 μE m^{-2} s^{-1}. Over the same irradiance range, there was a five-fold decrease in PS II reaction centers. They further examined whether the changes in the slow turnover time were correlated with changes in the stoichiometry of electron transport components between PS II and PS I. The results of that study revealed that the proportion of PS II reaction centers to plastoquinone to cytochrome f to PS I reaction centers was independent of growth irradiance. Thus, all measured components of electron transport increased concordantly with decreased irradiance. They called the ensemble of these components an "electron transport chain." However, the cellular level of the carboxylation enzyme, RuBisCO was independent of growth irradiance. When the rate constant, the reciprocal

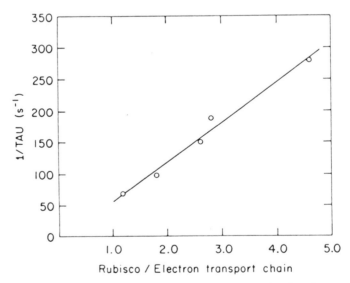

Fig. 5.   Relationship between the ratio of Ribulose 1,5-bis-phosphate carboxylase-oxygenase (Rubisco) to electron-transport chain and the maximal turnover rate of the photosynthetic apparatus in *Dunaliella tertiolecta* grown in nutrient-replete conditions. Note the linear dependence of the maximal rate of photosynthetic electron transport on the RuBisCO/ETC ratio (after Sukenik et al., 1987).

of the slow turnover time, was plotted as a function of the ratio of RuBisCO/electron transport chain for each irradiance, a linear relationship was found (Fig. 5). These results strongly suggested that carboxylation, or some process associated with carbon fixation, is the rate-limiting factor at light saturation, and not electron flow between the two photosystems. Similar results were subsequently obtained by Fischer et al. (1989).

The slow turnover time does not always vary with growth irradiance. For example, in *Isochrysis galbana* grown under nutrient-replete conditions, Falkowski et al. (1985) found that the turnover time was independent of growth irradiance, averaging *ca.* 4 ms. They did not examine the relationship between electron-transport components and Calvin cycle enzymes, so it is unclear if the electron-transport components were maintained in constant proportion to carbon-fixing components. In nitrogen-limited *Isochrysis*, however, Herzig and Falkowski (1989) found that the turnover time increased from 4.5 to 6.3 ms, as cells became increasingly nitrogen-limited. Interestingly, the relationship between RuBisCO/RCII was not a linear function of the maximum rate of whole chain electron transport under these conditions, suggesting that RuBisCO *per se* is not rate limiting.

The basic consensus is that Calvin cycle processes limit the maximum photosynthetic rate; however, the physiological literature is rich with discussion and debate as to which specific reaction(s) in the cycle is rate limiting. The activation of RuBisCO by the enzyme RuBisCO activase (Portis et al., 1986) has been implicated, although the regeneration of ribulose 1,5-bisphosphate also has been suggested (Farquahar et al., 1980). It is interesting that most of the key enzymes involved in the dark reactions of carbon fixation are soluble proteins located in the chloroplast stroma, while the electron-transport components of the light reactions are hydrophobic molecules found in the thylakoid membranes. The addition of surface area to the thylakoid membrane accompanying adaptation to low irradiance is not accompanied by an increase in the volume of the chloroplast (Berner et al., 1989). Thus, potentially one factor limiting P^B_{max} might be the surface area to volume ratio of within the chloroplasts, in which the diffusion of reductant and ATP from the thylakoid membranes to the soluble proteins becomes rate determining. Whatever the exact causes, however, it is clear that proper parameterization of the slow turnover time is essential for estimating P^B_{max} from chlorophyll and irradiance.

The Calculation of τ

At subsaturating irradiance, the initial slope of the photosynthesis-irradiance curve, α, can be described by:

$$\alpha = \sigma_{PSII} * n \tag{1}$$

where n is the concentration of PS II reaction centers (usually expressed as O_2/chlorophyll). At saturating irradiance levels:

$$P_{max} = n/\tau \tag{2}$$

where τ is the slow turnover time (Herron and Mauzerall, 1972; Falkowski, 1981). Recall that:

$$I_k = \alpha/P_{max} \tag{3}$$

Thus, the slow turnover time can be calculated from knowledge of the absorption cross-section of PS II and from I_k:

$$\tau = 1/(I_k * \sigma_{PSII}) \tag{4}$$

Both I_k and σ_{PSII} can be measured with the pump-and-probe fluorometer (Falkowski and Kolber, 1990), and thus τ is derived (Kolber and Falkowski, unpublished). Calculated values of τ in the ocean appear to average about 4 ms, although we have data for few samples. We expect that over the next decade the biological oceanographic community will have a much more extensive data set related to the climatology of τ, the relationship between τ and oceanographic regimes, and a more complete understanding of the biochemical and molecular biological factors controlling τ.

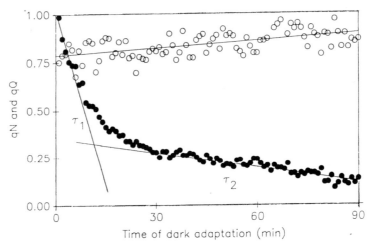

Fig. 6. Changes is non-photochemical quenching (qN) and photochemical quenching (qQ) in a natural phytoplankton community sampled near the surface at mid-day in summer. Note the rapid decrease in qN (τ_1) followed by a slow decrease (τ_2). The slow decrease in qN is mirrored by a proportional increase in qQ. We infer that τ_1 corresponds to "true" non-photochemical quenching, while τ_2 is related to photoinhibitory damage to PS II (data courtesy of Z. Kolber).

PHOTOINHIBITION AND NONPHOTOCHEMICAL QUENCHING

In most vertical profiles of phytoplankton photosynthesis, measured in bottles by either following radiocarbon incorporation or oxygen evolution, there is a depression in P^B_{max} near the surface (e.g., McAllister, 1961; see Neale, 1987 for a review). This effect can be simulated in cultures by increased irradiance levels, and has operationally been called photoinhibiton (Myers and Burr, 1940). Photoinhibition has been correlated with damage to both PS I (Gerber and Burris, 1981) and PS II (Kyle et al., 1985), although most of the experimental phenomena implicate PS II. The exact mechanism of photoinhibitory damage to PS II reaction centers is under some debate (Demmig-Adams, 1990). It has been suggested that photoinhibition is due to the degradation of the reaction-center protein, D1 (Ohad et al., 1990). Specifically, based on molecular models of the folded protein, hydropathy plots, and by analogy to the L subunit of *Rhodopseudomonas viridis*, D1 contains 5 transmembrane spans. Photoinhibition leads to cleavage between a residues 238 and 248, between the fourth and fifth transmembrane regions, and in a region where Q_a and DCMU bind. It has been suggested that the exact site of cleavage is at tyrosine residue. Alternatively, it has been suggested that photoinhibition is due to damage to the primary donor and acceptors within the reaction centers. Operationally, both types of mechanisms lead to the same basic phenomenology, namely a loss of photosynthetic capacity (Osmond, 1981), and a decrease in variable fluorescence (Critchley and Smillie, 1981; Vincent et al., 1984).

Functionally, the loss of activity within the ensemble of PS II reaction centers results in a decrease in the maximum quantum yield of photosynthesis and an increase in τ. Both the induction and recovery from photoinhibition are time-dependent. The induction is non-linearly dependent on photon dose (Takahashi et al., 1971), which is the product of photon fluence rate and time. The recovery occurs over several hours and is correlated with the synthesis of D1 (Kyle et al., 1985), a chloroplast-encoded protein which can be blocked by chloramphenicol. It should be pointed out that D1 turnover occurs under low as well as high irradiance levels; supra-optimal irradiance merely increases the rate of degradation of the protein. While field measurements of changes in the dark-adapted quantum yields of fluorescence (Fig. 6) suggest that near-surface irradiance levels may cause a loss of functional PS II reaction centers in natural phytoplankton communities (a contention which supports the diel variations in DCMU-enhanced fluorescence as shown by Vincent et al., 1984), there is no evidence that reaction center proteins such as D1 are damaged in natural phytoplankton communities. One possible means of assessing the effect of irradiance and mixing on PS II reaction centers would be to follow the pool size and turnover of D1 with a combination of Western blots and pulse labeling autoradiography in natural phytoplankton under a range of irradiance levels.

During the day, *in vivo* fluorescence per chlorophyll is generally lower than at night. This so-called fluorescence quenching may appear to be due to a loss of PS II activity resulting from photoinhibition. However, a significant fraction may also be due to a phenomenon called non-photochemical quenching.

The change in the quantum yield of fluorescence due to non-photochemical quenching is relatively easy to measure (Kolber et al., 1990). In chlorophytes and other green algae, the quenching process is correlated with the light-dependent conversion of violaxanthin to zeaxanthin through the intermediate antheroxanthin. In chromophyte algae such as diatoms, diadinoxanthin is photochemically converted to diatoxanthin. In both such types of algae, the back reactions are enzyme mediated and, therefore, temperature sensitive (Table 2). However, the kinetics of the changes in non-photochemical quenching due to the xanthophyll cycle or other relatively reversible processes (half time of a few minutes) can be readily distinguished from repair of PS II reaction centers, which occurs on much longer time scales (half times of a few hours) (Fig. 6).

Much remains to be understood about the mechanisms and significance of non-photochemical quenching, especially in natural phytoplankton. If non-photochemical

Table 2.   First-order rate constants (h^{-1}) for changes in the xanthophylls diadinoxanthin (DD) and diatoxanthin (DT) after a shift from low light to high light (LL/HL) or from high light to low light (HL/LL); (-) indicates a decrease whereas (+) indicates an increase. Note that the photo-induced conversion of DD to DT is temperature insensitive, while the reverse reaction is temperature dependent (Mortain-Bertrand, unpublished).

		LL/HL	HL/LL
18°C	Diadinoxanthin	-15.4	+6.6
	Diatoxanthin	+17.2	-7.5
	DD/DT	-42.0	+9.5
	ΔΦ	-20.6	+7.5
6°C	Diadinoxanthin	-14.4	+0.6
	Diatoxanthin	+16.9	-3.1
	DD/DT	-40.8	+0.7
	ΔΦ	-16.8	+3.2

quenching were only due to energy dissipation within the antenna, the process should dissipate more absorbed photons as heat, presumably reducing the overexcitation of PS II (Kiefer and Reynolds, this volume). If the quenching process is located in the antenna serving PS II reaction centers, the process should be observed as a decrease in the effective absorption cross-section for PS II. There does appear to be some reduction in PS II cross-section due to non-photochemical quenching (Genty et al., 1990); however the reduction is relatively small, on the order of 10%.

It should be pointed out that the rate of nutrient supply can affect both the degree and recovery from photoinhibition, as well as non-photochemical quenching. Nutrient limitation promotes an increase in the effective absorption cross-section of PS II. Thus, nutrient-limited cells will tend to become photoinhibited at lower photon fluence rates than cells growing in nutrient-rich regimes. Moreover, as previously discussed, limitation by such elements as iron or nitrogen, tends to lead to a selective loss of reaction center proteins, and potentially reduces the ability of cells to repair photoinhibitory damage.

CONCLUSIONS

The simultaneous development of molecular biological techniques and biophysical tools over the past decade has allowed biological oceanographers to investigate the mechanisms responsible for the photophysiological responses of phytoplankton in a variety of oceanic regimes. Casting empirical equations relating photosynthesis to irradiance in the biophysical terms of quantum yields, absorption cross-sections, and turnover times allows molecular models to be constructed which relate to the mathematical model abstractions. For example, if, as a first approximation, the functional absorption cross-section of PS II is related to the ratio of LHC II to reaction center proteins, such as D1, then it becomes of interest to understand the factors regulating the synthesis and destruction of these protein elements. If the quantum yield is expressed as a ratio of two cross-sections (see Dubinsky, this volume), then the factors controlling the synthesis of specific proteins and prosthetic groups can be related to variability in quantum yields. The objective of these investigations is to understand the biological "rules" that phytoplankton follow in adjusting the photosynthetic apparatus to variations in the external world.

Over a decade ago, I began to investigate phytoplankton productivity by focussing on physiological acclimation to irradiance. My personal belief was that productivity was probably much more related to irradiance than any other factor, while the distribution of biomass was much more related to physics, as it controls the distribution of nutrients. However, the vertical distribution of the photochemical energy conversion ($\Delta\phi_{sat}$), or of σ_{PSII} in the oceans cannot be explained based on irradiance alone. At present, the only factor which we know can modify these parameters, and which is consistent with the oceanographic observations, is nutrient supply. It therefore seems that nutrients modify photosynthetic responses and that light-chlorophyll models must mathematically incorporate these modifications to accurately predict primary productivity in the sea.

ACKNOWLEDGEMENTS

I thank Dick Eppley for his quiet support and inspiration, and many colleagues, including John Bennett, Zvy Dubinsky, Richard Geider, Ronny Herzig, Dale Kiefer, Zbigniew Kolber, Julie LaRoche, David Mauzerall, Anne Mortain-Bertrand, Assaf Sukenik, Kevin Wyman, and Jon Zehr for discussions leading to mutual growth. This research was supported by the U.S. Department of Energy and NASA.

REFERENCES

Allen, J. F., Bennett, J., Steinback, K. E., and Arntzen, C. J., 1981, Chloroplast protein phosphorylation couples plastoquinone redox state to distribution of excitation energy between photosystems, *Nature,* 291:25.

Bannister, T. T. and Laws, E. A., 1980, Modeling phytoplankton carbon metabolism, *in:* "Primary Productivity in the Sea," P.G. Falkowski, ed., Plenum Press, New York.

Beale, S. I. and Appleman, D., 1971, Regulation by degree of light limitation of growth, *Plant Physiol.,* 59:230.

Bennett, J., 1991, Protein phosphorylation in green plant chloroplasts, *Annu. Rev. Plant Physiol. Plant Mol. Biol. 1991,* 42:281.

Berner, T., Dubinsky, Z., Wyman, K., and Falkowski, P.G., 1989, Photoadaptation and the "package effect" in *Dunaliella tertiolecta* (Chlorophyceae), *J. Phycol.,* 25:70.

Bidigare, R. R., this volume.

Bonaventura, C. and Meyers, J., 1969, Fluorescence and oxygen evolution from *Chlorella pyrenoidosa, Biochim. Biophys. Acta,* 189:366.

Butler, W. L., and Kitajima, M., 1975, Fluorescence quenching of in photosystem II of chloroplasts, *Biochim. Biophys. Acta,* 376:116.

Caperon, J., 1968, Population growth response of *Isochrysis galbana* to nitrate variation at limiting concentrations, *Ecol.,* 49:866.

Cleveland, J. S., Bidigare, R., and Perry, M. J., 1989, Maximum quantum yield of photosynthesis in the northwestern Sargasso Sea, *J. Mar. Res.,* in press.

Critchley, C. and Smillie, R. M., 1981, Leaf chlorophyll fluorescence as an indicator of photoinhibition in *Cucumis sativus* L., *Aust. J. Plant Physiol.,* 8:133.

Cullen, J. J., 1990, On models of growth and photosynthesis in phytoplankton, *Deep-Sea Res.,* 37:667.

Cullen, J. J., Yang, X., and MacIntyre, H. L., this volume.

Demmig-Adams, B., 1990, Carotenoids and photoprotection in plants: A role for xanthophyll zeaxanthin, *Biochim. Biophys. Acta,* 1020:1.

Dubinsky, Z., this volume.

Dubinsky, Z., Falkowski, P. G., and Wyman, K., 1986, Light-harvesting and utilization by phytoplankton, *Plant Cell Physiol.,* 27:1335.

Dugdale, R. C., 1967, Nutrient limitation in the sea: Dynamics, identification, and significance, *Limnol. Oceanogr.,* 12(4):685.

Eppley, R. W., Stewart, E., Abbot, R. M., and Heyman, V., 1985, Estimating ocean primary production from satellite chlorophyll, introduction to regional differences and statistics for the Southern California Bight, *J. Plankton Res.,* 7:57.

Falkowski, P. G., 1981, Light-shade adaptation and assimilation numbers, *J. Plankton Res.,* 3:203.

Falkowski, P. G., Owens, T. G., Ley, A. C., and Mauzerall, D. C., 1981, Effects of growth irradiance levels on the ratio of reaction centers in two species of marine phytoplankton, *Plant Physiol.,* 68:969.

Falkowski, P. G., 1984, Physiological responses of phytoplankton to natural light regimes, *Limnol. Oceanogr.,* 39:311.

Falkowski, P. G., Dubinsky, Z., and Wyman, K., 1985, Growth-irradiance relationships in phytoplankton, *Limnol. Oceanogr.,* 39:311.

Falkowski, P. G., Wyman, K., Ley, A. C., and Mauzerall, D. C., 1986, Relationship of steady state photosynthesis to fluorescence in eucaryotic algae, *Biochim. Biophys. Acta,* 849:183.

Falkowski, P. G., Kolber, Z., and Fujita, Y., 1988, Effect of redox state on the dynamics of Photosystem II during steady-state photosynthesis in eucaryotic algae, *Biochim. Biophys. Acta,* 933:432.

Falkowski, P. G., Sukenik, A., and Herzig, R., 1989, Nitrogen limitation in *Isochrysis galbana* (Haptophyceae), II. Relative abundance of chloroplast proteins, *J. Phycol.,* 25:471.

Falkowski, P. G. and Kolber, Z., 1990, Phytoplankton photosynthesis in the Atlantic Ocean as measured from a submersible pump and probe fluorometer *in situ, in:* "Current Research in Photosynthesis IV," M. Baltscheffsky, ed., Kluwer, London.

Falkowski, P. G. and LaRoche, J., 1991, Acclimation to spectral irradiance in algae, *J. Phycol.,* 27:8.

Falkowski, P. G., Ziemann, D., Kolber, Z., and Bienfang, P. K., 1991, Role of eddy pumping in enhancing primary production in the ocean, *Nature,* 352:55.

Farquhar, G., Von Caemmerer, S., and Berry, J., 1980, A biochemical model of photosynthetic CO_2 assimilation in leaves of C_3 species, *Planta,* 149:78.

Fischer, T., Shurtz-Swirski, R., Gepstein, S., and Duzinsky, Z., 1989, Changes in the levels of ribulose-1,5-bisphosphate carboxylase/oxygenase (Rubisco) in *Tetraedon minimum* (Chlorophyta) during light and shade adaptation, *Plant Cell Physiol.,* 30:221.

Geider, R. J., Platt, T., and Raven, J. A., 1986, Size dependence of growth and photosynthesis in diatoms: A synthesis, *Mar. Ecol. Prog. Ser.,* 30:93.

Genty, B., Harbinson, J., Briantais, J. -M., and Baker, N. R., 1990, The relationship between non-photochemical quenching of chlorophyll fluorescence and the rate of photosystem 2 photochemistry in leaves, *Photosynthesis Res.,* 25:249.

Gerber, D. W. and Burris, J. E., 1981, Photoinhibition and P700 in the marine diatom *Amphora* sp., *Plant Physiol.,* 68:699.

Gibbs, P. B. and Biggins, J., 1991, *In vivo* and *in vitro* protein phosphorylation studies on *Ochromonas danica*, an alga with a chlorophyll *a/c* fucoxanthin binding protein, *Plant Physiol.*, 97:388.

Goldman, J. C., 1980, Physiological processes, nutrient availability, and the concept of relative growth rate in marine phytoplankton ecology, *in* "Primary Productivity in the Sea," P.G. Falkowski, ed., Plenum Press, New York.

Greenbaum, E., 1977, The photosynthetic unit of hydrogen evolution, *Science*, 196:879.

Greenbaum, N. L., Ley, A. C., and Mauzerall, D. C., 1987, Use of a light-induced respiratory transient to measure the optical cross-section of photosystem I in *Chlorella*, *Plant. Physiol.*, 84:879.

Greenbaum, N. L. and Mauzerall, D., 1991, Effects of irradiance level on distribution of chlorophylls between PSII and PSI as determined from optical cross-sections, *Biochim. Biophys. Acta*, 1057:195.

Greene, R. M., Geider, R. J., and Falkowski, P. G., 1991, Effect of iron limitation on photosynthesis in a marine diatom, *Limnol. Oceanogr.* (in press).

Herzig, R. and Falkowski, P. G., 1989, Nitrogen limitation in *Isochrysis galbana* (Haptophyceae), I. Photosynthetic energy conversion and growth efficiencies, *J. Phycol.*, 25:462.

Jassby, A. D. and Platt, T., 1976, Mathematical formulation of the relationship between photosynthesis and light for phytoplankton, *Limnol. Oceanogr.*, 21:540.

Kiefer, D. A. and Mitchell, B. G., 1983, A simple, steady-state description of phytoplankton growth based on absorption cross-section and quantum efficiency, *Limnol. Oceanogr.*, 28:770.

Kiefer, D. A. and Reynolds, R., this volume.

Kiefer, D. A., Chamberlain, W. S., and Booth, C. R., 1989, Natural fluorescence of chlorophyll a: Relationship to photosynthesis and chlorophyll concentration in the western South Pacific gyre, *Limnol. Oceanogr.*, 34:868.

Kirk, J. T. O., 1983, "Light and Photosynthesis in Aquatic Ecosystems," Cambridge University Press, New York.

Kirk, J. T. O., this volume.

Kolber, Z., Zehr, J., and Falkowski, P. G., 1988, Effects of growth irradiance and nitrogen limitation on photosynthetic energy conversion in Photosystem II, *Plant Physiol.*, 88:923.

Kolber, Z., Wyman, K. D., and Falkowski, P. G., 1990, Natural variability in photosynthetic energy conversion efficiency: A field study in the Gulf of Maine, *Limnol. Oceanogr.*, 35:72.

Kuhlbrandt, W. and Wang, D. N., 1991, Three-dimensional structure of plant light-harvesting complex determined by electron crystallography, *Nature*, 351:130.

Kyle, D. J., Ohad, I., and Arntzen, C. J., 1985, Membrane protein damage and repair: Selective loss of quinone protein function in chloroplast membranes, *Proc. Nat. Acad. Sci. USA*, 81:4070.

LaRoche, J., Mortain-Bertrand, A., and Falkowski, P. G., 1991, Light intensity-induced changes in *cab* mRNA and light-harvesting complex II apoprotein levels in the unicellular chlorophyte *Dunaliella tertiolecta*, *Plant Physiol.*, 97:147.

Laws, E. A., DiTullio, G. R., and Redalje, D. G., 1987, High phytoplankton growth and production rates in the North Pacific subtropical gyre, *Limnol. Oceanogr.*, 32:905.

Lewis, M., this volume.

Ley, A. C., 1980, The distribution of absorbed light energy for algal photosynthesis, *in:*, "Primary Productivity in the Sea," P.G. Falkowski, ed., Plenum, Press, New York.

Ley, A. C. and Mauzerall, D., 1982, Absolute absorption cross-sections for photosystem II and the minimum quantum requirements for photosynthesis in *Chlorella vulgaris*, *Biochim. Biophys. Acta,* 680:95.

Mauzerall, D. and Greenbaum, N. L., 1989, The absolute size of a photosynthetic unit, *Biochim. Biophys. Acta,* 974:119.

McAllister, C. D., 1961, Observations on the variation of planktonic photosynthesis with light intensity, using both the O_2 and ^{14}C-methods, *Limnol. Oceanogr.,* 6:483.

Michel, H. -P. and Deisenhofer, J., 1988, Relevance of the photosynthetic reaction center from purple bacteria to the structure of photosystem II, *Biochem.,* 27:1.

Michel, H. -P., Tellenbach, M., and Boschetti, A., 1983, A chlorophyll *b*-less mutant of *Chlamydomonas reinhardtii* lacking in the light-harvesting chlorophyll *a/b*-protein complex but not in its apoproteins, *Biochim. Biophys. Acta,* 725:417.

Morel, A., 1991, Light and marine photosynthesis: A spectral model with geochemical and climatological implications, *Prog. Oceanog.,* 26:263.

Mortain-Bertrand, A., Bennett, J., and Falkowski, P. G., 1990, Photoregulation of the light-harvesting chlorophyll protein complex associated with Photosystem II in *Dunaliella tertiolecta, Plant Physiol.,* 94:304.

Myers, J. and Burr, G. O., 1940, Studies on photosynthesis: Some effects of light of high intensity on *Chlorella, J. Gen. Physiol.,* 24:45.

Myers, J., 1980, On the algae: Thoughts about physiology and measurements of efficiency, *in:* "Primary Productivity in the Sea," P.G. Falkowski, ed., Plenum Press, New York.

Myers, J. and Graham, J. -R, 1971, The photosynthetic unit in *Chlorella* measured by repetitive short flashes, *Plant Physiol.,* 48:282.

Neale, P. J., 1987, Algal photoinhibition and photosynthesis in the aquatic environment, *in* "Photoinhibition," D. Kyle, C.J. Arntzen, and B. Osmond, eds., pp. 39-65, Elsevier, Amsterdam.

Ohad, I., Adir, N., Koike, H., Kyles, D. J., and Inoue, Y., 1990, Mechanism of photoinhibition *in vivo, J. Biol. Chem.,* 265:1972.

Osmond, C. B., 1981, Photorespiration and photoinhibition: Implications for the energetics of photosynthesis, *Biophys. Biochim. Acta,* 639:77.

Osborne, B. A., and Geider, R. J., 1986, Effect of nitrate-nitrogen limitation on photosynthesis of the diatom *Phaeodactylum tricornutum* Bohlin (Bacillariophyceae), *Plant Cell Environ.,* 9:617.

Owens, T. G., 1986, Light-harvesting function in the diatom *Phaeodactylum tricornutum,* II. Distribution of excitation energy between the photosystems, *Plant Physiol.,* 80:739.

Platt, T., 1986, Primary production of the ocean water column as a function of surface light intensity: algorithms for remote sensing, *Deep-Sea Res.,* 33:149.

Platt, T. and Sathyendranath S., 1988, Oceanic primary production: estimation by remote sensing at local and regional scales, *Science,* 241:1613.

Portis, A. R., Salvucci, M. E., and Ogren, W. L., 1986, Activation of ribulose bisphosphate carboxylase/oxygenase at physiological CO_2 and ribulose bisphosphate concentrations by rubisco activase, *Plant Physiol.,* 82:967.

Ryther, J. H. and Yentsch, C. S., 1957, The estimation of phytoplankton production in the ocean from chlorophyll and light data, *Limnol. Oceanogr.,* 2:281.

Sakshaug, E., Kiefer, D. A., and Andersen, K., 1989, A steady state description of growth and light absorption in the marine planktonic diatom *Skeletonema costatum, Limnol. Oceanogr.,* 34:198.

Sakshaug, E., Johnson, G., Andresen, K., and Vernet, M., 1991, Modeling of light-dependent algal photosynthesis and growth: Experiments with the Barents Sea diatoms *Thalassiosira nordenskioeldii* and *Chaeotoceros furcellatus, Deep-Sea Res.,* 38:415.

Schatz, G. H., Brock, H., and Holzwarth, A. R., 1988, Kinetic and energetic model for the primary processes in photosystem II, *Biophys. J.,* 54:397.

Sukenik, A., Bennett, J., and Falkowski, P. G., 1987, Light-saturated photosynthesis: limitation by electron transport or carbon fixation?, *Biochim. Biophys. Acta,* 891:205.

Sukenik, A., Bennett, J., and Falkowski, P. G., 1988, Changes in the abundance of individual apoproteins of light-harvesting chlorophyll *a/b*-protein complexes of Photosystem I and II with growth irradiance in the marine chlorophyte *Dunaliella tertiolecta, Biochim. Biophys. Acta,* 932:206.

Takahashi, M., Shimura, S., Yamaguchi, Y., and Fujita, Y., 1971, Photo-inhibition of phytoplankton photosynthesis as a function of exposure time, *J. Oceanogr. Soc. Japan,* 27:43.

Telfer, A., Bottin, H., Barber, J., and Mathis, P., 1984, The effect of magnesium and phosphorylation of the light-harvesting chlorophyll *a/b* protein on the yield of P700 photooxidation in pea chloroplasts, *Biochim. Biophys. Acta,* 764:324.

Thomas, W. H., 1970, On nitrogen deficiency in tropical Pacific Ocean phytoplankton: Photosynthetic parameters in poor and rich water, *Limnol. Oceanogr.,* 15:380.

Turpin, D. H., 1991, Effects of inorganic N availability on algal photosynthesis and carbon metabolism, *J. Phycol.,* 27:14.

Turpin, D. H. and Bruce, D., 1990, Regulation of photosynthetic light-harvesting by nitrogen assimilation in the green alga *Selenastrum minutum, FEBS Lett.,* 263:99.

Vincent, W. F., Neale, P. J., and Richerson, P. J., 1984, Photoinhibition: Alga responses to bright light during diel stratification and mixing in a tropical alpine lake, *J. Phycol.,* 20:201.

Weaver, E. C. and Weaver, H. E., 1969, Paramagnetic unit in spinach subchloroplast particles; estimation of size, *Science (Wash., D.C.),* 165:906.

Welschmeyer, M. A. and Lorenzen, C. J., 1981, Chlorophyll-specific photosynthesis and quantum efficiency at subsaturating light intensities, *J. Phycol.,* 17:283.

Zipfel, W. and Owens, T. G., 1991, Calculation of absolute Photosystem I absorption cross-sections from P_{700} photooxidation kinetics, *Photosyn. Res.,* 29:23.

NUTRIENT LIMITATION OF MARINE PHOTOSYNTHESIS

John J. Cullen*, Xiaolong Yang, and Hugh L. MacIntyre

Bigelow Laboratory for Ocean Sciences
McKown Point
West Boothbay Harbor, ME 04575

INTRODUCTION

Guided by insightful presentations at the previous Brookhaven Symposium (Bannister and Laws, 1980; Eppley, 1980; Goldman, 1980) we address here three questions that have challenged oceanographers for decades: 1) Can photosynthetic performance be used to diagnose the nutritional status of phytoplankton? 2) Should nutrients be incorporated into models of oceanic photosynthesis as a function of chlorophyll and light? and 3) How might we assess nutrient limitation of the specific growth rates or standing crop of phytoplankton in the ocean? We find that ambiguities thwart attempts to formulate robust generalizations. Accordingly, when it comes to nutrient limitation of marine photosynthesis, a good paradigm is hard to find.

NUTRIENTS AND PHOTOSYNTHETIC PERFORMANCE

The growth and photosynthesis of plants depends on nutrients. It is reasonable to expect, then, that some aspects of photosynthetic performance will reflect the nutritional status of phytoplankton. Here we examine the rate of photosynthesis normalized to chlorophyll a (P^B), asking if P^B tells us anything about the nutritional status or specific growth rates of phytoplankton.

Regional Differences in Normalized Photosynthesis

Measurements from different regions of the ocean suggest that P^B may indeed reflect the supply of nutrients to the euphotic zone (Barber and Chavez, in press). Consider a comparison of observations from the central equatorial Pacific and the Gulf of Mexico (Fig. 1). For these two environments, temperature and light conditions were similar, yet normalized photosynthesis (g C g Chl^{-1} d^{-1}) was much higher over the Texas shelf, where nitrate concentrations were low, but where trace elements, such as iron, were much more abundant than in the equatorial Pacific (cf. Martin et al, 1989). If the

---

*Address for correspondence: Department of Oceanography, Dalhousie University, Halifax, Nova Scotia B3H 4J1, Canada

Primary Productivity and Biogeochemical Cycles in the Sea
Edited by P.G. Falkowski and A.D. Woodhead, Plenum Press, New York, 1992

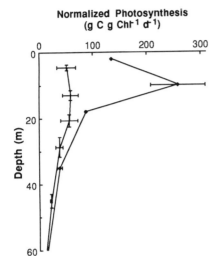

Normalized Photosynthesis
(g C g Chl⁻¹ d⁻¹)

Depth (m)

TEXAS SHELF
(water depth 70 m)

Surface $[NO_3] < 0.3 \ \mu M$
Surface Chl = 0.08 mg m⁻³
Surface Temperature = 28°C
1% surface irradiance at 70 m

EQUATORIAL PACIFIC
(150° W)

Surface $[NO_3] > 3 \ \mu M$
Surface Chl > 0.2 mg m⁻³
Surface Temperature = 26°C
1% surface irradiance at 70 m

Fig. 1.   Vertical profiles of normalized photosynthesis (g C g Chl⁻¹ d⁻¹) from the Texas Shelf (higher values; 4 h simulated *in situ* incubations, hourly rates multiplied by 10 ± range of duplicates; Cullen et al., unpubl.) and from the equatorial Pacific (lower values; 24 h simulated *in situ* incubations ± s.d. for 6 days; data of R. Barber and F. Chavez presented in Cullen et al., in press).

higher P^B off Texas corresponded to higher specific growth rates of phytoplankton, one might infer that the specific growth rates of the equatorial phytoplankton were limited by nutrient supply, possibly iron. However, specific growth rate, μ, (d⁻¹) is determined by both P^B (expressed as net photosynthesis over 24 h) and the carbon/chlorophyll ratio of phytoplankton (C:Chl):

$$\mu = \frac{P^B}{C:Chl} \tag{1},$$

so that the relationship between P^B and μ can be determined only if C:Chl is specified. Unfortunately, C:Chl, which is quite variable under physiological control, cannot be measured directly and accurately on natural samples (Eppley, 1972, 1980; but see Redalje and Laws, 1981), so geographical patterns in P^B cannot be related directly to μ unless more information is provided.

Laboratory Models of Nutrient Limitation

Pertinent information on the relationship between P^B and μ comes from models of photosynthesis, nutrition, and growth of phytoplankton. These models, which will not be reviewed here, are based on results from laboratory experiments. The experiments are usually performed either by allowing a culture to deplete a limiting nutrient (nutrient-starved batch culture) or by growing phytoplankton in nutrient-limited continuous culture. The two experimental regimes are fundamentally different, and the differences should be appreciated.

Batch culture. When a batch culture of microalgae runs out of a limiting nutrient, growth is unbalanced (Eppley, 1981) because photosynthesis proceeds, but synthesis of critical cellular constituents is restricted by lack of the limiting nutrient. Biochemical composition (Strickland et al., 1969; Sakshaug and Holm-Hansen, 1977; Flynn, 1990) and physiological capabilities (Horrigan and McCarthy, 1981; Cleveland and Perry, 1987) change, as does the rate of cell division. Both the assimilation number (P^B at optimal irradiance; Glover, 1980) and photosynthetic efficiency in subsaturating light (Welschmeyer and Lorenzen, 1981; Cleveland and Perry, 1987) decline in what appears to be an unavoidable consequence of nutrient starvation.

Continuous culture. In nutrient-limited continuous culture, physiological regulation culminates in balanced growth of phytoplankton: cellular concentrations of the compounds that require the limiting nutrient decline and storage products (such as carbohydrate or lipid) are accumulated until the amounts of all cellular constituents increase exponentially at the same nutrient-limited rate, averaged over the photocycle (Shuter, 1979; Eppley, 1981). That is, cellular carbon increases until "the growth rate of algal carbon is reduced to match the growth rate allowed by the nutrient supply" (Bannister and Laws, 1980). In response to nitrogen limitation, for example, cellular chlorophyll concentration is regulated so that changes in C:Chl are principally responsible for the variation in growth rate (e.g., Laws and Bannister, 1980). As a result, P^B at growth irradiance can be independent of nutrient-limited growth rate (Bannister and Laws, 1980; Sakshaug et al., 1989; Cullen, 1990).

Thus, the results for nutrient starvation in batch culture are not interchangeable with the results for continuous cultures. When relating experimental results to the ocean, it is useful to consider which laboratory model is likely to apply in a given oceanographic situation (Yentsch et al., 1977).

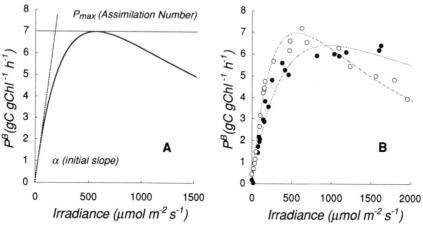

Fig. 2. The relationship between photosynthesis and irradiance. (A) A P versus I curve following the model of Platt et al. (1980), with initial slope α, and maximal rate P_{max}. Susceptibility to photoinhibition is designated by the parameter β. (B) P versus I curves for the diatom *Thalassiosira pseudonana* (clone 3H) grown at 200 μmol photons m^{-2} s^{-1} in nutrient-replete media (filled circles are from semi-continuous cultures, 12 h: 12 h light:dark cycle; $\mu = 0.85$ d^{-1}) and in nitrate-limited continuous culture at 0.3 d^{-1} (open circles; Yang et al., see Table 1). In this particular comparison, P_{max} is similar, but the nitrogen-limited culture has a higher α and greater susceptibility to photoinhibition. The generality of this comparison is examined below and found to be lacking.

When phytoplankton are grown at steady state in nutrient-limited continuous culture, the growth rate equals the dilution rate, which is under precise experimental control. The influence of nutrient limitation on photosynthesis can then be examined by measuring C:Chl and calculating P^B at growth irradiance by rearranging eq. 1 (e.g., Laws and Bannister, 1980). Also, photosynthesis versus irradiance (P versus I) can be measured directly on subsamples. More sophisticated biochemical or biophysical measurements also can be made (Falkowski, this volume), but the P versus I relationship is examined here because it is relatively easy to measure, and it contains information on photosynthetic efficiency, photosynthetic capacity, and the susceptibility of the alga to photoinhibition (Fig. 2A). Each of these facets of photosynthetic performance might respond independently to nutrient limitation (Fig. 2B). Environmental influences on P versus I of natural phytoplankton have been examined (e.g., Platt and Jassby, 1976; Harrison and Platt, 1980, 1986; Falkowski, 1981; Malone and Neale, 1981), but as the following discussion will show, the effects of nutrient-limitation on P versus I merit review.

Assimilation number. The assimilation number is P^B at light saturation, equivalent to P_{max} in the P versus I relationship. Assimilation number is perhaps the easiest photosynthetic parameter to use to compare measurements between the laboratory and the field, because it is relatively insensitive to differences in light quality, its calculation does not rely on accurate measurement of irradiance, and it can reasonably be compared to the maximum P^B measured in vertical profiles. The variability of assimilation number has been studied for years (e.g., Curl and Small, 1965; Eppley, 1972; Yentsch et al., 1974; Platt and Jassby, 1976; Glover, 1980; Harrison and Platt, 1980; Falkowski, 1981). Here, we examine one aspect, the relationship between nitrogen-limited growth rate and assimilation number in continuous culture.

Results spanning nearly 20 years are compiled in Fig. 3. The basic experimental design is straightforward, although details differ (Table 1). Simply, cultures were grown at a series of nitrogen-limited growth rates, and photosynthesis was determined experimentally, either by measuring O_2 evolution or the uptake of ^{14}C-bicarbonate. For comparison, we converted published results to plots of P_{max} (g C g Chl^{-1} h^{-1}) versus relative growth rate (μ/μ_{max}; Goldman, 1980). Also, we included our own results from studies of nitrate-limited growth of the neritic diatom, Thalassiosira pseudonana (clone 3H). Our cultures were grown on light:dark cycles, and the diel variation of P versus I was examined.

A pattern can be seen in each experimental series, but patterns differ between experiments. Several studies show that P_{max} is strongly depressed at low growth rates (Thomas and Dodson, 1972; Glover, 1980; Osborne and Geider, 1986; Kolber et al., 1988; Chalup and Laws, 1990), whereas others indicate that P_{max} is largely independent of nitrogen-limited growth rate (Eppley and Renger, 1974; Herzig and Falkowski, 1989; Yang et al., Fig. 3H). Experimental conditions and cultured species differed (Table 1), but we find no factor or set of factors that is clearly associated with either result. For example, one might look for a contrast between neritic and oceanic species (Sakshaug et al., 1987). However, both patterns have been observed in oceanic diatoms (Thomas and Dodson, 1972; and Eppley and Renger, 1974), and different patterns have been observed in the same clone of a neritic diatom (Kolber et al., 1988; and Yang et al., unpublished, Fig. 3H). Perhaps the experimental photocycle is important: assimilation number was little affected by nitrogen-limitation in two studies that employed light:dark cycles (Eppley and Renger, 1974; Yang et al., Fig. 3H), but it was strongly a function of growth rate in most studies using continuous light. However, in the thorough study made by Herzig and

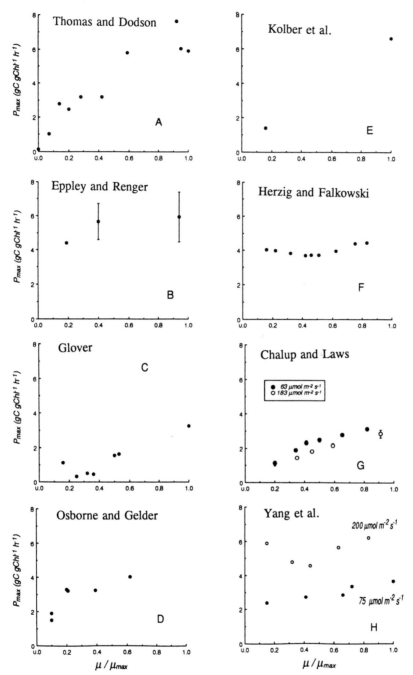

Fig. 3. Relationships between P_{max} and nitrogen-limited growth rate (converted to relative growth rate by normalizing to the nutrient saturated rate at that temperature and irradiance, Goldman, 1980): nitrogen-limited continuous cultures. Oxygen-based measurements of gross photosynthesis were converted using photosynthetic quotients in the original publications (but 0.8 mol C mol O_2^{-1} was assumed for Kolber et al., 1988). References and experimental details are presented in Table 1.

Table 1. Description of experiments described in Fig. 3. Irradiance measurements have been converted for comparison.

Reference	Species	Nutrient	Illumination	Other	Steady-state	Method	Result
Thomas and Dodson (1972)	Chaetoceros gracilis (oceanic)	10 µM NH$_4$ modified F	Continuous cool-white 160 µmol m^{-2} s^{-1}	Aerated and stirred 25°C	Successive cell counts	3 h ^{14}C uptake maximum about 320 µmol m^{-2} s^{-1}	P$_{max}$ was a strong function of nitrate-limited growth rate
Eppley and Renger (1974)	Thalassiosira oceanica (13-1) (oceanic)	25 µM NH$_4$ + 25 µM NO$_3$ IMR/2	12h:12h quartz-iodide 225 µmol m^{-2} s^{-1}	Aerated 20°C	Run for many days prior to measurements, some variation noted	1 h ^{14}C-uptake	P$_{max}$ was not clearly a function of nitrate-limited growth rate
Glover (1980)	Phaeodactylum tricornutum (Cambridge)	20 µM NO$_3$ F-1	Continuous warm-white ~240 µmol m^{-2} s^{-1}	Stirred 20°C	48 h constant cell density	^{14}C uptake over 3 min after resuspension in NO$_3$ free medium	P$_{max}$ was a strong function of nitrate-limited growth rate, with one higher point at the lowest growth rate
Osborne and Geider (1986)	Phaeodactylum tricornutum (SMBA strain 14)	100 µM NO$_3$ modified Aquil	Continuous cool-white 160 µmol m^{-2} s^{-1}	Aerated 23-25°C	Cell count ± 10% for 3 days	O$_2$ electrode	P$_{max}$ was a strong function of nitrate-limited growth rate, especially at low growth rates
Kolber et al. (1988)	Thalassiosira pseudonana (3H) (neritic)	75 µM NH$_4$ F/2	Continuous cool white 150 µmol m^{-2} s^{-1}	Aerated 18°C	One week	O$_2$ electrode	P$_{max}$ was much lower in the nitrate-limited culture
Herzig and Falkowski (1989)	Isochrysis galbana (SERI ISOCHI)	150 µM NO$_3$ F/2	Continuous cool-white 175 µmol m^{-2} s^{-1}	Aerated 18°C	At least a week, measurements made on three successive days	O$_2$ electrode	P$_{max}$ remained high at low nitrate-limited growth rates
Chalup and Laws (1990)	Pavlova lutheri (SIO MONO L)	25 µM NO$_3$ IMR medium	Continuous cool-white, 63 and 189 µmol m^{-2} s^{-1}	Aerated and stirred 22°C	Cell count ±5% 2 or more days, measurements repeated 3 days	30 min ^{14}C-uptake	P$_{max}$ was a strong function of nitrate-limited growth rate
Yang, MacIntyre and Cullen (unpubl.)	Thalassiosira pseudonana (3H) (neritic)	50 µM NO$_3$ modified F/2	12h:12h vita-light, 75 and 200 µmol m^{-2} s^{-1}	Aerated and stirred 20°C	Several parameters steady for several days - measurements repeated twice	20 min ^{14}C-uptake, 5 times during photocycle	P$_{max}$ remained high at low nitrate-limited growth rates. Possibly some nutrient effect for lower-irradiance series

Falkowski (1989), assimilation number was largely independent of nutrient-limited growth rate in continuous light. The largest discrepancies between experiments are at low dilution rates, when it is difficult to establish steady-state conditions.

We feel that no set of results can be considered "right" or "wrong"; rather, they differ for reasons that have not been identified or explicitly studied. It is curious and unsettling to discover that the laboratory experiments central to our understanding of nitrogen limitation and photosynthetic physiology can have such disparate results, for no obvious reason. A systematic examination of this problem is warranted.

Because P^B can remain high even when growth rates are severely limited by nitrogen supply, we conclude that assimilation number is an unreliable diagnostic of nitrogen limitation (Laws and Bannister, 1980; Herzig and Falkowski, 1989). That is, geographical uniformity of assimilation number cannot be assumed to reflect uniform nutritional status. However, when temperature and light are held constant in laboratory

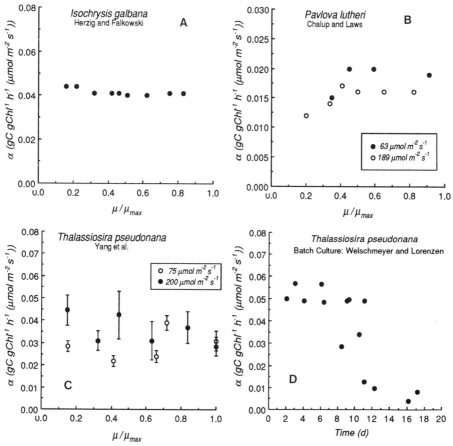

Fig. 4. Effects of nutrition on the initial slope of the P versus I curve, α. Results from continuous cultures described in Table 1: (A) Herzig and Falkowski (1989); (B) Chalup and Laws (1990); and (C) Yang et al. (unpubl.). The error bars represent diel variation of α: they are averages of the standard deviations of replicate series of 5 measurements during the light period. (D) An example from a batch culture (Welschmeyer and Lorenzen, 1981), showing the decline of α upon depletion of nitrate.

studies of cultures, low assimilation numbers implicate nutritional deficiency. Thus, regional differences of assimilation number at similar temperatures and irradiance (Fig. 1) are consistent with nutritional differences, but rigorous demonstration of causality would require more information.

Initial slope, α. The initial slope of the P versus I curve, [α: commonly used units, g C (g Chl)$^{-1}$ h^{-1} (μmol photons m^{-2} s^{-1})$^{-1}$)], is the product of the specific absorption coefficient for chlorophyll a, a_p (m^2 mg Chl^{-1}), and the quantum yield for photosynthesis ϕ_{max} [mol C (mol photons)$^{-1}$] (Falkowski, 1980; Dubinsky, this volume). It is a measure of photosynthetic efficiency, but it is not equivalent to quantum yield. The initial slope is more difficult to measure accurately than is maximum photosynthesis, and its magnitude is sensitive to the spectral quality of the light source and the absorption characteristics of the phytoplankton. Nonetheless, variability in α has been studied systematically in the field and in the laboratory.

Measurements on continuous cultures indicate that α is not very sensitive to nitrogen limitation (Fig. 4A, B, C). This result contrasts with observations on batch cultures subjected to nitrogen starvation (Welschmeyer and Lorenzen, 1981; Fig. 4D), illustrating the fundamental difference between nitrogen limitation in continuous cultures and nitrogen starvation in batch cultures. The relative constancy of α over a broad range of nitrogen-limited growth rates does not mean that photosynthetic efficiency stays high during nutrient limitation. In fact, it has been observed that the quantum yield for photosynthesis is depressed at low nitrogen-limited growth rates; however, this effect on α is more-or-less compensated by a concomitant increase in the specific absorption coefficient as a consequence of reduced chlorophyll per cell (Herzig and Falkowski, 1989; Chalup and Laws, 1990; Dubinsky, this volume). Thus, α is a poor diagnostic of nitrogen limitation, but published results indicate that quantum yield (hence, photosynthetic energy conversion efficiency; see Kolber et al., 1988) is low when nitrogen limits growth rate, whether in balanced or unbalanced growth. The efficiencies of photosynthetic energy

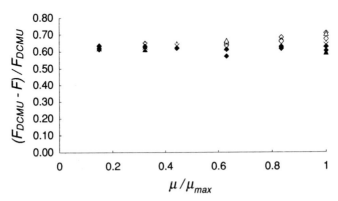

Fig. 5.   *Thalassiosira pseudonana* (3H): variation of DCMU-enhanced fluorescence, (F$_{DCMU}$-F)/F$_{DCMU}$, with nitrate-limited relative growth rate at 200 μmol photons m^{-2} s^{-1} (Yang et al., Table 1). Symbols represent the 5 different sampling periods during the light period with filled symbols indicating lights on and lights off.

conversion of natural phytoplankton, measured with a pump-and-probe fluorometer, were strongly related to the supplies of dissolved inorganic nitrogen in the Gulf of Maine (Kolber et al., 1990), leading to the suggestions that rates of photosynthesis were nitrogen-limited and that non-invasive measurements could be used to assess relative growth rates of phytoplankton *in situ*.

An apparently anomalous set of observations complicates our understanding of the relationship between nitrogen limitation and photosynthetic energy conversion. Dark-adapted fluorescence enhancement with 3-(3,4-dichlorophenyl)-1,1-dimethylurea (DCMU) is a crude measure of the photochemical conversion efficiency of photosystem II (Vincent et al., 1984). Relative enhancement, expressed as $(F_{DCMU} - F)/F_{DCMU}$, reliably reflects the degradation of photosynthetic quantum yield during nitrogen starvation (Welschmeyer and Lorenzen, 1981; Cleveland and Perry, 1987). In contrast, the magnitude of this parameter was unaffected by nitrogen limitation in continuous cultures of *Thalassiosira pseudonana* (Fig. 5). It would be very interesting to make the simple measurements of DCMU-enhanced fluorescence concurrent with pump-and-probe fluorometry to find out if culture conditions, species, or the type of measurement are responsible for the discrepant results.

Interpreting Spatial Patterns of Normalized Photosynthesis

The data reviewed here show that the photosynthetic characteristics of phytoplankton, normalized to chlorophyll, can be insensitive to large variations in nutrient-limited specific growth rates of phytoplankton. Yet, in nature, assimilation number can vary substantially in patterns that seem strongly related to nutrient supply (e.g., Curl and Small, 1965; Barber and Chavez, 1991; Fig. 1). How should these patterns be interpreted?

We recommend a cautious approach. It is instructive to compare patterns in assimilation number to inferred patterns in nutrient supply. For example, if there is a strong correlation between inferred nutrient supply and assimilation number, nutrient limitation of primary productivity is suggested, even though nutrient-limited specific growth rates cannot be specified. However, normalized photosynthesis can be insensitive to nutrition, and, as we show in the following section, little or no pattern might be observed over a strong gradient of nutrient supply, so that measurements of photosynthesis do little to resolve questions about the specific growth rates of phytoplankton.

Our review has a practical message: there is no experimental justification for converting measured values of P^B directly into estimated growth rates by assuming a constant C:Chl over an environmental gradient. Simply, P^B is not a reliable proxy for relative growth rate.

PHOTOSYNTHESIS - LIGHT MODELS

Having established that in some circumstances P^B can be nearly independent of nutrient-limited growth rate, we ask if nutrients need be incorporated into models of photosynthesis as a function of chlorophyll and light. The application considered here is remote sensing to estimate primary productivity. Bio-optical models and growth models will not be discussed. The approach is empirical; we will examine P versus I relationships in natural phytoplankton of the northwest Atlantic Ocean to see if inferred differences in nutrient supply influence P^B. If the effects of nutrient supply on P^B are small, the utility of geographical representations of P versus I (Platt et al., 1988) is enhanced.

Photosynthesis versus Irradiance near the Gulf Stream

We measured P versus I on samples from the vicinity of the Gulf Stream, including "green water" (surface chlorophyll $\geq$ 0.3 mg m^{-3}) and "blue water" (surface chlorophyll <0.3 mg m^{-3}). The arbitrary distinction reflects differences in nutrient supply to the euphotic zone associated with vertical mixing and sloping isopycnals across the Gulf Stream (Yentsch, 1974). If nutrient supply strongly influenced P versus I (as it does photosynthetic energy conversion efficiency; Kolber et al., 1990), then P versus I on samples of "green water" would differ from that in samples of "blue water".

Temperature can have a strong influence on seasonal (Harrison and Platt, 1980) and latitudinal (Harrison and Platt, 1986) comparisons of P versus I. However, during this cruise, most of the samples came from water in a narrow temperature range (22°C-26°C), so temperature was not an important factor.

A composite presentation (Fig. 6A) may look like a jumble, but much of the variability in P versus I is easily explained by invoking light regime as the principal influence. The curves with low maximum rates and pronounced susceptibility to photoinhibition are from deep samples; shallow samples have higher P_{max} and less susceptibility to photoinhibition (cf. Harrison and Platt, 1986). The pattern at a station with near-surface stratification (Fig. 6B) is consistent with photoacclimation, essentially the same as previously described (e.g., Falkowski, 1980; Richardson et al., 1983; Harding

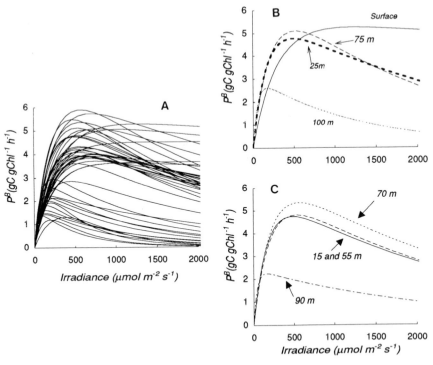

Fig. 6.   Photosynthesis versus irradiance in and near the Gulf Stream. Curve-fits of 24-point P versus I curves (Lewis and Smith, 1983) to the model of Platt et al. (1980). A) A composite of all results from "blue water" during the Biosynop cruise, 12-21 October 1988. B) Station 68, where weak stratification permitted photoacclimation of the phytoplankton assemblage. C) Station 71, with an apparently active mixed layer to 75 m.

et al., 1987), whereas P versus I in an actively mixing surface layer (Fig. 6C) is uniform because the rate of vertical mixing exceeds the rate of photoacclimation (Steemann Nielsen and Hansen, 1959; Falkowski, 1983; Lewis et al., 1984; Cullen and Lewis, 1988).

For cultures of phytoplankton, if P^B is little influenced by nutrition, then a plot of P^B against growth irradiance will conform to a saturating function of irradiance (Bannister and Laws, 1980; see Fig. 2 in Cullen, 1990). For samples from the field, we do not know growth irradiance, but we can compare nutrient regimes by plotting P^B as a function of *in situ* irradiance. That is, for each curve, P^B is calculated at the irradiance corresponding to the depth of sampling, estimated from measured extinction coefficients assuming surface irradiance of 2000 μmol m^{-2} s^{-1}. The result is gratifying if not remarkable (Fig. 7): P^B is clearly a function of irradiance, with very little scatter around a saturation curve. In fact, the data are consistent with the model of Ryther and Yentsch (1957) as reformulated by Cullen (1990). Rates for "green water" are higher than those for "blue water", but the differences are not great, and the sample size for the more eutrophic stations is small.

Apparently, in this region, nutrition had little influence on P versus I. This finding does not mean that growth rates were the same along the gradient of nutrient supply. Perhaps the more important observation is that a single P versus I relationship could describe the data with good precision. Such good agreement between measurements and chlorophyll-light models are not always obtained, however (Campbell and O'Reilly, 1988; Balch et al., 1989). Besides the possible physiological and ecological explanations for

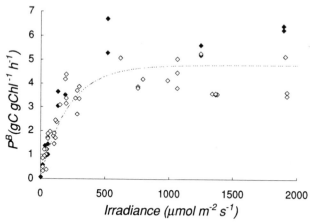

Fig. 7.　Calculated P^B at *in situ* irradiance for samples from the Biosynop cruise in and near the Gulf Stream (see text). Closed symbols are from stations with surface chlorophyll > 0.3 mg m^{-3}, open symbols from stations with surface chlorophyll < 0.3 mg m^{-3}. The line is the chlorophyll-light model of Ryther and Yentsch (1957) as reformulated by Cullen (1990). Estimates from "green water" tend to be higher than the others, but the effect is small compared to the variability that might be expected from inspection of the individual P versus I curves (Fig. 6). This result suggests that the phytoplankton were adapted to their irradiance regimes and that nutrition had a small influence on P^B at growth irradiance (cf. Cullen 1990).

variable photosynthesis-light relationships for natural phytoplankton (Cullen, 1990), results from conventional *in situ* or simulated *in situ* incubations might be more variable than estimates from short-term P versus I estimates for methodological reasons, such as longer incubation times and fluctuating solar irradiance during conventional incubations.

EXPERIMENTAL ASSESSMENT OF NUTRIENT LIMITATION

Nutrients can control primary productivity in several ways. For example, nutrient availability can regulate rate processes, such as photosynthesis (Blackman, 1905), or the final yield of a plant crop (Liebig limitation). The two types of limitation are conceptually distinct (Browne, 1942), but they are not mutually exclusive in ecological systems. Perhaps that is why the fundamental distinction between the Liebig limitation of yield and Blackman's rate-limiting factor has been blurred in ecological studies (Odum, 1971). Consequently, the term "limitation of phytoplankton growth" has assumed many meanings, including limitation of the specific growth rates of phytoplankton or limitation of standing crop (Cullen, in press). To be effective, hypotheses should specify which type of limitation is acting. Our discussion, so far, has focussed on rate processes, and how certain short-term rate measurements on natural phytoplankton might be influenced by nutrient supply. Now, we discuss how nutrients might control marine photosynthesis by limiting the standing crop of phytoplankton.

The "Iron Hypothesis" and Liebig Limitation of Standing Crop

Recently, it was suggested that iron limits the growth of phytoplankton in large parts of the ocean (Martin, 1990). Enrichment experiments have been used to assess the Liebig-limitation of phytoplankton standing crop by iron (Martin, in press). The results of these experiments, along with supporting measurements of dissolved iron concentrations, a variety of provocative paleoceanographic observations, and the suggestion that fertilization of the Southern Ocean with iron might mitigate the increase of atmospheric CO_2 (Martin, 1990), have generated intense interest and discussion (Chisholm and Morel, in press). The broad issues of the "iron hypothesis" (Martin, 1990, in press; Cullen, in press) are well beyond the scope of this presentation, but the interpretation of enrichment experiments is included as one more example of the ambiguity that can be encountered when trying to assess nutrient limitation in the ocean. Other aspects of nutrient-enrichment experiments are reviewed by Hecky and Kilham (1988).

The design of Martin's enrichment experiments is simple: samples of natural plankton are incubated in bottles and changes in chlorophyll, nitrate and other parameters are measured over several days. Samples enriched with small amounts of iron are compared to unenriched controls, and differences between experimental treatments and controls are attributed to iron. A particularly noteworthy achievement is in executing the experiments free from contamination. The results are consistent: in iron-poor waters, more phytoplankton accumulate in the iron-enriched bottles than in the unenriched controls. Ambient nitrate is depleted in the enriched samples, whereas some residual nitrate persists in the controls. Although the results are reported as growth rates (a contentious approach: cf. Banse, 1990; Martin et al., 1990), Martin interprets the experiments in the context of the Liebig limitation (Martin, 1990), with the fundamental observation being that there is not enough iron in the water to support the accumulation of phytoplankton and assimilation of residual nitrate.

More complete nutrient-enrichment experiments, with factorial design, were performed several times in the past (e.g., Ryther and Guillard, 1959; Menzel et al., 1963;

Thomas, 1969). It is now recognized that these experiments were compromised by contamination (Huntsman and Sunda, 1980), but their design bears examination. The standard of proof for such an experiment is stringent: to demonstrate Liebig-limitation, growth (relative to an unenriched control) must be stimulated by addition of the purported limiting nutrient and that nutrient alone, and growth should be unaffected when the limiting nutrient is omitted from a complete nutrient addition. The technical obstacles to performing such experiments without contamination are formidable, and the nature of oceanic trace-element nutrition (Morel et al., in press; Bruland et al., in press) is such that such a rigorous demonstration is unlikely.

There is another standard for evaluating the results of iron-enrichment experiments. Simply, the results should be consistent with the hypothesis that the availability of iron limits the standing crop of phytoplankton. Strictly interpreted, this means that there must be no increase of standing crop in the unenriched control. After all, if iron limited standing crop *in situ*, no increase in biomass would be observed during incubation of an ideal uncontaminated sample. Suppose that another factor, such as grazing or the balance between light-limited growth and grazing (cf. Banse, 1991), regulated the concentration of phytoplankton at some concentration below that which could be supported by iron. If grazing were diminished in bottles or if light-limited growth rates were increased during on-deck incubations, growth would be observed in controls until the nutrient in shortest supply was depleted. In fact, this is what regularly occurs: controls grow some, until iron is depleted, but iron-enriched samples grow much more (growth was indicated by disappearance of nitrate; Martin, in press). At face value, this result is a falsification of the null hypothesis that available iron limits the standing crop of phytoplankton *in situ*. Experimental artifacts, iron contamination, and second-order effects must be considered in a more rigorous interpretation.

Some control samples seemed not to grow (measurements of Chl: equatorial Pacific results of Martin, in press). It should be noted that the large initial decreases in chlorophyll were very likely the result of light-shock during on-deck incubations at supra-optimal irradiance: uncontaminated samples protected by neutral-density screens (Price et al., in press) showed increases of chlorophyll.

Interpretation of published results. Iron-enrichment experiments described to date show that in large regions of the ocean, iron is in short supply and that, isolated from allochthonous sources, iron would run out before nitrate was depleted. The experiments do not demonstrate that iron limits standing crop *in situ*, however. Also, these results do not exclude the possibility that iron limits the specific growth rates of larger phytoplankton (especially diatoms) and that an enhanced supply of iron would stimulate primary productivity and the utilization of nitrate (Cullen, in press). Simply, the bottle experiments are inconclusive.

Assessing Nitrogen Limitation of Standing Crop with a Factorial Experiment

Nitrogen is much easier to work with than iron, and Liebig-limitation can be examined with a factorial experiment. Results from a class exercise illustrate nitrogen limitation in Texas coastal waters. A sample from the Aransas Pass, Port Aransas, Texas was obtained near midday on an incoming tide. Sampling isolated the phytoplankton from the benthos, where much of the grazing pressure is thought to occur. A series of enrichments was prepared, and changes in chlorophyll *a*, particulate protein, particulate carbohydrate, nitrate, nitrite, ammonium, and phosphate were recorded during a 24-h incubation in 8-l polycarbonate bottles, water-cooled, and covered with neutral-density screen to simulate mid-depth irradiance. Changes of chlorophyll were consistent with the other measurements (Fig. 8). There was no increase of chlorophyll in the control, which was not surprising because nitrogen is required for net chlorophyll synthesis and dissolved inorganic nitrogen was very low to begin with. In response to a complete nutrient addition (major and minor nutrients, nitrogen added as ammonium, 16 μM), chlorophyll increased more than threefold over 24 h. Ammonium alone (16 μM) stimulated a substantial increase of chlorophyll, but phosphate was depleted within 12 h, and the +N treatment did not stimulate the same increase of biomass as the complete nutrient addition. The nutrient-omission treatment, complete minus N, was indistinguishable from the control, indicating clearly that N limited the standing crop of phytoplankton. It is not simple to relate these results of small-scale experiments to the functioning of the ecological system (Hecky and Kilham, 1988). It would be harder yet if the results were not so clear.

The limitation of standing crop by nitrogen was not necessarily accompanied by severe limitation of specific growth rates. In fact, the rapid increase in the enriched sample, with little lag period, suggests that specific growth rates were quite high. Thus, physiological diagnostics of nitrogen limitation might not reveal the control that nitrogen supply exerts on standing crop, hence primary productivity (Flynn, 1990). Clearly, a variety of approaches should be used when studying nutrient limitation of marine primary productivity.

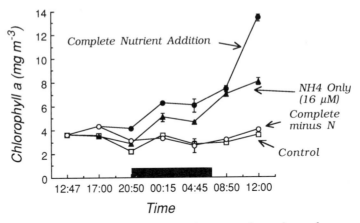

Fig. 8.   A factorial nutrient-enrichment experiment, performed as a class exercise using water from the Aransas Pass, Texas (see text for details). Error bars are the range of duplicates. The solid bar indicates darkness.

FROM LABORATORY TO NATURE: PHYSIOLOGICAL VERSUS GENETIC VARIABILITY

We have discussed here studies performed in the laboratory and in nature. When the results of laboratory experiments on individual clones of phytoplankton are used to predict the effects of environmental variability on the photosynthesis, growth, or chemical composition of phytoplankton in the sea, an implicit assumption is made: that for complex natural assemblages of phytoplankton responding to environmental variability, combined physiological responses (acclimation) and genotypic responses (including species succession and intraspecific changes in genotype frequency) will resemble the physiological responses of a phytoplankton clone in culture. So far, this assumption has served us fairly well. General trends have been described (e.g., Eppley and Renger, 1974; Goldman, 1980; Laws and Bannister, 1980; Kiefer and Mitchell, 1983; Geider, 1987) and different adaptational patterns have been identified (e.g., Richardson et al., 1983; Brand et al., 1983; Sakshaug et al., 1987). For some studies of marine processes, genetic responses need not be considered. In studies of diel variability or short-term physiological responses to vertical mixing, changes of species composition should be unimportant, so one need worry only about how well the physiological responses of cultured phytoplankton represent those in the natural setting. But, if geographical or longer-term temporal patterns are to be predicted or explained, we must recognize that in reaction to environmental change, species succession can dominate the responses of phytoplankton, and even if species persist, physiologically distinct genotypes might exchange dominance (Gallagher, 1982; Wood, 1988). For adaptation to temperature, a genetic component of phytoplankton response has been clearly recognized (Eppley, 1972). With respect to nutrient-limitation of photosynthesis, it seems warranted to examine in detail the possible genetic contributions to relationships between photosynthetic performance and nutrient-limited growth rate.

In a nutrient-limited system in which natural selection has had time to operate, the better competitors survive, and those that are most nutrient-limited become rare. Those rare species would probably exhibit the most pronounced physiological characteristics of nutrient limitation, and they would be likely to respond disproportionately to nutrient enrichment (cf. Martin, in press). Conversely, the dominant species would be least likely to show severe nutrient limitation. It would seem profitable, then, to use single cell methods, such as fluorescence microscopy or flow cytometry, to look for biochemical or physiological diagnostics of nutrient limitation of specific growth rates in the sea.

CONCLUSIONS

1. Models of nutrient limitation are based on laboratory experiments. There is a fundamental distinction between nutrient starvation (batch culture, unbalanced growth) and nutrient limitation (continuous culture, balanced growth over a photocycle), so the results of experiments should be interpreted and models should be generated with clear recognition of the stress imposed, and how it relates to oceanographic processes.

2. It is tempting to interpret geographical variability of P^B as a manifestation of nutrient limitation. However, in continuous culture, P^B can remain high despite nutrient limitation because phytoplankton adapt by regulating cellular chlorophyll concentration. Thus, although low values of P^B may, indeed, reflect nutrient stress, P^B is not a robust diagnostic of nutrient limitation and cannot be related directly to growth rate unless C:Chl is known.

3. Nutrient-limited algal cultures are used routinely to search for new diagnostics of nutritional status. With respect to assimilation number, the responses of phytoplankton cultures to nutrient limitation vary widely, and the source of this variation cannot be resolved. Because we do not know how the sources of variability would influence other potentially useful diagnostic parameters, it is essential to find the reason for the wide range of results.

4. Within some oceanic regions, P^B at ambient irradiance appears to be only weakly influenced by nutrient supply. Thus, models based on a regional P^B versus I relationship should have some utility.

5. The results of bioassays for iron-limitation, when interpreted strictly, reject the hypothesis of iron-limitation of standing crop and are inconclusive with respect to iron limitation of the specific growth rates of phytoplankton.

6. The responses of phytoplankton to environmental variability are often generalized on the basis of physiological responses to experimental conditions, i.e., how individual clones respond in the laboratory. In nature, though, genetic responses are undoubtedly important (natural selection). It may be instructive to question the conventional physiological approach. In the search for diagnostics of nutrient limitation in the sea, work should focus on the physiological characteristics and growth responses of rare species, which would flourish if purported nutrient limitation were alleviated.

ACKNOWLEDGEMENTS

This paper is dedicated to Dick Eppley, on the occasion of his retirement, with personal and professional gratitude. We thank Mary-Elena Carr for comments on the manuscript, A. Michelle Wood for assistance with the section on genetics, and Eric Mills for help with historical aspects of Liebig limitation. Supported by the Office of Naval Research (Oceanic Biology, Remote Sensing), NASA, and an NSERC Canada International Scientific Exchange Award. Bigelow Laboratory Contribution No. 91019.

REFERENCES

Balch, W. M., Abbott, M. R., and Eppley, R. W., 1989, Remote sensing of primary production - I. A comparison of empirical and semi-analytical algorithms, *Deep-Sea Res.*, 36:281.

Bannister, T. T., and Laws, E. A., 1980, Modeling phytoplankton carbon metabolism, *in*: "Primary Productivity in the Sea," P. G. Falkowski, ed., Plenum Press, New York.

Banse, K., 1990, Does iron really limit phytoplankton production in the offshore subarctic Pacific?, *Limnol. Oceanogr.*, 35:772.

Banse, K., 1991, Rates of phytoplankton growth, *Limnol. Oceanogr.*, Spec. Symp. Vol.: (in press).

Barber, R. T., and Chavez, F. P., 1991, Regulation of primary productivity rate in the equatorial Pacific Ocean, *Limnol. Oceanogr.*, Spec. Symp. Vol.: (in press).

Blackman, F. F., 1905, Optima and limiting factors, *Ann. Bot.*, 19:281.

Brand, L. E., Sunda, W. G., and Guillard, R. R. L., 1983, Limitation of marine phytoplankton reproductive rates by zinc, manganese, and iron, *Limnol. Oceanogr.*, 28:1182.

Browne, C.A., 1942, Liebig and the law of the minimum, *in*: "Liebig and after Liebig: A Century of Papers in Agricultural Chemistry," F.R. Moulton, ed., AAAS, Washington, D.C.

Bruland, K., et. al., 1991, Interactive influences of bioactive trace metals on biological production in the ocean, *Limnol. Oceanogr.*, Spec. Symp. Vol.: (in press).

Campbell, J. W., and O'Reilly, J. E., 1988, Role of satellites in estimating primary productivity on the northwest Atlantic continental shelf, *Cont. Shelf Res.*, 8:179.

Chalup, M. S., and Laws, E. A., 1990, A test of the assumptions and predictions of recent microalgal growth models with the marine phytoplankter *Pavlova lutheri*, *Limnol. Oceanogr.*, 35:583.

Chisholm, S. W., and Morel, F. M. M., 1991, What controls phytoplankton production in nutrient-rich areas of the open sea?, *Limnol. Oceanogr.*, Spec. Symp. Vol.: (in press).

Cleveland, J. S., and Perry, M. J., 1987, Quantum yield, relative specific absorption and fluorescence in nitrogen-limited *Chaetoceros gracilis*, *Mar. Biol.*, 94:489.

Cullen, J. J., and Lewis, M.R., 1988, The kinetics of algal photoadaptation in the context of vertical mixing, *J. Plankton Res.*, 10:1039.

Cullen, J. J., 1990, On models of growth and photosynthesis in phytoplankton, *Deep-Sea Res.*, 37:667.

Cullen, J. J., 1991, Hypotheses to explain high-nutrient conditions in the open sea, *Limnol. Oceanogr.*, Spec. Symp. Vol.: (in press).

Cullen, J. J., Lewis, M. R., Davis, C. O., and Barber, R. T., 1991, Photosynthetic characteristics and estimated growth rates indicate grazing is the proximate control of primary production in the equatorial Pacific, *J. Geophys. Res.*, (in press).

Curl, H., and Small, L. R., 1965, Variations in photosynthetic assimilation ratios in natural phytoplankton communities, *Limnol. Oceanogr.*, 10:R67.

Dubinsky, Z., this volume.

Eppley, R. W., 1972, Temperature and phytoplankton growth in the sea, *Fish. Bull.*, 70:1063.

Eppley, R. W., 1980, Estimating phytoplankton growth rates in the central oligotrophic oceans, *in*: "Primary Productivity in the Sea," P. G. Falkowski, ed., Plenum Press, New York.

Eppley, R. W., 1981, Relationship between nutrient assimilation and growth rate in phytoplankton with a brief view of estimates growth rate in the ocean., *in*: "Physiological Bases of Phytoplankton Ecology," T. Platt, ed., Ottawa.

Eppley, R. W., and Renger, E. H., 1974, Nitrogen assimilation of an oceanic diatom in nitrogen-limited continuous culture, *J. Phycol.*, 10:15.

Falkowski, P. G., 1980, Light-shade adaptation in marine phytoplankton, *in*: "Primary Productivity in the Sea," P. G. Falkowski, ed., Plenum Press, New York.

Falkowski, P. G., 1981, Light-shade adaptation and assimilation numbers, *J. Plankton Res.*, 3:203.

Falkowski, P. G., 1983, Light-shade adaptation and vertical mixing of marine phytoplankton: A comparative field study, *J. Mar. Res.*, 41:215.

Falkowski, P. G., 1991, Molecular ecology of phytoplankton photosynthesis, *in*: "Primary Productivity and biogeochemical Cycles in the Sea," P.G. Falkowski and A. Woodhead, eds., Plenum, New York.

Flynn, K. J., 1990, The determination of nitrogen status in microalgae, *Mar. Ecol. Prog. Ser.*, 61:297.

Gallagher, J. C., 1982, Physiological variation and electrophoretic banding patterns of genetically different seasonal populations of *Skeletonema costatum* (Bacillariophyceae), *J. Phycol.*, 18:148.

Geider, R. J., 1987, Light and temperature dependence of the carbon to chlorophyll *a* ratio in microalgae and cyanobacteria: implications for physiology and growth of phytoplankton, *New Phytol.*, 106:1.

Glover, H. E., 1980, Assimilation numbers in cultures of marine phytoplankton, *J. Plankton Res.*, 2:69.

Goldman, J. C., 1980, Physiological processes, nutrient availability, and concept of relative growth rate in marine phytoplankton ecology., *in*: "Primary Productivity in the Sea.," P. G. Falkowski, ed., Plenum Press, New York.

Harding, L. W. J., Fisher, T. R. J., and Tyler, M. A., 1987, Adaptive responses of photosynthesis in phytoplankton: specificity to time-scale of change in light, *Biol. Oceanogr.*, 4:403.

Harrison, W. G., and Platt, T., 1980, Variations in assimilation number of coastal marine phytoplankton: Effects of environmental covariates, *J. Plankton Res.*, 2:249.

Harrison, W. G., and Platt, T., 1986, Photosynthesis-irradiance relationships in polar and temperate phytoplankton populations, *Polar Biol.*, 5:153.

Hecky, R. E., and Kilham, P., 1988, Nutrient limitation of phytoplankton in freshwater and marine environments: A review of recent evidence on the effects of enrichment, *Limnol. Oceanogr.*, 33:796.

Herzig, R., and Falkowski, P. G., 1989, Nitrogen limitation in *Isochrysis galbana* (Haptophyceae). I. Photosynthetic energy conversion and growth efficiencies, *J. Phycol.*, 25:462.

Horrigan, S. G., and McCarthy, J. J., 1981, Urea uptake by phytoplankton at various stages of nutrient depletion, *J. Plankton Res.*, 3:403.

Huntsman, S. A., and Sunda, W. G., 1980, The role of trace metals in regulating phytoplankton growth, *in*: "The Physiological Ecology of Pphytoplankton," I. Morris, ed., University of California, Berkeley.

Kiefer, D. A., and Mitchell, D. G., 1983, A simple, steady state description of phytoplankton growth based on absorption cross section and quantum efficiency, *Limnol. Oceanogr.*, 28:770.

Kolber, Z., Zehr, J. R., and Falkowski, P. G., 1988, Effects of growth irradiance and nitrogen limitation on photosynthetic energy conversion in photosystem II, *Plant Physiol.*, 88:923.

Kolber, Z., Wyman, K. D., and Falkowski, P. G., 1990, Natural variability in photosynthetic energy conversion efficiency: A field study in the Gulf of Maine, *Limnol. Oceanogr.*, 35:72.

Laws, E. A., and Bannister, T. T., 1980, Nutrient- and light-limited growth of *Thalassiosira fluviatilis* in continuous culture, with implications for phytoplankton growth in the ocean, *Limnol. Oceanogr.*, 25:457.

Lewis, M. R., and Smith, J. C., 1983, A small volume, short-incubation-time method for measurement of photosynthesis as a function of incident irradiance, *Mar. Ecol. Prog. Ser.*, 13:99.

Lewis, M. R., Cullen, J. J., and Platt, T., 1984, Relationships between vertical mixing and photoadaptation of phytoplankton: Similarity criteria, *Mar. Ecol. Prog. Ser.*, 15:141.

Malone, T. C., and Neale, P. J., 1981, Parameters of light-dependent photosynthesis for phytoplankton size fractions in temperate estuarine and coastal environments, *Mar. Biol.*, 61:289.

Martin, J. H., 1990, Glacial-interglacial CO_2 change: The iron hypothesis, *Paleoceanography*, 5:1.

Martin, J. H., this volume.

Martin, J. H., Broenkow, W. W., Fitzwater, S. E., and Gordon, R. M., 1990, Yes it does: A reply to the comment by Banse, *Limnol. Oceanogr.*, 35:775.

Martin, J. H., Gordon, R. M., Fitzwater, S., and Broenkow, W. W., 1989, VERTEX: phytoplankton/iron studies in the Gulf of Alaska, *Deep-Sea Res.*, 36:649.

Menzel, D. W., Hulbert, E. M., and Ryther, J. H., 1963, The effects of enriching Sargasso Sea water on the production and species composition of the phytoplankton, *Deep-Sea Res.*, 10:209.

Morel, F. M. M., Hudson, R. J. M., and Price, N. M., 1991, Trace metal limitation in the sea, *Limnol. Oceanogr.*, Spec. Symp. Vol. (in press).

Odum, E. P., 1971, "Fundamentals of Ecology," W.B. Saunders Co., Philadelphia.

Osborne, B. A. and Geider, R. J., 1986, Effect of nitrate-nitrogen limitation on photosynthesis of the diatom Phaeodactylum tricornutum Bohlin (Bacillariophyceae), *Plant Cell Environ.*, 9:617.

Platt, T., Gallegos, C. L., and Harrison, W. G., 1980, Photoinhibition of photosynthesis in natural assemblages of marine phytoplankton, *J. Mar. Res.*, 38:687.

Platt, T., and Jassby, A. D., 1976, The relationship between photosynthesis and light for natural assemblages of coastal marine phytoplankton, *J. Phycol.*, 12:421.

Platt, T., Sathyendrenath, S., Caverhill, C. M., and Lewis, M. R., 1988, Oceanic primary production and available light: Further algorithms for remote sensing, *Deep-Sea Res.*, 35:855.

Price, N. M., Andersen, L. F., and Morel, F. M. M., 1991, Iron and nitrogen nutrition of equatorial Pacific plankton, *Deep-Sea Res.*, (in press).

Redalje, D. G., and Laws, E. A., 1981, A new method for estimating phytoplankton growth rates and carbon biomass, *Mar. Biol.*, 62:73.

Richardson, K., Beardall, J., and Raven, J. A., 1983, Adaptation of unicellular algae to irradiance: an analysis of strategies, *New Phytol.*, 93:157.

Ryther, J. H., and Guillard, R. R. L., 1959, Enrichment experiments as a means of studying nutrients limiting to phytoplankton populations, *Deep-Sea Res.*, 6:65.

Ryther, J. H., and Yentsch, C. S., 1957, The estimation of phytoplankton production in the ocean from chlorophyll and light data, *Limnol. Oceanogr.*, 2:281.

Sakshaug, E., Demers, S., and Yentsch, C. M., 1987, *Thalassiosira oceanica* and *T. pseudonana*: Two different photoadaptational responses, *Mar. Ecol. Prog. Ser.*, 41:275.

Sakshaug, E., and Holm-Hansen, O., 1977, Chemical composition of *Skeletonema costatum* (Grev.) Cleve and *Pavlova* (*Monochrysis*) *lutheri* (Droop) Green as a function of nitrate-, phosphate-, and iron-limited growth., *J. Exp. Mar. Biol. Ecol.*, 29:1.

Sakshaug, E., Kiefer, D. A., and Andresen, K., 1989, A steady state description of growth and light absorption in the marine planktonic diatom *Skeletonema costatum*, *Limnol. Oceanogr.*, 34:198.

Shuter, B., 1979, A model of physiological adaptation in unicellular algae, *J. Theor. Biol.*, 78:519.

Steemann Nielsen, E., and Hansen, V. K., 1959, Light adaptation in marine phytoplankton populations and its interrelation with temperature, *Physiol. Plant.*, 12:353.

Strickland, J. D. H., Holm-Hansen, O., Eppley, R. W., and Linn, R. J., 1969, The use of a deep tank in plankton ecology. I. Studies of the growth and composition of phytoplankton crops at low nutrient levels., *Limnol. Oceanogr.*, 14:23.

Thomas, W. H., 1969, Phytoplankton nutrient enrichment experiments off Baja California and in the eastern equatorial Pacific Ocean, *J. Fish. Res. Board Can.*, 26:1133.

Thomas, W. H., and Dodson, A. N., 1972, On nitrogen deficiency in tropical Pacific oceanic phytoplankton. II. Photosynthetic and cellular characteristics of a chemostat-grown diatom, *Limnol. Oceanogr.*, 17:515.

Vincent, W. F., Neale, P. J., and Richerson, P. J., 1984, Photoinhibition : algal responses to bright light during diel stratification and mixing in a tropical alpine lake, *J. Phycol.*, 20:201.

Welschmeyer, N. A., and Lorenzen, C. J., 1981, Chlorophyll-specific photosynthesis and quantum efficiency at subsaturating light intensities, *J. Phycol.*, 17:283.

Wood, A. M., 1988, Molecular Biology, single cell analysis, and quantitative genetics: new evolutionary genetic approaches in phytoplankton ecology, *in*: "Immunochemical approaches to coastal, estuarine, and oceanographic questions," C.M.Yentsch, F.C. Mague, and P.K. Horan, eds., Springer-Verlag.

Yentsch, C. M., Yentsch, C. S., and Strube, L. R., 1977, Variations in ammonium enhancement, and indication of nitrogen deficiency in New England coastal phytoplankton populations, *J. Mar. Res.*, 35:537.

Yentsch, C. S., 1974, The influence of geostrophy on primary production, *Tethys*, 6:111.

Yentsch, C. S., Yentsch, C. M., Strube, L. R., and Morris, I., 1974, The influence of temperature on the efficiency of photosynthesis in natural populations of marine phytoplankton, *in*: "Thermal Ecology," J. W. Gibbons and R. R. Sharitz, eds., AEC, Oak Ridge, Tenn.

GEOLOGIC AND CLIMATIC TIME SCALES OF NUTRIENT VARIABILITY

Richard T. Barber

Duke University Marine Laboratory
Beaufort, NC 28516

INTRODUCTION

Most ocean scientists are convinced that oceanic primary production plays some role in the global fluxes of biologically active elements such as carbon, nitrogen, oxygen, and phosphorus (Berger and Herguera, this volume). That conviction leads to an interesting question, "What processes regulate primary producers on the large time- and space-scales characteristic of the geologic and climatic scales?"

This question is not new. At the Brookhaven Symposium on Primary Productivity in the Sea a decade ago, Yentsch (1980) began his contribution with these words, "In other discussions I have expressed the opinion that the introduction of the experimental approach, an approach that uses the techniques derived largely from biochemistry and microbiology, has not greatly changed our thinking on how numbers of phytoplankton are regulated in time and space." Yentsch (1980) went on to express the need for coalescence of physical oceanography with biology to understand how phytoplankton growth in the sea is regulated. In a comment on the paper, Morris (p. 31, Yentsch, 1980) emphasized Medawar's (1967) sentiment that science deals with the art of the solvable and that it makes no sense to engage in scientific activity on problems that have no obvious means of approach and, therefore, are not solvable. In the context of Morris's comment, it appears that there has been some progress in the decade since that exchange. How phytoplankton are regulated in time and space in the ocean is now a problem with an obvious means of approach, if not a solution. The approach is to examine large-scale and low-frequency ocean patterns and determine what changes in the processes regulating productivity are coherent with productivity over the appropriate time and space scales.

To justify this approach, I use an exercise in logic that is based on three uncontroversial observations. One observation is that first principles argue that the regulation of primary production is, to first approximation, a function of light, temperature, and nutrients. The second observation, supported by a rich lode of paleoproductivity work, is that ocean productivity has varied over recent geologic time (Berger et al., 1989, and the references therein). The third and final observation is that presentations of the global pattern of primary productivity, from the first effort by Sverdrup (1955) to a recent effort by Berger (1989), all show a coherent spatial pattern of variability in the contemporary ocean. Together, these three observations lead one to

Primary Productivity and Biogeochemical Cycles in the Sea
Edited by P.G. Falkowski and A.D. Woodhead, Plenum Press, New York, 1992

ask what is the primary driver of this pattern of variability over large space scales and geologic time scales: light, temperature, or nutrients?

We are easily convinced that variations in primary production over the recent geologic record were not driven by variations in photosynthetically active radiation (PAR) from the sun. There have been insolation variations driven by the orbital Milankovitch cycle (Hays et al., 1976; Imbrie and Imbrie, 1980). However, the changes in insolation were relatively small, certainly less than 10 percent, and they occurred mostly in high latitudes; and the sign of the insolation changes (less during glacial periods) was opposite to the sign of the productivity changes (more during the glacial periods), at least in the low-latitude ocean (Arrhenius, 1952; Mix, 1989). In parallel reasoning, although the eastern sides of low-latitude contemporary ocean basins have much higher primary productivity than waters of the western portions of the basins (Barber, 1988), the input of light is not significantly asymmetric from west to east across the ocean basins. In fact, over most of the world's ocean there is no coherence between the spatial pattern of annual primary productivity and annual light supply. The lack of coherence of productivity and light supply in large-scale and low-frequency variations does not contradict our recognition that primary productivity clearly is regulated by light supply daily and seasonally. The explanation is simply that at different time-and space-scales, different processes structure the biological response to environmental conditions (Powell, 1989).

In summary, evidence from both the sediment record and the contemporary ocean indicates that changes in the supply of light for photosynthesis are not the cause of the geologic variability or global spatial pattern of productivity. One interpretation of this information is that over most, but not all, of the ocean there is presently, and there has been over the recent geologic past, an excess of the annual supply of light for photosynthesis relative to the annual supply of nutrients.

If temperature were the primary regulator of primary production through a physiological or biochemical interaction, the well-defined Arrhenius relationship between temperature and photosynthesis (Li, 1980) would lead us to expect an increase in temperature to cause an increase in primary production. Evidence from the sediment record and the contemporary ocean does not indicate such an association on geologic scales and, in fact, there is abundant evidence for the opposite association. Orbital cycles, which resulted in cooling ocean temperatures (CLIMAP, 1976), were associated with productivity increases (Berger et al., 1989; Mix, 1989). In a parallel manner, in the contemporary low-latitude and middle-latitude ocean basins, cooler eastern boundary regions have consistently higher primary productivity than warmer western boundary regions (Berger, 1989). The reason for this east versus west productivity asymmetry lies in the east versus west asymmetry of nutricline topography (Barber, 1988); here, I emphasize that geologic and oceanographic evidence is that temperature does not directly regulate oceanic primary production through physiological or biochemical means.

Logical elimination leaves nutrients, then, as the prime factor to account for the variations in primary productivity observed in the sediment record and for the large-scale pattern of spatial heterogeneity of the contemporary ocean. If nutrients are the regulating factor, how can we gain some appreciation of the processes that control the supply of nutrients to the sunlit surface layer of the ocean over the large time-and space-domains characteristic of the geologic and climatic scales? One approach is to examine the processes responsible for existing large-scale patterns in nutrient abundance and the other is to examine the processes responsible for low frequency (interannual) variability in nutrient availability in the contemporary ocean. Such an analysis of spatial and interannual variability provides a glimpse of the processes at work on the geologic and climatic scales.

To provide that glimpse, using examples from recent oceanographic and limnological studies at a variety of sites, is the goal of this paper.

SOURCES OF NUTRIENT VARIABILITY

The domains of geologic and climatic scales of variability considered here are shown by shading in the Stommel (1963) diagram of the spectral distribution of sea level variability (Fig. 1). The spatial domain involves gradients over 1000 km and in the time domain, periods over one year.

How can the supply of nutrients to the euphotic layer vary over these space and time domains? One potential source of variability involves changes in the nutrient content of the deepwater nutrient reservoir. Paleoceanographers have shown convincingly that the nutrient content of deep water varies on the glacial-interglacial time scale (Boyle, 1990), but contemporary oceanographers have rarely seen evidence of processes that would drive such changes. Dugdale et al. (1977) observed complete denitrification (Fig. 2) in a layer of water from 150 m to 250 m off the coast of Peru in 1976. However, the size of the region affected was less than 1000 km in any dimension and the duration was less than a month. Denitrification, particularly in eastern boundary regions, has the potential for significantly reducing the concentration of inorganic nitrogen and shifting the N to P ratio of subsurface waters towards a higher atom ratio excess of P relative to N (Codispoti, 1989), but we have meager evidence that this is now occurring (Codispoti and Christensen, 1985). Apparently, basinwide changes in deepwater nutrient content, such as those discussed by Boyle (1990), occur over such a long time that little insight into the phenomenon is provided by process studies in the modern ocean.

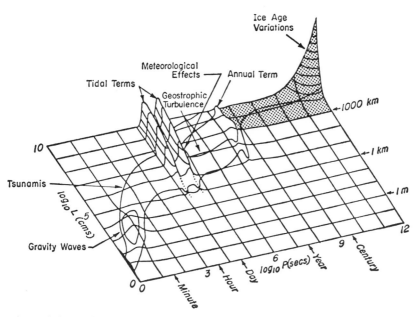

Fig. 1.   Schematic diagram of the spectral distribution of sea level with shading to show the climatic and geologic domain. From Stommel (1963).

Another potential source of variability involves changes in the eolian transport of material from the continents to the sea. Martin (1990) hypothesized that higher primary productivity in the glacial ocean was due to enhanced eolian supply of the limiting micronutrient, iron. In the contemporary ocean there is considerable spatial heterogeneity in the flux of eolian iron to surface waters (Duce and Tindale, 1991), and Martin and his coauthors believe this spatially variable iron flux is responsible for modulating productivity in the relatively high ambient nutrient concentrations of the North Pacific, equatorial Pacific and Southern Ocean (Martin and Gordon, 1988; Martin et al., 1989; Martin et al., 1990). Independent evidence that spatial variability in the eolian flux limits the rate of chlorophyll-specific primary productivity has been obtained from the equatorial Pacific Ocean (Barber and Chavez, 1991). Martin (this volume) discusses further the role of eolian processes.

The focus of this paper is another source of nutrient variability: processes that significantly alter the vertical flux of nutrients into the euphotic zone. This focus resulted from observing large, spatially coherent nutrient changes in the eastern Pacific during the 1982-83 El Niño (Halpern et al., 1983; Barber and Chavez, 1983) and in the western Pacific during the 1986-87 El Niño (Barnett et al., 1988).

THE ENSO PHENOMENON

During the 1980s the low latitude Atlantic and Pacific underwent a series of climate variations accompanied by coherent basinwide changes in surface nutrients (Barber and Chavez, 1983; Barber and Kogelschatz, 1990; Chavez et al., 1990). Two lines of new understanding emerged from analysis of this natural variability. One deals with processes responsible for the climatological mean pattern of nutrient abundance on the basin space scale, or, what makes the eastern tropical Atlantic and Pacific so rich in nutrients and the western tropical regions of the two oceans so poor? The other line deals with mechanisms responsible for the climatic scale (interannual) variability in nutrients at a given region or location.

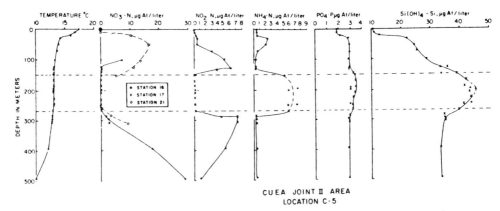

Fig. 2.   Profiles of temperature and nutrients from a station seaward of the continental shelf at 15°S along the coast of Peru in April 1976. The denitrified water also contained detectable concentrations of hydrogen sulfide. From Dugdale et al. (1977).

The episodes of climate variation driven by the El Niño Southern Oscillation (ENSO) in the 1980s have contributed to an understanding of the processes responsible for the basinwide nutrient pattern by showing how the mean pattern of nutrient abundance can be modified. (See Philander, 1990, for a physical description of the ENSO cycle.) Before the episodes of variability in the 1980s, basinwide interannual nutrient variability with excursions as large as the climatological mean concentration were not recognized. From observations during the 1982-83 (Halpern et al., 1983) and 1986-87 (Barnett et al., 1988) ENSO events, the existence and magnitude of interannual nutrient variability have been documented for the eastern and western Pacific, and the processes responsible for this low-frequency variability as well as the climatological mean condition have begun to be understood (Barber and Chavez, 1983).

Attention was focused on remotely forced variability in nutricline depth when it became obvious that during El Niño, local upwelling favorable winds did not cease and, in fact, they sometimes increased (Enfield, 1981). Current meter observations and models of both coastal and equatorial upwelling have found that water entrained by wind-driven

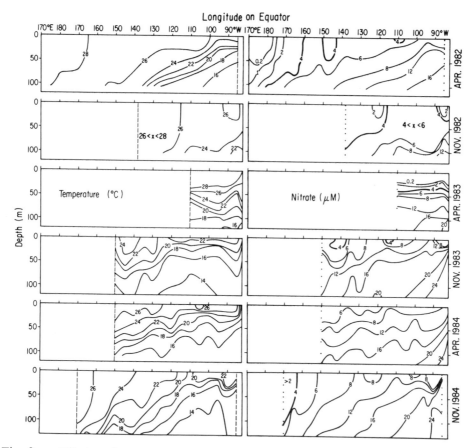

Fig. 3.   Nitrate and temperature profiles on an equatorial transect from 80°W to 170°E. April 1982, November 1983, April 1984 and November 1984 show normal thermocline and nutricline structure from the coast of South America to the dateline. The November 1982 and April 1983 sections show the 1982-83 El Niño thermocline and nutricline anomalies. From Barber and Chavez (1986).

upwelling comes from relatively shallow depths, usually 25 to 100 m (see Fig. 5 in Barber and Smith, 1981, for evidence of coastal upwelling and Fig. 4 in Halpern and Freitag, 1987, for equatorial data). If the thermocline and nutricline are deeper than the entrainment depth, then local processes upwell, or mix, nutrient-depleted water to the surface. How this scale interaction can work is illustrated by Fig. 3, which shows

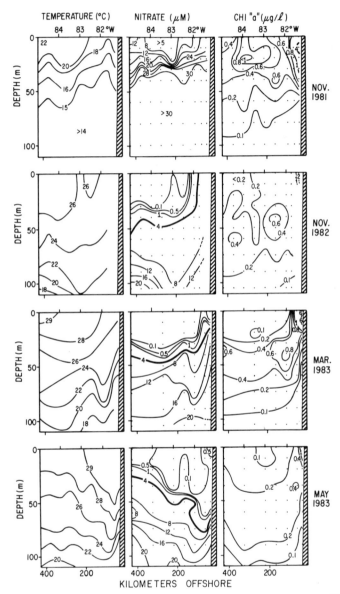

Fig. 4. Cross-shelf profiles of temperature, nitrate, and chlorophyll off the coast of Peru on a transect from 85°W to the coast. November 1981 shows normal austral spring conditions with no nitrate concentrations below 5 μM. In the lower three profiles, the 4 μM nitrate isopleth is bold to show the top of the nutricline. November 1982 is during onset of the 1982-83 anomaly, March 1983 is the mature phase, and May 1983 is the peak of the mature phase. From Barber et al. (1985).

variability driven by El Niño in basinwide nutricline and thermocline topography, and Fig. 4 which shows coastal temperature, nitrate, and chlorophyll conditions before and during the 1982-83 El Niño.

In Fig. 3, the April 1982 and April 1984 profiles show the normal thermocline and nutricline during austral fall with the 20°C isotherm and 12 μM nitrate isopleth shoaling from 100 m at 130°W to about 25 m at the coast of South America. During austral spring (November 1983 and November 1984), the thermocline and nutricline were shallower but essentially similar to the fall condition in that at the 25 to 100 m depth, where coastal upwelling entrains water, the temperature was less than 20°C and the nitrate concentration was higher than 12 μM.

The downward displacement of the thermocline (and nutricline) appeared to be 50 to 100 m in Fig. 3; using long records from moored instruments McPhaden and Hayes (1990) determined that during the 1982-83 event the seasonally averaged depth anomaly of the 20°C isotherm was 97 m. During the August to December 1982 period, when sea surface temperature was increasing, nutrients were decreasing and the thermocline and nutricline were sinking, Halpern (1987) measured zonal winds at 95°W and 110°W that were westward (upwelling favorable) with speeds similar to those of the previous year. The anomalous increase in the heat storage of the upper ocean was not locally wind forced, but remotely forced by an advective phenomenon that was caused by wind changes in the western Pacific (Halpern, 1987).

Observations on coastal winds established that in previous El Niño events (Enfield, 1981) and in the 1982-83 event (Smith, 1983; Huyer et al, 1987; Huyer et al., 1991) upwelling favorable winds continued blowing equatorward along the coast of central Peru during most of the event. In the 1982-83 event some locally forced upwelling was occurring from the beginning of the anomaly in September 1982 until April 1983, but anomalous warming and decreases in nutrients continued in the presence of upwelling as a result of anomalies in the thermocline and nutricline (Barber and Chavez, 1983), the Peru Undercurrent (Huyer et al., 1991), and the coastal pressure field (Huyer et al., 1987). The nutricline was progressively depressed so that in November 1982 (Fig. 4) water entrained into the upwelling circulation at 25 to 100 m depth was 10 to 20 μM lower in nitrate than in November 1981. Upward tilt of isopleths of nitrate and the continued onshore/offshore gradient in nitrate concentration in March 1983 show that upwelling continued, but the nutrient concentrations continued to decrease. The May 1983 section in Fig. 4 shows the strongest anomaly of the 1982-83 El Niño, with the surface temperatures about 12°C above the climatological monthly mean temperature of 17°C and the 4 μM isopleth of nitrate depressed below 50 m close to the coast of Peru. In the May 1983 section, the isotherms and isopleths of nitrate slope downward towards the coast indicating that flow was onshore and poleward in the coastal region and upwelling stopped due to the cross-shelf pressure gradient (Huyer et al., 1987).

A similar interaction of local and large-scale processes resulted in nutrient decreases on the equator (Barber and Chavez, 1986) despite continued favorable winds and upwelling along most of the equatorial Pacific during the 1982-83 event (Halpern, 1987). The area normally occupied by the equatorial cold tongue reaches from the coast of South America at about 80°W westward to the dateline at 180° (Wyrtki, 1981); the climatological mean nutrient condition of this cold tongue had surface nitrate concentrations of between 4 and 8 μM (Fig. 3) and moderate levels of primary productivity (Chavez and Barber, 1987). The ocean region occupied by the cold tongue reached over about a quarter of the circumference of the earth, roughly 90 degrees of longitude, or about 10,000 km in zonal extent; satellite observations of temperature (see

Fig. 9.5 in Barber, 1988) indicate that the entire equatorial cold tongue was occupied by anomalously warm water during the mature phase of the anomaly in May and June 1983. Hydrographic observations taken during the 1982-83 event indicate that for about 210 days, from December 1982 through June 1983, the surface layer concentration of nitrate was very low, often below the detection limit of 0.2 μM. Hydrographic observations of nutrient concentrations were made of only the eastern third of the equatorial Pacific (Fig. 3), but the well-developed temperature-nitrate relationship evident in Fig. 3 allows the satellite temperature data to be used to estimate that an enormous expanse of the region had very low nutrient concentration during the 1982-83 anomaly (Hayes et al., 1987). Estimates of equatorial productivity (Chavez and Barber, 1987) and the 1982-83 productivity anomaly summarized in Table 2 of Barber and Kogelschatz (1990) indicate that the rate of total primary production during the mature phase of El Niño was 21 to 26 percent of the normal conditions, while the rate of new production was five to six percent of the normal rate. Figure 3 shows that the nutrient-depleted layer was only about 30 m in depth during April 1983, but this was sufficient to cause a 20-fold reduction in new primary production. The magnitude of the biological anomaly emphasizes the relationship between nutricline topography and productivity.

To this point, I have discussed climatic-scale nutrient variability involving responses to El Niño along the eastern boundary and equatorial cold tongue where advective increases in heat storage forced the thermocline and nutricline down and reduced the flux of nutrients to the surface layer by locally forced processes. Philander (1990) emphasizes that cool anomalies forced by stronger than average trade winds are also an integral part of the ENSO cycle. The cool anomaly, commonly referred to as La Niña in

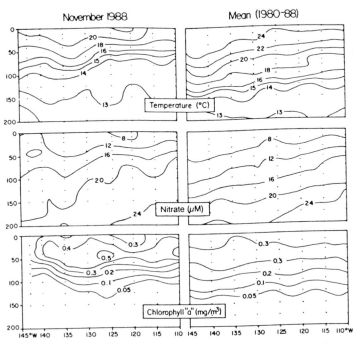

Fig. 5.   Temperature, nitrate, and chlorophyll profiles on a transect along the equator from 110°W to 145°W. Profiles from November 1988 are contrasted against mean profiles of the same properties derived from multiple occupations of the transect and show the cooler temperature and higher nitrate concentrations during November 1988. From Chavez et al. (1990).

Table 1. Enhancement of primary production in the warm pool of the western Pacific at 165°E over the region from 10°N to 6°S during the El Niño of 1986-87 (Barnett et al., 1988). The warm pool heat storage anomaly is a calculation provided by Klaus Wyrtki in the form of an upper ocean volume anomaly. The upper ocean volume relative to its mean value is calculated for the western Pacific from 170°E to 130°E over the region from 15°N to 15°S. See Wyrtki (1985) for a description of the general concept and calculation, but note that the anomaly given here is only for the warm pool (170°E to 130°E). Nutricline depth is depth of the 4 μM nitrate concentration (see Fig. 6).

Condition	Sample Size (n)	Zonal Wind Anomaly (s.d.)	Heat Storage Anomaly ($10^{14} m^3$)	Nutricline Depth (m)	Primary Production (mgC/m²/d)
Normal (Feb 86)	20	-0.8	+0.5	140	221±18
El Niño (Oct 87)	23	+0.0	-1.8	90	326±26

opposition to the warm anomaly called El Niño, clearly should increase nutrient concentration in surface waters of the cold tongue of the equatorial Pacific. During La Niña of 1988 (Kerr, 1988) when trade winds over the western Pacific intensified (McPhaden and Hayes, 1990), the isotherms and isopleths of the central and eastern Pacific were raised as shown in Fig. 5 from Chavez et al. (1990). In November 1988, the 12 μM isopleth of nitrate was elevated from 25 to 100 m and even intercepted the surface around 140°W, where continuously recording meteorological moorings showed that the westward zonal winds were not anomalously strong (McPhaden and Hayes, 1990).

WARM POOL AND GYRE VARIABILITY

When normal trade winds persist for several years, they accumulate surface water in the western equatorial Pacific and the upper layer heat storage, or upper ocean volume, is increased in a hydrographic region called the warm pool (Wyrtki, 1975). Conceptually, the warm pool is simply the ocean site where heat advected out of the equatorial cold tongue is stored. To quantitatively evaluate changes in heat storage of the low latitude ocean, Wyrtki (1985) developed an upper-layer volume index to track water displacement in the Pacific and better understand the genesis of El Niño cycles. The index calculates the volume of water above the 14°C isotherm between 15°N and 15°S across the Pacific basin.

What triggers the release or relaxation of potential energy stored as excess upper ocean water or excess heat content in the warm pool? Current understanding involves the generation by warm surface waters of anomalous atmospheric convection, which weakens trade winds in the western Pacific (McPhaden and Picaut, 1990). As the trades weaken, surface currents accelerate eastward and large-scale equatorial Kelvin and Rossby waves radiate and spread the warm sea surface anomaly eastward across the basin. As this displacement of water takes place, the thermocline (and nutricline) rises in the western

Pacific (and descends in the eastern cool tongue) to compensate for the upper ocean water drained from the western Pacific.

To investigate warm pool thermal dynamics, the United States and the People's Republic of China began a bilateral air-sea interaction program in 1986, with a mooring program, and two cruises a year to the region. The principal biological hypothesis tested was that El Niño events cause significant increases in primary productivity in the warm pool by elevating the nutricline (Barber and Kogelschatz, 1990). In late 1986, and most of 1987, a moderate strength event occurred (Barnett et al., 1988). The thermocline responded as the Wyrtki model predicted (McPhaden et al., 1990), the nutricline was elevated as predicted (Fig. 6), and there was a 30 to 40 percent increase in primary production in the region along 165°E from 10°N to 6°S (Table 1).

Natural variability of the 1986-87 El Niño demonstrated that a decrease in upper ocean heat storage can drive a large-scale increase in primary productivity. This natural variability made it clear that the east versus west asymmetry in mean annual productivity across ocean basins in both the Atlantic and Pacific is, to a first approximation, maintained by the ocean/atmosphere processes that drive asymmetry in heat storage.

Another issue concerns latitudinal extent of the east versus west asymmetry in upper ocean volume set up by large-scale winds and, in turn, setting up a mean basin wide tilt in the thermocline and nutricline. Figure 7 from Reid (1965), a composite of zonal sections from 35°N to 27°N between Japan and California, shows the concentration of phosphate in the mid-latitude North Pacific. The strong cross basin tilt of phosphate

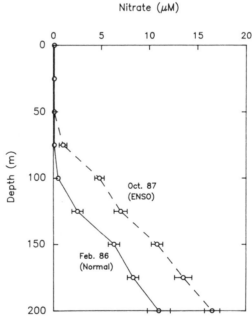

Fig. 6.   Mean vertical profiles of nitrate from stations at 165°E from 10°N to 6°S during a normal period (February 86) and an ENSO event (October 87). See McPhaden et al. (1990) for evidence that during February 1986 conditions along 165°E were close to the long-term climatological mean. October 1987 was close to the minimum in upper ocean volume anomaly during the 1986-87 event (n = 23 for the ENSO period; n = 20 for the normal period).

concentration isopleths emphasizes that east versus west asymmetry in nutrient availability exists in the permanently stratified mid-latitude regions as well as in low latitudes.

Turning to the central gyre of the North Pacific, I cite the time series by workers at Scripps Institution of Oceanography (Hayward, 1987) to show that permanently stratified regions removed from the equator show a surprisingly large amount of interannual variability in thermocline and nutricline depth. Figure 8 shows the interannual variability of nitrate depth profiles from the North Pacific central gyre in the region of 28°N and 155°W (Hayward, 1987). Hayward (1987) reports that there is no consistent relationship between nutrient distribution and productivity pattern, but these observations from 1973 to 1983 show a 40 m range in the depth of the top of the nutricline during this 10-year record. For vertical nutrient transport processes that are forced from the surface, interannual variability in nutricline depth will significantly alter the flux of nutrients to the euphotic layer.

SEASONALLY MIXED REGIONS

Heat storage also plays a primary role in determining the mean pattern and interannual variability of the nutrient supply in seasonally mixed waters, because the magnitude of the annual nutrient pulse is determined by the depth of the annual winter mixing (Lewis et al., 1988). Deeper winter mixing entrains higher nutrient water into the surface layer that will be isolated by the seasonal thermocline. Lewis et al. (1988) show that the difference between the depth of the summer and winter mixed layer increases sharply between 30°N and 40°N in the Atlantic and Pacific (see results from the Pacific in Fig. 9). The position of the steep gradient in the depth of winter mixing coincides with

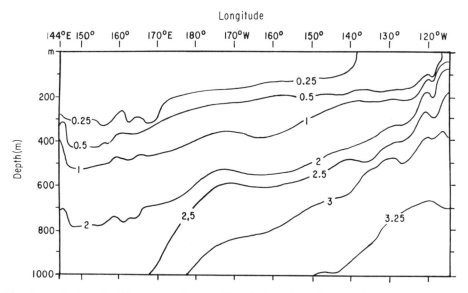

Fig. 7.   A transPacific composite profile of phosphate in μM from Japan to North America. The latitudinal domain of the zonal profile was from 35°N to 27°N. from Reid (1965).

a steep gradient in phytoplankton chlorophyll abundance. The climatological mean position and the magnitude of the chlorophyll front can be accounted for by one controlling process, depth of winter mixing. Lewis et al. (1988) used these observations to explain the mean spatial pattern in annual mean nitrate flux and phytoplankton biomass, but this process could be a major source of interannual variability, particularly when low latitude thermal perturbations export heat to high latitudes (Wyrtki, 1985).

Although there are few time series showing interannual variability in nutrient abundance in high latitudes, Fig. 10 from Miller et al. (1991) shows differences in concentrations of mixed layer nitrate that were observed during spring and summer cruises in 1984, 1987, and 1988 at Ocean Station Poppa at 55°N and 140°W. Strong year to year variations in the nitrate supply are related to the constrained vertical mixing that is the key to ecosystem function in this ocean region (Miller et al., 1991).

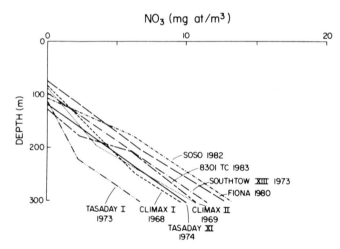

Fig. 8. Interannual variability in the vertical nitrate distribution at the CLIMAX site in the North Pacific. Cruises 8301 TC, SOUTHTOW XII, and TASADAY were made in winter; the rest in summer. From Hayward (1987).

SUMMARY

Observations of interannual nutrient and thermal variability of the 1980s have led to the generic suggestion that nutrient transport to the euphotic layer is regulated by two independent processes that are forced at different time scales: locally driven vertical transport processes and large-scale processes that set up basinwide properties (Barber, 1988). The latter appear to be primarily responsible for the climatic scale nutrient variability.

A decade ago in the first symposium in this series, Yentsch (1980) emphasized examples of geostrophic and baroclinic large-scale processes from a mid-latitude western boundary region (Figs. 3, 4, and 5 in Yentsch, 1980). Experience in low latitudes leads me to use examples that emphasize wind-driven advective large-scale processes, but both examples describe processes that regulate thermocline and nutricline topography over large time and space scales.

Local upwelling, or mixing, is a necessary but not sufficient condition for nutrient transport to the surface layer. The second necessary condition deals with the nutrient content of the subsurface water that is available to be advected, or mixed, into the euphotic layer. The connection between variability in the upper ocean thermal processes and nutrient variability over large space and time scales depends on the relationship between upper ocean heat content, thermocline depth, and nutricline depth. When the upper ocean heat content increases, that is, when the volume of warm water in the surface layer increases, sea level rises while the thermocline and nutricline are depressed (Wyrtki, 1985). This depression pushes the subsurface nutrient reservoir deeper (Barber, 1988); so local processes that drive vertical nutrient transport, such as wind driven upwelling,

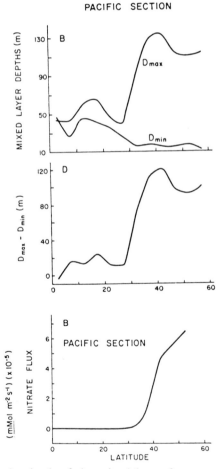

Fig. 9. Difference in the depth of the mixed layer from summer to winter and the computed annual nitrate flux resulting from winter convective mixing. From Lewis et al. (1988).

tidal mixing, island wake mixing or breaking internal waves, transport fewer nutrients to the sunlit zone when an increase in heat storage in the surface layer depresses the thermocline and nutricline.

Two important generalizations that follow are (1) this property, the mean upper ocean heat content, or mean upper ocean volume anomaly in Wyrtki's (1985) terminology (synonymous with the depth of the thermocline and nutricline), is the primary physical property responsible for the basinwide east-west asymmetry in nutrient abundance; and (2) changes in upper ocean heat storage or upper ocean volume in a particular ocean region provide a mechanism that can drive large-scale nutrient variability regardless of whether or not local forcing is changed.

In a parallel manner, although the argument has not been developed in detail here, I propose that (1) depth of winter mixing, a property determined in part by upper ocean heat content, is the primary determinant of the mean north versus south asymmetry in the annual nutrient flux to the mixed layer and (2) interannual changes in upper ocean heat storage provide a mechanism that will necessarily change the annual flux of nutrients available for the spring bloom.

The importance of these generalizations is that they operate for the most part at the climatic time and space scale and that they are easily observed at that scale by coastal stations, *in situ* arrays, and satellites. Logical progression suggested by Stommel's (1963) spectral distribution (Fig. 1) is that the climatic scale smoothly grades into the geologic

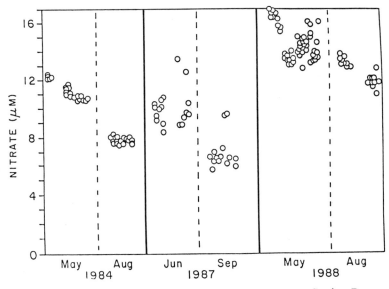

Fig. 10. Interannual variability in nitrate concentration at Ocean Station Poppa at 55°N and 140°W. From Miller et al. (1991).

scale of variability. The climatic scale gives insight into the processes that operate at the geologic scale. If this is true, then the heat storage scenario provides an approach, in the sense of Medawar (1967), to the problem of what regulates ocean primary production over the geologic time scale.

That Stommel (1963) chose sea level as the property for discussion in his classic paper, "Varieties of oceanographic experience," is fortunate for my purposes. Sea level has been measured over the climatic time scale (Marmer, 1925) and very large space scales (Wyrtki, 1984), and in the near future it will be measured continuously on a global basis by the TOPEX-Poseidon satellite. The spectral distribution of sea level variability is well known, but is this knowledge exploitable by a biogeochemist? Wyrtki (1985) points out that in the low latitude ocean there is a close relationship between sea level and the depth of the thermocline (or nutricline) as given by the depth of a selected isotherm (or nutrient isopleth). The correlation between sea level and the depth of the 20°C isotherm was 0.90 in the eastern Pacific at the Galapagos Islands, 0.80 in the central equatorial Pacific at Christmas Island, and 0.90 in the western Pacific at Truk (Wyrtki, 1985). Huyer (1980) found along the Peru coast a well-defined relationship between sea level variations and subsurface temperature and density structure over the shelf and slope, and Chavez et al. (1989) showed that along the Peru coast sea level and the nitrate concentration at 60 m were better correlated than 60 m temperature and 60 m nitrate, or 60 m density and 60 m nitrate. Observations of sea level variability, then, provide ocean biogeochemists with an approach to the problem of nutrient variability at climatic and geologic scales.

ACKNOWLEDGEMENTS

This research was supported by National Science Foundation Grants OCE-8613759 and OCE-8901929 and by National Oceanic and Atmospheric Administration Grant NA85AA-D-AC137. The Equatorial Pacific Ocean Climate Study (EPOCS) and the Tropical Ocean Global Atmosphere (TOGA) program provided space on cruises and the EPOCS and TOGA scientists provided insight into the physical processes.

REFERENCES

Arrhenius, G. O. S., 1952, Sediment cores from the east Pacific, *Rep. Swed. Deep Sea Exped. 1947-1948*, Vol. 5.

Barber, R. T., 1988, The ocean basin ecosystem, *in*: "Concepts of Ecosystem Ecology," J.J. Alberts and L.R. Pomeroy, eds., Springer-Verlag, New York.

Barber, R. T., and Chavez, F. P., 1983, Biological consequences of El Niño, *Science*, 222:1203.

Barber, R. T., and Chavez, F. P., 1986, Ocean variability in relation to living resources during the 1982-83 El Niño, *Nature*, 319:279-285.

Barber, R. T., and Chavez, F. P., 1991, Regulation of primary productivity rate in the equatorial Pacific Ocean, *Limnol. Oceanogr.*, in press.

Barber, R. T., and Kogelschatz, J. E., 1990, Nutrients and productivity during the 1982/83 El Niño, *in*: "Global Ecological Consequences of the 1982-83 El Niño - Southern Oscillation," P. Glynn, ed., Elsevier, New York.

Barber, R. T., and Smith, R. L., 1981, Coastal upwelling ecosystems, *in*: "Analysis of Marine Ecosystems," A. Longhurst, ed., Academic Press, New York.

Barber, R. T., Kogelschatz, J. E., and Chavez, F. P., 1985, Origin of productivity anomalies during the 1982/83 El Niño, *California Cooperative Oceanic Fisheries Investigation Reports*, 26:65.

Barnett, T. P., Graham, N., Cane, M., Zebiak, S., Dolan, S., O'Brien, J., and Legler, D., 1988, On the prediction of El Niño of 1986-1987, *Science*, 241:192.

Berger, W. H., 1989, Global maps of ocean productivity, *in*: "Productivity of the Ocean: Present and Past," W.H. Berger, V.S. Smetacek, and G. Wefer, eds., John Wiley, New York.

Berger, W. H., Smetacek, V. S., and Wefer, G., 1989, "Productivity of the Ocean: Present and Past," John Wiley, New York.

Boyle, E. A., 1990, Quaternary deepwater paleoceanography, *Science*, 249:863-870.

Chavez, F. P., and Barber, R. T., 1987, An estimate of new production in the equatorial Pacific, *Deep-Sea Res.*, 34:1229.

Chavez, F. P., Barber, R. T., and Sanderson, M. P., 1989, The potential primary production of the Peruvian upwelling ecosystem: 1953-1984, *in*: "The Peruvian Upwelling Ecosystem: Dynamics and Interactions," D. Pauly, P. Muck, J. Mendo, and I. Tsukayama, eds., ICLARM Conference Proceedings 18.

Chavez, F. P., Buck, K. R., and Barber, R. T., 1990, Phytoplankton taxa in relation to primary production in the equatorial Pacific, *Deep-Sea Res.*, 37:1733.

CLIMAP, 1976, The surface of the ice-age earth, *Science*, 191:1131.

Codispoti, L. A., 1989, Phosphorus vs. nitrogen limitation of new and export production, *in*: "Productivity of the Ocean: Present and Past," W.H. Berger, V.S. Smetacek, and G. Wefer, eds., John Wiley, New York.

Codispoti, L. A., and Christensen, J. P., 1985, Nitrification, denitrification and nitrous oxide cycling in the eastern tropical South Pacific Ocean, *Mar. Chem.*, 16:277.

Duce, R. A., and Tindale, N. W., 1991, The atmospheric transport of iron and its deposition in the ocean, *Limnol. Oceanogr.*, in press.

Dugdale, R. C., Goering, J. J., Barber, R. T., Smith, R. L., and Packard, T. T., 1977, Denitrification and hydrogen sulfide in Peru upwelling during 1976, *Deep-Sea Res.*, 24:601.

Enfield, D. B., 1981, Thermally driven wind variability in the planetary boundary layer above Lima, *J. Geophys. Res.*, 86:2005.

Halpern, D., 1987, Observations of annual and El Niño thermal and flow variations at 0°, 110°W and 0°, 95°W during 1980-1985, *J. Geophys. Res.*, 92:8197.

Halpern, D., and Freitag, H. P., 1987, Vertical motion in the upper ocean of the equatorial Eastern Pacific, *Oceanologica Acta*, 6:19.

Halpern, D., Hayes, S. P., Leetmaa, A., Hansen, D. V., and Philander, S. G. H., 1983, Oceanographic observations of the 1982 warming of the tropical Eastern Pacific, *Science*, 221:1173.

Hayes, S. P., Mangum, L. J., Barber, R. T., Huyer, A., and Smith, R. L., 1987, Hydrographic variability west of the Galapagos Islands during the 1982/83 El Niño, *Progress in Oceanography*, 17:137.

Hays, J. D., Imbrie, J., and Shackleton, N. J., 1976, Variations in the earth's orbit: pacemaker of the ice ages, *Science*, 194:1121.

Hayward, T. L., 1987, The nutrient distribution and primary production in the central North Pacific, *Deep-Sea Res.*, 34:1593.

Huyer, A., 1980, The offshore structure and subsurface expression of sea level variations off Peru, 1976-1977, *J. Phys. Oceanogr.*, 10:1755.

Huyer, A., Smith, R. L., and Paluszkiewicz, T., 1987, Coastal upwelling off Peru during normal and El Niño times, 1981-1984, *J. Geophys. Res.*, 92:14,297.

Huyer, A., Knoll, M., Paluszkiewicz, T., and Smith, R. L., 1991, The Peru Undercurrent: a study in variability, *Deep-Sea Res.*, 38:S247.

Imbrie, J., and Imbrie, J. Z., 1980, Modelling the climatic response to orbital variations, *Science*, 207:943.

Kerr, R. A., 1988, La Niña's big chill replaces El Niño, *Science*, 241:1037.

Lewis, M. R., Kuring, N., and Yentsch, C., 1988, Global patterns of ocean transparency: implications for the new production of the open ocean, *J. Geophys. Res.*, 93:6847.

Li, W. K. W., 1980, Temperature adaptation in phytoplankton: cellular and photosynthetic characteristics, *in*: "Primary Productivity in the Sea," P.G. Falkowski, ed., Plenum Press, New York.

Marmer, H. A., 1925, Variability of sea level along the Atlantic coast of the United States, *Geograph. Rev.*, 15:438.

Martin, J. H., 1990, Glacial-interglacial CO_2 change: the iron hypothesis, *Paleoceanogr.*, 5:1.

Martin, J. H., and Gordon, R. M., 1988, Northeast Pacific iron distributions in relation to phytoplankton productivity, *Deep-Sea Res.*, 35:177.

Martin, J. H., Fitzwater, S. E., and Gordon, R. M., 1990, Iron deficiency limits phytoplankton growth in Antarctic waters, *Global Biogeochem. Cycles*, 4:5.

Martin, J. H., Gordon, R. M., Fitzwater, S., and Broenkow, W. W., 1989, VERTEX: phytoplankton/iron studies in the Gulf of Alaska, *Deep-Sea Res.*, 36:649.

McPhaden, M. J., and Hayes, S. P., 1990, Variability in the eastern equatorial Pacific Ocean during 1986-1988, *J. Geophys. Res.*, 95:13,195.

McPhaden, M. J., and Picaut, J., 1990, El Niño-Southern Oscillation displacements of the western equatorial Pacific warm pool, *Science*, 250:1385-1388.

McPhaden, M. J., Hayes, S. P., Mangum, L. J., and Toole, J. M., 1990, Variability in the western equatorial Pacific Ocean during the 1986-87 El Niño/Southern Oscillation event, *J. Phys. Oceanogr.*, 20:190.

Medawar, P. B., 1967, "The Art of the Soluble," Methuen, London.

Miller, C. B., Frost, B. W., Wheeler, P. A., Landry, M. R., Welschmeyer, N., and Powell, T. M., 1991, Ecological dynamics in the subarctic Pacific, a possibly iron-limited ecosystem, *Limnol. Oceanogr.*, in press.

Mix, A. C., 1989, Pleistocene paleoproductivity: evidence from organic carbon and foraminiferal species, *in*: "Productivity of the Ocean: Present and Past," W.H. Berger, V.S. Smetacek, and G. Wefer, eds., John Wiley, New York.

Philander, S. G., 1990, "El Niño, La Niña, and the Southern Oscillation," Academic Press, New York.

Powell, T. M., 1989, Physical and biological scales of variability in lakes, estuaries, and the coastal ocean, *in*: "Perspectives in Ecological Theory," J. Roughgarden, R. M. May, and S. A. Levin, eds., Princeton University Press, Princeton, NJ.

Reid, J. L., 1965, "Intermediate Waters of the Pacific Ocean," Johns Hopkins Press, Baltimore.

Smith, R. L., 1983, Peru coastal currents during El Niño: 1976 and 1982, *Science*, 221:1397.

Stommel, H., 1963, Varieties of oceanographic experience, *Science*, 139:572.

Sverdrup, H. U., 1955, The place of physical oceanography in oceanographic research, *J. Mar. Res.*, 14:287.

Wyrtki, K., 1975, El Niño--The dynamic response of the equatorial Pacific Ocean to atmospheric forcing, *J. Phys. Oceanogr.*, 5:572.

Wyrtki, K., 1981, An estimate of equatorial upwelling in the Pacific, *J. Phys. Oceanogr.*, 11:1205.

Wyrtki, K., 1984, The slope of sea level along the equator during the 1982/1983 El Niño, *J. Geophys. Res.*, 89:10,419.

Wyrtki, K., 1985, Water displacements in the Pacific and the genesis of El Niño cycles, *J. Geophys. Res.*, 90:7129.

Yentsch, C. S., 1980, Phytoplankton growth in the sea - a coalescence of disciplines, *in*: "Primary Productivity in the Sea," P.G. Falkowski, ed., Plenum Press, New York.

NUTRIENT LIMITATION OF NEW PRODUCTION IN THE SEA

Richard Dugdale and Frances Wilkerson

Department of Biological Sciences and
Hancock Institute for Marine Studies
University of Southern California
Los Angeles, CA 90089-0371

INTRODUCTION

Light and nutrients are the two well-known basic requirements for primary production in the sea and are usually supplied in opposing vertical gradients. When radiant energy of correct wavelengths is available, the vertical advection of nutrients from below the euphotic zone sets the maximum rate of absorption of these new nutrients and the ensuing primary production is considered to be new production (Dugdale and Goering, 1967) in contrast to the regenerated production based upon recirculating nutrients. When Sverdrup (1955) presented a map of world ocean primary production based upon his understanding of the vertical advective regimes of nutrients in various regions, he was actually providing the first global map of new production. Dinitrogen fixation also fuels new production, but this source of new nutrient (along with atmospheric and terrestrial inputs of nitrate and ammonium) is relatively minor compared to the advection of nitrate (e.g., Carpenter, 1983). The fractionation of nitrogen species from new into regenerated forms, through grazing and bacterial activity, and the existence of a practical tracer for nitrogen, the stable isotope ^{15}N (e.g., Dugdale and Wilkerson, 1986), has made it possible to investigate the new and regenerated pathways in the marine ecosystem. The flow of nitrogen through the euphotic zone ecosystem has proved more complex than originally suggested (Dugdale and Goering, 1967) and is more realistically described by several possible schemes, including that of Michaels and Silver (1988), where the size fractions of the phytoplankton and bacteria are included explicitly along with other elements of the microbial loop. Their analysis showed that the microautotrophs provided the major source of new nitrogen for sinking particle formation, with the picoplankton participating primarily within the microbial grazing loop. However, both nanoplankton and picoplankton are capable of new production. For example, unicellular cyanobacteria (Glibert and Ray, 1990) and bacteria (Brown et al., 1975) may use nitrate and were shown to fix dinitrogen (e.g., Mitsui et al., 1986; Martinez et al., 1983).

The trophic status of a marine ecosystem is generally determined by the rate of ingestion of new nitrogen, which maintains and pumps up both the primary and higher production levels. Consequently, regions of the ocean where vertical advection of primary nutrients and ample irradiance occur are highly productive, i.e., eutrophic, and areas with poor nutrient supplies are poorly productive, i.e., oligotrophic. Regions also exist with high surface nutrients, but low productivity, whose functioning is poorly understood.

Primary Productivity and Biogeochemical Cycles in the Sea
Edited by P.G. Falkowski and A.D. Woodhead, Plenum Press, New York, 1992

Several new production models have been developed relating new production to other oceanic variables. Dugdale et al. (1989) describe a model for predicting new production from remotely sensed sea-surface temperature in upwelling areas. Eppley et al. (1979) and Eppley and Peterson (1979) related total production (expressed as carbon) to the proportion of new to new plus regenerated nitrogen uptake, which they referred to as f, equivalent to the percent new production of Dugdale and Goering (1967). High carbon uptake was a function of high values of f in data from several coastal and oceanic areas. Platt and Harrison (1985) proposed a hyperbolic model for the computation of f from nitrate concentrations at low levels of ammonium and showed that it worked well for the oligotrophic Sargasso Sea near Bermuda. However, that model failed in the equatorial Pacific upwelling system (Dugdale et al., in press; Wilkerson and Dugdale, in press), predicting high saturating values of f at relatively too low concentrations of nitrate. Elevated values of f are not always correlated with high total production, e.g., in the Antarctic seas where little biomass accumulation occurs and productivity remains low in most regions throughout the year. Obviously, there are factors that influence the uptake of nitrate beyond the mere presence of nitrate, and we shall examine the effects of nutrient concentration and related factors in limiting new production in both rich and poorly productive areas.

EUTROPHIC AND OLIGOTROPHIC AREAS

Concentrations of maximum surface nitrate and mean surface values of nitrate uptake for a series of eutrophic, oligotrophic, and high-nutrient, low productivity regions are shown in Table 1. Eutrophic areas, as represented by coastal upwelling regions, exhibit high concentrations of nitrate, high values of biomass (as particulate nitrogen) specific nitrate uptake rate (VNO_3), high new production rates (i.e. nitrate uptake, ρNO_3) and high f values. Oligotrophic regions have concentrations of surface nitrate that are usually below detection, using conventional AutoAnalyzer procedures that measure nitrate in the micromolar range. However, the use of a chemoluminescent technique (Cox, 1980; Garside, 1982) shows nitrate may be available at the nanomolar level (e.g., Glover et al., 1988). These nitrate-poor areas have low VNO_3 values, low new production rates, and low f values. There are some high-nutrient, low productivity areas with new production characteristics that place them functionally with oligotrophic low nutrient regimes, although there is abundant surface nitrate. These areas often have values of f that are intermediate between the truly oligotrophic and eutrophic regimes. The range in values of nitrate uptake shown in Table 1 is about two orders of magnitude, and although the oligotrophic values of VNO_3 are undoubtedly diluted more by non-phytoplankton nitrogen than those for the eutrophic regions (Dortch and Packard, 1989), there is a clear distinction between measured values of VNO_3 in the two regimes.

Another way to compare the eutrophic and oligotrophic regimes is to plot f versus VNO_3 (Figure 1). Increased values of VNO_3 lead to increased values of f, following a saturation type of curve with the initial slope containing the data points from the oligotrophic and high-nutrient, low production systems. The implication is that in the most productive, eutrophic regions of the ocean, the phytoplankton import a large proportion of new nitrogen as compared to the consumption of recycled nitrogen.

There is an upward gradient in concentration of chlorophyll between oligotrophic and eutrophic conditions, associated with a gradient in cell size of the phytoplankton. Chisholm (this volume) has shown that each increase of chlorophyll from oligotrophic concentrations of about 0.05 μg l^{-1} is accompanied by an increase in cell size. Malone

Table 1. Representative Values of Surface Maximum Nitrate and Mean Nitrogen Uptake

Region	Reference	NO_3 μM	VNO_3 d^{-1}	ρNO_3 nmol l^{-1} d^{-1}	f %
EUTROPHIC					
15°S, Peru	MacIsaac et al., 1985 Wilkerson et al., 1987	26.68	0.840	4200.0	82
Cap Blanc, NW Africa	Codispoti et al., 1982 Dugdale, 1985	12.00	0.230	400.0	70[*]
Pt. Conception, CA	OPUS 83 Prod. data Wilkerson et al., 1987	32.52	0.28	555.4	57
Baja Calif.	Codispoti et al., 1982 Wilkerson et al., 1987	13.28	0.280	1000.0	78[*]
OLIGOTROPHIC					
Fieberling Guyot	Kopczak et al., 1990[+] Wilkerson et al., 1990	0.03	0.011	4.2	9
Mediterranean Sea	MacIsaac & Dugdale, 1972	0.05	0.035	7.0	21
Sargasso Sea	Glibert et al., 1988	b.d.	0.022	2.0	4
Gulf Stream	Glibert et al., 1988	b.d.	0.006	0.6	3
HIGH-NUTRIENT, LOW PRODUCTIVITY					
Antarctic	Glibert et al., 1982 Olson, 1980	31.00	0.031	32.1	37
Equatorial Pacific, 1°N-1°S,150°W	WEC 88 Pro data	7.00	0.027	19.2	36
Station P, Northeast Pacific	Wheeler & Kokkinakis, 1990 Miller et al., 1988	17.00	0.056	84.2	48

Daily values were obtained from hourly data by multipling NO_3 uptake by 12 and NH_4 by 18.

f calculated from mean surface values of ρNO_3 and ρNH_4 except [*] calculated from values of depth integrated ρN.

[+] nanomolar measurement using chemoluminescent technique.

b.d. below detection using AutoAnalyzer

(1980) suggested that new production is carried out primarily by larger cell-sized organisms (e.g., chain-forming diatoms) that predominate in eutrophic regimes, while open ocean productivity is carried out by regenerated production by nanoplankton and picoplankton. Probyn (1985) measured nitrate uptake in all size fractions of phytoplankton in the Benguela upwelling system. The distribution of new production between the various size fractions depends upon both the nitrate uptake activity and biomass of each fraction. We calculated the proportion of new production in the shelf region for each of Probyn's size fractions from the maximum nitrate uptake (Table 2). The largest proportion of nitrate uptake is by the microplankton (64.3%). The nanoplankton plus the picoplankton account for 35.7%, which includes the very small contribution of the picoplankton (3.6%). As noted by Bigelow (1926), "all fish is diatoms", and the highest new production leading to highest yields are likely to occur with bloom-forming diatoms. The relatively low amount of nitrate uptake by picoplankton is the result of low values of biomass specific nitrate uptake, VNO_3 (Figure 2), and of the characteristic low biomass of this fraction. Figure 2 shows that the low values of VNO_3 reported by Probyn (1990) for the picoplankton are not the result of high $PON/Chl-a$ values, because nitrate uptake normalized to chlorophyll-a plotted versus those normalized to PON fall close to the 1:1 slope. The same is true for the other size fractions. These measurements were made in aged upwelled waters.

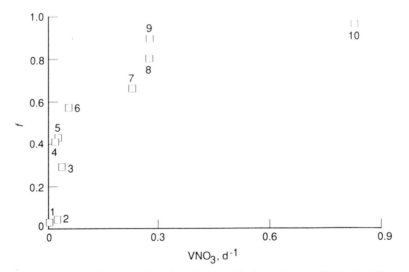

Fig. 1. Mean surface f values plotted against VNO_3 from data in Table 1. Oligotrophic: 1 = Gulf Stream; 2 = Sargasso Sea; 3 = Mediterranean Sea; High-nutrient, low productivity: 4 = Equatorial Pacific; 5 = Antarctic Ocean; 6 = Station P, northeast Pacific; Eutrophic: 7 = Cap Blanc, northwest Africa; 8 = Point Conception, CA; 9 = Baja, California; 10 = 15°S, Peru.

NEW PRODUCTION IN HIGH-NUTRIENT AREAS

New production (ρNO_3, units: $\mu mol\ l^{-1}\ h^{-1}$) can be expressed as the product of specific nitrate uptake rates (VNO_3, units: h^{-1}) and particulate nitrogen (PON, units: μg-at l^{-1}):

$$\rho NO_3 = VNO_3 * PON$$

Table 2. Contribution of Different Size Fractions to Total Nitrate Uptake in the Shelf Area of the Southern Benguela Upwelling System (from Probyn, 1985)

Size fraction	Maximum NO_3 uptake $\mu mol\ l^{-1}\ h^{-1}$	% of total NO_3 uptake
$<1\mu m$	0.015	3.6
$<10\mu m$	0.150	35.7
$>10\mu m<212\mu m$		64.3
$<212\mu m$	0.420	100.0

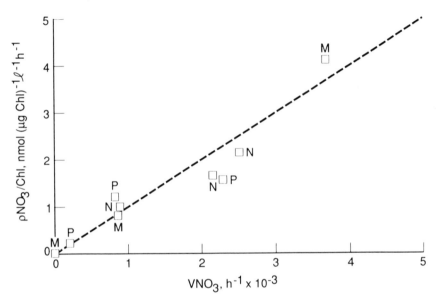

Fig. 2.   Nitrate uptake normalized to chlorophyll-*a* plotted against nitrate uptake normalized to particulate nitrogen for micro- (M), nano- (N), and picoplankton (P) from the Benguela upwelling system. Taken from data in Probyn (1990).

This relationship shows that high new production rates can result from either high VNO_3 or from the accumulation of active PON biomass, or both. The positive re-inforcement of these parameters can be seen in Figure 3, where ρNO_3 is plotted against VNO_3 for the upwelling center at 15°S, Peru. If there was no accumulation of PON, the data would lie along the straight line with the slope equal to the concentration of PON. However, although there is a general tendency for the points to follow a straight line with slope equal to a constant PON concentration up to VNO_3 of about 0.025 h^{-1}, the data begin to spread out above this, suggesting that some minimum value of VNO_3 must be reached to overcome the losses by sinking and predation. Then biomass as PON can accumulate and data fall more along the upward curves of biomass accumulation predicted for different drifter-following experiments from the model of Dugdale et al., (1990).

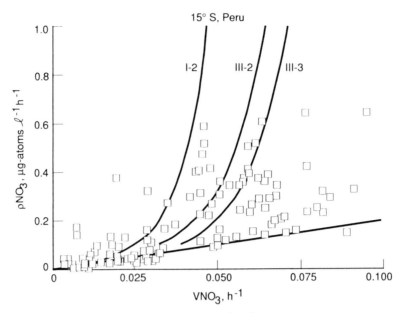

Fig. 3. New production rate (ρNO_3, $\mu mol\ l^{-1}\ h^{-1}$) versus specific nitrate uptake (VNO_3, h^{-1}) for the upwelling center at 15°S, Peru. Squares are ^{15}N measurements collected during Joint 2, 1977. The straight line is when PON remains constant and equal to the initial PON. The curved lines are theoretical relationships calculated from initial conditions, assuming all nitrate taken up is converted and retained as PON. Taken from Dugdale et al. (1990).

The nitrogen productivity cycle in high nutrient upwelling areas can be interpreted according to a conveyer belt hypothesis in which newly upwelled phytoplankton, with a history of light deprivation, adapt to advantageous surface conditions as they drift away from the upwelling center (MacIsaac et al., 1985; Wilkerson and Dugdale, 1987). There is a shift-up in phytoplankton metabolism, and first, VNO_3 increases with time, followed by carbon fixation. The acceleration of VNO_3 is a function of both the initial concentration of surface nitrate at the beginning of the upwelling cycle and the time elapsed since upwelling (Dugdale et al., 1990; Zimmerman et al., 1987). Reduced incoming irradiance may lower the rate of acceleration. The interpretation that VNO_3 acceleration is a physiological response may be challenged on the basis that the increasing VNO_3 is either an artefact due to decreasing detrital particulate nitrogen or due to changes in phytoplankton species composition (such that low nitrate uptake species are outcompeted by high nitrate uptake species), or both. Species shifts were not apparent during shipboard enclosure experiments carried out with water from the upwelling center at Point Conception, California (Dugdale and Wilkerson, unpublished). Typically, the population was dominated in numbers by picoplankton and in volume by diatoms. For example, in one enclosure, *Thalassiosira rotula* dominated the total cell volume from the beginning of the experiment through the period when acceleration of VNO_3 was measured. To examine whether VNO_3 changes were caused by changes in detrital PON, the ratio of VNO_3 to VNH_4 (both using saturated concentrations of ^{15}N innoculum) was examined as an alternative to VNO_3 alone. This parameter, which is not affected by the detrital or non-phytoplankton particulate nitrogen, was found to increase from 0.3 to 1.8 when acceleration was measured.

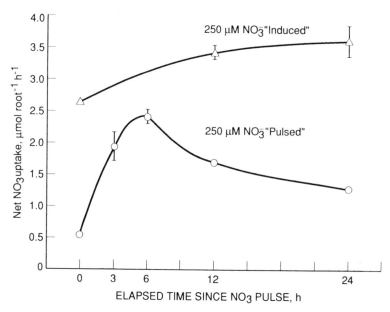

Fig. 4. Net nitrate uptake of maize roots given 250 μM nitrate either continuously (triangles) or in a one-hour pulse (circles). Details in MacKown and McClure (1988).

The observation in the enclosure experiments made at Point Conception, California of $VNO_{3(max)}$ values higher than $VNH_{4(max)}$, i.e. the ratio of saturated VNO_3/saturated VNH_4 being greater than one, is supported by the results of Dortch et al. (1991). *Thalassiosira pseudonana* grown under nitrate limitation at $\mu/\mu_{max} = 0.51$ developed a higher short-term rate of uptake of nitrate than for ammonium in cultures grown under ammonium limitation at about the same value of μ/μ_{max}. These results also are consistent with those of Demanche et al. (1979), which demonstrated the ability of *Skeletonema costatum* to take up nitrate and form internal storage pools at very high initial rates.

Several key features of the shift-up scenario described here for reaching high new production in high-nutrient regimes are supported by laboratory studies. Higher plant systems have been shown to develop an accelerated rate of nitrate uptake following initial exposure to nitrate concentrations in the 10 μM range (Jackson et al., 1986). MacKown and McClure (1988) measured 4 to 6-fold increases in net nitrate uptake of maize roots within 12 hours of exposure to 10 μM nitrate, and a 3-fold increase, when the inducing concentration was increased to 250 μM. Inhibition of accelerated nitrate uptake by cyclohexamide (an inhibitor of protein synthesis at the translation level) occurred, in agreement with previous observations by Jackson et al. (1986) that for development of accelerated nitrate uptake protein synthesis was required. Figure 4, taken from MacKown and McClure (1988), shows nitrate uptake to increase and then decrease when pulsed nitrate was provided at a concentration of 250 μM, the uptake being modulated by internal nitrate levels. When nitrate was provided continuously (the induced curve), the increase of nitrate uptake continued, i.e., acceleration of nitrate uptake was sustained. This parallels our results from ship-board enclosures filled with upwelled water from Point Conception, California, in which acceleration of VNO_3 was driven by ambient nitrate in the upwelled water; when nutrients were added, this acceleration was maintained (Figure 5).

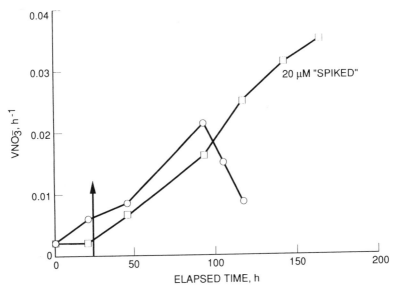

Fig. 5. Time course of nitrate uptake in shipboard enclosures after filling at the upwelling center at Point Conception, CA. Squares = Barrel 58B2; circles = Barrel 58B3. Arrow indicates addition of 20 μM nitrate to 58B2. Taken from Wilkerson and Dugdale (1987).

Shift-up and shift-down of nitrate metabolism by upwelled phytoplankton populations also showed a similar time response as maize roots when exposed to suitable conditions that initiated shift-up. Our upwelling studies showed peak VNO_3 (i.e., maximum shift-up) to be reached 5 to 7 days since upwelling (Figure 6). Bowsher et al. (1991) showed nitrate reductase (NR) and nitrite reductase (NiR) activity in maize shoots and roots to reach a maximum 5-6 days after transferring the seedlings from plates to a hydroponic system with a 16 hour light/8 hour dark photoperiod, and providing 10 mM KNO_3, 24 hours before harvesting (Figure 7). Assimilation of nitrate taken up requires the action of both NR and NiR, and the synthesis of these enzymes requires light and nitrate in higher plants (Beevers and Hageman, 1980; Gupta and Beevers, 1984). Induction of the system occurs at the transcription level, and in the study by Bowsher et al. (1991), synthesis of NR mRNA and NR activity correlated well under a light-dark regime after 6 days, but in darkness declined to zero within 8 hours. These observations fit well with those of Smith (pers. comm) for *Skeletonema costatum* in a laboratory study specifically designed to mimic conditions in upwelling production cycles. *S. costatum* cells responded to light and nitrate shifts by an increase in nitrate reductase transcript followed by nitrogen reductase protein. The change in cellular constituents was in an ordered fashion much as Schaechter (1968) had proposed for *Escherichia coli* in an early description of the shift-up phenomenon in bacteria. The acceleration rates for VNO_3 in this laboratory study were similar to those reported for natural upwelling populations (Table 3).

The uptake and assimilation of nitrate by micro-algae (especially diatoms) in steady state culture is described by Michaelis-Menten kinetics with a K_s of about 1 μM (e.g., Dugdale, 1976; Eppley et al., 1969) and concentrations of nitrate somewhat above the K_s value of 1 μM, e.g., 5 μM, should be sufficient to saturate the uptake and assimilation systems of phytoplankton. However, as more data become available from

Table 3.   Acceleration of VNO_3 in Upwelling and Simulated Upwelling Situations

Location/ Organism	Acceleration Rate $h^{-2} \times 10^{-4}$	Comments	Reference
15°S, Peru 1977	5.5	Drogue I-2, 100%LPD	Dugdale et al., 1990
	3.2	Drogue I-2, 50%LPD	Dugdale et al., 1990
	15.8	Drogue III-2, 50%LPD	Dugdale et al., 1990
	12.4	Drogue III-3, 100%LPD	Dugdale et al., 1990
Point Conception, CA 1983	3.8	Drifter S77, 100%LPD	Dugdale et al., 1990
	1.2	Drifter S239, 100%LPD	Dugdale et al., 1990
	12.0	Barrel 74B1	Wilkerson & Dugdale, 1987
Skeletonema costatum	7.4	Laboratory culture	Smith, G.J (pers. comm.)

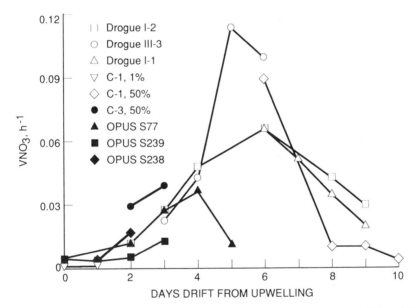

Fig. 6.   Time course of nitrate uptake in a series of hold-over and drifter-following experiments carried out at 15°S, Peru and Point Conception, CA. For details see Dugdale et al. (1990).

various upwelling areas, it has become apparent that for high VNO_3 and ρNO_3 values, nitrate concentrations of greater than 5 μM may be necessary (e.g., Dugdale, 1985). Further, when the maximum surface VNO_3 observed during a cruise to a particular upwelling area was plotted against the maximum surface nitrate concentration, a straight line with a positive slope was found (Figure 8) with an intercept on the nitrate axis above 5 μM.

This relationship is consistent with induction kinetics described in the experiments of Robertson and Button (1987), which show concentration and time-

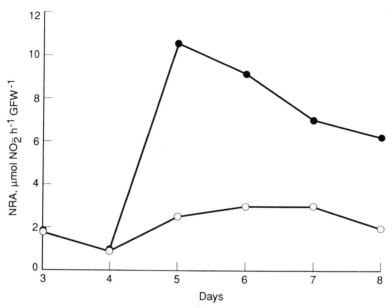

Fig. 7. Time course of nitrate reductase activity in maize shoots (closed circles) and roots (open circles) after induction with 16 hour light/8 hour dark photoperiod and 10 mM KNO_3. For details see Bowsher et al. (1991).

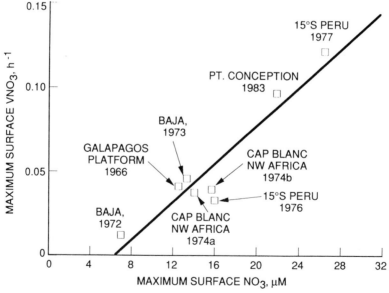

Fig. 8. Regression of maximum surface nitrate uptake versus maximum surface nitrate for the following coastal upwelling cruises ($r^2 = 0.9$): Baja, 1972 = Mescal 1; Baja, 1973 = Mescal 2; Galapagos Platform, 1966 = R/V Anton Bruun 15; Cap Blanc NW Africa, 1974a = Joint 1; Cap Blanc NW Africa, 1974b = Cineca V; 15°S Peru, 1976 = Joint 2-76; 15°S Peru, 1977 = Joint 2-77; Pt. Conception = OPUS83.

dependent induction of toluene uptake and metabolism by bacteria. The pattern for toluene induction by strongly repressed *Pseudomonas* is similar (Figure 9) to that for nitrate uptake (Figure 8) by phytoplankton in different upwelling systems. In both data sets, a 'foot' region appears in which the inducing concentration of substrate (either toluene as in Figure 9 or nitrate as in Figure 8) is too low to induce increased enzymatic activity, enzyme synthesis or nitrate uptake, and assimilation. In higher plants, which commonly show two nitrate transport systems, one with a high maximal uptake rate and low affinity for nitrate, and the other with a low maximal uptake rate and high affinity for nitrate, the initial nitrate concentration determines the pattern of induction of nitrate uptake, possibly due to the different responses of the various uptake systems (Larsson and Ingemarsson, 1989). Serra et al. (1978) showed a transition from hyperbolic, carrier-mediated kinetics to linear kinetics in *Skeletonema costatum* at 6 μM, virtually the threshold value that was found in upwelling areas (Figure 8). Other examples of nutrient concentration thresholds already known, include a minimum silicate concentration for initiation of silicate uptake in diatoms (Paasche, 1973), and an intercept on the nitrate axis for the regression of assimilation number versus nitrate concentration in the upwelling regions of northwest Africa and Peru (Huntsman and Barber, 1977).

LOW NEW PRODUCTION IN HIGH-NUTRIENT REGIONS

Certain areas exhibit surface nutrients well above detectable levels and high relative to oligotrophic regimes, but do not attain high new production rates (Table 1). Three such areas are the equatorial Pacific, the Southern Ocean, and the northeast Pacific near weather station Papa. These areas fail to achieve high new production as a result of low values of VNO_3, i.e., these regions fail to shift-up nitrate uptake to expected levels, and have low particulate nitrogen biomass. The model of Zimmerman et al. (1987) suggests that the mixed layer depth is too great for shift-up to occur in these regions (Dugdale and Wilkerson, in press). Martin and Fitzwater (1988) and Martin et al. (1990) suggest that lack of Fe from atmospheric fallout is a common cause for such low new production. However, there are few field observations of Fe concentrations, and there is a lack of agreement on the interpretation of experimental results (Banse, 1990; Dugdale and Wilkerson, 1990). Experimental additions of Fe, chelators, and nitrate to equatorial waters showed the most response to added nitrate and no effect of Fe additions on VNO_3 (Yang and Dugdale, 1990). Other explanations for new production anomalies in these areas exist, mostly based on sparse data but worthy of consideration. An explanation favored by Chavez (1989) for the eastern equatorial Pacific is a lack of seed diatom populations. Grazing control has been proposed for Station P by Frost (1987) and Miller et al. (1988), and for the eastern equatorial Pacific (Cullen et al., in press). Chronic low temperature and weak stratification (Dugdale and Wilkerson, 1989; in press) are unique features of the Southern Ocean that will influence phytoplankton physiology. The Q_{10} values for nitrate uptake and other parameters are very high, and the slope of VNO_3 versus temperature in sea-ice algae has been shown to be very steep in the range of -1 to +3°C by Priscu et al. (1989), suggesting strong temperature control on new production in the Southern Ocean.

An important factor affecting new production in all of these areas may be the concentration of silicate. If we accept that the high new production condition is equivalent to diatom productivity, the potential role of silicate can not be ignored. In the equatorial Pacific, Pena et al. (1990) show a low concentration of silicate and a low silicate to nitrate ratio at 135°W that supports this. There are few investigators working in the field of silicate effects on primary productivity, perhaps due to the lack of an easy method for measuring rates of transfer through the ecosystem. Nevertheless, silicate is a key factor,

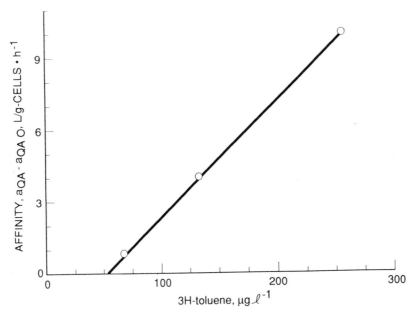

Fig. 9. Induction (increase in partial specific affinity) of toluene uptake by toluene-deprived *Pseudomonas* sp. strain T2 following exposure to toluene. For details see Robertson and Button (1987).

and has some interesting and complex features. For example, Paasche (1973) showed that silicate uptake by 5 planktonic diatoms required a threshold concentration that had to be subtracted from ambient values before computing kinetic parameters from Michaelis-Menten kinetics. In that study, dissolution of silicate simultaneously with uptake was considered a likely cause of the phenomenon, suggesting the possibility of a threshold concentration linked to temperature. Kinetics of silicate uptake in 5 Antarctic diatoms at 0°C were radically different from nitrate uptake kinetics in experiments of Sommer (1986). The values for K_s for nitrate ranged from 0.3 to 4.2 μM, within normal expectations, while the K_s for Si ranged from 5.7 μM to the extraordinarily high value of 88.7 μM. Values for VSi have been calculated for Sommer's data using a minimum value of silicate of 5 μM for the Drake Passage (Dugdale and Wilkerson, 1989; Sommer and Stabel, 1986) in an unmodified Michaelis-Menten expression. An estimate of the corresponding VNO_3 was obtained assuming balanced growth (i.e. VSi = VN = VC = ...), and a value of $f = 0.8$, taken as characteristic of a fully functioning eutrophic system. The computed minimum silicate-limited values of VNO_3 were 0.024 d^{-1} for *Corethon criophylum* and *Nitzchia kerguelensis*, the two diatoms with highest values of K_s for silicate of 60.1 and 88.7 μM respectively. The mean value of VNO_3 for the Scotia Sea was 0.031 d^{-1} (Table 1). The similarity in mean VNO_3 and silicate-limited estimate of VNO_3 are not inconsistent with silicate limitation in some regions of the Southern Ocean.

CONCLUSIONS

Two regimes of new production have been identified; one shows high VNO_3-high new production and is characteristic of coastal upwelling regions. The other has low VNO_3 values, low new production, and is characteristic of oligotrophic and some low production, high-nutrient regions.

Achievement of high new production depends upon several factors enabling development of high nitrate uptake rates and build-up of biomass. Nitrate and light are required for induction of nitrate uptake and for transcription and synthesis of assimilatory enzymes. The rate of induction is dependent upon the concentration of nitrate and a threshold concentration, or critical concentration of nitrate of about 6 μM appears in the field data from coastal upwelling areas to represent the minimum concentration for the initiation of the nitrate uptake and assimilation system. Next, there must be sufficient time in the high nitrate (and typically high light) situation for integration of induction to high VNO_3 values and high f conditions. As nitrate uptake proceeds, and if uptake is sufficiently rapid to overcome grazing and sinking losses, biomass will accumulate. This situation is greatly enhanced if there is a seed stock of bloom-forming diatom species that can become the major new producers and accumulate biomass readily. The product of high VNO_3 rates and high accumulation of biomass results in high new production. The sequence of events described above occurs cyclically in upwelling areas as upwelling favorable winds develop, and newly upwelled water drifts away from the upwelling center. In another mode, highest new production occurs as a result of sustained high nitrate concentrations, resulting in continuously shifted-up populations of bloom-forming diatoms. Both modes were observed by Dortch and Postel (1989) along the Washington coast, where hydrographic conditions favor retention of high concentrations of phytoplankton.

In areas where there is either too little nitrate for induction processes to occur at high rates, or because the seed populations are all nanoplankton or picoplankton, VNO_3 will not be able to increase rapidly enough to overcome the grazing pressure, and biomass will not build-up, in part, because of the small cell-size of the population. High rates of new production can not be attained as a result of both low VNO_3 and low biomass, and these systems remain locked in a grazing controlled loop. Low VNO_3 and new production in the presence of high nutrients may be the result of several limitations, including nutrient concentrations below threshold levels, inadequate levels of micro-nutrients, lack of seed stock of bloom-forming diatoms, low temperature, unfavorable physical conditions, and grazing.

The nitrate-based framework presented here does not explain every aspect of regional new production, but should be helpful in planning and executing the experiments and observations necessary to achieve an understanding of the key control mechanisms operating in the various regions of the ocean. When these key mechanisms are embedded in general circulation models, improved global new production estimates can be made, and a better understanding of the role of the marine phytoplankton in global atmospheric processes can be achieved.

ACKNOWLEDGMENTS

This study was supported by the following grants: National Science Foundation grant OCE-8710774, Office of Naval Research grant N00014-89-J-1423, and NASA grant 161-30-33 to C.O. Davis.

REFERENCES

Banse, K., 1990, Does iron really limit phytoplankton growth in the offshore subarctic Pacific? *Limnol. Oceanogr.*, 35: 772.

Beevers, L., and Hageman, R.H., 1980, Nitrate and nitrite reduction, *in*: "The Biochemistry of Plants," Vol 5, B.J. Miflin, ed., Academic Press, New York.

Bigelow, H.B., 1926, Plankton of the offshore waters of the Gulf of Maine, *U.S. Bur. Fish. Bull.*, 40: 1.

Bowsher, C.G., Long, D.M., Oaks, A., and Rothstein, S.J., 1991, Effect of light/dark cycles on expression of nitrate assimilatory genes in maize shoots and roots, *Plant Physiol.*, 95: 281.

Brown, C.M., MacDonald-Brown, D.S., and Stanley, S.O., 1975, Inorganic nitrogen metabolism in marine bacteria: nitrate uptake and reduction in a marine pseudomonad, *Mar. Biol.*, 31: 7.

Carpenter, E.J., 1983, Nitrogen fixation by marine Oscillatoria (*Trichodesmium*) in the world's oceans, *in*: "Nitrogen in the Marine Environment", E.J. Carpenter and D.G. Capone, eds., Academic Press.

Chavez, F.P., 1989, Size distribution of phytoplankton in the central and eastern tropical Pacific, *Global Biogeochemical Cycles*, 3: 27

Chisholm, S., This volume.

Codispoti, L.A., Dugdale, R.C., and Minas, H.J., 1982, A comparison of the nutrient regimes off Northwest Africa, Peru and Baja California, *Rapport et Proces-verbaux des reunions. Conseil permanent International pour l'Exploration de la Mer*, 180: 184.

Cox, R.D., 1980, Determination of nitrate and nitrite at the parts per billion level by chemoluminescence, *Anal. Chem.*, 52: 332.

Cullen, J.J., Lewis, M.R., Davis, C.O., and Barber, R.T., Photosynthetic characteristics and estimated growth rates indicate grazing is the proximate control of primary production in the equatorial Pacific. *J. Geophys. Res.*, in press.

Demanche, J.M., Curl, H.C., Lundy, D.W., and Donaghay, P.L., 1979, The rapid response of the marine diatom *Skeletonema costatum* to changes in external and internal nutrient concentration, *Mar. Biol.*, 53: 323.

Dortch, Q., Thompson, P.A., and Harrison, P.J., 1991, Variability in nitrate uptake kinetics in *Thalassiosira pseudonana* (Baccillariophyceae), *J.Phycol.*, 27: 35.

Dortch, Q., and Packard, T.T., 1989, Differences in biomass structure between oligotrophic and eutrophic marine ecosystems, *Deep-Sea Res.*, 36: 223.

Dortch, Q., and Postel, J.R., 1989, Biochemical indicators of N utilization by phytoplankton during upwelling off the Washington coast, *Limnol. Oceanogr.*, 34: 758.

Dugdale, R.C., 1976, Nutrient cycles, *in*: "The Ecology of the Sea", D.H. Cushing and J.J. Walsh, eds., Blackwells, London.

Dugdale, R.C., 1985, The effects of varying nutrient concentration on biological production in upwelling regions, *CalCOFI Rep.*, 26: 93.

Dugdale, R.C., and Goering, J.J., 1967, Uptake of new and regenerated forms of nitrogen in primary productivity, *Limnol. Oceanogr.*, 12: 196.

Dugdale, R.C., Morel, A., Bricaud, A., and Wilkerson, F.P., 1989, Modeling new production in upwelling centers: II A case-study of modeling new production from remotely sensed temperature and color, *J. Geophys. Res.*, 94: 18119.

Dugdale, R.C., and Wilkerson, F.P., 1986, The use of ^{15}N to measure nitrogen uptake in eutrophic oceans; experimental considerations, *Limnol. Oceanogr.*, 31: 673.

Dugdale, R.C., and Wilkerson, F.P., 1989, Regional Perspectives in Global New Production, *in*: "Oceanologie acualité et prospective," M.M. Denis, ed., Centre d'Oceanologie de Marseille, France.

Dugdale, R.C., and Wilkerson, F.P., 1990, Iron addition experiments in the Antarctic: a re-analysis, *Global Biogeochem. Cycles*, 4: 13.

Dugdale, R.C., and Wilkerson, F.P., 1991, Low specific nitrate uptake rate: a common feature of high nutrient, low chlorophyll marine ecosystems, *Limnol. Oceanogr.*, in press.

Dugdale, R.C., Wilkerson, F.P., Barber, R.T., and Chavez, F.P., 1991, Estimating new production in the equatorial Pacific at 150°W, *J. Geophys. Res.*, in press.

Dugdale, R.C., Wilkerson, F.P., and Morel, A., 1990, Realization of new production in coastal upwelling areas: a means to compare relative performance, *Limnol. Oceanogr.*, 35: 822.

Eppley, R.W., and Peterson, B.J., 1979, Particulate organic matter flux and planktonic new production in the deep ocean, *Nature*, 282: 677.

Eppley, R.W., Renger, E.H., and Harrison, W.G., 1979, Nitrate and phytoplankton production in southern California coastal waters, *Limnol. Oceanogr.*, 24: 483.

Eppley, R.W., Rogers, J.N., and McCarthy, J.J., 1969, Half stauration constants for uptake of nitrate and ammonium by marine phytoplankton, *Limnol. Oceanogr.*, 14: 912.

Frost, B.W., 1987, Grazing control of phytoplankton stock in the open subarctic Pacific Ocean: a model assessing the role of mesoplankton, particularly the large calanoid *Neocalanus* spp, *Mar. Ecol. Prog. Ser.*, 39: 49.

Garside, C., 1982, A chemoluminescent technique for the determination of nanomolar concentrations of nitrate and nitrite in seawater, *Mar. Chem.*, 11: 159.

Glibert, P.M., Biggs, D.C., and McCarthy, J.J., 1982, Utilization of ammonium and nitrate during the austral summer in the Scotia Sea, *Deep-Sea Res.*, 29: 837.

Glibert, P.M., Dennett, M.R., and Caron, D.A., 1988, Nitrogen uptake and NH_4 regeneration by pelagic microplankton and marine snow from the North Atlantic, *J. Mar. Res.*, 46: 837.

Glibert P.M. and Ray, R.T., 1990, Different patterns of growth and nitrogen uptake in two clones of marine *Synechoccus* spp., *Mar. Biol.*, 107: 273.

Glover, H.E., Prézelin, B.B. Campbell, L., Wyman, M., and Garside, C., 1988, A nitrate-dependent *Synechococcus* bloom in surface Sargasso Sea water, *Nature*, 331: 161.

Gupta, S.C., and Beevers, L., 1984, Synthesis and degradation of nitrite reductase in pea leaves, *Plant Physiol.*, 75: 251.

Huntsman, S.A., and Barber, R.T., 1977, Primary production off northwest Africa: the relationship to wind and nutrient conditions, *Deep-Sea Res.*, 24: 25.

Jackson, W.A., Pan, W.L., Moll, R.H., and Kamprath, E.J., 1986, Uptake, translocation and reduction of nitrate, *in*: "Biochemical Basis of Plant Breeding, Vol II Nitrogen Metabolism," V.A. Neyra, ed., CRC Press, Boca Raton.

Kopczak, C.D., Wilkerson, F.P., and Dugdale, R.C., 1990, Structure of the nutricline near Fieberling Guyot (32° 25′N, 127° 47′ W) in the eastern North Pacific, *Eos*, 71: 173.

Larsson, C-M., and Ingemarsson, B., 1989, Molecular aspects of nitrate uptake in higher plants, *in*: "Molecular and Genetic Aspects of Nitrate Assimilation", J.L. Wray and J.R. Kinghorn, eds., Oxford Science Publ.

MacIsaac, J.J., Dugdale, R.C., Barber, R.T., Blasco, D., and Packard, T.T., 1985, Primary production cycle in an upwelling center, *Deep-Sea Res.*, 32: 503.

MacKown, C.T., and McClure, P.R., 1988, Development of accelerated net nitrate uptake, *Plant Physiol.*, 87: 162.

Malone T.C., 1980, Size-fractionated primary productivity of marine phytoplankton, *in*: "Primary Productivity in the Sea", P.G. Falkowski, ed., Plenum Press, New York.

Martin, J.H., and Fitzwater, S., 1988, Iron deficiency limits phytoplankton growth in the north-east Pacific subarctic, *Nature*, 331: 341.

Martin, J.H., Fitzwater, S., and Gordon, R.M., 1990, Iron deficiency limits phytoplankton growth in Antarctic waters, *Global Biogeochem. Cycles*, 4: 5.

Martinez, L., Silver, M.W., King, J.M., and Alldredge, A., 1983, Nitrogen fixation by floating diatom mats, *Science*, 221: 152.

Michaels, A.F., and Silver, M.W., 1988, Primary production, sinking fluxes and the microbial food-web, *Deep-Sea Res.*, 35: 473.

Miller, C.B., and SUPER Group., 1988, Lower trophic level production dynamics in the oceanic subarctic Pacific Ocean, *Bull. Ocean Res. Inst., Univ Tokyo*, 26: 1.

Mitsui, A., Kumazawa, S., Takahashi, A., Ikemoto, H., Cao, S., and Arai, T., 1986, Strategy by which nitrogen-fixing unicellular cyanobacteria grow photoautotrophically, *Nature*, 323: 720.

Olson, R.J., 1980, Nitrate and ammonium uptake in Antarctic waters, *Limnol. Oceanogr.*, 25: 1064.

Paasche, E., 1973, Silicon and the ecology of marine plankton diatoms. II Silicate-uptake kinetics in five diatom species, *Mar. Biol.*, 19: 262.

Pena, M.A., Lewis, M.R., and Harrison, W.G., 1990, Primary productivity and size structure of phytoplankton biomass on a transect of the equator at 135°W in the Pacific, *Deep-Sea Res.*, 37: 295.

Platt, T., and Harrison, W.G., 1985, Biogenic fluxes of carbon and oxygen in the ocean, *Nature*, 318: 55.

Priscu, J.C., Palmisano, A.C., Priscu, L.R., and Sullivan, C.W., 1989, Temperature dependence of inorganic nitrogen uptake and assimilation in Antarctic sea-ice micro-algae, *Polar Biol.*, 9: 443.

Probyn, T.A., 1985, Nitrogen uptake by size fractionated phytoplankton in the southern Benguela upwelling system, *Mar. Ecol. Prog. Ser.*, 22: 249.

Probyn, T.A., 1990, Size-fractionated measurements of nitrogen uptake in aged upwelled waters: implications for pelagic food webs, *Limnol. Oceanogr.*, 35:202.

Robertson, B.R., and Button, D.K., 1987, Toluene induction and uptake kinetics and their inclusion in the specific affinity relationship for describing rates of hydrocarbon metabolism, *Appl. Environ. Microbiol.*, 53: 2193.

Serra, J.L., Llama, M.J., and Cadenas, E., 1978, Nitrate utilization by the diatom *Skeletonema costatum*. I. Kinetics of nitrate uptake, *Plant Physiol.*, 62: 987.

Schaechter, M., 1968, Growth: Cells and populations, *in*: "Biochemistry of Bacterial Growth," J. Mandelstam and K. McQuillen, eds., Wiley.

Sommer, U., 1986, Nitrate- and silicate-competition among Antarctic phytoplankton, *Mar. Biol.*, 91: 345.

Sommer, U., and Stabel, H-H., 1986, Near surface nutrient and phytoplankton distribution in the Drake Passage during early December, *Polar Biol.*, 6: 107.

Svederup, H.J., 1955, The place of physical oceanography in oceanographic research, *J. Mar. Res.*, 14: 287.

Wheeler, P.A., and Kokkinakis, S.A., 1990, Ammonium recycling limits nitrate use in the oceanic subarctic Pacific, *Limnol. Oceanogr.*, 35: 1267.

Wilkerson, F.P., and Dugdale, R.C., 1987, The use of large shipboard barrels and drifters to study the effects of coastal upwelling on phytoplankton nutrient dynamics, *Limnol. Oceanogr.*, 32: 368.

Wilkerson, F.P., and Dugdale, R.C., 1991, Measurements of nitrogen productivity in the equatorial Pacific, *J. Geophys. Res.*, in press.

Wilkerson, F.P., Dugdale, R.C., and Barber, R.T., 1987, Effects of El Niño on new, regenerated and total production in eastern boundary upwelling systems, *J. Geophys. Res.*, 92: 14347.

Wilkerson, F.P., Dugdale, R.C., and Kopczak, C.D., 1990, Measurements of new and regenerated production in the water column over Fieberling Guyot, *Eos*, 71: 173.

Yang, S.R., and Dugdale, R.C., 1990, Effect of iron on primary production in the equatorial Pacific, *Abst. ASLO*, Virginia, USA.

Zimmerman, R.C., Kremer, J.N., and Dugdale, R.C., 1987, Acceleration of uptake by phytoplankton in a coastal upwelling ecosystem: a modeling analysis, *Limnol. Oceanogr.*, 32: 359.

IRON AS A LIMITING FACTOR IN OCEANIC PRODUCTIVITY

John H. Martin

Moss Landing Marine Laboratories
Moss Landing, CA 95039

INTRODUCTION

Iron has been hypothesized to be a factor limiting phytoplankton standing crop in the ocean for decades. For example, in 1931, Gran suggested that, "If the productivity of the coastal waters is dependent on any factor of a chemical nature acting as a minimum factor, it must be an element which in its circulation does not follow the nitrates and phosphates accumulating in solution in the deep sea and reaching the surface again by vertical circulation of any kind. If such minimum stuffs exist, they must irreversibly go out of circulation in the sea, so that they can only be renewed from the land." Based on growth in culture solution, Gran (1931, p.41) concluded that lack or low concentration of iron probably limited plant growth at times and in areas of the sea where it was not replenished by land drainage. Previous data, and those obtained by Braarud & Klem (1931) off the Norwegian coast at his instigation, showed the iron content of sea water to be very small, ranging from 3 to 21 mg. Fe per m^3 (Harvey, 1938, p.205)

Three years later, in 1934, Hart suggested, "Among the . . . chemical constituents of sea water . . . possibly limiting phytoplankton production, iron may be mentioned . . . it may help to explain the observed richness of the neritic plankton . . . the land being regarded as a source of iron."

The above quotes remind us that iron availability is not a new addition to the list of basic factors regulating the plant life in the sea. Its importance has long been recognized, but "Our knowledge of the iron content is rather limited, as the quantities present are so minute that it is difficult to get the reagents sufficiently free from iron."(Gran, 1931). Only in the last few years have we learned that the iron content of sea water is very minute (Fig. 1). However, instead of the 3-21 mg m^{-3} noted by Harvey, the offshore-inshore range is more on the order of 0.001 to 0.5 mg m^{-3} (Martin and Gordon, 1988).

In view of the high loads of iron-rich sediments near shore, a confirmation of large values in coastal water for dissolved (e.g., 0.5 mg m^{-3} or about 10 nmol Fe kg^{-1}) and particulate iron was expected. What was really surprising was how low the offshore quantities were. For example, the Fe concentration for offshore surface water in Drake Passage is about 0.16 nmol kg^{-1} (Martin et al., 1990). Depending on the phytoplankton's

Primary Productivity and Biogeochemical Cycles in the Sea
Edited by P.G. Falkowski and A.D. Woodhead, Plenum Press, New York, 1992

C:Fe ratio (10,000 to 100,000C:1Fe; Morel and Hudson, 1985; Anderson and Morel, 1982), this would be enough Fe to produce a maximum of 16 μmol of phytoplankton C. Along with the Fe, there was 24.8 μmol NO_3 kg^{-1} i.e., assuming the Redfield ratio of 6.6C:1N, enough N to support the production of 160 μmol of C. Thus, based on Fe abundance only 10% of the NO_3 would be used by the phytoplankton (Open ocean C:Fe ratios may be as high as 500,000:1; see below).

These observations and those in the Gulf of Alaska (Martin et al., 1989) lead to the disturbing conclusion that open ocean water is basically infertile. Major nutrient-rich waters mixing up into the photic zone have scarcely any Fe, and, to support maximal growth via the use of major nutrients, supplemental Fe must be made available from other sources.

This argument leads to a second disturbing conclusion. Since we are talking about open ocean upwelling regions far removed from Fe-rich continental margins, the only way phytoplankton can obtain this essential element is via long-range, wind-blown transport and fallout of Fe-rich atmospheric dust originally derived from terrestrial arid regions (Duce, 1986). Scientists trying to understand phytoplankton production are used to thinking about major nutrient amounts and rates of supply, available light levels, and cell removal via zooplankton grazing (e.g., Harvey, 1938). If our current understanding is correct, oceanographers must also concern themselves with the amounts of dust falling out on these surface waters.

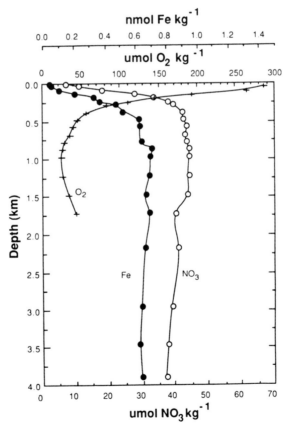

Fig. 1.   Vertical distribution of dissolved iron at Gulf of Alaska station, "P" together with oxygen and nitrate data (from Martin et al., 1989).

The recent knowledge gained about ocean Fe distributions and sources provides very strong indirect evidence supporting the hypothesis that iron is a limiting nutrient. Excess nutrients do not occur in Fe-rich coastal waters. For example, normal nutrient depletion occurs as the Alaskan continental margin is approached (Martin et al., 1989). As pointed out by Hart (1934) and El-Sayed (1988) among others, maximum populations of Antarctic phytoplankton is in Fe-rich neritic waters or along receding ice edges that also represent important Fe sources (Martin et al., 1990). Atmospheric dust loads in the Antarctic and equatorial Pacific are the lowest in the world (Prospero, 1981; Uematsu, 1987). It is noteworthy that excess nutrients do not occur in the equatorial Atlantic where large amounts of Saharan dust fall out on the sea surface (Prospero, 1981).

Direct evidence that open ocean phytoplankton benefit from atmospheric dust was recently obtained in a study in the equatorial Pacific. Open ocean aerosol was resuspended in clean seawater; growth rates in bottles with aliquots of the filtered leachate were 2-4 times higher than those in the controls without Fe (Martin et al., 1991). In general, these results were similar to those obtained in the Gulf of Alaska and Antarctic (Fig. 2).

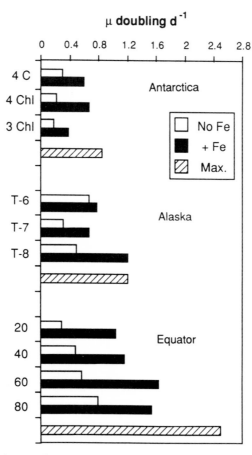

Fig. 2.   Comparisons of phytoplankton doubling rates with and without added iron. Theoretical maxima also are shown (from Martin et al., 1991).

Present-day atmospheric dust loads in the Antarctic are the lowest in the world (Prospero, 1981). However, this was not always so. The many active dune fields (Sarnthein, 1978) during the last glacial maximum (18,000 years ago) indicate that tropical arid areas were 5 times larger, wind speeds 1.3 to 1.6 times higher, and atmospheric dust loads 10-20 times greater than now (Petit et al., 1981). Furthermore, ice core data (De Angelis et al., 1987) show that large amounts of this dust were reaching the Antarctic; i.e., about 50 times as much windborne Fe was reaching the Antarctic during the last ice age than it is today.

When these data are compared with glacial-interglacial CO_2 data (Barnola et al., 1987), a striking inverse relationship is observed (Fig. 3). This leads to the hypothesis (Martin, 1990) that southern ocean phytoplankton received essential iron during the glacials from increased atmospheric dust input; the phytoplankton bloomed, the "biological pump" turned on, and CO_2 was withdrawn from the atmosphere. In contrast, present-day southern ocean phytoplankton are not receiving essential iron, the biological pump is turned off, and relatively little CO_2 is being removed from the atmosphere. Although this is an attractive hypothesis, there is little direct evidence supporting it. In fact, southern ocean glacial sediment opal levels are lower than those for the interglacial (see below).

Thoughts follow about how much Fe would be required to turn on the biological pump. Back-of-the-envelope calculations suggest that the number is surprisingly small, on the order of a few hundred thousand tons, which suggests that large-scale (100s of km) Fe-enrichment experiments could be performed. How large these experiments might become is a subject of much debate (Sarmiento, 1991; Lloyd, 1991).

Although much remains to be learned about Fe, especially in major nutrient-rich seas, the powerful effects of Fe have been documented in several independent field studies: e.g., in reporting work performed in the Gulf of Alaska, Coale (1991) states that "...Dramatic increases in phytoplankton productivity, chlorophyll-*a* and cell densities

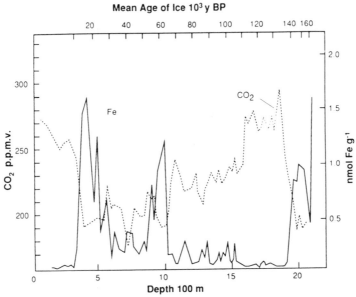

Fig. 3.   Iron and carbon dioxide versus age in Vostok ice core. Figure from Martin et al., (1989), based on original data of De Angelis et al., (1987) and Barnola et al., (1987).

occurred after the addition of 0.89 nM Fe..." de Baar et al., (1990) "...always observed Fe to stimulate chlorophyll-*a* synthesis and nutrient assimilation..." in the Scotia and Weddell Seas. In the central equatorial Pacific, "Addition of 1 nM Fe to seawater samples increased the final concentration of chlorophyll-*a* and particulate organic carbon and nitrogen..." (Price et al., 1991), while in the Drake Passage, "Fe addition...increased chlorophyll-*a* concentrations by a factor of 4 to 7 times..." (Helbling et al., 1991).

The above discussion summarizes the current situation in iron research in relation to other field studies, and I will not repeat detailed descriptions here. Instead, I will discuss my work in relation to some recent related research results; i.e., the very low C:Fe ratios of Sunda et al., (1991) and the lack of support for the glacial-interglacial CO_2 change iron hypothesis (Mortlock et al., 1991). I will also present some preliminary thoughts about a prototype large-scale enrichment experiment.

New C:Fe ratios

The most important factor in the iron-limitation hypothesis, that Fe levels in offshore sea water are so low that this essential element has to be provided via the fallout of Fe-rich dust, has recently been questioned. In laboratory experiments, Sunda et al., (1991) found an open ocean phytoplankton species to have extremely low Fe requirements. It was capable of rapid growth with Fe amounts resulting in C:Fe ratios of 500,000:1.

In the past, we and others have attempted to assign C:Fe ratios to better define the phytoplankton's needs for iron. For example, we used a range of 10,000:1 (Fe-replete, Morel and Hudson, 1985) to 100,000:1 (Fe-deplete, Anderson and Morel, 1982). We also used our estimate of 33,000 C:1 Fe, based on our field observations (Martin et al., 1989). Sunda and co-workers argue that previous C:Fe ratios were too low because the coastal species *Thalassiosira weissflogii* was used in the laboratory experiments; also, our field estimates were inaccurate because of a variety of errors. On the other hand, when using sophisticated techniques (100s of nmols of Fe, Cu, Zn, Mn, and Co are buffered with 0.1 mM EDTA to yield free ion Fe concentrations of 0.05 to 10 nM), the oceanic diatom *T. oceanica* achieved specific growth rates of ~ 1.0 (d^{-1}). These results indicate that this oceanic diatom can obtain near-maximum growth rates with much lower than the minimum amounts of Fe thought necessary to meet the metabolic needs of plant cells (Raven, 1988), i.e. less than 10% of the calculated amounts needed for growth, based on Fe enzymatic requirements in photosynthesis, respiration, and NO_3 reduction.

Sunda et al., (1991) believe that our assumption and that of Duce (1986), namely, that amounts of Fe in upwelling open ocean water were 10-100 times too low and that Fe had to be supplied via the atmosphere, should be reassessed. Accordingly, we re-examined our equatorial Pacific data (Martin et al., 1991) in light of their findings and in light of our own recent data on Fe distribution for the equatorial Pacific.

Under experimental conditions that we tried to keep as natural as possible (only ambient nutrients, phytoplankton species, and no chelaters), equatorial Pacific (0°; 140°W) phytoplankton obtained doubling growth rates of only 0.2 to 0.5 (d^{-1}) with the Fe available (Fig. 2). However, when we simulated a dust event by adding atmospheric-dust-sea-water leachate, these rates increased to 70-100% of those reported by Sunda et al., (1991).

We now have environmental Fe concentration data to better interpret our experimental findings for the equator (Fig. 4). When Fe amounts are compared with NO_3 profiles at the equatorial station, there is a subsurface "Redieldian" N:Fe ratio relationship commented on by Sunda et al.; i.e., converting to C by multiplying the NO_3 concentration by a 6.6 C:N ratio and dividing by dissolved Fe, gives ratios of 400,000-600,000 at 125-250 m depth. However, when approaching the euphotic zone ($\sim$ 80 m), the amounts of Fe become very low (<0.02-0.05 nmol Fe kg[-1]), while the NO_3 concentrations remain relatively high (Fig. 4). This results in ratios on the order of 800,000-2,700,000 C:Fe at the equator. The situation is worse at 3°S; near-surface ratios usually fall within the range of 1,200,000 to 5,200,000 C:Fe. These findings suggest that the dissolved Fe is removed preferentially, with the result that potential C:Fe ratios increase far beyond those reported by Sunda et al.

In our Fe enrichment experiments we always looked for the apparent C:Fe ratio but met with little success. Our results, instead, suggest a threshold effect. Below a certain level, growth is severely limited because diffusive rates to the cell surface are so low that the plants are unable to concentrate the Fe needed for maximal growth rates. This condition is reversed when the threshold level for diffusion to the cell surface is exceeded. Additional Fe has little effect. Our results suggest that in North Atlantic and Ross Sea experiments, the threshold is around 0.5 nmols while at the equator the threshold is around 0.3 nmols. These are rough estimates, because they are based on total Fe levels (aluminosilicate Fe, Fe oxides, extracellular Fe as well as biologically concentrated Fe); how much is in the cell is unknown.

These very low values remind me of Hudson and Morel's (1990) observations which I paraphrase here: open ocean Fe levels are so low that physical limits on diffusive transport are reached, only marginal Fe transport into the cell is achieved, and organisms cannot grow at maximal rates. A very sharp cut-off exists between organisms that can grow maximally at oceanic Fe levels (like the ones studied by Sunda et al., with C:Fe ratios of 500,000: 1) and those that must wait for periodic elevated Fe levels in order to grow (i.e., the ones that grow when we add Fe to bottles or the ones that grow after a dust event). I believe that this situation is reversed when a dust event occurs, Fe levels are raised, the phytoplankton can obtain sufficient Fe, and productivity rates increase. (e.g., see DiTullio and Laws, 1991; Young et al., 1991)

We have reassessed our previous conclusions in light of the findings of Sunda et al., (1991). We maintain that amounts of Fe upwelling to the surface in the open ocean are too low for maximal growth even if the phytoplankton C:Fe ratio is on the order of 500,000:1. We conclude that maximal phytoplankton growth cannot occur in the open ocean without supplemental Fe provided by the fall out of wind-blown atmospheric dust.

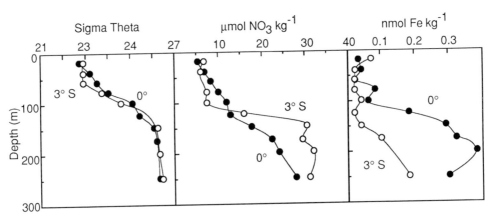

Fig. 4.   Comparison of density, nitrate, and dissolved iron on the equator and at 3°S;140°W .

Equatorial Pacific Productivity

 In the equatorial Pacific, it is almost always observed that primary productivity rates are highest in the immediate vicinity of the equator, in spite of the fact that nutrient levels are about the same a few degrees to the north and south. Chavez et al., (1991) suggested that the enhanced rates might be due to greater Fe availability. Barber and Chavez (1991) also argue convincingly that eastern equatorial Pacific phytoplankton benefit from enhanced Fe input. As mentioned above, we measured dissolved and particulate Fe levels on a section along 140°W (9°N to 3°S) in June/July of 1990. In reviewing the results, we compared the equator station with one at 3°S. The latter was selected because of the high NO_3 values and because of the very low input rates of atmospheric dust south of the equator (Uematsu et al., 1983).

 Although density and major nutrient levels (e.g., NO_3) were similar at the two stations, dissolved Fe amounts were substantially higher on the equator (Fig. 4). Fe began to increase with depth at the bottom of the photic zone and peaked at 200 m. In contrast, no increase in Fe was observed at 3°S shallower than 150 m. Furthermore, concentrations increased very slowly with depth in comparison to the equator. As noted above, a fairly good correlation between Fe:NO_3 was found at the equator (similar to that in the Gulf of Alaska), while no relationship was noted at 3°S.

 In addition to dissolved phases, suspended particulates were also analyzed for various trace elements. A pronounced maximum was observed for particulate aluminum at about 150m depth (Fig. 5) at the core of the equatorial counter current (ECC). The high particulate Al and Fe levels in the core suggest that small amounts of fine alumino-silicate particles may be transported away from the western Pacific margin out into the central equatorial Pacific via the rapidly flowing ECC. This source may also contribute to the larger dissolved Fe concentrations found here. In addition to the greater amounts of Fe, vigorous upwelling and shear mixing must also result in elevated Fe fluxes to the surface.

 I believe that the enhanced Fe supply enables phytoplankton at the equator to produce more chlorophyll than those at 3°S (Fig. 6). With the extra chlorophyll, more C is fixed in the water column although amounts of C fixed/unit chlorophyll are about the same (Fig. 6). Thus, with the extra Fe and chlorophyll, integrated productivities at the

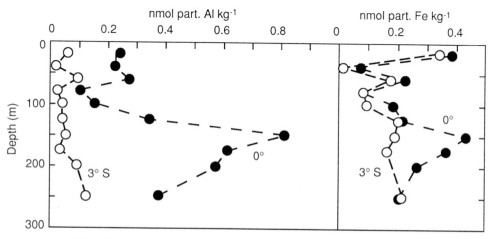

Fig. 5.   Particulate Al and Fe at the equator and 3°S; 140°W.

equator are twice as high as those 3°S (1280 vs 640 mg C m^{-2} day^{-1}). Nevertheless, these chlorophylls are very low in view of the levels of the major nutrients. As before, I contend that even right on the equator there is insufficient Fe in upwelling water, and that normally, open ocean phytoplankton have to obtain their Fe from the fallout of dust.

The Glacial-interglacial CO$_2$ Change Iron Hypothesis

The discovery that glacial atmospheric CO$_2$ concentrations were lower than those in the interglacials has led to many hypotheses. Three similar ones (Sarmiento and Toggweiler, 1984; Knox and McElroy, 1984; Siegenthaller and Wenk, 1984) all suggest that if the abundant nutrients in the southern ocean were used up, the consequent bloom could have resulted in the glacial drawdown in CO$_2$. In keeping with the iron deficiency hypothesis, I looked for evidence that Fe might have been more available during the ice ages. The answer was provided by the Vostok ice-core data of De Angelis et al. (1987). Because winds were stronger, and tropical arid areas larger, global atmospheric dust loads were 10-20 times higher during glacial periods. It is evident that much of this material was swept into the Antarctic because dust concentrations in the ice cores were 50 times higher than they were (and are) during past and present interglacials (Fig. 3). This finding suggests that Fe-rich dust blew into the Antarctic, the phytoplankton bloomed, the biological pump turned on, and CO$_2$ was withdrawn from the atmosphere (Martin, 1990).

One of the drawbacks of this hypothesis is that there is no direct evidence that ice-age productivity in the southern ocean was higher. Recently, Mortlock et al., (1991) analyzed cores from the South Atlantic near the present Polar Frontal Zone ($\sim$ 50°S). They concluded that productivity was not higher, but, on the contrary, was lower during glacial periods than it is today. One of their main lines of evidence involves the abundance of opal; south of the present polar frontal zone, opal in modern sediments is on the order of 70-75%, while during the last glacial, the opal content was around 40-55%.

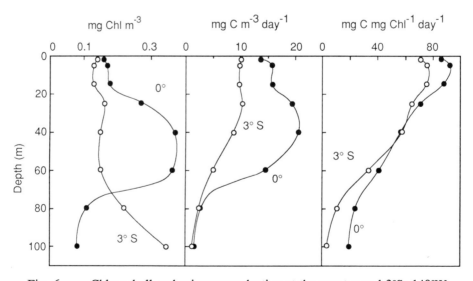

Fig. 6.   Chlorophyll and primary production at the equator and 3°S; 140°W.

In conversations with one of the authors (P.N. Froelich) of the paper, I pointed out that there is wide variability in diatom C:Si ratios (Fig. 7). One extreme is the North Atlantic, where there is very little silica in the water column. *Chaetoceros atlanticus* produced 95 umol C with only 3 umol Si; this gives a ratio of 28C:1Si. On the other hand, the mother load of opal and Si occurs in the Ross Sea (e.g., Dunbar et al., 1989), where phytoplankton C:Si ratios of 1-6.7 were reported (Nelson and Smith, 1986). Nevertheless, the addition of Fe appears to affect this ratio. By and large, Si is concentrated at about the same rates with and without Fe while the accumulation of organic C doubles or triples with Fe (Martin et al., 1991): this leads to C:Si ratios of 2.5 without Fe and 5.0 with Fe. Thus, if we assume that twice as much C was transported along with the Si found in the cores of Mortlock et al. (1991), it would mean that more C was being removed during the Fe-rich glacial periods in spite of larger percentages of opal accumulating in the sediments on the sea floor during the interglacials.

We also pointed out that one of the major southern ocean species, *Phaeocystis pouchetii*, has no opal. Perhaps this species bloomed during the ice ages and no hard-part record was left behind. Support for this idea is provided by the sulphur (originally dimethyl sulphide, DMS) data in the Vostok core (Legrand et al., 1991). In Fig. 8, I overlay their data with the Vostok core dust values of De Angelis et al., (1987). The fit is striking. Legrand et al. discuss the possibility that the sulphur might have the same terrestrial origin as the dust, but, after examining the evidence, they state that "..our data suggest enhanced productivity from the DMS-producing portion of the oceanic biota between 18 and 70 kyr BP (before present)." However, they go on to say that, "Whether this increase represents a real increase in productivity or simply an ecological shift favoring DMS producers is not known."

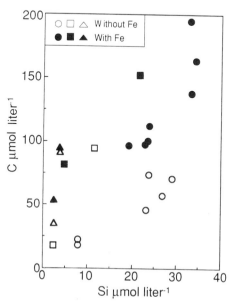

Fig. 7.   Silica versus carbon from the North Atlantic (triangles), the Ross Sea (circles), and the equatorial Pacific (squares).

Phaeocystis is known to be one of the major DMS producers (Barnard et al., 1984; Crocker et al., 1991; McTaggard, 1991). It also can rapidly transport large amounts of organic matter away from the surface via the formation of marine snow and subsequent rapid sinking (Wassmann et al., 1990). All in all, the glacial interglacial CO_2 change Fe hypothesis has neither been proved or disproved; i.e., quoting Berger (1991), "...Mortlock et al., stage an impressive attack on the problem of Antarctic productivity changes; it should severely damage conventional belief in Antarctic modulation of atmospheric carbon dioxide. Yet, the basic question is not resolved. The Antarctic has not been eliminated from contention as co-regulator of CO_2 and Martin's iron hypothesis has not been discredited."

Large-Scale Iron Enrichment Experiments

Although much has been learned, and there is still more to learn from Fe bottle experiments, the basic criticism remains that large grazers are excluded. Hence, it is time to start thinking about large-scale enrichment experiments that can reflect the use of both natural and artificial Fe sources. For example, the marked plumes of color streaming off the Galapagos (Feldman, 1986) suggest that these islands may represent a discrete Fe source. If this is confirmed, much could be learned via upstream/downstream Galapagos studies (Minas et al., 1990) without having to introduce iron artificially.

Nevertheless, the time has come to begin serious planning of a relatively large-scale Fe enrichment experiment. Several questions must be addressed. Where will the experiment be conducted? What form of Fe will be used? How will it be added? How much will it cost?

In thinking about where to do the experiment, several factors must be considered; logistical convenience, weather, international politics, availability of background information and other limiting factors such as light, temperature, and major nutrients. This list more or less eliminates the Antarctic; logistics are difficult, the seas are rough, work would have to be done with approval of the signatories of the Antarctic Treaty (at least south of 56°S). Ambient light levels and low temperatures can result in slow growth,

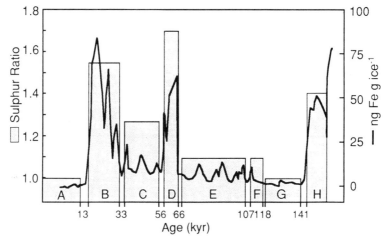

Fig. 8.   Iron and sulphur data from the Vostok ice core. Iron data are based on Al data of De Angelis et al., (1987). Sulphur data are ratios between various glacial stages and the present glacial stage "A" (from Legrand et al., 1991).

which would necessitate following a fertilized patch for a long period. Furthermore, there is relatively little background information to enable a decision to be made on exactly where the experiment should be done.

In contrast, the Gulf of Alaska has many advantages; convenient logistics, fairly good weather in late summer, no territorial problems, intermediate light and temperature levels, and abundant background information from work by the SUPER group and others at Ocean Station P. On the other hand, all of these factors also are true for the Equatorial Pacific. If the work was done south of the Galapagos ($\sim 5^\circ$S; 90°W), steaming time from Ecuador would be about two days, and if the weather is good, rapid growth should result from high light levels and temperatures, the experiment would be in international waters, and there is abundant background information on the physical oceanography of this area. As noted above, there is the additional opportunity of the natural Fe enrichment that appears to be occurring at the Galapagos Island. For these reasons, the equatorial Pacific merits serious consideration for the first enrichment experiment.

The area south of the Galapagos and much of the South Equatorial Current ($\sim 5^\circ$S) appears to have nearly linear flow. This is indicated by drifting buoys drogued to 15m set out by Don Hansen who informs me (pers. commun.) that the 90° - 100° W SEC region appears to be ideal since the drifters go due west or southwest throughout the year. This finding suggests that there is surface laminar flow and that buoys placed on the four corners of an enriched patch might facilitate tracking. On the other hand, there is a shear problem because the wind direction is opposite that of the pressure field (Wanninkhof, pers. commun.) and a surface patch would "smear out" about 10 km each day.

Currents are very swift here; e.g., Hansen (pers. commun.) reports that average SEC current speeds in March, 1991 were 65 cm sec^{-1}, so that the patch of water in a week-long experiment would move about 400 km to the west. Nevertheless, the drifting buoys suggest that the patch would maintain its integrity and the Fe should stay with it; this would have to be verified.

To learn more, we will propose seeding a patch of water with Fe and a purposeful tracer such as sulphur hexaflouride (e.g., Wanninkhof et al., 1991). Drifters would be placed at the four corners of a patch 20 km long and 5 km wide (long axis perpendicular to the equator). A ship steaming at 12 knots could cover the long axis about once each hour and 24 passes would be made each day.

How much Fe would be needed to seed this patch? The NO_3 concentration here is about 6-7 umol liter^{-1}. Assuming a N:Fe ratio of 5,000N:1Fe, then, 7000 nmol NO_3/ 5000 = 1.4 nmol. To add a safety factor, we double this amount and by converting from nmol to ng we estimate a requirement of ~ 150 ng/liter or 150 mg m^{-3}. The area of the patch (5,000 X 20,000m) is 1×10^8 m^2 ; thus, for each meter of downward mixing, 15 kg of Fe will be required; e.g., if the Fe mixes down to 20 m, the total volume will be 20 m X 1×10^8 m^2 = 2×10^9 m^3. This volume times the Fe concentration is 3×10^{11} ug or 300 kg of Fe. This is a very small amount of Fe and demonstrates the feasibility of Fe fertilization.

How would this Fe be applied? Airplanes are very difficult to deal with in remote ocean regions, and, in general, they are cost-prohibitive. It would be best if the Fe could be applied from a ship. If the Fe was injected into the wash from a ship's propellor (assume the wash is 3 m deep and 10 m wide) as the ship is steaming at 12 knots (about 20,000 m/hr), Fe could be injected into 600,000 m^3 of water each hr, or 1.4×10^7 m^3 each day. This would cover about 5% of the upper 3 m of the 5 X 20 km patch. To put in the 300 kg of Fe in a day (assuming a 20 m patch depth) would require an Fe concentration of about 20 mg Fe m^{-3}, which leads to problems about Fe solubility.

The solubility of Fe is thought to be about 1.1 mg m^{-3} (Byrne and Kester, 1975). As the Fe is introduced, it will be rapidly oxidized from the soluble $+2$ to insoluble $+3$ state, forming Fe(OH)$_3$. This hydroxy complex may remain in the water column and go back into solution as the fertilized waters mix with the Fe-poor background waters. If, on the other hand, the hydroxy-complexed Fe sinks out of the surface water, only about 10% of the introduced Fe would be available to the phytoplankton. If this occurs, the Fe will have to be introduced at a much slower rate, or a chelated Fe source will have to be used. Adding Fe to the patch at the solubility maximum, 1.1 mg m^{-3}, would require 19 days; this is unacceptable. The patch would have moved over 1000 km by the time fertilization was completed. An alternative Fe source that would stay in solution would have to be used.

An inorganic Fe source might be FeCl$_3$ or FeNO$_3$. FeCl$_3$ costs about \$150/drum. The volume of five 55 gallon drums is $\sim$ 1000 liters. Depending on dilution, the five drums would contain the required 300 kg of Fe. Thus, the Fe for this experiment might cost as little as \$750. This estimate is probably overly optimistic, but even if I am off by an order of magnitude, the price of the Fe for such an experiment is very reasonable.

Thus, I envision a preliminary experiment where Fe and a purposeful tracer would be spread over a 100 km^2 patch in one day. The ship would steam back and forth over the patch between the drogue buoys located at each corner of the rectangle. The mixing of the Fe and tracer away from surface and within the patch would be observed, as well as the apparent loss from the patch. The hexaflouride tracer would be monitored via a flow-through gas chromatograph. Fe would be measured in real time via the chemoluminescent technique of Elrod et al., (1991). Chlorophyll production and NO$_3$ uptake also would be monitored. The entire experiment should take about a week. The modest cost of this preliminary experiment would enable better design to be made of larger efforts, on scales of 100s of km, which would, in turn, enable a fuller understanding of the role of Fe as a limiting factor in the sea.

ACKNOWLEDGMENTS

Our research is supported by grants from the ONR Ocean Chemistry Program (N 000 14-84-C-0619), the NSF Marine Chemistry Program (OCE 8813565), and the Biology and Medicine Section of NSF Polar Programs (DPP 8716460).

REFERENCES

Anderson, G. C., and Morel, F. M. M., 1982, The influence of aqueous iron chemistry on the uptake of iron by the coastal diatom *Thalassiosira weissflogii, Limnol. Oceanogr.*, 27:789.

Barber, R. T., and Chavez, F. P., 1991, Regulation of primary productivity rate in the equatorial Pacific Ocean, *in*: "What Controls Phytoplankton Production in Nutrient-rich Areas of the Open Sea?", ASLO Symposium, Lake San Marcos, California, February 22-24, 1991, S.W. Chisholm and F.M.M. Morel, eds., Allen Press, Lawrence, in press.

Barnard, W.R., Andreae, M. O., and Iverson, R. L., 1984, Dimethyl sulphide and *Phaeocystis pouchetii* in the southeastern Bering Sea, *Cont. Shelf Res.*, 3:103.

Barnola, J. M., Raynaud, D., Korotkevich, Y. S., and Lorius, C., 1987, Vostok ice core provides 160,000 year record of atmospheric CO_2, *Nature*, 329:408.

Berger, W. H., 1991, No change down under, *Nature*, 351:186.

Braarud, T., and Klem, A., 1931, Hydrological and chemical investigations in the sea off More. Skr. *Uttgift av Hvalradets ved Univ. Biol. Lab. Oslo*, No. 1.

Byrne, R. H., and Kester, D. R., 1975, Solubility of hydrous ferric oxide and iron speciation in seawater, *Mar. Chem.*, 4:255.

Chavez, F. P., Buck, K.R., and Barber, R. T., 1990, Phytoplankton taxa in relation to primary production in the equatorial Pacific, *Deep-Sea Res.*, 37:1733.

Coale, K. H., 1991, Effects of iron, managnese, copper and zinc enrichments on productivity and biomass in the subarctic Pacific, *in*: "What Controls Phytoplankton Production in Nutrient-rich Areas of the Open Sea?", ASLO Symposium, Lake San Marcos, California, February 22-24, 1991, S.W. Chisholm and F.M.M. Morel, eds., Allen Press, Lawrence, in press.

Crocker, K., Ondrescek, M., and Petty, R., 1991, Dimethylsulphide production from a *Phaeocystis* bloom in the Bellingshausen Sea, Antarctica. Abstract, in press.

de Baar, H. J. W., Buma, A. G. J., Nolting, R. F., Cadee, G. C., Jaques, G., and Treguer, 1990, On iron limitation in the Southern Ocean: experimental observations in the Weddell and Scotia Seas, *Mar. Ecol. Prog. Ser.*, 65:105.

De Angelis, M., Barkov, N. I., and Petrov, V.N., 1987, Aerosol concentrations over the last climatic cycle (160 kyr) from an Antarctic ice core, *Nature*, 325:318.

DiTullio, G. R., and Laws, E. A., 1991, Impact of an atmospheric-oceanic disturbance on phytoplankton community dynamics in the north Pacific central gyre, *Deep-Sea Res.*, (in press).

Duce, R. A., 1986, The impact of atmospheric nitrogen, phosphorus, and iron species on marine biological productivity, *in*: "The Role of Air-sea Exchange in Geochemical Cycling," P. Buat-Menard, ed., D. Reidel Publishing Company, Dordrecht.

Dunbar, R. B., Leventer, A. R., and Stockton, W. L., 1989, Biogenic sedimentation in McMurdo Sound, Antarctica, *Mar Geol.*, 85:155.

Elrod, V. A., Johnson, K.S., and Coale, K. H., 1991, Determination of subnanomolar levels of iron(II) and total dissolved iron in seawater by flow injection analysis with chemiluminescence detection, *Anal. Chem.*, 63:893.

El-Sayed, S. Z., 1988, Productivity of the Southern Ocean: A closer look, *Compar. Biochem. Physiol.*, 90B:489.

Feldman, G. C., 1986, Patterns of phytoplankton production around the Galapagos Islands, *in*: "Tidal Mixing and Plankton Dynamics; Lecture Notes on Coastal and Estraurine Studies, Vol. 17, J. Bowman, M. Yentsch, and W.T. Peterson, eds., Springer-Verlag, Berlin.

Gran, H. H., 1931, On the conditions for the production of plankton in the sea, *Rapp. Proc. Verb. Cons. Int. Explor. Mer.*, 75:37.

Hart, T. J., 1934, On the phytoplankton of the south-west Atlantic and the Bellingshausen Sea, 1929-31, *Discovery Reports, Vol. VIII*.

Harvey, H. W., 1938, The supply of iron to diatoms. *J. Mar. Biol. Assoc., U.K.*, 22:205.

Helbling, E. W., Villafane, V., and Holm-Hansen, O., 1991, Effect of Fe on productivity and size distribution of Antarctic phytoplankton, *in*: "What Controls Phytoplankton Production in Nutrient-rich Areas of the Open Sea?", ASLO Symposium, Lake San Marcos, California, February 22-24, 1991, S.W. Chisholm and F.M.M. Morel, eds., Allen Press, Lawrence, in press.

Hudson, J. M., and Morel, F. M. M., 1990, Iron transport in marine phytoplankton: Kinetics of cellular and medium coordination reactions, *Limnol. Oceanogr.*, 35:1002.

Knox, F., and McElroy, M. B., 1984, Changes in atmospheric carbon dioxide: Influence of the marine biota at high latitude, *J. Geophys. Res.*, 89:4629.

Legrand, M., Feniet-Saigne, C., Saltzman, E. S., Germain, C., Barkov, N.I., and Petrov, V.N., 1991, Ice-core record of oceanic emissions of dimethylsulphide during the last climate cycle, *Nature*, 350:144.

Lloyd, P., 1991, Iron determinations, *Nature*, 350:19.

Martin, J. H., 1990, Glacial-Interglacial CO_2 change: The iron hypothesis, *Paleoceanogr*, 5:1.

Martin, J. H., and Gordon, R. M., 1988, Northeast Pacific iron distributions in relation to phytoplankton productivity, *Deep-Sea Res.*, 35:177.

Martin, J. H., Gordon, R. M., Fitzwater, S., and Broenkow, W. W., 1989, VERTEX: Phytoplankton/iron studies in the Gulf of Alaska, *Deep-Sea Res.*, 36:649.

Martin, J. H., Gordon, R. M., and Fitzwater, S.E., 1990, Iron in Antarctic waters, *Nature*, 345:156.

Martin, J. H., Gordon, R. M., and Fitzwater, S. E., 1991, The case for iron, *in*: "What Controls Phytoplankton Production in Nutrient-rich Areas of the Open Sea?", ASLO Symposium, Lake San Marcos, California, February 22-24, 1991, S.W. Chisholm and F.M.M. Morel, eds., Allen Press, Lawrence, in press.

McTaggert, A., 1991, The biogeochemistry of dimethylsulphide in Antarctic coastal seawater, Abstract, this volume.

Minas, H. J., Coste, B., Minas, M., and Raimbault, P., 1990, Conditions hydrologiques, chimiques et production primaire dans les upwellings du Perou et des iles Galapagos, en regime d'hiver austral (campagne Paciprod), *Oceanologia Acta*, 10:383.

Morel, F. M. M., and Hudson, R. J. M., 1985, The cycle of trace elements in aquatic systems: Redfield revisited, *in*: "Chemical Processes in Lakes," W. Stumm, ed., Wiley, New York.

Mortlock, R. A., Charles, C. D., Froelich, P. N., Zibello, M. A., Saltzman, J., Hays, J.D., and Burckle, L. H., 1991, Evidence for lower productivity in the Antarctic Ocean during the last glaciation, *Nature*, 351:220.

Nelson, D. M., and Smith, W. O., Jr., 1986, Phytoplankton bloom dynamics of the western Ross Sea ice edge - II. Mesoscale cycling of nitrogen and silicon, *Deep-Sea Res.*, 33:1389.

Petit, J. R., Briat, M., and Royer, A., 1981, Ice age aerosol content from East Antarctic ice core samples and past wind strength, *Nature*, 293:391.

Price, N. M., Andersen, L.F., and Morel, F. M. M., 1991, Iron and nitrogen nutrition of equatorial Pacific plankton, *Deep-Sea Res.*, (in press).

Prospero, J. M., 1981, Eolian transport to the world ocean, *in*: "The Sea," Vol. 7, C. Emiliani, ed., Wiley, N.Y.

Raven, J. A., 1988, The iron and molybdenum use efficiencies of plant growth with different energy, carbon and nitrogen sources, *New Phytol.*, 109:279.

Sarmiento, J. L., 1991, Slowing the buildup of fossil CO_2 in the atmosphere by iron fertilization: A comment, *Global Biogeochem. Cycles*, 5:1.

Sarmiento, J. L., and Toggweiler, L. R., 1984, A new model for the role of the oceans in determining atmospheric carbon dioxide levels, *Nature*, 308:621.

Sarnthein, M., 1978, Sand deserts during glacial maximum and climatic optimum, *Nature*, 272:43.

Siegenthaler, U., and Wenk, T. H., 1984, Rapid atmospheric CO_2 variations and ocean circulation, *Nature*, 308:624.

Sunda, W. G., Swift, D. G., and Huntsman, S. A., 1991, Low iron requirement in oceanic phytoplankton, *Nature*, 351:55.

Uematsu, M., 1987, Study of the continental material transported through the atmosphere to the ocean, *J. Oceanogr. Soc. Japan*, 43:395.

Uematsu, M., Duce, R. A., Prospero, J. M., Chen, L., Merrill, J. T., and McDonald, R. L., 1983, Transport of mineral aerosol from Asia over the North Pacific Ocean, *J. Geophys. Res.*, 88:5343.

Wanninkhof, R., Ledwell, J. R., and Watson, A. J., 1991, Analysis of sulphur hexaflouride in seawater, *J. Geophys. Res.*, 96:8733.

Wassmann, P., Vernet, M., Mitchell, B. G., and Rey, F., 1990, Mass sedimentation of *Phaeocystis pouchetii* in the Barents Sea, *Mar. Ecol. Prog. Series*, 66:183.

Young, R. W., Carder, K. L., Betzer, P. R., Costello, D. K., Duce, R. A., DiTullio, G. R., Tindale, N. W., Laws, E. A., Uematsu, M., Merrill, J. T., and Feely, R. A., 1991, Atmospheric iron inputs and primary productivity: Phytoplankton responses in the north Pacific, *Global Biogeochem. Cycles* (in press).

SATELLITE OCEAN COLOR OBSERVATIONS OF GLOBAL BIOGEOCHEMICAL CYCLES

Marlon R. Lewis

Department of Oceanography
Dalhousie University
Halifax, Nova Scotia B3H 4J1 Canada

INTRODUCTION

Variability in the optical properties of the upper ocean influences the color of the ocean as seen from space. For most of the ocean, the optical properties are controlled by the concentration of biogenic particles and dissolved matter -- phytoplankton, bacteria, and their degradation products. Variations in the optical properties modify the spectral and geometrical distribution of the underwater light field, and thereby alter the color of the sea. Biologically rich and productive waters are characterized by green water; the relatively depauperate open ocean regions are blue.

The prime oceanographic motivation for observing the color of the oceans from space is to permit better understanding of the role of the ocean in the global carbon cycle. Presently, prediction of the concentrations of carbon dioxide in the atmosphere, and hence the development of the greenhouse effect, are hampered by uncertainties in the ocean's role (e.g. Sarmiento and Toggweiler, 1984). The biological processes of the organisms that are principally responsible for variations in the color of the sea, the microscopic phytoplankton, may play a significant part in altering the exchange of carbon between atmosphere and oceans. Photosynthesis by phytoplankton can reduce the partial pressure of carbon dioxide in the surface waters; a portion of the carbon sequestered as organic matter then sinks below the layer of the ocean that actively interacts with the atmosphere. The net effect is a flux of carbon from the atmosphere to the deep sea - "the biological pump".

The concentrations of phytoplankton in the upper ocean can be determined by their effect on the color of the ocean as seen from space. Observations made throughout the 8-year lifetime of the Coastal Zone Color Scanner (CZCS), which orbited the Earth on NASA's Nimbus-7 spacecraft, have revolutionized the field of biological oceanography. By providing a synoptic view of phytoplankton pigment concentrations over global scales at high spatial resolution and with roughly daily repeat coverage, upcoming ocean color missions, such as the SEASTAR/SeaWiFS project, will provide data to determine the magnitude, and spatial and temporal variability, in both the concentration of pigment and the flux of carbon between the atmosphere and ocean on regional to global scales (e.g. Gordon et al., 1980; Esaias, 1981; Feldman et al., 1984; Brown et al., 1985; Evans et al.,

Primary Productivity and Biogeochemical Cycles in the Sea
Edited by P.G. Falkowski and A.D. Woodhead, Plenum Press, New York, 1992

1985; Eslinger et al., 1989; Lewis, 1989; McClain et al., 1990; Baker, 1990; Mitchell et al., in press; Abbott and Chelton, 1991).

In addition to modifying the global fluxes of carbon, oceanic phytoplankton also may play a more direct role in moderating the heat budget of the Earth in two ways. First, it was recently postulated that a byproduct of algal metabolism, dimethyl sulphide (DMS), may enhance the formation of low-lying clouds over the ocean (Charlson et al., 1987; Bates et al., 1987). DMS transported to the atmosphere may chemically and physically interact in the lower atmosphere and form nuclei that result in condensation, thus leading to cloud formation. The enhanced albedo that results cools the Earth and provides a potentially negative feedback to warming associated with enhanced concentrations of carbon dioxide. Second, phytoplankton themselves absorb solar radiation at visible frequencies. High concentrations of phytoplankton trap solar energy near the ocean surface. The buoyancy flux generated stabilizes the ocean, and retains solar energy in a depth horizon that actively interacts with the atmosphere over relatively short time-scales (e.g. Denman, 1973; Lewis et al., 1983; Woods et al., 1984; Lewis, 1987). For the equatorial Pacific, variability in such energy exchange may influence processes that result in the well-known El Nino-Southern Oscillation (ENSO) events (Lewis et al., 1990). For the tropical Arabian sea, inclusion of absorption by phytoplankton increases the modelled sea-surface temperature over models that assume no absorption other than that associated with seawater alone (Sathyendranath et al., 1991).

Synoptic measurements of the phytoplankton concentration are required to adequately test these hypotheses over the spatial scales of interest. Phytoplankton have a short lifespan, about 1-10 days, and are subject to the vagaries of ocean currents at a range of spatial and temporal scales. Consequently, the ocean pigment field decorrelates with itself on time-scales of about 1-10 days and on spatial scales about 1-5 km, depending on environmental conditions (Abbott and Zion, 1987). Satellite observations are the only means to measure the phytoplankton concentration on the global scales required, and also to sample with sufficient spatial and temporal resolution to avoid aliasing the long-term and large-scale estimates. Satellite observations also are the only means to "revisit" large areas of the oceans on a precise (i.e. to 1 km) grid to monitor any changes in the pigment content, whether due to biological processes such as growth and grazing, or to physical transport by ocean currents.

In summary, some of the research uses of the ocean color data set will be:

* To specify quantitatively the ocean's role in the global carbon cycle and other major biogeochemical cycles;

* To determine the magnitude and variability of annual primary production by marine phytoplankton on both local and global scales, and their role in modifying the chemical and physical processes of the ocean and atmosphere;

* To acquire global data on marine optical properties for analyzing ocean transparency and local heating rates.

HERITAGE AND FUTURE

The heritage of current efforts to acquire satellite ocean color data, and the basis for developing both applications and a data system, is the Coastal Zone Color Scanner (CZCS), which operated onboard the NIMBUS-7 from 1978 to 1986 (Gordon et al.,

Table 1.   SeaWiFS Radiance Bands

Band Number	Wavelength Center (nm)	Purpose
1	412	In-water detritus correction
2	443	Blue/Green Pigment algorithm
3	490	Blue/Green Pigment algorithm
4	510	Blue/Green Pigment algorithm
5	555	Blue/Green Pigment algorithm
6	670	Atmospheric Correction
7	765	Atmospheric Correction
8	865	Atmospheric Correction

1980). The CZCS, in turn, was based on aircraft observations over the ocean (Clarke et al., 1970). Over 65,000, two-minute scenes, which resulted from the CZCS, each of which represented roughly 2,200 by 800 kilometers, have allowed an unprecedented view of biological modification of the global ocean. The data system that evolved with the CZCS (Feldman et al., 1989) has been a model for other developing data systems, notably the EOS Pathfinder exercise, which will put the global AVHRR sea-surface temperature and land vegetation index data sets into the hands of researchers.

Unfortunately, the CZCS ceased operations in 1986, and because it was never designed to be an operational instrument, coverage was uneven in space and time. Consequently, considerable uncertainty still exists in the estimation of the role of the ocean in global biogeochemical cycles, and hence, in the prediction of future climate.

The next opportunity for the oceanographic community to have access to global ocean color data will be the SEASTAR/SeaWiFS mission. This dedicated sensor will be launched in late summer, 1993, and will provide data on a global basis every two days for five years, with a spatial resolution of 4 km globally and 1 km locally. The instrument includes increased radiometric performance, additional bands for new applications, and operational performance (Table 1).

In the longer term, Japan plans to launch the Ocean Color and Temperature Sensor (OCTS) onboard the ADEOS satellite in 1995. The international Mission to Planet Earth includes ocean color sensors provided by the European nations (MERIS), and NASA's EOS/MODIS, and EOS/HIRIS sensors for use in the International Geosphere/Biosphere Program (IGBP), and other international global change research programs now under consideration.

However, much work needs to be done now to advance the state of the field to be able to make adequate use of the very high data rate associated with the future ocean color sensors. This is as true for developing atmospheric and bio-optical algorithms as it is for developing a data and information system, which can make the huge data volume useful in addressing the scientific questions outlined above.

REQUIRED SENSOR SUITE AND RATIONALE

The research uses described above impose stringent requirements on the quality and delivery of the ocean color data set, and consequently, on the instrument, spacecraft, launch, ground operations, and data system employed. The required wavebands and radiometric requirements, such as accuracy, precision, polarization sensitivity, and signal to noise characteristics of the sensor are driven by current and future atmospheric and in-water algorithms for determining surface pigment (Table 1). The sensor must be capable of avoiding sun glint, which would mask any signal received from the ocean interior. Many of the problems are global in nature; global coverage must result. Since the sea-surface pigment field decorrelates over time scales about 1-10 days, and over spatial scales about 1-5 km, spatial and temporal resolution of 1 km and 2 days are required to avoid aliasing the long-term and global pigment fields. Many of the applications rely on accurate georeferencing, for example, determining surface currents from displacements of the surface pigment field from day to day. Hence, pixel knowledge must be within 1 km of its actual location for accurate retrievals. For detecting long-term changes in the ocean pigment field, many of which may be relatively small, long-term calibration stability is of paramount importance, as is the reliability of the entire system for the duration of the mission.

The global ocean color data set to be acquired with SEASTAR/SeaWiFS will be of sufficient quality to permit as required by the above: a) developing atmospheric correction algorithms that will allow the derivation of water-leaving radiances from the satellite data, which are accurate to within plus or minus 5 percent on a routine basis; and b) developing bio-optical algorithms for derivation of chlorophyll concentration from water-leaving radiances, which are routinely accurate to within plus or minus 35 percent for Case I (Morel, 1978) optical waters over a range of 0.05 to 50 mg m^{-3}, and to within a factor of 2 everywhere. The spatial resolution to be achieved will be 4 km globally and 1 km over selected areas, with a capability to georeference the data without ground control points to within 1 km; global coverage with a temporal resolution of 2 days will be delivered.

The necessity to achieve these objectives to the desired accuracy has driven the technical development; the data system must not impose any additional constraints. The guiding principal has been to ensure that SEASTAR remains a small source of error in the retrieval of the concentration and geolocation of surface pigment, both in the case of current, as well as for more advanced, algorithmic determination of pigment concentrations from satellite-viewed spectral radiances. To achieve this, SEASTAR/SeaWiFS improves upon the CZCS for long-term calibration stability, where there was a severe problem with CZCS (by using a lunar calibration target), for reducing cloud-ringing problems (by reducing the transient response characteristics), for additional wavebands to permit better atmospheric and bio-optical algorithms, and for improvements to both absolute and relative accuracies and band to band stability (by significantly increasing the radiometric sensitivity and by increasing the bit quantization from 8 to 10 bits). To avoid sun-glint contamination of the scene, the sensor is capable of tilting fore and aft by 20 degrees. Polarization sensitivity is reduced for all tilts and view angles to below 2%. The requirement for precision georeferencing will be achieved by sufficient pointing knowledge of the spacecraft and sensor to permit 1-km ground referencing. A data quality assurance program, based on advanced pre-launch and post-launch calibration techniques and characterization methods, was developed to ensure that the sensor will not limit the accurate retrieval of pigment for the entire 5-year lifetime of the mission.

The benchmark for the technical development has been the estimation of surface pigment concentration from a measurement of spectral radiance at the level of the spacecraft; the sensor noise introduced must be sufficiently small so that future improvements in atmospheric correction and bio-optical algorithms will not be constrained by the sensor. In a general sense, error in determining surface pigment results from errors in a) the bio-optical algorithms relating water-leaving radiances to surface pigment, b) determining the atmospheric scattering and absorption, and c) the sensor itself.

ALGORITHMS FOR RETREIVAL OF OCEANOGRAPHIC VARIABLES

Satellite ocean color instruments measure radiances entering the aperture of the sensor in space. To be useful, these radiances must be corrected to retrieve information on both the optical and biological properties of the oceans. Algorithms for achieving this goal remain under active development; in what follows, a brief synopsis of current and futures efforts are given.

Radiances observed at the level of the satellite are not derived solely from the ocean; as much as 95-99% of the radiance at the level of the spacecraft is derived from light scattered into the viewing angle of the sensor by the atmosphere:

$$L_t\ (\lambda,\theta,\theta_0,\Phi) = L_{atm}\ (\lambda,\theta,\theta_0,\Phi) + t\ (\lambda)L_w(\lambda), \qquad (1)$$

where L_t is the radiance viewed at the level of spacecraft; L_{atm} is the radiance derived from the atmosphere; L_w is the radiance exiting the surface of the ocean; and t is the atmospheric transmission, all of which are functions of wavelength, λ (e.g. Gordon and Morel, 1983). The dependencies indicated on the viewing angle of the spacecraft relative to the water (θ), on the Sun zenith angle relative to the water surface (θ_0), and on the azimuthal angle difference between the vertical planes of the spacecraft and Sun (Φ) will be suppressed in what follows.

Atmospheric Correction

Given an accurate (5% or better) calibration of the instrument itself, the next task will be to correct for atmospheric influences. This is a key step; the atmospheric signal exceeds that from the ocean by an order of magnitude. The baseline for the atmospheric correction will be that currently used with the CZCS, developed primarily by Howard Gordon at the University of Miami (e.g. Gordon, 1978; Gordon and Clark 1981; Gordon and Castano, 1987; Gordon et al., 1988; 1989; Gordon and Castano, 1989) and European researchers (Viollier et al, 1980; Bricaud and Morel, 1987; Andre and Morel, 1989; 1991). There are three components to this correction, presently treated separately. The first is to correct for molecular (Rayleigh) scattering by the atmosphere; the second is to correct for scattering by aerosols. Finally, the diffuse transmittance of the atmosphere must be computed to propagate the water-leaving signal to the level of the spacecraft. The Rayleigh scattering typically makes the largest contribution to the signal. Initially, it was believed that a single-scattering correction would suffice for this term. For large look angles, i.e. at high latitudes or on the wings of the scan, this turned out to be a poor assumption and current processing includes a detailed multiple scattering correction (Gordon and Castano, 1987; Gordon et al., 1988). Aerosol correction currently relies on the concept of "clear water radiances", where the water-leaving radiances in the red channels are presumed known for concentration of pigment less than 0.25 mg m^{-3}, and the spectral characteristics of the aerosol scattering are presumed constant over the scene of interest. Finally, the diffuse transmittance of the atmosphere must be determined; it may be estimated with acceptable accuracy from surface pressure fields and ozone optical thicknesses. Presently, often climatological fields are used (zonally and seasonally averaged values), although it may be useful to incorporate measured fields in the future derived from the Total Ozone Mapper (TOMS) and wind field derived from geostrophy.

Future atmospheric correction models may employ iterative schemes, which convolute the in-water radiance models with atmospheric models on a pixel-by-pixel basis (Smith and Wilson, 1981; Gordon et al., 1988; Bricaud and Morel, 1987; Andre and Morel, 1991), although this will be problematic for regions of high pigment concentration.

The atmospheric correction problem will be made easier with the SeaWiFS instrument. Three, rather than one, channel will be available in the red/near infrared portions of the spectrum, which will significantly improve skill in accounting for aerosol effects. The derivation of the parameters of atmospheric radiative transfer models used to correct the atmosphere will depend on radiances in these bands; a complete discussion on the atmospheric correction and the range of expected errors, can be found in Gordon (1978), Gordon and Clark (1981), Gordon and Morel (1983), Bricaud and Morel (1987), Gordon et al. (1988), and Andre and Morel (1989). It will be useful, and perhaps required, to derive the components of the atmosphere, the optical thickness and aerosol concentration, and place these in archives for the future.

Bio-optical Algorithms

Given that the atmospheric radiances can be eliminated from Eq.1 entirely, it remains to interpret variations in L_w in terms of first, the optical properties of the upper ocean, and second, more biologically relevant quantities, such as the chlorophyll concentration or primary production. Following Gordon et al. (1988), define the normalized water-leaving radiance:

$$L_w|_n = [(1-\rho)\ (1-\rho')F_oR]/[m^2Q(1-rR)] \tag{2}$$

where $L_w|_n$ is the water-leaving radiance that would occur under conditions where the sun is at zenith and there was no atmosphere. F_o is the solar flux; ρ and ρ' refer to components of the Fresnel reflectance at the sea-surface; R is the reflectance ($\equiv E_u/E_d$, where E_u and E_d are the upwelling and downwelling irradiances, respectively); m is the index of refraction; Q is a parameter defining the distribution of radiance with respect to angle about the vertical ($\equiv E_u/L_u$, where L_u is the upwelling radiance); and the term $(1-rR)$ takes account of internal reflection at the sea surface by the upwelling radiance stream.

The normalized water-leaving radiances are directly related to the concentration of surface pigment, C, through its influence on the inherent optical properties of the upper ocean. The predictive relationships developed for use with the CZCS archive are empirical representations of the link between upwelled radiance and the concentration of pigment in the upper ocean, for example (Gordon et al., 1988):

$$C = 1.15 \ (L_w(443)|_n/L_w(560)|_n)^{-1.42}, \ C < 1 \ \text{mg m}^{-3} \tag{3}$$

and

$$C = 3.64 \ (L_w(500)|_n/L_w(560)|_n)^{-2.62}, \ C < 1 \ \text{mg m}^{-3} \tag{4}$$

where the coefficient of determination is greater than 0.95, and the relative error is approximately 20% for Eq. 3 and 30% for Eq. 4. More analytical approaches are under development (Gordon et al., 1988; Morel, 1988), which will take advantage of the enhanced spectral resolution which will be provided in the blue-green part of the spectrum in determining the concentration of pigment in the ocean (Table 1).

Part of the additional error in the "green-water" algorithm (Eq. 4) over that for "blue-water" (Eq. 3) is due to the presence of high concentrations of detrital material and dissolved organic matter or gelbstoffe in these, usually coastal, waters. For SEASTAR/SeaWiFS, a new spectral band at 412 nm has been added, which will be useful in correcting for this effect (Carder and Steward, 1985; Carder et al., 1989; Table 1).

Another important source of error in current pigment retrievals is the very high reflectances associated with blooms of a certain type of marine algae, the coccolithophorids. These organisms lay down a calcium carbonate skeleton; when these are shed in large numbers as often observed in the North Atlantic, high reflectance or "milky water" results (Balch et al., 1989b). Efforts are underway to better understand the nature of this interference and to develop correction schemes for pigment retrieval.

The long-term accuracy of both the atmospheric and bio-optical algorithms depends on reliable calibration of the sensor. In addition to a rigorous pre-launch calibration and characterization of the sensor, and routine on-orbit use of solar diffuser calibrations, monthly lunar views through the entire optical train are planned to maintain calibration throughout the mission.

In the end, the problem reduces to inversion of Eq. 1 to estimate the concentration of pigment in the upper ocean which, in conjunction with a knowledge of the ground location of the pixels, permits the production of the global pigment field. For a uniform pigment distribution, error in the estimate will result from natural variability in the coefficients of the bio-optical algorithms (Eqs. 3,4), errors in the atmospheric correction, and instrument error as well as any covariances. The object of the instrument and mission design is to ensure that the contribution of the sensor to this variance decomposition will be sufficiently small so that improvements in atmospheric and bio-optical algorithms over those used with CZCS will result in significant improvement in predicting pigment in the upper ocean.

Primary Productivity

The base properties to be derived from past and future ocean color satellites will include concentration of pigment and attenuation coefficients (K) at least at one wavelength. Further developments will likely result in desire to derive additional parameters such as the phytoplankton absorption coefficient, the concentration of coccoliths, and ultimately, the primary productivity of the ocean.

Current efforts have made a good start towards estimating the total rate of primary productivity from space. All methods to date are based, either implicitly or explicitly, on the photosynthesis-irradiance relationship of natural phytoplankton communities, and how this relationship is modified by environmental variation (c.f. Bidigare, this volume). Early efforts (e.g. Smith et al., 1982; Eppley et al., 1985) made use of empirical relationships between surface chlorophyll concentrations derived from either ships or satellites, and ship-board measured rates of primary production. The empirical estimators can be recast in terms of parameters of the photosynthesis-irradiance relationships (Lewis et al., 1986), and more recent work has attempted to ground the remote estimation of primary production more firmly on the basis of physiological first principles (Perry, 1986; Platt, 1986; Platt and Lewis, 1987; Platt et al., 1988; Balch et al., 1989a; Morel and Berthon, 1989; Smith et al., 1989; Platt and Sathyendranath, 1989; Morel, 1991; Balch et al., in press).

Assuming that the satellite measures the surface concentration of chlorophyll without error, that the pigment concentration, and the optical and physiological properties, are uniformly distributed with respect to depth, and that the surface solar irradiance and the attenuation of the seawater is given, then the integrated total productivity, P, is:

$$P = (P_m^B/K) \, f(E_o/E_k) \tag{5}$$

where P_m^B is the (light-saturated) maximum photosynthetic rate; E_o is the surface irradiance; and E_k is the so-called light-adaptation parameter which, depending on the parameterization, is more or less equivalent to the irradiance level at which the photosynthetic rate is half of the maximum. Attempts to estimate primary production from space have used highly parameterized versions of these relationships (e.g. Eppley et al., 1985; Banse and Yong, 1990; Balch et al., in press) while others have included much more detail, encompassing non-uniform distribution of the pigment field, a fuller accounting of the geometrical distribution of the radiance field with depth, spectral decomposition of the terms in Eq. 5, and horizontal and vertical variations in the parameters of the photosynthesis-irradiance relationships (e.g. Platt and Sathyendranath, 1989). Experience has suggested that the latter is responsible for much of the variance, and has led to a suggestion of establishing "bio-optical provinces" (Mueller and Lange, 1989) in the world oceans, where these parameters can be considered relatively uniform (Platt and Sathyendranath, 1989). The high variability in physiological parameters, even over short time and space scales (particularly in the vertical) will continue to limit these methods unless some means is found to parameterize the variabililty in terms of observations that can be made remotely.

Given an exact estimation of chlorophyll, sea-surface irradiance, and the attenuation coefficient (not possible from space-borne observations alone), some of these relationships may account for as much as 75% of the variability in primary productivity on a global scale; when the additional uncertainties in estimating the forcing from satellite observations are considered, and propagated through, the error explained is much less, perhaps 20-30% (e.g. Kuring et al., 1990; Balch et al., 1989a; in press). Nonetheless, this is a significant improvement over current estimates of global productivity based on a severely undersampled population from ship observations, and leads to much encouragement for future work in this area.

NEW PRODUCTION FROM SPACE

The efforts described here concern estimating total primary production from space-borne observations. Of considerable interest, however, is the proportion of the total production, which is exported from the surface ocean, which interacts most actively with the atmosphere. This portion, termed the new production (Dugdale and Goering, 1967; Eppley and Peterson, 1979), represents the fraction of the production which can be lost in steady-state. The loss of carbon as organic export is compensated by a supply of inorganic carbon (in a one-dimensional sense) by vertical transport from below the surface, and by a flux of carbon from atmosphere to oceans. Hence, new production would seem to be of interest in predicting future concentrations of atmospheric carbon dioxide.

However, there is no straightforward correspondence between any electromagnetic signal that can be observed from space, and either the rate of new production, or its percentage of the total. Indirect means are called for.

Lewis et al. (1988) used the global marine transparency field derived from historical Secchi disk observations to estimate the new production rate of the Atlantic and Pacific Ocean Basins. Simple models of ecosystem function were combined with ship-based observations of nitrate input from seasonal oscillations in the mixed layer depth to estimate new production rates for meridional transcects along the middle of both basins. Nitrate fluxes, generally in the vertical, are associated with new production because they represent an external source of nutrient which, in the steady-state, must be balanced in a stoichiometric sense with the export flux of carbon (Dugdale and Goering, 1967; Eppley and Peterson, 1979; Lewis et al., 1986; Lewis et al., 1988). The model accounted for much of the variation in surface chlorophyll as derived from the Secchi disk observations, particularly the optical front associated with the subtropical convergence, and provided an estimate of the basin-scale new production rate.

A recent attempt to estimate the new production rate on local scales used the correlation between surface temperature and nitrate concentrations in upwelling regions of the upper ocean (Dugdale et al., 1989). Areas of high concentration of nitrate would be expected to have high percentages of new production (see Eppley and Peterson, 1979). Temperature and nitrate, while both non-conservative in surface waters because of surface heat fluxes and biological uptake respectively, are correlated in deeper waters. The temperature of the sea-surface in many areas can be related to the concentration of local nitrate through empirical relationships, and the concentration of nitrate can be related to the fraction of the total production which can be considered "new". A similar computation for the Georges Bank coastal region was recently reported (Sathyendranath et al., 1991).

Another means to estimate the new production rate is more direct, but also relies on the relationship between nitrate and temperature derived from historical observations. In a one-dimensional sense, and in steady-state, the net heat flux at the ocean's surface (sum of latent, solar, sensible and turbulent) must be balanced by a subsurface, say turbulent flux of heat,

$$\overline{w'T'} = (1/C_p\rho)Q_o \tag{6}$$

where w' represents the 'turbulent' velocity fluctuations on time scales less than that used for the mean (typically annually or greater), and T' is likewise the fluctuations in the local temperature. The net surface heat flux, Q_o, modifies the temperature through the specific heat, C_p, and the local density, ρ. The fluid transport term is evaluated at the base of the upper layer. Considerable progress has been made in the remote estimation of the various components of the surface heat fluxes (e.g. Liu, 1988; Liu and Niiler, 1984; Liu and Gautier, 1990).

It remains to estimate the upward nitrate flux, and hence, the stoichiometrically equivalent downward vertical carbon flux. As above, the vertical turbulent nitrate flux can be related in the steady-state with the biological uptake (Eppley and Peterson, 1979; Lewis et al., 1986):

$$\overline{w'N'} = P_n \qquad (7)$$

where, as in Eq.6, the primed terms refer to fluctuations, and Pn represents the depth-integrated uptake of nitrate or new production. Using historical data on the relationship between nitrate and temperature, one can compute the slope of this relationship at a given reference temperature, $\partial T'/\partial N'$. Substitution of this quantity into the above leads to an estimate of the new production rate as a function of the surface heat fluxes, determined solely from satellite observations of the surface heat fluxes, physical constants and historical data on the relationship between nitrate and temperature:

$$P_n = (1/C_p\rho) \, (\partial N'/\partial T') \, Q_o \qquad (8)$$

Applications of this technique to the Equatorial Pacific Ocean result in an estimate very similar to that derived on the basis of a full-scale, general ocean-atmosphere circulation model, including biological fluxes.

New Production and Carbon Flux

However interesting determining the rate of new production is in developing an understanding of marine ecosystem function, it is not equivalent, and may bear little relation to, the net flux of carbon across the air-sea interface. The principal reason for this involves the vertical transport of inorganic carbon from the deep sea to the surface ocean. New production is defined in terms of the supply of inorganic nitrate to the sea-surface; if this is accompanied by a stoichiometrically equivalent (i.e. Redfield ratio) flux of carbon, the net effect is no transport across the air-sea interface apart from that associated with temperature and alkalinity induced solubility variations.

Temporal variations in total inorganic carbon can be shown in the same simple manner as above:

$$\frac{\partial C}{\partial t} = \frac{\overline{\partial w'C'}}{\partial z} - P_c \qquad (9)$$

where C is the concentration of total inorganic carbon, and P_c represents the net, local community production in carbon units. Again, for convenience but with no loss of generality, all vertical fluxes are considered turbulent. Unlike nitrate, however, the boundary condition at the sea surface permits exchange of inorganic carbon between atmosphere and ocean:

$$\text{Air-sea Carbon Flux} = k \, (\Delta_p CO_2) \qquad (10)$$

where $\Delta_p CO_2$ is the partial pressure differential across the air-sea interface, and k is the gas exchange coefficient. To make the boundary layer formulation in Eq. 10 consistent with Eq. 9, we replace $\Delta_p CO_2$ with $(C_a - C)$, where C_a is the equivalent concentration of carbon in the overlying atmosphere corrected for surface temperature and alkalinity, and C is now the concentration of total inorganic carbon in the surface ocean. In the differential form of Eq. 7 above, we can replace the local new production rate with $f\mu B_n$, where B_n is the organic nitrogen concentration; μ is the specific productivity rate (T^{-1}); and f is the fraction of the total production which can be attributable to new production.

Likewise, the net community uptake of carbon in Eq. 9 is expressed as $\zeta \mu B_c$, where B_c is the organic carbon content, and ζ is the fraction of the total production which represents net community production. The Redfield ratio of the organic matter, R, is given as B_c/B_n. Combining Eqs. 7 and 9, substituting in the Redfield ratio, and evaluating the integrals from the surface to some arbitrary depth and rearranging leads to:

$$\frac{k\ (C_a-C)}{(Rw'N')} = \frac{\zeta}{f} - \frac{\Delta C}{R\Delta N} \tag{11}$$

where the term on the left-hand side refers to the ratio of the air-sea exchange of carbon to the flux of carbon that would be stoichiometrically equivalent to the nitrate flux in the vertical. The two terms on the right-hand side refer to the ratio of the net community production to the new, nitrogen-based production and the stoichiometric ratio of the vertical gradients in inorganic carbon and nitrogen at the base of the layer under consideration. The usual assumption has been made that the turbulent transport of scalars can be expressed as a product of an eddy diffusivity and the local gradient, and that the eddy diffusivity for carbon and nitrate are equivalent. In a steady-state, time-and space-averaged sense, both terms on the right would be expected to equal one; the conclusion is that the air-sea flux of carbon associated with the biological pump is vanishingly small in the steady-state.

This conclusion regarding the relationship between new production and the air-sea carbon flux is not new: "Since dissolved inorganic carbon moves upward along with the vertically transported nitrate, in approximately the Redfield ratio of 106 C atoms: 16 N atoms, only the sinking flux due to new production associated with nitrogen fixation and nutrient inputs from terrestrial and atmospheric sources can be identified as a biologically-mediated transport of atmospheric carbon dioxide to the deep ocean" (Eppley and Petersen, 1979). For most of the oceans, the new production associated with nitrogen fixation and external inputs is a small fraction of the new production rate as generally measured.

The caution is that this represents the steady-state. Time and space variations in the Redfield ratio for the upward inorganic fluxes, and in the ratio of new production to net community production, will cause deviations from the steady-state in the flux ratio represented on the left-hand side. Local net fluxes of carbon from atmosphere to ocean associated with the biological pump may result (Watson et al., 1991). Nonetheless, it is clear that the upward flux of nitrate and the new production rate, cannot be equated nor even related in a straightforward way to the air-sea flux of carbon. Better oceanic observations of the variability (and covariability) in both the vertical profiles of inorganic carbon and nitrogen, and the relative magnitudes of the new production and net community production rates will go a long way towards better evaluating the biologically mediated net flux of carbon between the atmosphere and the ocean.

CONCLUSIONS

Satellite observations of the color of the ocean's surface have dramatically changed the scope of biological oceanography. Many of the intellectual challenges that will face the community in the future will require the global perspective provided by the space-borne observations. Many of the problems, particularly that of estimating the role of the ocean in the global carbon cycle, will require not only global observations of ocean color, but concurrent satellite observations of wind stress, sea-surface temperature, and air-sea heat fluxes. These observations will be assimilated, along with *in situ* data, into new generations of ocean and atmosphere models to predict biogeochemical processes and fluxes in the Earth system. As the cost of accessing and using the quantities of data provided by the satellite diminish, and the availability of well-calibrated data sets increases, we look forward to a day when satellite observations are as routine as CTD measurements.

ACKNOWLEDGEMENTS

I thank John Cullen and Doug Wallace for critical coments, and NASA for their support.

REFERENCES

Abbott, M. R., and Chelton, D. B., 1991, Advances in passive remote sensing of the ocean, *Rev. Geophysics*, in press.

Abbott, M. R., and Zion, P. M., 1987, Spatial and temporal variability of phytoplankton pigment off northern California during Coastal Ocean Dynamics Experiment 1, *J. Geophys. Res.*, 92:1745.

Andre, J. M., and Morel, A., 1989, Simulated effects of barometric pressure and ozone content upon the estimate of marine phytoplankton from space, *J. Geophys. Res.*, 94:1029.

Andre, J. M., and Morel, A., 1991, Atmospheric corrections and interpretation of marine radiances in CZCS imagery, revisited, *Oceanol. Acta.*, 14:3.

Baker, D. J., 1990, Planet Earth: The view from space, Harvard University Press, Cambridge, Massachusettes.

Balch, W. M., Abbott, M. R., and Eppley, R. W., 1989a, Remote sensing of primary production - I. A comparison of empirical and semi-analytical algorithms, *Deep Sea Res.*, 36:281.

Balch, W. M., Eppley, R. W., Abbott, M. R., and Reid, R. M. H., 1989b, Bias in satellite-derived pigment measurements due to coccolithophores and dinoflagellates, *J. Plankton Res.*, 11:575.

Balch, W. M., Evans, R., and Brown, J., 1991, The remote sensing of ocean primary productivity - use of a new data compilation to test satellite algorithms, *J. Geophys. Res.*, in press.

Banse, K., and Yong, M., 1990, Sources of variability in satellite-derived estimates of phytoplankton production in the eastern tropical Pacific, *J. Geophys. Res.*, 95:7201.

Bates, T. S., Charlson, R. J., and Gammon, R. H., 1987, Evidence for the climatic role of marine biogenic sulphur, *Nature*, 329:319.

Bidigare, R. R., this volume.

Bricaud, A., and Morel, A., 1987, Atmospheric corrections and interpretation of marine radiances in CZCS imagery: use of a reflectance model, *Oceanol. Acta* 7:33.

Brown, O. B., Evans, R. H., Brown, J. W., Gordon, H. R., Smith, R. C., and Baker, K. S., 1985, Phytoplankton blooming off the U.S. East coast: A satellite description, *Science*, 229:163.

Carder, K. L., and Steward, R. G., 1985, A remote-sensing reflectance model of a red-tide dinoflagellate off West Florida, *Limnol. Oceanogr.*, 30:286.

Carder, K. L., Steward, R. G., Harvey, G. R., and Ortner, P. B., 1989, Marine humic and fulvic acids: Their effects on remote sensing of ocean chlorophyll, *Limnol. Oceanogr.*, 34:68.

Charlson, J. E. Lovelock, Andreae, M. O., and Warren, S. G., 1987, Oceanic phytoplankton, atmospheric sulphur, cloud albedo and climate, *Nature*, 326:655.

Clarke, G. L., Ewing, G. C., and Lorenzen, C. J., 1970, Spectra of backscattered light from the sea obtained from aircraft as a measure of chlorophyll concentration, *Science*, 167:1119.

Denman, K. L., 1973, A time-dependent model of the upper ocean, *J. Phys. Oceanogr.*, 3:173.

Dugdale, R. C., and Goering, J. J., 1967, Uptake of new and regenerated forms of nitrogen in primary productivity, *Limnol. Oceanogr.*, 12:196.

Dugdale, R. C., Morel, A., Bricaud, A., and Wilkerson, F. P., 1989, Modeling new production in upwelling centers: A case study of modeling new production from remotely sensed temperaure and color, *J. Geophys. Res.*, 94:18119.

Eppley, R. W., and Peterson, B. J., 1979, Particulate organic matter flux and planktonic new production in the deep ocean, *Nature*, 282:677.

Eppley, R. W., Stewart, E., Abbott, M. R., and Heyman, U., 1985, Estimated ocean primary production from satellite chlorophyll, Introduction to regional differences and statistics for the Southern California Bight, *J. Plankton Res.*, 7:57.

Esaias, W. E., 1981, Remote sensing in biological oceanography, *Oceanus*, 24:33.

Eslinger, D. L., O'Brien, J. J., and Iverson, R. L., 1989, Empirical orthogonal function analysis of cloud-containing Coastal Zone Color Scanner images of northeastern North American coastal waters, *J. Geophys. Res.*, 94:10884.

Evans, R. H., Baker, K. S., Brown, O. B., and Smith, R. C., 1985, Chronology of Warm-Core ring 82B, *J. Geophys. Res.*, 90:8803.

Feldman, G. C., Clark, D., and Halpern, D., 1984, Satellite color observations of the phytoplankton distribution in the Eastern Equatorial Pacific during the 1982-1983 El Niño, *Science*, 226:1069.

Feldman, G. C., Kuring, N., Ng, C., Esaias, W., McClain, C., Elrod, J., Maynard, N., Endres, D., Evans, R., Brown, J., Walsh, S., Carle, M., and Podesta, G., 1989, Ocean Color: Availability of the global data set, *EOS*, 70:634.

Gordon, H. R., 1978, Removal of atmospheric effects from satellite imagery of the oceans, *Appl. Opt.*, 17:1631.

Gordon, H. R., Brown, J. W., Evans, R. H., 1989, Exact Rayleigh scattering calculations for use with the Nimbus-7 Coastal Zone Color Scanner, *Appl. Optics*, 27:862.

Gordon, H. R., Brown, O. B., Evans, R. H., Brown, J. W., Smith, R. C., Baker, K. S., and Clark, D. K., 1988, A semianalytic radiance model of ocean color, *J. Geophys. Res.*, 93:10909.

Gordon, H. R., and Castano, D. J., 1987, Coastal Zone Color Scanner atmospheric correction algorithm: multiple scattering effects, *Appl. Optics*, 26:2111.

Gordon, H. R., and Castano, D. J., 1989, Aerosol analysis with the Coastal Zone Color Scanner: A simple method for including multiple scattering effects, *Appl. Optics*, 28:1320.

Gordon, H. R., and Clark, D. K., 1981, Clear-water radiances for atmospheric correction of Coastal Zone Color Scanner imagery, *Appl. Opt.*, 20:4175.

Gordon, H. R., Clark, D. K., Mueller, J. L., and Hovis, W. A., 1980, Phytoplankton pigments derived from the Nimbus-7 CZCS: Initial comparisons with surface measurements, *Science*, 210:63.

Gordon, H. R., and Morel, A., 1983, Remote Assessment of Ocean Color for Interpretation of Satellite Visible Imagery: A Review, Springer-Verlag, New York.

Kuring, N., Lewis, M. R., Platt, T., and O'Reilly, J. F., 1990, Satellite-derived estimates of primary production in the Northwestern Atlantic, *Cont. Shelf Res.*, 10:461.

Lewis, M.R., 1987, Phytoplankton and thermal structure in the tropical ocean, *Oceanolgica Acta*, SP:91.

Lewis, M. R., 1989, The variegated ocean: A view from space, *New Scientist*, 1685.

Lewis, M. R., Carr, M. E., Feldman, G. C., Esaias, W., and McClain, C., 1990, Influence of penetrating solar radiation on the heat budget of the equatorial Pacific Ocean, *Nature*, 347:543.

Lewis, M. R., Cullen, J. J., and Platt, T., 1983, Phytoplankton and thermal structure in the upper ocean: Consequences of nonuniformity in chlorophyll profile, *J. Geophys. Res.*, 88:2565.

Lewis, M. R., Harrison, W. G., Oakey, N. S., Hebert, D., and Platt, T., 1986, Vertical nitrate fluxes in the oligotrophic ocean, *Science*, 234:870.

Lewis, M. R., Kuring, N., and Yentsch, C. S., 1988, Global patterns of ocean transparency, Implications for the new production of the open ocean, *J. Geophys. Res.*, 93:6847.

Lewis, M. R. and Platt, T., 1987, Remote observation of ocean colour for prediction of upper ocean heating rates, *Adv. Space Res.*, 7:6.

Liu, W. T., 1988, Moisture and latent heat flux variabilities in the tropical Pacific derived from satellite data, *J. Geophys. Res.*, 93:6749.

Liu, W. T., and Gautier, C., 1990, Thermal forcing on the tropical Pacific from satellite data, *J. Geophys. Res.*, 95:13209.

Liu, W. T., and Niiler, P. P., 1984, Determination of monthly mean humidity in the atmospheric surface layer over oceans from satellite data, *J. Phys. Oceanogr.*, 14:1451.

McClain, C. R., Esaias, W. E., Feldman, G. C., Elrod, J., Endres, D., Firestone, J., Darzi, M., Evans, R., and Brown, J., 1990, Physical and biological processes in the north Atlantic during the First GARP Global Experiment, *J. Geophys. Res.*, 95:18027.

Mitchell, B. G., Esaias, W. E., Feldman, G., Kirk, R. G., McClain, C. R, and Lewis, M.R., Satellite ocean color data for studying oceanic biogeochemical cycles, IEEE Pub., in press.

Morel, A., 1978, Available, usable, and stored radiant energy in relation to marine photosynthesis, *Deep Sea Res.*, 25:673.

Morel, A., 1988, Optical modelling of the upper ocean in relations to its biogenous matter content (Case I waters), *J. Geophys. Res.*, 93:10749.

Morel, A., 1991, Light and marine photosynthesis: a spectral model with geochemical and climatological implications, *Prog. Oceanogr.*, 26:263.

Morel, A., and Berthon, J.-F., 1989, Surface pigments, algal biomass profiles and potential production of the euphotic layer: Relationships reinvestigated in view of remote-sensing applications, *Limnol. Oceanogr.*, 34:1545.

Mueller, J. L., and Lange, R. E., 1989, Bio-optical provinces of the Northeast Pacific Ocean: A provisional analysis, *Limnol. Oceanogr.*, 34:1572.

Perry, M. J., 1986, Assessing marine primary production from space, *BioScience*, 36:461.

Platt, T., 1986, Primary production of the ocean water column as a function of surface light intensity: algorithms for remote sensing, *Deep Sea Res.*, 33:149.

Platt, T., and Lewis, M. R., 1987, Estimation of phytoplankton production by remote sensing, *Adv. Space Res.*, 7:10.

Platt, T., and Sathyendranath, S., 1989, Oceanic primary production: estimation by remote sensing at regional and larger scales, *Science*, 241:1613.

Platt, T., Sathyendranath, S., Caverhill, C. M., and Lewis, M. R., 1988, Oceanic primary production and available light: further algorithms for remote sensing, *Deep Sea Res.*, 35:855.

Sarmiento, J. L., and Toggweiler, R. R., 1984, A new model for the role of the oceans in determining atmospheric pCO2, *Nature*, 308:621.

Sathyendranath, S., Platt, T., Horne, E. P. W., et al., 1991, Estimation of new production in the ocean by compound remote sensing, *Nature*, 353:129.

Smith, R. C., Eppley, R. W., Baker, K. S., 1982, Correlation of primary production as measured aboard ship on Southern California coastal waters and as estimated "from" satellite chlorophyll images, *Mar. Biol.*, 66:281.

Smith, R. C., and Wilson, W. H., 1981, Ship and satellite bio-optical research in the California Bight., *In*: "Oceanography from Space," J. Gower, ed., Plenum, New York.

Smith, R. C., Prezelin, B. B., Bidigare, R. R., and Baker, K. S., 1989, Bio-optical modelling of photosynthetic production in coastal waters, *Limnol. Oceanogr.*, 34:1524.

Violler, M., Tanre, D., and Deschamps, P. Y., 1980, An algorithm for remote sensing of water color from space, *Boundary-Layer Meterol.*, 18:247.

Watson, A. J., Robinson, C., Robinson, J. E., Williams, P. J. leB., and Fasham, M. J. R., 1991, Spatial variability in the sink for atmospheric carbon dioxide in the North Atlantic, *Nature*, 350:50.

Woods, J. D., Barkman, W., and Horch, A., 1984, Solar heating of the oceans - diurnal, seasonal, and meridional variation, *Quart. J.R. Meterol. Soc.*, 110:633.

ADVANCES IN UNDERSTANDING PHYTOPLANKTON FLUORESCENCE AND PHOTOSYNTHESIS

Dale A. Kiefer and Rick A. Reynolds

Department of Biological Sciences
University of Southern California
University Park
Los Angeles, CA 90089-0371

INTRODUCTION

Significant technological and scientific advances were made during the last decade in the measurement of fluorescence from the photosynthetic pigments of natural marine populations of phytoplankton and cyanobacteria. Field studies have begun to employ a diverse array of fluorescence sensors. Airborne Lidar has been used to obtain synoptic, one-dimensional transects of the concentration of chlorophyll *a* and phycoerythrin. Towed fluorometers have provided rapid, two-dimensional transects of the distribution of chlorophyll *a*. Moored active and passive fluorometers have given continuous, long-term records of the concentration of chlorophyll with unprecedented temporal detail. Flow cytometers have measured the fluorescence and scattering cross-sections of individual cells. As conventional spectrofluorometers provided fluorescence excitation and emission spectra for entire assemblages of cells and particles, microspectrophotometers provided such data for individual cells.

Such studies have increased our understanding of the spatial distribution, temporal variability, and taxonomic composition of phytoplanktonic and cyanobacterial populations. However, because of interest in using fluorescence measurements as indicators of photosynthetic activity, we focus our discussion on advances in the understanding of the relationship between the fluorescence of chlorophyll *a* and photosynthetic rate. In particular, we will consider three types of studies: (1) A model of primary photochemistry, based upon an analysis of the absorption and fluorescence of isolated photosynthetic units induced by picosecond flashes of light, (2) An analysis of the fluorescence in cultures and at sea induced by microsecond pulses of variable intensity (pump-probe fluorescence), and (3) An analysis of solar-induced fluorescence in the field. We have selected for discussion studies that we consider relevant to marine planktonic photoautotrophs. Falkowski and Kiefer, 1985; Kraus and Weis, 1991; and Owens, in press, give a more comprehensive treatment of fluorescence and photosynthesis.

Primary Productivity and Biogeochemical Cycles in the Sea
Edited by P.G. Falkowski and A.D. Woodhead, Plenum Press, New York, 1992

155

Table 1. Definitions and Units

Symbol	Description	Units
λ	light wavelength	nm
$E_o(\lambda)$	scalar irradiance	$molm^{-2}s^{-1}$
I	flash intensity	$mol\ m^{-2}flash^{-1}$
I_k	irradiance where photosynthesis approaches saturation	$mol\ m^{-2}s^{-1}$
k_{cf}	value of irradiance at $1/2\ (\Phi_c/\Phi_F)max$	$mol\ m^{-2}s^{-1}$
$a(\lambda)$	spectral absorption coefficient of PSII	m^{-1}
σ_a	absorption cross-section pf PSII	$m^{-2}mol^{-1}$
σ_{po}	cross-section for photochemistry of open PSII reaction centers	$m^{-2}mol^{-1}$
J_a	rate of light absorption by PSII	$mol\ m^{-3}s^{-1}$
J_F	rate of fluorescence emission	$mol\ m^{-3}s^{-1}$
J_e	rate of photosynthetic electron transport	$mol\ m^{-3}s^{-1}$
J_c	rate of photosynthetic carbon assimilation	$mol\ m^{-3}s^{-1}$
Φ_p	quantum yield of primary photochemistry	mol e/mol quanta
Φ_c	quantum yield of carbon fixation	mol C/mol quanta
Φ_F	quantum yield of fluorescence	quanta/quanta
A	fraction of open reaction centers	dimensionless (0-1)
F	fluorescence intensity induced by flash	relative units
τ	turnover time of PSII reaction centers	s
Ψ_x	probability of occurrence of x	dimensionless (0-1)
n	concentration of PSII in suspension	$mol\ m^{-3}$

BASIC RELATIONSHIPS

We begin by considering two types of fluorescence measurements used in field studies of marine phytoplankton. The first involves passive measurement of the solar-induced fluorescence of chlorophyll *a* within the water column; this is often referred to as "natural" fluorescence. Natural fluorescence can be accurately measured within the euphotic zone, except just below the surface where solar irradiance in the red is large (Kiefer et al., 1989). The second measurement involves analysis of the fluorescence of phytoplankton exposed to repetitive flashes of light; this is often called "pump-probe" fluorescence (Falkowski et al., 1986).

Natural Fluorescence

The relationship between the fluorescence of chlorophyll *a* and photosynthetic rate of planktonic photoautotrophs is most easily described in terms of the quantum yields of the two processes. Because 95% or more of the *in vivo* fluorescence originates from the chlorophyll of the antenna of photosystem II (PSII), we will discuss this photosystem specifically. Consider an optically thin lamina of cells: the rate of light absorption by photosystem II (J_a) in the suspension is simply the integral over the spectrum of visible light of the product of the ambient scalar irradiance, $E_o(\lambda)$, and the absorption coefficient of PSII, $a(\lambda)$ (see Table 1 for a description of symbols and units):

$$J_a = \int_{400}^{700} a(\lambda)E_o(\lambda)d\lambda \tag{1}$$

$a(\lambda)$ is equal to the product of the concentration of photosystem II found in the cell suspension, n, and the absorption cross section of the photosystem, $\sigma_a(\lambda)$:

$$a(\lambda) = n \; \sigma_a(\lambda) \tag{2}$$

$a(\lambda)$ can be estimated from measurements of cellular spectra for fluorescence excitation and absorption (Sakshaug et al., 1991). If we accept the errors caused by ignoring spectral variations in ambient irradiance and the absorption coefficient of the cells, Eq 1 can be approximated as the product of the scalar irradiance of photosynthetically available radiation (PAR), and the mean absorption coefficient of the population of photosystem units:

$$J_a = a \; E_o \tag{3}$$

In the case of solar-induced natural fluorescence, the instantaneous rates of fluorescence, J_F (moles of photons m^{-3} s^{-1}), and gross photosynthetic carbon fixation J_c (moles carbon m^{-3} s^{-1}) are, by definition, the product of the rate of light absorption and the respective quantum yields of fluorescence (Φ_F) and photosynthesis (Φ_c) for the photosystem:

$$J_c = \Phi_c(E_o) \; J_a \tag{4}$$

$$J_F = \Phi_F(E_o) \; J_a \tag{5}$$

Φ_c and Φ_F are assumed to be independent of wavelength but dependent upon light intensity. Φ_c and Φ_F may also vary with other factors, such as nutrient concentration, temperature, and the species composition of the crop. Much of the variability in both photosynthetic rate and fluorescence in the sea is caused by variations in the rate of cellular light absorption and, thus, by variability in $E_o(\lambda)$ and $a(\lambda)$.

Substituting Eq 5 into 4 yields:

$$J_c = \frac{\Phi_c(E_o)}{\Phi_F(E_o)} \; J_F \tag{6}$$

Equation 6 is the basis for predicting photosynthetic rate from measurements of natural fluorescence. The accuracy of predicting gross photosynthetic rate from natural fluorescence will be independent of spectral variations in either irradiance or the absorption coefficient, and will depend largely upon the accuracy with which the value of the ratio of quantum yields of photosynthesis to fluorescence can be predicted. This ratio is predicted from measurements of ambient irradiance.

Pump-Probe Fluorescence

When the fluorescence emitted by chlorophyll a in a suspension of cells is excited by a flash of light that is sufficiently short, infrequent, and low in intensity, the redox poise of the photosynthetic electron transport system will be unaltered by the flash. The fluorescence induced by such a flash is called the probe fluorescence, and we will represent the dose of exciting photons as I (photons m^{-2} $flash^{-1}$). The fluorescence quantum yield of such a flash will be similar to that of natural fluorescence, Φ_F. Because the dose of the probe flash is much smaller than the ambient irradiance, flash-induced fluorescence (F) can be represented as:

$$F = \Phi_F \, Ia \tag{7}$$

Φ_F will vary with the redox poise of the photosynthetic electron transport system, which itself is determined by the ambient irradiance, E_o. If the probe flash is preceded by a "pump flash" whose dose (I_s) is sufficiently large to perturb the redox poise of the photosynthetic electron transport system, the quantum yield of fluorescence induced by the probe flash, $\Phi_F(E_o, I_s)$, will differ from its natural value $\Phi_F(E_o)$. The ability to systematically perturb the electron transport system with a pump flash and monitor the fluorescence response is the basis of the pump-probe measurement.

Much theoretical and experimental work has been devoted to formulating the functional relationship between the quantum yield of photosynthesis and the flash-induced fluorescence under ambient and perturbed conditions. We will show that the quantum yield of photosynthesis may be predicted from several measurements of pump-probe fluorescence in ambient light and in the dark.

A MODEL OF ENERGY TRANSFORMATION IN PHOTOSYSTEM II

General Description

Recently, Schatz et al. (1988) have developed a model of energy transformations in photosystem II that is based upon measurements of fluorescence and absorption changes in photosystem II preparations from a thermophilic species of *Synechococcus*. The particles included the chlorophyll *a* core of the photosystem and the reaction center, but not the bilisomes that are the principal components of light capture in the complete photosystem. The fluorescence and absorption signals, which are induced by flashes from a laser, were resolved to the picosecond range and the data were analyzed by a Laplace transformation of the signals. The model itself was presented in terms of the first-order rate constants of the key transformations.

Figure 1 is a diagram of the five components of their model and the five stages or conditions in which these components may be found during photon capture, charge separation, and subsequent charge stabilization. Each transformation is represented by an arrow. One of the five components is the core of the antenna, which consists of about eight molecules of chlorophyll *a* that comprise the core of the antenna chlorophyll. These molecules are the sole source of measurable fluorescence. The remaining four components are subunits of the reaction center. A dimer of chlorophyll *a* (P) is initially excited to P^* when the exciton reaches the reaction center. Upon charge separation, it then becomes oxidized to P^+. Changes in the redox state of P are monitored by differential absorption. The initial charge separation also produces a short-lived, reduced phaeophytin a intermediate (I^-) which is proximate to P. This phaeophytin molecule is, in turn, oxidized by a distil quinone molecule, Q. Ultimately, P^+ is reduced by the primary electron donor Z, which is part of the oxygen-evolving complex.

In the figure, Stage A, which is the unexcited photosynthetic unit, is shown in the lower left-hand corner. Upon absorption of light, Stage A is converted to Stage B. This stage is shown as two forms; one, in which the exciton is found in the antenna, and the other in which the exciton is localized in the reaction center. Since the difference in free energy between the two forms is small (the values for the rate constants k_T and k_{-T} are similar), the exciton may cycle between the reaction center and antenna several times before another stage is reached. Stage C is produced by actual charge separation; P is oxidized by the neighboring phaeophytin. Stage C may be considered unstable because the probability of charge recombination is large. Stage D, which is very stable, is the result of charge stabilization; I is oxidized by Q. Stage E is attained when P^+ is reduced by the primary donor Z. Both stages D and E are characterized by the presence of Q^-, and, at these stages, the reaction center is said to be "closed." When the unit is in Stage A, the reaction center is "open". The presence in closed reaction centers of an electrostatic field produced by the anion Q^- causes large changes in the value of the probabilities of several of the transformations shown in the figure.

While the transformations between adjacent stages symbolized by the arrows were characterized by Schatz et al. (1988) in terms of the first-order rate constants, we chose instead to characterize the processes by their "single step" probabilities of occurrence, Ψ. Where appropriate, two values for Ψ are given. When the reaction centers are open (Q oxidized), the subscript of the probability will end in "o", and when the reaction center is closed (Q reduced), the subscript will end in "s".

In Fig. 1, we note that the fate of the excited photosynthetic unit (Stage B) is largely determined by whether the reaction center is opened or closed. If the reaction center is open, the probability, Ψ_{cso}, that charge separation will occur (Stage B to Stage C) is predicted to be 0.91, while the probability that de-excitation (Stage B to Stage A) will occur is 0.09. The probability that de-excitation will be by fluorescence, Ψ_{Fo}, is 0.011, and the probability that de-excitation will be "radiationless", Ψ_{Do}, is 0.079. On the other hand, if the reaction center is closed, the probability, Ψ_{css}, that charge separation will occur (Stage B to Stage C) will decrease to only 0.58, while the probability that de-excitation (Stage B to Stage A) will occur increases to 0.42. The probability that de-

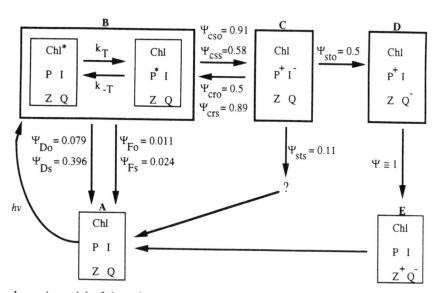

Fig. 1.   A model of the primary reactions occuring in PSII (after Schatz et al., 1988). The arrows represent transformations between adjacent states, and Ψ denotes the "single step probability" of occurrence for each transition. See text for further details.

excitation will be by fluorescence, Ψ_{Fs}, is 0.024, and the probability that de-excitation will be "radiationless", Ψ_{Ds}, is 0.396. (We estimated the probabilities of fluorescence and "radiationless" de-excitation by assuming that the quantum yields of fluorescence are 0.02 and 0.05 for open and closed reaction centers, respectively.)

The fate of Stage C also depends upon the whether the reaction center is opened or closed. If the reaction center is open, the probability, Ψ_{sto}, that charge stabilization will occur (Stage C to Stage D) is 0.5, and the probability that charge recombination, Ψ_{cro}, (Stage C to Stage B) will occur is 0.5. If the reaction center is closed, the probability that charge separation will occur (Stage C to Stage D) is 0; instead, either an unknown product will be formed with a probability of 0.11 (Ψ_{sts}), or charges will recombine (Stage C to Stage B), Ψ_{crs}, with a probability of 0.89.

The values of these probabilities show that there is considerable photochemical cycling by charge separation and recombination between Stages B and C. In this regard, the Schatz model is similar to the bipartite model proposed by Butler (1978). However, there are several important differences between the two models. First, Butler's model changes in the value of rate constants and single-step probabilities were not considered. Second, in the earlier model, cycling occurs between the antenna and the reaction center rather than between charge separated and charge recombined states of the reaction center. Furthermore, cycling occurs in the Butler model only when reaction centers are closed; in the Schatz model, cycling occurs with comparable probability whether the reaction centers are opened or closed. Third, in the Schatz model, the primary cause of increased fluorescence with the closing of reaction centers is the increase in the value of Ψ_{Fs} over Ψ_{Fo}. In the earlier model, increased fluorescence is caused by increased cycling between the reaction center and the antenna.

<u>Derivations</u>

The information shown in Fig. 1 is sufficient to formulate the quantum yields of primary photochemistry and fluorescence under steady-state conditions. Following the nomenclature and derivation of Butler and Kitajima (1975), we let A represent the fraction of a population of photosystems that are open. When all reaction centers are open ($A = 1$), the quantum yield of primary photochemistry is maximal, and the quantum yield of fluorescence is minimal. The value of A changes rapidly (milliseconds) with changes in light intensity, and we will see that the ability to rapidly effect the value of A (and, thus, the fluorescence quantum yield) with exposure of the cells to flashes of light of variable intensity is the key feature of the pump-probe measurement. When all reaction centers are closed ($A = 0$), the opposite is true. The yield of primary photochemistry (Φ_p) is:

$$\Phi_p = A \ (\Psi_{cso} \ \Psi_{sto} + \Psi_{cso} \ \Psi_{cro} \ \Psi_{sto} + \Psi_{cso} \ (\Psi_{cro} \ \Psi_{sto})^2 + \Psi_{cso} \ (\Psi_{cro} \ \Psi_{sto})^3 + \ldots)$$

The quantum yield of fluorescence (Φ_F) is:

$$\Phi_F = A \ (\Psi_{Fo} + \Psi_{Fo} \ \Psi_{cso} \ \Psi_{sro} + \Psi_{Fo} \ (\Psi_{cso} \ \Psi_{sro})^2 + \Psi_{Fo} \ (\Psi_{cso} \ \Psi_{cro})^3 + \ldots) +$$
$$(1\text{-}A) \ (\Psi_{Fs} + \Psi_{Fs} \ \Psi_{css} \ \Psi_{crs} + \Psi_{Fs} \ (\Psi_{css} \ \Psi_{crs})^2 + \Psi_{Fs} \ (\Psi_{css} \ \Psi_{crs})^3 + \ldots)$$

The exponent of the terms ($\Psi_{cso} \ \Psi_{cro}$) and ($\Psi_{css} \ \Psi_{crs}$) is equal to the number of times an exciton has cycled through the charge separation stage. Because the values for probabilities are greater than 0 but less than 1, these series converge to:

$$\Phi_p = \frac{\Psi_{cso} \, \Psi_{sto} \, A}{1 - \Psi_{cso} \, \Psi_{cro}} \tag{8}$$

$$\Phi_F = \frac{\Psi_{Fo} \, A}{1 - \Psi_{cso} \, \Psi_{cro}} + \frac{\Psi_{Fs} \, (1-A)}{1 - \Psi_{css} \, \Psi_{crs}} \tag{9}$$

The first term of Eq 9 identifies the quantum yield of fluorescence when all reactions centers are open ($A = 1$):

$$F_{Fo} = \frac{\Psi_{Fo}}{1 - \Psi_{cso} \, \Psi_{cro}} \tag{10}$$

The second term of the equation identifies the quantum yield of fluorescence when all reaction centers are closed ($A = 0$):

$$\Phi_{Fs} = \frac{\Psi_{Fs}}{1 - \Psi_{css} \, \Psi_{crs}} \tag{11}$$

We note from these equations that when all the reaction centers are either closed or open, the quantum yields of fluorescence depend only upon the values of the probability coefficients. Also, the quantum yield of gross photosynthesis (carbon fixation), Φ_c, is simply the product of the quantum yield of primary photochemistry and the stoichiometric coefficient of steady state carbon fixation to electron flow, J_c/J_e:

$$\Phi_c = \frac{J_c}{J_e} \, \Phi_p = \frac{J_c}{J_e} \, \frac{\Psi_{cso} \, \Psi_{sto} \, A}{1 - \Psi_{cso} \, \Psi_{cro}} \tag{12}$$

At this point, it is worth noting that the quantum yields for the transformations we have discussed also can be expressed in terms of cross sections. For example, the quantum yield of primary photochemistry (charge separation and stabilization), Φ_p, of Eq 12 can be also written in terms of A, σ_{po} (the cross section for photochemistry of photosystems with open reaction centers), and σ_a (the cross section for light absorption by the photosystem):

$$\Phi p = \frac{\sigma_{po}}{\sigma_a} \, A = \frac{\Psi_{cso} \, \Psi_{sto}}{1 - \Psi_{cso} \, \Psi_{cro}} \, A \tag{13}$$

Photochemical Quenching

We consider Eqs 9 and 12 to be the bases for understanding the relationship between fluorescence and photosynthesis. Variation in the yield of *in vivo* fluorescence caused by the opening and closing of reactions centers has been called photochemical quenching. The value of A can be determined from measurements of fluorescence quantum yields under ambient light (Φ_F), when all reaction centers are open (Φ_{Fo}), and when all reaction centers are closed (Φ_{Fs}). Substitution of Eqs 10 and 11 into 9 yields:

$$A = \frac{\Phi_{Fs} - \Phi_F}{\Phi_{Fs} - \Phi_{Fo}} \tag{14}$$

The value of A can be manipulated and monitored by measuring the quantum yield of fluorescence immediately after a pump flash of variable intensity, $\Phi_F(I_s)$:

$$A\,(I_s) = \frac{\Phi_{Fs} - \Phi_F(I_s)}{\Phi_{Fs} - \Phi_{Fo}} \tag{15}$$

Furthermore, substitution of Eq 14 into 12 yields a relationship between the quantum yields of fluorescence and gross photosynthetic rate:

$$\Phi_c = \frac{J_c}{J_e}\,\Phi_p = \frac{J_c}{J_e}\,\frac{\Psi_{cso}\,\Psi_{sto}\,(\Phi_{Fs} - \Phi_{Fa})}{(1 - \Psi_{cso}\,\Psi_{cro})\,(\Phi_{Fs} - \Phi_{Fo})} \tag{16}$$

Equation 16 can also be expressed in terms of the values for fluorescence rather than for fluorescence yields (see Eq 7):

$$\Phi_c = \frac{J_c}{J_e}\,\frac{\Psi_{cs}\,\Psi_{st}\,(F_s - F)}{(1 - \Psi_{cr}\,\Psi_{cs})(F_s - F_o)} \tag{17}$$

Equation 17 provides the basis of predicting photosynthetic rates from measurements of pump-probe fluorescence. The term $F_s - F$ is called variable fluorescence, and the term $F_s - F_o$ is called the maximum variable fluorescence. If the probability coefficients in this equation remain constant in the ocean, we need measure only F, F_o, and F_s to predict the quantum yield of photosynthesis. Under such conditions, the quantum yields of fluorescence and photosynthesis would be inversely proportional to each other.

Figure 2 is a plot of Φ_p and Φ_F for variations in the fraction of open reaction centers. When all reaction centers are open, Φ_p is 0.83 and Φ_F is 0.02. These values linearly decrease and increase with decreases in A, respectively. When all reaction centers are closed, Φ_p is 0 and Φ_F is 0.5. As presented by Falkowski and Kiefer (1985), the value

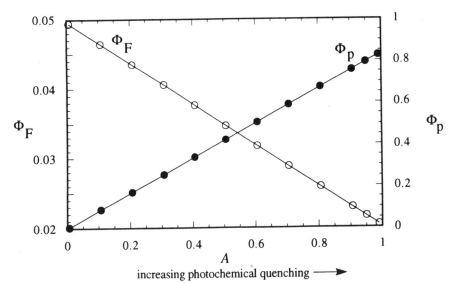

Fig. 2. Predicted changes in the quantum yields of flourescence (Φ_F) and photochemistry (Φ_p) as a function of the fraction of open PSII reaction centers. Φ_p and Φ_F were calculated from Eqs 8 and 9, respectively, using the probability coefficients depicted in Fig. 1.

of A will be a function of irradiance and the I_k value of the instantaneous photosynthetic response curve:

$$A = e^{-\sigma_{po}\tau E_o} = e^{-E_o/I_k} \tag{18}$$

where τ is the minimum steady-state turnover time of the reaction centers. In Fig. 3a we have applied Eq 18 and plotted the quantum yield of photosynthesis (Eq 8) and fluorescence (Eq 9) as a function of E_o/I_k. These functions are curvilinear with respect to light intensity. We also note that quantum yields are predicted to be half maximal when irradiance is about two-thirds that of I_k. In Fig. 3b, we have estimated the quantum yield of carbon-fixation using a value of 0.22 for J_c/J_e (Kok, 1952) and determined the ratio Φ_c/Φ_F as a function of normalized irradiance. The function shown in this figure may be compared to the empirical function shown in Fig. 9 that is applied to the prediction of photosynthetic rate from measurements of natural fluorescence.

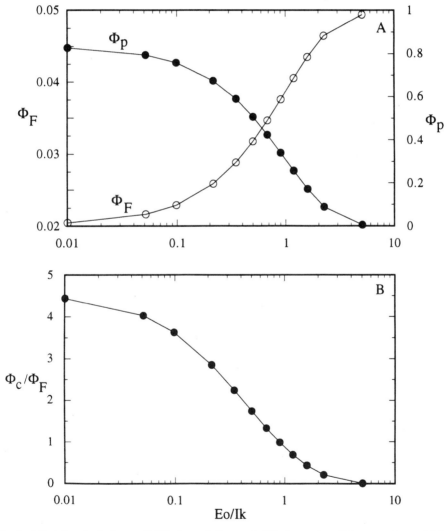

Fig 3. Model calculations of (A) Φ_p and Φ_F, and (B) the ratio of the quantum yields of carbon-fixation (Φ_c) to fluorescence (Φ_F) as a function of normalized irradiance. Φ_c was calculated from Eq 12 assuming a value of 0.22 for J_c/J_e.

Eliminating A from Eqs 14 and 18, we obtain a formulation of I_k in terms of variable fluorescence:

$$I_k = - \frac{E_o}{\ln \dfrac{F_s - F}{F_s - F_o}} \qquad (19)$$

In practice, estimates of I_k are made by a best-fit analysis of measurements made over a range of ambient irradiance. The form of Eq 18 also is valid if the rate of light absorption by a cell suspension that has been placed in the dark is perturbed by a pump flash of variable intensity:

$$A = e^{-\sigma_{po} I_s} \qquad (20)$$

Under such conditions, the effective cross section of the photosynthetic unit can be determined by eliminating A from Eqs 15 and 20:

$$\sigma_{po} = - \frac{\ln A}{I_s} = - \frac{\ln \dfrac{F_s - F(I_s)}{F_s - F_o}}{I_s} \qquad (21)$$

Equations 19-21 show how I_k, σ_{po} and τ can be determined from pump-probe fluorescence (Falkowski et al., 1991).

Nonphotochemical Quenching

Measurements in the field and laboratory indicate that at least some of the probability coefficients, Ψ, in these equations vary with light intensity. Measurements at the sea surface of the ratio of fluorescence to the concentration of chlorophyll a often show

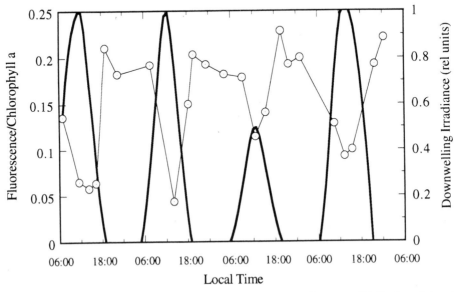

Fig. 4.   Diel measurements of downwelling surface irradiance (solid line) and *in vivo* fluorescence per chlorophyll a (circles) for a natural phytoplankton community in the N. Central Pacific (29° 0.0'N, 122° 31.2'W). (From Kiefer, 1973b).

164

decreases in the middle of the day when light intensities are highest. For example, Fig. 4 depicts a record of fluorescence/chl for sea water (measured with a flow-through Turner fluorometer) and irradiance for 4 days in the north Central Pacific (Kiefer, 1973a). The decrease in the fluorescence with increasing irradiance is the opposite response to what would be expected if increases in irradiance only caused a decrease in the fraction of open reaction centers. Because the irradiance within the Turner Fluorometer is low, the fraction of open reaction centers was probably large, and thus, according to Eq 10, the decrease in fluorescence quantum yield were most likely caused by decreases in Ψ_{Fo}, Ψ_{cso}, and or Ψ_{cro} induced by exposure to high irradiance. Such a light-induced decrease in the quantum yield of fluorescence has been called nonphotochemical quenching.

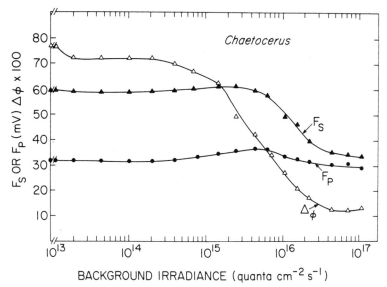

Fig. 5.   Photosynthetic rates and pump-probe fluorescence measurements for a culture of *Chaetoceros gracilis* under continuous background irradiance. Fluorescence was measured with a weak probe flash before (F_p) and following (F_s) a saturating pump flash. $\Delta\Phi$ is the variable fluorescence normalised to F_p. F_b is the fluorescence from the background source. (Falkowski, unpublished).

Figure 5 shows photosynthetic rates and fluorescence as a function of light intensity for a culture of *Chaetoceros gracilis* (Falkowski, unpublished). Photosynthesis was measured with an oxygen electrode, and fluorescence was measured with a pump-probe system. Variations in the yield of induced fluorescence are caused by both photochemical and nonphotochemical quenching. The increases in F with increases in irradiance above 10^{15} q cm^{-2} s^{-1} is largely caused by the closing of reaction centers (decrease in photochemical quenching), while the decrease in F at high irradiance is caused by nonphotochemical quenching. Again, one notes that nonphotochemical quenching begins near I_k values of irradiance.

While photochemical quenching (oxidation and reduction of Q) responds rapidly (μsec to msec) to changes in irradiance, nonphotochemical quenching responds relatively slowly (seconds to minutes) (Horton and Hague, 1988). Such temporal differences allows us to distinguish between the two processes: all reaction centers can be opened or closed within milliseconds of perturbations to the irradiance, allowing the changes in nonphotochemical quenching to be monitored by measuring either Φ_{Fo} or Φ_{Fs}. Although both photochemical and nonphotochemical quenching are being extensively studied in higher plants, there have been few studies of common marine phytoplankton.

Several types of evidence suggest that much of the nonphotochemical quenching results from increased heat production within the antenna of photosystem II (Genty et al., 1990; Rees et al., 1990). In particular, studies on higher plants have shown that increases in nonphotochemical quenching are accompanied by the rapid interconversion of xanthophyll pigments within the antenna bed of photosystem II (reviewed by Demmig-Adams, 1990). However, there are indications that nonphotochemical quenching also may be associated with inactivation of reaction centers (Neale, 1987) and increased heat production within the centers themselves (Weis and Berry, 1987).

Although there are few detailed studies, the information we have discussed allows us to speculate on how nonphotochemical quenching may affect the relationship between fluorescence and photosynthesis. If the principle consequence of increased nonphotochemical quenching is an increase in the rate constant for heat production in the antenna and all other rate constants remain constant, then only the probability coefficients for fluorescence and charge separation (Ψ_{Fo}, Ψ_{Fs}, Ψ_{cso}, Ψ_{css}) will vary. Furthermore, it can be shown that under such conditions the ratios $\dfrac{\Psi_{Fo}}{\Psi_{cso}}$ and $\dfrac{\Psi_{Fs}}{\Psi_{css}}$ remain constant. Fluorescence can be measured with all reaction centers open both in the dark, $F_o(D)$, when there is no nonphotochemical quenching and under ambient light, F_o (E_o), when nonphotochemical quenching is variable. In the dark, the probability coefficient, $\Psi_{cso}(D)$,

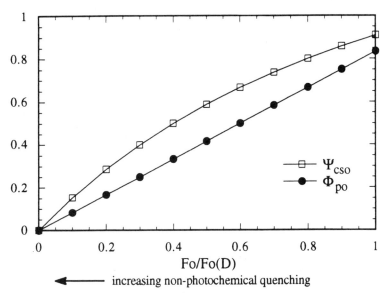

increasing non-photochemical quenching

Fig. 6.   The probability of charge separation in open reaction centers (Ψ_{cso}, Eq 22) and the quantum yield of open reaction centers (Φ_{op}, Eq 16) as a function of non-photochemical quenching. Non-photochemical quenching is represented as the ratio of F_o in dark-adapted samples.

is constant and known. With this information, one can calculate the value of Ψ_{cso}, presumably the only probability coefficient in Eq 16 that varies with nonphotochemical quenching:

$$\Psi_{cso}(E_o) = \frac{Y_{cso}(D) \, F_o(E_o)}{F_o(D) \, (1 - \Psi_{cso}(D) \, \Psi_{cro}) + F_o(E_o) \, \Psi_{cso}(D) \, \Psi_{cro}} \qquad (22)$$

Figure 6 is a plot of Eq 22; it shows that increases in nonphotochemical quenching will cause decreases in the ratio of F_o of cells in ambient light, $F_o(E_o)$, to F_o in dark adapted cells, $F_o(D)$. Such decreases will cause curvilinear decreases in the probability of charge separation in open reaction centers, Ψ_{cso}, and linear decreases in the quantum yield of primary photochemistry in photosystems that have open reaction centers, Φ_{po}. If phytoplankton maintained constant values of A in different irradiances by merely varying the rate of heat dissipation in the antenna, the quantum yields of natural fluorescence and photosynthesis would vary proportionately.

Field and laboratory studies suggest that both photochemical and nonphotochemical quenching occur in the sea. The calculation of the quantum yield of photosynthesis from pump-probe measurements can now be written explicitly stating the light dependence of some of the terms in Eq 17:

$$\Phi_C = \frac{J_c}{J_e} \, \frac{\Psi_{cso}(E_o) \, \Psi_{sto} \, (F_s(E_o) - F(E_o))}{(1 - \Psi_{cso} \, \Psi_{cro}(E_o)) \, (F_s(E_o) - F_o(E_o))} \qquad (23)$$

Equations 22 and 23 are proposed as the basis for measuring gross photosynthetic rate from measurements of fluorescence under variable photochemical and nonphotochemical quenching. Because measurements of F_o and $F_o(D)$ have not been routinely made in the sea, the validity of these hypotheses has not been tested.

A second approach to monitor non-photochemical quenching is based upon measurements by the pump-probe technique of the photochemical cross-section of open photosystems (σ_{po}). According to Eq 21, σ_{po} can be determined by measuring the fraction of reaction centers closed by a pump flash of dose I_s. It then follows from Eq 13 that:

$$\sigma_{po} = \frac{\Psi_{cso} \, \Psi_{sto} \, \sigma_a}{1 - Y_{cso} Y_{cro}} \qquad (24)$$

Because σ_a, Ψ_{sto}, and Ψ_{cro} are, presumably, not affected by non-photochemical quenching, changes in σ_{po} are affected only by changes in Ψ_{cso}. By eliminating σ_a from Eqs 2 and 13, we obtain:

$$J_c = \frac{J_c}{J_e} \, \sigma_{po} \, A \, E_o \, n \qquad (25)$$

Unfortunately, there is no simple means to measure the concentration of photosystems in a seawater suspension. Falkowski et al. (1991) used a similar approach to estimate P^B ($= J_c/$chl) by assuming that n is proportional to the concentration of chlorophyll.

A FIELD STUDY OF FLUORESCENCE AND PHOTOSYNTHESIS

In March 1988 and 1989, measurements of natural fluorescence, variable fluorescence, and photosynthesis were made in the mid-Atlantic Bight. Vertical profiles of solar-stimulated fluorescence were obtained with a Biospherical Instruments profiling natural fluorometer (Chamberlin et al., 1990). Measurements of $F_s(E)$ and $F(E)$ were made *in situ* with a custom-built, submersible pump-probe fluorometer (Falkowski et al., 1991); $F_o(D)$ was measured at discrete depths on dark-adapted (30 min) water samples.

For each station, water samples were collected in 5-liter Niskin bottles attached to a CTD rosette. Pigment samples were filtered onto Whatman GF/F filters and extracted in 90% acetone; chlorophyll *a* was determined fluorometrically on a Turner designs Model 10 fluorometer previously calibrated with pure chlorophyll *a*. Productivity rates were determined in an onboard incubator, using ^{14}C-bicarbonate as a tracer. Water samples from each depth were incubated 4-6 hours under surface irradiance attenuated with neutral density filters, then filtered onto Millipore HA filters. Filters were fumed for 60 seconds over concentrated HCl and the remaining radioactivity was assayed by liquid scintillation counting.

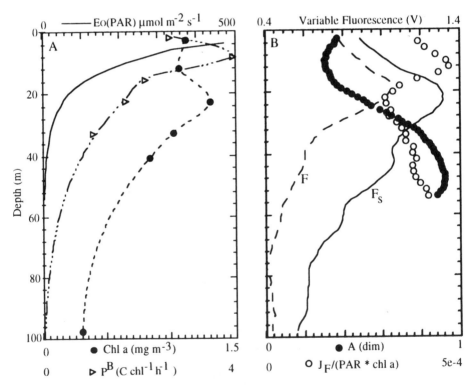

Fig. 7.   Vertical profiles from station EN89312 on March 18, 1989 [37° 41.37'N, 74° 20.55'W: 19:45 local time]. (A) The distribution of scalar irradiance (E_oPAR), chlorophyll *a* concentration (Chl a), and photosynthetic assimilation numbers (P^B). (B) The relative *in situ* yield of fluorescence excited by a weak probe flash before (F) and following (F_s) a saturating pump flash. The fraction of open reaction centers (A, dark circles) was calculated from Eq 14 using F_o from dark-adapted samples. The ratio $J_F/(PAR * chl a)$ is directly proportional to the quantum yield of natural fluorescence (open circles).

Figure 7 shows a profile obtained in the late afternoon. Stratification resulted from the presence of a thermocline extending to a depth of 28m. The chlorophyll *a* maximum was located above the bottom of the thermocline, corresponding to the 10% light level (Fig. 7A). Maximal photosynthetic assimilation occurred at 10m, with photoinhibition evident at the surface.

In general, both F_s and F covaried with the distribution of chlorophyll *a* (Fig. 7B). F increased relative to F_s towards the surface, reflecting increases in photosynthetic activity. At depths less than 11m, both F_s and F deviate from the distribution of chlorophyll and decline, indicative of non-photochemical quenching. The decay in the natural fluorescence signal approximated the attenuation of the exciting scalar irradiance, although at depths of 20m (near the chlorophyll maximum) and 9m (near the photosynthetic maximum) small changes in the slope could be discerned.

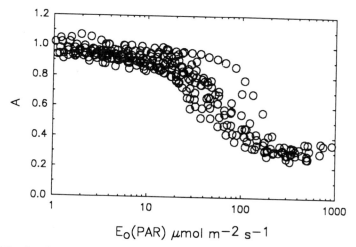

$E_0(PAR)$ μmol m^{-2} s^{-1}

Fig. 8.   The fraction of open reaction centers (A) as a function of incident irradiance for several profiles in the mid-Atlantic Bight.

We estimated the fraction of open reaction centers (*A*) from the magnitude of photochemical quenching of variable fluorescence (Genty et al., 1989, Eq 14). At depth, reaction centers are open and photochemical quenching is maximal (Fig. 7B). With increasing irradiance towards the surface, photochemical quenching diminishes as the photosynthetic rates approach maximum. At a light level of approximately 180 μmol m^{-2} s^{-1}, photosynthetic activity declines and there is an apparent increase in photochemical quenching. The depth where this inflection occurs is interpreted to represent the irradiance where photosynthetic activity approaches saturation (I_k of the photosynthesis vs irradiance curve).

The natural fluorescence signal, normalized to the exciting irradiance (PAR) and chlorophyll *a* concentration, varied 1.5 fold within the euphotic zone. According to Eq 5, this value is equal to the product of the absorption coefficient (a) and the quantum yield of fluorescence (Φ_F). If changes in the absorption coefficient are small, then the observed changes are predominantly caused by changes in Φ_F. Below the 10% light level, the fluorescence yield appears to vary little, although a slight increase is observed below the mixed layer which may indicate increasing contributions of detrital pigments to fluorescence. Towards the surface, the *in vivo* quantum yield increases as a greater proportion of reaction centers are closed by ambient irradiance (decreased photochemical quenching). At depths less than 11m, photosynthetic reactions approach saturation and the onset of non-photochemical quenching processes initiate a decline in the natural fluorescence yield.

It is interesting to note that at the surface, where non-photochemical quenching processes dominate the fluorescence yield, an increasing fraction of photosystem units appear to open. Weis and Berry (1987) suggested that some non-photochemical quenching processes have a photoprotective role, decreasing the rate of excitation energy to PSII reaction centers at supersaturating irradiance. If a large proportion of non-photochemical quenching results from increased rates of thermal dissipation within the antenna bed (as suggested), then this would cause a decrease in the photochemical cross section of open photosystems (σ_{po}) (Genty et al., 1990, Eq 13). Relative estimates of σ_{po} can be obtained with the pump-probe fluorometer by varying the intensity of the pump flash (I_s) and measuring the fluorescence response (see Eq 21). At some stations, σ_{po} decreased at the surface. This decrease was not consistently observed, however, presumably because σ_{po} is also influenced by other conditions of growth, such as nutrient supply and growth irradiance (Kolber et al., 1988; Kolber et al., 1990).

In Fig. 8, we have plotted A as a function of incident irradiance for several vertical profiles. Despite considerable variability between stations, the general shape agrees well with the predicted function of Eq 18. Even at the highest light levels, A did not decrease

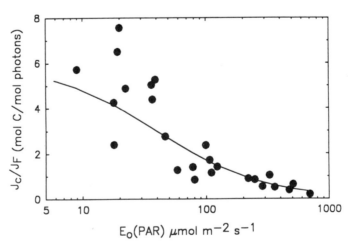

Fig. 9.   The ratio of the rates of photosynthetic carbon assimilation (J_c) to rates of solar-simulated natural fluorescence (J_F) as a function of ambient irradiance. The solid line represents the empirical equation proposed by Chamberlin et al., 1990 (Eq 26).

below 0.3. A fit of this data to Eq 18 yields a mean value for I_k of ~ 165 μmol m^{-2} s^{-1}. This value is similar to the I_k obtained by fitting the measured radiocarbon productivity rates as a function of irradiance using the traditional P vs I approach (Jassby and Platt, 1976).

As expressed in Eq 6, the link between rates of natural fluorescence and photosynthetic carbon assimilation is the ratio of the quantum yields of the two processes.

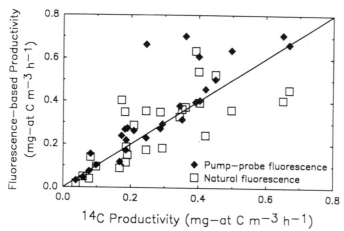

Fig. 10. Productivity rates estimated from pump-probe fluorescence and natural fluorescence compared with rates of radiocarbon assimilation. The line represents 1:1 correspondence.

As we have seen, both quantum yields depend upon the state of the PSII reaction centers, which, in turn, are a function of the ambient light field experienced by the cells and their photoadaptive state. In Fig. 10, the ratio of photosynthesis to natural fluorescence (= Φ_c/Φ_F, see Eqs 4 and 5) is depicted as a function of incident irradiance. A decrease in this ratio with increasing irradiance is apparent. Chamberlin et al. (1990) proposed an empirical equation to describe this dependence:

$$\frac{J_c}{J_F} = \frac{\Phi_c}{\Phi_F} = \frac{k_{cf}}{k_{cf} + E_o(PAR)} [\Phi_c/\Phi_F] \tag{26}$$

where $[\Phi_c/\Phi_F]$ represents the maximum value of the quantum yield ratio and k_{cf} is the irradiance at which this value is one-half the maximum. The general shape of this relationship is comparable with the prediction from the Schatz model (Fig. 3b), suggesting that decreases in the photosynthetic quantum yield with increasing irradiance typically exceed decreases in the fluorescence quantum yield. These results indicate that the quantum yield ratio may vary considerably with increasing depth in the euphotic zone.

Productivity rates were estimated from variable fluorescence measurements according to Falkowski and Kolber (1990). The rate of photosynthetic electron transport was calculated with a kinetic model of PSII, introducing measured values of E_o, σ_{po}, *in situ* quenching of variable fluorescence yields, and the turnover time of photosynthetic electron transport. Carbon assimilation rates then were estimated, assuming eight electrons are required to fix one carbon molecule. Productivity estimates from natural fluorescence were determined by multiplying measured values of natural fluorescence by the predicted ratio of the quantum yields (Eq 26).

Figure 10 compares fluorescence-derived predictions of photosynthetic rate with radiocarbon assimilation from all stations at discrete depths; the results show significant correlations between these estimates of production. Estimates of natural fluorescence explained 74% of the variance in primary production. Measurements of variable fluorescence appeared to be more robust, accounting for 86% of the variability in the radiocarbon measurement.

We point out that discrepancies between ^{14}C and fluorescence-based production are expected from fundamental differences in the methodology. Fluorescence provides an instantaneous measure of gross productivity, while ^{14}C incubations includes the time-integrated effects of irradiance fluctuations and respiratory losses of carbon. In addition, fluorescence measurements are non-intrusive and eliminate possible artifacts arising from sample collection and enclosure (so-called bottle effects). A regression of natural fluorescence on variable fluorescence predictions yielded excellent agreement, with a slope of 1.08 and correlation coefficient of 0.87.

CONCLUSIONS

Since the initial measurements of phytoplankton *in vivo* fluorescence by Lorenzen in 1966, significant advances have been made in the technology and instrumentation for measuring fluorescence at sea. Concomitantly, our understanding of *in vivo* fluorescence of chlorophyll *a* has also progressed. The conceptual basis for interpreting the relationship between chlorophyll *a* fluorescence and photosynthesis appear to be well described by models such as those of Butler and the one discussed in this paper. Numerous laboratory studies have demonstrated the validity of these models in interpreting fluorescence data. Increasingly, studies of fluorescence in the ocean indicate that these concepts can also be applied to mixed populations of marine phytoplankton under natural conditions. However, there is little information on common species of marine phytoplankton; the most detailed studies have been made on higher plants and green algae (i.e., chl a/b systems). Studies are particularly needed on the differences between algal classes (chl a/c and cyanobacteria), and the variability induced by differing conditions of irradiance, temperature, and nutrient supply.

Traditionally, fluorescence in biological oceanography has been used as an indicator of biomass. As our understanding of the variety of processes which control *in vivo* fluorescence yields has progressed, it has become apparent that much more information can be obtained from these measurements. Already, fluorescence has been used as a diagnostic test for photoinhibition (Neale, 1987) and nutrient limitation (Kolber et al., 1990). Considerable progress has been made in understanding how light is transmitted and subsequently absorbed by phytoplankton within the ocean, and now we are able to monitor the fate of this absorbed energy within photosynthetic systems.

REFERENCES

Butler, W. L., and Kitajima, M., 1975, Fluorescence quenching in Photosystem II of chloroplasts, *Biochim. Biophys. Acta.*, 376: 116.

Butler, W. L., 1978, Energy distribution in the photochemical apparatus of photosynthesis, *Ann. Rev. Plant Physiol.*, 29: 345.

Chamberlin, W. S., Booth, C. R., Murphy, R. C., Morrow, J. H., and Kiefer, D. A., 1990, Evidence for a simple relationship between natural fluorescence and photosynthesis in the Sea, *Deep-Sea Res.*, 37(6): 951.

Demmig-Adams, B., 1990, Carotenoids and photoprotection in plants: A role for the xanthophyll zeaxanthin, *Biochim. Biophys. Acta.*, 1020: 1.

Falkowski, P. G., and Kiefer, D. A., 1985, Chlorophyll *a* fluorescence in phytoplankton: Relationship to photosynthesis and biomass, *J. Plankton Res.*, 7 (5): 715.

Falkowski, P. G., and Kolber, Z., 1990, *Current Research in Photosynthesis IV*, M. Baltscheffsky, ed., Kluwer, London.

Falkowski, P. G., Wyman, K., Ley, A. C., and Mauzerall, D. C., 1986, Relationship of steady state photosynthesis to fluorescence in eucaryotic algae, *Biochim. Biophys. Acta.*, 849: 183.

Falkowski, P. G., Ziemann, D., Kolber, Z., and Bienfang, P. K., 1991, Role of eddy pumping in enhancing primary production in the ocean, *Nature*, 352:55.

Genty, B., Briantais J. M., and Baker, N. R., 1989, The relationship between the quantum yield of photosynthetic electron transport and quenching of chlorophyll fluorescence, *Biochim. Biophys. Acta.*, 990: 87.

Genty, B., Harbinson, J., Briantais, J. M., and Baker, N. R., 1990, The relationship between non-photochemical quenching of chlorophyll fluorescence and the rate of photosystem 2 photochemistry in leaves, *Photosynth. Res.*, 25: 249.

Horton, P., and Hague, A., 1988, Studies on the induction of chlorophyll fluorescence in isolated barley protoplasts. IV. Resolution of non-photochemical quenching. *Biochim. Biophys. Acta*, 932: 107.

Jassby, A. D., and Platt, T., 1976, Mathematical formulation of the relationship between photosynthesis and light for phytoplankton, *Limnol. Oceanogr.*, 21 (4): 540.

Kiefer, D. A., 1973a, Fluorescence properties of natural phytoplankton populations, *Mar. Biol.*, 23: 263.

Kiefer, D. A., 1973b, Chlorophyll *a* fluorescence in marine centric diatoms: Responses of chloroplasts to light and nutrient stress, *Mar. Biol.*, 23: 39.

Kiefer, D. A., Chamberlin, W. S., and Booth, C. R., 1989, Natural fluorescence of chlorophyll *a*: Relationship to photosynthesis and chlorophyll concentration in the western south pacific gyre, *Limnol. Oceanogr.*, 34 (5): 868.

Kolber, Z., Wyman, K. D., and Falkowski, P. G., 1990, Natural variability in photosynthetic energy conversion efficiency: A field study in the Gulf of Maine, *Limnol. Oceanogr.*, 35(1): 72.

Kolber, Z. , Zehr, J., and Falkowski, P. G., 1988, Effects of growth irradiance and nitrogen limitation on photosynthetic energy conversion in photosystem II, *Plant Physiol.*, 88: 923.

Kok, B., 1952, On the efficiency of Chlorella growth, *Acta Bot. Neerl.*, I: 445.

Lorenzen, C. J., 1966, A method for the continuous measurement of *in vivo* chlorophyll concentration, *Deep-Sea Res.*, 13:223.

Mitchell, B. G., and Kiefer, D. A., 1988, Chlorophyll *a* specific absorption and fluorescence excitation spectra for light-limited phytoplankton, *Deep Sea Res.*, 35 (5): 639.

Neale, P. J., 1987, Algal photoinhibition and photosynthesis in the aquatic environment, *in:*, "Photoinhibition," D. Kyle, C.B. Osmond, and C.J. Arntzen, eds., Elsevier, Amsterdam.

Owens, T. G., 1991, Energy transformation and fluorescence inphotosynthesis, NATO-Advanced Study Insitute Workshop Series, in press.

Rees, D., Noctor, G. D., and Horton, P., 1990, The effect of high energy-state excitation quenching on maximum and dark level chlorophyll fluorescence yield, *Photosynth. Res.*, 25: 199.

Sakshaug, E., Johnsen, G., Andresen, K., and Vernet, M., 1991, Modeling of light-dependent algal photosynthesis and growth: Experiments with the Barents Sea diatoms *Thalassiosira nordenskioeldii* and *Chaeotoceros furcellatus*, *Deep-Sea Res.*, **38**(4):415.

Schatz, G. H., Brock, H., and Holzwarth, A. R., 1988, Kinetic and energetic model for the primary processes in Photosystem II, *Biophys. J.*, 54: 397.

Weis, E., and Berry, 1987, Quantum efficiency of Photosystem II in relation to "energy"-dependent quenching of chlorophyll fluorescence, *Biochim. Biophys. Acta.*, 894: 198.

BIO-OPTICAL MODELS AND THE PROBLEMS OF SCALING

Robert R. Bidigare[1], Barbara B. Prézelin[2] and Raymond C. Smith[3]

[1]Department of Oceanography
University of Hawaii at Manoa
Honolulu, HI 96822

[2]Department of Biological Sciences
Marine Science Institute
University of California at Santa Barbara
Santa Barbara, CA 93106

[3]Center for Remote Sensing and Environmental Optics
University of California at Santa Barbara
Santa Barbara, CA 93106

INTRODUCTION

Historically, the construction of global maps of ocean productivity has been a difficult task (Berger, 1989). Representative, precise, and accurate measurements of carbon fixation rates have been hampered by the errors associated with methodological problems and sampling limitations (Jahnke, 1990). The frustration of biological oceanographers in dealing with this issue was best summarized by Eppley (1980) during the first 'Primary Productivity in the Sea' symposium over a decade ago: "These disparate results beg for reconciliation as they suggest an order of magnitude uncertainty in the rate of primary production in the central oceans. Is it of the order 50-150 mg C m^{-2} d^{-1} as the standard ^{14}C data have suggested for twenty years or is it 1-2 g C m^{-2} d^{-1} as the diel oxygen and POC changes and the PIT collections imply?" While this issue has yet to be completely resolved, considerable progress has been made during the last decade towards the reconciliation of differences in primary productivity estimates based on the standard ^{14}C-labeling technique (Steeman Nielsen, 1952) and those based on geochemical tracer distributions (Jenkins and Goldman, 1985; Williams and Robertson, 1991). It appears that systematic errors in the different methodologies used to determine rates of oxygen production and carbon fixation contribute to observations of high photosynthetic quotients (mol O_2 evolved per CO_2 fixed) for phytoplankton communities (Laws, 1991; Prézelin and Glover, 1991; Williams and Robertson, 1991). In addition, when care is taken to minimize trace metal contamination (Fitzwater et al., 1982), avoid the toxic effects of latex and neoprene rubber closure mechanisms of Niskin® bottles (Price et al., 1986; Williams and Robertson, 1989), and incubate seawater samples under the appropriate spectral distribution of light (Laws et al., 1990), then ^{14}C-measured rates of

Primary Productivity and Biogeochemical Cycles in the Sea
Edited by P.G. Falkowski and A.D. Woodhead, Plenum Press, New York, 1992

primary productivity for the central Pacific Ocean are several fold higher than historical values (i.e., 428 $\pm$ 249 mg C m^{-2} d^{-1}, n = 11, 24 hour simulated *in situ* incubations, Station ALOHA, October 1988 to November 1989, HOT program, 1990).

Furthermore, it has been established that the base of the euphotic zone for primary production may lie well below the traditionally defined 1% light depth (cf. Harris, 1986) and can account for up to 15-20% of total integrated water column productivity in the Sargasso Sea (Prézelin and Glover, 1991). Likewise, the discovery of diverse and often fragile photosynthetic picoplankton as ubiquitous and dominant components of oligotrophic water masses has led to the development of methods in the last decade to define their biomass and include their photosynthetic activity in productivity estimates (Glover, 1986; Hooks et al., 1988; Chisholm, this volume). Given the potential for <1 μm phytoplankton cells to contribute more than half of the productivity of oligotrophic water columns (cf. Glover, 1986; Platt and Li, 1986), it is not surprising that recent ^{14}C measurements have resulted in large increases in areal production estimates for the Sargasso Sea (i.e., 218 $\pm$ 101 mg C m^{-2} d^{-1}, n = 25, spar buoy and shipboard estimates for two stations 284 km apart during summer 1986, Prézelin and Glover, 1991; also see Lohrenz et al., 1991).

In addition to the methodological problems associated with primary production measurements, there are also problems of scaling (cf. Harris, 1986). Current estimates of global ocean production are biased by the errors associated with our inability to sample on the appropriate time and space scales. Given our changing view of the stratified upper ocean as a constant 'steady-state' system, some of the discrepancies between oxygen and ^{14}C estimates of primary production may reflect differences in the time scales over which data were gathered (Kerr, 1986). As recently stated by Platt and co-workers (1989), "the incompatibility of intrinsic time scales, and the difficulty of extrapolating one component of primary production to estimate another, limit the resolving power of the comparison..." While tracer distributions (e.g., oxygen anomaly, oxygen utilization rate, and ^{3}He excess) integrate biological and physical processes over seasonal to annual time scales, *in vitro* ^{14}C uptake and oxygen production measurements are typically short in duration (hours to one day). Recent physical models (Klein and Coste, 1984) and direct nanomolar nitrate measurements (Garside, 1985; Eppley and Renger, 1988; Glover et al., 1988a,b) suggest that the euphotic oligotrophic nitrate environment may be perturbed on biological time scales and be reflected in large transient fluctuations in carbon fixation rates (Glover et al., 1988b; Prézelin and Glover, 1991). Under such conditions, the rate processes controlling carbon fixation and oxygen production may become decoupled from one another (Williams and Robertson, 1991). Improvements in establishing regional- to global-scale oceanic production will be realized when both temporally and spatially dependent variations are included in the integral.

Lastly, it is important to ensure that the sampling strategy and the methodology employed match the scale of the scientific question being addressed (Prézelin et al., 1991). For instance, the initial imagery collected with the Nimbus-7 coastal zone color scanner (CZCS, Hovis et al., 1980) revealed that distributions of phytoplankton in the upper ocean are quite complex, hence, poorly sampled by classical shipboard operations. This has led to the development of remote sensing methods to estimate pigment biomass (Gordon et al., 1980; Smith and Baker, 1982; Shannon et al., 1983; McLain et al., 1984; Gordon et al., 1988; Carder et al., person. commun.) and primary production (Smith et al., 1982, 1987a; Brown et al., 1985; Eppley et al., 1985; Collins et al., 1986; Platt, 1986; Lohrenz et al., 1988; Platt et al., 1988; Platt and Sathyendranath, 1988; Balch et al., 1989a, 1989b; Sathyendranath et al., 1989; Platt et al., 1990; Morel, 1991). These approaches have advanced recent investigations of biological variability in the world's ocean. However, those studies that are completely dependent on atmospheric remote sensing measurements are limited to providing information only for the upper 25% of the euphotic zone (Smith et al., 1987a) over long time scales ($\geq$ day, depending on cloud cover which often negates observations in many regions). Recognizing these limitations, there are now compelling reasons for the development of multiplatform sampling strategies where ships, moorings, and drifters provide complementary information on the bio-optical and hydrographic characteristics of diverse water columns over shorter time scales ($\geq$ minute to hours).

The present discussion will address the application of 'light-pigment' models for estimating primary production profiles from bio-optical data provided by drifter, mooring and shipboard platforms. Descriptions of other types of bio-optical production models based on satellite imagery (M.R. Lewis) and fluorescence (P. Falkowski, D.A. Kiefer, and R. Reynolds) are treated separately in this volume. We begin with a discussion of the variability of relevant oceanic processes and the sampling strategies required for their resolution. Next, we present a summary of the evolution of 'light-pigment' bio-optical models over the last three decades. Attention is given to the variability of the required input parameters and how this variability limits the accuracy of bio-optical models when applied to larger spatial scales. Finally, we conclude with a list of requirements necessary for the future use of bio-optical models for addressing key questions on global change.

TEMPORAL-SPATIAL VARIABILITY OF OCEANIC PROCESSES

The abundance and distribution of marine phytoplankton are dependent primarily upon mixing processes in the upper ocean, photon and nutrient fluxes controlling algal growth rate, and the grazing pressure and sedimentation processes driving loss rates. It is clear that near synoptic and continuous knowledge of the magnitude and variability of these forcing functions is of critical importance to successful local, regional and global predictions of phytoplankton distribution, abundance and activities.

A schematic drawing depicting time and space scales of variability of relevant oceanic processes in the upper ocean is shown in Fig. 1. Since the variability spans many orders-of-magnitude and since any one sampling approach has inherent limits of resolution, there is now a general concensus that a multiplatform, multidisciplinary sampling strategy is required for the determination of phytoplankton distributions in space and time (Esaias, 1980; Campbell et al., 1986; Harris, 1986; Smith et al., 1987a; Dickey, 1988, 1990, 1991; Prézelin et al., 1991). Once accurate assessments of phytoplankton distributions are available, then concurrently measured and modeled light distributions can potentially be used in conjunction with light-pigment models to estimate rates of primary productivity. It is important to emphasize that physical observations be done on time and space scales consistent with bio-optical observations in order to relate forcing functions with biological responses. For example, the complexity of the distributions of phytoplankton originates in large part to the ambiguity of local versus advective processes and scales of physical mixing.

Resolution of Time-dependent Vertical Variability

Vertical distributions of phytoplankton, or more precisely pigment biomass, are best measured from shipboard and mooring platforms (Table 1 and Fig. 2). Depth-dependent variations in pigment biomass are poorly resolved from satellites and planes since these measurements represent depth-weighted averages of pigments in near-surface waters, i.e., they are limited by the optical attenuation depth (Gordon and McCluney, 1975; Smith, 1981). Because ship availability and cruise duration are limited, shipboard measurements

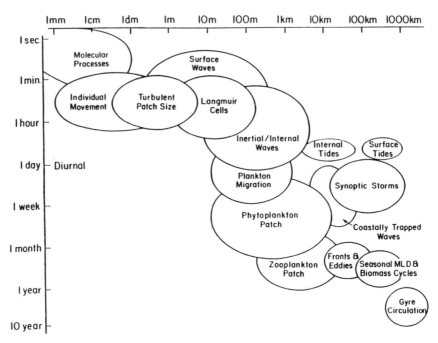

Fig. 1.   Relevant time and space scales characteristic of biological and physical processes in the upper ocean (Dickey, 1991).

Table 1.   Time and Space Sampling Domains of Various Research Platforms

Platform Description	Sampling Domains		
	Time	Horizontal	Vertical
Moorings	1 min - years	cms	At fixed depths: 10 m - 100 m
Ship on station	<1 h - 4 weeks	On station: cms Inter-station: kms	<1 m - 100 m
Ship mapping/fixed depth	1.5 days - 4 weeks	<1 m - 100 km	± few meters upper 100 m
Ship tow-yos	1.5 days - 4 weeks	0.5 km - 100 km	<1 m - 100 m
Drifters/fixed depth	1 min - 6 months	<1 m - 1000 km	± few meters upper 10 m
Planes	1 day - 1 week	10 m - 1000 km	Upper few meters (opt. atten. depth)[a]
Satellites	1 day - years	1 km - global	Upper few meters (opt. atten. depth)

[a]optical attenuation depth

in the upper 100's m of the upper water column are constrained to time scales of 1 hour to several weeks. Moorings, on the other hand, can be sequentially deployed at 3-6 month intervals to provide high resolution measurements (minutes to hours) over time scales of several years (Dickey, 1991). It should be noted that periodic shipboard 'ground truthing' of mooring sensors is critical for obtaining calibration and complimentary data sets. During 1987, the Office of Naval Research (ONR) sponsored the deployment of the first open-ocean bio-optical mooring as part of the Biowatt advanced research initiative. The mooring was located at 34°N, 70°W in the Sargasso Sea and collected physical, biological and optical data in the upper 160 m during three sequential three-month intervals. The multi-variable moored system (MVMS, which measures currents, temperature, conductivity beam transmission, stimulated fluorescence, dissolved oxygen, and photosynthetically active radiation; Dickey et al., 1991) and bio-optical moored system (BOMS, which measures depth, temperature, downwelling spectral irradiance, upwelling spectral radiance, and tilt; Smith et al., 1991a) instrument packages deployed on the Biowatt mooring recorded the springtime stratification of the upper ocean and a phytoplankton bloom event, which occurred shortly after stabilization of the water column. Interestingly, the bloom (> 1 mg chlorophyll m^{-3}) lasted approximately 2 days and would have undoubtedly been missed by the infrequent shipboard sampling in this region. Unfortunately, the lack of concurrently collected color satellite images precluded an estimate of the horizontal length of this bloom.

Resolution of Time-dependent Horizontal Variability

Horizontal distributions of pigment biomass can be measured using a variety of platform types, including, drifters, planes, ships and satellites (Table 1 and Fig. 3). Since moorings and 'ships on station' represent fixed point sampling strategies, they are

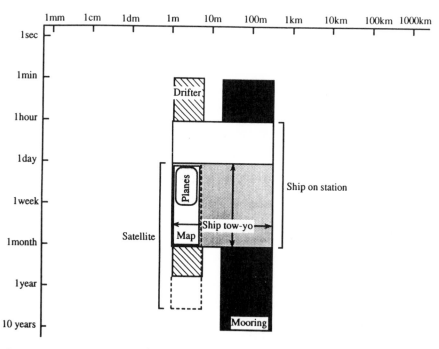

Fig. 2.   Time and vertical space sampling domains of the various platforms used in oceanographic research.

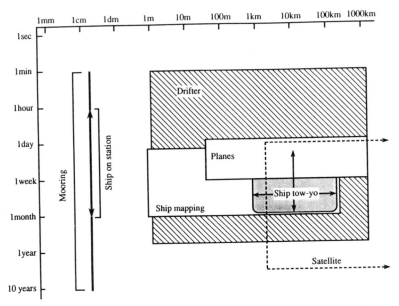

Fig. 3.　Time and horizontal space sampling domains of the various platforms used in oceanographic research.

inefficient in providing information on the horizontal gradients of pigment biomass. The use of drifters, planes and ships have limitations in providing synoptic data on basin and global scales due to their restricted areal coverage, as well as the associated cost and time constraints; these sampling strategies are best suited for regional scale studies. Satellites are an excellent platform for sampling horizontal pigment gradients. Global coverage is provided on a daily basis (depending on cloud cover) on length scales of $\geq$ kms. Unfortunately, as of 1986 when the CZCS failed, the oceanographic community has not had access to color satellite images. The next launch of an ocean color imaging system, Sea-WiFS, is currently scheduled for August 1993. Sea-WiFS is an improved version of the original CZCS and will have additional channels (i.e., 412, 443, 490, 520, 565, 665, 765 and 865 nm) to provide chlorophyll biomass estimates, accurate atmospheric corrections, and information on the attenuation of light by colored dissolved organic matter.

LIGHT-PIGMENT PRODUCTION MODELS

Evolution of Bio-optical Production Models

In this section the evolution of models for 'in water' light-pigment production models is described. This is not a comprehensive review, but merely summarizes some of the important models which have been formulated over the last three decades. Discussion of the physiological assumptions underlying many of the models is presented in following sections. For comparison, all of the models have been cast in similar notation and units. The notation used in the bio-optical models is summarized in Table 2 and

Table 2. Summary of Notation Used in the Bio-optical Production Models. These Symbols are Defined when Initially Used in the Text and Have Been Summarized for Convenience.

Symbol	Definition (units)
z	Water column depth (m)
λ	Wavelength (nm)
Q_o (λ, z)	Quantum scalar irradiance (4π, μEin m^{-2} nm^{-1} s^{-1})
Q_{opar} (z)	Quantum scalar photosynthetically active radiation (par, 4π, μEin m^{-2} s^{-1})
Q_d (λ, z)	Quantum downwelling irradiance (2π, μEin m^{-2} nm^{-1} s^{-1})
Q_{par} (0^-)	Downwelling par measured just below the surface (2π, μEin m^{-2} s^{-1})
Q_{par} (z)	Downwelling par at depth z (2π, μEin m^{-2} s^{-1})
$Q_{par}(z)/Q_{par}(0^-)$	Fractional light depth (dimensionless)
K_{Qpar} (z)	Diffuse attenuation coefficient for Q_{par} at depth z (m^{-1})
K_d (λ, z)	Diffuse attenuation coefficient for downwelling irradiance (m^{-1})
Chl (z)	Concentration of chlorophyll a at depth z (mg m^{-3})
C_i (z)	HPLC measured concentration of pigment group i at depth z (mg m^{-3})
a^*_i (λ)	Weight-specific absorption coefficient for pigment group i (m^2 mg^{-1})
a_{ph} (λ, z)	Phytoplankton absorption coefficient (m^{-1})
a_{ph} Chl^{-1} (λ, z)	Chl-specific phytoplankton absorption coefficient (m^2 mg^{-1})
$\overline{a_{ph}}$ (z)	Spectrally weighted phytoplankton absorption coefficient (m^{-1})
$\overline{a_{ph}}$ Chl^{-1} (z)	Spectrally weighted Chl-specific phytoplankton absorption coefficient (m^2 mg^{-1})
k_c	Spectrally averaged Chl-specific vertical attenuation coefficient for phytoplankton (m^2 mg^{-1})
AQ_{ph} (z)	Integrated quanta absorbed by phytoplankton (400-700 nm, μEin m^{-3} s^{-1})
AQ_{acc} (z)	Integrated quanta absorbed by accessory pigments (400-700 nm)
AQ_i (z)	Integrated quanta absorbed by pigment group i (400-700 nm)
Φ_{max}	Maximum quantum efficiency of photosynthesis ($\equiv \alpha/\overline{a_{ph}}$, mol C Ein^{-1})
Φ_c (z)	Quantum efficiency of photosynthesis at depth z (mol C Ein^{-1})
K_ϕ	Irradiance at which quantum efficiency is equal to $\Phi_{max}/2$
P_{max}	Maximum rate of photosynthesis (mg C m^{-3} h^{-1})
P_{max} Chl^{-1}	Chl-specific maximum rate of photosynthesis (mg C mg Chl^{-1} h^{-1})
α	Rate of light limited photosynthesis (mg C m^{-3} h^{-1} (μEin m^{-2} s^{-1})$^{-1}$)
α Chl^{-1}	Chl-specific alpha (mg C mg Chl^{-1} h^{-1} (μEin m^{-2} s^{-1})$^{-1}$)
I_k	Saturation parameter of photosynthesis ($\equiv P_{max} / \alpha$, μEin m^{-2} s^{-1})
P_s Chl^{-1}	Chl-specific maximum rate of photosynthesis in the absence of photo-inhibition (mg C mg Chl^{-1} h^{-1})
β Chl^{-1}	Chl-specific photoinhibition parameter (mg C mg Chl^{-1} h^{-1} (μEin m^{-2} s^{-1})$^{-1}$)
P (z)	Calculated hourly rate of photosynthesis at depth z (mg C m^{-3} h^{-1})
ΣP (t)	Integral photosynthesis over time interval t (mg C m^{-2} t^{-1})
R	Relative photosynthesis parameter (cf. Ryther and Yentsch, 1957)
R_d (z)	Relative photosynthesis at depth z (cf. Ryther and Yentsch, 1957)
P_d	Calculated daily rate of photosynthesis at depth z (mg C m^{-3} d^{-1}, cf. Ryther and Yentsch, 1957)

represents a subset of bio-optical notation recommendations which will appear elsewhere (U.S. JGOFS, 1991).

Talling (1957, 1965) treated phytoplankton as a "compound photosynthetic system" and described a mathematical model to estimate integral photosynthesis (ΣP):

$$\Sigma P \text{ (hourly)} = \frac{Chl \cdot (P_{max} \, Chl^{-1})}{K_{Qpar}} \ln \frac{Q_{par} \, (0-)}{0.5 \, I_k} \tag{1}$$

where, Chl is chlorophyll concentration; $P_{max} \, Chl^{-1}$ is the chlorophyll-specific maximum rate of photosynthesis; $Q_{par}(0-)$ is downwelling photosynthetically active radiation (par) measured just below the surface; K_{Qpar} is the diffuse attenuation coefficient for Q_{par}; and I_k is the saturation parameter for photosynthesis (i.e., the minimum irradiance to sustain light-saturated rates of photosynthesis). In this form, the half-saturation constant for photosynthesis (i.e., $0.5 \, I_k$) is comparable to the K_m ($= 0.5 \, V_{max}$) for enzyme saturation kinetics and provided one of the first approaches towards linking the natural variability in the photo-physiological state of phytoplankton to the surrounding light environment. The Talling (1957) model for estimating integral photosynthesis is based on the graphical representation of a rectangle (ABCD), where one side (AB) is equal to P_{max} and the other side (BC) is equal to the depth ($z = 0.5 \, I_k$) at which the onset of saturation of photosynthesis begins. This approach assumes I_k does not vary with light depth. The product of these two parameters equals integral photosynthesis:

$$\Sigma P \text{ (hourly)} = Z_{0.5Ik} \cdot Chl \cdot (P_{max} \, Chl^{-1}) \tag{2}$$

Rodhe (1965) further simplified this model based on field measurements which revealed that the depth corresponding to $0.5 \, I_k$ is approximately equal to the depth of the 10% light level (i.e., $Z_{0.1Qpar(0-)}$, the depth at which $_{Qpar}(z)_{Qpar}(0^-) = 0.1$):

$$\Sigma P \text{ (hourly)} = Z_{0.1Qpar(0-)} \cdot Chl \cdot (P_{max} \, Chl^{-1}) \tag{3}$$

We note that the application of these models is restricted to situations where the phytoplankton, or more specifically chlorophyll *a*, is uniformly distributed in the upper water column and photosynthesis-irradiance (P-I) parameters do not vary with light depth - i.e., mixing regimes where rates of photoadaptation are significantly slower than rates of mixing (Lewis et al., 1984; Cullen and Lewis, 1988). Rodhe (1965) found an excellent agreement between calculated (Eq 2) and measured rates of photosynthesis for several different lakes, and obtained a slightly poorer correlation between measured and calculated primary production rates when Eq (3) was applied.

Ryther and Yentsch (1957) developed an analogous model based on P-I relationships determined for marine phytoplankton cultures (Ryther, 1956):

$$\Sigma P \text{ (daily)} = \frac{R}{K_{Qpar}} \cdot Chl \cdot (P_{max} \, Chl^{-1}) \tag{4}$$

where, R equals a relative photosynthesis parameter which varies as a function of total daily surface radiation (see Fig. 1 of Ryther and Yentsch, 1957). While this formulation has the same restrictions as the Talling and Rodhe models, Ryther and Yentsch (1957) also presented an equation which allows the estimation of depth-dependent daily rates of photosynthesis (P_d) when phytoplankton are non-uniformly distributed in the upper water column:

$$P_d = R_d \cdot \text{Chl} (z) \cdot (P_{max} \text{ Chl}^{-1})$$ (5)

where, R_d equals a depth-dependent relative photosynthesis parameter which varies as a function of total daily surface radiation and fractional light depth (see Fig. 1 of Ryther and Yentsch, 1957). These investigators found good agreement between calculated (Eq 4) and measured rates of photosynthesis for a variety of oceanic water types.

Webb et al. (1974) formulated an exponential expression for predicting photosynthetic rates as a function of growth irradiance. This model was recast by Jassby and Platt (1976) into a form containing commonly measured P-I parameters:

$$P (z) = \text{Chl} (z) \cdot (P_{max} \text{ Chl}^{-1}) \cdot (1 - e^{-Q_{par}(z)/I_k})$$ (6)

where, $Q_{par}(z)$ is downwelling par at depth z. Geider et al. (1986) modified this basic equation to develop an energy-balance model of micro-algal physiology which relates photosynthesis to algal growth and respiration rates. Using this energy-balance approach, Geider et al. (1986) were able to examine the size dependence of several metabolic variables in diatoms, including maximum quantum yield, maximum growth rate, the Chl:C ratio, and the Chl-specific absorption coefficient.

Platt et al. (1980) expanded Eq (6) to include a term to correct for the effects of photoinhibition at high growth irradiances:

$$P (z) = \text{Chl} (z) \cdot (P_s \text{ Chl}^{-1}) \cdot (1 - e^{-a}) \cdot e^{-b}$$ (7)

where, $a = [(\text{alpha Chl}^{-1}) \cdot Q_{par} (z) \cdot (P_s \text{ Chl}^{-1})^{-1}]$; $b = [(\text{beta Chl}^{-1}) \cdot Q_{par}(z) \cdot (P_s \text{Chl}^{-1})^{-1}]$; alpha ($\alpha$) Chl^{-1} is the Chl-specific rate of light limited photosynthesis; beta (β) Chl^{-1} is the Chl-specific photoinhibition parameter (same units as α Chl^{-1}); and P_s Chl^{-1} is the chlorophyll-specific maximum rate of photosynthesis in the absence of photoinhibition. This model (Eq 7) has been used to estimate time- and depth-dependent rates of primary production for phytoplankton populations sampled from arctic and temperate marine waters (Harrison et al., 1985), an oligotrophic basin and warm core eddy in the Northwest Atlantic (Prézelin et al., 1986), as well as the equatorial Pacific (Cullen et al., 1991).

Jassby and Platt (1976) first introduced the hyperbolic tangent model for describing P-I relationships:

$$P (z) = \text{Chl} (z) \cdot (P_{max} \text{ Chl}^{-1}) \cdot \tanh (Q_{par}(z) / I_k)$$ (8)

Since its inception, it has become the most widely used model for predicting the photosynthetic rates of natural phytoplankton populations (Harrison et al., 1985; Herman and Platt, 1986; Smith et al., 1987b, 1989; Prézelin and Glover, 1991). Pahl-Wostl and Imboden (1990) transformed Eq (8) into a 'DYnamic model for the PHOtosynthetic Rate of Algae (DYPHORA) for application to lake systems where the combination of photoinhibition by bright light and mixing time scales can dominate the kinetics of primary productivity over the time-course of a day. The authors suggested that this model may be useful for testing hypotheses regarding the selective role of vertical mixing in the competition between algal species. This model describes the dynamic response of the rate of photosynthesis to changing growth irradiances using two characteristic times: the response time of P_{max} for increasing light (τ_r) and the light inhibition decay time (τ_i). Results of simulations performed with DYPHORA agreed with experimental data when τ_r and τ_i values of 0.5-5 and 30-120 minutes, respectively, were used in the model.

Kiefer and Mitchell (1983) described a simple, steady state model of (gross) primary production:

$$P(z) = \Phi_c(z) \cdot (1.2 \times 10^4) \cdot Q_{par}(z) \cdot Chl(z) \cdot (\overline{a_{ph}} Chl^{-1}(z)) \qquad (9)$$

where, $\Phi_c(z)$ is the quantum efficiency of photosynthesis at depth z; 1.2×10^4 is a factor to convert mol C to mg C; and $\overline{a_{ph}} Chl^{-1}(z)$ is the spectrally weighted chlorophyll-specific phytoplankton absorption coefficient at depth z. It should be noted that $Q_{par}(z)$ is used in place of quantum scalar par $(Q_{opar}(z)$; Kiefer and Mitchell, 1983) to keep consistent notation between equations. Tyler (1975) has shown that Φ_c varies inversely with growth irradiance. Kiefer and Mitchell (1983) chose a Michaelis-Menton (rectangular hyperbolic) function to describe the dependence of Φ_c on growth irradiance:

$$\Phi_c(z) = \Phi_{max} \frac{K_\Phi}{K_\Phi + Q_{par}(z)} \qquad (10)$$

where, Φ_{max} is the maximum quantum efficiency of photosynthesis and K_Φ is the irradiance at which the quantum efficiency is equal to $\Phi_{max}/2$. As such, this model of Φ_c is not wavelength-dependent and was only validated for 'white-light' grown cells. It should also be noted that the Kiefer-Mitchell model differs from the P-I expressions described above (Eqs 6-8) in that Eqs (9) and (10) refer only to the 'adapted' rate of photosynthesis at a given light level (cf. Geider et al., 1986). With the exception of near-surface waters, Marra and Heinemann (1987) found a good agreement between calculated (Eqs 9 and 10, $\Phi_{max} = 0.06$ mol C Ein^{-1}, and $\overline{a_{ph}} Chl^{-1} = 0.017$ m^2 mg^{-1}) and directly measured rates of primary production in the North Pacific Central Gyre.

Bidigare, Smith and colleagues (1987) modified Eq (9) by transforming it into a spectral model which takes into account depth- and wavelength-dependent variations in irradiance and phytoplankton absorption properties (also see full spectral production model recently described by Morel, 1991):

$$P(z) = \Phi_c(z) \cdot (1.2 \times 10^4) \cdot AQ_{ph}(z) \qquad (11)$$

where, $AQ_{ph}(z)$ is the integrated quanta (400-700 nm) absorbed by the phytoplankton population at depth z:

$$AQ_{ph}(z) = \int_{400}^{700} Q_d(\lambda, z) \cdot a_{ph}(\lambda, z) \, d\lambda \qquad (12)$$

where, $Q_d(\lambda, z)$ is the quantum downwelling irradiance at depth z and wavelength λ, and $a_{ph}(\lambda, z)$ is the phytoplankton absorption coefficient at depth z and wavelength λ. Since the estimation of $a_{ph}(\lambda, z)$ is confounded by the presence of detritus (see below), Bidigare et al. (1987, 1990a) developed a spectral reconstruction technique which allows an estimation of $a_{ph}(\lambda, z)$ based on pigment concentrations and their *in vivo* absorption properties determined by high-performance liquid chromatography (HPLC):

$$a_{ph} (\lambda, z) = \sum_{i=1}^{n} a^*_i (\lambda) \cdot C_i (z) \tag{13}$$

where, $a^*_i (\lambda)$ is the weight-specific absorption coefficient for pigment group i and $C_i (z)$ is the HPLC determined concentration of pigment group i at depth z. Since spectral quality changes with depth, a new formulation of $\Phi_c(z)$ was required for use in Eq (11) (Bidigare et al., 1987):

$$\Phi_c(z) = \Phi_{max} \frac{P_{max}/\Phi_{max}}{P_{max}/\Phi_{max} + AQ_{ph}(z)} \tag{14}$$

This equation differs from the Kiefer-Mitchell formulation (Eq 10) in that the wavelength-dependent ($AQ_{ph}(z)$) and the wavelength-independent (P_{max} and Φ_{max}) terms are treated separately. For comparisons made in open-oceanic waters, primary production estimates based on the 'Photosynthetically absorbed radiation' (Phar) model (Eqs 11-14) agreed to within 30% (or better) of the rates determined by *in situ* incubation in the Sargasso Sea (Bidigare et al., 1987).

Predictive accuracy improved further for production estimates of diverse water types of the Southern California Bight, when Smith et al. (1989) further derived a variant of Eq (14) by replacing the P_{max}/Φ_{max} term with its mathematically equivalent form, $I_k \cdot k_c \cdot Chl$, where k_c is the spectrally averaged Chl-specific vertical attenuation coefficient for phytoplankton. As such, Eq (14) is transformed into a fully spectral equation for estimating Φ_c as a function of I_k, k_c and AQ_{ph}. In addition, for laboratory studies performed with the diatom *Chaetoceros gracile* and the prymnesiophyte *Emiliania huxleyi*, Schofield et al. (1990) reported a close agreement between photosynthetic rates determined by Eq (8), 'enhanced' carbon-action spectra and the absorption-based model. Similar correspondence between spectral and non-spectral components of production for natural phytoplankton communities from diverse water types has also been documented by Schofield and coworkers (1991).

Comparison of Bio-optical Production Models

The absorption-based bio-optical models of production (Eqs 9 and 11) described above approximate gross photosynthesis (i.e., net production *plus* autotrophic respiration), whereas the predictions resulting from the P-I models (Eqs 6 and 8) approach net production (i.e., gross production *minus* autotrophic respiration) since they contain parameters determined by short-term ^{14}C incubations (Jassby and Platt, 1976; Kiefer and Mitchell, 1983). Furthermore, primary production estimates based on 24 h *in situ* (IS) or simulated *in situ* (SIS) ^{14}C measurements approach net community production (i.e., gross production *minus* autotrophic and microscopic heterotrophic respiration; cf. Platt et al., 1989; Prézelin and Glover, 1991) because of phytoplankton-zooplankton interactions in the incubation bottles. Thus, intercomparisons of these different approaches for estimating primary productivity are not strictly valid and often biased.

In addition to these 'operational' differences, there are 'mathematical' differences between Eqs (6), (8), and (9, 14) resulting from the form of the expression used (i.e., exponential (Exp), hyperbolic tangent (Tanh) or absorption-based rectangular hyperbolic (Phar) function, respectively). Chalker (1980) presented a derivation of Eq (6) and its mathematical relationship to Eq (8) by performing a Taylor series expansion on the expression $dP/dI = F(P)$. Model simulations using identical (or mathematically equivalent) input parameters (cf. Jassby and Platt, 1976) were performed with Eqs (6), (8), and (9, 14) in order to examine between-model differences (Fig. 4). It should be noted that these comparisons do not correct for autotrophic respiration and photorespiration. The results obtained reveal that P (Tanh) > P (Exp) > P (Phar). Interestingly, when comparisons have been made in the field, P (Tanh) > (S)IS > P (Exp), P (Phar) (Harrison et al., 1985; Smith et al., 1989; Cullen et al., 1991; Prézelin and Glover, 1991). To arrive at the 'best' mathematical formulation for predicting P-I responses, Jassby and Platt (1976) tested 8 different equations using empirical data consisting of 188 duplicate light saturation experiments. These authors found that the hyperbolic tangent function provided the best fit to the empirical data. With these results in mind, we derive a hyperbolic tangent expression for estimating Φ_c (for use in Eq 9 or Eq 11) by setting Eq (9) equal to Eq (8), and solving for Φ_c as follows (also see similar derivations by Platt and Jassby, 1976):

$$\Phi_c(z) \cdot Q_{par}(z) \cdot a_{ph}(z) \overline{=} P_{max} \cdot \tanh(Q_{par}(z) / I_k) \qquad (15)$$

Multiplying the left-hand side of the equation by Φ_{max} / Φ_{max} yields:

$$\frac{\Phi_c(z)}{\Phi_{max}} \cdot \Phi_{max} \cdot Q_{par}(z) \cdot a_{ph}\overline{(z)} = P_{max} \cdot \tanh(Q_{par}(z) / I_k) \qquad (16)$$

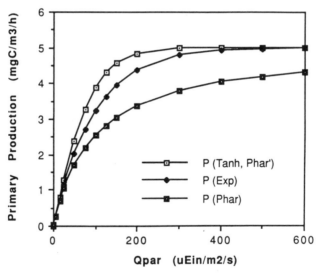

Fig. 4. Primary production estimates based on the hyperbolic tangent model of Jassby and Platt (1976; Tanh model, Eq 8); the exponential model of Webb et al. (1974; Exp model, Eq 6); and the absorption model of Kiefer and Mitchell (1983; Eq 9), where Φ_c is calculated using a modified rectangular hyperbolic (Bidigare et al., 1987; Phar model, Eq 14) or a hyperbolic tangent function (Phar' model, Eq 21). Production was estimated for 'white light' irradiances ranging from 0 to 600 μEin m^{-2} s^{-1} (Chl = 1 mg m^{-3}; Φ_{max} = 0.08 mol C Ein^{-1}; P_{max} Chl^{-1} = 5 mg C mg Chl^{-1} h^{-1}; chlorophyll-specific phytoplankton absorption coefficient = 0.015 m^2 mg^{-1}; and I_k = 96.5 μEin m^{-2} s^{-1}). Note results from the hyperbolic tangent model (Eq 8) and the absorption-based model (Eqs 9 and 21) are mathematically identical for irradiances greater than 0 μEin m^{-2} s^{-1}.

Since,

$$\Phi_{max} = \alpha \cdot (\overline{a_{ph}}(z))^{-1}, \text{ and} \qquad (17)$$

$$P_{max} = I_k \cdot \alpha \qquad (18)$$

Eq (16) can be rewritten as:

$$\frac{\Phi_c(z)}{\Phi_{max}} \cdot \frac{\alpha}{\overline{a_{ph}}(z)} \cdot Q_{par}(z) \cdot \overline{a_{ph}}(z) = I_k \cdot \alpha \cdot \tanh(Q_{par}(z) / I_k) \qquad (19)$$

Which is equivalent to:

$$\frac{\Phi_c(z)}{\Phi_{max}} \cdot Q_{par}(z) = I_k \cdot \tanh(Q_{par}(z) / I_k) \qquad (20)$$

Thus, Φ_c can be defined by the following expression:

$$\Phi_c = \Phi_{max} \cdot (I_k/Q_{par}(z)) \cdot \tanh(Q_{par}(z)/I_k) \qquad (21)$$

By means of the derivation used, which ignores autotrophic respiration and photorespiration, the hyperbolic tangent model (Tanh, Eq 8) is mathematically identical to the absorption-based model (Phar', Eqs 9 and 21) for $Q_{par}(z)$ values $> 0 \ \mu$Ein m^{-2} s^{-1} (Fig. 4). The authors refer the readers to Platt et al. (1988) and Sathyendranath et al. (1989) for a more complete discussion of the equivalence of different photosynthesis and growth equations.

INPUT PARAMETERS FOR BIO-OPTICAL PRODUCTION MODELS

This section addresses the input parameters which drive the bio-optical production models presented above. Our ability to measure these parameters and their variability has greatly improved since the initial 'Primary Productivity in the Sea' symposium held in 1980. These improvements are highlighted below.

Light

Light in the marine environment is extremely variable and its availability in the upper ocean is the most important factor controlling rates of carbon fixation by resident phytoplankton. Energy available for in-water photoprocesses, with photosynthesis being most important, is modulated by solar zenith angle, cloud fraction and optical thickness, atmospheric chemistry (aerosols, ozone, water vapor, oxygen content) and sea state. This variability determines the amount, spectral composition, and angular distribution of the light field just below the air-water interface. The consequent distributions of radiant energy as a function of depth depend upon this surface input plus the absorption and scattering properties of seawater itself and the dissolved and suspended particulate material within the water column.

Instruments and methods developed within the past decade now permit spectral irradiance to be continuously estimated as a function of depth over space and time scales previously not possible in spite of the large environmental variability. These advances include:

(1) satellite imagery that allows the spatial and temporal surface solar irradiance to be accurately estimated (Gautier et al., 1980; Bishop and Rossow, 1991);

(2) models linking total irradiance to Q_{par} (Frouin et al., 1989) and spectral irradiance (Smith et al., 1991b);

(3) improved ocean color algorithms linking in-water spectral properties to the spectral water leaving radiance (Gordon et al., 1988);

(4) satellite estimates of ocean optical properties, especially the diffuse attenuation coefficient for irradiance;

(5) ship deployed bio-optical profiling systems for rapid optical and biological observations (Aiken, 1981; Smith et al., 1984);

(6) optical sensors on untended moorings for the continuous determination of in-water spectral irradiance and the estimation of both optical and biological parameters (Booth and Smith, 1988; Smith et al., 1991a);

(7) analysis methods (Smith and Baker, 1986) and instrument techniques (Waters et al., 1991) that permit correction or collection of optical data without ship perturbation of the light field;

(8) methods of optical data analysis (Smith and Baker, 1984, 1986) and in-water bio-optical algorithms (Baker and Smith, 1982; Smith and Baker, 1978, 1981; Morel, 1988); and

(9) the increased use of optical methods for the study of biological processes (cf., special Hydrologic Optics issue of *Limnology and Oceanography*, vol. 8, no.8, 1989).

Items (1) and (2) potentially provide input data necessary for regional and global production estimates; (3) provides input data for regional and global estimates of ocean optical water types; (4-7) permit the continuous estimation of the in-water spectral irradiance field and the consequent application to a range of biological problems (Bidigare et al., 1987; Smith et al., 1987b; Smith et al., 1989). Thus, radiant energy can be used as the input driver for accurately estimating phytoplankton production rates over a wide range of space and time scales. A further discussion of the nature and measurement of the light environment in the ocean is given by J. T. O. Kirk in this volume.

Pigments

In 1983, Mantoura and Llewellyn described a rapid, routine HPLC technique for the determination of algal pigments and their degradation products in seawater. Since then, numerous papers have been published which document the distribution and detailed composition of photosynthetic pigments in algal cultures and oceanic waters (Gieskes and Kraay, 1983a,b, 1984, 1986a,b; Liaaen-Jensen, 1985; Bidigare et al., 1986, 1989a, 1990b,c, 1991; Smith et al., 1987b; 1990; Gieskes et al., 1988; Hooks et al., 1988; Whitledge et al., 1988; Bidigare, 1989; Siegel et al., 1990; Ondrusek et al., 1991). Most importantly, these studies have revealed that accessory pigments are (1) structurally diverse; (2) heterogeneously distributed with respect to depth; and (3) highly variable on time scales of $\geq$ days to weeks. During the last decade, at least 10 novel chlorophyll and carotenoid pigments have been described for marine phytoplankton and bacterioplankton (Table 3). In addition, a number of novel phycoerythrins have been reported for marine coccoid cyanobacteria (Alberte et al., 1984; Ong et al., 1984). Many of these pigments have yet to be structurally characterized and little, if anything, is known about their *in vivo* absorption properties and photosynthetic function. Clearly, this is an important area of future study. The introduction of analytical methods providing greater resolution (supercritical fluid chromatography/mass spectrometry, Frew et al., 1988; liquid chromatography/mass spectrometry, Eckardt et al., 1990) of complex pigment mixtures and specific detection should lead to the discovery of an even more diverse suite of phytoplankton pigments.

Phytoplankton Absorption Properties

The rate of phytoplankton photosynthesis is ultimately dependent upon the rate of photon capture. The effectiveness of utilizing photons is controlled by two coefficients: the absorption coefficient (a_{ph} (λ, m^{-1})), which represents the ability of photosynthetic

Table 3. Occurrence of Novel Chlorophyll and Carotenoid Pigments in Marine Phytoplankton and Bacterioplankton.

Pigment	Occurrence[a]	Reference
Chlorophylls		
Phytyl-chlorophyll c-like	Prymnesiophytes	Nelson and Wakeham (1989)
Chlorophyll c(CS-170)[b]	*Micromonas pusilla*	Jeffrey (1989)
Chlorophyll c(MPPAV)[c]	*Pavlova gyrans*	Fawley (1989)
Divinyl chlorophyll a	Prochlorophytes	Chisholm et al. (1988)
Divinyl chlorophyll b	Prochlorophytes	Goericke (1990)
Chlorophyll c_3	Prymnesiophytes & chrysophytes	Fookes and Jeffrey (1989)
Bacteriochlorophylls		
Geranyl-4-isobutylbacterio- chlorophyll e	Photosynthetic bacteria	Repeta et al. (1989)
Carotenoids		
19'-Butanoyloxyfucoxanthin	Chrysophytes	Bjørnland et al. (1984)
Eutreptiellanone	Euglenophytes	Fiksdahl et al. (1984)
Prasinoxanthin	Prasinophytes	Foss et al. (1984)

[a] denotes the occurrence of novel pigments in *some* members of the algal and bacterial groups specified (i.e., they are *not* present in all species of each group)
[b] chlorophyll c-like pigment from a tropical strain of *Micromonas pusilla* (clone CS-170)
[c] chlorophyll c-like pigment from *Pavlova gyrans* (clone MPPAV)

pigments to absorb photons, and the quantum yield of photosynthesis (Φ_c), which represents the efficiency with which absorbed light is used to fix carbon. The former, a_{ph} (λ), is wavelength dependent and its magnitude is strongly influenced by several factors, including cell size and shape, intracellular pigment concentration and composition, chloroplast geometry, and organization of the thylakoid membrane (Platt and Jassby, 1976; Morel and Bricaud, 1981, 1986; Falkowski et al., 1985; Post et al., 1985; Haardt and Maske, 1987; Morel et al., 1987; Yentsch and Phinney, 1988; Berner et al., 1989; Nelson and Prézelin, 1990). Determination of a_{ph} (λ) for natural phytoplankton populations has historically been difficult. This difficulty arises from the problems associated with performing absorption measurements on particles, as well as the co-occurrence of detrital matter in natural suspended particulate samples (Yentsch, 1962; Kiefer and SooHoo, 1982; Kirk, 1983; Kishino et al., 1984, 1985, 1986; Lewis et al., 1985a; Maske and Haardt, 1987; Mitchell and Kiefer, 1988a). During the last decade several new methods have been introduced for the determination of a_{ph} (λ). These techniques are based on 4 different approaches, quantitative glass fiber filter spectrophotometry (Kiefer and SooHoo, 1982; Mitchell and Kiefer, 1988b; Mitchell, 1990), quantitative spectrophotometry of resuspended Nuclepore® filtered cells (Nelson et al., 1991; Schofield et al., 1991), microphotometry (Iturriaga et al., 1988), and spectral reconstruction (Bidigare et al., 1987, 1989a,b, 1990a). The methods which are routinely used for for estimating a_{ph} (λ) in natural waters are summarized in Table 4 and discussed below.

Table 4. Summary of the Techniques Used to Estimate the Phytoplankton Absorption Coefficient (a_{ph} (λ, z), m^{-1}) in Natural Waters

Technique and Detrital Correction Method	Reference
Quantitative glass fiber filter spectrophotometry	
Multivariate analysis of absorption spectra	Kiefer and SooHoo (1982)
MeOH extraction of glass fiber filters	Kishino et al. (1985)
Spectral fluorescence excitation ratio ("F-ratio") correction	Mitchell (1987)
Multiple linear regression analysis of absorption spectra	Morrow et al. (1989)
Quantitative spectrophotometry of resuspended cells concentrated by Nuclepore® filtration	
Spectral analysis of suspensions of MeOH extracted cell detritus concentrated by Nuclepore® filtration	Nelson et al. (1991) Schofield et al. (1991)
Microphotometry	
Statistical analysis of cellular and detrital absorption spectra	Iturriaga and Siegel (1989)
Spectral reconstruction	
Spectral reconstruction based on HPLC-determined pigment concentrations and their *in vivo* absorption properties	Bidigare et al. (1989; 1990a) Nelson and Prézelin (1990)

Kiefer and SooHoo (1982) described a quantitative method for determining the spectral absorption coefficient for natural suspended particles. This technique has subsequently been modified by Mitchell and Kiefer (1988b). Initially, the diffuse transmittance of the filter is measured spectrophotometrically (400-750 nm), using a wetted filter as a blank. The diffuse transmittance spectrum is transformed into an absorption spectrum by normalization at 750 nm and applying a correction for (1) the volume of seawater filtered, (2) the clearance area of the filter, and (3) the multiple pathlength effect (Mitchell and Kiefer, 1988b). Mitchell (1990) recently published new algorithms for determining spectral absorption coefficients by this method. The resulting spectra represent the absorption (m^{-1}) due to living and detrital particles in a collimated light field. Deconvolution of these spectra into phytoplankton and detrital components has been achieved by multivariate analysis (Kiefer and SooHoo, 1982), MeOH extracted difference spectra (Kishino et al., 1985), multiple linear regression analysis (Morrow et al., 1989), and a detrital correction factor based on the spectral fluorescence excitation ratio ("F-ratio", Mitchell, 1987). Iturriaga et al. (1988) developed a microphotometric technique for the determination of the absorption efficiency factor ($Q_a (\lambda)$, 400-750 nm) of individual phytoplankton cells and detrital particles. A statistical comparison of $Q_a (\lambda)$ values with the total particulate absorption coefficient enabled Iturriaga and Siegel (1989) to partition absorption contributions by phytoplankton cells and detrital particles.

For quantitative spectrophotometry of resuspended Nuclepore® filtered cells (Nelson et al., 1991; Schofield et al., 1991), phytoplankton absorption spectra [$a_{ph} (\Delta\lambda_i, z)$] were measured using the opal glass technique on whole water samples that had been gently filtered on 0.4 μm Nuclepore® polyester filters and resuspended by gentle agitation in the seawater filtrate. Resuspended particle absorption [$a_p(\lambda)$] was measured using the sample filtrate as the blank (Nelson and Prézelin, 1990). Replicate absorption measurements were made on serially diluted samples (optical densities values <0.05) in order to confirm that Beer's law applied and that multiple scattering effects were insignificant (Kirk, 1983; Sathyendranath et al., 1987; Morel et al., 1987). Phytoplankton losses during filtration and resuspension were quantified. After absorption spectra were recorded, suspensions were refiltered and the extracted Chl concentration was determined. HPLC analyses of selected filters showed no systematic bias toward the retention of particular pigments on the Nuclepore® filters. Refiltered extracts were compared to chlorophyll a concentrations of whole water suspensions, in order to correct for phytoplankton losses and to allow for direct estimates of volume-specific absorption parameters. The chlorophyll concentrations of refiltered particle suspensions were used to determine Chl-specific absorption parameters reported here and to reconstruct spectral absorption reported elsewhere (Nelson et al., 1991). Direct measurements of total particle absorption were "detrital"-corrected following the procedure of Kishino et al. (1985). Aliquots of the particle suspension were extracted overnight in cold methanol and refiltered to remove extracted pigments and to collect extracted particles. The extracted particles were resuspended in distilled water and the spectral absorption by detritus absorption [$ad(\lambda)$] was quantified. These "detrital" spectra were subtracted from the total particle absorption to yield "detrital"-corrected phytoplankton absorption spectra [$a_{ph}(\lambda) = a_p(\lambda) - a_d(\lambda)$].

A spectral reconstruction technique for indirectly estimating $a_{ph}(\lambda)$ was described by Bidigare et al. (1990a). In this approach, $a_{ph}(\lambda)$ spectra are reconstructed from the *in vivo* absorption coefficients derived for the major pigment groups (chlorophyll a, chlorophyll b, chlorophyll c, photosynthetic carotenoids (psc), and photoprotectant carotenoids (ppc)) and their volume-based concentrations determined by HPLC (Eq 13). This calculation assumes that the $a_{ph} (\lambda)$ spectrum is a linear combination of absorption contributions provided by "unpackaged" pigment chromophores located within phytoplankton cells. As a consequence, the accuracy of this method is limited by

191

(1) incomplete pigment extraction, (2) pigment "package" effects, and (3) differences in the intracellular absorption properties of the pigments classified within each of the pigment groups. Thus, the accuracy of this technique should be greatest where phytoplankton cell size is small (cf. Bidigare et al., 1989b; Stramski and Morel, 1990) and/or where pigment "package" effects are minimal. This approach technique has provided reasonable approximations of a_{ph} (λ) for cyanobacteria cultures (Bidigare et al., 1989b) and natural phytoplankton communities sampled from the chlorophyll maximum layer of the Sargasso Sea (Bidigare et al., 1990a), diverse waters in the Southern California Bight (Nelson et al., 1991) and diatom-dominated waters in Antarctic waters (Nelson, Schofield and Ondrusek, pers. comm.). In cultures of red tide dinoflagellates, where differences can arise between directly measured and spectrally reconstructed a_{ph}, limits to the spectral dependency of pigment packaging effects have been described (Nelson and Prézelin, 1990) Similar exceptions were seen upon occasion in chlorophyll maximum communities of the Southern California Bight (Nelson et al., 1991) and often in *Phaeocystis*-dominated communities of the Antarctic marginal ice zone where pigment "package" effects appear to be evident (Nelson, Schofield and Ondrusek, pers. comm.).

The spectral reconstruction technique has provided new insight into the role of accessory pigmentation in light absorption. For the Sargasso Sea, accessory pigments accounted for 60% and 90% of the light absorbed by resident phytoplankton at the sea surface and the base of the euphotic zone, respectively (Fig. 5). The optical depth-dependent absorption contribution provided by accessory pigmentation followed the same pattern when winter and summer Biowatt II cruises are compared. This result is surprising since chlorophyll *a* distributions and water column optical properties measured during these cruises were markedly different. Absorption contributions provided by individual accessory pigment groups (chlorophyll *b*, chlorophyll *c*, photosynthetic carotenoids, and photoprotectant carotenoids), however, showed large between-cruise differences (Fig. 6). These results document the physiological plasticity of marine phytoplankton and provide strong evidence which supports the notion that marine phytoplankton are chromatically adapted to their light environment. In addition, the spectral reconstruction technique allows an estimation of phytoplankton absorption which is not biased by the photoprotectant accessory pigments which have poor energy transfer efficiencies (Fig. 7).

Absorption-based Photosynthetic Parameters

Over the last decade, advancements in sea-going optical instrumentation, sampling strategies and physiological-based bio-optical models have led to a surge in multi-disciplinary field programs specifically designed to quantify the time and space scales of phytoplankton variability. Together, these achievements have brought abilities to monitor, understand and predict changing patterns of photosynthesis in the sea to its present state where, among other possibilities:

1) *in situ* spectral and non-spectral parameters of photosynthesis can be determined accurately over a wide range of time and space scales (Figs. 7-10);

2) *in situ* spatial gradients in photosynthetic parameters can be employed to delineate phytoplankton communities that differ in composition and/or photophysiology (Smith et al., 1987b; 1991c; Prézelin et al.1991; Prézelin and Glover, 1991);

3) the variability in photosynthetic parameters on time scales >1 day, vertical space scales >5m, and horizontal space scales >10 km can be linked to variability in hydrographic properties and their regulatory effects on phytoplankton community dynamics (cf. Harris, 1986; Prézelin et al., 1986, 1987);

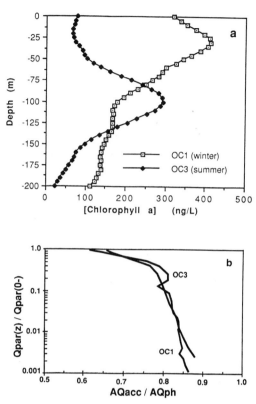

Fig. 5. Profiles of (a) chlorophyll a (ng L^{-1}) vs depth and (b) AQ_{acc}/AQ_{ph} vs optical depth for Biowatt II cruises OC1 (March 1987, winter) and OC3 (August 1987, summer). Stations were occupied at the Biowatt mooring site (34°N, 70°W).

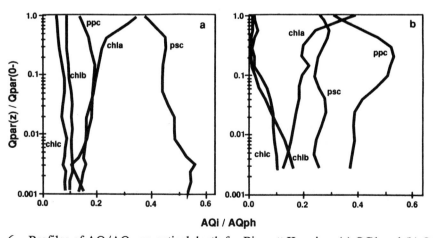

Fig. 6. Profiles of AQ_i/AQ_{ph} vs optical depth for Biowatt II cruises (a) OC1 and (b) OC3.

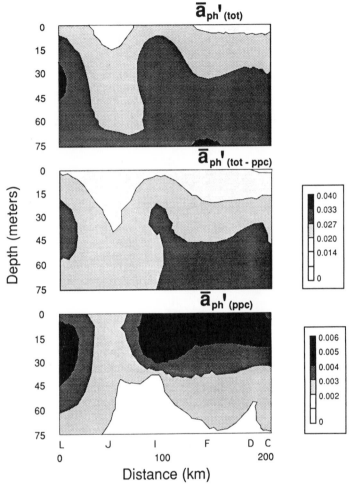

Fig. 7. Comparisons of spatial variations in the spectrally-weighted coefficients for absorption by all phytoplankton pigments (top), photosynthetic pigments (middle) and photoprotective pigments (bottom) in the Southern California Bight during Watercolors '88 (Prézelin et al., 1991; Prézelin, unpubl. data). Between dawn and dusk of a single day, the 200 km transect crossed California Current waters flowing to the south at Sta. L, through post-upwelling water subducting offshore at Sta. J, over the Continental Shelf Break at Sta. I and into hydrographically related Southern California Counter Current waters flowing to the north at Stas. C-I.

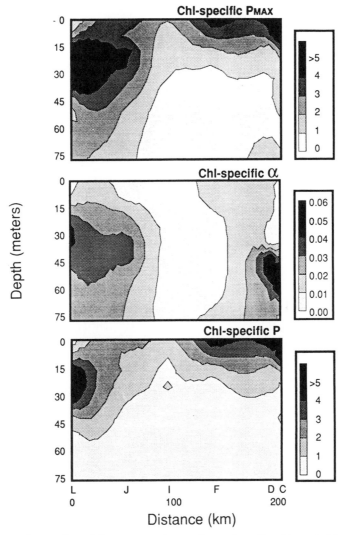

Fig. 8. Comparisons of spatial contours of noontime values for chlorophyll-specific P_{max}, α and *in situ* productivity in the Southern California Bight during Watercolors '88 (Prézelin, unpubl. data; see Fig. 7 for station details).

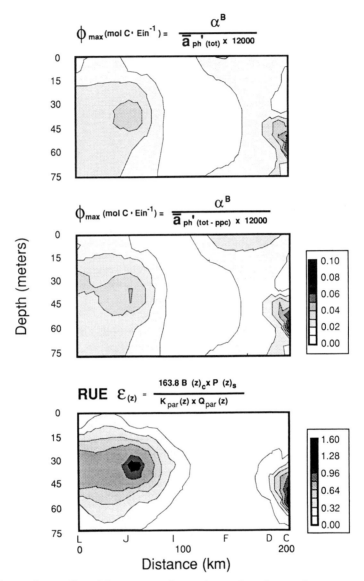

Fig. 9. Comparisons of spatial contours of noontime values for maximum quantum yield, based on absorption by total phytoplankton pigments (top), photosynthetic pigments alone (middle), and radiation utilization efficiency (bottom) in the Southern California Bight during Watercolors '88 (Prézelin, unpubl.; see Fig. 7 for details).

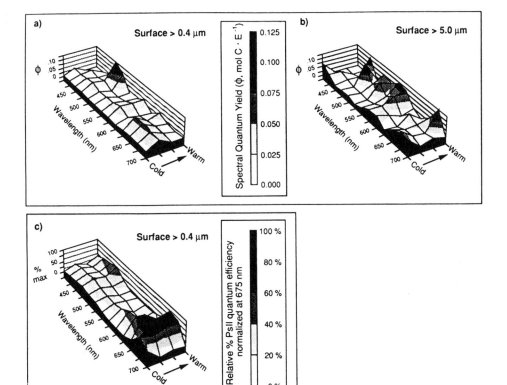

Fig. 10. Comparison of 3-dimensional surface plots of absorption-based estimates of spectral quantum yield of carbon fixation for (a) >0.4 μm and (b) >5.0 μm phytoplankton communities sampled from surface waters of the Southern California Bight during Watercolors '88. Fluorescence-based estimates of the spatial variability in PsII quantum efficiency for > 0.4 μm phytoplankton communities is presented in (c) (Prézelin, Schofield and Nelson; Prézelin, unpubl. data)

4)   *in situ* spatial gradients in photoadaptive parameters can be used to estimate minimum rate constants for hydrographic events, including rates of turbulent mixing (Lewis et al., 1984; Cullen and Lewis, 1988), as well as scales of vertical stratification within the upper mixed layer (Prézelin et al., 1986, 1987);

5)   on the time scale of 1 day, periodicity in production parameters can be linked to intrinsic and, for the most part, environmentally-independent biological clocks and/or environmentally-mediated cell cycle events (cf. Prezelin, 1991) and lastly;

6)   the mechanistic and predictive linkages between spectral and non-spectral parameters of photosynthesis can be derived (Fig. 11), used to test assumptions implicit in earlier bio-optical models, and provide direction for improving existing models.

Measurements of light-dependent rates of photosynthesis provides significant information on the photophysiology of phytoplankton (cf. Prézelin, 1981, 1991; Richardson et al., 1983; Prézelin et al., 1991). Useful parameters derived directly from curve fitting P-I data include α, the light-limited slope and an index of the 'relative' quantum efficiency of photosynthesis; P_{max}, the light-saturated rate of photosynthesis and a measure of photosynthetic capacity of cells; I_k ($\equiv P_{max}/\alpha$, the Talling constant) which is an index of the onset of light-saturated photosynthesis and photoadaptive state; β, the slope of the negative decline in photosynthesis at photoinhibiting light intensities; and I_b ($\equiv P_{max}/\beta$) the parameter which quantifies the onset of photoinhibition. With sufficient data and attention to technique, the standard deviation about the mean of these parameters can be maintained within 3-10% for P_{max}; 5-30% for α and β; and 10-40% for I_k and I_b (Zimmerman et al., 1987; Schofield et al., 1990). Much of the improvement in field efforts to derive these parameters is attributable to the development of non-spectral, spectral, and spectrally enhanced "photosynthetron" systems which allow for quick, accurate, and numerous shipboard determinations of radiolabeled carbon uptake rates (Lewis and Smith, 1983; Lewis et al., 1985a,b; Prézelin et al., 1986, 1987; Schofield et al., 1991).

The dynamics of physical processes determine most, but not all, of the temporal and spatial variability in phytoplankton productivity. Physical factors also largely determine how long phytoplankton communities have to adapt to a given set of growth conditions. Adaptability of phytoplankton communities to new growth conditions, and hence the constancy of photophysiological indices of primary production, depend on the

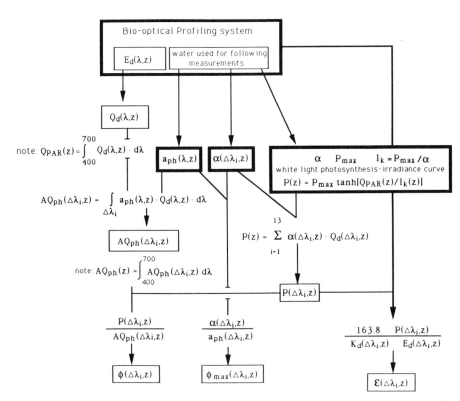

Fig. 11. Flow chart of spectral and non-spectral derivations of absorption-based production parameters from spectral and non-spectral field measurements (in bold boxes) carried out during Watercolors '88 (Schofield et al., 1991).

nature of the environmental change as well as the growth requirements and physiological state of the species present. On times scales of a few days or less, phytoplankton communities tend to respond to hydrographic variability via physiological adaptations (Harris, 1986). Over longer times scales, the composition of phytoplankton communities tends to change. Those species best suited to the direction of hydrographic change will grow and divide, while those less suited may grow too slowly to maintain a significant presence and/or be more susceptible to grazing, senescence and sedimentation (cf. Morris, 1981; Harris, 1986). For those interested in documenting how variability in a single environmental factor may drive variability in algal biology and *vice versa*, it is advisable to choose a biological parameter which responds on a similar time scale and with sufficient sensitivity to track the environmental change. It is best if the proxy measure chosen is one least affected by other environmental or biological variability that might be occurring on overlapping time scales (cf. Prézelin, 1991; Prézelin et al., 1991).

A perturbation in the ambient light field can induce changes in multiple phytoplankton properties simultaneously, with the individual photoadaptive response times varying by several orders of magnitude (cf. Prézelin et al., 1991). For instance, a sustained change in irradiance can induce changes in Chl fluorescence on time scales of sec to min while changes in Chl concentrations, cell absorption and P-I parameters generally will be evident on time scales of hours or longer. Given that wind forcing largely determines the vertical location of phytoplankton with the euphotic zone of most water columns, it is to be expected that a relationship should exist between vertical gradients in photoadaptive indices (i.e., I_k or Chl-specific α) and turbulent motion. If the time scale response of photoadaptation is shorter than that for vertical mixing, phytoplankton within the mixed layer will exhibit a vertical gradient associated with adaptation to ambient light intensities. However, if mixing occurs at a rate which is faster than that of the photoadaptation time scale, then photophysiological parameters will be uniform within the mixed layer (Lewis et al., 1984; Cullen and Lewis, 1988). This can be demonstrated with changes in the fluorescence ratio (i.e., *in vivo* fluorescence with DCMU: *in vivo* fluorescence without DCMU) which is an index of the coupling between light-harvesting pigments and photosynthetic reaction centers of PsII. The fluorescence ratio will respond to changes in the underwater light field on the order of 30 minutes (Harris, 1986), and, when the water column is strongly stratified (Richardson number > 10,000), cells have sufficient time to adapt photosynthetic machinery to the ambient light field. When vertical mixing dominates (Richarson number < 500), there is no change in the fluorescence ratio as mixing scales are shorter than the response time of the fluorescence ratio.

The difficulties in simply inversing the expected relationship in order to utilize observed gradients in photoadaptive parameters (and associated rate constants, γ) to predict *in situ* turbulent mixing rates or scales of stratification is two-fold. First, photoadaptive rate constants are not universal constants whether one considers different photoadaptive parameters, the same parameter within different phytoplankton species, or even the same parameter in a single species growing at different rates (cf. Prézelin et al., 1991). Second, *in situ* photoadaptive rate constants, and resulting vertical profiles of the photoadaptive parameter, are subject to differential regulation by environmental factors that are uncoupled or only loosely coupled to rates of turbulent mixing or scales of vertical mixing. For example, as an uniformly mixed water column begins to stratify, even small vertical differences in absolute temperature or nutrient flux (i.e., zooplankton swarms) could drive the photoadaptative rate of phytoplankton faster at one depth to another depth. As a result, the vertical profile for a given photoadaptive trait more likely reflects the differential rates of photoadaptation at different depths which, in turn, is largely but not exclusively dependent upon the recent history of vertical mixing in the water column. But even with these limitations, it is likely that defining gradients in photoadaptive parameters can provide most field researchers with general insights into mixing regimes that would not otherwise be available.

Of the vast quantities of radiant energy impinging onto the surface of water bodies only a small proportion is converted into chemically bound energy and even a smaller quantity is transferred to the consumer food web. The efficiency with which AQ_{ph} is photochemically converted to photosynthate is termed quantum yield, while the efficiency with which Q_{par} is photochemically converted to photosynthate is termed radiation utilization efficiency (RUE, ϵ) (Fig. 9). Sensitivity analyses indicate that the associated estimations of in situ Φ_{max} (Fig. 9), par-averaged Φ and $\Phi_{max}(\lambda)$ (Fig. 11) is the most significant source of error in present attempts to bio-optically estimate in situ production (Bannister and Weidemann, 1984; Smith et al., 1989; Schofield et al., 1991). For instance, it has been assumed to varying degrees that P-I parameters and related measures of quantum yield are invariant, maximal or show a definite depth-dependent profile that is broadly applicable over different time and space scales. That such generalities can not be universally applied without significantly affecting the predictive accuracy of modeling attempts has become apparent in the last few years. An example of the time/space variability in P-I parameters, maximum quantum yield and radiation utilization efficiency (RUE) measured during the Watercolors '88 cruise in the Southern California Bight is illustrated in Figs. 7-10. Among other things, these findings (Prézelin et al., 1991 and unpubl. data) document that in situ Φ_{max} values (1) are highly variable on the daily time scale and spatial scales of a few hundred kilometers; (2) do not routinely increase with light depth; (3) are not correlated with any single environmental variable but, rather, appear to reflect the synergistic interaction of different environmental variables on algal photophysiology; (4) were routinely less than half the theoretical Φ_{max} values (i.e., 0.08 - 0.10 mol C Ein^{-1}) commonly used in model calculations where quantum yield estimates are not available; and (5) that operational (in situ) Φ is often a small fraction of in situ Φ_{max} (Prézelin et al., 1991; Prézelin, unpubl. data). The spectral dependency of in situ quantum yield and RUE is evident and can be associated with different phytoplankton size fractions within a community, as well a photophysiological differences related to hydrographic and optical variability on different space scales (Fig. 10, Schofield et al., 1991). The potential variability in spectral quantum yield and RUE are sufficient to impact the predictive linkages between spectrally-weighted and nonspectral estimates of these parameters and bio-optical models of productivity dependent upon them (Fig. 11).

FIELD APPLICATIONS AND SCALING CONSTRAINTS

The bio-optical models outlined above can be classified either as P-I based (Eqs 6-8) or absorption-based (Eqs 9 and 11). When these models are used in conjunction with data collected from ships, moorings and drifters, it is important that depth-dependent photosynthesis rates (mg C m^{-3} h^{-1}) be computed over short time intervals (≤ 2 h) and then integrated with respect to time to obtain the daily primary production profile (mg C m^{-3} d^{-1}). Integration of the resulting profile over depth yields the areal daily production rate (mg C m^{-2} d^{-1}). Due to the non-linearity of the P-I response, use of the average daily irradiance to compute daily primary production will lead to a significant overestimation ($\sim 30\%$) of the actual rate (Platt et al., 1984).

Shipboard application of the P-I based models requires knowledge of Q_{par} (z, t); Chl (z, t); and the appropriate P-I parameters (z, t). Specifically, the latter parameters include P_{max} Chl^{-1} and I_k or P_{max} Chl^{-1} and α Chl^{-1}. Further, if the effects of photoinhibition are to be accounted for as in Eq (7), then knowledge of β Chl^{-1} and P_s Chl^{-1} is also required.

Since P-I parameters and Q_{par} are known to vary significantly with respect to depth and time of day (see above), their variability must be assessed in the field to ensure accurate model predictions (Smith et al., 1987b, 1989; Cullen et al., 1991; Prézelin et al., 1991). In general, over short time scales ($\leq$ days), chlorophyll concentrations and water column optical properties at a given location and depth are considerably less variable than Chl-specific P-I parameters or Q_{par}. Smith et al. (1987b, 1989) took advantage of this and developed a "quasi-synoptic" shipboard sampling strategy for bio-optically modeling primary production rates in highly variable oceanic regions. Initially, the diurnal variability of P-I parameters of phytoplankton sampled at different optical depths across a coastal front was established (Prézelin et al., 1987). Next, a rapid frontal transect ($\sim$ day) was performed to measure depth-dependent variations in P-I responses, pigment concentrations and diffuse attenuation coefficients (K_d (λ, z)). The water column optical data was used in conjunction with continuous shipboard solar radiation measurements to compute Q_{par} (z, t). Finally, the resulting data set was input into Eq (8) to create a "spreadsheet" which described daytime changes in depth-dependent *in situ* production for every 2 h interval of the day. The uniqueness of the sampling approach described by Smith et al. (1987b, 1989) is that while the ship took a full day to perform the transect, the resulting production estimates for each station are time-corrected for diurnal variations in P-I parameters and Q_{par}. This "quasi-synoptic" shipboard sampling strategy was recently used during the Watercolors '88 cruise to investigate spatial and temporal variations in primary productivity in the Southern California Bight (Prézelin et al., 1991; Fig. 8).

Shipboard application of the full spectral absorption-based bio-optical model (Eqs 11 and 12) requires knowledge of Φ_c (z, t); Q_d (λ, z, t); and a_{ph} (λ, z, t) (cf. Bidigare et al., 1987; Smith et al., 1989). The latter term can be determined by any of the approaches described above (e.g., Kishino et al., 1985; Morrow et al., 1989; Bidigare et al., 1990a). *In situ* quantum yield (Φ_c) can be computed by means of Eq (21) which requires additional knowledge of I_k (see above) and Φ_{max}. If direct measurements of Φ_{max} are not available, this parameter can be estimated mathematically as follows and treated as a constant (however, see the limitations of this assumption described above). The theoretical limit of Φ_{O_2} predicted by the Z-scheme of photosynthesis is 1 O_2 evolved per 8 quanta absorbed ($1/8 = 0.125$). However, energy losses caused by cycling of Photoreaction I and non-perfect energy transfers yields an actual limit of ~ 0.10 (Myers, 1980). Since Eq (11) is a carbon-based model of photosynthesis, an actual limit for Φ_{CO_2} ($\equiv \Phi_{max}$) is required for use in Eq (21). Φ_{max} is dependent upon the nitrogen source assimilated during photosynthesis (Myers, 1980). The photosynthetic quotients calculated by Laws (1991) can be used to estimate Φ_{max} for phytoplankton assimilating NO_3^- (PQ = 1.4), NO_3^- plus NH_4^+ (PQ = 1.25) and NH_4^+ (PQ = 1.1); these PQs translate into Φ_{max} values of 0.07, 0.08, and 0.09 mol C Ein^{-1}, respectively.

Despite the attractiveness of light-pigment production models, their use has been restricted to regional spatial scales over time intervals of $\leq$ weeks (Harrison et al., 1985; Herman and Platt, 1986; Bidigare et al., 1987; Smith et al., 1987b, 1989; Prézelin and Glover, 1991). Factors contributing to their limited use include the (1) variability associated with P-I parameters and Φ_{max} (Cullen, 1990; Falkowski, this volume); and (2) lack of high resolution pigment and light data needed to drive the models. Bidigare et al. (1987) suggested that the full spectral absorption-based model may be useful for predicting production profiles from bio-optical data (Q_d (λ, z, t) and Chl (z, t)) provided by untended moorings. Since it is now technically feasible to perform such measurements (Dickey et al., 1991; Smith et al., 1991a), it should be possible to optically estimate (Eqs 11, 12 and 21) primary production profiles on time scales ranging from hours to months. Periodic shipboard measurements, however, would be required to provide the additional parameters (i.e., a_{ph} Chl^{-1} (λ, z), Φ_{max}, and I_k (z, t)) needed for computing production rates. A distinct advantage of the absorption-based model is that photosynthetic rates can be described as a function of only one P-I parameter, I_k, as opposed to the two required (i.e., P_{max} Chl^{-1} and I_k <u>or</u> P_{max} Chl^{-1} and α Chl^{-1}) for the P-I models, assuming Φ_{max} is a constant dependent upon nitrogenous nutrient quality not quantity.

PROSPECTS AND CONCLUSIONS

If in the future, photosynthetic pigment concentrations can be optically determined (cf. Smith and Baker, 1978; Baker and Smith, 1982) and the dependence of I_k on growth irradiance established (Smith, Baker and Prézelin, pers. comm.), then it may be possible to estimate photosynthetic rates from spectral irradiance distributions alone. The areal extent to which primary production rates could be estimated by this approach will ultimately be limited by the cost constraints associated with the deployment of bio-optical moorings. Thus, it is obvious that a multiplatform sampling strategy involving satellites, ships, and moorings will be required to provide realistic estimates of global production (Fig. 12). For example, satellite-derived estimates of chlorophyll a concentration (Sea-WiFS) and global solar surface irradiance (Bishop and Rossow, 1991) could be used in conjunction with the absorption-based remote sensing models (Collins et al., 1986; Morel, 1991) to extrapolate mooring-derived production profiles to three dimensions (i.e., P (x, y,z,t).

In summary, while the absolute predictive ability of production models remains about the same as those formulated three decades ago, our ability to determine important

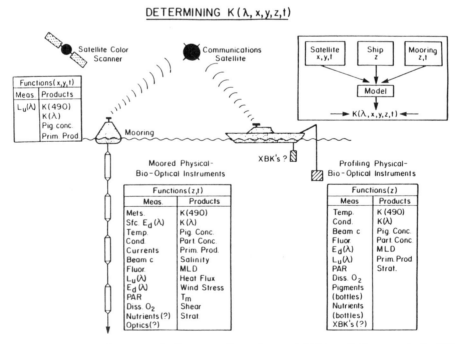

Fig. 12. Conceptual approach for sampling the variability of the spectral diffuse attenuation coefficient ($K_d(\lambda)$, Dickey, 1991).

input parameters to these models has improved significantly. Methods for the accurate determination of the mean and variance and the space/time variabilty of the parameters which drive bio-optical models are now available. This is significant since this space/time variabilty has been the greatest source of uncertainity in the estimation of regional and global productivity. Further, the models now encompass full spectral information thus properly accounting for changes in the in-water light field. In the next decade, special emphasis should be placed on the development of multiplatform techniques for determining full spectral information for bio-optical model parameters, including *in situ* distributions of quantum scalar irradiance (Q_o (λ, z), 4π, μEin m^{-2} nm^{-1} s^{-1}). It is envisioned that the availability of high resolution optical data collected from moorings and satellites would allow improved chlorophyll *a* estimates as well as predictions of accessory pigment concentrations. Future research efforts should also be directed toward determining the structures and *in vivo* absorption properties of novel algal pigments. This knowledge will undoubtedly improve the predictive abilities of bio-optical models as well as provide new insight into the photoadaptive behavior of marine phytoplankton.

ACKNOWLEDGEMENTS

Research support was provided by National Aeronautics and Space Administration grant NAGW-290 (RCS); National Science Foundation grants OCE 88-13727 (RRB), OCE 88-00099 (BBP) and OCE 88-13728 (RCS); and Office of Naval Research contract N00014-89-J-1691 (RRB). We thank Tommy Dickey (University of Southern California) for his valuable contributions on time-space variability; John Cullen (Bigelow Laboratory for Ocean Sciences and Dalhousie University) for discussions regarding derivation of Eq (21); Oscar Schofield (University of California, Santa Barbara) for discussions of spectral versus non-spectral measurements of quantum yield and primary production; and Karen S. Baker (Scripps Institution of Oceanography) for the bio-optical computations she provided for Figures 5B and 6. Biowatt contribution No. 46 and Watercolors contribution No. 15.

REFERENCES

Aiken, J., 1981, The undulating oceanographic recorder Mark 2, *J. Plankton Res.*, 3:551.

Alberte, R. S., Wood, A. M., Kursar, T. A., and Guillard, R. R. L., 1984, Novel phyco-erythrins in marine *Synechococcus* spp., *Plant Physiol.*, 75:732.

Baker, K. S., and Smith, R. C., 1982, Bio-optical classification and model of natural waters. 2, *Limnol. Oceanogr.*, 27:500.

Balch, W. M., Abbott, M. R., and Eppley, R. W., 1989a, Remote sensing of primary production, I, A comparison of empirical and semi-analytical algorithms, *Deep-Sea Res.*, 36:281.

Balch, W. M., Eppley, R. W., and Abbott, M. R., 1989b, Remote sensing of primary production, II, A semi-analytical algorithm based on pigments, temperature and light, *Deep-Sea Res.*, 36:1201.

Bannister, T. T., and Weidemann, A. D., 1984, The maximum quantum yield of phytoplankton photosynthesis *in situ*, *J. Plankton Res.*, 6:275.

Berger, W. H., 1989, Global maps of ocean productivity, *in*: "Productivity of the Ocean: Present and Past," W.H. Berger, V.S. Smetacek and G. Wefer, eds., John Wiley & Sons, Chichester.

Berner, T., Dubinsky, Z., Wyman, K., and Falkowski, P. G., 1989, Photoadaptation and the "package" effect in *Dunaliella tertiolecta* (Chlorophyceae), *J. Phycol.*, 25:70.

Bidigare, R. R., 1989, Photosynthetic pigment composition of the brown tide alga: Unique chlorophyll and carotenoid derivatives, *in*: "Novel Phytoplankton Blooms," E. Cosper, E.J. Carpenter and M. Bricelj, eds., Coastal and Estuarine Studies, Vol. 35, Springer-Verlag, Berlin.

Bidigare, R. R., Frank, T. J., Zastrow, C., and Brooks, J. M., 1986, The distribution of algal chlorophylls and their degradation products in the Southern Ocean, *Deep-Sea Res.*, 33:923.

Bidigare, R. R., Kennicutt II, M. C., Ondrusek, M. E., Keller, M. D., and Guillard, R. R. L., 1990c, Novel chlorophyll-related compounds in marine phytoplankton: Distributions and geochemical implications, *Energy & Fuels*, 4:653.

Bidigare, R. R., Marra, J., Dickey, T. D., Iturriaga, R., Baker, K. S., Smith, R. C., and Pak, H., 1990b, Evidence for phytoplankton succession and chromatic adaptation in the Sargasso Sea during springtime 1985, *Mar. Ecol. Prog. Ser.*, 60:113.

Bidigare, R.R., Ondrusek, M.E., and Brooks, J.M., 1991, Influence of the Orinoco River outflow on distributions of algal pigments in the Caribbean Sea, *J. Geophys. Res.* (in press).

Bidigare, R. R., Ondrusek, M. E., Morrow, J. H., and Kiefer, D. A., 1990a, *In vivo* absorption properties of algal pigments, *Proc. SPIE Ocean Opt. X*, 1302:290.

Bidigare, R. R., Morrow, J. H., and Kiefer, D. A., 1989a, Derivative analysis of spectral absorption by photosynthetic pigments in the western Sargasso Sea, *J. Mar. Res.*, 47:323.

Bidigare, R. R., Schofield, O., and Prézelin, B. B., 1989b, Influence of zeaxanthin on quantum yield of photosynthesis of *Synechococcus* clone WH7803 (DC2), *Mar. Ecol. Prog. Ser.*, 56:177.

Bidigare, R. R., Smith, R. C., Baker, K. S., and Marra, J., 1987, Oceanic primary production estimates from measurements of spectral irradiance and pigment concentrations, *Global Biogeochem. Cycles*, 1:171.

Bishop, J. K. B., and Rossow, W. B., 1991, Spatial and temporal variability of global surface solar irradiance, *J. Geophys. Res.* (in press).

Bjørnland, T., Liaaen-Jensen, S., and Throndsen, J., 1989, Carotenoids of the marine chrysophyte *Pelagococcus subviridis*, *Phytochemistry*, 28:3347.

Booth, C. R., and Smith, R. C., 1988, Moorable spectroradiometer in the Biowatt experiment, *Proc. SPIE Ocean Opt. IX*, 925:176.

Brown, O. B., Evans, R. H., Gordon, H. R., Smith, R. C., and Baker, K. S., 1985, Blooming off the U.S. coast: A satellite description, *Science*, 229:163.

Campbell, J. W., Yentsch, C. S., and Esaias, W. E., 1986, Dynamics of phytoplankton patches on Nantucket Shoals: An experiment involving aircraft, ships and buoys, *in*: "Lecture Notes on Coastal and Estuarine Studies," Vol. 17, J. Bowman, M. Yentsch and W.T. Peterson, eds., Springer-Verlag, Berlin.

Chalker, B. E., 1980, Modelling light saturation curves for photosynthesis: An exponential function, *J. Theor. Biol.*, 84:205.

Chisholm, S. W., this volume.

Chisholm, S. W., Olson, R. J., Zettler, E. R., Goericke, R., Waterbury, J. B., and Welschmeyer, N. A., 1988, A novel free-living prochlorophyte abundant in the oceanic euphotic zone, *Nature*, 334:340.

Collins, D. J., Kiefer, D. A., SooHoo, J. B., Stallings, C., and Yang, W., 1986, A model for the use of satellite remote sensing for the measurement of primary production in the ocean, *Proc. SPIE Ocean Opt. VIII*, 637:335.

Cullen, J. J., 1990, On models of growth and photosynthesis in phytoplankton, *Deep-Sea Res.*, 37:667.

Cullen, J. J., and Lewis, M. R., 1988, The kinetics of algal photoadaptation in the context of vertical mixing, *J. Plankton Res.*, 10:1039.

Cullen, J. J., Lewis, M. R., Davis, C. O., and Barber, R. T., 1991, Photosynthetic characteristics and estimated growth rates indicate grazing is the proximate control of primary production in the equatorial Pacific, *J. Geophys. Res.,* in press.

Dickey, T. D., 1988, Recent advances and future directions in multi-disciplinary in situ oceanographic measurement systems, *in:* "Towards a Theory on Biological-Physical Interactions in the World Ocean," B. J. Rothschild, ed., Kluwer Academic, Dordrecht.

Dickey, T.D., 1990, Physical-optical-biological scales relevant to recruitment in large marine ecosystems, *in:* "Large Marine Ecosystems: Patterns, Processes, and Yields," K. Sherman, L. M. Alexander, and B. D. Gold, eds., AAAS, Washington, D. C.

Dickey, T. D., 1991, The emergence of concurrent high resolution physical and bio-optical measurements in the upper ocean and their applications, *Rev. of Geophys.,* in press.

Dickey, T. D., Marra, J., Granata, T., Langdon, C., Hamilton, M., Wiggert, J., Siegel, D. A., and Bratkovich, A., 1991, Concurrent high resolution bio-optical and physical time series observations in the Sargasso Sea during the spring of 1987, *J. Geophys. Res.,* 96:8643.

Dubinsky, Z., Falkowski, P. G., and Wyman. K., 1986, Light harvesting and utilization by phytoplankton, *Plant Cell Physiol.,* 27:1335.

Eckardt, C. B., Carter, J. F., and Maxwell, J. R., 1990, Combined liquid chromatography/mass spectrometry of tetrapyrroles of sedimentary significance, *Energy & Fuels,* 4:741.

Eppley, R. W., 1980, Estimating phytoplankton growth rates in the central oligotrophic oceans, *in:* "Primary Productivity in the Sea," P.G. Falkowski, ed., Plenum Press, New York.

Eppley, R. W., and Renger, E. H., 1988, Nanomolar increases in surface layer nitrate concentration following a small wind event, *Deep-Sea Res.,* 35:1119.

Eppley, R. W., Stewart, E., Abbott, M. R., and Heyman, U., 1985, Estimating ocean production from satellite chlorophyll: Introduction to regional differences and statistics for the southern California bight, *J. Plankton Res.,* 7:57.

Esaias, W. E., 1980, Remote sensing of oceanic phytoplankton: Present capabilities and future goals, *in:* "Primary Productivity in the Sea," P.G. Falkowski, ed., Plenum Press, New York.

Falkowski, P. G., this volume.

Falkowski, P. G., Dubinsky, Z., and Wyman, K., 1985, Growth-irradiance relationships in phytoplankton, *Limnol. Oceanogr.,* 30:311.

Fawley, M. W., 1989, A new form of chlorophyll *c* involved in light-harvesting, *Plant. Physiol.,* 91:727.

Fiksdahl, A., Bjørnland, T., and Liaaen-Jensen, S., 1984, Algal carotenoids with novel end groups, *Phytochemistry,* 23:649.

Fitzwater, S. E., Knauer, G. A., and Martin, J. H., 1982, Metal contamination and its effect on primary production measurements, *Limnol. Oceanogr.,* 27:544.

Fookes, C. J. R., and Jeffrey, S. W., 1989, The structure of chlorophyll c_3, a novel marine photosynthetic pigment, *J. Chem. Soc., Chem. Commun.,* 18:27.

Foss, P., Guillard, R. R. L., and Liaaen-Jensen, S., 1984, Prasinoxanthin - A chemosystematic marker for algae, *Phytochemistry,* 23:1629.

Frew, N. M., Johnson, C. G., and Bromund, R. H., 1988, Supercritical fluid chromatography - Mass spectrometry of carotenoid pigments, *in:* "Supercritical Fluid Extraction and Chromatography," B. A. Charpentier and M. R. Sevenants, eds., ACS Symposium Series 366, American Chemical Society, Washington, D.C.

Frouin, R., Lingner, D. W., Gautier, C., Baker, K. S., and Smith, R. C., 1989, A simple analytical formula to compute clear sky total and photosynthetically available solar irradiance at the ocean surface, *J. Geophys. Res.*, 94:9731.

Garside, C., 1985, The vertical distribution of nitrate in open ocean surface water, *Deep-Sea Res.*, 32:723.

Gautier, C., Diak, G. R., and Masse, S., 1980, A simple physical model to estimate incident solar radiation at the sea surface from GOES satellite data, *J. Appl. Meteor.*, 19:1005.

Geider, R. J., Platt, T., and Raven, J. A., 1986, Size dependence of growth and photosynthesis in diatoms: A synthesis, *Mar. Ecol. Prog. Ser.*, 30:93.

Gieskes, W. W. C., and Kraay, G. W., 1983a, Unknown chlorophyll *a* derivatives in the North Sea and tropical Atlantic Ocean revealed by HPLC analysis, *Limnol. Oceanogr.*, 28:757.

Gieskes, W. W. C., and Kraay, G. W., 1983b, Dominance of Cryptophyceae during the phytoplankton spring bloom in the central North Sea detected by HPLC analysis of pigments, *Mar. Biol.*, 75:179.

Gieskes, W. W. C., and Kraay, G. W., 1984, Phytoplankton, its pigments, and primary production at a central North Sea station in May, July and September 1981, *Neth. J. Sea Res.*, 18:51.

Gieskes, W. W. C., and Kraay, G. W., 1986a, Floristic and physiological differences between the shallow and the deep nanophytoplankton community in the euphotic zone of the open tropical Atlantic revealed by HPLC analysis of pigments, *Mar. Biol.*, 91:567.

Gieskes, W. W. C., and Kraay, G. W., 1986b, Analysis of phytoplankton pigments by HPLC before, during and after mass occurrence of the microflagellate *Corymbellus aureus* during spring bloom in the open northern North Sea in 1983, *Mar. Biol.*, 92:45.

Gieskes, W. W. C., Kraay, G. W., Nontji, A., Setiapermana, D., and Sutomo, 1988, Monsoonal alteration of a mixed and a layered structure in the phytoplankton of the euphotic zone of the Banda Sea (Indonesia): A mathematical analysis of algal pigment fingerprints, *Neth. J. Sea Res.*, 22:123.

Glover, H. E., 1985, The physiology and ecology of the marine cyanobacterial genus Synechococcus, *in*: "Advances in Aquatic Microbiology," Vol. 3, H. W. Jannasch and P. J. le B. Williams, eds., Academic Press, London.

Glover, H. E., Prézelin, B. B., Campbell, L., and Wyman, M., 1988a, Pico- and ultraplankton Sargasso Sea communities: variability and comparative distribution of *Synechococcus* spp. and algae, *Mar. Ecol. Prog. Ser.*, 49:127.

Glover, H. E., Prézelin, B. B., Campbell, L., Wyman, M., and Garside, C., 1988b, A nitrate-dependent *Synechococcus* bloom in surface Sargasso Sea water, *Nature*, 331:161.

Goericke, R., 1990, "Pigments as ecological tracers for the study of the abundance and growth of marine phytoplankton," Ph. D. dissertation, Harvard University, Cambridge.

Gordon, H. R., Brown, O. B., Evans, R. H., Brown, J. W., Smith, R. C., Baker, K. S., and Clark, D. K., 1988, A semianalytic radiance model of ocean color, 1988, *J. Geophys. Res.*, 93:10,909.

Gordon, H. R., Clark, D., Mueller, J. L., and Hovis, W. A., 1980, Phytoplankton pigments from Nimbus-7 coastal zone color scanner: Comparison with surface measurements, *Science*, 210:63.

Gordon, H. R., and McCluney, W. R., 1975, Estimation of the depth of sunlight penetration in the sea for remote sensing, *Appl. Opt.*, 14:413.

Gregg, W. W., and Carder, K. L., 1990, A simple spectral solar irradiance model for cloudless maritime atmospheres, *Limnol. Oceanogr.*, 35:1657.

Haardt, H., and Maske, H., 1987, Specific *in vivo* absorption coefficient of chlorophyll *a* at 675 nm, *Limnol. Oceanogr.*, 32:608.

Harris, G. P., 1986, "Phytoplankton Ecology: Structure, Function and Fluctuation," Chapman and Hill, London.

Harrison, W. G., Platt, T., and Lewis, M. R., 1985, The utility of light-saturation models for estimating marine productivity in the field: A comparison with conventional "simulated" *in situ* methods, *Can. J. Fish. Aquat. Sci.*, 42:864.

Herman, A. W., and Platt, T., 1986, Primary production profiles in the ocean: Estimation from a chlorophyll/light model, *Oceanol. Acta, 9*:31.

Hooks, C. E., Bidigare, R. R., Keller, M. D., and Guillard, R. R. L., 1988, Coccoid eukaryotic marine ultraplankters with four different HPLC pigment signatures, *J. Phycol.*, 24:571.

HOT Program, 1990, "Data Report 1," School of Ocean and Earth Science and Technology, University of Hawaii, Honolulu.

Hovis, W. A., Clark, D. K., Anderson, F. P., Austin, R. W., Wilson, W. H., Baker, E. T., Ball, D., Gordon, H. R., Mueller, J. L., El-Sayed, S. Z., Strun, B., Wrigley, R. C., and Yentsch, C. S., 1980, Nimbus-7 coastal zone color scanner: System description and initial imagery, *Science,* 210:60.

Iturriaga, R., Mitchell, B. G., and Kiefer, D. A., 1988, Microphotometric analysis of individual particle absorption spectra, *Limnol. Oceanogr.*, 33:128.

Iturriaga, R., and Siegel, D. A., 1989, Microphotometric characterization of phytoplankton and detrital absorption properties in the Sargasso Sea, *Limnol. Oceanogr.*, 34:1706.

Jahnke, R. A., 1990, Ocean flux studies: A status report, *Rev. of Geophys.*, 28:381.

Jassby, A. D., and Platt, T., 1976, Mathematical formulation of the relationship between photosynthesis and light for phytoplankton, *Limnol. Oceanogr.*, 21:540.

Jeffrey, S. W., 1989, Chlorophyll *c* pigments and their distribution in the chromophyte algae, *in*: "The Chromophyte Algae: Problems and Perspectives," J.C. Green, B.S.C. Leadbeater, and W.L. Diver, eds., Clarendon Press, Oxford.

Jenkins, W. J., and Goldman, J. C., 1985, Seasonal oxygen cycling and primary production in the Sargasso Sea, *J. Mar. Res.*, 43:465.

Kerr, R. A., 1986, The ocean's deserts are blooming, *Science,* 232:1345.

Kiefer, D. A., and Mitchell, B. G., 1983, A simple, steady state description of phytoplankton growth based on absorption cross section and quantum efficiency, *Limnol. Oceanogr.*, 28:770.

Kiefer, D. A., and SooHoo, J. B., 1982, Spectral absorption by marine particles of coastal waters of Baja California, *Limnol. Oceanogr.*, 27:492.

Kirk, J. T. O., 1983, "Light and Photosynthesis in Aquatic Ecosystems", Cambridge University Press, New York.

Kishino, M., Booth, C. R., and Okami, N., 1984, Underwater radiant energy absorbed by phytoplankton, detritus, dissolved organic matter, and pure water, *Limnol. Oceanogr.*, 29:340.

Kishino, M., Okami, N., Takahashi, M., and Ichimura, S., 1986, Light utilization efficiency and quantum yield of phytoplankton in a thermally stratified sea, *Limnol. Oceanogr.*, 31:557.

Kishino, M., Takahashi, N., Okami, N., and Ichimura, S., 1985, Estimation of the spectral absorption coefficients of phytoplankton in the sea, *Bull. Mar. Sci.*, 37:634.

Klein, P., and Coste, B., 1984, Effects of wind stress variability on nutrient transport into the mixed layer, *Deep-Sea Res.*, 4 31:21.

Laws, E. A., 1991, Photosynthetic quotients, new production and net community production in the open ocean, *Deep-Sea Res.*, 38:143.

Laws, E. A., DiTullio, G. R., Carder, K. L., Betzer, P. R., and Hawes, S., 1990, Primary production in the deep blue sea, *Deep-Sea Res.*, 37:715.

Lewis, M. R., Cullen, J. J., and Platt, T., 1984, Relationships between vertical mixing and photoadaptation of phytoplankton: Similarity criteria, *Mar. Ecol. Prog. Ser.*, 15:141.

Lewis, M. R., Warnock, R. E., Irwin, B., and Platt, T, 1985b, Measuring photosynthetic action spectra of natural phytoplankton populations, *J. Phycol.* 21: 310.

Lewis, M. R., Warnock, R. E., and Platt, T., 1985a, Absorption and photosynthetic action spectra for natural phytoplankton populations, *Limnol. Oceanogr.*, 30:794.

Lewis, M. R., and Smith, J. C., 1983, "Photosynthetron": a small volume, short-incubation time method for measurement of photosynthesis as a function of incident irradiance, *Mar. Ecol. Prog. Ser.*, 13: 99.

Liaaen-Jensen, S., 1985, Carotenoids of lower plants - Recent Progress, *Pure Appl. Chem.*, 57:649.

Lohrenz, S. E., Arnone, R. A., Wiesenburg, D. A., and DePalma, I. P., 1988, Satellite detection of transient enhanced primary production in the western Mediterranean Sea, *Nature*, 335:245.

Lohrenz, S. E., Knauer, G. A., Asper, V. L., Tuel, M., Knapp, A. H., and Michaels, A. F., 1991, Seasonal variability in primary production and particle flux in the northwestern Sargasso Sea: U.S. JGOFS Bermuda Atlantic time series, *Deep-Sea Res.*, in press.

Mantoura, R. F. C., and Llewellyn, C. A., 1983, The rapid determination of algal chlorophyll and carotenoid pigments and their breakdown products in natural waters by reverse-phase high-performance liquid chromatography, *Anal. Chim. Acta*, 151:297.

Marra, J., and Heinemann, K. R., 1987, Primary production in the North Pacific Central Gyre: Some new measurements based on ^{14}C, *Deep-Sea Res.*, 34:1821.

Maske, H., and Haardt, H., 1987, Quantitative *in vivo* absorption spectra of phytoplankton: Detrital absorption and comparison with fluorescence excitation spectra, *Limnol. Oceanogr.*, 32:620.

McClain, C. R., Pietrafesa, L. J., and Yoder, J. A., 1984, Observations of Gulf Stream induced and wind driven upwelling in the Georgia bight using ocean color and infrared imagery, *J. Geophys. Res.*, 89:3705.

Mitchell, B. G., 1987, "Ecological implications of variability in marine particulate absorption and fluorescence excitation spectra," Ph. D. dissertation, University of Southern California, Los Angeles.

Mitchell, B. G., 1990, Algorithms for determining the absorption coefficient for aquatic particles using the quantitative filter technique, *Proc. SPIE Ocean Opt. X*, 1302:137

Mitchell, B. G., and Kiefer, D. A., 1988a, Variability in pigment specific particulate fluorescence and absorption spectra in the northeastern Pacific Ocean, *Deep-Sea Res.*, 35:665.

Mitchell, B. G., and Kiefer, D. A., 1988b, Chlorophyll *a* specific absorption and fluorescence excitation spectra for light-limited phytoplankton, *Deep-Sea Res.*, 35:639.

Morel, A., 1988, Optical modeling of the upper ocean in relation to its biogenous matter content (case I waters), *J. Geophys. Res.*, 93:10,749.

Morel, A., 1991, Light and marine photosynthesis: A spectral model with geochemical and climatological implications, *Prog. Oceanogr.*, 26:263.

Morel, A., and Bricaud, A., 1981, Theoretical results concerning light absorption in a discrete medium, and application to specific absorption of phytoplankton, *Deep-Sea Res.,* 28:1375.

Morel, A., and Bricaud, A., 1986, Inherent optical properties of algal cells including picoplankton: Theoretical and experimental results, *in:* "Photosynthetic Picoplankton," T. Platt and W. K. W. Li, eds., *Can. Bull. Fish. Aquat. Sci.,* 214:521.

Morel, A., Lazzara, L., and Gostan, J., 1987, Growth rate and quantum yield time response for a diatom to changing irradiances (energy and color), *Limnol. Oceanogr.,* 32:1066.

Morris, I., 1981, "The Physiological Ecology of Phytoplankton," Blackwell Sc. Publ. Berkley.

Morrow, J. H., Chamberlin, W. S., and Kiefer, D. A., 1989, A two-component description of spectral absorption by marine particles, *Limnol. Oceanogr.,* 34:1500.

Myers, J., 1980, On the algae: Thoughts about physiology and measurements of efficiency, *in:* "Primary Productivity in the Sea," P.G. Falkowski, ed., Plenum Press, New York.

Nelson, J. R., and Wakeham, S. G., 1989, A phytol-substituted chlorophyll *c* from *Emiliania huxleyi* (Prymnesiophyceae), *J. Phycol.,* 25:761.

Nelson, N. B., and Prézelin, B. B., 1990, Chromatic light effects and physiological modeling of absorption properties of *Heterocapsa pygmaea* (= *Glenodinium sp.*), *Mar. Ecol. Prog. Ser.,* 63:37.

Nelson, N. B., Prézelin, B. B., Bidigare, R. R., Smith, R. C., and Baker, K. S., 1991, Spatial and temporal variability of phytoplankton spectral absorption properties in the Southern California Bight, *Deep-Sea Res.,* in press.

Ondrusek, M. E., Bidigare, R. R., Sweet, S. T., DeFreitas, D. A., and Brooks, J. M., 1991, Distribution of algal pigments in the North Pacific Ocean in relation to physical and optical variability, *Deep-Sea Res.,* 38:243.

Ong, L. J., Glazer, A. N., and Waterbury, J. B., 1984, An unusual phycoerythrin from a marine cyanobacterium, *Science,* 224:80.

Pahl-Wostl, C., and Imboden, D. M., 1990, DYPHORA - A dynamic model for the rate of photosynthesis of algae, *J. Plankton Res.,* 12:1207.

Platt, T., 1986, Primary production of the ocean water column as a function of surface light intensity: Algorithms for remote sensing, *Deep-Sea Res.,* 33:149.

Platt, T., Gallegos, C. L., and Harrison, W. G., 1980, Photoinhibition of photosynthesis in natural assemblages of marine phytoplankton, *J. Mar. Res.,* 38:687.

Platt, T., Harrison, W. G., Lewis, M. R., Li, W. K., Sathyendranath, S., Smith, R. E., and Vezina, A. F., 1989, Biological production of the oceans: The case for a consensus, *Mar. Ecol. Prog. Ser.,* 52:77.

Platt, T., and Jassby, A. D., 1976, The relationship between photosynthesis and light for natural assemblages of coastal marine phytoplankton, *J. Phycol.,* 12:421.

Platt, T., Lewis, M. R., and Geider, R., 1984, Thermodynamics of the pelagic ecosystem: Elementary closure conditions for biological production in the open ocean, *in:* "Flows of Energy and Materials in Marine Ecosystems," M. J. R. Fasham, ed., Plenum Press, New York.

Platt, T., and Li, W. K. W., eds., 1986, "Photosynthetic Picoplankton," *Can. Bull. Fish. Aquat. Sci.,* 214:583.

Platt, T., and Sathyendranath, S., 1988, Oceanic primary production: Estimation by remote sensing at local and regional scales, *Science,* 241:1613.

Platt, T., Sathyendranath, S., and Ravindran, P., 1990, Primary production by phytoplankton: Analytic solutions for daily rates per unit area of water surface, *Proc. R. Soc. Lond. B,* 241:101.

Platt, T., Sathyendranath, S., Caverhill, C. M., and Lewis, M. R., 1988, Ocean primary production and available light: Further algorithms for remote sensing, *Deep-Sea Res.*, 35:855.

Post, A. F., Dubinsky, Z., Wyman, K., and Falkowski, P. G., 1985, Physiological responses of a marine planktonic diatom to transitions in growth irradiance, *Mar. Ecol. Prog. Ser.*, 25:141.

Prézelin, B. B., 1991, Diel periodicity in phytoplankton productivity, *Hydrobiol.* (in press).

Prézelin, B. B., Bidigare, R. R., Matlick, A., Putt, M., and Ver Hoven, B., 1987, Diurnal patterns of size-fractioned primary productivity across a coastal front, *Mar. Biol.*, 96:591.

Prézelin, B. B., and Glover, H. E., 1991, Variability in time/space estimates of phytoplankton, biomass and productivity in the Sargasso Sea, *J. Plankton Res.*, 13S:45.

Prézelin, B. B., Putt, M., and Glover, H. E., 1986, Diurnal patterns in photosynthetic capacity and depth-dependent photosynthesis-irradiance relationships in *Synechococcus* spp. and larger phytoplankton in three water masses in the Northwest Atlantic Ocean, *Mar. Biol.*, 91:205.

Prézelin, B. B., Tilzer, M. M., Schofield, O., and Haese, C., 1991, Review: Control of the production process of phytoplankton by the physical structure of the aquatic environment, *Hydrobiol.*, in press.

Price, N. M., Harrison, P. J., Landry, M. R., Azam, F., and Hall, K. J. F., 1986, Toxic effect of latex and Tygon tubing on phytoplankton, zooplankton and bacteria, *Mar. Ecol. Prog. Ser.*, 34:41.

Repeta, D. J., Simpson, D. J., Jorgensen, B. B., and Jannasch, H. W., 1989, Evidence for anoxygenic photosynthesis from the distribution of bacteriochlorophylls in the Black Sea, *Nature*, 342:69.

Rodhe, W., 1965, Standard correlations between pelagic photosynthesis and light, *in*: "Primary Productivity in Aquatic Environments," C. R. Goldman, ed., University of California Press, Berkeley.

Ryther, J. H., 1956, Photosynthesis in the ocean as a function of light intensity, *Limnol. Oceanogr.*, 1:61.

Ryther, J. H., and Yentsch, C. S., 1957, The estimation of phytoplankton production in the ocean from chlorophyll and light data, *Limnol. Oceanogr.*, 2:281.

Sathyendranath, S., Lazzara, L., and Prieur, L., 1987, Variations in the spectral values of specific absorption of phytoplankton, *Limnol. Oceanogr.*, 32:403.

Sathyendranath, S., and Platt, T., 1989, Computation of aquatic primary production: Extended formalism to include effect of angular and spectral distribution of light, *Limnol. Oceanogr.*, 34:188.

Sathyendranath, S., Platt, T., Caverhill, C. M., Warnock, R. E., and Lewis, M. R., 1989, Remote sensing of oceanic primary production: Computations using a spectral model, *Deep-Sea Res.*, 36:431.

Schofield, O., Bidigare, R. R., and Prézelin, B. B., 1990, Chromatic photoadaptation and enhancement effects on wavelength-dependent quantum yield and productivity in the diatom *Chaetoceros gracile* and the prymnesiophyte *Emiliania huxleyi*, *Mar. Ecol. Prog. Ser.*, 64:175.

Schofield, O., Prézelin, B. B., Smith, R. C., Stegmann, P., Nelson, N. B., Lewis, M. R., and Baker, K.S., 1991, Spectral photosynthesis, quantum yield, and radiation utilization efficiency across the Southern California Bight, *Mar. Ecol. Prog. Ser.*, in press.

Shannon, L. V., Mostert, S. A., Walters, N. M., and Anderson, F. P., 1983, Chlorophyll concentrations in the southern Benguel current region as determined by satellite (Nimbus-7 CZCS), *J. Plankton Res.*, 5:565.

Siegel, D. A., Iturriaga, R., Bidigare, R. R., Smith, R. C., Pak, H., Dickey, T. D., Marra, J., and Baker, K. S., 1990, Meridional variations of the springtime phytoplankton community in the Sargasso Sea, *J. Mar. Res.*, 48:379.

Smith, R. C., 1981, Remote sensing and depth distribution of ocean chlorophyll, *Mar. Ecol.*, 5:359.

Smith, R. C., and Baker, K. S., 1978, Optical classification of natural waters, *Limnol. Oceanogr.*, 23:260.

Smith, R. C., and Baker, K. S., 1982, Oceanic chlorophyll concentrations determined by satellite (Nimbus-7 coastal zone color scanner), *Mar. Biol.*, 66:269.

Smith, R. C., and Baker, K. S., 1984, The analysis of ocean optical data, *Proc. SPIE Ocean Opt. VII*, 489:119.

Smith, R. C. and Baker, K. S., 1986, The analysis of ocean optical data II, *Proc. SPIE Ocean Opt. VIII*, 637:95.

Smith, R. C., Bidigare, R. R., Prézelin, B. B., Baker, K. S., and Brooks, J. M., 1987b, Optical characterization of primary productivity across a coastal front, *Mar. Biol.*, 96:563.

Smith, R. C., Booth, C. R., and Star, J. L., 1984, Oceanographic bio-optical profiling system, *Appl. Opt.*, 23:2791.

Smith, R. C., Brown, O. B., Hoge, F. E., Baker, K. S., Evans, R. H., Swift, R. N., and Esaias, W. E., 1987a, Multiplatform sampling (ship, aircraft, and satellite) of a Gulf Stream warm core ring, *Appl. Opt.*, 26:2068.

Smith, R. C., Eppley, R. W., and Baker, K. S., 1982, Correlation of primary production as measured aboard ship in southern California coastal waters and as estimated from satellite chlorophyll images, *Mar. Biol.*, 66:281.

Smith, R. C., Prézelin, B. B., Baker, K. S., Bidigare, R. R., Boucher, N. P., Coley, T., Karentz, D., MacIntyre, S., Matlick, H. A., Menzies, D., Ondrusek, M., Wan, Z., and Waters, K., 1991c, Ozone depletion: Ultraviolet radiation and phytoplankton biology in Antarctic waters, *Science*, in press.

Smith, R. C., Prézelin, B. B., Bidigare, R. R., and Baker, K. S., 1989, Bio-optical modeling of photosynthetic production in coastal waters, *Limnol. Oceanogr.*, 34:1526.

Smith, R. C., Wan, Z., and Baker, K. S., 1991b, Ozone depletion in Antarctica: Satellite and ground measurements, and modeling under clear-sky conditions, *J. Geophys. Res.*, in press.

Smith, R. C., Waters, K. J., and Baker, K. S., 1991a, Optical variability and pigment biomass in the Sargasso Sea as determined using deep-sea optical mooring data, *J. Geophys. Res.*, 96:8665.

Steemann Nielsen, E., 1952, The use of radioactive carbon (^{14}C) for measuring organic production in the sea, *J. Cons. Cons. Int. Explor. Mer*, 18:117.

Stramski, D., and Morel, A., 1990, Optical properties of photosynthetic picoplankton in different physiological states as affected by growth irradiance, *Deep-Sea Res.*, 37:245.

Talling, J. F., 1957, The phytoplankton population as a compound photosynthetic system, *New Phytol.*, 56:133.

Talling, J. F., 1965, Comparative problems of phytoplankton production and photosynthetic productivity in a tropical and temperate lake, *in*: "Primary Productivity in Aquatic Environments," C. R. Goldman, ed., University of California Press, Berkeley.

Tyler, J. E., 1975, The *in situ* quantum efficiency of natural phytoplankton populations, *Limnol. Oceanogr.,* 18:442.

U.S. JGOFS, 1991, "Bio-optics in JGOFS," U.S. JGOFS Planning Report No. 13, Woods Hole Oceanographic Institution, Woods Hole, in press.

Waters, K. J., Smith, R. C., and Lewis, M. L., 1990, Avoiding ship-induced light-field perturbation in the determination of oceanic optical properties, *Oceanography,* 3:18.

Webb, W. L., Newton, M., and Starr, D., 1974, Carbon dioxide exchange of Alnus rubra: A mathematical model, *Oecologia,* 17:281.

Whitledge T.E., Bidigare, R. R., Zeeman, S., Sambrotto, R.N., Roscigno, P.F., Jensen, P.R., Brooks, J.M., Trees, C.C., and Veidt, D.M, 1988, Biological measurements and related chemical features in Soviet and U.S. regions of the Bering Sea, *Continental Shelf Res.,* 8:1299.

Williams, P. J. le B., and Robertson, J. E., 1989, A serious inhibition problem from a Niskin sampler during plankton productivity studies, *Limnol. Oceanogr.,* 34:1300.

Williams, P. J. le B., and Robertson, J. E., 1991, Overall planktonic oxygen and carbon dioxide metabolism: The problem of reconciling observations and calculations of photosynthetic quotients, *J. Plankton Res.,* 13S:153.

Yentsch, C. S., 1962, Measurement of visible light absorption by particulate matter in the ocean, *Limnol. Oceanogr.,* 9:207.

Yentsch, C. S., and Phinney, D. A., 1988, Relationship between cross-sectional absorption and chlorophyll content in natural populations of marine phytoplankton, *Proc. SPIE Ocean Opt. X,* 1302:109.

Zimmerman, R. C., SooHoo, J. B., Kremer, J. N., and D'Argenio, D. Z, 1987, Evaluation of variance approximation techniques for non-linear photosynthesis-irradiance models, *Mar. Biol.,* 95:209.

PHYTOPLANKTON SIZE

Sallie W. Chisholm

48-425 Ralph M. Parsons Laboratory
Massachusetts Institute of Technology
Cambridge, MA 02139

INTRODUCTION

In reviewing this subject, it became clear to me that plankton ecologists fall out into two groups: Those who delight in finding the patterns in nature that can be explained by size, and those who delight in finding exceptions to the established size-dependent rules. I came to appreciate the degree to which the satisfaction of both groups is equally justified. The mechanisms underlying the size-dependent patterns have undoubtedly steered the general course of phytoplankton evolution, but the organisms that do not abide by the rules reveal the wonderful diversity of ways in which cells have managed to disobey the "laws" scripted for them. The simplicity of the general relationships serves as a stable backdrop against which the exceptions can shine. By understanding the forces that have driven the design of these exceptions, we can begin to understand the ecology that has shaped past and present planktonic ecosystems.

This is not a comprehensive review, since all aspects of the life of a phytoplankton cell are influenced, more or less, by its size. Instead, I have chosen to explore specific dimensions of the topic which I find to be particularly provocative, or which appear to be ripe for new advances and approaches.

ALLOMETRIC RELATIONSHIPS

Size-Dependence of Growth and Metabolism

Allometry is the study of the correlates of body size. Although its origins can be traced back to Elton (1927), the field has undergone a resurgence in the last decade (Peters, 1983; Calder, 1984). Plankton ecologists have participated in this resurgence, stimulated largely by the works of Fenchel (1974), Banse (1976; 1982), Peters (1978) and Platt and co-workers (e.g. Platt and Denman, 1977; 1978; Platt and Silvert, 1981; Silvert and Platt, 1978; 1980; Geider et al., 1986).

Primary Productivity and Biogeochemical Cycles in the Sea
Edited by P.G. Falkowski and A.D. Woodhead, Plenum Press, New York, 1992

According to allometric theory, a diverse set of characteristics of organisms scale with body size such that:

$$R = a\, W^b \tag{1}$$

i.e. $$\log R = \log a + b \log W \tag{2}$$

where R is a specific rate process (t^{-1}), e.g. respiration or growth rate, W is some measure of body mass, and a and b are constants. Indeed, this relationship seems to hold for a diversity of processes among unrelated organisms (Peters, 1983). The mass-specific value of the exponent, b, is relatively constant (-0.25) for large data sets covering broad size ranges. The value of a, on the other hand, is variable, differentiating major groups of organisms such as homeotherms, heterotherms, and unicells (Fenchel, 1974). The fact that b is dimensionless and relatively constant for a variety of physiological processes, has elevated it nearly to the status of a "natural law" in the biological sciences (Peters, 1978; Platt and Silvert, 1981).

How well does this relationship apply within the restricted realm of phytoplankton cells? The answer is not straightforward. Eppley and Sloan (1965) were among the first to reveal the dependency of respiration rate on cell size in marine phytoplankton. Applying allometric theory to Eppley and Sloan's data, Laws (1975) and Banse (1976) found that respiration rate scaled roughly as the -0.25 power of cell mass, supporting the universality of the "-0.25 rule." Subsequently, however, Falkowski and Owens (1978) and Langdon (1987; 1988) found no relationship whatsoever between mass and respiration. One is left with the impression that there simply is not enough data to say anything conclusive about this relationship.

Eppley and Sloan (1966) also were among the first to look at the relationship between cell size and maximum specific growth rate in phytoplankton. Here, a reasonably consistent picture has emerged (albeit, after several rounds of revision). For large data sets spanning broad size ranges (10^1 - 10^5 pg C cell-1), the allometric equation has been

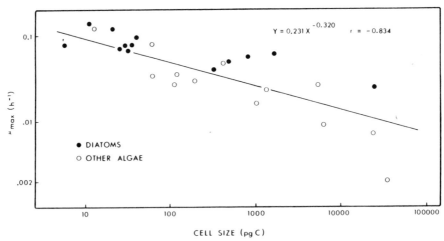

Fig. 1.   Relationship between phytoplankton cell size and the maximum specific growth rate of species grown under optimal conditions of light and temperature. Growth rates were normalized to 20°C when necessary using a Q_{10} of 1.88. From Schlesinger et al., (1981). Note that the slope for diatoms is significantly less negative than that of the data set as a whole.

shown to apply for eukaryotic phytoplankton grown under optimum conditions (Fig. 1). The value of b is closer to -0.30 than -0.25, however, (Langdon, 1988; Schlesinger et al., 1981), as was predicted by Platt and Silvert (1981) for aquatic organisms on theoretical grounds.

The error associated with data sets such as that shown in Fig. 1 is large, however. When one systematically examines specific groups of phytoplankton under identical experimental conditions, one finds that the value of b is consistently less negative than -0.30, i.e. the mass-dependency of growth rate is weaker. For diatoms and dinoflagellates, for example, the value of b is -0.13 and -0.15, respectively (Blasco et al., 1982; Banse, 1982; Chan, 1978), and it decreases significantly (to -0.08) for low-temperature species (Sommer, 1989). More importantly, although b is roughly the same for diatoms and dinoflagellates, the value of a for these two groups is very different: 0.14 for dinoflagellates vs. 0.48 for diatoms. In other words, diatoms grow three times faster than dinoflagellates of equal size. Therefore, in many ways, it is the properties of the value of a rather than b that are of most interest to phytoplankton ecologists (Platt, 1985).

Thus, although the allometric equations derived by Schlesinger et al. (1981) and Langdon (1988) are valid for the particular data sets they collected, they are not of great use in predicting the growth rate of a given phytoplankton cell from its size (or carbon content). This is obvious from the discussion of diatoms and dinoflagellates above, and is also apparent if one tries to apply the relationship to prokaryotic picoplankton. For example, the use of the Schlesinger et al. (1981) equation:

$$\mu_{max} = 5.53 \ W^{-0.32} \tag{3}$$

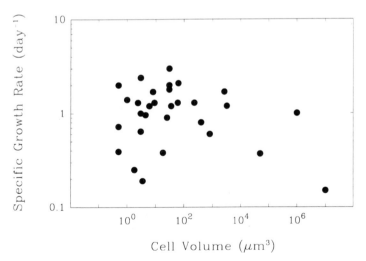

Fig. 2.   Relationship between phytoplankton cell size and the maximum specific growth rate of species grown under optimal conditions of light and temperature. Data includes all of the autotrophic organisms in Table 4 from Raven (1986). Growth rates were normalized to 20°C using a Q_{10} of 2.0. Note, for comparison with (a), that a cell 75 μm^3 in volume (5 μm in diameter) has roughly 10 pg C. The organisms smaller than this deviate significantly from the general relationship.

(where μ is expressed in day[-1]) to predict the maximum growth rates of *Synechococcus* (which have roughly 0.25 pg C cell[-1]) yields a μ_{max} in excess of 8 day[-1], which is unrealistic. This result is not surprising considering that the maximum growth rates for diatoms are around 2.5 day[-1] (Blasco et al., 1982) whereas those for *Synechococcus* are around 1.5 day[-1] (Kana and Glibert, 1987), and that picoplankton were not considered in the derivation of the Schlesinger et al. (1981) relationship. The deviation of very small cells from the Schlesinger/Langdon relationships is revealed even more dramatically in a compilation of maximum growth rates presented by Raven (1986), which includes data points for prokaryotes and other small-sized species (Fig. 2).

To further complicate this picture, even within a given species of diatom, the maximum growth rate can vary by a factor of 2, *increasing* with size as cells are transformed from pre- to post-auxospore cells. Similarly, when mean cell size decreases in diatoms because of the reduction of frustule size over succeeding generations, growth rates decrease (Chisholm and Costello, 1980). This goes against all allometric reasoning, but is a central characteristic of diatoms, and speaks to the influence of individual species characteristics on intrinsic growth rates. To quote LaBarbera (1989), "Scaling studies paint nature with a very broad brush; they are more akin to the gas laws of physics than to Newton's laws. They do not afford much precision in their predictions for individuals or perhaps even species..."

To sum up, size-dependence of growth rate and metabolic processes in phytoplankton as a group is weak. The ecological differences between the taxonomic groups appear to override the influence of size on growth potential.

Size and Maximum Abundance

Anyone who has cultured organisms, or observed them in the field, has an intuitive sense of the inverse relationship between the size of an organism and its local abundance. This relationship formed the conceptual foundation for the "ecological pyramid" of Elton, and can have implications in the analysis of food web dynamics (Platt and Denman, 1977, 1978). In a rather provocative paper, Duarte et al. (1987) showed that the relationship is highly conserved for aquatic organisms in culture: the maximum density achievable in

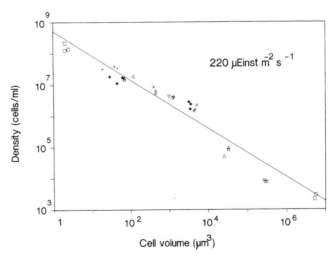

Fig. 3.   Relationship between maximal cell density (D_{max}) and cell volume (V) in phytoplankton cultures grown under identical conditions. Different symbols represent different species (from Agusti and Kalff, 1989).

culture (D_{max}), of organisms ranging in size from bacteria to fish, is inversely related to body size (V), according to the equation:

$$\log (D_{max}) = 8.53 - 0.95 \log (V) \tag{4}$$

Because the slope of this line is close to unity, the maximum biomass (V * D_{max}) achievable in cultures of aquatic organisms is a constant (6 x 10^8 μm^3 ml^{-1}), regardless of the size of individual organisms. At these biomass levels, organisms are using 0.1 % of the volume available to them and are separated by 10-20 equivalent spherical diameters. It is probably not coincidental that the region around an organism defined by a radius of 10 spherical diameters is the region within which the cell can influence the concentration of nutrients relative to that of the bulk fluid. Under severe nutrient limitation, the transport of nutrients to the cell through this region can be limited by molecular diffusion (see below).

Again, the question is, do these relationships hold when phytoplankton alone are considered? By reviewing the literature, Agusti et al. (1987) found the relationship between the maximum density in phytoplankton cultures and cell size to be:

$$\log D_{max} = 9.04 - 1.27 \log V \tag{5}$$

In subsequent work, Agusti and Kalff (1989) tested the size-density relationship directly by growing cells ranging from 2 to 5 x 10^6 μm^3 under identical conditions in culture (Fig. 3). They found that the relationship between maximum density and cell volume was:

$$\log D_{max} = 8.7 - 0.79 \log V \tag{6}$$

In this case, the slope is significantly less than one, the opposite of the results from the literature survey (Eq. 5). If one "splits the difference", however, it is not difficult to be convinced that on average, phytoplankton reach a constant maximum biomass in culture, regardless of their size.

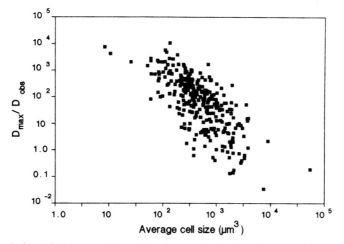

Fig. 4. Relationship between the ratio of the predicted maximal cell densities (D_{max}) and the observed densities (D_{obs}) and the mean cell size of phytoplankton communities from 165 Florida lakes (from Augusti et al., 1990).

It appears that there is a "...fundamental regularity that determines how organisms use the available space..." (Duarte et al., 1987). No clear picture has emerged as to what this "regularity" is for organisms in general, nor phytoplankton in particular, but for the latter, light and nutrient supply would be the likely regulators. Agusti and Kalff (1989) showed that the size dependency of D_{max} (Eq. 6) is quantitatively conserved under light-limited conditions in cultures, however, and evidence from Florida lakes suggests that at D_{max}, the phytoplankton community is regulated by autogenic factors rather than nutrients (Agusti et al., 1990). Furthermore, there is an inverse relationship between density and average cell size in these lakes (Fig. 4), consistent with the culture work described above, and patterns observed in marine ecosystems (see below). Finally, Duarte et al. (1990) also showed that the range of population densities achievable by different phytoplankton genera in the Florida lakes is a function of the size plasticity in the genera, again pointing to a fundamental regulatory role of size in community dynamics.

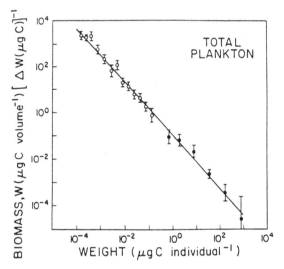

Fig. 5. Normalized biomass spectrum of the data of Beers et al. (1982) for euphotic zone plankton in the North Pacific Central Gyre (from Rodriguez and Mullin, 1986; see also Platt et al., 1984).

Community Size Spectra

Particulate matter in seawater is distributed among size classes in a manner that is reminiscent of those described above. Small particles are numerically more abundant than large ones, such that there are roughly equal amounts of material in logarithmically equal size intervals (Sheldon and Parsons, 1967; Sheldon et al., 1972). The situation is more complex here, however, because the diversity of species and trophic levels are contributing to the biomass in each size class, and the flow of energy and matter through the food web leaves its imprint on the spectrum. In the face of all of this complexity, however, normalized biomass spectra (Platt and Denman, 1977; 1978) constructed with the living component of the particulate matter in pelagic ecosystems (e.g. Sprules and Munawar,

1986; Sprules et al., 1983; Platt et al., 1984; Rodriguez and Mullin, 1986) have surprisingly conservative properties which conform to the equation:

$$\log B_w = a + b \log W \qquad (7)$$

where B_w is the total biomass of organisms in weight class W, divided by W (Fig. 5). Since B_w is equal to the numerical abundance of organisms in a given size class, Eq 7 is identical in form to Eqs 4, 5, and 6, with weight substituted for volume. Although the shape of this spectrum can be predicted from simple assumptions about ecological efficiencies and food web structure (Kerr, 1974; Platt and Denman, 1977; 1978), random encounter models also are successful (Kiefer and Berwald, 1991), leaving mechanistic interpretations ambiguous. Much like the V/D_{max} spectrum, the slope of the normalized

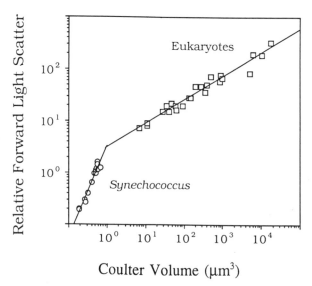

Coulter Volume (μm^3)

Fig. 6.   Relationship between Coulter volume and forward angle light scatter measured using flow cytometry. Several species of eukaryotic phytoplankton were used to construct the upper portion of the curve, and different strains of *Synechococcus* were used for the lower portion. (unpublished data of R. Olson, M. DuRand, and E. Zettler).

biomass spectrum is usually close to -1. It has been shown to vary with the productivity and size of ecosystems, however, decreasing with increasing eutrophy (Sprules and Munawar, 1986). As such, these spectra can be used with some success for the prediction of fish stocks (e.g. Sheldon et al., 1967; Maloney and Field, 1985; Borgmann, 1982). I believe that biomass spectra are underutilized as tools for understanding planktonic ecosystems, particularly in the oceans (cf. Agusti et al., 1990; 1991; Duarte et al., 1990). This is undoubtedly because of sampling problems, and the tedium involved in building the data sets required for the analyses.

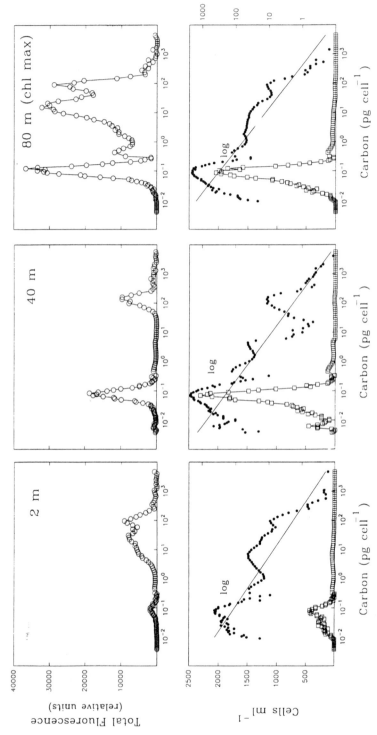

Fig. 7. Distributions of total fluorescence and cell concentration among size classes of fluorescent particles at three depths from the OFP station (31°50'N, 64°10'W) off Bermuda. The top graphs show the total fluorescence in a given size category, and the bottom shows the distribution of the numerical abundance of the cells on a log (solid dots) and linear (squares) scale. Numerically dominant cells in the small size classes are *Synechococcus* and *Prochlorococcus*.

Given the promise of size spectra as synoptic measures of the structure and function of planktonic ecosystems, we have begun to explore the utility of flow cytometric measurements for their automated construction (see also Yentsch and Phinney, 1989). Present technology limits us to particles no larger than 150 μm in diameter, but high sensitivity flow cytometry (Button and Robertson, 1989; Olson et al., 1990b) allows us to push the lower limits of spectra down to about 0.3 μm diameter, which is difficult to achieve using conventional techniques. Although flow cytometers measure light scatter rather than equivalent spherical diameter, the former can be converted to the latter using a reasonably simple calibration curve (Fig. 6), and volume can be converted to particulate carbon using appropriate equations (Strathmann, 1967; DuRand, unpublished).

Examples of flow cytometrically derived size spectra of particles from three depths in the Sargasso Sea are shown in Fig. 7. Although only the autofluorescent particles were analyzed in this particular example, it is technically feasible to collect data on all particles to construct a full spectrum analogous to that shown in Fig. 5. Our presumption, based on the theory above, is that in doing so the "valleys" in the log transformed data in Fig. 7 would disappear because of the contribution of heterotrophic organisms[1]. Besides highlighting the potential of flow cytometric analysis for the automated generation of pelagic size spectra, the examples in Fig. 7 illustrate an interesting point: Although the size distribution of phytoplankton appears to be unimodal and dramatically skewed toward the smaller size classes (cf. Yentsch and Phinney, 1989), the distribution of total fluorescence among size classes is bimodal, with a sizable fraction of the total fluorescence coming from the larger cells. This is not easily reconciled with the size-fractionated chlorophyll generalizations discussed below, but we cannot rule out the possibility that the picoplankton have a lower fluorescence yield than do the larger cells. Further work is needed, but the potential utility of these types of spectra for analyzing the planktonic community is clear.

Samples in Fig. 7 were analyzed using an Epics V flow cytometer, modified for high sensitivity (Olson et al., 1990) for the ultraplankton, and rapid sample throughput (Olson et al., 1991) for the larger cells. Cell volumes were calculated from light scatter according to the calibration curve in Fig. 6, and then converted to particulate carbon according to the Strathmann (1967) equations for the eukaryotic phytoplankton, and similar equations derived experimentally for the picoplankton (DuRand, unpublished).

SIZE-FRACTIONATED CHLOROPHYLL

One thing that can be stated unequivocally about phytoplankton size is that the fractional contribution of small cells to the standing crop increases as total chlorophyll decreases. If size-fractionated chlorophyll measurements are pooled from all over the oceans, regardless of climate or depth, we find that there is an envelope which defines the maximum fraction of the chlorophyll that will be found in cells less than 1 μm in diameter at a given total chlorophyll concentration (Fig. 8). The upper bound fraction amounts to roughly 0.50 μg l^{-1}, regardless of the total chlorophyll concentration. In extremely oligotrophic waters, the picoplankton chlorophyll is often much less than this, but usually comprises as much as 90% of the total chlorophyll biomass; in nutrient rich (high chlorophyll) areas they usually realize their maximum "potential biomass" (i.e. 0.50 μg l^{1}),

---

[1]At present, we cannot distinguish between detritus and living non-fluorescent organisms, but this is theoretically possible through the use of fluorescent vital stains.

which under these conditions is a relatively small percentage of the total phytoplankton chlorophyll.

Using an extensive and coherent data set from the Mediterranean Sea, Raimbault et al. (1988) extended this general analysis to other size classes of cells, revealing a general pattern in the size structure of phytoplankton communities. The total amount of chlorophyll in each size fraction has an upper limit, corresponding to roughly 0.5, 1, and 2 μg l^{-1} for the <1, <3, and <10 μm size fractions, respectively (Fig. 9). Thus, beyond certain thresholds, chlorophyll can only be added to a system by adding a larger size class

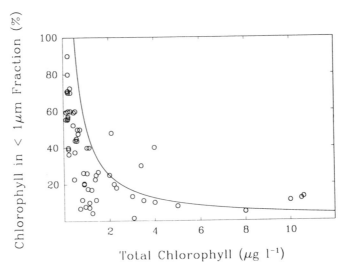

Fig. 8.   Fraction of total chlorophyll that will pass through a 1 μm pore size filter as a function of total chlorophyll for a variety of locations and depths throughout the world's oceans. The solid line describes the set of theoretical values where the concentration of chlorophyll in the less than 1 μm fraction is 0.50 μg l^{-1}. This appears to be the maximum amount found in this fraction, regardless of total chlorophyll concentration. Data were compiled from: Takahashi and Bienfang, 1983; Raimbault et al., 1988; Herbland et al., 1985; Chavez, 1989; Herbland and LeBouteiller, 1981; Smith et al., 1985; Harrison and Wood, 1988; Platt et al., 1983; Furnas and Mitchell, 1988; and Hopcroft and Roff, 1990. In cases where a single reference had too many data points to include in the figure, a random sample of points was selected.

of cells. Diatoms play a particularly important role in this "additional" chlorophyll. In a transect across the western north Pacific, Odate and Maita (1988) showed that diatoms were responsible for virtually all of the variation in the >10 μm fraction. Similarly, Chavez (1989) showed that variability in chlorophyll off the coast of Peru was directly related to numbers of diatoms present.

The patterns that emerge from studies of size fractionated chlorophyll biomass leave us with a fundamental question: Why are the oligotrophic oceans dominated by small cells? The answer is not clear-cut, but we will explore some ideas in the next two sections.

SIZE, NITROGEN PREFERENCE, AND OLIGOTROPHY

In oligotrophic oceans, which are dominated by small cells, the majority of the primary production is believed to be driven by reduced (remineralized) nitrogen (NH_4^+ and urea). Conversely, large cells dominate in areas which have relatively high supply rates of NO_3^- ("new" nitrogen) and can support a relatively large phytoplankton biomass (Dugdale and Goering, 1967; Eppley and Peterson, 1979; Malone, 1980a,b). This rather universal trend has contributed to sustaining the hypothesis that large cells use primarily NO_3^- as their nitrogen source, and small cells use primarily NH_4^+. The origin of this hypothesis is usually attributed to Malone (1971,1980b); however, he was the first to insist that high levels of net plankton productivity in NO_3^- rich areas need not result from a causal relationship between size and NO_3^- assimilation (Malone, 1975), which is the case developed below.

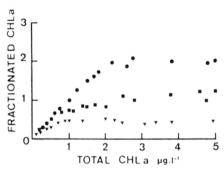

Fig. 9.   Relationships between total chlorophyll a and the maximum amount of chlorophyll *a* in the <1 μm (triangles), <3 μm (squares) and <10 μm (circles) fraction in the Mediterranean Sea (from Raimbault et al., 1988).

One should be able to test this hypothesis by examining the forms of nitrogen assimilated by different size classes using N^{15} tracer experiments. Indeed, much of the evidence in support of it (Nalewajko and Garside, 1983; Probyn, 1985; Probyn and Painting, 1985; Koike et al., 1986; Harrison and Wood, 1988; Bienfang and Takahashi, 1983) has emerged from such measurements. It is difficult to do these measurements without ambiguity, however, because the small size fraction contains heterotrophic bacteria, which can be responsible for a significant fraction of the total NH_4^+ assimilation in the euphotic zone (Eppley et al., 1977; Laws et al., 1984; 1985; and most notably Wheeler and Kirchman, 1986). Thus, measured nitrogen assimilation by the small size fraction is most likely biased in the direction of NH_4^+ preference.

Even with this bias, however, an overview of the literature does not lead to exclusive support for the hypothesis that small cells preferentially assimilate reduced nitrogen. Sherr et al. (1982) found, for example, that smaller cells had a greater affinity for both NH_4^+ and NO_3^- than did larger cells in Lake Kinneret. Similarly, Furnas (1983) found that ammonium uptake averaged between 50 and 67% of the total uptake for both

Table 1. Fraction of the Total Nitrogen Assimilated as Nitrate (f-ratio) for Different Size Fractions of Phytoplankton from Various Regions of the World's Oceans (courtesy of S. Carler, Bedford Institute).

Location and Water Type	Size (μm)	F-Ratio	Reference
Narragansett Bay (USA) "COASTAL"	TOTAL < 10	0.29 0.27	Furnas (1983)
N.W. Atlantic "COASTAL" "OCEANIC"	TOTAL < 1 TOTAL < 1	0.48 0.34 0.02 0.02	Harrison & Wood (1988)
Antarctic "COASTAL"	TOTAL < 15 < 1	0.58 0.53 0.57	Probyn & Painting (1985)
Antarctic "COASTAL"	TOTAL < 20 < 10	0.27 0.14 0.22	Koike et al. (1986)
Antarctic "COASTAL"	TOTAL < 100 < 10	0.28 0.19 0.21	Koike et al. (1981)
South Benguela Upwelling (South Africa) "COASTAL" "OCEANIC"	TOTAL < 10 < 1 TOTAL < 10 < 1	0.71 0.58 0.44 0.17 0.13 0.33	Probyn (1985)
Sapelo Island (USA) "COASTAL"	< 1	0.03	Wheeler & Kirchman (1986)
Lake Kinneret (Israel)	TOTAL < 10	0.18 0.16	Sherr et al. (1982)

the total plankton and < 10μm fractions, even though the standing crop was distributed evenly between the large (> 10 μm) and small fractions. Ronner et al. (1983) found similar trends for the Scotia Sea. Although Harrison and Wood (1988) found that NH_4^+ was favored disproportionately by the picoplankton fraction in a variety of oceanographic regimes, the differences in the fractional uptake were not large. Finally, if one examines the f-ratio (i.e. the fraction of the total nitrogen uptake accounted for by NO_3^- uptake) of total and fractionated populations from a variety of locations, strong patterns are lacking (Table 1).

Thus, the dominance of larger cells in areas enriched with NO_3^- does not appear to be a result of a causal link between cell size and preference for either nitrate or ammonium. Indeed, Taylor and Joint (1990) showed through simulation that the relative biomass of picoplankton and larger phytoplankton is a function of the total nitrogen concentration below the mixed layer, and not whether the nitrogen is in the form of nitrate or ammonium. They showed further that intense mixing favors larger cells, and that the change to picoplankton dominance (and low f-ratios) under low mixing regimes occurs even when the physiology of the large and small size fractions is modelled identically. Chavez (1989) postulated a similar mechanism.

DIFFUSION LIMITATION OF NUTRIENT TRANSPORT

The dominance of picoplankton in oligotrophic oceans also might result from the fact that the acquisition of nutrients by large cells can be limited by molecular diffusion at the very low nutrient concentrations. Munk and Riley (1952) were the first to analyze the potential for phytoplankton to be limited by the rate at which molecular diffusion can supply nutrients to the surface of the cell. The problem was reconsidered by Pasciak and Gavis (1974), who concluded that diffusion limited transport could play a role in limiting the growth of very large cells in the sea. These studies pre-date the discovery of picoplankton, however, and also did not have the benefit of accurate measurements of the extremely low concentrations for both macro- and micronutrients in the oligotrophic oceans (Garside, 1982; Bruland, 1983).

We now know that typical concentrations of NO_3^- and NH_4^+ in the surface waters of oligotrophic oceans are about 10-20 nM (e.g. Eppley and Koeve, 1990). To appreciate exactly how dilute this environment is, it helps to create a mental picture: There are roughly 10^{16} molecules of NO_3^- in a liter of seawater at this concentration, and the molecules are roughly 1 μm apart. (To complete the image, in typical oligotrophic surface waters individual cyanobacteria are roughly 1 mm apart from one another, and large net plankton are 3 cm apart). Thus, a single *Synechococcus* cell, would have at most a few NO_3^- molecules near its surface membrane at any instant in time. Molecules of trace elements are even more rare on the horizon.

With this new picture in mind, at what values of cell size, nutrient concentration, and growth rate would diffusion limit the supply of nutrients to a cell? Using the approach of Hudson and Morel (1991) and Morel et al. (1991), consider the limiting case, i.e. a spherical cell which is a perfect sink for nutrients, such that the uptake rate is fast enough that the concentration of nutrients is zero at the cell surface. For a cell with a cell quota of limiting nutrient equal to Q (mole cell^{-1}), and a specific growth rate, μ (sec^{-1}), the supply of nutrients to the cell, J, must be greater than or equal to $\mu * Q$ if the cell is to survive. Using the formulation for steady diffusion to a sphere with zero boundary conditions, one can calculate the value of J as follows:

$$J = 4\pi r D S \tag{8}$$

where r is the radius of the cell, D is the molecular diffusion coefficient ($= 2 \times 10^{-5}$ cm^2 sec^{-1}), and S is the ambient concentration of the nutrient in question. One can calculate the carbon content of a cell of radius r using the equation of Strathmann (1967), and the value of Q using the Redfield ratio. Knowing Q, one can calculate, for a range of cell sizes and growth rates, the ambient concentration of nutrient at which diffusion limitation would limit the growth of phytoplankton cells of different sizes (Fig. 10).

The advantages of being small in low nutrient environments can be seen clearly in Fig. 10. At ambient nitrogen concentrations of 10 nM, a cell of radius 0.35 μm can grow at a rate of 1 day^{-1}, which is not uncommon for these regions (e.g., Iturriaga and Mitchell, 1986; Bienfang and Takahashi, 1983; Douglas, 1984; Iturriaga and Marra, 1988), without being near diffusion limitation. A cell of radius 1 μm, however, could not grow at this rate at 10 nM N, because diffusion could not supply it fast enough. In other words, in this environment a 10 μM radius cell would be at the threshold for diffusion limitation when growing at about 0.1 day^{-1}, whereas a 0.5 μm cell could grow at 20 times this rate and not exceed the diffusion limitation threshold. Hudson and Morel (1991) and Morel et al. (1991) did a similar analysis for iron and zinc availability in oligotrophic oceans, and conclude that diffusion limitation imposes constraints on cell size and/or cell quotas of these two trace elements in oligotrophic oceans. Indeed, Sunda et al. (1991) have shown that open ocean strains of diatoms have much smaller cell quotas for iron than do their coastal counterparts.

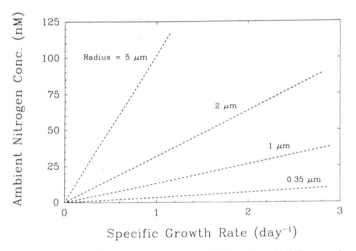

Fig. 10. Ambient concentration of nitrogen at which cells of different sizes would be diffusion limited as a function of growth rate. A cell of radius 5 μm growing at 1 day^{-1}, for example, would be diffusion limited at nitrogen concentrations less than 100 nM, whereas a cell of radius 0.35 μm growing the same rate would not be diffusion limited until concentrations dropped below 5 nM.

I have ignored sinking, swimming, and turbulent shear in this simple analysis, each of which can reduce the extent of diffusion limitation for larger cells to some extent (Pasciak and Gavis, 1975; Gavis, 1976). The influence is not large, however. To use an extreme case to illustrate the outer bound, a 200 μm diameter cell swimming at maximal swimming rates (300 μm sec^{-1}) can, at most, double its nutrient uptake rate over the diffusion limited rate; motions because of shear and sinking would less than double it (Gavis, 1976). Even large flocs of diatoms, which are centimeters in diameter, can, at most, double their nutrient uptake rate through rapid settling (Logan and Alldredge, 1989). Although these effects are not insignificant, they do not change the general message of Fig. 10. The size dependencies depicted here are logarithmic, thus, doubling does not greatly change the results. I also am ignoring the fact that the cell quota, Q, decreases with growth rate in phytoplankton, thus the nutrient requirement would be lower for lower growth rates. Although I have not done a quantitative analysis of how Fig. 10 would change with a variable Q, the effect should not be dramatic.

According to the analysis shown in Fig. 10, at ambient nitrogen concentrations typical of oligotrophic oceans, a cell of radius 5 μm would be diffusion limited at growth rates greater than 0.1 day^{-1}. Since there is reasonable evidence that cells grow at rates significantly faster than this (e.g., Laws et al., 1984), one cannot help but wonder: Why are there any large cells in the oligotrophic oceans at all? How do they cope with this handicap? Although this is one of those "whys" that has no real answer, we can identify some mechanisms that have evolved in large cells which help them overcome their disadvantage. For example, shape can be very important. A spherical shape is not the optimum configuration for transport of nutrients (Munk and Riley, 1952). In fact, over certain size ranges, prolate spheroidal cells have a greater nutrient acquisition ability than spherical cells of equivalent volume (Grover, 1989). It is probably not coincidental, therefore, that pennate diatoms are found in great abundance in the equatorial Pacific (Chavez et al., 1991), where iron availability has been hypothesized to limit phytoplankton growth (Martin et al., 1991).

Nitrogen fixation, either autonomous or through symbioses, is another mechanism that allows large cells to thrive in nutrient poor waters. The colonial, nitrogen-fixing cyanobacterium, *Trichodesmium*, for example, plays a more significant role in the total productivity in oligotrophic oceans than previously thought, largely because of novel mechanisms for nutrient acquisition (e.g. Karl et al., 1991a,b). Analogously, large diatoms have enhanced nitrogen availability through symbioses with nitrogen-fixing cyanobacteria (Venrick, 1974; Mague et al., 1974; Heinbokel, 1986), which in some cases can supply 100% of the nitrogen requirement (Martinez et al., 1983). Diatoms may have other gimmicks: Goldman (1988) built a relatively strong theoretical case for a "two layer" model of oligotrophic systems, in which large diatoms occupy the deep euphotic zone and can grow at relatively rapid rates driven by periodic inputs of nitrate from below the mixed layer. Moreover, evidence suggests that large diatoms can regulate their buoyancy enough to facilitate daily excursions between the nitricline and the surface waters (Villareal, 1988; Villareal and Carpenter, 1989).

Therefore, extremely low nutrient concentrations are probably a powerful selective force against large cells in the oligotrophic oceans. The fact that small cells also dominate nutrient-rich areas of the open sea, such as the equatorial Pacific (Chavez, 1989), is not inconsistent with this hypothesis if we accept the evidence that iron limits production in these regions (Martin et al., 1991; Hudson and Morel, 1991; Morel et al., 1991). In fact, the dominance of small cells in these areas could be used as circumstantial support for the hypothesis that some form of trace nutrient limitation (most probably iron) plays a role in regulating the size of phytoplankton crops in nutrient rich oceanic zones.

SYNECHOCOCCUS AND *PROCHLOROCOCCUS*: SAME SIZE, COMPLEMENTARY NICHES

Up to this point we have examined the phytoplankton community as an assemblage of cells of different sizes and examined the properties of the community which can be explained on the basis of the size distribution. I would like now to examine the properties of two species belonging to one size class, such that, by definition, all differences between the species must be size-independent.

Table 2. The Pigment Composition of Typical Marine *Prochlorococcus* and *Synechococcus* (from Chisholm et al., 1988; 1991; Goericke and Repeta, 1991).

SYNECHOCOCCUS	PROCHLOROCOCCUS
Chlorophyll *a*	Divinyl Chlorophyll *a*
Phycoerythrin	Divinyl Chlorophyll *b*
ß-carotene	α-carotene
xeaxanthin	xeaxanthin
	Chlorophyll *c* - like pigment

In oligotrophic oceans the dominant species that pass through a 1 μm filter are *Synechococcus* and the marine prochlorophyte, *Prochlorococcus* (Chisholm et al., 1988; 1991). These cells are quite similar in size, although *Prochlorococcus* is almost always slightly smaller than *Synechococcus* (Olson et al., 1990b). Recent DNA sequence analyses (Urbach et al., 1991; Palenik and Haselkorn, 1991) suggest a relatively close taxonomic affinity between these two groups, yet they differ dramatically in their pigment composition (Table 2). Unlike typical cyanobacteria, *Prochlorococcus* lack phycobili-proteins and contain chlorophyll *b*, α-carotene, xeaxanthin, and a chlorophyll *c* - like pigment, possibly Mg 3,8-divinylpheoporphyrin a_5. Moreover, both their chlorophyll *a* and *b* are divinyl chlorophylls (Goericke and Repeta, 1991).

Although the two types of cells are nearly identical in size, are closely related, and almost always co-occur, their relative abundance in time and space is different (Olson et al., 1990b). For several years now, we have been following the two populations at monthly intervals at the OFP station off Bermuda. The numerical abundance of *Prochlorococcus* is always greater than that of *Synechococcus* (Fig. 11), and the median depth of these cells is always equal to or deeper than that of *Synechococcus*. When surface temperatures are high (Fig. 11) and the mixed layer is shallow, *Prochlorococcus* forms a sizeable sub-surface maximum layer (July in Fig. 12), and is an order of magnitude more abundant than *Synechococcus*. As the mixed layer deepens in the fall, and nutrients are entrained into the mixed layer from below the thermocline, *Prochlorococcus* "blooms" in the surface waters, where it greatly outnumbers *Synechococcus* (Nov. in Fig. 12). When surface temperatures are cold, however, and the water column is well mixed (Feb. in Fig. 12), the two types of cells have roughly the same abundance in the water column. The same dynamics depicted in Figs. 11 and 12 can be seen in the spatial dimension for various transects in the Atlantic and Pacific oceans (Olson et al., 1990b).

The general picture that emerges suggests that *Prochlorococcus* is more efficient than *Synechococcus* at using low light, as would be expected from their pigment compositions (Glover et al., 1986; Waterbury et al., 1986; Olson et al., 1988, 1990a,b). Evidence also suggests that *Prochlorococcus* may have a different temperature optimum than does *Synechococcus*, and may be less efficient at using the low nutrient concentrations characteristic of the surface waters of the Sargasso Sea in summer (Olson et al., 1990b).

An interesting outcome of the dynamics of these two populations is that the balance of the advantages and disadvantages for the two types of cells results in a fairly constant total cell biovolume when the two populations are considered as one (faint dotted line in Fig. 11). The two populations are complementary: They fill the space "allotted" to 1 μm cells in the euphotic zone throughout the year. Although our data are not as resolved for the larger cells (Olson et al., 1990b), their numerical abundance does not display this type of constancy through the seasons. This is consistent with the image created above, i.e. that the small cells are a lawn upon which larger cells can be found in varying amounts depending on the available light and nutrients.

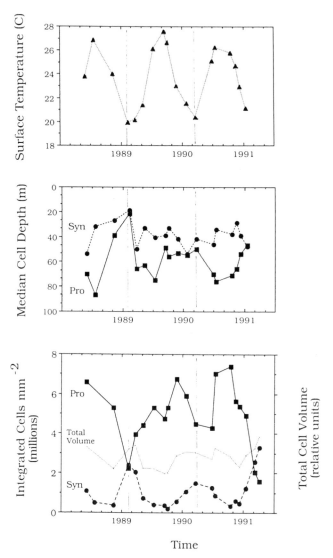

Fig. 11. Seasonal dynamics of *Synechococcus* (Syn) and *Prochlorococcus* (Pro) in the Sargasso Sea off Bermuda ("OFP" 31°50'N, 64°10'W).   (A) Surface temperature, (B) Median depth occupied by the two populations, (C)Intergrated numerical abundance of the individual species, and the summed biovolume (faint dotted line) of the two species calculated using a diameter of 0.7 and 1.0 μm for Pro and Syn, respectively. (adapted from Olson et al., 1990, and unpublished data of Olson and Chisholm).

EPILOGUE

It is sobering to realize how much our picture of the size structure of phytoplankton communities has changed in the past decade. In the early years the community was separated into "large and small" categories (net- and nannoplankton) by a 20 μm mesh net. When marine *Synechococcus* were discovered a decade ago, followed by the discovery of *Prochlorococcus* and eukaryotic ultraplankton, our image of the median size category in phytoplankton communities gradually shifted such that the modal size is now around 2-3 μm. One cannot help but wonder if there is an even smaller group waiting to be discovered. We can take some comfort, however, in the fact that the size of *Prochlorococcus* is near the theoretical limit for the smallest possible autotroph (Raven, 1986).

I have not included some very important size-dependent properties of phytoplankton in this review, which should at least be noted. First, there is a pigment "package effect" in cells, which is an increasing function of cell size and influences the efficiency with which cells absorb light (e.g. Bricaud et al., 1988; Yentsch and Phinney, 1989). Second, I have ignored the influence of size-selective grazing, grazing thresholds, and differential

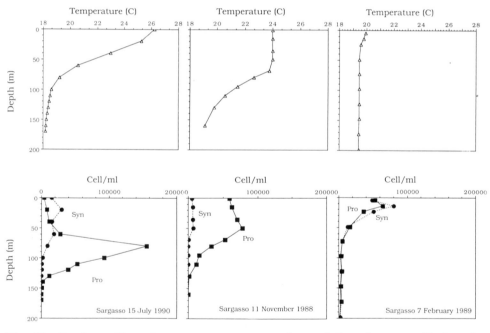

Fig. 12. Depth profiles of temperature, concentrations of *Synechococcus* (Syn) and *Prochlorococcus* (Pro) typical in the summer, winter, and fall in the Sargasso Sea. Under highly stratified summer conditions, *Synechococcus* dominates in the surface waters, and *Prochlorococcus* forms a large sub-surface maximum, which sits below the thermocline and just above the nitricline. As the mixed layer deepens and nutrients are entrained into the surface waters in the fall, *Prochlorococcus* "blooms" in the mixed layer, significantly outnumbering *Synechococcus*. Under the deep mixing conditions of winter, both organisms have similar distributions (adapted from Olson et al., 1990, and unpublished data of Olson and Chisholm).

settling rates on phytoplankton size distributions, which must influence the size distribution we measure in any snapshot of the pelagic community. Third, I have not addressed the puzzling dependence of DNA content on cell size in phytoplankton (Holm-Hansen, 1969), the adaptive significance of which has been the subject of much speculation over the years (e.g., Cavalier-Smith, 1978; 1980; Lewis, 1985). Each of these topics is worthy of a review in and of itself.

ACKNOWLEDGEMENTS

I am deeply indebted to Dick Eppley for his support, encouragement, and inspiration over the years. The following people were immensely helpful to me in preparing this paper: Sophie Carler, Michele DuRand, Jeff Dusenberry, Paul Falkowski, Sheila Frankel, Bob Hudson, Francois Morel, Rob Olson, Keith Stolzenbach, Christina Ware, and Erik Zettler. I also thank Brian Binder, Lisa West-Johnsrud, and Ena Urbach for comments on the manuscript. The work was supported in part by the following grants: NSF OCE9000043, OCE9012117, DSR9020254, and ONR N00014-87-K-0007.

REFERENCES

Agusti, S., and Kalff, J., 1989, The influence of growth conditions on the size dependence of maximal algal density and biomass, *Limnol. Oceanogr.*, 34:1104.

Agusti, S., Duarte, C. M., and Kalff, J., 1987, Algal cell size and the maximum density and biomass of phytoplankton, *Limnol. Oceanogr.*, 32:983.

Agusti, S., Duarte, D. M., and Canfield, D. E., 1990, Phytoplankton abundance in Florida lakes: Evidence for frequent lack of nutrient limitation, *Limnol. Oceanogr.*, 35-181.

Agusti, S., Duarte, D. M., and Canfield, D. E., 1991, Biomass partitioning in Florida phytoplankton communities, *J. Plank. Res.*, 13:239.

Banse, K., 1982, Cell volumes, maximal growth rates of unicellular algae and ciliates, and the role of ciliates in the marine pelagial, *Limnol. Oceanogr.*, 1059-1071.

Banse, K., 1976, Rates of growth, respiration, and photosynthesis of unicellular algae as related to cell size--A review, *J. Phycol.*, 12:135.

Beers, J. R., Reid, F. M. H., and Stewart, G. L., 1982, Seasonal abundance of the microplankton population in the N. Pacific central gyre, *Deep-Sea Res.*, 29:217.

Bienfang, P. K., and Takahashi, M., 1983, Ultraplankton growth rates in a subtropical ecosystem, *Mar. Biol.*, 76:213.

Blasco, D., Packard, T. T., and Garfield, P. C., 1982, Size dependence of growth rate, respiratory electron transport system activity and chemical composition of marine diatoms in the laboratory, *J. Phycol.*, 18:58.

Borgmann, U., 1982, Particle-size-conversion efficiency and total animal production in pelagic ecosystems, *Can. J. Fish. Aquat. Sci.*, 39:668.

Bricaud, A., Bedhomme, A. -L., and Morel, A., 1988, Optical properties of diverse phytoplanktonic species: Experimental results and theoretical interpretation, *J. Plank. Res.*, 10:851.

Bruland, K. W., 1983, Trace elements in sea-water, *in*: "Chemical Oceanography, Vol. 8, J.P. Riley and R. Chester, eds., Academic Press, London.

Button, D., and Robertson, B., 1989, Kinetics of bacterial processes in natural aquatic systems based on biomass as determined by high-resolution flow cytometry, *Cytometry*, 10:558.

Calder, W. A. III, 1984, "Function and Life History," Harvard University Press, Cambridge.

Cavalier-Smith, T., 1980, r- and K-tactics in the evolution of protist developmental systems: Cell and genome size, phenotype diversifying selection, and cell cycle patterns, *Biosystems*, 12:43.

Cavalier-Smith, T., 1978, Nuclear volume control by nucleoskeletal DNA, selection for cell volume and cell growth rate and the solution of the DNA C-value paradox, *J. Cell. Sci.*, 34:247.

Chan, A. T., 1978, Comparative physiological study of marine diatoms and dinoflagellates in relation to irradiance and cell size, I. Growth under continuous light, *J. Phycol.*, 14:396.

Chavez, F. P., 1989, Size distribution of phytoplankton in the central and eastern tropical Pacific, *Global Biogeochem. Cycles*, 3:27.

Chavez, F. P., Buck, K. R., Coale, K., Martin, J. H., DiTullio, G. R., Welshmeyer, N. A., Jacobson, A. C., and Barber, R. T., 1991, Growth rates, grazing, sinking and iron limitation of equatorial Pacific phytoplankton, *Limnol. Oceanogr.*, in press.

Chisholm, S. W., and Costello, J. C., 1980, Influence of environmental factors and population composition on the timing of cell division in *Thalassiosira fluviatilis* (Bacillariophyceae) grown on light/dark cycles, *J. Phycol.*, 16:375.

Chisholm, S. W., Olson, R. J., Zettler, E. R., Goericke, R., Waterbury, J., and Welschmeyer, N., 1988, A novel free-living prochlorophyte abundant in the oceanic euphotic zone, *Nature*, 334:340.

Chisholm, S. W., Frankel, S. L., Goericke, R., Olson, R. J., Palenik, B., Waterbury, J. B., West-Johnsrud, L., and Zettler, E. R., 1991, *Prochlorococcus* marinus nov. gen. nov. sp.: A oxyphototrophic marine prokaryote containing divinyl chlorophyll *a* and *b*, *Archiv. Microbiol.*, in press.

Costello, J. C., and Chisholm, S. W., 1981, The influence of cell size on the growth rate of *Thalassiosira weissflogii*, *J. Plank. Res.*, 3:415.

Douglas, D. J., 1984, Microautoradiography-based enumeration of photosynthetic picoplankton with estimates of carbon-specific growth rates, *Mar Ecol. Prog. Ser.*, 14:223.

Duarte, C. M., Agusti, S., and Peters, H., 1987, An upper limit to the abundance of aquatic organisms, *Oecologia* (Berlin), 74:272.

Duarte, D. M., Agusti, S., and Canfield, D. E., 1990, Size plasticity of freshwater phytoplankton: Implications for community structure, *Limnol. Oceanogr.*, 35:1846.

Dugdale, R. C., and Goering, J. J., 1967, Uptake of new and regenerated forms of nitrogen in primary productivity, *Limnol. Oceanogr.*, 12:196.

Elton, C., 1927, "Animal Ecology," Macmillan, New York.

Eppley, R. W., and Sloan, P. R., 1965, Carbon balance experiments with marine phytoplankton, *J. Fish. Res. Bd. Can.*, 22:1083.

Eppley, R. W., and Peterson, B. J., 1979, Particulate organic matter flux and planktonic new production in the deep ocean, *Nature*, 282:677.

Eppley, R. W., and Sloan, P. R., 1966, Growth rates of marine phytoplankton: Correlation with light absorption by cell chlorophyll *a*, *Physiol. Plant.*, 19:47.

Eppley, R. W., and Koeve, W., 1990, Nitrate use by plankton in the eastern subtropical North Atlantic, March-April 1989, *Limnol. Oceanogr.*, 35:1781.

Eppley, R. W., Sharp, J. H., Renger, E. H., Perry, M. J., and Harrison, W. G., 1977, Nitrogen assimilation by phytoplankton and other microorganisms in the surface waters of the central North Pacific Ocean, *Mar. Biol.*, 39:111.

Falkowski, P. G., and Owens, T. G., 1978, Effects of light intensity on photosynthesis and dark respiration in six species of marine phytoplankton, *Mar. Biol.*, 45:289.

Fenchel, T., 1974, Intrinsic rate of natural increase: The relationship with body size, *Oecologia* (Berlin), 14:317.

Furnas, M. J., 1983, Nitrogen dynamics in lower Narragansett Bay, Rhode Island, 1. Uptake by size-fractionated phytoplankton populations, *J. Plank. Res.*, 5:657.

Furnas, M. J., and Mitchell, A. W., 1988, Photosynthetic characteristics of Choral Sea Picoplankton (<2 μm size fraction), *Biol. Oceanogr.*, 5:163.

Garside, C., 1982, A chemiluminescent technique for the determination of nanomolar concentrations of nitrate and nitrate, or nitrite alone in seawater, *Mar. Chem.*, 11:159.

Gavis, J., 1976, Munk and Riley revisited: Nutrient diffusion transport and rates of phytoplankton growth, *J. Mar. Res.*, 34:161.

Geider, R. J., Platt, T., and Raven, J. A., 1986, Size dependence of growth and photosynthesis in diatoms: A synthesis, *Mar. Ecol. Prog. Ser.*, 30:93.

Glover, H. E., Campbell, L., and Prezelin, B. B., 1986, Contribution of Synechococcus to size-fractionated primary productivity in three water masses in the Northwest Atlantic Ocean, *Mar. Biol.*, 91:193.

Goericke, R., and Repeta, D., 1991, The pigments of Prochlorococcus marinus: The presence of divinyl-chlorophyll *a* and *b* in a marine cyanobacterium, *Limnol. Oceanogr.*, in press.

Goldman, J. C., 1988, Spatial and temporal discontinuities of biological processes in pelagic surface waters, *in*: "Toward a Theory on Biological Physical Interactions in the World Ocean," B.J. Rothschild, ed., Kluwer Academic Publishers, New York.

Grover, J. P, 1989, Influence of cell shape and size on algal competitive ability, *J. Phycol.*, 25:402.

Harrison, W. G., and Wood, L. J. E., 1988, Inorganic nitrogen uptake by marine phytoplankton, *Limnol. Oceanogr.*, 33:468.

Heinbokel, J. F., 1986, Occurrence of Richelia intracellularis (Cyanophyta) within the diatoms Hemiaulus haukii and H. membranaceus off Hawaii, *J. Phycol.*, 22:399.

Herbland, A., Le Bouteiller, A., and Raimbault, P. L., 1985, Size structure of phytoplankton in the equatorial Atlantic Ocean, *Deep-Sea Res.*, 32:819.

Herbland, A., and Le Bouteiller, A., 1981, The size distribution of phytoplankton and particulate organic matter in the Equatorial Atlantic Ocean, importance of ultraseston and consequences, *J. Plank. Res.*, 3:6659.

Hopcroft, R. R., and Roff, J. C., 1990, Phytoplankton size fractions in a tropical neritic ecosystem near Kingston Jamaica, *J. Plank. Res.*, 12:1069.

Hudson, R. J., and Morel, F. M. M., 1991, Trace metal transport by marine microorganisms: Implications of metal coordination kinetics, *Deep-Sea Res.*, in press.

Holm-Hansen, O., 1969, Algae: Amounts of DNA and organic carbon in single cells, *Science*, 163:87.

Iturriaga, R., and Mitchell, B. G., 1986, Chroococcoid cyanobacteria: A significant component of the food web dynamics of the open ocean, *Mar. Ecol. Prog. Ser.*, 28:291.

Iturriaga, R., and Marra, J., 1988, Temporal and spatial variability of chroococcoid *cyanobacteria Synechococcus spp.* specific growth rates and their contribution to primary production in the Sargasso Sea, *Mar. Ecol. Prog. Ser.*, 44:175.

Kana, T. M., and Glibert, P. M., 1987, Effect of irradiances up to 2000 μE m-2 sec-1 on marine *Synechococcus* WH7803 - I. Growth, pigmentation, and cell composition, *Deep-Sea Res.*, 34:479.

Karl, D. M., Bird, D. F., Hebel, D. V., Letelier, R., Sabine, C., and Winn, C. D., 1991b, Nitrogen fixation contributes to new production in the oligotrophic North Pacific Gyre, unpublished.

Karl, D. M., Hebel, D. V., Bird, D. F., Letelier, R., and Winn, C. D., 1991a, Trichodesmium blooms and new nitrogen in the North Pacific Gyre, in: "Biology and Ecology of Diazotrophic Marine Organisms: Trichodesmium and Other Species," E.J. Carpenter, D.G. Capone, and J.G. Rueter, eds., Kluwer Academic Publishers, New York.

Kerr, S. R., 1974, Theory of size distribution in ecological communities, J. Fish. Res. Bd. Can., 31:1859.

Kiefer, D. A., and Berwald, J., 1992, A random encounter model for the microbial planktonic community, Limnol. Oceanogr., in press.

Koike, I., Ronner, U., and Holm-Hansen, O., 1981, Microbial nitrogen metabolism in the Scotia Sea, Antarctic J., 16:165.

Koike, I., Holm-Hansen, O., and Biggs, D. C., 1986, Inorganic nitrogen metabolism by Antarctic phytoplankton with special reference to ammonia cycling, Mar. Ecol. Prog. Ser., 30:105.

LaBarbera, M., 1989, Analyzing body size as a factor in ecology and evolution, Ann. Rev. Ecol. Syst., 20:97.

Langdon, C., 1987, On the causes of interspecific differences in the growth-irradiance relationship for phytoplankton, I. A comparative study of the growth-irradiance relationship of three marine phytoplankton species: Skeletonema costatum, Olisthodiscus luteus and Gonyaulax tamarensis, J. Plank. Res., 9:459.

Langdon, C., 1988, On the causes of interspecific differences in the growth-irradiance relationship for phytoplankton, II. A general review, J. Plank. Res., 10:1291.

Laws, E. A., 1975, The importance of respiration losses in controlling the size distribution of marine phytoplankton, Ecology, 56:419.

Laws, E. A., Redalje, D. G., Haas, L. W., Bienfang, P. K., Eppley, R. W., Harrison, W. G., Karl, D. M., and Marra, J., 1984, High phytoplankton growth and production rates in oligotrophic Hawaiian coastal waters, Limnol. Oceanogr., 29:1161.

Laws, E. A., Harrison, W. G., and DiTullio, G. R., 1985, A comparison of nitrogen assimilation rates based on N-15 uptake and autotrophic protein synthesis, Deep-Sea Res., 32:85.

Lewis, W. M., 1985, Nutrient scarcity as an evolutionary cause of haploidy, Amer. Nat., 125:692.

Logan, B. E., and Alldredge, A. L., 1989, Potential for increased nutrient uptake by flocculating diatoms, Mar. Biol., 101:433.

Mague, T. H., Weare, N. M., and Holm-Hansen, O., 1974, Nitrogen fixation in the north Pacific Ocean, Mar Biol., 24:109.

Malone, T., 1975, Environmental control of phytoplankton cell size, Limnol. Oceanogr., 20:490.

Malone, T., 1971, The relative importance of nannoplankton and netplankton as primary producers in the California current system, Fish. Bull., 69:799.

Malone, T. C., 1980a, Algal size, in: "The Physiological Ecology of Phytoplankton," I. Morris, ed., U. Calif. Press, Berkeley and Los Angeles.

Malone, T. C., 1980b, Size-fractionated primary productivity of marine phytoplankton, in: "Primary Productivity in the Sea," P.G. Falkowski, ed., Brookhaven Symposium in Biology, Plenum, New York.

Maloney, C. L., and Field, J. G., 1985, Use of particle-size data to predict potential pelagic-fish yield of some South African areas, S. Afr. J. Mar Sci., 3:119.

Martin, J. H., Gordon, R. M., and Fitzwater, S. E., 1991, The case for iron, *in*: "What Controls Phytoplankton Production in Nutirent Rich Areas of the Open Sea?", S.W. Chisholm and F.M.M. Morel, eds., *Limnol. Oceanogr.* (Special issue), in press.

Martinez, L., Silver, M. W., King, J. M., and Alldredge, A. L., 1983, Nitrogen fixation by floating diatom mats: A source of new nitrogen to oligotrophic ocean waters, *Science*, 221:152.

Morel, F. M. M., Hudson, R. J., and Price, N. M., 1991, Trace metal limitation in the sea, *in*: "What Controls Phytoplankton Production in Nutirent Rich Areas of the Open Sea?", S.W. Chisholm and F.M.M. Morel, eds., *Limnol. Oceanogr.* (Special Issue), in press.

Munk, W. H., and Riley, G. A., 1952, Absorption of nutrients by aquatic plants, *J. Mar. Res.*, 11:215.

Murphy, L. S., and Haugen, E. M., 1985, The distribution and abundance of phototrophic ultraplankton in the N. Atlantic, *Limnol. Oceanogr.*, 30:47.

Nalewajko, C., and Garside, C., 1983, Methodological problems in the simultaneous assessment of photosynthesis and nutrient uptake in phytoplankton as functions of light intensity and cell size, *Limnol. Oceanogr.*, 28:591.

Odate, T., and Maita, Y., 1988, Regional variation in the size composition of phytoplankton communities in the Western North Pacific Ocean, Spring 1985, *Biol. Oceanogr.*, 6:65.

Olson, R. J., Zettler, E. R., Dusenberry, J., and Chisholm, S. W., 1991, Advances in oceanography through flow cytometry, *in*: "Individual Cell and Particle Analysis in Oceanography, S. Demers and M. Lewis, eds., in press.

Olson, R. J., Chisholm, S. W., Zettler, E. R., and Armbrust, E. V., 1988, Analysis of *Synechococccus* pigment types in the sea using single and dual beam flow cytometry, *Deep-Sea Res.*, 35:425.

Olson, R.J., Chisholm, S.W., Zettler, E.R., and Armbrust, E.V., 1990a, Pigments, size, and distribution of *Synechococcus* in the North Atlantic and Pacific Oceans, *Limnol. Oceanogr.*, 35:45.

Olson, R.J., Chisholm, S.W., Zettler, E.R., Altabet, M.A., and Dusenberry, J.A., 1990b, Spatial and temporal distributions of prochlorophyte picoplankton in the North Atlantic Ocean, *Deep-Sea Res.*, 37:1033.

Palenik, B.P., and Haselkorn, R., 1991, Multiple evolutionary origins of prochlorophytes, the chlorophyll *b*-containing prokaryotes, *Nature*, in press.

Pasciak, W. J., and Gavis, J., 1974, Transport limitation of nutrient uptake in phytoplankton, *Limnol. Oceanogr.*, 19:881.

Peters, R. H., 1978, Empirical physiological models of ecosystem processes, *Verh. Int. Ver. Theor. Angew. Limnol.*, 20:110.

Peters, R. H., 1983, "The Ecological Implications of Body Size," Cambridge University Press, Cambridge.

Platt, T., 1985, Structure of the marine ecosystem: Its allometric basis, *in*: "Ecosystem Theory for Biological Oceanography," R.E Ulanowicz and T. Platt, eds., *Can. Bull. Fish. Aquat. Sci.*, 213:55.

Platt, T., and Denman, K. L., 1977, Organization in the pelagic ecosystem, *Helgolander wiss. Meeresunters*, 30:575.

Platt, T., and Denman, K. L., 1978, The structure of pelagic marine ecosystems, *Rapp. P.-V. Reun. Cons. Perm. Int. Explor. Mer.*, 173:60.

Platt, T., and Silvert, W., 1981, Ecology, physiology, allometry and dimensionality, *J. Theor. Biol.*, 93:885.

Platt, T., Subba Rao, D. V., and Irwin, B., 1983, Photosynthesis of picoplankton in the oligotrophic ocean, *Nature*, 301:702.

Platt, T., Lewis, M., and Geider, R., 1984, Thermodynamics of the pelagic ecosystem: Elementary closure conditions for biological production in the open ocean, *in*: "Flows of Energy and Materials in Marine Ecosystems," M.J.R. Fasham, ed., Plenum, New York.

Probyn, T. A., 1985, Nitrogen uptake by size-fractionated phytoplankton populations in the southern Benguela upwelling system, *Mar. Ecol. Prog. Ser.*, 22:249.

Probyn, T. A., and Painting, S. J., 1985, Nitrogen uptake by size-fractionated phytoplankton populations in Antarctic surface waters, *Limnol. Oceanogr.*, 30:1327.

Raimbault, P., Rodier, M., and Taupier-Letage, I., 1988, Size fraction of phytoplankton in the Ligurian Sea and the Algerian Basin (Mediterranean Sea): Size distribution versus total concentration, *Mar. Microb. Food Webs*, 3:1.

Raven, J. A., 1986, Physiological consequences of extremely small size for autotrophic organisms in the sea, *in*: "Photosynthetic Picoplankton," T. Platt and W.K.W. Li., eds., *Can. Bull. Fish. Aquat. Sci.*, 214:583.

Rodriguez, J., and Mullin, M. M., 1986, Relation between biomass and body weight of plankton in a steady-state oceanic ecosystem, *Limnol. Oceanogr.*, 31:316.

Ronner, U., Sorennsson, F., and Holm-Hansen, O., 1983, Nitrogen assimilation by phytoplankton in the Scotia Sea, *Polar Biol.*, 2:137.

Schlesinger, D. A., Molot, L. A., and Shuter, B. J., 1981, Specific growth rates of freshwater algae in relation to cell size and light intensity, *Can. J. Fish. Aquat. Sci.*, 38:1052.

Sheldon, R. W., Prakash, A., and Sutcliffe, W. H., 1972, The size distribution of particles in the ocean, *Limnol. Oceanogr.*, 17:327.

Sheldon, R. W., and Parsons, T. R., 1967, A continuous size spectrum for particulate matter in the sea, *J. Fish. Res. Bd. Can.*, 24:909.

Sherr, E. B., Sherr, B. F., Berman, T., and McCarthy, J. J., 1982, Differences in nitrate and ammonia uptake among components of a phytoplankton population, *J. Plankton Res.*, 4:961.

Silvert, W., and Platt, T., 1978, Energy flux in the pelagic ecosystem: A time-dependent equation, *Limnol. Oceanogr.*, 23:813.

Silvert, W., and Platt, T., 1980, Dynamic energy flow model of the particle size distribution in pelagic ecosystems, *in*: "Evolution and Ecology of Zooplankton Communities," W. Charles Kerfoot, ed., The University Press of New England, N.H.

Smith, J. C., Platt, T., Li, W. W. K., Horne, E. H. P., Harrison, W. G., Subba Rao, D. U., and Irwin, B. P., 1985, Arctic marine photoautrotophic picoplankton, *Mar. Ecol. Prog. Ser.*, 20:207.

Sommer, U., 1989, Maximal growth rates of Antarctic phytoplankton: Only weak dependence on cell size, *Limnol. Oceanogr.*, 34:1109.

Sprules, W. G., and Munawar, M., 1986, Plankton size spectra in relation to ecosystem productivity, size, and perturbation, *Can. J. Fish. Aquat. Sci.*, 43:1789.

Sprules, W. G., Casselman, J. M., and Shuter, B. J., 1983, Size distribution of pelagic particles in lakes, *Can. J. Fish. Aquat. Sci.*, 40:1761.

Strathmann, R. R., 1967, Estimating the organic carbon content of phytoplankton from cell volume or plasma volume, *Limnol. Oceanogr.*, 12:411.

Sunda, W. G., Swift, D. G., and Huntsman, S. A., 1991, Low iron requirement in oceanic phytoplankton, *Nature*, 351:55.

Takahashi, M., and Bienfang, P. K., 1983, Size structure of phytoplankton biomass and photosynthesis in subtropical Hawaiian waters, *Mar. Biol.*, 76:203.

Taylor, A. H., and Joint, I., 1990, A steady state analysis of the 'microbial loop' in stratified systems, *Mar. Ecol. Prog. Ser.*, 59:1.

Urbach, E. Robertson, D., and Chisholm, S. W., 1991, Multiple evolutionary origins of prochlorophytes within the cyanobacterial radiation, *Nature*, in press.

Venrick, E. L., 1974, The distribution and significance of Richelia intracellularis Schmidt in the North Pacific Central Gyre, *Limnol Oceanogr.*, 19:437.

Villareal, T. A., and Carpenter, E. J., 1989, Nitrogen fixation, suspension characteristics and chemical composition of Rhizosolenia mats in the central N. Pacific Gyre, *Biol. Oceanogr.*, 6:327.

Villareal, T. A., 1988, Positive buoyancy in the oceanic diatom Rhizosolenia debyana H. Peragallo, *Deep-Sea. Res.*, 35:1037.

Waterbury, J. B., Watson, S. W., Valois, F. W., and Franks, D. G., 1986, Biological and ecological characterization of the marine unicellular cyanobacterium *Synechococcus*, *in*: "Photosynthetic Picoplankton," T. Platt and W.K.W. Li., eds., *Can. Bull. Fish. Aquat. Sci.*, 214:583.

Wheeler, P. A., and Kirchman, D. L., 1986, Utilization of inorganic and organic nitrogen by bacteria in marine systems, *Limnol. Oceanogr.*, 31:998.

Yentsch, C. S., and Phinney, D. A., 1989, A bridge between ocean optics and microbial ecology, *Limnol. Oceanogr.*, 34:1694.

PRODUCTIVITY OF SEAWEEDS

J. Ramus

Duke University, Marine Laboratory
Beaufort, NC 28516

INTRODUCTION

The goal of this manuscript is to assess this field of research for the 1980s, and to highlight the progress and the needs with specific examples. The goal is not an exhaustive review of the literature; thus, many fine contributions are not cited here. The topic is seaweed productivity *sensu strictu*, from the ecological perspective rather than the physiological perspective. The subject is seaweed biomass and net carbon (C) fixation rates, the regulation of those parameters by environmental properties, and techniques for assessing those parameters. Seaweeds here are marine macroalgae, mostly the chlorophytes (greens), phaeophytes (browns), and rhodophytes (reds).

THE SEAWEED SCALE

Whittaker and Likens (1975) estimated the net annual primary production of the biosphere to be 172.5 x 10⁹ metric tons of dry matter, of which 32% or 55.2 x 10⁹ tons is marine. Converting dry matter to C (x 0.45), the number approximates 24.8 x 10¹⁵ gC/y for the marine environment. The estimate included the coastal zone, but did not distinguish between pelagic and benthic productivities. Bunt (1975), in the same text, attempted an estimate for the coastal fringe based on the extent of the benthic environment lying within the photic zone, and gave an order of magnitude range of 0.65 - 6.5 x 10¹⁵ gC/y, or 2.6% - 26% of the total marine productivity and 0.8% - 8% of global productivity. In a treatise on marine macrophytes as a global carbon sink, Smith (1981) estimated the areal standing biomass to be 400 times that of phytoplankton, and annual productivity to be 1 x 10¹⁵ gC. Marine macrophytes here included benthic submerged plants, both seagrasses and seaweeds, from estuaries, algal beds, and reefs. Macroalgae *per se* are not partitioned from any of the above coastal or macrophyte estimates. Thus, for global macroalgal productivity, the range of uncertainly is sufficiently large to warrant considerable attention in the future.

The uncertainty in productivity estimates has as its root cause the enormous range in dimensional scale encompassed by seaweed biology, for estimating standing crop (*B*),

Primary Productivity and Biogeochemical Cycles in the Sea
Edited by P.G. Falkowski and A.D. Woodhead, Plenum Press, New York, 1992

productivity (*P*), and turnover rates (*P/B*). Linear dimensions range from mm to tens of m, mass dimensions from mg to kg, and temporal dimensions from hours to months. Further, the dimensions often escape the scale of physics important to phytoplankton productivity. This range in scale can be usefully bounded by contrasting two habitats, namely those with transportable sediments and low relief, and those with rocky substrates and high relief. As I will show, there are long-term data bases from both habitats, at least at one geographic location each.

Habitats with transportable sediments and low relief include the wetlands, estuaries, and mudflats of the temperate and subtropical latitudes of eastern North America. One such system is the 30 km^2 North Inlet system, South Carolina, an NSF Long Term Ecological Research (LTER) site for which data has been produced for more than a decade (Dame et al., 1986). Here, the water is turbid, and thus, the vertical distribution of seaweeds is confined to 1-2 m. Substrate for attachment is spatially rare, and includes peat banks, shell litter, polychaete worm tubes, seagrass blades, and the bases of marshgrass culms. The physical environment is harsh because of shallow waters, moving sediments, and long exposures. Zonation is inconspicuous. Small grazers abound, e.g., snails, amphipods, and shrimp, which rapidly consume blooms of seaweed. Biomass and diversity are low, confined primarily to small ruderal species as *Enteromorpha*, *Ectocarpus*, and *Porphyra*. Unattached *Ulva* thalli a meter or more in length are carried horizontally through the shallows, oscillating with the tides, and bathed in a nutrient- and light-abundant environment.

Based on measurements of O_2 exchange rates on whole plants, seaweed productivities ranged from a few gC/m^2•y for the high marsh to greater than 1000 gC/m^2•y for the flood tide delta (Coutinho and Zingmark, pers. comm.). Area-averaged net annual primary production was near 300 gC/m^2 for benthic macroalgae, near 400 gC/m^2 for marshgrasses, and near 300 gC/m^2 for benthic microalgae compared with 55 gC/m^2 for phytoplankton (Dame et al., 1986).

Estimates have been made for a similar but temperate latitude system, notably Flax Pond, New York (Woodwell et al., 1979). Here, area-averaged net annual primary production for benthic macroalgae was 75 gC/m^2, near 300 gC/m^2 for marshgrasses, and 50 gC/m^2 for benthic microalgae compared with 12 gC/m^2 for phytoplankton. The Flax Pond system is less than half (429 gC/m^2) as productive as the North Inlet system (1059 gC/m^2), the latter being productive in the cold season as well as in the warm season.

The sublittoral seaweed communities of temperate-latitude, rocky coasts are best developed, especially where episodic upwelling is a feature, e.g., southern California. There are distinct patch types composed of species of algae that can be categorized into vegetation layers, distinguished by distinct morphological adaptations (Dayton et al., 1984). These layers include (1) a floating canopy (*Macrocystis, Nereocystis*) supported at or near the surface by floats; (2) a stipitate, erect understory in which the fronds are supported well above the substratum by stipes (*Pterygophora, Eisenia, Laminaria*); (3) a prostrate canopy in which the fronds lie on or immediately above the substratum (*Laminaria, Cystoseira, Dictyoneurum*); (4) a densely packed algal turf of articulated coralline algae (*Calliarthron*) and many species of foliose and siphonous red algae; and (5) encrusting coralline algae (*Lithothamnion and Lithophyllum*). The physical environment is relatively stable, as is patch dynamics. Biomass is the highest of all seaweed communities, as is diversity. The kelp patch actually affects the physical and chemical environment, by slowing and directing longshore currents (Jackson and Winant, 1983), drawing down DIN and increasing DOC loads.

Despite the conspicuousness and inferred ecological importance of these kelp systems, data on biomass and productivity are scanty. Although these five vegetation layers have been identified, data are generally available for only one layer, namely, for the floating canopy layer. For example, Mann (1982) cites data for kelp ecosystems around the globe, and includes *Macrocystis pyrifera* systems from southern California. Based on measurements of instantaneous photosynthetic rate and growth rate, P_n has a maximum value of 6 % of biomass/d, but sustained rates are more on the order of 1-4 % of biomass/d. Taking 4 kg/m^2 as an average biomass, this gives a P_n of 40-160 g fresh weight/m^2•d, about 1-4 gC/m^2•d, or 350-1500 gC/m^2•y. Mann concludes that a rough estimate of average P_n for *Macrocystis* beds might be taken as 800-1000 gC/m^2•y, depending on latitude and nutrient supply, and the assumption of $P/B \approx 1$.

Mann (1982) also cited his own work on the *Laminaria*-dominated systems of Nova Scotia, specifically for St. Margaret's Bay which measures about 10 x 14 km. From transects, total biomass averaged almost 1500 kg/m shoreline, 84% of which was *Laminaria* and *Agarum* and about 9% rockweed. Productivity of this community was measured by determining the amount of new tissue added to the blades of these short kelps by marking the blade next to the meristem with a hole at monthly intervals. Averaged over the seaweed zone, this new growth amounted to 1750 gC/m^2•y. The area-averaged seaweed productivity for the Bay was about 600 gC/m^2•y, compared with 200 gC/m^2•y for phytoplankton.

The productivity data from the above systems are extrapolated largely from measurements of segments of plants, and, on occasion, from whole plants. The measurements usually include only the most conspicuous biomass dominants in the community. Further, the elapsed time of measurements is usually hours, some days and months, over several seasons at most. Yet these communities contain tiny filamentous epiphytic species with biomass turnover times of about a day, to species which are the spatial equivalents of forest trees with turnover times of about a year. Further, as I shall document in the following two examples, productivities are regulated by temporal events of the frequency of sunflecks to those of the El Niño-Southern Oscillation (ENSO) cycle. Relative to phytoplankton, seaweed populations are much more anisotropic in time and space.

Studying the light environment within a giant kelp (*Macrocystis pyrifera*) forest, Gerard (1984a) concluded that the greatest reduction in the vertical gradient of irradiance occurred in the uppermost 1 m of the water column, where the floating portions of the kelp fronds formed surface canopies with high blade densities. Average irradiances low enough to limit kelp photosynthesis (<200 μmol photons/m^2•s) were recorded at 1 m depth below dense canopies under sunny surface conditions. Light penetration was exponentially related to canopy density, but was higher than predicted from transmission through individual kelp blades because of the heterogenous distribution of canopy tissue. Light penetration was considerably lower immediately adjacent to an individual kelp plant than in spaces between plants, indicating that there was significant self-shading. Spatial and temporal variability in both irradiance and light penetration were highest just below the canopy and decreased with increasing depth. High-frequency light fluctuations due to wave-focusing are a prominent feature of shallow marine waters. These sunflecks (>200 μmol photons/m^2•s) range from <1 to >60s in duration and are a large proportion of total irradiance at 1 m depth, even during recordings with low average irradiances. Gerard concludes that these fluctuations affect the photosynthetic efficiency of kelp tissues within or immediately below the canopy, and, thus, productivity estimates in general. Greene and Gerard (1990) determined that growth rates of the rhodophyte *Chondrus crispus* were higher under fluctuating light regimes than in constant light regimes with equivalent daily irradiances. Growth was enhanced by fluctuations of O.1 and 1 Hz, but not by O.01 Hz.

To have an inventory of a harvestable resource, an historical record has been kept of the surface canopy cover of the kelp forest at Pt. Loma, California, dating from 1911, a crude but effective measure of biomass. The resistance and resilience stability of these kelp patches were characterized from 1970-1981 (Dayton et al., 1984). In the winter of 1982-1983, 11 unusually powerful storms struck the southern California coastline. For 15 days, wave heights exceeding 3m were measured, some of these waves were the largest measured in eight years. Seven of the 11 storms had wave periods of more than 20 seconds, the first such occurrence in over a decade. These storms occurred during an El Niño event, which, by many indications, appeared to be the most extreme ever measured (Dayton and Tegner, 1984).

The winter storms reduced the area of the surface canopy from over 600 to less than 40 ha, and there were periods when essentially there was no canopy. Mortality, measured by densities of extant plants and recently killed holdfasts or fresh holdfast scars, was highest (66%) in the shallow (12 m) inner margin of the forest, lower (47%) at mid depths (15 m) of the central part of the forest, and lowest (13%) in the deeper (18 m) central outer margin. Mortality at the northern and southern ends of the forest, 40 and 41% respectively, was higher than that at the central outer station even though these regions were all at the same depth (18 m). Two- and three-year-old plants survived better than others.

Besides the catastrophic physical effects of the winter storms, the subsequent high water temperatures ($>20°C$) and low nutrient concentrations associated with the El Niño event adversely effected the physiology and growth of *Macrocystis* (Gerard, 1984b). The kelp canopy deteriorated rapidly following the onset of high ambient surface temperatures and the depletion of internal nitrogen reserves. Reductions in chlorophyll content, photosynthetic capacity, and growth rates of canopy fronds were all attributable to N-starvation. Although N-reserves were depleted more slowly in smaller kelp fronds on the same plants, elongation rates of all fronds decreased concurrently.

REMOTE SENSING, MAPPING AND MODELLING

Recognizing the extremely patchy distribution of seaweeds in time and space, new techniques have been brought to bear on the determinations of biomass and productivity. The first is the remote sensing of sunlight backscattered in the visible and near-infrared spectral distributions by seaweed biomass, and is particularly applicable to the mapping of standing crop. There are two caveats, however: 1) the seaweed must be emergent or near emergent, and 2) the remote images require rigorous ground calibration for quantification. Estuarine macrophytobenthos biomass over a 900 ha area was assessed using aerial (small, fixed-wing airplane) photography and infrared false-color film with considerable accuracy (10%) (Meulstee et al., 1986). Biomass estimates were partitioned over exposed mudflat seagrasses, chlorophytes, and phaeophytes. Densities of colored emulsion layers on the film were measured with an HP86 computer-coupled transmission densitometer and correlated with field samples. A color-density ratio algorithm (green/red) was correlated with field areal dry biomass (correlation coefficient = 95%).

Littler and Littler (1987) similarly quantified an exposed rocky intertidal from a low-flying helicopter using a film for daylight-color projection slides. Photographic samples were taken obliquely at approximate right angles to the shoreline in a continuous format. Highly reproducible quantitative information (% cover) on dominant populations and major community types were produced by photographimetric techniques. Here, the images were projected onto a mechanically randomized pattern of fine red dots, and values for the relative percentage cover were expressed as the number of "hits" for each species.

The potential application of remote sensing to seaweed biomass mapping by a satellite-borne spectroradiometer was presented by Belsher et al. (1988). A salutary limitation of current instruments is their resolution. The so-called second generation instruments as the French SPOT satellite have design resolutions of 10m with panchromatic channels, and 20m with multispectral channels, which when used in combination, may provide very accurate data of the kelp forest surface canopy. To simulate SPOT-type satellite imagery, remote sensing of the Bay of San Cyprianu, Corsica, was carried out using a *Daedalus* radiometer mounted in an aircraft. The simulated image was calibrated using a chart of the benthos produced by SCUBA diving. The exercise showed that remote sensing by satellite could resolve certain bottom types and benthic communities in water down to 12 m.

The ability to distinguish seaweed classes by fluorescence emission may allow the type and abundance of subtidal seaweeds to be characterized by existing laser-induced fluorescence methodology (LIDAR) from low-flying aircraft. In a survey of 20 species, Topinka et al. (1990) used narrow waveband light to excite suites of light-harvesting pigment protein complexes (LHPPs) characteristic of brown, green and red seaweeds and measured chlorophyll *a* fluorescence emission at 685 nm. Fluorescence induced by 540 and 465 nm wavelengths gave excitation ratios of 0.28 ± 0.07 for 6 green seaweeds, 0.59 ± 0.07 for 7 brown seaweeds, and 3.67 ± 0.56 for 7 red seaweeds. These data suggest that fluorescence excitation signatures are relatively uniform within phylogenetic class but differ substantially between classes. The phaeophyte *Ectocarpus siliculosus* was cultured over a large range of light and nitrate regimes, and the 540:465 fluorescence excitation ratio showed little variation. Thus, the stage is set for field trials of this active remote sensing technique, which has an enhanced ability to detect biomass in the z dimension.

Modeling seaweed growth presents challenges not encountered with phytoplankton because of the enormous range in spatial and temporal scale encompassed by their biology. G.A. Jackson (pers. comm.) notes that models of phytoplankton growth usually assume average properties of large numbers of individuals. Nutrient distributions and light fields operate at scales larger than those of the individual cell. Models of phytoplankton growth and spatial distributions must include terms for physical processes such as advective transport, turbulent mixing, light absorption and scattering, and biological processes such as cellular nutrient uptake and release. The high concentrations of phytoplankton cells, typically 10^2-10^5cells/l, make it possible to conduct laboratory experiments that average over the responses of many individuals. The fast specific growth rates of microalgae, on the order of 1/d, can cause rapid changes in the environment that are relatively easy to monitor. Models are frequently used to relate biological responses measured in the laboratory to physical processes to predict the temporal and spatial distributions of phytoplankton.

By contrast (G.A. Jackson, pers. comm.), individual seaweeds can be large compared to important physical scales. Further, slower specific growth rates, typically <0.05/d, and longer turnover times, on the order of months to a year, give seaweeds the capacity to integrate environmental properties. For example, many kelp species can store nitrogen taken from seawater at times of high-nutrient and low-light availability, and use the nitrogen to meet growth demands during times of low-nutrient, high-light availability

(Mann, 1982). Similarly, a single plant can simultaneously have tissue at the surface exposed to full sunlight in water depleted of nutrients and tissue below the pycnocline at irradiances $<I_c$ in water with high nutrient concentrations. Seaweed tissues can be shaded by the surface segments of the same individual, by adjacent plants, and by suspended materials. Thus, an individual plant must simultaneously react to a range of conditions; the net response is the sum of these reactions. Laboratory experiments on large seaweeds are typically performed on separated parts of the plant because of the difficulty of working with the whole organism. Advection and turbulence are not important in distributing seaweeds but are important in renewing nutrients, enhancing boundary-layer transport, tearing seaweeds from their benthic attachments and in pulling seaweeds away from the ocean surface. A prime task of a seaweed model is to relate the workings of the different parts of a plant, especially under the range of environmental conditions that an individual plant experiences.

Jackson (1987) developed a bio-optical production model which describes the biomass and production of *Macrocystis pyrifera* over a solar year as a function of environmental properties which affect primarily the irradiance available for photosynthesis. These properties include solar angle, turbidity, spacing between plants, bottom depth, latitude, and photosynthetic response (P_{max} & I_k). The model considers the kelp plant as the sum of its fronds, and each frond is characterized by its length and its age. At the beginning of a model day, the biomass distribution of each frond is calculated along its length.

Results for a standard set of conditions (latitude 33°N, 3 m plant spacing, k = 0.115/m, and 12 m depth) yield a peak daily P_g of almost 6 gC/m^2•d, a peak daily P_n of almost 3 gC/m^2•d, and a peak μ of about O.022/d. Annual P_g for this case is 1567 gC/m^2; annual P_n is 537 gC/m^2. These values are comparable to those from field measurements. The size and timing of biomass and production peaks are affected by changes in the terms describing the light field, with peaks usually occurring later in the year for more adverse circumstances. In higher latitudes, the seasonal variation is so extreme that the plant could not last the year at 53°N in 12 m of water, although it can survive in shallower water.

Ramus (1990) modelled a bio-optical production parameter called photon growth yield (*PGY*), the growth (or the time-integrated) analog of quantum yield for photosynthesis (ϕ). Here, $PGY = \mu/I_a$, where μ is the C-specific growth rate and I_a is the light absorbed. I_a is the product of I_o, the irradiance incident to the plant, and a_c, the optical absorption cross section in this case normalized to C biomass. The term a_c is a mathematical construct, i.e., C has no real absorption cross section. Nevertheless, the term is very useful ecologically, namely in predicting light capture efficiency for growth, i.e., for *PGY*. For example, the *PGY* was predicted as a function of an increasing scale of available light for two chlorophytes, *Codium* and *Ulva*, and the values were compared. *Ulva* has a relatively low but uniform *PGY* for all I_o, whereas *Codium* has a relatively high *PGY* at low I_o which decreases rapidly thereafter. Thus, the two species utilize the natural irradiance scale in very different ways.

EXPLORATION

Seaweed productivity is known largely from temperate latitude coasts which are accessible to marine biologists. Habitats which have been inaccessible to marine biologists in the past, e.g., polar seas, the deep benthos, and subtropical gyres, are poorly known, but now are gaining increasing attention.

Large areas of kelp associated with a diverse invertebrate fauna have been discovered in the arctic, notably the north coast of Alaska and the Canadian High Arctic. These waters have prolonged periods of darkness because of high latitude, ice cover, and turbidity during open water periods. I_o seldom exceeds 5 μmol photons/m$^2 \bullet$s during ice-cover and 200 μmol photons/m$^2 \bullet$s during open water months. Water temperatures vary from 2°C to -2°C. The biomass dominant (>90%) in these areas is *Laminaria solidungula*, which lends itself to measurements of whole-plant photosynthetic rates (^{14}C-incubations in plastic bags) or measurements of linear growth rates (the hole- punching technique). In the Canadian High Arctic, Chapman and Lindley (1981) determined annual productivities of about 20 gC/m^2, roughly equivalent to phytoplankton production in open water. Growth was greatest in late winter and early spring, and was correlated with the concentration of DIN.

In an area known as the Boulder Patch in the Alaskan Beaufort Sea, Dunton et al. (1982) estimated annual productivities to be about 7 gC/m^2, but under substantially different light regimes than above. Here, kelps experience nine months of complete darkness caused by a turbid ice canopy, and they rely on carbohydrate reserves to complete over 90% of their annual linear growth. Production was best predicted by H_{sat} (time during which $I_o \geq I_k$) rather than by the total quantity of photons received over a growing season. This conclusion is consistent with the highly significant relationship between H_{sat} with linear growth and percent carbon (Dunton, 1990) as originally proposed by Dennison and Alberte (1985).

The use of manned deep submersibles for research (Earle, 1985) has dramatically extended bathymetric records for seaweed distibutions and productivities. Perhaps the most salutary example comes from an expedition of the Harbor Branch Oceanographic Institution's submersible *Johnson-Sea-Link I* to a seamount off San Salvador Island, Bahamas (Littler et al., 1985; Littler et al., 1986). The submarine traced a vertical transect up the seamount from a trench at 520 m depth to the top at 81 m depth. The optical properties of the water column were those of Jerlov type I, the clearest oceanic water, and the diffuse attenuation coefficient (k) was determined to be 0.0453/m. During the upward transect, a crustose coralline rhodophyte was found in abundance growing attached at a depth of 268 m. This plant represents the deepest known macrophyte population that has been directly observed, photographed, and collected.

Percent cover of this undescribed crustose coralline as well as other species in this transect were determined from videotape and still photographs. I_{268m} was determined to be 0.0005 full sunlight or 0.015 - 0.025 μmol photons/m$^2 \bullet$s. Autotrophic competency for the crustose coralline rhodophyte was established from samples brought to the surface and O_2-exchange rates measured at 20 μmol photons/m$^2 \bullet$s, or approximately 1% surface irradiance. At this irradiance, productivity rates were 0.43 mgC/g organic dry wt$\bullet$h, comparable to rates measured for shallow-water crustose corallines at the same irradiance.

The seaweed vegetation of the North Atlantic subtropical gyre includes two holopelagic phaeophyte species, *Sargassum natans* and *S. fluitans*. Although these plants are characteristic of the Sargasso Sea, they are broadly distributed by surface currents between the Sargasso Sea, the Caribbean, the Gulf of Mexico, and the Gulf Stream. An estimated 7-10 million tons of *Sargassum* occurs in the Sargasso Sea alone, of which 90% is *S. natans* and 10% is *S. fluitans* (Butler and Stoner, 1984). The historical notion is that the *Sargassum* vegetation propagates indefinitely by vegetative fragmentation within the bounds of the Sargasso Sea. Primary productivity of the gyre is characteristically low due to limited vertical flux of nutrients to surface waters. Biomass turnover times for *Sargassum* have been estimated to be 10-100 years, which seems an appropriate value for these nutrient-poor waters.

LaPointe (1986) conducted growth enrichment studies utilizing *in situ* cage cultures and a shipboard flowing seawater culture system with whole-plant populations of pelagic *Sargassum* in the western Sargasso Sea and in the Straits of Florida. Control plants showed surprisingly high growth rates, in the range of 0.031-0.045 biomass doublings/d, or 11-17 turnovers/y. Enrichment with 0.2 mM NO_3^- or NH_4^+ did not increase growth rates; however, enrichment with 0.2mM PO_4^{3-} nearly doubled growth rates. These data indicate that pelagic *Sargassum* could account for at least 10% of the total productivity for the Sargasso Sea, much higher than previous estimates. Also, these data suggesting P-limited growth and productivity of *Sargassum*, are particularly interesting as they support the view held by most geochemists that P availability limits net organic production in the sea.

LaPointe (pers. comm.) demonstrated that boundary current circulation of pelagic *Sargassum* through neritic waters provides a significant source of "new" nutrients which support transient periods of high primary production and biomass production during transport through the Sargasso Sea. Enhanced productivity occurs from enrichment in the Florida Straits and Gulf Stream where the highest levels of tissue N and P also occurred. Mesoscale nutrient sources generally available in neritic waters include those associated with shelf-break upwelling, riverine inputs to coastal waters, benthic nutrient regeneration, and cold-core rings.

NUTRIENT LIMITATION AND CULTURAL EUTROPHICATION

Ecosystems of the world's shorelines are receiving elevated loadings of nutrients as a consequence of human activities; hence, the term "cultural eutrophication." These ecosystems, which include estuaries, shallow bays, sounds and lagoons, function as nursery areas for shellfish and finfish, stopovers for migrating birds, and sites of biogeochemical cycling. The sources of nutrients are the application of fertilizers and wastewater disposal in watersheds, and industrial emissions to the atmosphere. The loading of N and P to coastal aquatic environments even exceed those to fertilized agroecosystems (Nixon et al., 1986). This increased nutrient loading from anthropogenic sources is pervasive around the globe, and and is probably changing the structure and function of shallow coastal ecosystems.

Blooms of seaweeds in coastal waters have become more commonplace; however, there is not always conclusive evidence linking the blooms to anthropogenic nutrient enrichment. A body of evidence is, nevertheless, accumulating. For example, the proliferation of the chlorophyte *Cladophora prolifera* is a dramatic change in the oligotrophic inshore waters of Bermuda during the last 25 years, and today forms extensive, unattached, drift mats 5-100 cm thick over tens to hundreds of acres (LaPointe and O'Connell, 1989). The *Cladophora* mats are comprised of individual, spherical plants up to 7 cm in diameter, which fragment easily and are dispersed through Bermuda's inshore waters by wind-driven and tidal currents. Decomposition at the bottom of dense mats results in anoxia, which reduces the diversity of the infaunal and epifaunal benthic species, including the commercial calico clam.

Both N- and P-enrichments as nighttime pulses decreased the biomass doubling times of *Cladophora* from 100 days in control environments to 14 days in N+P-enrichments. Nutrient enrichment also increased P_{max} from 0.5 mgC/g dry wt•h in controls to 1.0 mgC/g dry wt•h in N+P-enrichments. Tissue C:N, C:P, and N:P ratios of unenriched *Cladophora* were elevated levels (25, 942, and 49, respectively) that suggest limitation by both N and P but primary limitation by P. Pore-waters under *Cladophora* mats had reduced salinities, elevated concentrations of NH_4^+, and high N:P ratios (= 85), suggesting that N-rich groundwater seepage enriches the algal mats. The alkaline phosphatase capacity of *Cladophora* was high compared to other macroalgae in shallow Bermudan waters, and its capacity was enhanced by N-enrichment and suppressed by P-enrichment. Because the productivity of *Cladophora* is nutrient-limited in the shallow Bermudan waters, enhanced growth and increased biomass of the species result from cumulative seepage of N-rich groundwaters coupled with efficient utilization and recycling of dissolved organophosphorous compounds.

A similar occurrence was documented for a temperate latitude system, notably Waquoit Bay on the south shore of Cape Cod, MA (Valiella et al., 1990). Fast growing opportunist seaweeds, e.g., *Cladophora* and *Gracilaria*, thought to be favored by high nutrient loadings, have taken over much of the bay floor that was previously dominated by eelgrass meadows. In shallow embayments, such as Waquoit Bay, seaweeds can grow profusely at shoal depths where light is not limiting, and may carry out as much as 60% of the primary production in these systems.

Layers of *Cladophora* up to 75 cm thick were measured in Waquoit Bay. Such layers contain large pools of nutrients, estimated very conservatively at 0.2 - 11.2 x 10^4 kg N for the entire Bay. The estimate does not include N from other species, and brackets the estimate for annual N loading to the entire Bay, i.e., 2.7 x 10^4 kg N/y, mostly from anthropogenic sources. The explanation for low residual concentrations of dissolved nutrients in this and similar systems may be that nutrients are stored in macroalgal biomass. The nutrients may be released at the time of year when macroalgae senesce and decay. Nutrient enrichment has so fostered macroalgal growth in Waquoit Bay that macroalgae, in turn, may have become the governors of nutrient dynamics.

Increased nutrient loading may cause fundamental changes in the trophic-dynamic structure of shallow coastal waters as Waquoit Bay (Valiella et al., 1990). Phytoplankton and seaweeds may compete seasonally for nutrients. If the seaweeds control nutrient availability, phytoplankton abundance and productivity may be depressed during times when macroalgae grow, and phytoplankton ought to bloom when macroalgae senesce; this hypothesis needs testing. The eelgrass *Zostera marina* and seaweeds such as *Polysiphonia* were formerly widely distributed in Waquoit Bay, but aerial photographs from the past several decades show that there has been a sharp decrease in the distribution of eelgrass beds in recent years. Nutrient loading increases growth of light-intercepting epiphytes on leaves of eelgrass, many of which are macroalgae. Eelgrass production is not nutrient-limited in these shallow systems, but rather light-limited, so these epiphytes have serious consequences on seagrass growth by intercepting the light. The changes in vegetation brought about by nutrient loading have thoroughly altered the rest of the trophic web in Waquoit Bay. The large macroalgal biomass itself respires, and also produces large quantities of DOM and POM. Both result in high rates of BOD, hypoxia, and anoxia, and increased fish and invertebrate mortality. Not only are animal abundances reduced, but the species composition is radically altered to favor those which can survive on top of the macroalgal canopy or in reduced O_2 conditions.

Research on vectors for the loading of nutrients to coastal waters has focused primarily on deeper estuaries in which flow from rivers and estuaries dominate water budgets. For example, rivers and streams contribute 74% of the N to the Baltic, while atmospheric deposition, sewers, and groundwater contribute only 13%, 10%, and 3%, respectively. For shallow coastal systems which are underlain by coarse, unconsolidated sands of glacial or marine origin, groundwater flow is being recognized as a major vector of nutrient loading (Valiella et al., 1990). Most of the fresh groundwater discharging into coastal waters does so very near the shore. The importance of groundwater flow is not so much the magnitude of flow rates, but rather the high nutrient concentrations of groundwater relative to those of the receiving seawater, ranging to 2 - 3 orders of magnitude greater. For coastal bays and lagoons of New England, groundwater accounts for 71 - 97% of the total N inputs.

Anthropogenic activity accounts for a large percentage of nutrient loadings into groundwater. For example, LaPointe et al. (1990) performed a one-year study to determine the effects of on-site sewage disposal systems (septic tanks) on nutrient relations of limestone groundwaters and nearshore surface waters in the Florida Keys. Wells were installed on inhabited lots and a control site wildlife refuge, and concentrations of dissolved inorganic nitrogen (DIN) and soluble reactive phosphate (SRP), temperature and salinity were monitored in the groundwaters. Significant nutrient enrichment (up to 5000-fold) occurred in groundwaters contiguous to septic tanks; DIN was enriched an average of 400-fold and SRP an average of 70-fold compared to control groundwaters. Ammonium was the dominant nitrogenous species and its concentration ranged from a low of 0.77 μM in control groundwaters to 2.75 mM in septic-tank-enriched groundwaters. Nitrate + nitrite ranged from 0.05 μM in control groundwaters to 2.89 mM in enriched groundwaters. SRP ranged from 30 nM to 107 μM under similar tests. N:P ratios of enriched groundwaters were consistently >100 and increased with increasing distance from the septic tank, suggesting significant, but incomplete, adsorption of SRP by subsurface flow through carbonate substrata. Nutrient concentrations of groundwaters also varied seasonally and were approximately two-fold higher in winter compared to summer. In contrast, nutrient concentrations in surface water was two-fold higher in summer than in winter.

Direct measurement of the flow rate of subsurface groundwater indicated that tides and increased groundwater recharge enhanced flow some two-fold and six-fold, respectively. Accordingly, the observed seasonal coupling of septic-tank derived nutrients from groundwaters to surface waters is maximum during summer because of seasonally maximum tides and increased hydraulic head during the summer wet season. The yearly average benthic flux of anthropogenic DIN into contiguous canal surface waters is 55 mmol/m^2•d, a value some five-fold greater than the highest rate of benthic N-fixation measured in carbonate-rich tropical marine waters.

There is an active debate over whether N or P limits primary production in marine systems (summarized by Howarth, 1988), including seaweed productivity. The potential importance of P relative to N is implicated in global surveys of tissue C:N:P data. For example, Atkinson and Smith (1983) reported a mean C:N:P ratio of 700:35:1 for benthic marine plants. To a first approximation, N:P ratios <10 indicate N-limitation and N:P ratios >30 indicate P-limitation. LaPointe et al. (1991) surveyed populations of frondose epilithic seaweeds from tropical waters over carbonate sediments and from temperate waters over siliciclastic sediments. Those seaweeds from carbonate-rich tropical waters were significantly depleted in P relative to C and N when compared to seaweeds from temperate siliciclastic waters. Percent C and N dry weight contents were similar between tissues from the siliciclastic and carbonate-rich waters (means of 22.6% vs 20.1%, and

1.0% vs 1.2%, respectively), but P levels were two-fold lower (0.15% vs 0.07%) in the carbonate-rich waters. Accordingly, the molar C:N tissue ratios were comparable between seaweeds from the siliciclastic and carbonate-rich waters (mean of 29.2 vs 23.1), whereas large differences were observed for the C:P (mean of 430 vs 976) and N:P ratios (mean of 14.9 vs 43.4). In addition, alkaline phosphatase activity was low and often undetectable in the seaweeds from siliciclastic waters (mean 7.3 μmol PO_4^{3-} released/g dry wt•h) compared to seven-fold higher rates (52.5 μmol PO_4^{3-}/g dry wt•h) for seaweeds from carbonate-rich waters. Seawater taken adjacent to seaweeds from the carbonate-rich waters contained relatively high concentrations of DIN with low concentrations of SRP, and showed elevated N:P ratios (mean = 36), compared to siliciclastic waters (mean <3). These data support the notion that N availability limits seaweed productivity in temperate latitude siliciclastic waters, and, by contrast, that P availability limits seaweed productivity in tropical latitude carbonate-rich waters.

In general, seaweeds have been regarded as obligate photoautotrophs, although there has been speculation that some species are capable of heterotrophic growth, and that the heterotrophic growth is ecologically meaningful. The effect of external glucose (51 mM) and acetate (17 mM) on growth and photosynthetic capacity of the chlorophyte *Ulva* was tested in laboratory cultures over 41 days in the dark and dim light ($\approx$ 1 μmol photons/m^2•s) at 7-8°C (Markager and Sand-Jensen, 1990). Both organic-C sources had a significant positive effect on growth rate, chlorophyll content, and ϕ, both in the dark and in dim light. The C-gain from hetertorophic uptake was low and only allowed *Ulva* to maintain a μ of 0.005/d compared to 0.06 - 0.1/d at higher irradiances. Plants without added glucose and acetate lost pigmentation and photosynthetic competence after 41 days in the dark or in dim light. The data suggest that the ecological significance of heterotrophic uptake is to allow *Ulva* to survive prolonged low-light conditions with an intact photosynthetic apparatus. This ability becomes functional during burial in sediments or high latitude winters, during which time the plant can perform heterotrophic uptake from the sediments.

The temporal patchiness of nutrient availability has been shown to affect growth rates in seaweeds (Ramus and Venable, 1987). The growth rates of two chlorophytes, *Codium* and *Ulva*, were compared in response to varied, but periodic, NH_4^+ enrichments (pulses). The species were chosen to contrast radically different morphology but similar photochemistry. Pulse frequency and pulse duration were varied independently; however, the mass balance of nutrient supply was equivalent in each treatment. The growth rate of *Codium* varied neither as a function of pulse frequency nor duration; the growth rate of *Ulva* varied with pulse frequency but not duration. The data were related to the morphology and physiological characteristics of the species, and discussed in the context of the *function form* hypothesis (Littler and Littler, 1980). *Ulva* can use transiently high NH_4^+ concentrations and translate them into transiently high growth rates; such attributes contribute to its role as ruderal species or a variability tracker. In contrast, *Codium's* life form allows storage of transient pulses of NH_4^+ but relatively low growth rates, thereby contributing to its role as a persistent species or a variability integrator.

THE FATE OF SEAWEED CARBON

It is currently thought that more energy and materials flow through detritus food webs than through grazer food webs (reviewed by Mann, 1988). Of the total primary production of an ecosystem, more is transmitted to other trophic levels from dead decomposing plant tissue than from living tissue consumed by the grazer. These general principles yield to questions of the relative importance of plant detritus derived from seagrasses, seaweeds, and phytoplankton in coastal systems. An intuitive observation might be based on an index of food quality, e.g., tissue C:N atom ratio, where the ratio is inversely proportional to food quality. The greater is the tissue N-content relative to C, the greater is the quality, and the more likely the tissue will enter the grazing food web. The food quality index partitions between C-rich structural materials *vs* N-rich photosynthetic apparatus. Phytoplankton have a low C:N (ideally near the Redfield ratio of 6.7) and are mostly grazed. Seagrass and seaweed tissues have higher C:N atom ratios than phytoplankton, although the range is enormous (C:N > 50, Atkinson and Smith, 1983). The detritus food chain implies reworking of the tissue by microbes and microzooplankton, solubilizing the C and enriching it with N. Historically, great emphasis has been placed on studying the fate of vascular plant detritus, and the fate of macroalgal detritus is less well understood. The most complete picture to date comes from the study of kelp ecosystems.

The kelp beds of the Cape Peninsula, South Africa, are dominated by *Ecklonia maxima* and *Laminaria palida*. They produce about 500 gC/m²•y of POM and 250 gC/m²•y of DOM. Phytoplankton production in the area also was about 500 gC/m²•y (Newell and Field, 1983; Wulff and Field, 1983). If the system were closed, the combined phytoplankton and kelp production would just about supply the nutritional needs of the filter-feeding invertebrates. The mussels had enzyme systems capable of digesting the carbohydrates contained in the kelps. Using a double-labeling technique, Stuart et al. (1982) showed that the mussels in the kelp bed derived much of their tissue C from kelp material rather than from colonizing bacteria. These Cape Peninsula beds are adjacent to the Benguela upwelling system. During upwelling the filter feeders receive predominantly kelp detritus, and during downwelling they receive mostly phytoplankton C. A simulation model showed that the relative importance of phytoplankton and detritus to the filter feeders depends on the frequency of upwelling and on the rate of water movement through the kelp beds (Wulff and Field, 1983).

A separate study of the fate of kelp biomass cast up on beaches near Cape Town (Koop et al., 1982*a,b*) showed that invertebrates consumed 74% of the kelp within eight days and bacteria consumed the remaining 26%. Invertebrate feces amounted to 67% of the kelp C, and, in turn, were metabolized by bacteria in the sand. Overall, 100 gC of kelp yielded 23-28 g of bacterial C available for higher trophic levels, but the startling finding was that this bacterial biomass contained 94% of the N originally present in the kelp.

The trophic role of kelp-derived C in a wide range of marine organisms was assessed by a natural experiment in the Aleutian Islands (Duggins et al., 1989). Here, the mid- and low-intertidal zones are dominated by kelps belonging to the genera *Laminaria* and *Alaria*, which exist in a refuge above and below the foraging range of urchins. Rocky subtidal habitats at islands with sea otters are characterized by a low biomass of urchins and large stands of understory and surface-canopy kelps; however, islands where sea otters have not become reestablished are characterized by large biomasses of urchins and few kelps. This nearshore community variation allowed an assessment (by comparison) of the importance of kelp C to secondary production. Two analyses were used: 1) the introduction of two suspension feeders from a common source, a mussel and a barnacle, and 2) age-size relations from naturally occurring mussels. Mussels in kelp-dominated habitats grew from two (subtidal) to four (intertidal) times as fast as those in urchin-dominated habitats. Likewise, barnacles grew up to five times as fast in kelp-dominated habitats. For year classes 2-5, mussels were significantly larger at islands with substantial subtidal kelp forests.

Analyses of stable C isotope ratios ($\delta^{13}C$) revealed the extent to which consumers were utilizing kelp-derived organic C, and if such use differed among islands as predicted. In these arctic systems, there are fundamentally only two sources of C, namely phytoplankton and kelp; thus, stable C isotope signature analyses are much less ambiguous than they are in temperate latitude systems where the range of C sources is much greater. The mean $\delta^{13}C$ values were $-24\%_{oo}$ for phytoplankton and $-17.7\%_{oo}$ for kelps, the kelps being enriched for $\delta^{13}C$ relative to phytoplankton. Eleven benthic particle consumers were analyzed; six suspension feeders, two detritivores and three predator taxa. The consumer animals from kelp-dominated islands were consistently more enriched in ^{13}C than those animals from urchin-dominated islands, thus supporting the hypothesis. On the basis of a simple mixing model, primary consumers at kelp-dominated islands average, conservatively, $58.3\%_{oo}$ kelp-derived C, whereas at urchin-dominated islands they average only $32\%_{oo}$, despite all consumers being collected for $\delta^{13}C$ analyses in midsummer, the peak period of phytoplankton abundance. Further, isotopically enriched signatures were obtained for organisms at kelp-dominated islands other than the benthic particle feeders, including mysids, rock greenling and pelagic cormorants. Thus, there is a strong trophic link between kelps and a wide range of organisms of varied feeding strategies and trophic levels, extending beyond the obvious kelp-grazer-predator food chain.

In a similar, natural arctic experiment, Dunton and Schell (1987) showed the importance of kelp C (-13.6 to $-16.5\%_{oo}$) relative to phytoplankton C (-25.5 to $-26.5\%_{oo}$). Animals that showed the greatest assimilation of kelp C ($\geq 50\%$) included macroalgal herbivores (gastropods and chitons, -16.9 to $-18.2\%_{oo}$), a non-selective suspension feeder (an ascidian, $-19\%_{oo}$) and a predatory gastropod ($-17.6\%_{oo}$). Animals which showed the least assimilation of kelp C into body tissues ($\leq 7\%_{oo}$) included selective suspension-feeders (hydroids, soft corals, and bryozoans, -22.8 to $-25.1\%_{oo}$). Distinct seasonal changes in the $\delta^{13}C$ values of several animals indicated an increased dependence on kelp C during the dark winter period when phytoplankton were absent. Up to 50% of the body C of mysid crustaceans, which are key prey species for birds, fishes, and marine mammals, was composed of C derived from kelp detritus during the ice-covered period.

In the seagrass meadows of the northwest Gulf of Mexico, small epiphytic seaweeds were shown to have higher productivites, greater palatability and a more important trophic role than the seagrasses on which they grow (Kitting et al., 1984). Epiphytic seaweeds had mean productivities of 7 mgC/g•h, 5 mgC/g•h for seagrass and the epiphytic seaweeds accounted for 50 - 126% of the seagrass areal productivity. Feeding behavior of common invertebrates was quantified by remote sensing, i.e., by recording underwater camera and hydrophones. Some species have distinctive feeding sounds, and feeding rates can be quantified from acoustic recordings. These techniques established intensive nighttime feeding on epiphytic seaweeds, by several taxa of shrimp and snails. Stable C isotope ratio analyses ($\delta^{13}C$) established that the shrimp and snails were consuming the epiphytic seaweed C rather than seagrass C.

It is well known that pelagic *Sargassum* plays a similar role in the North Atlantic, serving as an bioreactor surface for small epiphytic seaweeds which are grazed by associated fauna. Pelagic *Sargassum* is also important to the material flux to the deep sea (Fowler and Knauer, 1986). Patches of *Sargassum* were observed from the research submersible *ALVIN* at station Deep Ocean Station 2, at a depth of 3600 m on the continental rise and beneath the axis of the Gulf Stream, south of Cape Cod, MA (Grassle and Morse-Porteous, 1987). These patches are a form of disturbance favoring relatively opportunistic species of polychaetes and spionids. Clearly seaweed C is very important in many trophic-dynamic relations.

CONCLUSIONS

Considerable progress has been made in the exploration of habitats for which little data on the dynamics of seaweed production had previously existed. These habitats include subtropical gyres, polar seas and the euphotic limits of the sublittoral zone. The latter has been redefined by excursions of a manned deep submersible to a seamount in clear oceanic water.

Eutrophication via groundwater discharge to coastal waters has increased nuisance blooms of seaweeds on the global scale. The blooms are usually monotypic and are composed of species with high intrinsic growth rates. In these blooms, seaweed production exceeds herbivore consumption, which results in major changes in community structure and in the control of biogeochemical cycles. Seaweed productivity in waters over siliciclastic sediments is usually controlled by N-nutrient availability. However, a major departure from conventional wisdom is that seaweed productivity in waters over carbonate-rich sediments is usually controlled by P-nutrient availability.

The fate of seaweed mass and carbon in trophic-dynamic transfers is quite dependent on tissue quality, as indexed by C:N atom ratio. Some biomass is consumed directly, but most is reworked by microbes and protozoans, and enters the food web as DOM and enriched detritus. Stable C isotope analyses have revealed that in some coastal systems as much as 60% of the C consumed is seaweed-C during peak abundances of phytoplankton.

Global productivity estimates for seaweeds remain sketchy due to the anisotropy of the seaweed scale, i.e., in the linear, mass, and temporal dimensions. Properties such as emersion, water turbidity, community structure, and substrate relief further complicate the scale. The anisotropy in scale introduces large quantitative uncertainties in the measurement of standing crops and C-fixation rates using standard techniques, i.e., those which were developed for terrestrial plants and phytoplankton. Sampling by remote radiometers which collect either reflected sunlight or induced fluorescence has met with some success, but needs further refinement. Modeling has met with success as well, but the number of applications is few. Clearly, the anisotropy of the seaweed scale requires the development of sampling tools designed specifically for the seaweeds. The critical need is to think of seaweeds in functional terms other than as terrestrial plants or as phytoplankton.

REFERENCES

Atkinson, M.J., and Smith, S.V., 1983, C:N:P ratios of benthic marine plants, *Limnol. Oceanogr.*, 28:568.

Belsher, T., Meinesz, A., Lefevre, J.L., and Boudouresque, C-F., 1988, Simulation of SPOT satellite imagery for charting shallow-water benthic communities in the Mediterranean, *P.S.Z.N.I.: Mar. Ecol.*, 9:157.

Bunt, J.S., 1975, Primary productivity of marine ecosystems, *in* "Primary Productivity of the Biosphere," H. Leith and R.H. Whittaker eds., Springer Verlag, New York.

Butler, J.N., and Stoner, A.W., 1984, Pelagic *Sargassum*: has its biomass changed in the last 50 years? *Deep-Sea Res.*, 31:1259.

Chapman, A.R.O., and Lindley, J.E., 1981, Productivity of *Laminaria solidungula* J. Ag. in the Canadian High Arctic: A year round study, *in* "Proc. 10th International Seaweed Symposium," T. Levring ed., Walter de Gruyter, Berlin.

Dame, R., Chrzanowski, T., Bildstein, K., Kjerfve, B., McKellar, H., Nelson, D., Spurrier, J., Stancyk, S., Stevenson, H., Vernberg, J., and Zingmark, R., 1986, The outwelling hypothesis and North Inlet, South Carolina, *Mar. Ecol. Prog. Ser.*, 33:217.

Dayton, P.K., and Tegner, M.J. 1984, Catastrophic storms, El Niño, and patch stability in a southern California kelp community, *Science*, 224:283.

Dayton, P.K., Currie, V., Gerrodette, T., Keller, B.D., Rosenthal, R., and Ven Tresca, D., 1984, Patch dynamics and stability of some California kelp communities, *Ecol. Monogr.*, 54:253.

Dennison, W.C., and Alberte, R.S. 1985, Role of daily light period in the depth distribution of *Zostera marina* (eelgrass), *Mar. Ecol. Prog. Ser.*, 25:51.

Duggins, D.O., Simenstad, C.A., and Estes, J.A., 1989, Magnification of secondary production by kelp detritus in the coastal marine ecosystems, *Science*, 245:170.

Dunton, K.H., 1990, Growth and production in *Laminaria solidungula*: Relation to continuous underwater light levels in the Alaskan High Arctic, *Mar. Biol.*, 106:297.

Dunton, K.H., Reimnitz, E., and Schonberg, S., 1982, An arctic kelp community in the Alaskan Beaufort Sea, *Arctic*, 35:465.

Dunton, K.H., and Schell, D.M., 1987, Dependence of consumers on macroalgal (*Laminaria solidungula*) carbon in an arctic kelp community: $\delta^{13}C$ evidence, *Mar. Biol.*, 93:615.

Earle, S.A., 1985, Equipment for conducting research in deep water, *in*: "Handbook of Phycological Methods. Ecological Field Methods: Macroalgae," M.M. Littler, and D.S. Littler, eds., Cambridge University Press, Cambridge.

Fowler, S.W., and Knauer, G.A., 1986, Role of large particles in the transport of elements and organic compounds through the oceanic water column, *Prog. Oceanogr.*, 16:147.

Gerard, V.A., 1984a, The light environment in a giant kelp forest: Influence of *Macrocystis pyrifera* on spatial and temporal variability, *Mar. Biol.*, 84:189.

Gerard, V.A., 1984b, Physiological effects of El Niño on giant kelp in southern California, *Mar. Biol. Lett.*, 5:317.

Grassle, J.F., and Morse-Porteous, L.S., 1987, Macrofaunal colonization of disturbed deep-sea environments and the structure of deep-sea benthic communities, *Deep-Sea Res.*, 12:1911

Greene, R.M., and Gerard, V.A., 1990, Effects of high-frequency light fluctuations on growth and photoacclimation of the red alga *Chondrus crispus*, *Mar. Biol.*, 105:337.

Howarth, R.W., 1988, Nutrient limitation of net primary production in marine ecosystems, *Ann. Rev. Ecol.*, 19:89.

Jackson, G.A., 1987, Modelling the growth and harvest yield of the giant kelp *Macrocystis pyrifera*, *Mar. Biol.*, 95:611.

Jackson, G.A., and Winant, C.D., 1983, Effect of a kelp forest on coastal currents, *Cont. Shelf Res.*, 2:75.

Kitting, C.L., Fry, B., and Morgan, M.D., 1984, Detection of inconspicuous epiphytic algae supporting food webs in seagrass meadows, *Oecologia*, 62:145.

Koop, K., Newell, R.C., and Lucas, M.I., 1982a, Biodegradation and carbon flow based on kelp (*Ecklonia maxima*) debris in a sandy beach microcosm, *Mar. Ecol. Prog. Ser.*, 7::315.

Koop, K., Newell, R.C., and Lucas, M.I., 1982b, Microbial regeneration of nutrients from the decomposition of macrophyte debris on the shore, *Mar. Ecol. Prog. Ser.*, 9:91.

LaPointe, B.E., 1986, Phosphorous-limited photosynthesis and growth of *Sargassum natans* and *Sargassum fluitans* (Phaeophyceae) in the western North Atlantic, *Deep-Sea Res.*, 33:391.

LaPointe, B.E., and O'Connell, J., 1989, Nutrient-enhanced growth of *Cladophora prolifera* in Harrington Sound, Bermuda: Eutrophication of a confined phosphorous-limited marine ecosystem, *Est., Coast. Shelf Sci.*, 28:347.

LaPointe, B.E., O'Connell, J., and Garrett, G.S., 1990, Nutrient couplings between on-site sewage disposal systems, groundwaters, and nearshore surface waters of the Florida Keys, *Biogeochem.*, 10:289.

LaPointe, B.E., Littler, M.M., and Littler, D.S., 1991, N:P availability to marine macroalgae in siliciclastic versus carbonate-rich coastal waters, *Estuaries*, in press.

Littler, M.M., and Littler, D.S., 1980, The evolution of thallus form and survival strategies in benthic marine macroalgae: Field and laboratory tests of a functional form model, *Am. Nat.*, 116:25.

Littler, M.M., Littler, D.S., Blair, S.M., and Norris, J.M., 1985, Deepest known plant life discovered on an uncharted seamount, *Science*, 227:57.

Littler, M.M., Littler, D.S., Blair, S.M., and Norris, J.B., 1986, Deep-water plant communities from an uncharted seamount off San Salvador Island, Bahamas: Distribution, abundance, and primary productivity, *Deep-Sea Res.*, 33:881.

Littler, D.S., and Littler, M.M., 1987, Rocky intertidal aerial survey methods using helicopters, *Photo-Interprétation*, 1:31.

Mann, K.H., 1982, Ecology of Coastal Waters, A Systems Approach, U. Calif. Press, Berkeley.

Mann, K.H., 1988, Production and use of detritus in various freshwater, estuarine, and coastal marine ecosystems, *Limnol. Oceanogr.*, 33:910.

Markager, S., and Sand-Jensen, K., 1990, Heterotrophic growth of *Ulva lactuca* (Chlorophyceae), *J. Phycol.*, 26:670.

Meulstee, C., Nienhuis, P.H., and Van Stokkom, H.T.C., 1986, Biomass assessment of estuarine macrophytobenthos using aerial photography, *Mar. Biol.*, 91:331.

Newell, R.C., and Field, J.G., 1983, The contribution of bacteria and detritus to carbon and nitrogen flow in a benthic community, *Mar. Biol. Lett.*, 4:23.

Nixon, S.W., Oviatt, C.A., Firthsen, J., and Sullivan, B., 1986, Nutrients and the productivity of estuarine and coastal ecosystems, *J. Limnol. Soc. South Africa*, 12:43.

Ramus, J., 1990, A form-function analysis of photon capture for seaweeds, *Hydrobiologia* 204/205:65.

Ramus, J., and Venable, M., 1987, Temporal ammonium patchiness and growth rate in *Codium* and *Ulva* (Ulvophyceae), *J. Phycol.*, 23:518.

Smith, S.V., 1981, Marine macrophytes as a global carbon sink, *Science*, 211:838.

Stuart, V., Field, J.G., and Newell, R.C., 1982, Evidence for the absorption of kelp detritus by the ribbed mussel *Aulacomya* after using a new ^{13}C-labelled microsphere technique, *Mar. Ecol. Prog. Ser.*, 9:263.

Topinka, J.A., Bellows, W.K., and Yentsch, C.S., 1990, Characterization of marine macroalgae by fluorescence signatures, *Int. J. Remote Sensing*, 11:2329.

Valiella, I., Costa, J., Foreman, K., Teal, J.M., Howes, B., and Aubrey, D., 1990, Transport of groundwater-borne nutrients from watersheds and their effects on coastal waters, *Biogeochem*, 10:177.

Whittaker, R.H., and Likens, G.E., 1975, The biosphere and man, in: "Primary Productivity of the Biosphere," H. Leith and R.H. Whittaker, eds., Springer Verlag, New York.

Woodwell, G.M., Houghton, R.A., Hall, C.A.S., Whitney, D.E., Moll, R.A., and Juers, D.W., 1979, The Flax Pond ecosystem study: The annual metabolism and nutrient budgets of a salt marsh, *in*: "Ecological Processes in Coastal Environments," R.L. Jeffries, ed., Blackwell Scientific, Oxford.

Wulff, F.Y., and Field, J.G., 1983, Importance of different trophic pathways in a nearshore benthic community under upwelling and downwelling conditions, *Mar. Ecol. Prog. Ser.*, 12:217.

PRODUCTIVITY OF ZOOXANTHELLAE AND BIOGEOCHEMICAL CYCLES

Leonard Muscatine and Virginia Weis

Department of Biology
University of California
Los Angeles, CA 90024

INTRODUCTION

Symbiotic dinoflagellates (zooxanthellae) are dominant primary producers in tropical reef communities along with benthic algae (macrophytes), unicellular and filamentous sand algae, turf algae, sea grasses, and phytoplankton (Larkum, 1983). Zooxanthellae are widely distributed and abundant in the cells of foraminiferans, radiolarians, sponges, cnidarians and molluscs. Among the cnidarians, they inhabit true stony corals, soft corals, gorgonians, sea anemones, milleporines, zoanthids, and hydrozoans. Although all of these taxa are represented on coral reefs and contribute to reef productivity, corals are most often used as models for productivity of zooxanthellae. This is because zooxanthellae population densities often exceed 10^6 cells per cm^2 of the surface area of the coral (Muscatine, 1980), corals cover from 10% to 50% of the projected surface area of many reefs (Larkum, 1983), and coral reef communities cover 6×10^5 km^2 of the world's oceans (Smith, 1978). Corals emerge as the source of the most detailed information. Moreover, measurement of coral productivity has now achieved sufficient precision and standardization so that results from a wide range of studies can easily be compared.

There are several contemporary reviews of zooxanthellae productivity (Hatcher, 1988; Barnes and Chalker, 1990; Muscatine, 1990) which cover broad aspects of the subject, but touch only briefly on the role of zooxanthellae in biogeochemical cycles. In this chapter, we delve more deeply into this topic and discuss qualitative and quantitative aspects of daily budgets of photosynthetically fixed carbon, the fate of this fixed carbon in the coral host and in the community, the role of nutrients, and novel adaptations manifested by the host coral animal for increased light harvesting and biogeochemical cycling.

Whereas the earliest Paleozoic corals were solitary and believed to be cosmopolitan (Tasch, 1980), present-day reef building corals are predominantly colonial and exhibit a much narrower global distribution, governed by light and temperature. They are most abundant in the euphotic zone of warm, shallow tropical seas in Western ocean basins between 30° N. Lat. and 30° S. Lat., where the mean annual temperature is greater than 18°C. As a result of this narrow distribution, zooxanthellae productivity affects biogeochemical cycles primarily on a microscale at the level of the host animal, and secondarily on a mesoscale at the level of coral reef communities.

Primary Productivity and Biogeochemical Cycles in the Sea
Edited by P.G. Falkowski and A.D. Woodhead, Plenum Press, New York, 1992

MICROSCALE EFFECTS OF ZOOXANTHELLAE PHOTOSYNTHESIS ON BIOGEOCHEMICAL CYCLES OF CARBON IN REEF CORALS

The capacity of zooxanthellae to influence biogeochemical cycles of carbon and nutrients is potentially the same as that of free-living algae. But, whereas the habitat of free-living phytoplankton is a column of sea water, the habitat of zooxanthellae is a "column" of animal cells. As noted by Hatcher (1988) "...corals combine the advantages of microalgal kinetics and macrophyte structure..." Because of this novel spatial and functional relationship with a heterotroph, important differences emerge.

The customary path of carbon and nutrients (e.g., nitrogen) in an aquatic heterotroph is shown in Fig. 1(a-h). Particulate and dissolved organic carbon and nitrogen are taken up by holozoic feeding or by transport from solution (a). Carbon dioxide and ammonia and some particulate and dissolved organic materials are released via catabolism (b,c). When the fluxes of particulate and dissolved carbon and nitrogen in a non-nitrogen fixing autotroph are intimately merged with those in the heterotroph, new routes of flux emerge. The heterotroph now becomes a sink for inorganic carbon and nitrogen (d), receives organics from the autotroph via translocation (e), retains excretory nitrogen (f), and releases dissolved organic material which is qualitatively and quantitatively modified (g). Moreoever, in the calcifying corals (h), the photosynthesis by symbiotic dinoflagellates increases the rate of calcification by the heterotroph up to 15-fold (Goreau, 1964).

In the past decade, attempts to quantify the various fluxes depicted in Fig. 1, especially those for carbon, and to a certain extent for nitrogen (see reviews of Muscatine, 1990; D'Elia and Wiebe, 1990) have been reasonably successful. Efforts to determine the fate of fixed carbon have resulted in the construction of daily carbon budgets. A representative daily budget of photosynthetically fixed carbon is shown in Fig. 2. The diagram is scaled, and the flow emulates gross carbon fixed and respired, and net carbon used for zooxanthellae growth. The balance is translocated to the animal, respired, and used for the animal's growth. The pool of carbon remaining is used variously in the skeletal carbonate and skeletal organic matrix, and export of DOC, mucus, egesta, and reproductive products. Budgets from representative studies for corals from different habitats are given in Table 1. The character of these independently produced data sets is astonishingly similar. The budgets are normalized to percent of total daily fixed carbon, and those for shallow water corals are averaged to facilitate generalization.

Daily Budgets of Photosynthetically Fixed Carbon in Reef Corals

Carbon fixation by zooxanthellae. Gross photosynthesis for corals in shallow water, reported as daily rates on an areal basis ($X \pm S.D.$) average 221.69 ± 82.5 ug C cm^{-2}d^{-1} (n=5) (Davies, 1977; Porter et al., 1984; Muscatine et al., 1984; McCloskey and Muscatine, 1984; Porter, 1985). This figure is equivalent to about 809 g C m^{-2}y^{-1}, and is similar to values of 774 and 960 g C m^{-2}y^{-1} reported by Muscatine (1990) and Davies (1977). These values, normalized to plane surface area of colonies, would be significantly higher if normalized to their projected surface area. Taking the total area of coral reefs as 6×10^5 km^2, and conservatively estimating the areal coverage of zooxanthellae in corals as about 10% of this area, then at an annual rate of production of 1000 g C m^{-2}y^{-1}, total annual production by coral zooxanthellae amounts to 0.6×10^8 metric tons y^{-1}; this is about 10% of Bunt's (1975) estimate of total global benthic production. If zooxanthellae in all symbiotic taxa are included, this figure could easily double. In any case, zooxanthellae productivity is substantial.

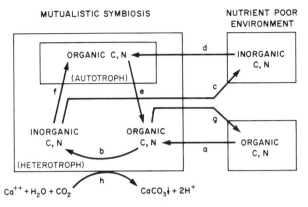

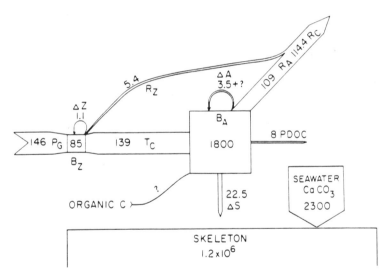

Fig. 1. A model of the potential pathways of carbon and nitrogen flux in a symbiotic reef coral (Modified from Lewis and Smith, 1971).

Fig. 2. Scaled flow path of carbon in light-adapted *Stylophora pistillata*. Fluxes are in units of ug C m^{-2}d^{-1}; biomass values are in units of ug C cm^{-2}. P$_g$ = gross photosynthesis of the coral; ΔZ = growth of zooxanthellae; R$_{z,a,c}$ = respiration of zooxanthellae, animal, or coral; ΔA = growth of animal; ΔS = skeletal growth; P/DOC = particulate and/or dissolved organic carbon; B$_{z,a}$ = biomass of zooxanthellae, animal. (After Falkowski et al., 1984).

Table 1.   Daily Budgets of Photosynthetically Fixed Carbon in Selected Reef Corals

A. Shallow Water

Species	Location	Zooxanthellae				Animal			References
		Photosynthesis	Respiration	Growth	Translocation	Respiration	Growth	Losses	
Stylophora pistillata	Red Sea*	100	3.7	0.7	95.2	74.6	2.4	18.5	Falkowski et al., 1984
Stylophora pistillata	Red Sea*	100	1.2	1.9	96.7	61.3	1.9	33.7	McCloskey and Muscatine, 1984
Pocillopora eydouxi	Guam	100	9.8	0.1	90.1	41.3	0.8	47.9	Davies, 1984
Porites porites	Jamaica	100	21.2	0.8	78.0	26.3	6.3	45.4	Edmunds and Davies, 1986
Pocillopora damicornis	Hawaii	100	8.3	0.3	91.4	67.9	3.5	20.0	Davies, 1991
Montipora verrucosa	Hawaii	100	13.1	0.4	86.5	49.9	3.7	32.4	Davies, 1991
Porites lobata	Hawaii	100	32.9	1.6	65.5	37.7	8.5	19.3	Davies, 1991
"Average" coral shallow water x̄		100	12.8	0.82	86.2	51.2	3.87	31.0	
± S.D.			10.9	0.67	11.0	17.4	2.64	12.3	

B. Shade, deep water

Species	Location	Zooxanthellae				Animal			References
		Photosynthesis	Respiration	Growth	Translocation	Respiration	Growth	Losses	
Stylophora pistillata (3 m. shade)	Red Sea*	100	6.5	2.2	91.2	195.44	0.6	15.5	Falkowski et al., 1984
Stylophora pistillata (35 m.)	Red Sea**	100	1.5	4.0	94.4	131.7	?	?	McCloskey and Muscatine, 1984
Pocillopora damicornis (overcast day)	Hawaii	100	13.1	0.5	86.4	107.3	5.5	-26.4	Davies, 1991

*Nabek, Israel; **Elat, Israel

Respiration by zooxanthellae. The respiration rate of zooxanthellae *in hospite* is difficult to assess as it is coincident with respiration by the animal. Rates have been estimated either from respiration of zooxanthellae *in vitro* (Davies, 1984; 1991), or as a proportion of total respiration based on the proportion of zooxanthellae:animal biomass (Muscatine et al., 1981), or by the difference between the respiration of symbiotic and aposymbiotic specimens (Smith, 1984). Each approach carries with it a particular uncertainty (see Muscatine, 1990; also Edmunds and Davies, 1986). Table 1A shows the values obtained by the first two methods. Zooxanthellae respire from 1.2% to 32.9% of the total daily carbon fixed, with a mean of 12.8%.

Growth of zooxanthellae. The specific growth rate of zooxanthellae *in hospite* can be estimated from zooxanthellae mitotic index or from the increase in surface area of the coral skeleton (Wilkerson et al., 1983; Davies, 1984; Muscatine et al., 1984). On average, less than 1% of the daily carbon fixed by coral zooxanthellae is used to make new cells (Table 1A).

In general, the growth rate of zooxanthellae *in hospite* is relatively low (Wilkerson et al., 1988), while the rate of photosynthesis is relatively high. For example, the specific growth rate (u_z) of zooxanthellae in *S. pistillata* is 0.013 d^{-1}, corresponding to a doubling time of 53 days, but the carbon-specific growth rate (u_c) is 1.36 d^{-1} (Muscatine et al., 1984). These data reveal a central dogma of algae-invertebrate endosymbioses; namely, that zooxanthellae photosynthesis is uncoupled from cell growth. This leaves a substantial portion of fixed carbon available for export to the animal, and, since growth is unbalanced, the standing stock of zooxanthellae is maintained at a constant level. This situation is in sharp contrast to that in free-living microalgae in which photosynthesis and growth are coupled, and growth is balanced.

Loss of fixed carbon via expulsion of zooxanthellae from the coral amounts to about 0.01% of the total daily carbon fixed (Hoegh-Guldberg et al., 1987). Loss of zooxanthellae by digestion or autolysis has been investigated in several algae-invertebrate associations (Fitt and Trench, 1985; Colley and Trench, 1985), but is largely unstudied in corals. As these two processes are the means by which the animal may acquire algal biomass, any loss may be accounted for by considering it as part of the pool of translocated carbon.

Translocation of fixed carbon to the animal. Given the large differences between u_z and u_c, it is not surprising that 65.5% to 96.7% of the carbon fixed daily by zooxanthellae is released to the host (Table 1A). The products of fixation are discussed elsewhere (Schmitz and Kremer, 1977; Hofmann and Kremer, 1981). Glycerol, lipid, and a few amino acids are among the major products released by zooxanthellae (Schlichter et al., 1983, 1984; Battey and Patton, 1984, 1986; Crossland et al., 1980a,b; Szmant-Froelich, 1981; Patton and Burris, 1983; Kellogg and Patton, 1983). Translocation is customarily viewed in terms of the movement of carbon from algae to animal. However, it is really the net effect of both the forward (algae to animal) and reverse (animal to algae) movement of fixed carbon. Little is known of reverse translocation in corals. It may function largely in instances of facultative heterotrophy (Steen, 1986).

Respiration by the animal. Shallow water corals respire at least half of the carbon acquired by translocation (Table 1A), and, in doing so, subsidize 100% or more of their demand for carbon for respiration and growth. Such corals are said to be phototrophic with respect to carbon. Mass-specific respiration decreases with depth (Davies, 1977; McCloskey and Muscatine, 1984), offsetting the tendency for a decline in the ratio of photosynthesis to respiration.

Growth of the animal.   Coral animal tissues grow relatively slowly.   Surface area-specific growth rates range from 0.0014 d^{-1} to 0.0047 d^{-1} (Lewis, 1981; Davies, 1984), and tend to decrease with increasing colony size (Muscatine et al., 1985).   On average, about 4% of the translocated carbon is used to synthesize new animal biomass (Table 1A), so the carbon available may be as much as two orders of magnitude greater than needed.

"Losses" of fixed carbon; assimilation efficiencies.   After budgeting for respiration and growth, a fraction of the daily net fixed carbon is invariably "left over" for dissemination to skeletal carbonate, skeletal organic matrix, lipid stores, exported mucus, egesta, DOC, and reproductive products.   How the carbon is apportioned is uncertain as only a few of these sinks have been quantified. Independent assessment of efflux of DOC alone shows that it is substantial, and may reach 50% of the daily fixed carbon, depending on the species and technique of assessment (Cooksey and Cooksey, 1972; Crossland, 1980; Crossland et al., 1980a, b; Davies, 1984, 1991; Muscatine et al., 1984).   Efflux of this magnitude has persuaded investigators that release of DOC from corals is a major new carbon and energy sink that may significantly affect biogeochemical cycles in reef communities by providing a substrate for bacteria and microheterotrophs.   It appears to be a major route for transfer of zooxanthellae production to the reef communities.

On average, losses in Table 1A amount to about 33% of the daily fixed carbon. Dubinsky et al. (1984) reported that about 16% of the net radioactive carbon fixed by *S. pistillata* was incorporated into skeletal carbonate; this is inorganic carbon not measured as respiration.   If this percentage of carbon is subtracted from losses by corals in Table 1A, then losses of organic carbon to all other sinks would average about 15%.   Thus, organic carbon assimilated by the animal (animal growth + losses) would average about 19% of the gross carbon fixed.   Assimilation efficiencies in some corals are much lower. For example, even including DOC export, the assimilation efficiency in light-adapted *S. pistillata* is about 7-8% (Falkowski et al., 1984).   The prevailing view is that the zooxanthellae in *S. pistillata*, and probably a range of other species, are nitrogen-limited (Falkowski et al., 1984).   This idea explains how growth of zooxanthellae can be uncoupled from photosynthesis, and why material translocated from zooxanthellae to coral is nitrogen-deficient, largely respired, and deserving of its "junk food" label (Davies, 1984; Falkowski et al., 1984).   In this vein, it is noteworthy that exported DOC, such as coral mucus, and storage lipid are also nitrogen-poor.   Both contain a high proportion of hydrocarbon (Benson and Muscatine, 1974).

The Effect of Shade, Depth, and Season on Daily Carbon Budgets

Daily carbon budgets for shallow water corals reveal the high flux of fixed carbon through zooxanthellae and animal tissue.   There is sufficient carbon to support animal respiration and growth.   As irradiance decreases in shade, at depth, or in winter, the flux of carbon decreases, and is no longer sufficient to support the needs of the animal for respiration and growth.   These corals exhibit deficit budgets and are revealed as obligate heterotrophs, requiring carbon input from zooplankton or dissolved organic materials to meet their carbon needs.

Table 1B illustrates the effect of permanent shade and depth, and transient cloud cover on daily budgets of three corals.   Shade-adapted *S. pistillata* inhabits caves and crevices in shallow water.   It receives only 10-50 umole m^{-2}s^{-1} compared to the 650-1200 umole m^{-2}s^{-1} received by light-adapted conspecifics (Dubinsky et al., 1984). Despite its higher pigment content and increased light harvesting capacity, its quantum yields are low and it fixes only about 26 ug C cm^{-2}d^{-1}. This amount is about one-fifth of that fixed by the light-adapted conspecifics (Dubinsky et al., 1984) so that net carbon translocated does not meet the needs of the animal for respiration, growth, and export. Similarly, at 35 m. *S. pistillata* receives about 12% of surface irradiance, and concomitant with the decreased irradiance regime, fixes one-fourth to one-fifth of the carbon fixed by shallow water colonies (McCloskey and Muscatine, 1984). Both shade-adapted and deep water *S. pistillata* must supplement their carbon input by 50-60% from allochthonous sources. The same conclusion has been drawn for *Montastrea annularis* in shallow vs. deep water (Davies, 1977; Porter, 1985; see also Chalker et al., 1984). This pattern of carbon surplus and deficit also emerges in energy budgets for clear vs. overcast days. On clear days, *P. damicornis* exhibits an energy surplus. On overcast days, when total irradiance falls to about 43% of that on clear days, *P. damicornis* is predicted to incur an energy budget deficit. This deficit may be offset by drawing on lipid stores manufactured on clear days (Davies, 1991).

Budget deficits may be minimized by a variety of photoadaptations by zooxanthellae, summarized by Barnes and Chalker (1990). For example, *Acropora spp.* maintain a relatively constant P/R ratio above unity to a depth of 35 m. (5.8% of surface irradiance) by optimizing photokinetic parameters; these include an increase in chlorophyll *a* per cell, photosynthetic efficiency (α), I_c, and in respiration as growth irradiance decreases. Although photosynthesis decreases below 35 m., P/R is still ≥ 1 down to 60 m. A deficit of energy from photosynthesis and a supplement from heterotrophy would certainly be expected below this depth. However, Chalker et al. (1984) note that any heterotrophy which might be required to offset energy budget deficits does not extend the depth range of *Acropora granulosa*. This is in contrast to *M. annularis* which incurs obligate heterotrophy over most of its depth range to 50 m. These differences demonstrate the greater efficiency of light harvesting by the branching coral with high surface/volume ratio vs the massive coral with relatively low surface/volume (Porter, 1976).

The Role of Nutrients

Nutrient flux in corals and reef communities is reasonably well-defined. The emphasis has been on nitrogen, although interest in phosphorus is growing (reviewed by D'Elia and Wiebe, 1990; Miller and Yellowlees, 1989). Briefly, corals can take up inorganic nitrogen, particularly ammonium. The zooxanthellae are thought to be the primary agents of uptake. They assimilate ammonium via the GS-GOGAT pathway. They translocate organic nitrogen to the animal. ^{14}C-alanine is a large translocation component in studies using ^{14}C-labelled bicarbonate. Transamination by the animal is likely to yield other non-essential amino acids which may be incorporated into protein and then released by catabolism as ammonium. Corals normally excrete little ammonium as it is immediately scavenged by the zooxanthellae. Nitrogen acquired initially as ammonium may thus be recycled within a symbiosis, but it is "empty" nitrogen in the sense that it does not contribute to the animal's net protein synthesis.

The hypothesis that many corals, particularly those that live in nutrient-poor waters, are nitrogen limited, and that chronic nutrient limitation may explain why photosynthesis and growth of zooxanthellae are uncoupled, was recently tested. Investigators used the simple expedient of enriching the coral's maintenance medium with dissolved inorganic nitrogen (DIN) and phosphorus (DIP), and with daily rations of *Artemia* (Hoegh-Guldberg and Smith, 1989; Muscatine et al., 1989b; Dubinsky et al., 1989). After an 18-day incubation with either DIN, DIN+DIP, or *Artemia*, the population density of zooxanthellae in *S. pistillata* increased up to five-fold, depending on treatment (Fig. 3).

Many details remain to be explained, particularly in connection with feeding on *Artemia* and phosphorus enrichment. The significant feature is that environmental DIN affects the population density of zooxanthellae most dramatically. DIN not only re-couples photosynthesis and growth, but also alters the quality and quantity of both the translocated products and the products released by the corals (Muscatine et al., 1984).

Increased population densities of zooxanthellae do not necessarily increase the host's fitness. On the contrary, the densely packed zooxanthellae exhibit less photosynthesis per cell, suggesting that the dense populations must compete for CO_2 as it becomes a potential limiting factor. This limitation could initiate facultative heterotrophic utilization of the host's organic substrates (Steen, 1986).

Role of Host Animals in Light Harvesting, Carbon Dioxide Flux, and Biogeochemical Cycling

Light harvesting. Whereas zooxanthellae manifest adaptations for increased light harvesting, the host animals also show interesting adaptive emergent properties. The best known of these is an alteration in colony form with depth as an adaptation for light harvesting (see, for example, Fricke and Schuhmacher, 1983). Others include behavioral adaptations for light harvesting, by expansion of special zooxanthellae-laden "bubble" tentacles during the day, and plankton harvesting, by expansion of nematocyst-laden prehensile tentacles at night by the reef coral *Plerogyra sinuosa* (Fricke and Vareschi, 1982).

Some corals, such as *Leptoseris fragilis*, found at 145 m in the Red Sea, maintain zooxanthellae at extremely low light levels. The animal possesses a chromatophore system in the endoderm beneath the monolayer of zooxanthellae. Blue light at depth excites the chromatophores, and they fluoresce at longer wavelengths suitable for absorption by the zooxanthellae photosynthetic pigments. Thus, the chromatophore system is an adaptation for light harvesting by both the zooxanthellae and the animal host (Schlichter and Fricke, 1990; Schlichter et al., 1984, 1985, 1986).

Carbon dioxide flux. Coral zooxanthellae can obtain CO_2 for photosynthesis from animal respiration. When coral photosynthesis exceeds respiration, as it usually does in shallow water, the zooxanthellae must draw on the sea water bicarbonate reservoir. Until recently, this reservoir was viewed as unlimited. However, the animal's tissues are a

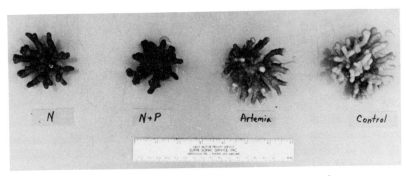

Fig. 3.   *Stylophora pistillata* after 18 days of treatment *in situ* with DIN, DIN + DIP, *Artemia*, control. (After Dubinsky et al., 1989).

barrier to diffusion and transport of bicarbonate, and so the delivery of CO_2 to zooxanthellae can be a problem, as is described below in several independent studies.

Burris et al. (1983) reported that when bicarbonate was added to sea water, the effect on photosynthesis by *Seratiopora hystrix* and *S. pistillata* was negligible. However, Dennison and Barnes (1987) observed a significant increase in rates of photosynthesis and calcification in *Acropora scandens* when the surrounding water was stirred. When photosynthesis was near compensation, stirring had no effect. Their observations suggested that photosynthesis could be limited by diffusion of substrate. This suggestion is supported by the results of stable carbon isotope studies. Coral zooxanthellae might be expected to have $\delta^{13}C$ around -25°/oo, similar to a range of marine phytoplankton. However, coral zooxanthellae have $\delta^{13}C$ values of around -13°/oo (Land et al., 1975; Goreau, 1977; Muscatine et al., 1989), probably because the supply of bicarbonate is limited by diffusion, and most of the available CO_2 inside the animal ($\delta^{13}C$ approximately -11.7 °/oo; Muscatine et al., 1989a) is fixed with minimal isotope fractionation.

Internal CO_2 depletion is exacerbated by high cell densities of zooxanthellae (Cummings and McCarty, 1982; Muscatine et al., 1984; Dubinsky et al., 1989).

Carbonic anhydrase (CA). Carbonic anhydrase catalyzes the interconversion of CO_2 and HCO_3^- in a wide variety of organisms and, with a turnover rate of $10^6/sec$, is the most efficient enzyme ever studied (Tashian, 1989). It has been widely reported (Lucas and Berry, 1985) that some microalgae and cyanobacteria possess a CO_2-concentrating mechanism which involves CA. The report by Graham and Smillie (1976) of CA activity, not only in zooxanthellae, but also in animal tissues of *Pocillopora damicornis*, and the observation by Goreau (1977) that CA could aid both in calcification and CO_2 delivery, prompted an investigation of CA in coral animal tissues. The results are worth noting in some detail.

Initial measurements of CA activity in 22 species of cnidarians, from Bermuda and Eilat, Israel led to two major observations. First, as CA activity in symbiotic species was, on average, 29 times higher than in non-symbiotic species (Weis et al., 1989), it seemed to be related to the presence of zooxanthellae. Weis and coworkers also found a positive correlation between CA activity in the Red Sea coral *Stylophora pistillata* and growth irradiance, suggesting a relationship between CA activity and photosynthesis. Further, $10^{-4}M$ acetazolamide, a noncompetitive inhibitor of CA, inhibited photosynthesis by 60 - 85% in three symbiotic cnidarian species *Agaricia fragilis*, *Millepora alcicornis*, and *Aiptasia pallida* from Bermuda (Weis et al., 1989). Moreover, $10^{-4}M$ acetazolamide inhibited photosynthesis by up to 80% in the tropical sea anemone *Aiptasia pulchella* even in the presence of added DIC at levels well above the ambient 2.2 mM (Weis, 1990). These findings strongly suggested that CA plays a role in dissolved inorganic carbon delivery to the algae.

The second observation was that CA activity in symbiotic cnidarians was not only 2 - 3 times higher in the animal tissue than in the zooxanthellae, but also there were spatial differences in CA activity within the same individual. Column tissue of the anemone *Condylactis gigantea*, which lacks zooxanthellae, had very low activity compared to the tentacle tissue which contains them. In addition, CA activity in aposymbiotic *A. pulchella* tissue was 2.5 times lower than that in control symbiotic animals and in aposymbiotic animals which had been repopulated with algae (Weis, 1991). These observations again suggest that CA activity in the animal tissue is directly related to the presence of zooxanthellae, and raise the questions of where CA originated in the symbiosis, in the animal or zooxanthellae or both, and where it was located in the animal tissue.

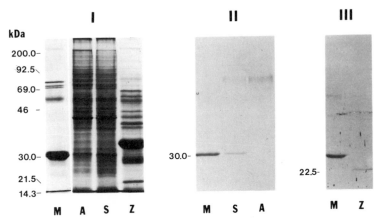

Fig. 4.   An SDS-polyacrylamide gel stained with Coomassie blue (I) and corresponding immunoblots (II and III). Blot II was probed with anti-human CA (Bioproducts for Science) and blot III with anti-human CA (ICN). M = control purified mammalian CA; S = symbiotic animal extract; A = aposymbiotic animal extract, and Z = cultured zooxanthellae extract (After Weis, 1991).

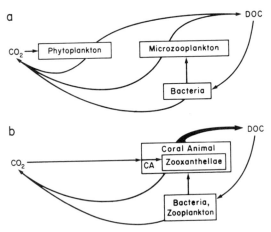

Fig. 5.   Contrasting roles of free-living and symbiotic microalgae in a biogeochemical cycle of carbon (Design after Eppley, 1980).

To address the question of the origin of CA activity, Weis (1991) probed for the presence of CA using polyclonal antisera against human CA in symbiotic and aposymbiotic *A. pulchella* and freshly isolated and cultured zooxanthellae. The resulting immunoblots (Fig. 4) showed one band with a molecular weight of 30 kDa in symbiotic animal tissue and control mammalian CA lanes, no bands in aposymbiotic animal lanes and one band at a molecular weight of 22.5 kDa in freshly isolated and cultured zooxanthellae lanes. Since there was no band at 22.5 kDa in the symbiotic animal tissue lanes, the high CA activity in symbiotic animal tissue was considered to be due to the induction of animal enzyme by the presence of zooxanthellae, not to the production and transfer of zooxanthellae CA into the animal tissue. This observation has important implications for genome interactions in these symbioses.

To determine the location of CA in the animal tissue at the ultrastructural level, Weis (1991) treated symbiotic *A. pulchella* with the same antibodies mentioned above, in conjunction with colloidal gold immunocytochemistry. CA was found on or near the anemone vacuolar membrane surrounding the zooxanthellae. Further, CA activity was measured in membrane and soluble fractions of anemone tissue. The majority of the CA activity appeared in the soluble fraction, suggesting that although CA is associated with the animal vacuolar membrane, it is not intrinsically bound to it. How the zooxanthellae induce CA activity in the animal tissue is a major new question for research in algae-cnidarian symbiosis.

In summary, it now appears that CA is an important element in inorganic carbon delivery to zooxanthellae in symbiosis with corals and other cnidarians.

MESOSCALE EFFECTS OF ZOOXANTHELLAE PHOTOSYNTHESIS ON BIOGEOCHEMICAL CYCLES OF CARBON IN REEF COMMUNITIES

Figure 5a shows an abbreviated scheme for a carbon cycle involving free-living microalgae. The algae fix CO_2. Microzooplankton graze on the algae and bacteria. All components contribute to the flux of DOC and to the mineralization of DOC to CO_2. In this scheme, growth of the algae is balanced, and population density varies with availability of nutrients.

The impact of symbiotic microalgae on such a cycle is shown in Fig. 5b. The cycle is now truncated, and a "short cut" is created by eliminating predation by microzooplankton. A carnivore, the coral, now takes on an additional trophic role as a facultative herbivore. The boundary conditions imposed on the zooxanthellae protect them from grazing, but they must somehow expend energy and resources to induce CA in the coral to ensure their supply of CO_2. The irony is that they still cannot grow due to chronic nitrogen limitation, which leads to uncoupling of photosynthesis and growth, and DOC export. The DOC, in turn, provides substrate for bacteria, which may cycle back to the coral via holozoic feeding. It would appear that release of DOC is one of the novel ways in which zooxanthellae influence biogeochemical cycles of carbon.

REFERENCES

Barnes, D.J., and Chalker, B.E., 1990, Calcification and photosynthesis in reef-building corals and algae, *in*: Coral Reefs, " Z. Dubinsky ed., Elsevier, Amsterdam.

Battey, J.F., and Patton, J.S., 1984, A reevaluation of the role of glycerol in carbon translocation in zooxanthellae-coelenterate symbiosis, *Mar. Biol.*, 79:27.

Battey, J.F., and Patton, J.S., 1986, Glycerol translocation in *Condylactis gigantea*, *Mar. Biol.*, 95:37.

Benson, A.A., and Muscatine, L., 1974, Wax in coral mucus: Energy transfer from corals to reef fishes, *Limnol. Oceanogr.*, 19:810.

Bunt, J., 1975, Primary productivity of marine ecosystems, *in*: "Primary Productivity of the Biosphere", H. Lieth and R.H. Whittaker eds., Springer-Verlag, New York.

Burris, J.E., Porter, J.W., and Laing, W.A., 1983, Effects of carbon dioxide concentration on coral photosynthesis, *Mar. Biol.*, 75:113.

Chalker, B.E., Cox, T., and Dunlap, W.C., 1984, Seasonal changes in primary production and photoadaptation by the reef-building coral *Acropora granulosa*, *in*: "Marine Phytoplankton and Productivity," O. Holm-Hansen, L. Bolis, and R. Giles eds., Springer-Verlag, New York.

Colley, N.J., and Trench, R.K., 1985, Cellular events in the reestablishment of symbiosis between a marine dinoflagellate and a coelenterate, *Cell Tissue Res.*, 239:93.

Cooksey, K., and Cooksey, B., 1972, Turnover of photosynthetically fixed carbon in reef corals, *Mar. Biol.*, 15:289.

Crossland, C.J., 1980, Release of photosynthetically-derived organic carbon from a hermatypic coral, *Acropora* cf. *acuminata*, *in*: "Endosymbiosis and Cell Biology," W. Schwemmler and H.E.A. Schenk eds., W. De Gruyter, Berlin.

Crossland, C.J., Barnes, D.J., and Borowitzka, M.A., 1980a, Diurnal lipid and mucus production in the staghorn coral *Acropora acuminata*, *Mar. Biol.*, 60:81.

Crossland, C.J., Barnes, D.J., Cox, T., and Devereaux, M., 1980b, Compartmentation and turnover of organic carbon in the staghorn coral *Acropora formosa*, *Mar. Biol.*, 59:181.

Cummings, C.E., and McCarty, H.B., 1982, Stable carbon isotope ratios in *Astrangia danae*: Evidence for algal modification of carbon pools used in calcifcation, *Geochem. et Cosmochim. Acta*, 46:1125.

Davies, P.S., 1977, Carbon budgets and vertical zonation of Atlantic reef corals, *Proc. 3rd Int. Reef Coral Symp.*, 1:392.

Davies, P.S., 1984, The role of zooxanthellae in the nutritional energy requirements of *Pocillopora eydouxi*, *Coral Reefs*, 2:181.

Davies, P.S., 1991, The effect of daylight variations on the energy budgets of shallow-water corals, *Mar. Biol.*, 108:137.

D'Elia, C.F., and Weibe, W.J., 1990, Biogeochemical nutrient cycles in coral-reef ecosystems, *in*: "Coral Reefs," Z. Dubinsky ed., Elsevier, Amsterdam.

Dennison, W.C., and Barnes, D.J., 1987, Effects of water motion on coral photosynthesis and calcification, *J. Exp. Mar. Biol. Ecol.*, 115: 67.

Dubinsky, Z., Falkowski, P.G., Porter, J.W., and Muscatine, L., 1984, Absorption and utilization of radiant energy by light- and shade-adapted colonies of the hermatypic coral *Stylophora pistillata*, *Proc. R. Soc. Lond.*, B 222:203.

Dubinsky, Z., Stambler, N., Ben-Zion, M., McCloskey, L., Muscatine, L., and Falkowski, P.G., 1989, The effects of external nutrient resources on the optical properties and photosynthetic efficiency of *Stylophora pistillata*, *Proc. R. Soc. Lond.*, B 239:231.

Edmunds, P.J., and Davies, P.S., 1986, An energy budget for *Porites porites* (Scleractinia), *Mar. Biol.*, 92:339.

Eppley, R.W., 1980, Estimating phytoplankton growth rates in the central oligotrophic oceans, *in*: "Primary Productivity in the Seas," P.G. Falkowski ed., Plenum Press, New York.

Falkowski, P.G., Dubinsky, Z., Muscatine, L. and Porter, J. W., 1984, Light and the bioenergetics of a symbiotic coral, *BioScience*, 34:705.

Fitt, W.K., and Trench, R.K., 1985, Endocytosis of the symbiotic dinoflagellate *Symbiodinium microcadriaticum* Freudenthal by endodermal cells of the scyphistomae of *Cassiopeia xamachana* and resistance of the algae to host digestion, *J. Cell. Sci.*, 64:195.

Fricke, H.W., and Vareschi, E., 1982, A scleractinian coral (*Plerogyra sinuosa*) with "photosynthetic organs," *Mar. Ecol.*, 7:273.

Fricke, H.W., and Schuhmacher, H., 1983, The depth limits of Red Sea stony corals: An ecophysiological problem (a deep diving survey by submersible). P.S.Z.N.I., *Mar. Ecol.*, 4:163.

Goreau, T.F., 1964, Mass expulsion of zooxanthellae from Jamaican reef communities after hurricane Flora, *Science*, 145:383.

Goreau, T.F., 1977, Coral skeletal chemistry: physiological and environmental regulation of stable isotopes and trace metals in *Montastrea annularis*, *Proc. R. Soc. Lond.*, B 196:291.

Graham, D., and Smillie, R.M., 1976, Carbonate dehydratase in marine organisms of the Great Barrier Reef, *Aust. J. Plant Physiol.*, 3:113.

Hatcher, B.G., 1988, Coral reef primary productivity: A beggar's banquet, *TREE*, 3:106.

Hoegh-Guldberg, O., McCloskey, L.R., and Muscatine, L., 1987, Expulsion of zooxanthellae from symbiotic cnidarians from the Red Sea, *Coral Reefs*, 7:113.

Hoegh-Guldberg, O., and Smith, G.J., 1989, Influence of the population density of zooxanthellae and supply of ammonium on the biomass and metabolic characteristics of the reef corals *Seriatopora hystrix* and *Stylophora pistillata*, *Mar. Ecol. Prog. Ser.*, 57:173.

Hofmann, D.K., and Kremer, B.P., 1981, Carbon metabolism and strobilation in *Cassiopea andromeda* (Cnidaria:Scyphozoa): significance of endosymbiotic dinoflagellates, *Mar. Biol.*, 65:25.

Kellogg, R.B., and Patton, J.S., 1983, Lipid droplets, medium of energy exchange in the symbiotic anemone *Condylactis gigantea*: A model coral polyp, *Mar. Biol.*, 75:137.

Land, L.S., Lang, J.C., and Smith, B.N., 1975, Preliminary observations on the carbon isotopic composition of some reef coral tissues and symbiotic zooxanthellae, *Limnol. Oceanogr.*, 20:283.

Larkum, A.W.D., 1983, The primary productivity of plant communities on coral reefs, *in*: "Perspectives on Coral Reefs," D.J. Barnes ed., B. Clouston, Australia.

Lewis, J.B., 1981, Estimates of secondary production of reef corals, *Proc. 4th Int. Coral Reef Symp.*, p. 369.

Lewis, D. and Smith, D.C., 1971, The autotrophic nutrition of symbiotic marine coelenterates with reference to hermatypic corals. I. Movement of photosynthetic products between the symbionts, *Proc. R. Soc, Lond.*, B 178:111.

Lucas, W.J., and Berry, J.A., 1985, Inorganic carbon transport in aquatic photosynthetic organisms, *Physiol. Plant*, 65:539.

McCloskey, L.R., and Muscatine, L., 1984, Production and respiration in the Red Sea coral *Stylophora pistillata* as a function of depth, *Proc. R. Soc. Lond.*, B 222:215.

Miller, D.J., and Yellowlees, D., 1989, Inorganic nitrogen uptake by symbiotic marine cnidarians: A critical review, *Proc. R. Soc. Lond.*, B 237:109.

Muscatine, L., 1980, Productivity of zooxanthellae, *in*: "Primary Productivity in the Sea," P.G. Falkowski, ed., Plenum, New York.

Muscatine, L., 1990, The role of symbiotic algae in carbon and energy flux in reef corals, *in*: "Coral Reefs," Z. Dubinsky, ed., Elsevier, Amsterdam.

Muscatine, L., Mccloskey, L. R., and Marian, R. E., 1981, Estimating the daily contribution of carbon from zooxanthellae to animal respiration, *Limnol. Oceanogr.*, 26:601.

Muscatine, L., Falkowski, P.G., Porter, J.W., and Dubinsky, Z., 1984, Fate of photosynthetic fixed carbon in light and shade-adapted colonies of the symbiotic coral *Stylophora pistillata*, *Proc. R. Soc. Lond.*, B 222:181.

Muscatine, L., McCloskey, L.R., and Loya, Y., 1985, A comparison of the growth rates of zooxanthellae and animal tissue in the Red Sea coral *Stylophora pistillata*, *Proc. 5th Int. Coral Reef Symp.*, 6:119.

Muscatine, L., Porter, J.W., and Kaplan, I.R., 1989a, Resource partitioning by reef corals as determined from stable isotope composition. I. $\delta^{13}C$ of zooxanthellae and animal tissue vs. depth, *Mar. Biol.*, 100:185.

Muscatine, L., Falkowski, P.G., Dubinsky, Z., Cook, P.A., and McCloskey, L.R., 1989b, The effect of external nutrient resources on the population dynamics of zooxanthellae in a reef coral, *Proc. R. Soc. Lond.*, B 236:311.

Patton, J.S., and Burris, J.E., 1983, Lipid synthesis and extrusion by freshly isolated zooxanthellae (symbiotic algae), *Mar. Biol.*, 75:131.

Porter, J.W., 1976, Autotrophy, heterotrophy and resource partitioning in Caribbean reef-building corals, *Am. Nat.*, 110:731.

Porter, J.W., 1985, The maritime weather of Jamaica: its effects on annual carbon budgets of the massive reef-building coral *Montastrea annularis*, *Proc. 5th Int. Coral Reef Symp.*, 6:363.

Porter, J.W., Muscatine, L., Dubinsky, Z., and Falkowski, P.G., 1984, Primary production and photoadaptation in light- and shade-adapted colonies of the symbiotic coral, *Stylophora pistillata*, *Proc. R. Soc. Lond.*, B 222:161.

Schlichter, D., Svoboda, A., and Kremer, B.P., 1983, Functional autotrophy of *Heteroxenia fuscescens* (Anthozoa:Alcyonaria): carbon assimilation and translocation of photosynthates from symbionts to host, *Mar. Biol.*, 78:29.

Schlichter, D., Kremer, D.P., and Svoboda, A., 1984, Zooxanthellae providing assimilatory power for the incorporation of exogenous acetate in *Heteroxenia fuscescens* (Cnidaria:Alcyonaria), *Mar. Biol.*, 83:277.

Schlichter, D., Fricke, H.W., and Weber, W., 1986, Light harvesting by wavelength transformation in a symbiotic coral of the Red Sea twilight zone, *Mar. Biol.*, 91:403.

Schlichter, D., Weber, W., and Fricke, H., 1985, A chromatophore system in the hermatypic, deep water coral *Leptoseris fragilis* (Anthozoa: Hexocorallia), *Mar. Biol.*, 89:143.

Schlichter, D., and Fricke, H.W., 1990, Coral host improves photosynthesis of endosymbiotic algae, *Naturwiss.*, 77:447.

Schmitz, K., and Kremer, B.P., 1977, Carbon fixation and analysis of assimilates in a coral-dinoflagellate symbiosis, *Mar. Biol.*, 43: 305.

Smith, G.J., 1984, Ontogenetic variation in the symbiotic associations between zooxanthellae (*Symbiodinium microadriaticum* (Freudenthal) and sea anemone (Anthozoa: Actiniaria) hosts, Ph. D. Dissertation, University of Georgia.

Smith, S.V., 1978, Coral reef area and the contributions of reefs to processes and resources of the world's oceans, *Nature*, 273:225.

Steen, R.G., 1986, Evidence for heterotrophy by zooxanthellae in symbiosis with *Aiptasia pulchella*, *Biol. Bull.*, 170:267.

Szmant-Froelich, A., 1981, Coral nutrition: Comparison of the fate of ^{14}C from ingested labeled brine shrimp and from the uptake of $NaH^{14}CO_3$ by zooxanthellae, *J. Exp. Mar. Biol. Ecol.*, 55:133.

Tasch, P., 1980, "Paleobiology of the Invertebrates", John Wiley, New York.

Tashian, R.E., 1989, The carbonic anhydrases: Widening perspectives on their evolution, expression and function, *Bioessays*, 10:186.

Weis, V.M. 1990, The role of carbonic anhydrase in symbiotic cnidarians, Ph.D. Thesis, University of California, Los Angeles.

Weis, V.M., 1991, The induction of carbonic anhydrase in the symbiotic sea anemone *Aiptasia pulchella*, *Biol. Bull.*, 180: (in press).

Weis, V.M., Smith, G.J., and Muscatine, L., 1989, A "CO_2 supply" mechanism in zooxanthellate cnidarians: Role of carbonic anydrase, *Mar. Biol.*, 100:195.

Wilkerson, F.P., Muller-Parker, G., and Muscatine, L., 1983, Temporal patterns of cell division in natural populations of endosymbiotic algae, *Limnol. Oceanogr.*, 28:1009.

Wilkerson, F.P., Kobayashi, D., and Muscatine, L., 1988, Growth of symbiotic algae in Caribbean reef corals. Mitotic index and size of Caribbean reef corals, *Coral Reefs*, 7:29.

THE IMPORTANCE AND MEASUREMENT OF NEW PRODUCTION

Trevor Platt[1], Pratima Jauhari[2], and Shubba Sathyendranath[1,3]

[1]Biological Oceanography Division
Bedford Institute of Oceanography
Box 1006, Dartmouth, Nova Scotia, Canada B2Y 4A2

[2]Geological Oceanography Division
National Institute of Oceanography
Dona Paula, Goa, India 403 004

[3]Department of Oceanography
Dalhousie University
Halifax, Nova Scotia, Canada B3H 4J1

INTRODUCTION

Public concern about the inadvertent interference by man in the processes that conspire to produce the earth's climate - the general problem of "global change" - has emphasised the need for scientists to gain a deeper understanding of the planet as a compound biogeochemical system. In particular it has become recognised that we must improve our knowledge of the global carbon cycle, including the part of it that lies in the oceans. It is in this context that a rather obscure concept in biological oceanography has been projected into the scientific limelight and has given a household word ("new production") to those institutions where research on global-change issues is carried out, written about, read about, or worried about. For understanding the oceanic carbon cycle, and the role of the oceans in the planetary carbon cycle, the concept of new production is central and fundamental.

For many concepts in ecology, of which we may consider biological oceanography to be a branch, there is a gap of enormous significance between basic definitions and operational definitions. Physical oceanographers do not face this difficulty: their questions are much better posed. For example, temperature is a property of the ocean that is of interest to physical oceanographers. We can proceed to measure temperature without debate about what temperature "really is", or what the chosen transducer is "really measuring", or whether temperature is what we "really should be measuring". Nor need the theoreticians concern themselves with such difficult questions. Temperature has a universal significance that transcends the methods used to measure it. Temperature data, carefully collected, can be incorporated without reservation into the conventional view of the ocean, and interpreted without fanfare by the theories of the day.

Primary Productivity and Biogeochemical Cycles in the Sea
Edited by P.G. Falkowski and A.D. Woodhead, Plenum Press, New York, 1992

If we are honest about it, the story in biological oceanography is a much less happy one. Many of the fundamental state variables have no universally-accepted definitions. And if the definitions could be agreed to, it would not follow that the variables so defined were, in fact, true observables of the system. Uncertainties arising from this intrinsic lack of clarity lie at the bottom of many of the pseudo-controversies in biological oceanography, such as that concerning the level of primary production in the oligotrophic oceans.

With respect to new production, we consider that the basic definition is acceptably unambiguous, within the limitation of its steady-state assumption. The operational definitions, however, generally do not respect the steady-state condition, and for this reason the results of different methods are difficult to reconcile.

BASIC DEFINITION

Both the concept and the definition of new production were given by Dugdale and Goering (1967). Discussing nutrient cycles in the pelagic ecosystem, and the capacity of a given region to support secondary and higher levels of production, they noted that the integrity of the autotrophic community would be maintained only if the export of an organically-bound nutrient from the system did not exceed the rate at which that nutrient was supplied from outside the system. The excess of supply over export would represent the capacity of the system to sustain secondary and higher levels of production. They recognised that any one of several nutrients could be used as common currency for this kind of discussion, but selected nitrogen on the grounds that it is thought often to limit (in the sense of Liebig) production in the sea; that it is the major structural component of cells; and that it stands in roughly constant ratio to two other important components, carbon and phosphorus.

Against this background, they defined new production as the primary production associated with newly-available nitrogen, such as nitrate and dinitrogen gas, as opposed to regenerated production, which would be the primary production associated with nitrogen recycled within the photic zone. In a companion paper, Dugdale (1967) elaborated this basic theme in a *steady-state context.*

Note that "regenerated" nutrients, in the usage of Dugdale and Goering, would necessarily be chemically reduced. Nutrients supplied from outside the system could be expected to be in a higher oxidation state, with possible exceptions supplied by horizontal advection. This was the basis for the *examples* of new nutrients, nitrate and nitrogen gas, given by Dugdale and Goering. They did not say that new production was production associated with nitrate. They *did* say that new production was production associated with newly-available nitrogen, *for example,* nitrate (our italics). We emphasise the distinction, having noticed an incipient growth industry among those who would misuse the definitions and then find them wanting.

Clearly, new production and regenerated production are complementary quantities: their sum is the total primary production. The ratio of new production to total primary production was called the *f*-ratio by Eppley and Peterson (1979). It has proved to be a useful index of trophic status.

Geologists, who are also interested to know about new production, prefer phosphorus to nitrogen as a bookkeeping nutrient on the grounds that, although nitrogen fixation is known to occur in the surface layer of the ocean, there is no analogous process

274

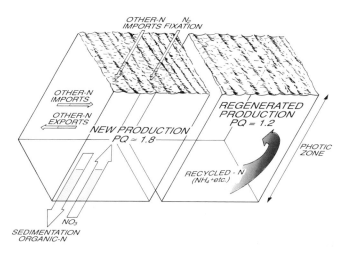

Fig. 1.   Schematic diagram showing the major fluxes for new and regenerated production, from Platt et al. (1989).   Regenerated production is entirely dependent on reduced nitrogen derived from excretion of organisms in the euphotic zone.   New production, on the other hand, depends on supply of nitrogen from the outside.   Nitrate from below the photic zone is one major source of this nitrogen.   In steady state, there is a balance between the nitrate supply, and the downward flux of nitrogen in the sedimenting particles.

for phosphorus: in this sense, phosphorus is seen as a more limiting nutrient, given the unlimited supply of nitrogen gas in the atmosphere. In other words, in the presence of unrestricted nitrogen fixation, the requirement for balanced growth would quickly cast phosphorus in the rôle of limiting nutrient. This view has attracted support in other branches of oceanography (Smith, 1984; Codispoti, 1989). However, Dugdale and Goering (1967) had selected nitrogen over phosphorus because of the rapid turnover rate of phosphorus in energy metabolism. Notwithstanding these distinctions, it is clear that in areas of the ocean, such as parts of the Antarctic and the north Pacific Oceans, where nutrients are believed not to regulate primary production, the entire formalism of new and regenerated production is called into question.

OPERATIONAL DEFINITIONS

With the exception of that applied to the uptake of N^{15}-labelled nitrate (*in vitro*) methods, the operational definitions of new production are all based on the idealised concept of a surface control volume in dynamic, steady state at some time scale. The relevant time scale is rarely specified explicitly, but by default, an implicit annual time scale is assumed. The operational definitions are best understood by reference to Fig. 1. When implemented, they translate into the methods listed in Table 1.

Nitrogen Inputs

The sum total of the inputs of biologically-usable nitrogen into the surface box is a measure of new production. Strictly, input through all faces of the box should be included. Fluxes through the vertically-oriented faces correspond to horizontal advection. Flux through the surface corresponds to atmospheric input, either of nitrogen gas or of other nitrogen compounds deposited by precipitation or as dryfall. Nitrogen gas can be considered as new nitrogen only if there are (nitrogen-fixing) organisms present that can use it. Here, then, the operational definition of new production becomes dependent upon the community structure of the pelagic zone.

Horizontal advection and atmospheric input are thought, at least in the open ocean, to be minor components of nitrogen supply. Under this assumption, the input of nitrogen to the surface box is dominated by the vertical flux from below. Being supplied from the deep water, this nitrogen is in the form of nitrate. New production can then be estimated as the rate of transport of nitrate through the lower face of the surface box. This is a diffusive flux computed as the action of a vertical diffusion coefficient on the vertical nitrate gradient (Lewis et al., 1986; Jenkins, 1988), or estimated by isotopic methods (Altabet and Deuser, 1985). The time scale of applicability of results obtained using this operational definition is the time scale over which the nitrate flux is averaged. It is not always demonstrated that this is a time scale on which the surface box can be assumed to be in steady state. Nevertheless, it is an essential requirement of the approach.

Moreover, the evidence that the vertical flux from below dominates the nitrate supply to the surface control volume is not strong, even for the open ocean. In particular, the quantitative role of nitrogen-fixing organisms still remains to be defined. Because nitrogen gas is so freely available, the active presence of large nitrogen-fixing communities would be sufficient to undermine this operational definition of new production.

Nitrogen Exports

The sum total of nitrogen exports from the surface box is another measure of new production: under the steady-state assumption, any exported nitrogen has to be balanced

Table 1. Methods for estimating primary production in the ocean and the nominal time-scales on which the results apply. The components P_g (gross primary production), P_n (net primary production), and P_c (net community production) of primary production refer to a scheme based on carbon; P_T (total primary production), P_r (regenerated production) and P_{new} (new production) to one based on nitrogen. Sedimentation rate refers to the gravitational flux of organic particles leaving the photic zone (= export production), not the (much smaller) flux arriving at the sediment surface. Based on Platt et al. (1989).

Method	Nominal component of production	Nominal time-scale
In vitro		
^{14}C assimilation†	P_T ($\equiv P_n$)	Hours to 1 d (duration of incubation)
O$_2$ evolution	P_T	Hours to 1 d (duration of incubation)
^{15}NO$_3$ assimilation	P_{new}	Hours to 1 d (duration of incubation)
^{15}NH$_4$ assimilation	P_r	Hours to 1 d (duration of incubation)
^{18}O$_2$ evolution††	P_{new} ($\equiv P_c$)	Hours to 1 d (duration of incubation)
Physical transport		
Sedimentation rate below photic zone	P_{new} ($\equiv P_c$)	Days to months (duration of trap deployment)
Bulk property		
NO$_3$ flux to photic zone‡	P_{new}	Hours to days
O$_2$ utilization rate OUR below photic zone	P_{new}	Seasonal to annual
Net O$_2$ accumulation in photic zone	P_{new}	Seasonal to annual
^{238}U/^{234}Th§	P_{new}	1d to 300d
^{3}H/^{3}He§§	P_{new}	Seasonal and longer
Upper and lower limits		
Optimal energy conversion of photons absorbed¶	P_T	Instantaneous to annual (upper limit)
Depletion of winter accumulation of NO$_3$	P_{new}	Seasonal (lower limit)
Other		
Remote Sensing¶¶	P_T, P_{new}	Days to weighted annual

† Here and elsewhere, P_{new} can be calculated from P_T if there is an independent measurement or estimate of the f-ratio.

‡ Altabet and Deuser (1985); Lewis et al. (1986); Jenkins (1988). †† Bender et al. (1987).

§ Coale and Bruland (1987). Isotopic disequilibrium in photic zone used to estimate export of particles from it.

§§ Jenkins (1982). Isotopic data used to establish a time scale for calculation of OUR.

¶ Platt et al. (1989).

¶¶ P_T from physiological model (Platt and Sathyendranath, 1988), f-ratio from temperature-nitrate correlations (Sathyendranath et al., in press).

by nitrogen input, which itself is a measure of new production. We have to take into account the horizontal advective fluxes, as in calculating the nitrogen input, but in the case of nitrogen export, the flux through the surface itself is less problematic, and, in this respect at least, the method is not sensitive to the pelagic community structure.

It is commonly believed that export, like import, is dominated by the vertical flux through the base of the surface box. The flux of nitrogen passing downwards through this lower boundary is then a measure of new production (Eppley and Peterson, 1979). The device used to measure it is the sediment trap. The time scale of applicability of results obtained with this method is the duration of trap deployment. Again, it is implicit in the application of the method that the duration of trap deployment is a time scale for which the steady-state assumption can be applied to the surface box. The introduction of the designation "export production" for the results obtained by this method (Berger et al., 1989) is a tacit admission that the steady-state assumption is not proven: if a genuine steady state obtains, export production and new production are one and the same thing.

Oxygen Consumption

Material that passes through the base of the surface box by sedimentation is attacked by microbial activity, and will be decomposed, with a corresponding consumption of oxygen. If there are no processes occurring in this subsurface layer that generate oxygen, the rate of consumption of oxygen in the layer is an index of the organic material being supplied to it, that is, of the new production (Jenkins, 1982), assuming, as always, that the steady state applies.

The loss rate of oxygen is known as the apparent oxygen utilization rate (AOU). It does not qualify directly as a measure of new production unless and until it has been converted to an equivalent nitrogen flux. To accomplish this conversion, it is necessary to extend the steady-state hypothesis to include the idea of stoichiometry. That is, we must assume that a given consumption of oxygen can be equated with microbial destruction of a known amount of organic nitrogen. This is an assumption that may be justified at the annual time scale, but may not be at the time scale on which the results apply, which is the time period over which the oxygen loss is calculated.

Uptake of Labelled Nitrogen

Direct measurement of the uptake rate of N^{15}-labelled nitrate by a sample of the autotrophic community enclosed in a bottle provides an operational definition of new production that comes closest to its conceptual definition (Dugdale and Goering, 1967). In this case, the steady-state hypothesis consists in the assumption that the rate of nitrate uptake in the (small) sample is a representative measure of the rate of nitrate uptake at the space and time scales to which the results are to be applied (Platt et al., 1989).

INTERPRETATION

Active Metabolism

Regenerated production (total production minus new production) is the production required to support the maintenance metabolism of the *entire* community in the surface box. The new production is then the production that can serve to support any metabolic processes, such as growth, over and above the maintenance requirement of the pelagic system. We may call this the active metabolism. It is the portion of primary production that can be used to fuel the net accumulation of biological tissue. Of course, under the steady-state hypothesis, tissue can accumulate only in the short term: on the time scale at which steady state is thought to hold, any net growth is removed from the control volume, leaving the biomass invariant.

Consideration of the new production in terms of system metabolism has led to a closed relationship between two important ecological coefficients: the f-ratio and the photosynthesis-to-respiration or P/R ratio (Quiñones and Platt, 1991).

Exploitable Production

Extending this idea further, the magnitude of the new production sets an upper limit to the rate at which biological tissue can be removed from an exploited marine ecosystem without impairing its long-term integrity (Platt et al., 1989; Horne et al., 1989). If material were removed at a rate that exceeded the new production, the regenerated production would decrease such that the maintenance requirements of the ecosystem could no longer be met. A similar argument applies to systems losing material by advection. Of course, other factors would intervene that would set the upper limit to exploitable production at a much lower figure than that defined by the level of new production; for example, destabilisation of the biomass flow structure by excessive removal of the biomass at one node.

Coupling Coefficient for the Water Column

Regenerated production depends upon processes occurring internal to the surface box. On the other hand, new production depends upon connections between the surface box and the rest of the water column. The ratio of new production to total production (the f-ratio) then contains information about the degree of coupling between the surface box and the layers below. In areas where the surface layer is strongly coupled to the subsurface one (frequent exchanges of water between the two), new production will be a large fraction of total production. Conversely, where the water column is stratified so strongly that the surface layer is effectively isolated from the water below, episodic new production will be correspondingly decreased.

Carbon Dioxide and the Greenhouse Effect

Regenerated production, being that part of the total production that is respired in the surface box, fixes carbon dioxide and respires it again within the layer that has access to the atmosphere on a time scale of hours to days (Platt et al., 1989). From the point of view of absorption of atmospheric carbon dioxide by the ocean, this part of the total production is therefore irrelevant. Conversely, new production is the part of the carbon fixed that is not respired until it has been removed from the surface box to the deeper water (under the steady-state hypothesis where the vertical fluxes dominate). Once transported below the surface box, any fixed carbon that is respired back to carbon dioxide has access to the atmosphere only on time scales very much longer than it would have had in the surface box itself, say tens of hundreds of years. Potentially, then, the process of new production can be seen as an avenue to move carbon dioxide from the atmosphere to the abyss. In this incarnation, new production is called the "biological pump", and is a mechanism by which man hopes that the pollution of the atmosphere will be swept under the rug of the thermocline. However, new production is not a measure of the air-sea flux of carbon dioxide, nor is it a measure of the anthropogenic perturbation in that flux.

Nonrenewable Resources

The new production is the part of total production that is removed from the surface box in the steady state, including material removed and transported towards the sediments. Not all the new production reaches the sediment, because much of it respires en route. Once arrived at the sediment, the material can accumulate: the respiration rate at the bottom of the deep oceans is low. The accumulated material may include structures that may be of interest for exploitation by man. In principle, maps of the distribution of such structures should bear some relation to the distribution of new production (Andrews et al., 1983).

THE QUESTION OF SCALE

Discussion of measurement procedures in oceanography should always give due consideration to the question of scale: in treating new production the matter of scale is of critical importance. Note that the dimensions of new production, $[M][L^{-2}][T^{-1}]$ or $[M][L^{-3}][T^{-1}]$ contain the time in the denominator. That is to say, new production is a *rate*. Hence the results of all measurements of new production will contain an intrinsic time scale determined by the method itself and by the way in which it is implemented. Because time and space scales are inextricably related for oceanographic phenomena, measurements of new production will also refer to a particular spatial scale (Platt et al., 1984; Platt, 1984; Platt and Harrison, 1986; Platt et al., 1989).

A further complication is that although the fundamental dimensions of new production are based on unit volume, $[M][L^{-3}][T^{-1}]$, only procedures for which the inherent time scale is short are capable of producing results in this form. Methods with longer intrinsic time scales, even though they may be invoked at discrete depths, give results with dimensions $[M][L^{-2}][T^{-1}]$, that is they are based on unit area for a layer of finite depth. This is because of the inevitable smearing in the vertical due to vertical diffusion, and the minimum thickness of the layer involved can be estimated as the result of the local vertical diffusion acting for a time equal to the intrinsic time scale of the method. It is convenient, however, and conventional, to express the results of such measurements in terms of unit surface area for the entire water column, or at least for that part of it enclosed by the surface box. This feature imposes a fundamental limitation on the comparison between measures of new production with different characteristic time scales (Platt et al., 1989).

The nominal time scales to which the various methods for measuring new production refer are listed in Table 1. Typically, the bulk property methods, as they are normally applied, produce results valid at longer time scales. This is because the observer is usually obliged to let the procedure run for a sufficiently long time to produce a measurable signal. Bulk property methods are integrative rather than instantaneous.

THE IMPORTANCE OF NEW PRODUCTION

Ecosystem Function

Part of the significance of the new production formalism is that it gives an additional dimension to the array of ideas we can use to analyse the structure and function of the marine ecosystem (Horne et al., 1989).

The concept of new production gives oceanographers the capability to distinguish, at the macroscopic level, between the energy required to maintain the metabolic integrity of the pelagic ecosystem and the energy available for growth in the pelagic ecosystem. Regenerated production satisfies the maintenance metabolic demands of the primary producers and of the rest of the pelagic community.

The f-ratio has been interpreted as an index of trophic status. This interpretation is correct only if the f-ratio has been referred to the annual time scale. It has been shown (Platt and Harrison, 1985) that a particular station can exhibit a range of trophic conditions, as judged by the local f-ratio, from eutrophic to oligotrophic, as the seasonal cycle progresses.

New production creates the possibility of secondary and higher levels of production. The balance between new and regenerated production (for example, the f-ratio) is then an index of the *efficiency* of ecosystem function. Another index of ecosystem efficiency, much discussed in the 1970s, is the ratio of primary production to total ecosystem and metabolism, and a formal connection between this index and the f-ratio has been made by Quiñones and Platt (1991). Ecosystem maturity is a property that increases as the degree of ecosystem perturbation decreases. Thus, the f-ratio is in some way diagnostic of the level of disturbance to which the system is subject. This idea is attractive in the sense that episodic nutrient supply to the surface layer arising from the passage of storms is thought to be an important contribution to new production.

It can be shown (Vézina and Platt, 1987) that the f-ratio is a property of the flow structure of the pelagic food web that can be modulated by the variations in the supply rate of nitrogen, provided that there is at least one nonlinear link in the flow structure. The question then arises whether the pelagic ecosystem, in its various manifestations in different oceanographic régimes, can be considered to have a common generic structure whose realisation in a particular location can be explained by the response of this generic structure to variations in nutrient supply. The formalism of new production allows such questions to be asked: without its conceptual background, the asking might not be worthwhile.

Finally, the f-ratio, by virtue of its value as an index of the degree of coupling between the surface layer and the rest of the water column, can tell us something, at least for shallow water stations, about the tightness of coupling between the pelagic and benthic ecosystems. Areas of high new production will be areas where the benthic ecosystem is better supplied with organically-bound carbon. Areas of high f-ratio will be areas where the benthic ecosystem receives a bigger share of the total primary production. The f-ratio is determined by the balance between physical and biological processes in regulating primary production.

Biogeochemical Cycles

As developed by Eppley and Peterson (1979), our understanding of the rôle of the ocean in the global carbon cycle has been much enhanced by the insights arising from the partition of primary production between new production and regenerated production.

Particularly in the context of the perceived climate change consequent upon the increasing concentration of carbon dioxide in the atmosphere, the concept of new production has allowed biological oceanographers to clarify the biogenic flux that is relevant as an avenue for sequestration of atmospheric carbon in the deep ocean circulation. That is to say, the biological analogue of the (physical) deep convective flux has been clearly defined. The rôle and significance of the oceanic microbiota in the great planetary cycles has thus been established and brought into focus. A profound consequence has been the general realisation that climate change issues in the ocean will not be resolved without the collaboration of all the subdisciplines of oceanography. This has been an important step in the maturity of the field and the impetus behind the emergence of oceanography as a unified discipline.

Cyclic Patterns of Glaciation

On a much longer time scale, the new production formalism has proved useful for interpretation of patterns in geological data from the Pleistocene. Following the principle of uniformitarianism - that the present is the key to the past - study of present-day sedimentation in relation to nutrient supply and new production provides a basis for understanding paleoclimate and the causes of its variation. Sediment trap deployments yield estimates of seasonality in the downward flux of carbon from the photic zone, and show how it depends on the time variation in delivery of nitrogen to the surface layer and the associated fluctuations in primary production. In turn, this leads to a deeper understanding of the empirical relationships between sedimentation rate in the water column (as measured by the traps) and the accumulation rate of material in surficial sediments.

Such work has attracted increased attention by the discovery of connections between fluctuations in new production and the cyclic pattern of glaciation over the past hundred thousand years (Sarnthein et al., 1987). Through the variable intensity of the biological pump, climatically-significant fluctuations in atmospheric carbon dioxide concentration on time scales of order 10^4 yr are believed to have been controlled by variations in oceanic nitrogen (McElroy, 1983). These perturbations represent major violations of the steady-state condition for the chemistry of the pelagic layer.

Shifts in the dominant wind patterns over the oceans, especially in lower latitudes, under the influence of fluctuations in the Earth's orbit around the Sun, have led to significant changes in the upwelling fluxes of nutrients to the photic zone. These changes in nutrient supply have caused the biological pump to transport varying quantities of carbon from the surface layer of the ocean and entrain it in the deep circulation. In response, the concentration of carbon dioxide in the atmosphere has been modulated, as evidenced by the data on carbon dioxide trapped in air bubbles sampled from polar ice cores (Barnola et al., 1987). New production may then be seen as the key to understanding the effect of global change on the marine ecosystem. The initial response of the ocean to climate change will be in the physical processes themselves: ecosystem processes will not respond unless and until these initial, physical reactions are manifest. The ecosystem property that will be affected is the new production.

CONCLUDING REMARKS

The importance of the new production formalism is that it both stated the need for, and provided, a conceptual basis for discussion of the relation between nutrient supply and ecosystem function, and how these relate to the major biogeochemical cycles. It is difficult to explain why the concept lay dormant for more than ten years after its introduction. Perhaps it is just an expression of its authors' being ahead of their time. Nevertheless, when the ideas were reintroduced by Eppley and Peterson, the community of biological oceanographers was more ready to pay attention. The biogeochemical paradigm was accepted, and, as its importance became understood, permeated to all branches of oceanography. Dissemination of these ideas was facilitated by the sense of urgency created by the growing realisation of the scale of man's perturbation of atmospheric carbon dioxide concentration.

As the ideas became rooted in oceanographic thinking, discussion of the magnitude and control of primary production moved to a more sophisticated lane than before. But residual confusion persisted. The difficulty revolved around failure by some to appreciate the central assumption of the steady state, which is at once the strength and the limitation of the formalism. Rigid application of the steady-state formalism of new production in circumstances where the steady state manifestly does not apply may lead to spurious conclusions. There is a sense in which a short time scale measurement of new production is not useful except in so far as it quantifies a contribution to the new production as defined and computed on a longer time scale through a weighted integral.

The surface layer of the ocean is best viewed as a system which is open, in the thermodynamic sense, and subject to frequent, but irregular, episodic or transient perturbations, in the intervals between which the dynamics of the ecosystem will respond to, and neutralise (in the sense of Le Chatelier) the effects of the disturbances. In this light, the biogeochemical state of the surface layer will be perpetually unsteady at the episodic time scale. In particular, there is very little likelihood that an oceanographic cruise of short duration will find its destination ecosystem in a locally-steady state.

It *is* possible that an oceanographic cruise would occupy a particular station long enough to observe the occurrence and effects of an episodic perturbation from start to finish. The input of nitrate nitrogen, the total production, an *in vitro* and a bulk property estimate of new production, and the export production, integrated over the duration of the disturbance, could all be measured. But the total ecosystem inventories at the beginning and end, necessary to measure net community production and therefore close the budget, would remain a formidable challenge.

ACKNOWLEDGEMENTS

We thank W.G. Harrison for helpful comments on the manuscript.

REFERENCES

Altabet, M. A., and Deuser, W. G., 1985, Seasonal variations in natural abundance of ^{15}N in particles sinking to the deep Sargasso Sea, *Nature*, 315:218.

Andrews, J., Friedrich, C., Pautot, G., Pluger, W., Renard, V., Melguen, M., Cronan, D., Craig, J., Hoffert, M., Stoffers, P., Shearme, S., Thijssen, T., Glasby, G., Note, L., and Saget, P., 1983, The Hawaii-Tahiti transect. The oceanographic environments of manganese nodule deposits in the Central Pacific, *Mar. Geol.*, 54:109.

Barnola, J. M., Reynaud, D., Korotkevich, Y .C., and Lorius, C., 1987, Vostok ice core provides 160,000-year record of atmospheric CO_2, *Nature*, 329:408.

Bender, M. L., Grande, K., Johnson, J., Marra, J., Williams, J. LeB., Sieburth, J., Pilson, M., Langdon, C., Hitchcock, G., Orchardo, J., Hunt, C., Donaghay, P., and Heinemann, K., 1987, A comparison of four methods for determining planktonic community production, *Limnol. Oceanogr.*, 32:1085.

Berger, W. H., Smetacek, V. S., and Wefer, G., 1989, Ocean productivity and paleoproductivity - an overview, in: "Productivity of the Ocean: Present and Past," W. H. Berger, V. S. Smetacek, and G. Wefer, eds., John Wiley and Sons Limited.

Coale, K. H., and Bruland, K. W., 1987, Oceanic stratified euphotic zone as elucidated by ^{234}Th:^{238}U disequilibria, Limnol. Oceanogr., 32:189.

Codispoti, L. A., 1989, Phosphorus vs. nitrogen limitation of new and export production, in: "Productivity of the Ocean: Present and Past," W. H. Berger, V. S. Smetacek, and G. Wefer, eds., John Wiley and Sons Limited.

Dugdale, R. C., 1967, Nutrient limitation in the sea: dynamics, identification, and significance, Limnol. Oceanogr., 12:685.

Dugdale, R. C., and Goering, J. J., 1967, Uptake of new and regenerated forms of nitrogen in primary productivity, Limnol. Oceanogr., 12:196.

Eppley, R. W., and Peterson, B. J., 1979, Particulate organic matter flux and planktonic new production in the deep ocean, Nature, 282:677.

Horne, E. P. W., Loder, J. W., Harrison, W. G., Mohn, R., Lewis, M. R., Irwin, B., and Platt, T., 1989, Nitrate supply and demand at the Georges Bank tidal front, Scient. Mar., 53:145.

Jenkins, W. J., 1982, Oxygen utilization rates in North Atlantic subtropical gyre and primary production in oligotrophic systems, Nature, 300:246.

Jenkins, W. J., 1988, Nitrate flux into the euphotic zone near Bermuda, Nature, 331:521.

Lewis, M. R., Harrison, W. G., Oakey, N. S., Herbert, D., and Platt, T., 1986, Vertical nitrate fluxes in the oligotrophic ocean, Science, 234:870.

McElroy, M. B., 1983, Marine biological controls on atmospheric CO_2 and climate, Nature, 302:328.

Platt, T., 1984, Primary productivity in the central North Pacific: comparison to oxygen and carbon fluxes, Deep Sea Res., 31:1311.

Platt, T., and Harrison, W. G., 1985, Biogenic fluxes of carbon and oxygen in the ocean, Nature, 318:55.

Platt, T., and Harrison, W. G., 1986, Reconciliation of carbon and oxygen fluxes in the upper ocean, Deep Sea Res., 33:273.

Platt, T., and Sathyendranath, S., 1988, Oceanic primary production: Estimation by remote sensing at local and regional scales, Science, 1613:1620.

Platt, T., Lewis, M. R., and Geider, R., 1984, Thermodynamics of the pelagic ecosystem: Elementary closure conditions for biological production in the open ocean, in: "Flows of Energy and Materials in Marine Ecosystems," M. J. R. Fasham, ed., Plenum Publishing Corporation, New York.

Platt, T., Harrison, W. G., Lewis, M. R., Li, W. K. W., Sathyendranath, S., Smith, R. E., and Vézina, A. F., 1989, Biological production of the oceans: The case for a consensus, Mar. Ecol. Prog. Ser., 52:77.

Quiñones, R. A., and Platt, T., 1991, The relationship between the f-ratio and the P:R ratio in the pelagic ecosystem, Limnol. Oceanogr., 36:211.

Sarnthein, M., Winn, K., and Zahn, R., 1987, Paleoproductivity or oceanic upwelling and the effect on atmospheric CO_2 and climate change during deglaciation times, in: "Abrupt Climatic Change - Evidence and Implications," W. H. Berger and L. D. Labeyrie, eds., Reidel, Dordrecht.

Sathyendranath, S., Platt, T., Horne, E. P. W., Harrison, W. G., Ulloa, O., Outerbridge, R., and Hoepffner, N., 1991, New production in the ocean: estimation by compound remote sensing, Nature, in press.

Smith, S.V., 1984, Phosphorus versus nitrogen limitation in the marine environment, Limnol. Oceanogr., 29:1149.

Vézina, A., and Platt, T., 1987, Small-scale variability of new production and particulate fluxes in the ocean, Can. J. Fish. Aquat. Sci., 44:198.

THE ROLE OF COASTAL HIGH LATITUDE ECOSYSTEMS IN GLOBAL EXPORT PRODUCTION

Paul K. Bienfang and David A. Ziemann

The Oceanic Institute
Makapuu Point
P. O. Box 25280
Honolulu, Hawaii 96825

INTRODUCTION

If one were to sit down and design a perfectly conservative pelagic ecosystem, it would likely have the following properties. First, it would be a steady-state system, in which the delivery of essential substrates would be constant over time to minimize transient oscillations of the food web to changing substrate supplies. Second, the biological components would be built upon small-sized primary producers (e.g., picoplankton), and consist of a complex network of numerous trophic levels of carefully balanced standing stocks, which are configured in a K-type strategy having numerous feedback loops for the regeneration of essential substrates. Third, the system would have negligible sinking rates to minimize vertical transport of material away from the photic influx at the surface. Fourth, the benthos would be located far away (i.e., in deep water) to maximize the time/opportunities for regeneration and, thus, minimize losses from the water column.

Conversely, a minimally conservative (i.e., a very leaky) pelagic system would be designed as follows. The supply of essential substrates would pulsate over a wide range of frequencies. This variability, ranging on time scales from episodic to seasonal, would promote variability in the state variables regulating biological rate processes, which, in turn, would promote uncouplings among trophic levels. The system would have strong gradients of physicochemical and nutrient parameters over space, time, and depth to maximize variability. The planktonic system would be based on large-celled phytoplankton and would consist of unbalanced trophic components organized as short, simple food chains, devoid of feedbacks. The potential for vertical material transport would be maximized by the large-cell primary trophic level having the highest potential sinking rates, and r-type growth strategies to augment the potential for continuous destabilization. The benthos would be located close to the photic zone to optimize the likelihood of the rapidly sinking material leaving the water column and actually reaching the bottom.

Table 1 summarizes the factors that influence the magnitude of organic production, the relative amounts of new versus regenerated production, and their relative values in pelagic oceanic areas and coastal high latitude areas.

Primary Productivity and Biogeochemical Cycles in the Sea
Edited by P.G. Falkowski and A.D. Woodhead, Plenum Press, New York, 1992

Table 1. Factors that Influence the Rates of Export Production in the Oceans, and their Characteristics in Areas of Low and High Export

	EXPORT PRODUCTION	
	LOW	HIGH
Physiographic Factors		
Water depth	deep	shallow
Photic zone depth	deep	shallow
Terrestrial proximity	distant	close
State Variables		
Temperature	high	low
Nutrient concentration	low	high
Light	constant	variable
Physical Factors		
Wind mixing	moderate	extreme
Currents	slow	rapid
Stratification	constant	intermittent
Variability		
Seasonality	moderate	strong
Episodicity	moderate	strong
Food Web Structure		
Phytoplankton	small cells	large cells
Zooplankton	constant	seasonal
Complexity	long food webs	short food chains
Degree of coupling	tightly coupled	loosely coupled
Strategy design	K-type	r-type

ANALYSIS AND DISCUSSION

Estimates of Global New Production

Estimation of total global ocean production is a critical first step when attempting to relate the importance of a particular region or process to the global scale. The estimation of the total production of ocean carbon has been attempted by several authors using a variety of data and assumptions. The results vary by more than a factor of 2, ranging from 23 to 56 gigatons (gt = 10^{15} g) carbon per year. Koblentz-Mishke et al. (1970) estimate global primary production at 23 gt C, based on various measurements of production rates. Platt and Subba Rao (1975) and Berger et al. (1987) estimate annual production at approximately 30 gt carbon, while Shushkina (1985) estimates global production at 56 gt, based on data collected from 130 stations between 1968 and 1982. This scale of variation results in large differences in the relative importance of particular areas for which direct measurements may be available. For example, a region which contributes 5 gt annual production constitutes from 9 to 22% of global production, depending on the global value used.

Processes Influencing New vs. Regenerated Production

New production is that portion of photosynthesis which is supported by inorganic nitrate supplied from outside the photic zone (either from below, as is generally the case in oceanic regions, or from terrigenous input, as may be the case in coastal areas), while regenerated production depends on recycling of nutrients within the photic zone. The proportion of new production to total production will be lowest in areas where the influx of inorganic nutrients is minimal, and highest where mixing, upwelling, or coastal influences are pronounced.

Where autotrophs and heterotrophs are in dynamic balance, the transfer of carbon and nitrogen from one trophic level to another is efficient. Recycling within the water column is the dominant process, and carbon fixed within the photic zone is rapidly consumed by heterotrophs within or just below the photic zone. Where natural variability in the physical environment dominates, however, dynamic balance often is not established. Instead, seasonal or episodic injections of inorganic nutrients into the photic zone result in rapid autotrophic growth which cannot be consumed by the heterotrophic population, and subsequently sinks out of the system.

There seems to be a lack of agreement on the fraction of total primary production which is new production. Eppley and Peterson (1979) estimate global new production at 3.4 to 4.7 gt, based on conversions from primary production to new production. Chavez and Barber (1987) and Berger et al. (1989) estimate that approximately 20% (6 gt) of total production is new production.

Processes Influencing Export

Some portion of the new production remains within the photic zone through heterotrophic consumption, and the nutrients contained therein are subsequently made available for recycling and regenerated production. That portion of new production which leaves the photic zone is known as export production. The rate of export production depends on the depth at which it is determined, since the action of continued grazing or bacterial action decreases the carbon as particulate material sinks (Seuss, 1980). For example, it is estimated that half the export production at 100 m depth is lost in transit between 100 m and 200 m depth.

Open Ocean Production/Sedimentation

This section summarizes the historical context of current perspectives on the sinking of phytoplankton in oceanic waters. Our interest in sedimentation began in the early 70s against a backdrop of working with continuous cultures. Early hallmark papers (e.g., Steele and Yentsch, 1960; Eppley et al., 1967; Smayda, 1970) had drawn attention to the importance of phytoplankton sinking and the overwhelming role of physiological state in controlling cell buoyancy. Chemostats offered the capability to control growth rates and provide steady-state cultures with various defined physiological states that could be used to unravel this association so important to material transfer in the sea. After describing the theory underlying the assessment of the mean sinking rate (Bienfang et al., 1977), we developed several analytical methods applicable to both laboratory and field settings (Bienfang, 1979; 1981). Several laboratory efforts then addressed the correlation between variability in phytoplankton sinking rate and specific growth rate, temperature, salinity, irradiance, various forms of nutrient stress, and taxonomic character (Bienfang, 1981; Bienfang et al., 1982; Bienfang and Harrison, 1984a; 1984b; Bienfang and Szyper, 1982; Smayda and Bienfang, 1983; Bienfang et al., 1983). Armed with this information, we then examined sinking rates in the field in various temperate, upwelling, subarctic, and subtropical environments (Bienfang, 1981; 1982; 1984; 1985a; 1985b; Laws et al., 1988). In addition to making direct measurements of the sinking rate of natural assemblages, we hoped that this information might be used to compare specific growth rates of these populations. In retrospect, our naivete seems appalling, but this was before there was any pervasive appreciation of the prevailing size structure of subtropical populations, when picoplankton and *Synechococcus* became household words.

The size structure of subtropical phytoplankton assemblages is the most significant characteristic influencing sedimentation in these oceanic environments. Ultraplankton ($< 3\mu$m), nanoplankton ($< 20\mu$m), and netplankton ($> 20\mu$m) account for about 80%, 98%, and 2% of the total subtropical chlorophyll standing stocks, respectively (Bienfang, 1980a; Bienfang and Szyper, 1981; Bienfang and Takahashi, 1983). Chlorophyll size structure among various marine environments are contrasted in Fig. 1; the size distributions of other microparticulate constituents are similar to those for chlorophyll. In subtropical waters, most of the biomass is within the $< 5\mu$m fraction and consists of small flagellates and coccoid cells (Takahashi and Bienfang, 1983; Bienfang et al., 1984; Bienfang, 1985a; 1985b). This is very different from temperate or subarctic systems where cells are larger and populations are composed primarily of centric diatoms.

The preponderance of small-celled biomass is characteristic of warm, well-stratified, oligotrophic ecosystems where low ambient nutrient levels prevail (see Fig. 2 in Bienfang, 1985b). This influences sedimentation losses in several ways: (1) For the vast majority of phytoplankton biomass, the sinking rates of intact cells are virtually zero. (2) The small size limits the maximum sinking rates that could result from events that reduce metabolic vigor and affect the physiological state to which buoyancy is coupled. The maximum sinking rate (ψ_{max}) of a cell is related primarily to its volume, whereas the relationship of sinking rate at any given time (ψ_t) to ψ_{max} depends primarily on the cell's physiological state. When growth is active, ψ_t approaches a minimum value (ψ_{min}). For large cells (where $\psi_{max} >> \psi_{min}$) such events can cause substantial episodic increases in downward flux, but for small-celled assemblages (where $\psi_{max} \approx \psi_{min}$) the resultant effects to sedimentation flux are much more modest, and the impact of episodic events is reduced. (3) Of the various types of nutrient limitation, the effects of silicon depletion elicit by far the greatest increase in sinking rate response. The predominance of small, non-diatomaceous components in the oceanic flora minimizes the sedimentary impact of this, the most pronounced form of nutrient stress to phytoplankton.

Phytoplankton sinking rates at the deep chlorophyll maximum are lower than those prevailing in the upper water column (Bienfang, 1980a; 1985b). This field observation has been related to physiological changes and to the lower light levels characteristic of this region (Bienfang et al., 1983). Though the size structure precludes a deceleration in sinking rate as the primary mechanism for chlorophyll accumulation in this layer, the observation reflects the presence of a physiological-based mechanism to retain phytoplankton biomass and reduce downward sedimentary transport in these waters.

The predominance of small phytoplankton in such oceanic systems means that the majority of herbivorous grazing is done by herbivores that do not produce distinct, rapidly sinking fecal pellets. The excreta of these microherbivores are small and amorphous, and have lower sinking rates than those of their filter-feeding counterparts in other systems. Earlier work (Bienfang, 1980b) showed that the composition of phytoplankton grazed upon by herbivores influences the net excess density and, therefore, the sinking rates of the fecal pellets produced. The predominance of non-diatom taxa in the oceanic assemblages available to grazers would indicate a lower silica content in the herbivore fecal material produced. Thus, even the intact fecal pellets produced by filter-feeding herbivores would be subject to reduced sedimentary transfer rates from the system.

Field analyses of subtropical oceanic microparticulates indicated that the sedimentation flux accounted for about 7% of the daily photosynthetic carbon production rates, and that particulates $> 20\mu m$ accounted for most of flux in the C, N, P, and Si. Constituent turnover rates from sedimentary losses alone comprise less than 1% of other biological rates processes (e.g., growth rates) that influence these material concentrations (Bienfang, 1985). This suggests that direct sedimentation of suspended microparticulates has relatively little effect on time-dependent changes of phytoplankton biomass in these waters.

Microparticulate fluxes of various constituents in a subtropical system are summarized in Table 2 of Bienfang (1985b). Since these analyses are based on discrete samples, it is improbable that this material includes the larger particles (e.g., fecal pellets, snow, aggregates) shown in trap studies to be important to total vertical flux. However, these data do reflect several significant processes. First, they reflect direct microparticulate fluxes for several constituents having both ecological and geochemical

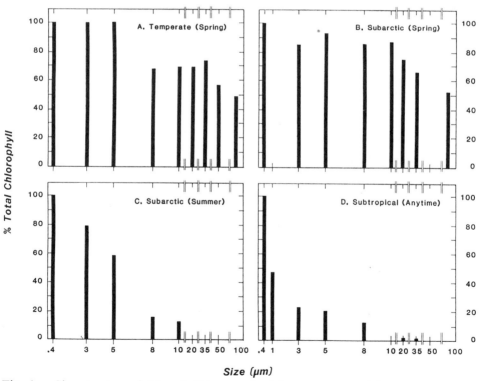

Fig. 1.   Size structure of chlorophyll in several marine ecosystems. The subarctic spring (B) and summer (C) data from Bienfang (1984) are compared with results from (A) temperate waters around Friday Harbor, Washington and (D) subtropical waters around Hawaii. Figure from Bienfang (1984).

importance; the elevated carbon content (C:N:Si:P = 544:69:12:1, by atoms) shows substantial detrital content, presumably relating to active reingestion and regeneration. The microparticulate flux values for carbon are about 20% of total carbon flux values ascertained by trap studies (e.g., Lorenzen et al., 1983) in these waters. Second, the extremely low chlorophyll flux reflects the dominance of very small cells having negligible sinking rates. Third, the relative amount of flux due to larger (i.e., $> 20\mu$m) particles ranges from 46 to 58% and is similar among the various constituents; the lowest value (22% for chlorophyll) results from the very low chlorophyll biomass in the $> 20\mu$m fraction. Low sedimentary loss is a characteristic feature of environments that have low rates of new production.

From these data, the calculation of turnover times, based solely on sedimentation losses, provides a basis for comparing loss rates with other biological rate processes, such as specific growth rates. Turnover estimates are derived by dividing flux values (F) by depth-integrated constituent levels (B*). For all constituents, turnover times for the >20 μm fractions are considerably larger (2 to 8 times) than those for total microparticulates (> 0.4 μm); this reflects the large relative amounts of particulates having negligible sinking rates. More importantly, these turnover rates for sedimentary losses are least 2 orders of magnitude lower than current estimates of specific growth rates. In addition to illustrating the conservative nature of these oligotrophic systems, such values indicate the comparatively minor importance of sinking to time-dependent changes of photoautotrophic biomass. Stated another way, the downward flux of (microparticulate) chlorophyll accounts for 0.03% of the total depth-integrated standing stock, or 0.25% of the depth-integrated standing stock smaller than 20μm. Lorenzen et al. (1983) estimated this proportion to be 0.8%; this value represents an upper bound for the relative chlorophyll loss due to sedimentation because their sampling methodology would include ingested/excreted chlorophyll. Such turnover estimates are small indeed when compared with more recent estimates of the specific growth rates of these assemblages (Laws et al., 1984). These results suggest that phytoplankton models do not need to incorporate a term for biomass loss resulting from direct sinking of intact cells until the precision of such efforts would be influenced by rate processes less than 1% those of phytoplankton growth. The modest importance of direct sinking losses is an important characteristic of oligotrophic environments which display comparatively little seasonality.

Coastal Production/Sedimentation

Export production is maximized when autotrophic production and heterotrophic consumption are essentially uncoupled. Generally speaking, spring blooms in high latitude coastal systems exhibit the greatest degree of uncoupling. Overwintering populations of herbivores are generally small and unable to reproduce with sufficient rapidity to consume the carbon fixed during the intense algal blooms which occur in these areas. The combination of high nutrient levels, sufficient light, a stable water column, and low grazing pressure result in algal blooms that yield extremely dense populations of phytoplankton before nutrient depletion occurs. Upon the onset of nutrient depletion, increases in sinking rates lead to rapid removal of algal material from the photic zone. The result is a "boom and bust" scenario in which algal production rapidly converts new nutrients to organic material which sediments out of the photic zone before it is consumed.

Interannual variations in the amounts of primary production and sedimentation during the spring bloom have been examined for a five-year period in Auke Bay, an embayment in southeastern Alaska (Ziemann et al., 1990). Cumulative production from mid-March to June of the five years ranged from 95 - 140 g C m^{-2}. All five years showed variations on the typical "boom and bust" bloom scenario. Interannual differences in the amounts of production were the result of differences in short-term weather patterns among the years. Nitrate concentrations in early spring for all five years were similar, at approximately 28 μM NO_3. Thus, in theory, the same amount of carbon production could have been supported each year. However, Auke Bay, like other high export systems, exhibits strong vertical stratification during the spring, which limits the depth over which nitrate is taken up in support of production. Interannual variations in the wind field resulted in variations in vertical mixing and, thus, variations in total carbon production.

The majority of carbon production during the primary spring bloom in Auke Bay was new production supported by nitrate. Much of that production was lost to the pelagic system via sedimentation. Kanda et al. (1990) showed that during the spring of 1986, new production constituted 62% of total production, and 77% of this new production was removed from the water column through sedimentation in the form of intact cells, fecal pellets, and detrital particles.

Variations in vertical mixing combined with interannual variations in the light field resulted in differences in species composition during the spring (see Fig. 10 in Ziemann et al., 1990), with resultant impacts to the production-sedimentation relationships. In years when the winds were light and the incident light could be characterized by 3-7 day periods of high light interspersed with brief periods of overcast, the spring bloom was dominated by the large celled species of *Thalassiosira*; in years when the light and wind fields were more highly variable, the blooms were dominated by small-celled diatoms of the genera *Skeletonema* and *Chaetoceros*. Experimental sinking rate studies determined that the *Thalassiosira* responded to nutrient depletion with significant increases in sinking rates,

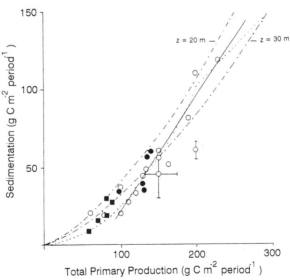

Fig. 2. Relationship between total primary production and sedimentation of particulate organic carbon (g C m^{-2} period^{-1}) from studies where both estimates have been measured simultaneously for more than 6 months of the productive part of the year (open circles). Predicted values from Betzer et al. (1984) ($-\cdot-\cdot-\cdot$) at two depths (20 and 30 m), Eppley and Peterson (1979) ($\cdots\cdots$), and the regression: J(z') = 0.733 PP - 51.6 (————) in Wassman (1988) are shown. Added to the graph are primary production/sedimentation data from the 1985-1988 spring periods (● symbols) and primary blooms (■ symbols) in Auke Bay, Alaska. Figure modified from Wassmann (1988).

and these species dominated the algal material collected in sediment traps. *Skeletonema* and *Chaetoceros*, on the other hand, showed less pronounced sinking rate responses to nutrient depletion, and were not found in abundance in the sediment traps even in years when they were dominant in the water column.

An examination of five years of data from Auke Bay suggests that carbon production and sedimentary carbon flux were correlated, and that about 40% of spring carbon production sediments out of the pelagic system. This appears to be typical of high latitude coastal systems. Wassman (1988) presents a summary of annual production and sedimentation estimates for primarily high latitude Atlantic coastal systems. The Auke Bay

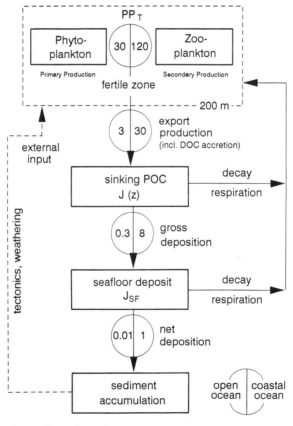

Fig. 3. Sketch of transfer of particulate organic carbon in the ocean, from primary production to burial in the sediment, showing the main elements of the "biological pump." Numbers are fluxes in g C m^{-2} y^{-1}; to the left in each circle is a typical open ocean value; to the right a typical coastal ocean value. Figure from Berger et al. (1989).

data fit well the Eppley and Peterson (1979) curve relating new production (sedimentation of carbon during steady state and long time intervals at the bottom of the photic zone) to total production (Fig. 2). Although collected over a period which was neither long nor in steady state, the Auke Bay data appear to fit the envelope of production and sedimentation data given by Wassman for depths of 20 - 30 m.

The fact that a single mathematical relationship adequately describes the relationship between total production and new production over a range of annual production levels, even for short periods and non-steady state conditions, suggests that the processes that control the fate of production in these high latitude coastal systems are closely linked. For example, very different phytoplankton communities developed during the spring blooms of several years in Auke Bay, with resultant differences in the sedimentation responses to nutrient depletion, and varied potential for losses due to grazing. In all years, however, the relationship between production and sedimentation remained relatively constant (at approx. 40%), suggesting that in years during which large losses occurred due to sinking of intact cells, the losses due to grazing were small, but that grazing losses were greater during years when sedimentation of intact cells was low.

Berger et al. (1989; Fig. 3) present pathways for the production and fate of organic carbon in the marine environment for "typical" oceanic (annual primary production = 30 g C m^{-2} y^{-1}) and coastal systems (primary production = 120 g C m^{-2} y^{-1}). Only about 10% of the production within the "fertile zone" of the pelagic ocean leaves the zone as export production, while 30% of annual production leaves the fertile zone in coastal systems. As a result, areal rates of export production are approximately ten times higher

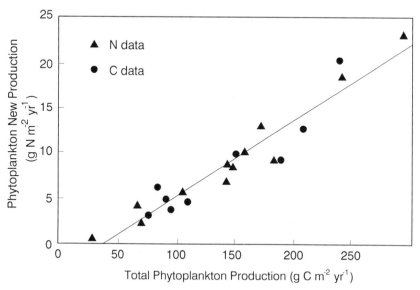

Fig. 4.   Phytoplankton N annual new production (ANNP) as a function of total phytoplankton annual C production (ACP). The linear regression yielded the equation: ANNP = -3.08 + 0.083 * ACP. The regression was significant (P < 0.01, F-test) and r^2 = 0.92. The regression intercept was significantly different from zero (P < 0.01, t-test). Figure from Iverson (1990).

for coastal areas than for oceanic regions. Although the oceanic waters constitute approximately 88% of the world ocean area and coastal waters constitute approximately 12%, Berger et al. (1989) estimate that both areas produce approximately 50% of the annual new production.

Coastal areas are characterized by the highest phytoplankton biomass, the largest benthic populations, and the largest removal of biomass, in the form of demersal fisheries. Coastal upwelling environments, for example, produce a large biomass of commercially important fishes, including many herbivorous species. These environments represent perhaps 1% of the total ocean surface area (Ryther, 1969) but produce more than 10% of global ocean fish production (Mann, 1984).

The Distribution of Export Production by Region

How large a role do coastal high latitude systems play in global biogeochemical fluxes? Iverson (1990) presents data (Fig. 4) relating the amount of annual new N production to annual C production for a variety of coastal and oceanic areas. The data show a strong linear relationship described by the equation:

$$ANNP = -3.08 + 0.083 \times ACP$$

where ANNP is annual new nitrogen production and ACP is annual carbon production. Annual new carbon production can be derived by multiplying ANNP by the Redfield ratio for C:N of 6:1. The annual carbon production level that is supported solely by regenerated N is found by solving the above equation for ANNP = 0; the resultant value is 37 g C m^{-2} y^{-1}.

We used this relationship and the summary data of Berger et al. (1987) to estimate annual amounts of both total and new production in various oceanic and coastal regions.

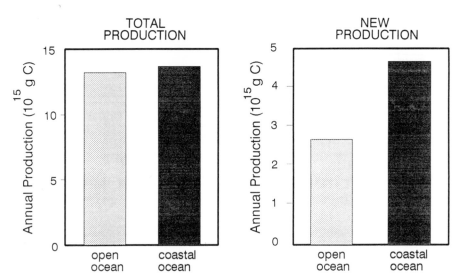

Fig. 5. Comparison of global carbon production in oceanic and coastal regions. Left: Total Carbon Production - oceanic and coastal regions are roughly equivalent; Right: New Production - coastal regions contribute almost twice as much as oceanic regions.

Following these computations for specific regions, the individual values were compiled to contrast the influence of all coastal and oceanic locales for both total production and new production (Fig. 5); the Arctic and Antarctic regions were included as coastal areas. This analysis shows an equivalent amount of annual carbon production for oceanic and coastal regions (13.2 vs 13.7 gt C y^{-1}). Likewise, the estimated distributions of new production showed 2.7 gt C y^{-1} for oceanic waters and 4.7 gt C y^{-1} for coastal regions. These analyses indicate that coastal regions support 64% of new production while oceanic regions contribute only 36%. This calculation emphasizes the importance of coastal high latitude environments in global new production.

CONCLUSIONS

Recent analyses suggest that coastal high latitude and open ocean systems provide roughly equal amounts of annual oceanic carbon production, but that coastal high latitude systems may provide almost twice the new production as the open ocean. The low rate of nutrient supply in oligotrophic, open ocean systems supports populations of small, slowly sinking cells. Little of this material leaves the photic zone directly through sinking, and microherbivore grazing generates fecal pellets that sink relatively slowly. In contrast, high latitude coastal systems are characterized by high nutrient supply rates and large, rapidly sinking cells. In this setting, herbivorous grazing produces fecal pellets with extremely fast sinking rates. Much of the biomass produced in high latitude coastal systems reaches the bottom quickly because of relatively shallow water and cold water temperatures, which tend to uncouple production from grazing. Thus, the role of high latitude coastal systems in global production and flux is perhaps more important than that of open ocean systems.

ACKNOWLEDGEMENTS

Production/sedimentation research in Auke Bay, Alaska was conducted as part of the APPRISE program, under Contract No. NA-85-ABH-0022 from the U.S. Department of Commerce, National Oceanic and Atmospheric Administration. Earlier research on sinking rates was supported by numerous grants from the National Science Foundation.

REFERENCES

Berger, W. H., Fischer, K., Lai, C., and Wu, G., 1987, Ocean productivity and carbon flux. Part I., Overview and maps of primary production and export production, *SIO Ref.*, 87-30.

Berger, W. H., Smetacek, V. S., and Wefer, G., 1989, Ocean productivity and paleoproductivity - an overview, *in*: "Productivity of the Ocean: Present and Past, Report of the Dahlem Workshop, Berlin," W. H. Berger, V. S. Smetacek, and G. Wefer, eds., Wiley, New York.

Beltzer, P.R., Howers, W.H.S., Laws, E.A., Win, C.D., Ditullio, G.R., and Kroopnick, P.M., 1984, Primary production and particulate fluxes on a transect of the equator at 153 degrees West in the Pacific Ocean, *Deep-Sea Res.*, 31:1.

Bienfang, P. K., 1979, Instruments and Methods. A new phytoplankton sinking rate method suitable for field use, *Deep-Sea Res.*, 26:719.

Bienfang, P. K., 1981, SETCOL - A technologically simple and reliable method for measuring phytoplankton sinking rates, *Can. J. Fish. Aquat. Sci.*, 38:1289.

Bienfang, P. K., Laws, E., and Johnson, W., 1977, Phytoplankton sinking rate determination: Technical and theoretical aspects, an improved methodology, *J. Exp. Mar. Biol. Ecol.*, 30:283.

Bienfang, P. K., 1980a, Phytoplankton sinking rates in oligotrophic waters off Hawaii, USA, *Mar. Biol.*, 61:69.

Bienfang, P. K., 1980b, Herbivore diet affects fecal pellet settling, *Can. J. Fish. Aquat. Sci.*, 37:1352.

Bienfang, P. K., and Szyper, J. P., 1981, Phytoplankton dynamics in the Subtropical Pacific Ocean off Hawaii, *Deep-Sea Res.*, 28:981.

Bienfang, P. K., 1982, Phytoplankton sinking-rate dynamics in enclosed experimental ecosystems, *in*: "Marine Mesocosms, Biological and Chemical Research in Experimental Ecosystems," G. D. Grice and M. R. Reeve, eds., Springer Verlag, New York.

Bienfang, P. K., Harrison, P. J., and Quarmby, L. M., 1982, Sinking rate response to depletion of nitrate, phosphate and silicate in four marine diatoms, *Mar. Biol.*, 67:295.

Bienfang, P. K., and Szyper, J. P., 1982, Effects of temperature and salinity on sinking rates of the centric diatom *Ditylum brightwelli*, *Biol. Oceanogr.*, 1:211.

Bienfang, P. K., and Takahashi, M., 1983, Ultraplankton growth rates in a subtropical ecosystem, *Mar. Biol.*, 76:213.

Bienfang, P. K., Szyper, J., and Laws, E., 1983, Sinking rate and pigment responses to light-limitation of a marine diatom: Implications to dynamics of chlorophyll maximum layers, *Oceanologica Acta*, 6:55.

Bienfang, P. K., Szyper, J. P., Okamoto, M. Y., and Noda, E. K., 1984, Temporal and spatial variability of phytoplankton in a subtropical ecosystem, *Limnol. Oceanogr.*, 29:527.

Bienfang, P. K., 1984, Size structure and sedimentation of biogenic microparticulates in a subarctic ecosystem, *Journal of Plankton Research*, 6:985.

Bienfang, P. K., 1985a, Sedimentation of suspended microparticulate material in the point conception upwelling ecosystem, *in*: "A Technical Report of Research Performed During the 1983 OPUS II Fieldwork," The Oceanic Institute, Waimanalo, Hawaii.

Bienfang, P. K., 1985b, Size structure and sinking rates of various microparticulate constituents in oligotrophic Hawaiian Waters, *Mar. Ecol. Pro. Ser.*, 23:143.

Bienfang, P. K., and Harrison, P. J., 1984a, Co-Variation of sinking rate and cell quota among nutrient replete marine phytoplankton, *Mar. Ecol. Pro. Ser.*, 14:297.

Bienfang, P. K., and Harrison, P. J., 1984b, Sinking-rate response of natural assemblages of temperate and subtropical phytoplankton to nutrient depletion, *Mar. Biol.*, 83:293.

Chavez, F. P., and Barber, R. T., 1987, An estimate of new production in the Equatorial Pacific, *Deep-Sea Res.*, 34:1229.

Eppley, R. W., Holmes, R. W., and Strickland, J. D. H., 1967, Sinking rates of marine phytoplankton measured with a fluorometer, *J. Exp. Mar. Biol. Ecol.*, 1:191.

Eppley, R. W., and Peterson, B. J., 1979, Particulate organic matter flux and planktonic new production in the deep ocean, *Nature*, 282:677.

Iverson, R. L., 1990, Control of marine fish production, *Limnol. Oceanogr.*, 35:1593.

Kanda, J., Ziemann, D. A., Conquest, L. D., and Bienfang, P. K., 1990, Nitrate and ammonium uptake by phytoplankton populations during the spring bloom in Auke Bay, Alaska, *Estuarine, Coastal and Shelf Science*, 30:509.

Koblentz-Mishke, O. I., Volkovinski, V. V., and Kabanova, J. G., 1970, Plankton primary production of the world ocean, *in*: "Scientific Exploration of the South Pacific," W. Wooster, ed., National Academy of Sciences, Washington, DC.

296

Laws, E. A., Redalje, D. G., Haas, L. W., Bienfang, P. K., Eppley, R. W., Harrison, W. G., Karl, D. M., and Jarra, J., 1984, High phytoplankton growth and production rates in oligotrophic Hawaiian coastal waters, *Limnol. Oceanogr.*, 29:1161.

Laws, E. A., Bienfang, P. K., Ziemann, D. A., and Conquest, L. D., 1988, Phytoplankton population dynamics and the fate of production during the spring bloom in Auke Bay, Alaska, *Limnol. Oceanogr.*, 33:57.

Lorenzen, C. J., Welschmeyer, N. A., and Copping, A. E., 1983, Particulate organic carbon flux in the subarctic Pacific, *Deep-Sea Res.*, 30:639.

Mann, K.H., 1984, Fish production in open ocean ecosystems, *in*: "Flows of Energy and Materials in Marine Ecosystems," *NATO Conf. Ser. 4. Mar. Sci., 13*, Plenum Press, New York.

Platt, T., and Subba Rao, D. V., 1975, Primary production of marine macrophytes, *in*: "Photosynthesis and Productivity of Different Environments," *International Biological Programme*, 3:249, Cambridge University Press.

Ryther, J. H., 1969, Photosynthesis and fish production in the sea, *Science*, 166:72.

Seuss, E., 1980, Particulate organic fluxes in the oceans - surface productivity and oxygen utilization, *Nature*, 228:260.

Shushkina, E. A., 1985, Production of principal ecological groups of plankton in the epipelagic zone of the ocean, *Oceanology*, 25:653.

Smayda, T. J., 1970, The suspension and sinking of phytoplankton in the sea, *Oceanogr. Mar. Biol. Ann. Rev.*, 8:353.

Smayda, T. J., and Bienfang, P. K., 1983, Suspension properties of various phyletic groups of phytoplankton and tintinnids in an oligotrophic subtropical system, *Mar. Ecol.*, 4:289.

Steele, J. H., and Yentsch, C. S., 1960, The vertical distribution of chlorophyll, *J. Mar. Biol. Assoc. U.K.*, 39:217.

Takahashi, M., and Bienfang, P. K., 1983, Size structure of phytoplankton biomass and photosynthesis in subtropical Hawaiian waters, *Mar. Biol.*, 76:203.

Wassman, P., 1988, Primary production and sedimentation, *in*: "Sediment Trap Studies in the Nordic Countries," P. Wassman, and A.-S. Heiskanen, eds., Ylio pisto-paino, Helsinki, Finland.

Wyrtki, K., 1963, The horizontal and vertical field of motion in the Peru Current, *Bulletin of the Scripps Institute of Oceanography*, 8:313.

Wyrtki, K., 1981, An estimate of equatorial upwelling in the Pacific, *J. Phys. Oceanogr.*, 11:1205.

Ziemann, D. A., Conquest, L. D., Fulton-Bennett, K. W., and Bienfang, P. K., 1990, Interannual variability in the Auke Bay phytoplankton, *in*: "APPRISE - Interannual Variability and Fisheries Recruitment," D.A. Ziemann and K.W. Fulton-Bennett, eds., The Oceanic Institute, Honolulu.

TRACER BASED INFERENCES OF NEW PRIMARY PRODUCTION IN THE SEA

W.J. Jenkins

Department of Chemistry
Woods Hole Oceanographic Institution
Woods Hole, MA 02543

D.W.R. Wallace

Department of Applied Science
Brookhaven National Laboratory
Upton, NY 11973

INTRODUCTION

Much of what we have learned about the oceans, particularly in the early decades of this century, stems from inferences drawn from the distribution of properties in the ocean. The pioneers of oceanography used observations of first, temperature and salinity, and then passive, nonconservative tracers, such as oxygen and dissolved nutrients to deduce the origins, pathways, and fates of water masses. The development of the dynamical methods further placed broad constraints on the rates of the processes involved. Recently, measurement of transient tracers, ie. the distributions of those substances resulting from human activities, which are changing with time, coupled with better knowledge of the "classical" tracer distributions and the advent of sophisticated numerical models, is leading to a more quantitative and complete understanding of the ocean circulation and ventilation. The quantification of these physical processes enables us to estimate the rates of biogeochemical processes in the ocean.

In this paper we summarize and review three geochemical techniques for estimating the "new" primary productivity in the sea. By "new" primary production we mean the export of organic carbon from the euphotic zone, which could be in either particulate or dissolved form. We discuss the interpretation of dissolved gas and transient tracer data, recognizing that additional tracers can be used to examine particle fluxes. The conclusions we draw from these studies are inferences, which are by nature model dependent. Wherever possible, we seek additional, independent evidence which bears on our individual conclusions, and as many different constraints as possible. The case that we build here is predicated on three independent lines of evidence, each of which provide mutually consistent results. The approaches are:

1. The consumption of oxygen in the aphotic zone results from the production and export of organic carbon from above. Apparent oxygen utilization rates have been estimated in several ways, primarily from large scale tracer balances.

Primary Productivity and Biogeochemical Cycles in the Sea
Edited by P.G. Falkowski and A.D. Woodhead, Plenum Press, New York, 1992

299

The vertical integral of the oxygen utilization rates (OUR) estimates new production. The advantage of this approach is that it provides long-term (seasonal to decadal), large-scale (*ca* 1000 km) average estimates of what is likely to be a spatially and temporally variable flux.

2. Oxygen production rates in the euphotic zone also provide an estimate of new production. In the subtropics, a subsurface oxygen excess in the summertime is produced as a result of a dominance of photosynthesis over respiration. By carefully accounting for the physical processes which can alter concentrations of oxygen, it is possible to estimate the seasonal "excess photosynthesis", which should be directly related to new production.

3. The flux of nitrate into the euphotic zone, which supports this new production, can be determined. The balance between the upward flux of tritiugenic ^{3}He from the thermocline and the loss-rate to the atmosphere from the mixed layer can be used to estimate the long-term average nitrate flux into the euphotic zone. This can be compared with nitrate flux estimates based on steady-state mixing processes.

Some implications of the results obtained from these geochemical approaches to estimating new production also need to be addressed. The vertical transport of nutrients estimated by the last approach makes demands on the physical processes involved. There are questions regarding the detailed oxygen:nutrient budgets associated with new production in the subtropics. Climatic variations in watermass renewal, and subsequent variations in the supply nutrients to the upper ocean via deep convection suggest a corresponding modulation of new production on interannual timescales. These issues are discussed in the last section.

APHOTIC ZONE OXYGEN UTILIZATION

The undersaturation of dissolved oxygen below the euphotic zone represents a balance between consumption because of the oxidation of sinking organic carbon, and the resupply of oxygen via advection and mixing. If the oxygen distribution is in steady-state, then the vertically integrated oxygen demand must be balanced by a stoichiometrically equivalent supply of organic carbon. Estimation of oxygen utilization rates (OUR) using large scale balances is particularly attractive because it yields averages of spatially and temporally variable processes.

Estimates of OUR based on tracers fall into two broad categories: box model (inventory) estimates, and advective-diffusive estimates. Box models represent the ultimate simplification of the processes at work, and hence are most subject to concerns regarding initial and boundary conditions, and the assumption of interior homogenization. Estimates, for example, of thermocline OURs based on tritium box models (Jenkins, 1980; Sarmiento et al., 1990) are consistently low (Doney and Jenkins, 1988; Sarmiento et al., 1990) because of differences in the upper ocean "boundary conditions" of tritium and oxygen. This is discussed in a later section. While it is difficult to generalize, such estimates may be as much as 2 to 3 times low (Doney and Jenkins, 1988).

Calculations of OURs based on advection-diffusion balances incorporate more complete physics, and hence are more attractive. The value of such estimates, however, depends on the integrity of the "calibration" of the advection-diffusion terms. For example, Riley (1951) used a three-dimensional advection-diffusion model of the North

Atlantic based on geostrophic velocity estimates. The results are thus sensitive to the choice of the depth of no motion: the shallow values, where velocities and OURs are greatest, are most robust, but the deeper values are subject to proportionally greater uncertainty. Studies using tritium-^{3}He age gradients in the eastern North Atlantic (Jenkins, 1982a; 1987) obtained values similar to Riley's shallow results, but were consistently greater below 250m depth. These results are particularly compelling, since the tritium-^{3}He age gradients agreed with the beta-spiral velocity estimates, and made no assumptions about initial conditions. Sarmiento et al. (1990) used ^{228}Ra versus apparent oxygen utilization (AOU) correlations in the main thermocline of the North Atlantic to estimate basin scale OURs. The values obtained were even higher, possibly due to a real geographic variation in OUR. Studies of CFC distributions also were used to estimate OURs (Wallace et al., 1987; Doney and Bullister, 1991).

^{3}H-^{3}He dating represents a powerful means by which to estimate OURs, by providing an apparent "age" for a water parcel with respect to contact with the sea surface. Further, the similarity between the behavior of ^{3}He and oxygen at the sea surface lends credence to the OUR estimates (eg. Jenkins, 1977; 1980), although the Age and AOU gradients are the most compelling constraints. Figures 1, 2, and 3 show maps of pressure, apparent oxygen utilization, and tritium-^{3}He age interpolated onto four density surfaces (data taken from the several cruises in 1981). The data show the effects of ventilation of the subtropical gyre along isopycnals outcropping in the north (with some rotation to the

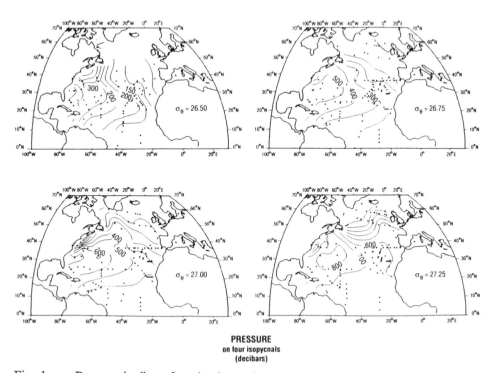

PRESSURE
on four isopycnals
(decibars)

Fig. 1.   Pressure in db on four density surfaces in the North Atlantic as sampled during the Transient Tracers in the Ocean survey of 1981. The dots indicate station positions used in the mapping.

NW for the deeper layers). AOU and ^{3}H-^{3}He age show roughly similar patterns because they share similar boundary conditions at the isopycnal outcrops. The effect of progressive carbon oxidation with time is clear from the maps.

Effects of Mixing on OUR Estimates

It might be considered straightforward to estimate oxygen utilization rates from these maps. One would be inclined to use a simple correlation between Age and AOU to estimate OUR. The concern, however, is over the influence of mixing in determining the distribution of tracers. Consider the advection-diffusion balance equation for oxygen (O) as:

$$0 = \kappa \nabla^2 O - \vec{u} \cdot \nabla O + J \qquad (1)$$

where κ is the eddy diffusivity, $\vec{u}$ is the velocity, O is the oxygen concentration, and J is the oxygen utilization rate (negative for consumption, positive for production). It is possible to estimate the magnitude of the error introduced by ignoring mixing and assuming an advective balance (ie. between the last two terms). We can compute a Peclet number based on typical upper thermocline velocities ($\sim 10^{-2}$ ms^{-1}), expected isopycnal

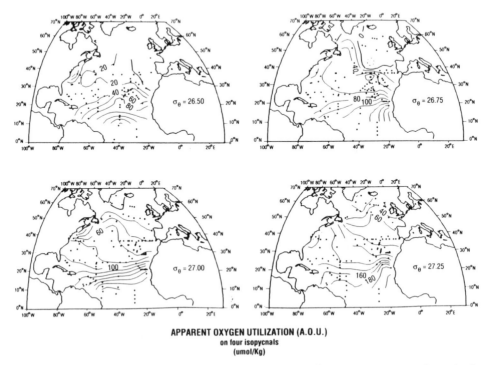

APPARENT OXYGEN UTILIZATION (A.O.U.)
on four isopycnals
(umol/Kg)

Fig. 2.   Apparent Oxygen Utilization (AOU) in μmol kg^{-1} on four density surfaces in the North Atlantic as sampled during the Transient Tracers in the Ocean survey of 1981. The dots indicate station positions used in the mapping.

eddy diffusivities ($\sim 10^3 m^2 s^{-1}$), and the horizontal length scale imposed by oxygen gradients ($\sim 10^6 m$). The value, about 10, implies that an error of about 10% is introduced throughout most of the subtropical gyre by ignoring mixing. This is borne out by direct determination of the first two terms in (1) in the Beta Triangle area in the North Atlantic (Jenkins, 1987). In regions of strong gradients or mixing, this approach needs to be more carefully examined. Further, these essentially qualitative arguments need to be supplemented with more rigorous numerical simulations.

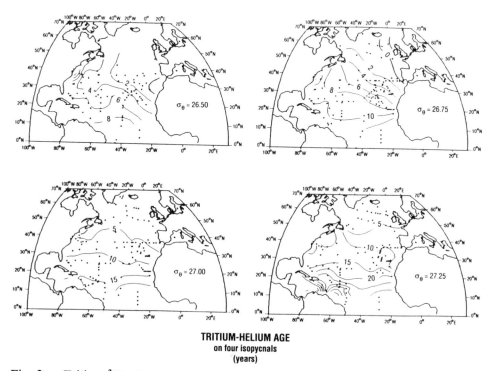

TRITIUM-HELIUM AGE
on four isopycnals
(years)

Fig. 3.   Tritium-^{3}He Age (years) on four density surfaces in the North Atlantic as sampled during the Transient Tracers in the Ocean survey of 1981. The dots indicate station positions used in the mapping.

The ^{3}H-^{3}He age may not be an ideal age tracer everywhere, because it is computed from the concentrations of two transient (i.e. non-steady state) tracers. Conceptually, the problem arises for mixtures of water parcels with two different ages: the resultant age of the mixture is usually biased towards the younger parcel's age, which is likely to have higher absolute ^{3}H concentrations. Jenkins (1987) has shown that the advection-diffusion equation for the tritium-^{3}He age can be written as:

$$\frac{\partial \tau}{\partial t} = \kappa \nabla^2 \tau - \vec{u} \cdot \nabla \tau + 1 + \kappa \left(\frac{\nabla \zeta}{\zeta} + \frac{\nabla \theta}{\theta} \right) \cdot \nabla \tau \qquad (2)$$

303

where τ is the tritium-helium age, θ is the tritium concentration, and ζ is the sum of the tritium and ^{3}He concentrations. Except for the last two terms and the unsteady term, this equation resembles that of an "ideal age tracer" (A):

$$0 = \kappa\nabla^2A - \vec{u}\cdot\nabla A + 1 \qquad (3)$$

The deviations from age ideality can be estimated in at least two ways, by:

1. directly estimating the magnitude of the non-linear terms in (2), using assumed eddy diffusion coefficients. In the upper 600-700 m in the eastern North Atlantic, the effects are about 10% or less (Fig. 4).

2. numerical model simulations: eg., of an idealized gyre circulation with reasonable volume transports and lateral diffusivities (Fig. 5).

Both approaches indicate that throughout much of the sub-tropical gyre, the ^{3}H-^{3}He age differs from the idealized age tracer by 10% or less. Consideration of all aspects of the problem, then, gives at most about 20% uncertainty to the OUR estimates made using tritium-^{3}He dating in the bulk of the North Atlantic subtropical gyre.

OUR and New Production Estimates

Despite the broad geographical coverage, and the differences in techniques, the above estimates, when vertically integrated, imply an oxygen demand of about 5 - 6 mol $(O_2)m^2y^{-1}$ (Fig. 6). Some of the scatter shown in the figure will be real (i.e., resulting from regional variations in OUR, or its distribution with depth), and some may be due to biases associated with the method for obtaining these estimates. Nonetheless, the general agreement despite rather different approaches (ranging from geostrophic to tracer mass balance calculations) builds considerable confidence in the results. Using an estimated Redfield oxygen:carbon ratio of 1.65 (Takahashi et al., 1985), the vertically integrated oxygen demand corresponds to a new productivity of about 3.0-3.5 mol(C)m^{-2}y^{-1} for the oligotrophic North Atlantic.

Arctic Ocean Shelf-Export Production

Tracer techniques are being applied to other ocean basins as well. Wallace et al. (1987) calculated the integrated OUR for a station in the central Arctic Ocean based on several highly simplified models calibrated with CFC (Freon) data. The approach was based on the vertical distribution of temperature, salinity, CCl_3F, and CCl_2F_2. The Arctic Ocean sub-surface temperature can be considered to be a conservative, dynamically-passive tracer which allows the effects of along-and across-isopycnal mixing processes on the ventilation of the Arctic Ocean halocline to be crudely separated using a 1.5-D model. In this remote region, there are insufficient data to address the appropriateness of the models used to calculate ventilation time scales, and the uncertainties of the OUR estimates are likely to be considerably larger than those derived from the sub-tropical North Atlantic ^{3}H-^{3}He ages. Nevertheless, a variety of simple models converged on an integrated AOUR for the central Arctic Ocean halocline of about 1mol (O_2) m^{-2}y^{-1}. The carbon equivalent of this apparent *new* production (approximately 0.6 mol(C) m^{-2}y^{-1}) is 5 to 10 times larger than the estimated *total* production in the ice-covered central Arctic basins, strongly suggesting that the OUR represents the effects of oxidation of carbon fixed over the continental shelves. This OUR estimate, therefore, represents a measure of export production from marginal

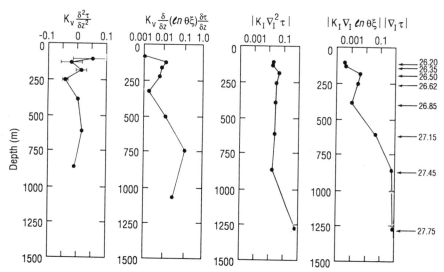

Fig. 4. Horizontal and vertical mixing terms in the tritium-^{3}He age equation for the eastern North Atlantic (from Jenkins, 1987). Note that these values should be compared with 1, and are less than 10% in the upper 600 m.

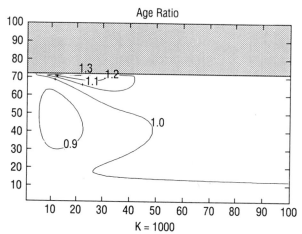

Fig. 5. The ratio of tritium-^{3}He age to "ideal" tracer age for a two-dimensional Stommel gyre with a total transport of 50×10^6 m^{3}s^{-1} and a horizontal diffusivity of 10^3 m^{2}s^{-1}. Note that the ratio is within 10% of unity throughout most of the gyre. This calculation is a replication of the experiment performed by Thiele and Sarmiento (1990).

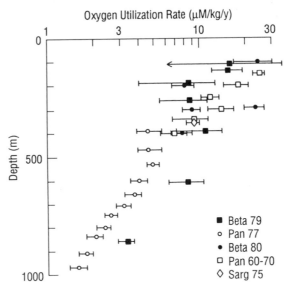

Fig. 6.   Various estimates of oxygen utilization rates (in μmol kg^{-1}) from several locales in the subtropical oligotrophic North Atlantic.

production from marginal seas to the basin interior. Hence, the extreme case of the Arctic Ocean, where production takes place almost exclusively at the basin margins, serves to emphasize the spatial averaging which is built into OUR-based estimates of new production.

OXYGEN PRODUCTION IN THE EUPHOTIC ZONE

The observation of seasonal, sub-surface oxygen maxima in the sub-tropics has led to attempts to relate the seasonal oxygen cycle to primary production. Shulenberger and Reid (1981) compared their estimates of primary production to ^{14}C incubation measurements and found roughly comparable results (see also Platt, 1984; Reid and Shulenberger, 1986). This is curious, however, since the latter are a determination of net productivity, whereas the accrual of oxygen in upper ocean waters must be a result of new production (nutrient regeneration entails oxygen consumption).

The problem is complicated by the need to account for physical processes in the upper water column: thermal cycling will produce apparent over- and under-saturations, air injection will supersaturate gases, and air-sea gas exchange will drive them towards equilibrium. Platt (1984) criticized Reid and Shulenberger's conclusions for many of these reasons. Craig and Hayward (1987) assessed the fraction of the observed oxygen excess, which was biological rather than physical, by using argon as an abiogenic analog of oxygen, but did not quantitatively estimate biological production.

Using the time series data obtained at station "S" near Bermuda, Jenkins and Goldman (1985) estimated new production with simple calculations, and obtained values of about 5 mol(O_2)m^{-2}y^{-1}. Chou (1985) and later Musgrave et al. (1988) used the Price et al. (1986) upper ocean model to simulate the seasonal cycle of oxygen at Bermuda, and obtained comparable values. Chou estimated an oxygen productivity almost identical to the estimate of Jenkins and Goldman, but the estimate by Musgrave et al. was lower. Part of the difference may be attributed to the formulation of gas transfer processes, in particular air injection, and part may be due to differences in the wind forcing used in the model. Such calculations, while physically well constrained (in particular by observations of the temperature and density fields), lack specific "calibration" of important processes such as air injection and gas exchange. Parameterization of those processes depended on other field experiments, and observations of the oxygen distributions could not be used to test them.

The above approaches were combined and augmented by measurement and modeling of a three year time series from 1985 to 1987 - again near Bermuda - of argon, helium, and oxygen measurements (Spitzer and Jenkins, 1989; Spitzer, 1989). The upper ocean's response to the seasonal heating and cooling cycle (Fig. 7) is to attempt to "breath" argon in and out (Fig. 8). The magnitude of the residual supersaturation (Fig.9) is a measure of vertical mixing and gas exchange rates. Vertical mixing rates also are tightly constrained by the seasonal temperature distribution. In the mixed layer, helium supersaturations are a strong function of air injection and gas exchange rates. Oxygen feels the combined forcing of both the physical and biological processes. These data were interpreted in the framework of an enhanced upper ocean model (Price et al., 1986), which incorporated gas exchange and two modes of air injection (partial vs complete bubble trapping). Sensitivities of several critical observational indices to variations in model parameters were established by Monte Carlo techniques, and a series of linearized indicial equations were constructed. Sufficient constraints were chosen to make the system overdetermined so that it could be solved in a least-squares sense. The intent was to test and evaluate model performance as well as to estimate the parameters. The indicial equations were then solved using the time series observations coupled with linear inverse methods to obtain both the optimal solutions and their uncertainties. The advantage of this approach is that it can also be used to explore which observations most tightly constrained the various processes. This has not only pedagogical value, but also permits the design of improved field experiments for the future.

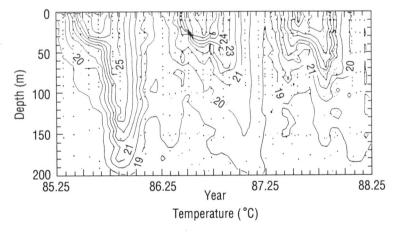

Fig. 7. The time series from 1985 to 1987 of temperature at station S near Bermuda (from Spitzer, 1989).

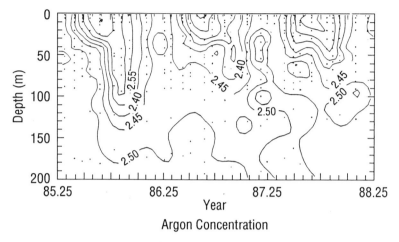

Argon Concentration

Fig. 8.   The time series from 1985 to 1987 of dissolved argon at station S near Bermuda (from Spitzer, 1989). Note the decreased concentrations during the summer warming period.

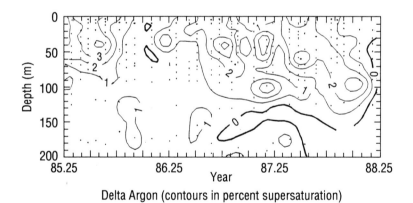

Delta Argon (contours in percent supersaturation)

Fig. 9.   The time series from 1985 to 1987 of argon percent supersaturation at station S near Bermuda (from Spitzer, 1989). Note the subsurface maximum due to thermal effects in the seasonal thermocline.

308

Table 1. Optimum Solutions for He- Ar- O_2 Time Series Near Bermuda

Parameter	Results	Units
Vertical Diffusivty	1.02 ± 0.06 x 10^{-4}	m^2 s^{-1}
Gas Exchange Rate	0.94 ± 0.11	rel. to Liss and Merlivat
Air Injection Amplitude	0.40 ± 0.10	non-dimensional
A.I. Wind Speed Exponent	2.10 ± 0.10	non-dimensional
A.I. Fraction:Trapping	0.50 ± 0.10	non-dimensional
New Production: Mixed Layer	9.80 ± 3.50	mol O_2 m^{-2} y^{-1}
New Production: at 75m	5.00 ± 1.00	mol O_2 m^{-2} y^{-1}
New Production: at 75m (ML=0)	6.20 ± 0.90	mol O_2 m^{-2} y^{-1}

Table 1 lists the optimum solutions derived from the three-year experiment (from Spitzer, 1989). The remarkable aspect of these results is that they provide relatively precise (about 10-20%) estimates of biogeochemically important processes on seasonal time scales. An important aspect of the study was the decomposition of the oxygen production function into a mixed layer and a sub-mixed layer component. This approach was required because the mixed layer production was poorly constrained, due to the dominance of air-sea gas exchange and the smallness of the resultant O_2 excesses over equilibrium. Estimates of the mixed-layer production were very large with correspondingly large systematic uncertainties. Not much faith can be placed in these mixed layer production estimates, because small errors in either observations or the air injection formulation could contribute to large uncertainties. The sub-mixed layer production was considerably better constrained, with an average for the three-year period of 5 ± 1 mol(0_2)m^{-2}y^{-1}. The character of the solutions was such that if the mixed layer productivity were forced to be zero, the sub-mixed layer component would rise by another 1 mol(0_2)m^{-2}y^{-1}. An estimate of 6.2 ± 1 mol(0_2)m^{-2}y^{-1} is, therefore, a lower limit to the new oxygen production. It is ironic that the two principle sources of uncertainty in the results turned out to be variance in the oxygen measurements, and poor knowledge of mean wind speed.

The results in Table 1 are from Spitzer (1989), and are slightly different from the results derived from a single year of data presented by Spitzer and Jenkins (1989), partly because of the longer period covered. A notable difference from the earlier treatment is that the gas exchange formulation now takes into account the climatological atmospheric pressure variability. This change brought the best-fit gas exchange coefficients into complete agreement with the Liss and Merlivat (1986) relationship. Further numerical experimentation using all of the noble gases indicates that both improved precision and better model testing and formulation could be attained by expanding the range of physical properties (molecular diffusivity, solubility, etc). This is particularly important if we are to successfully determine production in the mixed layer.

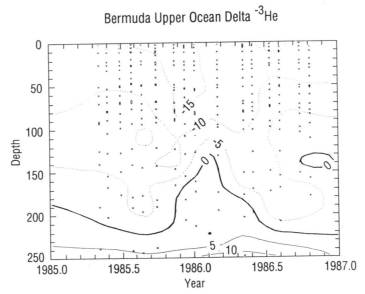

Fig. 10. A two-year record of the helium isotope ratio anomaly in the upper waters near Bermuda (from Jenkins, 1988a). Note that approximate equilibrium with the atmosphere is -17°/∞, a value affected by air injection and gas exchange.

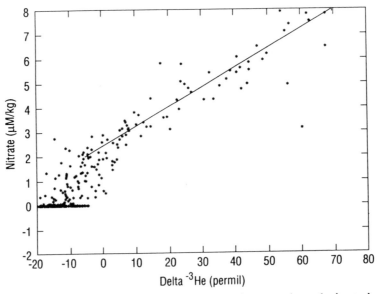

Fig. 11. The correlation between helium isotope ratio anomaly and nitrate in the main thermocline near Bermuda (from Jenkins, 1988a).

Oxygen Production and the New Production of Carbon

There is some uncertainty in the appropriate carbon:oxygen ratio to use in new production. The consensus is that the ratio is highest for the assimilation of nitrate (vs reduced forms of nitrogen). Values ranging from 1 to as high as 2 or more have been suggested (eg. Raine, 1983; Megard et al., 1985); convincing arguments based on biochemical mass balances (Laws, 1991) coupled with a reassessment of techniques (Williams and Roberston, 1991) strongly argue for a value of about 1.4 . This leads to a new production of carbon >3.6 or >4.4 mol(C)m^{-2}y^{-1} for the two scenarios outlined in the previous section.

ESTIMATING THE NITRATE FLUX INTO THE EUPHOTIC ZONE USING ^{3}He

Excess ^{3}He derived from tritium decay is observed in the mixed layer year-round (Jenkins, 1988a, Fig. 10). This excess is maintained in the presence of a gas exchange flux from ocean to atmosphere. From measurements of the excess ^{3}He and knowledge of the (wind-speed dependent) gas exchange coefficient for He, we can estimate the annual flux of ^{3}He. The magnitude of this flux is much larger (by an order of magnitude) than can be supported by tritium decay in the mixed-layer, and implies that it must ultimately come from tritium decay in the main thermocline. This tritiugenic ^{3}He flux must be transported upward by physical processes. In the main thermocline there is a strong correlation between nitrate and ^{3}He (Fig. 10) extending down to the ^{3}He and nutrient maxima at approximately 600m. Using this correlation, and our knowledge of the ^{3}He flux, we can estimate the nitrate flux into the euphotic zone (Jenkins, 1988b).

It is important to note that the upward ^{3}He flux into the euphotic zone from the main thermocline occurs year-round, and is not simply a result of winter convection. Note the summertime build up of ^{3}He in the seasonal pycnocline: i.e. beneath the mixed layer, but within the euphotic zone. This ^{3}He build-up should be associated with a nitrate flux (and primary production) but is not expressed in the summertime mixed-layer (and outgassed) because of restricted mixing across the seasonal pycnocline.

In calculating the annual loss of ^{3}He to the atmosphere, it is critical to assess the "dynamic equilibrium" concentration in the mixed layer. This dynamic equilibrium is a function of both the gas solubility and isotope fractionating effect of processes such as air-sea gas exchange and the amount and nature of air injection. Depending on the nature of the air injection process, it is possible to either elevate or depress the dynamic equilibrium He isotope ratio significantly relative to static solubility equilibrium. This effect has been addressed by modeling the distribution of argon (see above), and alternatively through consideration of the relationship between neon and helium saturation anomalies. In particular, the seasonal gas cycling model has been run for helium and neon, and the resulting range of saturation anomalies has been compared to surface water data collected during GEOSECS Atlantic cruises to determine the best-fit mode of air injection.

Nitrate Flux and New Production Estimates

The best-estimate ^{3}He fluxes resulting from such studies is $21 \pm 6°/_{\circ\circ}$ md^{-1}, of which only 10% can be attributed to the decay of tritium in the euphotic zone. The remainder of the ^{3}He flux must be derived from the main thermocline. This observed flux is consistent with the long-term decrease of ^{3}He and tritium from the main thermocline in this region although the agreement may be fortuitous: vertical fluxes of thermocline properties may occur laterally over significant distances, and it is unreasonable to demand a local, vertical balance. The agreement, nonetheless, is encouraging. Using the nitrate-^{3}He relationship, the associated flux of nitrate is 0.6 mol(NO_3)m^{-2}y^{-1}. Altabet (1989) determined a vertical particulate nitrogen flux of 0.33 Mm^{-2}y^{-1} from several short sediment trap deployments over one year. Considering temporal-spatial sampling issues, those results are consistent. Depending on the choice of C:N ratios, which range from 5.7 to 7.5 (eg. see Laws, 1991), the observed nitrate flux is stoichiometrically consistent with a new production of 3.4 - 4.5 mol(C)m^{-2}y^{-1}, with a relative uncertainty of about 30%.

It should be noted that the oceanic cycling of nutrients may not be fully understood. Sarmiento et al. (1990) constructed nitrate budgets for the subtropical North Atlantic, which argue for significant lateral fluxes of dissolved organic nitrogen and carbon. Such large scale balances can place important constraints on basin wide new production, but the calculations are singularly difficult with the available data.

SUMMARY OF RATE ESTIMATES

Three independent estimates of new production converge on a new production estimate of approximately 3-4 mol(C)m^{-2}y^{-1} for the oligotrophic subtropical gyre of the North Atlantic. These estimates are characterized by time scales ranging from at least seasonal to decadal, and are inherently averaged over spatial scales of about 1000 km to basin scale. This inherent spatial and temporal averaging must be considered when comparing to *in vitro* estimates of new production based on bottle incubations and nutrient uptake studies (eg. Platt et al., 1989), or to short deployment sediment trap results.

IMPLICATIONS AND CONUNDRUMS

The Vertical Transport of Nitrate

The estimates given above imply an upward nitrate flux of about 0.6 mol m^{-2}y^{-1}. However, the observed vertical nitrate gradients in the sub-tropical North Atlantic are about 2 to 3 x 10^{-5}mol(NO_3)m^{-4}, which together with an upper limit vertical diffusivity of about 10^{-4}m^{2}s^{-1} predicts a vertical flux 5 to 10 times lower. This estimate is conservative (i.e., high), because the diffusivity used, though consistent with seasonal temperature and argon budgets in the seasonal thermocline, is large compared to estimates for the main thermocline (cf. Jenkins, 1980; Schmitt, 1981), which is the ultimate source of nutrients. To insist on higher rates of diapycnal diffusivity would violate upper ocean thermal balances, aside from the fact that there is insufficient energy to support such high mixing rates (eg., Garrett, 1979). Further, short-term direct measurements of nitrate uptake and field estimates of nitrate flux support the lower flux (Lewis et al., 1986).

There is a possible explanation for the discrepancy. The only feasible mechanism for separation of the fluxes of heat and other properties is isopycnal transport. Although the large scale, climatological mean gradients in properties are insufficient to support the required fluxes with reasonable along-isopycnal diffusivities, one might propose smaller scale, eddy related processes. Jenkins (1988b) reported an episodic injection event associated with the passage of a mode-water eddy. The feature was responsible for lifting nutrient (and ^{3}He) bearing isopycnals into the euphotic zone. In addition, the high lateral velocity shears associated with the eddies (Brundage and Dugan, 1986) provide energy for mixing and transport. The existence of such features in the North Atlantic were documented (Brundage and Dugan, 1986). Other transient events have been reported (eg.

McGowan and Hayward, 1978; Falkowski et al., 1991) which have significant impact on local production. However, the question remains as to whether such features occur in sufficient quantity to provide the necessary flux. There are several critical unknowns: the amount of mixing or primary production associated with each feature, how it scales with the strength and evolution of the eddy, and their frequency of occurrence (see Spitzer, 1989). Falkowski et al. (1991) made an interesting estimate by scaling their observation of the primary production enhancement factor associated with eddies by the ratio of eddy to mean kinetic energy for the Pacific, suggesting an enhancement of only about 20% to the total gyre productivity. There are two aspects of this calculation which need to be thought about. First, Falkowski et al. may have underestimated the degree of local productivity enhancement, because their station line cut through the edge of the eddy. The degree to which this may be a problem is unclear, because the dependence of production on position in the eddy has not been mapped out. Second, the enhancement of nutrient flux, and hence, productivity may not scale simply with the kinetic energy ratio, which projects the energy available for mixing. The flux of nutrients also depends on the enhancement of gradients associated with the local isopycnal deformation.

We conclude, at this stage, that no convincing calculation has as yet been performed to prove or disprove the significance of eddies in augmenting new production. The evidence based on the large scale tracer studies suggests some kind of episodic, isopycnal process must be operating, but beyond this we have no quantitatively provable candidate.

The Oxygen-Nutrient Relationship

To fuel new production, nutrients are required; however, nutrients in the ocean are correlated with oxygen debt (AOU). If the stoichiometric ratio between AOU and nutrients is the same as or less than the ratio of O_2 production to nutrients consumed, then for a closed system, we would not expect to see any photosynthetic oxygen signal. However, as noted above, we observe a photosynthetically derived oxygen signal, which is comparable to the deeper AOU signal. It may be argued that very high photosynthetic quotients (O:C ratios) associated with nitrate assimilation at low light levels (eg. Megard et al., 1985) could account for the discrepancy; but recent arguments (Laws, 1991; Williams and Robertson, 1991) point to ratios of about 1.4, i.e., lower than the Redfield ratio of 1.65 estimated by Takahashi et al. (1985). This, therefore, is not an answer to the conundrum.

Another explanation may lie in the different physical boundary conditions for nutrients and oxygen. During the wintertime, deep convection occurs throughout the North Atlantic, the upper water column is charged with oxygen by gas exchange, negating the oxygen debt accrued during aphotic zone remineralization. The nutrients, on the other hand, can only be mixed vertically. If vertical mixing and gas exchange occur sufficiently rapidly to outstrip biological utilization, the oxygen inventory is enhanced. A simple calculation exemplifies this process: consider a 300m deep water column that contains a 100m euphotic zone. The upper 100m are assumed to be free of any nitrate or AOU. The deeper part contains AOU and nitrate in Redfield proportions, say 27μmol kg^{-1} and 3μmol kg^{-1}, respectively. In the late winter months, vertical mixing penetrates to 300m, and persists long enough to equilibrate the water column for oxygen. The resultant water column, upon stratification, is uniformly 0 μmol kg^{-1} and 1 μmol kg^{-1} for AOU and nitrate. Thus, the system would be primed for oxygen production, because there is a significant inventory of AOU free, nitrate laden water, both available for the spring bloom, and for subsequent upward mixing throughout the summer.

In the preceding, simplified discussion, we argue that oxygen has a "tense" boundary condition (i.e., it is held to atmospheric equilibrium), whereas nutrients have an "elastic" boundary condition (i.e., are only weakly affected by atmospheric contact). In the real ocean, the former is not completely true, in that gas exchange is not infinitely fast; however, the tracer separation does take place to a signficant extent. The ratio of "boundary condition tensions" is approximately the ratio of summer to winter mixed layer depths, which for the sub-tropical North Atlantic is about 2-4 (Doney and Jenkins, 1988). This implies that for tension ratios of about 3, the photosynthetic oxygen signal may be suppressed by approximately 30% because of the updraft of oxygen debt, but will certainly not be erased.

Climatic Variations in New Production

New production throughout the subtropics is predominantly nutrient limited. In the North Atlantic, at least, the important source of nutrients to the upper ocean is deep wintertime convection associated with watermass formation. Significant decade time scale variations in watermass formation rates, approaching a factor of two for some isopycnal classes, have occurred (Jenkins, 1982b). For example, subtropical mode water formation was dramatically curtailed in the mid 1970s (Jenkins, 1982b; Talley and Raymer, 1982). Venrick et al. (1987) also noted secular changes in chlorophyll-a in the central North Pacific, which they argued to be a response to changing wind stress. Such changes must surely affect regional new production rates.

The supply of nutrients to the upper ocean in the Atlantic, although ultimately driven by watermass renewal processes, will not linearly vary with variations in renewal rate. Boundary condition tension (Doney and Jenkins, 1988; see previous section) plays a role again. Watermass renewal variations, which occurred for gaseous tracers (oxygen and ^{3}He, Jenkins, 1982b), should be scaled down by the tension ratio for the North Atlantic, viz 2 to 4. Thus, corresponding variations in nutrient supply, and the resultant new production changes will only be about 20 to 50% for the subtropical North Atlantic. Presently, it would be inappropriate to extend this approach to the Pacific, for winter convection does not play the same role in vertical nutrient transport in that ocean. Variations in wind stress, as suggested by Venrick et al. (1987) play a more important role.

CONCLUSIONS

Tracer distributions have provided us with many estimates of basin scale, new production in oligotrophic waters. For the subtropical North Atlantic, several independent estimates yield mutually consistent results, giving us confidence in the overall approach. We must conclude from this that new production for the North Atlantic subtropical gyre is about 3-4 mol $m^{-2}y^{-1}$, a value substantially higher than was previously believed.

There are several implications of these results that need to be explored. Notably, the high vertical transport rates required to supply the nutrients to support this production suggest that poorly understood and quantified isopycnal eddy processes are at work. Consideration of boundary conditions of various tracers, particularly, nutrients and oxygen, resolve an apparent conundrum on nutrient:oxygen ratios, and allow us to set limits on the climatic variability of new production in the Atlantic subtropics.

ACKNOWLEDGEMENTS

W. J. Jenkins thanks Ted Packard for his support and encouragement, and Paul Falkowski for his patience and tolerance. This work was supported by the National Science Foundation, grant OCE-8911697.

REFERENCES

Altabet, M. A., 1989, Particulate new nitrogen fluxes in the Sargasso Sea, *J. Geophys. Res.*, 94:12771.

Brundage, W. L., and Dugan, J. P., 1986, Observations of an anticyclonic eddy of 18°C water in the Sargasso Sea, *J. Phys. Oceanogr.*, 16:717.

Chou, J. Z., 1985, Numerical modelling of oxygen cycling in the upper ocean, *WHOI-SSF, report No. 42.*

Craig, H., and Hayward, T., 1987, Oxygen supersaturation in the ocean: biological versus physical contributions, *Science*, 235:199.

Doney, S. C., and Bullister, J. L., 1991, A chlorofluorocarbon section in the eastern North Atlantic, *Deep-Sea Res.*, in press.

Doney, S. C., and Jenkins, W. J., 1988, The effect of boundary conditions on tracer estimates of thermocline ventilation rates, *J. Marine Res.*, 46:947.

Emerson, S., 1987, Seasonal oxygen cycles and biological new production in surface waters of the subarctic Pacific Ocean, *J. Geophys. Res.*, 92:6535.

Falkowski, P., Zeimann, D., Kolber, Z., and Bienfang, P. K., 1991, Role of eddy pumping in enhancing primary production in the ocean, *Nature*, 352:55.

Garrett, C., 1979, Mixing in the ocean interior, *Dyn. Atm. Oceans*, 3:239.

Jenkins, W. J., 1977, Tritium-helium dating in the Sargasso Sea: a measurement of oxygen utilization rates, *Science*, 196:291.

Jenkins, W. J., 1980, Tritium and ^{3}He in the Sargasso Sea, *J. Mar. Res.*, 38:533.

Jenkins, W. J., 1982a, Oxygen utilization rates in the North Atlantic Subtropical Gyre and primary production in oligotrophic systems, *Nature*, 300:246.

Jenkins, W. J., 1982b, On the climate of a subtropical ocean gyre: decade time-scale variations in water mass renewal in the Sargasso Sea, *J. Mar. Res.*, 40(supp):265.

Jenkins, W. J., 1987, ^{3}H and ^{3}He in the Beta Triangle: observations of gyre ventilation and oxygen utilization rates, *J. Phys. Oceanogr.*, 17:763.

Jenkins, W. J., 1988a, The use of anthropogenic tritium and helium-3 to study subtropical gyre ventilation and circulation, *Phil. Trans. R. Soc. Lond. A*, 325:43.

Jenkins, W. J., 1988b, Nitrate flux into the euphotic zone near Bermuda, *Nature*, 331:521.

Jenkins, W. J., and Goldman, J. C., 1985, Seasonal oxygen cycling and primary production in the Sargasso Sea, *J. Mar. Res.*, 43:465.

Laws, E. A., 1991, Photosynthetic quotients, new production and net community production in the open ocean, *Deep-Sea Res.*, 38:143.

Lewis, M. R., Harrison, W. G., Oakey, N. S., Hebert, D., and Platt, T., 1986, Vertical nitrate fluxes in the oligotrophic ocean, *Science*, 234:870.

Liss, P. S., and Merlivat, L., 1986, Air-sea gas exchange rates: introduction and synthesis, *in*: "The Role of Air-Sea Exchange in Geochemical Cycling," P. Buat-Menard, ed., Reidel Publ. Co.

McGowan, J. A., and Hayward, T. L., 1978, Mixing and oceanic productivity, *Deep-Sea Res.*, 25:771.

Megard, R. O., Berman, T., Curtis, P. J., and Vaughan, P. W., 1985, Dependence of phytoplankton assimilation quotients on light and nitrogen source: implications for oceanic primary productivity, *J. Plankt. Res.*, 7:691.

Musgrave, D. L., Chou, J., and Jenkins, W. J., 1988, Application of a model of upper-ocean physics for studying seasonal cycles of oxygen, *J. Geophys. Res.*, 93:15679.

Musgrave, D. L., 1990, Numerical studies of tritium and ^{3}He in the thermocline, *J. Phys. Oceanogr.*, 20:344.

Platt, T., 1984, Primary productivity in the central North Pacific: comparison of oxygen and carbon fluxes, *Deep Sea Res.*, 31:1311.

Platt, T., Harrison, W. G., Lewis, M. R.,Li, W. K. W., Sathyendranath, S., Smith, R. E., and Vezina, A. F., 1989, Biological production of the oceans: The case for a consensus, *Mar. Ecol. Prog. Ser.*, 52:77.

Price, J. F., Weller, R. A., and Pinkel, R., 1986, Diurnal cycling: observations and models of the upper ocean response to diurnal heating, cooling, and wind mixing, *J. Geophys. Res.*, 91:8411.

Raine, R. C. T., 1983, The effect of nitrogen supply on the photosynthetic quotient of natural phytoplankton assemblages, *Bot. Mar.*, 26:417.

Reid, J. L., and Shulenberger, E., 1986, Oxygen saturation and carbon uptake near 28N, 133W, *Deep-Sea Res.*, 33:267.

Riley, G. A., 1951, Oxygen, phosphate and nitrate in the Atlantic Ocean, *Bull. Bingham Oceanogr. Coll.*, 13:1.

Sarmiento, J. L., 1983, A tritium box model of the North Atlantic thermocline, *J. Phys. Oceanogr.*, 13:1269.

Sarmiento, J. L., Thiele, G., Key, R. M., and Moore, W. S., 1990, Oxygen and nitrate new production and remineralization in the North Atlantic Subtropical Gyre, *J. Geophys. Res.*, 95:18303.

Schmitt, R., 1981, Form of the temperature - salinity relationship in the Central Water: evidence for double diffusive mixing, *J. Phys. Oceanogr.*, 11:1015.

Shulenberger, E. and Reid, J. L., 1981, The Pacific shallow oxygen maximum, deep chlorophyll maximum, and primary productivity, reconsidered, *Deep-Sea Res.*, 28:901.

Spitzer, W. S., 1989, Rates of vertical mixing, gas exchange and new production: estimates from seasonal gas cycles in the upper ocean near Bermuda, PhD Thesis, *W.H.O.I.- M.I.T. Joint Program in Oceanography*.

Spitzer, W. S., and Jenkins, W. J., 1989, Rates of vertical mixing, gas exchange and new production: estimates from seasonal gas cycles in the upper ocean near Bermuda, *J. Mar. Res.*, 47:169.

Takahaski, T., Broecker, W. S., and Langer, S., 1985, Redfield ratio based on chemical data from isopycnal surfaces, *J. Geophys. Res.*, 90:6907.

Talley, L. D., and Raymer, M. E., 1982, Eighteen degree water variability, *J. Mar. Res.*, 40 (Suppl.),757.

Thiele, G., and Sarmiento, J. L., 1990, Tracer dating and ocean ventilation, *J. Geophys. Res.*, 95:9377.

Venrick, E. L., McGowan, J. A., Cayan, D. R., and Hayward, T. L., 1987, Climate and chlorophyll *a*: long-term trends in the central North Pacific Ocean, *Science*, 238:70.

Wallace, D. W. R., Moore, R. M., and Jones, E. P., 1987, Ventilation of the Arctic Ocean cold halocline: rates of diapycnal and isopycnal transport, oxygen utilization and primary production inferred using chlorofluoromethane distributions, *Deep-Sea Res.*, 34:1957.

Williams, P. J. leB., and Robertson, J. E., 1991, Overall planktonic oxygen and carbon dioxide metabolisms: the problem of reconciling observations and calculations of photosynthetic quotients, *J. Plankt. Res.*, 13:153.

NEW PRODUCTION AND THE GLOBAL CARBON CYCLE

Jorge L. Sarmiento

Program in Atmospheric and Ocean Sciences
Princeton University
Princeton, NJ

Ulrich Siegenthaler

Physics Institute
University of Bern
Bern, Switzerland

INTRODUCTION

The export of newly produced organic carbon from the surface ocean and its regeneration at depth account for an estimated three-quarters of the vertical ΣCO_2 gradient shown in Fig. 1 (Volk and Hoffert, 1985). If these processes, often referred to as the "biological pump," had ceased operating during the pre-industrial era, the increase in surface ΣCO_2 resulting from upward mixing of high ΣCO_2 deep waters would have raised atmospheric pCO_2 from 280 ppm to the order of 450 ppm (Sarmiento and Toggweiler, 1984) over a period of centuries. Vertical exchange, which gives an estimated upward flux of 100 GtC/yr (Fig. 2), works continuously to bring about just such a scenario. The biological pump prevents it by stripping out about 10 GtC/yr, so that the water arriving at the surface has a concentration equal to that which is already there.

The large flux of organic carbon out of the surface ocean has led many scientists to consider the possibility that biological processes might play an important role in determining the fate of the anthropogenic CO_2 transient. One possibility is that changes in new production which might result from variations in ocean circulation or mixing, or from a change in the ecology of the oceans, could have a significant impact on atmospheric pCO_2. The major problem that we address in this paper is what role biology plays in the fate of the anthropogenic CO_2 transient.

One of the properties which determines the chemical capacity of a parcel of water to take up additional CO_2 is the concentration of ΣCO_2 (Broecker and Peng, 1982), because the uptake capacity decreases with increasing ΣCO_2. The biological pump thus has, through its influence on the ΣCO_2 distribution, an indirect impact on the oceanic uptake of anthropogenic CO_2, which, however, is of academic interest only as long as there is no massive change of the biological pump. Similarly, as long as global marine biology continues to operate in the steady-state mode that was assumed in putting together

Primary Productivity and Biogeochemical Cycles in the Sea
Edited by P.G. Falkowski and A.D. Woodhead, Plenum Press, New York, 1992

317

the balanced pre-industrial carbon budget of Fig. 2, it will not contribute directly to the oceanic uptake of anthropogenic CO_2. The question arises as to whether it is reasonable to assume that the biological pump is in steady state. In the first section of this paper, we review evidence for the variability of the carbon cycle in the past, and discuss some of the mechanisms involving changes in new production which have been proposed to explain these variations. The observations suggest that the potential impact of natural variability on the CO_2 increase that has occurred in the last two centuries is negligible. Model studies of perturbations to the ocean carbon-cycle support this conclusion, showing that scenarios with atmospheric pCO_2 changes large enough to significantly affect the anthropogenic transient require extreme modifications of the biological pump or ocean circulation. Also discussed in this first section is the potential impact of alterations of the biological pump that may occur as a consequence of increasing UV radiation, and the direct effect of increasing CO_2 concentrations on the rate of photosynthesis.

The combined biological and solubility pumps (see Fig. 1 for a definition) give a complex spatial pattern of sea-surface pCO_2 which leads to such features as the net outward flux of 1 to 2 GtC/yr in the Equatorial region (e.g., Keeling, 1968; Tans et al., 1990). It is important to note that this net outward flux is balanced by inward fluxes at higher latitudes to give a total outward flux of only 0.6 GtC/yr required to balance the difference between the river inflow and sediment burial (Fig. 2). (The large 90 GtC/yr air-sea exchange fluxes shown in Fig. 2 are somewhat misleading in that by far the largest part represents simply *in situ* exchange of oceanic and atmospheric CO_2 molecules.) The second section of the paper is dedicated to a discussion of how anthropogenic CO_2 has modified the pre-industrial carbon budget of Fig. 2. The discussion will particularly focus on the latitudinal pattern of carbon fluxes.

NATURAL CARBON CYCLE

Since about 1910, the annual fossil fuel emissions of CO_2 have exceeded the observed rate of increase of atmospheric pCO_2. Thus, there can be no doubt that the

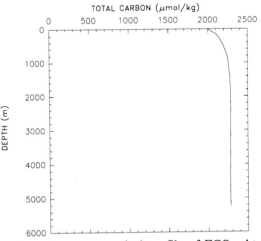

Fig. 1. The global horizontal average vertical profile of ΣCO_2 obtained from Geosecs observations. The surface ocean concentrations are of the order of 12% lower than concentrations in the deepest waters due, primarily, to the biological pump, which accounts for approximately three-quarters of the vertical gradient (Volk and Hoffert, 1985). The remainder of the vertical gradient is due to the "solubility pump," which results from the fact that CO_2 is more soluble in the cold waters which fill the deep ocean than it is in warm surface waters.

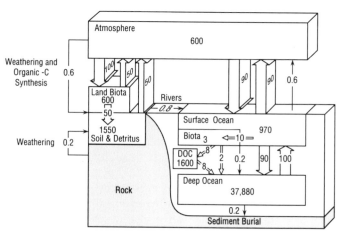

Fig. 2. An estimate of the pre-industrial carbon budget based on the Intergovernmental Panel on Climate Change (IPCC) budget for 1980-1989 published by Houghton et al. (1990). The pre-industrial inventories within the boxes were calculated from the IPCC inventories by subtracting estimates of the anthropogenic increases obtained from Siegenthaler and Oeschger's (1987) box-diffusion and outcrop-diffusion model studies, and the three-dimensional general circulation model of Sarmiento et al. (1991) (cf., the middle panel of Fig. 5). The fluxes shown in the Fig. are the same as those in the IPCC budget except for the following: (1) the anthropogenic component shown in the middle panel of Fig. 5 has been removed. (2) The river and sediment burial fluxes given in the IPCC budget have been closed by allowing a 0.2 GtC/yr flux of carbon from the surface to the deep ocean, a 0.6 GtC/yr efflux from the ocean to the atmosphere, and a combined terrestrial uptake from weathering and organic carbon synthesis of 0.8 GtC/yr (cf., Sabine and Mackenzie, 1991, who give a pre-industrial evasion of 0.46 GtC/yr). (3) The new production estimate has been increased from 4 GtC/yr to 10 GtC/yr based on general circulation models of the ocean by Najjar (1990) and Bacastow and Maier-Reimer (1991). New production is defined in the same sense as used by Eppley and Peterson (1979); as the net export of organic matter from the surface ocean. Both these models have difficulty in realistically simulating surface nutrient concentrations with estimates of new production as low as the value of 4 GtC/yr given in the IPCC study, which is based on estimates such as those summarized by Eppley (1989). (4) The new production has been split between a dissolved organic carbon (DOC) component of 8 GtC/yr, and a particulate component of 2 GtC/yr based on Toggweiler's (1989) analysis of Suzuki et al.'s (1985) dissolved organic nitrogen observations, and the model studies of Najjar (1990) and Bacastow and Maier-Reimer (1991). Neither model gave realistic interior-nutrient distributions with particle only simulations. (5) A DOC pool has been added with a magnitude estimated from the observations of Sugimura and Suzuki (1988) (Brewer, personal communication). (6) The upward vertical oceanic transport term has been increased in proportion to the new production term so that it still is about 10 times the new production. The magnitude of this term is very speculative. It involves not only such processes as deep water formation and upwelling, for which there are estimates, and which were most likely used in the IPCC estimate of 37 GtC/yr, but also diapycnic mixing and mixing along sloping isopycnal surfaces. The magnitude of vertical mixing resulting from such processes is poorly known, and cannot be separated into an upward and downward component. The basis for choosing 100 GtC/yr is the observation that the average surface ocean ΣCO_2 is about 10% lower than the average deep ocean ΣCO_2 (cf. Fig. 1). Once the new production is fixed at 10 GtC/yr, the maintenance of an average surface concentration 10% lower than the deep ocean average requires an upward flux of 100 GtC/yr. The downward flux of 90 GtC/yr is obtained from the difference between the upward flux and the new production.

increase in atmospheric pCO_2 which has occurred over most of this century is due to anthropogenic emissions, with all other carbon reservoirs combined having served as sinks during this time. However, this observation does not preclude the possibility that natural variability has played an important role in the observed *rate* of increase. To gain an understanding of natural variability, we turn to records of the behavior of atmospheric pCO_2 in the past.

The record which is most relevant to the anthropogenic transient is that obtained for the 10th to 18th centuries in trapped air bubbles from the South Pole ice-core (Fig. 3a; Siegenthaler et al., 1988). These measurements show that atmospheric pCO_2 varied by the order of 5 to 10 ppm around the baseline of 280 ppm, the generally accepted pre-industrial value. The rate of CO_2 increase in the 19th century shown, in Fig. 3b, is small enough that natural variability of about 5 to 10 ppm might have played a significant role. By contrast, the increase that has occurred during the 20th century would have overwhelmed any natural variability of such a magnitude. Figure 4 also shows natural variability resulting from seasonal changes and the El Niño, both of which occur on much shorter time scales and are smaller than the anthropogenic perturbations.

Delving further into the past shows much larger natural variability, as summarized in Fig. 4. Studies of Greenland ice-cores indicate rapid CO_2 variations of about 50 ppm on a century time scale during the last ice age (Stauffer et al., 1984). The variations are comparable in magnitude and timing to the rate of increase of atmospheric pCO_2 during the 20th century. Mechanisms for the pCO_2 changes associated with these episodes have not been studied directly, but we can infer the types of changes that may have occurred from studies of the ice-age reduction in CO_2 (e.g., Knox and McElroy, 1984; Sarmiento and Toggweiler, 1984; Siegenthaler and Wenk, 1984; Boyle, 1988; Broecker and Peng, 1989) and iron-fertilization scenarios (Peng and Broecker, 1991; Joos et al., 1991; Sarmiento and Orr, 1991).

The three-dimensional model of Sarmiento and Orr (1991) confirms earlier suppositions, based on box models, that the Southern Ocean surface carbon balance is most likely the dominant factor in oceanic control of atmospheric pCO_2. Table 1 summarizes a series of studies by Sarmiento and Orr in which the effect on atmospheric pCO_2 was determined of an extreme scenario of completely depleting nutrients in various regions of the ocean. These simulations were intended as an analysis of the possible impact of such changes on the anthropogenic transient. Therefore, the studies covered 100 years, beginning in 1990, with anthropogenic emissions prescribed according to the IPCC business-as-usual scenario (Houghton et al., 1990). The nutrient-depletion scenario of the Southern Ocean has the largest impact, because it is the region where the vast bulk of the deep ocean mixes up to the surface and is exposed to the atmosphere. The balance of carbon in the surface Southern Ocean is determined by the relative magnitude of the upward flux of carbon-rich deep waters and the stripping of carbon by the biological pump. The models show that one or both of these processes must change by a very large amount in order to explain the observed changes in pCO_2. For example, new production in this scenario increases by 12 GtC/yr, almost double the new production determined by the same model for the entire present ocean.

The only evidence of recent large-scale changes in the biological pump that we are aware of are the studies by Venrick et al. (1987), who documented an increase in total chlorophyll in the central North Pacific Ocean since 1968, and Falkowski and Wilson (personal communication), who showed an increase in Secchi depth in the North Pacific from 1900 to 1981. However, as suggested by the scenario on northern hemisphere nutrient depletion of Table 1, the North Pacific is not an area that has a significant impact

Table 1. Sensitivity of Atmospheric pCO$_2$ to a Complete Depletion of Nutrients for 100 Years in the Latitude Band Shown. These Results are Taken from the Three-dimensional Business-As-Usual Model Projection of Sarmiento and Orr (1991).

Latitude Band	Atmospheric pCO$_2$ perturbation (ppm)
North Atlantic (31.12°N to 80.06°N)	-12.7
North Pacific (31.12°N to 66.69°N)	- 6.9
Equatorial Region (17.78°S to 17.78°N)	- 2.8
Southern Ocean (South of 31.12°S)	-71.8

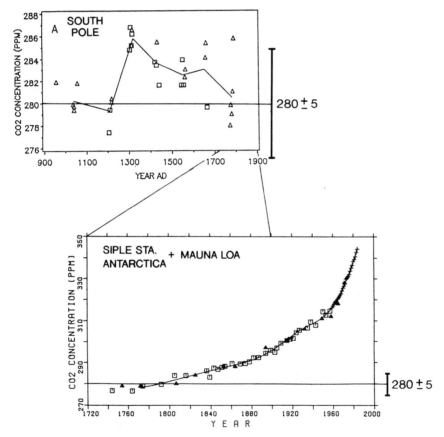

Fig. 3.   (a) Observations of pCO$_2$ in the South Pole ice core for 800 years before the beginning of the industrial revolution (Siegenthaler et al., 1988).   (b) Atmospheric pCO$_2$ from the 18th century until the present (Siegenthaler and Oeschger, 1987).   The solid line is a spline fit to the measurements shown in the figure, which came from trapped air bubbles in the Siple ice core (Neftel et al., 1985; Friedli et al., 1986), and from the Mauna Loa record of Keeling et al. (1989a).

on atmospheric pCO_2. The same is true of the North Atlantic, where changes in the process of deep water formation are suggested by evidence such as the freshening of deep water in the north of the basin over the 20 years prior to the TTO North Atlantic Study (e.g., Brewer et al., 1983).

Temporal variability of the biology and deep water formation of the Southern Ocean almost certainly occur (e.g., Gordon, 1988), but there is no direct evidence we are aware of that massive changes such as those postulated to explain the glacial and deglacial episodes have occurred in the Southern Ocean over the last few decades. However, given the difficulty of obtaining direct information of temporal variability, the strongest argument that can be made against the likelihood of such large changes is the one already mentioned above, that the 800-year pre-industrial record of Fig. 3a, coupled with longer records of interglacial pCO_2 (e.g., Barnola et al., 1987) show no atmospheric pCO_2 changes of such a magnitude under the interglacial climate regime. The rapid CO_2 variations during glacial time suggested by Greenland ice core studies occurred during climatic regimes which were very different from the interglacial climate of today, and were associated with the Dansgaard-Oeschger episodes of rapid climate change discussed by Broecker and Denton (1989). Even the 0.5°C increase in temperature of the last century, which is consistent with warming that would be expected from the greenhouse effect (Houghton et al., 1990),

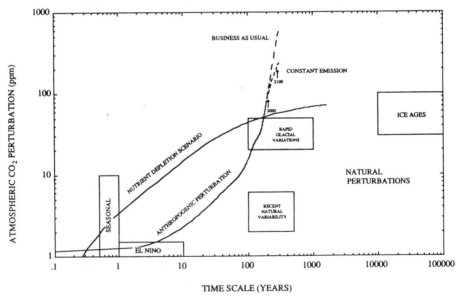

TIME SCALE (YEARS)

Fig. 4.   The magnitude of natural and anthropogenic atmospheric pCO_2 perturbations versus their estimated time scale. The anthropogenic perturbation up to 1990 is taken from Fig. 3b, wtih Time = actual year - 1800. The projection of the anthropogenic perturbation from 1990 to 2100 is from the model results of Sarmiento et al. (1991). The business-as-usual scenario is taken from the IPCC study (Houghton et al., 1990). The constant-emission scenario fixes anthropogenic emissions at the value obtained by the model in 1990 when forced with the observations in Fig. 3b. The figure also shows the reverse of the reduction in atmospheric pCO_2 which would occur in a scenario of a sudden depletion of Southern Ocean nutrients, beginning with an atmospheric pCO_2 of 280 ppm (the reduction is larger if we begin with the present pCO_2 of 355 ppm; Joos et al., 1991; Sarmiento and Orr, 1991).

is well within the range of natural climate changes that occurred during the interglacial time span shown in Fig. 3a, which includes the late Medieval warm period of the 13th century, and the culmination of the Little Ice Age in 1600 to 1860 A.D (Siegenthaler et al., 1988).

There has been considerable interest in the possible role of organic carbon export from the shelves to the deep waters of the slope as a sink for anthropogenic CO_2. The suggestion made by Walsh (1989) is that measurement-based estimates of CO_2 exchange at the air-sea interface, such as those reported on by Tans et al. (1990), are missing a large net sink of about 1 GtC/yr in the continental shelves due to carbon, which is taken up by photosynthesis and exported off the shelf before being regenerated. The existence of a net export from the shelf is controversial (Falkowski et al., 1988, reported that large zooplankton grazing and possible microbial sinks were missed in previous attempts to develop a budget for the New York Bight). However, even if there is such an export, Walsh's interpretation fails to recognize that the vast majority of the organic carbon exported from the shelves arrives there as inorganic carbon carried by water transport, not by air-sea exchange; this would continue to be the case even if the shelf ecosystem were altered by fishing, pollution, or other direct human intervention. A good analogy for the potential impact of such alterations is the scenario of equatorial nutrient depletion in Table 1, which shows a negligible change of atmospheric pCO_2.

Could the anthropogenic CO_2 perturbation itself lead to changes in oceanic processes? An analogy would be the effect of CO_2 levels on the photosynthesis of terrestrial C-3 plants (CO_2 fertilization). Photosynthesis in most phytoplankton occurs by the same Rubisco-catalyzed pathway used by C-3 plants on land. Oceanic concentrations of CO_2 are comparable to, or below, the half-saturation constant for Rubisco, but the evidence (most of it from near-shore phytoplankton) demonstrates that most organisms show very little sensitivity to oceanic CO_2 concentrations (Raven, 1991). Most phytoplankton have developed a means of concentrating CO_2 in the cell, perhaps by a bicarbonate pumping mechanism. Raven (1991) concludes that the increase in CO_2 concentration due to the anthropogenic transient will erode the competitive advantage of organisms with a CO_2 concentrating mechanism relative to those lacking the mechanism; the impact of this change on global marine productivity is difficult to judge. The model simulations discussed above suggest that even if marine productivity does change, it is unlikely to significantly affect the rate of increase of atmospheric pCO_2, unless the change is large enough to have a major impact on Southern Ocean surface nutrients.

Another possible feedback is through the effect on marine life of increased ultraviolet radiation resulting from ozone reduction (e.g., Smith and Baker, 1979; Hardy and Gucinski, 1989). The impact of decreased ozone will be greatest at high latitudes where deep water formation occurs. UV-B (280-320 nm) radiation causes significant damage to marine life. UV-B would increase by almost 50% at the ocean surface if ozone were reduced by just 16%. However, the vertical attenuation coefficient for UV-B weighted by DNA damage is about 0.4 m^{-1} (Smith and Baker, 1979), i.e., a factor of 10 reduction in 10 m. It is difficult to estimate exactly how this would affect marine life. It is somewhat less difficult to demonstrate that even a large impact on marine life would have a relatively small effect on atmospheric pCO_2. Peng (personal communication) used the ocean model of Peng and Broecker (1991) to show that even if increased UV radiation led to a cessation of all marine photosynthesis in the Southern Ocean, atmospheric pCO_2 would only increase by approximately 37 ppm over the next 100 years.

There are inconsistencies in our understanding of the carbon cycle which remain to be reconciled. The recent carbon-budget study of Tans et al. (1990) concluded that oceanic uptake of anthropogenic CO_2 is far smaller than required by ocean models which have been calibrated by tracer observations. The evidence reviewed above suggests that changes in the biological pump are unlikely to play a role. The following section discusses the Tans et al. study and suggests alternative mechanisms which may reconcile a large part of the difference.

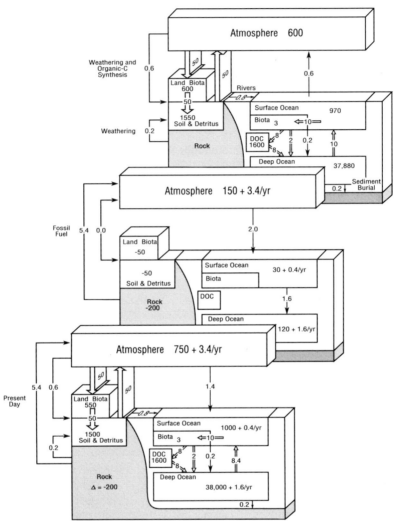

Fig. 5. Estimates of the pre-industrial carbon budget (upper panel), anthropogenic perturbation for 1980 to 1989 (middle panel) and present carbon budget for 1980 to 1989 (lower panel). The anthropogenic perturbation and present carbon budget are based on the IPCC report (Houghton et al., 1990) with modifications as discussed in the caption to Fig. 2. The upper panel is identical to Fig. 2 except that only the net flux across each interface is shown. The reduced carbon inventory of land biota and soil and detritus in the middle and lower panels is due to a net deforestation flux to the atmosphere which occurred prior to 1980. This deforestation flux is assumed to be balanced by a terrestrial biota uptake during the 1980 to 1989 decade, in order to account for what would otherwise be a missing sink of order 1.6 GtC/yr (Houghton et al., 1990).

(a) Inorganic Carbon Weathering and Sediment Burial

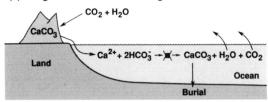

(b) Organic Carbon Weathering and Sediment Burial

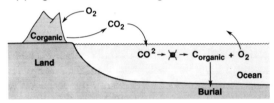

(c) Terrestrial Uptake of Organic Carbon and Export

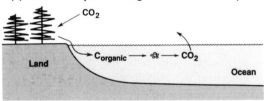

Fig. 6. (a) The upper panel is a scheme of the processes of weathering of $CaCO_3$, and the river, biological uptake, sediment burial, and air-sea fluxes which are required to balance the weathering. The burial of $CaCO_3$ equals the weathering rate in a steady state. The net effect of $CaCO_3$ weathering, as well as weathering of silicates and non-Ca carbonates, is to give a net air-sea CO_2 flux out of the ocean. (b) The weathering of organic carbon shown in the middle panel gives a net flux of CO_2 into the ocean. (c) The lower panel illustrates net uptake by terrestrial vegetation which is lost to the rivers, as well as the river flux, oceanic bacterial regeneration, and outward air-sea flux which are required to balance it in a steady state. Some of the river-borne organic carbon will be buried in sediments rather than remineralized, but a steady-state balance can only be maintained if the organic carbon burial in (b) is reduced by the same amount. Thus, the total organic carbon burial must equal organic carbon weathering on land, as shown in panel (b), and the flux of CO_2 out of the ocean must be large enough to balance exactly the uptake by terrestrial vegetation, as shown in panel (c). The overall impact of the three processes illustrated in this figure is to give a net outward flux of CO_2 from the ocean of about 0.6 GtC/yr (Sarmiento and Sundquist, 1991).

The upper panel of Fig. 5 simplifies the pre-anthropogenic steady-state carbon cycle of Fig. 1 to show only the net fluxes. The corresponding net air-sea exchange is added to the estimates of oceanic uptake of anthropogenic CO$_2$, shown in the middle panel of Fig. 5 to obtain the present air-sea exchange shown in the lower panel. The flux into the ocean in the lower panel is only 1.4 GtC/yr, rather than the 2 GtC/yr obtained by perturbation models (middle panel) because of the effect of the outward pre-industrial flux of 0.6 GtC/yr shown in the upper panel (Sarmiento and Sundquist, 1991; cf. Houghton et al., 1990; Sabine and Mackenzie, 1991). The 0.6 GtC/yr pre-industrial flux is a result of the processes illustrated in Fig. 6, with an important role for new production of organic carbon and CaCO$_3$ tests, as well as bacterial remineralization of organic matter. The inorganic carbon cycle of Fig. 6a leads to an outward flux of CO$_2$ from the ocean, as does the organic carbon transported by rivers into the ocean (Fig. 6c). Only the weathering of organic carbon results in a flux of CO$_2$ into the ocean (Fig. 6b). Thus, the ocean takes up approximately 2 GtC of anthropogenic carbon per year at present, but 0.6 GtC/yr of this flux is accounted for by the river flux minus sediment burial. The contribution from rivers helps to explain part of the difference between the large oceanic uptakes of the order of 2 GtC/yr obtained by ocean models (Houghton et al., 1990), and the far smaller air-sea exchange of < 1 GtC/yr obtained by the Tans et al. (1990) study of atmospheric transport, constrained by atmospheric pCO$_2$ observations and oceanic observations of air-sea CO$_2$ exchange.

The biological pump in the present carbon budget of Fig. 5c is identical to that in the pre-industrial budget of Fig. 5a. The only internal ocean flux that changes as a result of the anthropogenic transient is the net upward flux of dissolved inorganic carbon from the deep to the surface ocean, which is reduced from 10 GtC/yr in the pre-industrial budget to 8.4 GtC/yr at present. This reduction permits the net downward flux of 1.6 GtC/yr required by models of the anthropogenic perturbation.

The flux of CO$_2$ into the ocean occurs in response to the continuing anthropogenic addition of CO$_2$ to the atmosphere. If this perturbation were to cease, the ocean and atmosphere eventually would come into equilibrium. The ultimate fate of anthropogenic CO$_2$ in this case is perhaps best summarized by simulations of the response of the ocean-atmosphere system to an instantaneous pulse of CO$_2$ added to the atmosphere (Siegenthaler and Oeschger, 1978; Maier-Reimer and Hasselmann, 1987; Sarmiento et al.,1991). 83% of the CO$_2$ disappears into the ocean. The time scales of oceanic uptake get increasingly longer as the amount remaining in the atmosphere is reduced, and CO$_2$ begins to penetrate the more slowly circulating regions of the deep ocean. These simulations show that approximately 17% of the CO$_2$ will remain in the atmosphere forever. The simulations ignore dissolution of sedimentary CaCO$_3$ in the oceans, which will eventually take up a large fraction of the remaining atmospheric CO$_2$ perturbation through the reaction CO$_2$ + CaCO$_3$ + H$_2$O → Ca^{2+} + 2HCO$^-_3$ (Broecker and Takahashi, 1977).

Figure 7 shows various attempts to reconstruct the effect of the biological and solubility pumps on the source-sink distribution of the ocean. We will not discuss the details of these mechanisms, as this topic was covered by previous studies such as that of Volk and Liu (1988). What is of interest here is to demonstrate that recent atmospheric CO$_2$ transport studies, such as those of Keeling et al. (1989b) and Tans et al. (1990), place strong constraints on the latitudinal distribution of oceanic source-sink fluxes which make it far easier to distinguish between various scenarios for the uptake of anthropogenic CO$_2$. The ocean is broken down into an equatorial latitudinal band from 15°S to 15°N, and regions to the north and south of the Equatorial band.

Three scenarios are shown in Fig. 7. Scenario 1 is based on the measurements by Tans et al. (1990) of the oceanic air-sea CO_2 difference using their so-called empirical wind-speed-dependent, gas-exchange coefficient. The empirical gas-exchange coefficient is obtained by models of the oceanic uptake of atmospheric radiocarbon. The coefficient is approximately twice the magnitude of that of Liss and Merlivat (1986), which was based on laboratory measurements calibrated with lake observations. The use of the latter gas exchange coefficient would reduce the estimates of Scenario 1 by approximately a factor of 2, but is impossible to justify without totally ignoring the radiocarbon constraint (see further discussion in Sarmiento et al., 1991).

The total oceanic uptake in the Tans et al. observations is comparable to that obtained by perturbation model studies, such as that of Sarmiento et al. (1991) shown in

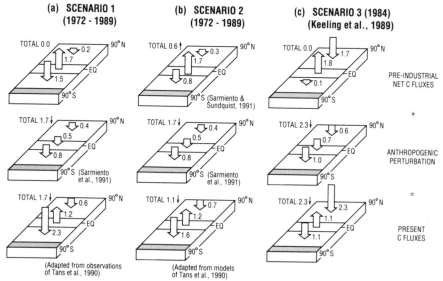

Fig. 7.   Three estimates of the present geographical distribution of air-sea CO_2 fluxes. The boundaries between the three latitudinal bands are at 15° N and 15° S for Scenarios 1 and 2, and 15.6° N and 15.6° S for Scenario 3. The upper set of panels shows estimated pre-anthropogenic fluxes, the middle panels show the anthropogenic perturbation, and lower panels show the sum of the upper two. (1) The present carbon cycle panel of Scenario 1 is the Tans et al. (1991) analysis of observations from 1972 to 1989; their recommended 0.3 GtC/yr reduction of the equatorial efflux accounts for the effect of El Niño. The total uptake is reduced from their El Niño corrected value of 1.9 GtC/yr to 1.7 GtC/yr to match the perturbation uptake of Sarmiento et al.'s (1991) model shown in the middle panel. Subtracting Sarmiento et al.'s anthropogenic perturbation gives the upper pre-industrial panel of Scenario 1. (2) The total flux of CO_2 from the ocean in the top panel of Scenario 2 is based on the Sarmiento and Sundquist (1991), who give a net pre-industrial oceanic efflux of 0.6 GtC/yr due to weathering, river transport, and the sediment burial cycle. Adding to this, Sarmiento et al.'s (1991) estimate for the oceanic uptake of anthropogenic CO_2 of 1.7 GtC/yr from 1972 to 1989 shown in the middle panel gives a total oceanic uptake of 1.1 GtC/yr shown in the lower panel (cf. Fig. 5). See text for an explanation of the latitudinal breakdown of the fluxes in the upper and lower panels. (3) Scenario 3 is obtained from the model of Keeling et al. (1989b).

the middle panel of Fig. 7a. The outstanding characteristic of Scenario 1 relative to Scenarios 2 and 3 is a very large oceanic sink in the southern hemisphere. Such a sink is inconsistent with atmospheric transport models constrained by measurements of atmospheric pCO_2, which do not permit enough southward atmospheric transport from anthropogenic sources primarily in the northern hemisphere (Keeling et al., 1989b; Tans et al., 1990). Thus, Scenarios 2 and 3, which do take the atmospheric transport models into consideration, both have smaller southern hemisphere sinks than does Scenario 1. Hence, Scenarios 2 and 3 are inconsistent with the southern hemisphere uptake estimate of Scenario 1, but the spatial and temporal coverage is so poor in this area that even Tans et al. consider it unreliable.

The major difference between Scenarios 2 and 3 is in the northern hemisphere sink. Scenario 3, obtained from Keeling et al. (1989b), has a total air-sea flux of anthropogenic CO_2 equal to that estimated by ocean perturbation models. Unable to take the CO_2 up in the southern hemisphere because of the atmospheric transport constraints in their model, Keeling et al. opt for putting a very large flux of anthropogenic CO_2 into the northern hemisphere ocean. Their estimate of anthropogenic uptake of 2.3 GtC/yr is larger than the 1.7 GtC/yr of Scenarios 1 and 2 because they used a different ocean model. There is also a small difference because Scenario 3 is for 1984, whereas Scenarios 1 and 2 are for 1972 to 1989.

Tans et al. (1990) go to the opposite extreme of ignoring the ocean model constraints and using only their northern hemisphere and equatorial ocean pCO_2 observations (where the data coverage is better than that in the southern hemisphere) to come up with a very small total oceanic uptake of substantially less than 1 GtC/yr (0.3 to 0.8 GtC/yr in their Scenarios 5 to 8). The anthropogenic carbon budget is closed by assuming that the terrestrial sink is larger than previously supposed. This is not a satisfactory solution. The 2 GtC/yr oceanic model estimates of anthropogenic CO_2 uptake are strongly constrained by tracer observations. Sarmiento et al. (1991) make the case that their estimate, which serves as the basis for the middle panel of Scenarios 1 and 2, is a lower limit to the anthropogenic CO_2 uptake.

A compromise between Scenarios 1 and 3, and Tans et al.'s scenario, is proposed in Scenario 2. Scenario 2 begins with the assumption that the total oceanic uptake of anthropogenic CO_2 is equal to the 1.7 GtC/yr obtained by Sarmiento et al. (1991), but includes in the overall budget the net pre-industrial riverine input of 0.6 GtC/yr discussed in connection with Fig. 5 (Sarmiento and Sundquist, 1991). Consequently, the oceanic uptake by air-sea exchange is required to be only 1.1 GtC/yr, in better agreement with the Tans et al. estimates of 0.3 to 0.8 GtC/yr. If Scenario 2 is to be consistent with the atmospheric transport constraint, terrestrial weathering and the photosynthetic sink, which provides the 0.6 GtC/yr to the rivers, must be primarily in the northern hemisphere, where most of the continental surface is.

The atmospheric transport models also place strong constraints on the oceanic uptake of CO_2 in the southern hemisphere. The value of 1.6 GtC/yr chosen for the bottom panel of Scenario 2 is the highest one obtained by Tans et al. in their Scenarios 5 to 8, i.e., 1.4 GtC/yr, with a 0.25 GtC/yr correction added in to account for the effect of carbon-monoxide transport and oxidation on the CO_2 budget (Enting and Mansbridge, 1991). The equatorial and northern hemisphere air-sea fluxes of Scenario 2 are adapted from the data-based estimates of Scenario 1. The sum of the two Scenario 1 values (1.2 - 0.6) gives a net outward flux of 0.6 GtC/yr, compared to the net outward flux of 0.5 GtC/yr required to balance the difference between the total air-sea uptake of 1.1 GtC/yr and the southern hemisphere uptake of 1.6 GtC/yr which we have already adopted for

Scenario 2. Thus, the data-based equatorial and northern hemisphere flux estimates are reduced by subtracting off a total area-weighted adjustment of 0.1 GtC/yr.

An adjustment of 0.1 GtC/yr can be readily justified on the basis of undersampling (Sarmiento et al., 1991; Watson et al., 1991), and may be due to Tans et al. not taking into consideration the fact that the skin temperature of the ocean may be about 0.11 to 0.30°C colder than the bulk water temperatures usually used in determining air-sea pCO_2 differences (e.g., Schluessel et al., 1990), so that these differences may be underestimated by 1 to 4 ppm (A. Watson, personal communication, UNESCO, 1991). Sarmiento and Sundquist (1991) estimate the correction due to this effect as an increased flux into the ocean of order 0.1 to 0.6 GtC/yr for the region north of 15°S.

CONCLUSIONS

The biological pump plays a major role in determining the distribution of carbon in the ocean and, through this, the atmospheric pCO_2. Thus, biological processes indirectly affect the capacity of the ocean to take up anthropogenic CO_2 because they influence the chemical composition of surface waters. However, the only way biological processes can play a direct role in the oceanic uptake of anthropogenic CO_2 is if they change in time. We have suggested that changes large enough to have had a significant impact on the anthropogenic transient are unlikely. Therefore, it is difficult to argue that an improved knowledge of the magnitude of the biological pump will have a significant impact on our present understanding of the anthropogenic transient. The most powerful tools for this are tracer-calibrated models of ocean circulation. On the other hand, an understanding of the biological pump is crucial if we are to understand the natural variability which is important on longer time scales, and may be of relevance as climate responds to the continued increase of greenhouse gases.

Perhaps the most interesting conclusion that can be drawn from our discussion pertains to the latitudinal pattern of air-sea exchange which arises from the influence of the biological and solubility pumps. The three scenarios shown in Fig. 7 have dramatically different implications for the anthropogenic carbon budget. They correspond to estimates of total oceanic uptake of anthropogenic carbon ranging from 1.1 GtC/yr to 2.3 GtC/yr. The difference between these two numbers is equivalent to more than 1/5th of the fossil-fuel source to the atmosphere. However, this difference in flux is equivalent to only a 4 to 5 ppm uncertainty in the measured air-sea CO_2 difference on a global scale. Temporal and spatial variability is so great as to make measurements of precision sufficient to resolve such an uncertainty extremely difficult (e.g., Watson et al., 1991).

However, when we consider the latitudinal distribution of the fluxes, and add to this the constraints given by atmospheric transport models, we come up with regional differences in oceanic fluxes which are associated with very large air-sea CO_2 differences that may very well be resolvable by measurements (Keeling et al., 1989b; Tans et al., 1990). Thus, the range in southern hemisphere uptakes of 1.1 to 2.3 GtC/yr in Fig. 7 corresponds to a range in the air-sea difference of ~11 ppm; the range in northern hemisphere uptakes of 0.6 to 2.3 GtC/yr corresponds to a range of ~23 ppm over that smaller ocean area. A breakdown into smaller latitudinal bands should be possible, and may show even greater sensitivity. We conclude that the measurement of air-sea CO_2 difference shows great promise of helping to constrain our knowledge of the anthropogenic carbon transient in the ocean. The difficulty in obtaining good temporal and spatial resolution of these measurements might be overcome by the use of temperature and chlorophyll estimates based on satellite observations, validated with *in situ* measurements, as suggested by the study of Watson et al. (1991). It is most likely that data assimilation models of ocean physics and biology will be required for this approach to succeed.

ACKNOWLEDGEMENTS

Support for this research has been provided by the Carbon Dioxide Research Division of the Department of Energy (DE-FG02-90ER61052 and DE-FG02-90ER61054). US received additional support from the Swiss National Science Foundation. We are grateful for the assistance of J. Olszewski in preparation of the manuscript, and appreciate the work of and helpful discussions with J. Orr and F. Joos.

REFERENCES

Bacastow, R., and Maier-Reimer, E., 1991, Dissolved organic carbon in modeling oceanic new production, *Global Biogeochem. Cycles*, 5:71.

Barnola, J. M., Raynaud, D., Kortkevich, Y.S., and Lorius, C., 1987, Vostok ice core provides 160,000-year record of atmospheric CO_2, *Nature*, 329:408.

Boyle, E. A., 1988, The role of vertical fractionation in controlling Late Quaternary atmospheric carbon dioxide, *J. Geophys. Res.*, 93:15701.

Brewer, P. G., Broecker, W. S., Jenkins, W. J., Rhines, P.B., Rooth, C.G., Swift, J.H., Takahashi, T., and Williams, R.T., 1983, A climatic freshening of the deep Atlantic north of 50°N over the past 20 years, *Science*, 222:1237.

Broecker, W. S., and Denton, G.H., 1989, The role of ocean-atmosphere reorganizations in glacial cycles, *Geochim. Cosmochim. Acta*, 53:2465.

Broecker, W. S., and Peng, T.-H., 1982, Tracers in the Sea, Eldigio Press, Palisades, New York.

Broecker, W. S., and Peng, T.-H., 1989, The cause of the glacial to interglacial atmospheric CO_2 change: A polar alkalinity hypothesis, *Global Biogeochem. Cycles*, 3:215.

Broecker, W. S., and Takahashi, T., 1977, Neutralization of fossil fuel CO_2 by marine calcium carbonate, *In*, "The Fate of Fossil Fuel CO_2 in the Oceans," N.R. Andersen and A. Malahoff, eds., Plenum Publishing Corp., New York.

Enting, I. G., and Mansbridge, J. V., 1991, Latitudinal distribution of sources and sinks of CO_2: Results of an inversion study, *Tellus*, 43B:156.

Eppley, R. W., 1989, New Production: History, Methods, Problems, *In*:, "Productivity of the Ocean: Present and Past", W. H. Berger, W.S. Smetacek, G. Wefer, and J. Wiley and Sons, eds., New York.

Eppley, R. W., and Peterson, B. J., 1979, Particulate organic matter flux and planktonic new production in the deep ocean, *Nature*, 282:677.

Falkowski, P. G., Flagg, C. N., Rowe, G. T., Smith, S. L., Whitledge, T.E., and Wirick, C.D., 1988, The fate of a spring phytoplankton bloom: Export or oxidation?, *Cont. Shelf Res.*, 8:457.

Friedli, H., Lötscher, H., Oeschger, H., Siegenthaler, U., and Stauffer, B., 1986, Ice core record of the $^{13}C/^{12}C$ ratio of atmospheric carbon dioxide in the past two centuries, *Nature*, 324:237.

Gordon, A. L. ,1988, Spatial and temporal variability within the Southern Ocean, *In*: "Antarctic Ocean and Resources Variability," D. Sahrhage, ed., Springer-Verlag Berlin Heidelberg.

Hardy, J., and Gucinski, H., 1989, Stratospheric ozone depletion: Implications for marine ecosystems, *Oceanography*, 2:18.

Houghton, J. T., Jenkins, G. J., and Ephraums, J. J., 1990, Climate Change, The IPCC Scientific Assessment, Cambridge U. Press.

Joos, F., Sarmiento, J. L., and Siegenthaler, U., 1991, Estimates of the effect of Southern Ocean iron fertilization on atmospheric CO_2 concentrations, *Nature*, 349:772.

Keeling, C.D., 1968, Carbon dioxide in surface ocean waters, 4, Global distribution, *J. Geophys. Res.*, 73:4543.

Keeling, C. D., Bacastow, R. B., Carter, A. F., Piper, S. C., Whorf, T. P., Heimann, M., Mook, W. G., and Roeloffzen, H., 1989a, A three dimensional model of atmospheric CO_2 transport based on observed winds: 1. Analysis of observational data, *In*: "Aspects of Climate Variability in the Pacific and the Western Americas," D.H. Peterson, ed., Geophysical Monograph 55, American Geophysical Union Washington (USA).

Keeling, C. D., Piper, S.C., and Heimann, M., 1989b, A three dimensional model of atmospheric CO_2 transport based on observed winds: 4. Mean annual gradients and interannual variations, *In*: "Aspects of Climate Variability in the Pacific and the Western Americas," D. H. Peterson, ed., Geophysical Monograph 55, American Geophysical Union Washington (USA), pp. 305-363.

Knox, F., and McElroy, M., 1984, Changes in atmospheric CO_2: Influence of the marine biota at high latitudes, *J. Geophys. Res.*, 89:4629.

Liss, P., and Merlivat, L., 1986, Air-sea exchange rates, introduction and synthesis, *In*, "The Role of Air-Sea Exchange in Geochemical Cycling," P. Buat-Menard, ed., D. Reidel Publ. Co., Dordrecht.

Maier-Reimer, E., and Hasselmann, K., 1987, Transport and storage of CO_2 in the ocean - an inorganic ocean-circulation cycle model, *Climate Dyn.*, 2:63.

Najjar, R. G., 1990, Simulations of the phosphorus and oxygen cycles in the world ocean using a general circulation model, Ph.D. Thesis, Princeton University, Princeton, New Jersey.

Neftel, A., E. Moor, H. Oeschger, and B. Stauffer, 1985. Evidence from polar ice cores for the increase in atmospheric CO_2 in the past two centuries, *Nature*, 315:45

Peng, T.-H., and Broecker, W. S.,1991, Dynamic limitations on the Antarctic iron fertilization strategy, *Nature*, 349:227.

Raven, J. A., 1991, Implications of inorganic C utilization: Ecology, evolution and geochemistry, *Can. J. Bot.*, 69:203.

Sabine, C. L., and Mackenzie, F. T., 1991, Oceanic sinks for anthropogenic CO_2, *International Journal of Energy Environment Economics*, 1:119.

Sarmiento, J. L., and Orr, J.C., 1991, Three dimensional ocean model simulations of the impact of Southern Ocean nutrient depletion on atmospheric CO_2 and ocean chemistry, *Limnol. Oceanogr.*, in press.

Sarmiento, J. L., and Sundquist, E., 1991, River and ocean sediment carbon fluxes play a major role in the oceanic anthropogenic CO_2 budget, In preparation.

Sarmiento, J. L., Orr, J. C., and Siegenthaler, U., 1991, A perturbation simulation of CO_2 uptake in an ocean general circulation model, *J. Geophys. Res.*, in press.

Sarmiento, J. L., and Toggweiler, J. R., 1984, A new model for the role of the oceans in determining atmospheric pCO_2, *Nature*, 308:621.

Schluessel, P., Emery, W. J., Grassl, H., and Mammen, T., 1990, On the bulk-skin temperature difference and its impact on satellite remote sensing of sea surface temperature, *J. Geophys. Res.*, 95:13341.

Siegenthaler, U., and Wenk, T., 1984, Rapid atmospheric CO_2 variations and ocean circulation, *Nature*, 308:624.

Siegenthaler, U., and Oeschger, H., 1978, Predicting future atmospheric carbon dioxide levels, *Science*, 199:388.

Siegenthaler, U., and Oeschger, H., 1987, Biospheric CO_2 emissions during the past 200 years reconstructed by deconvolution of ice core data, *Tellus*, 39B:140.

Siegenthaler, U., Friedli, H., LÜtscher, H., Moor, E., Neftel, A., Oeschger, H., and Stauffer, B., 1988, Stable-isotope ratios and concentration of CO_2 in air from polar ice cores, *Annals of Glaciology*, 10:1.

Smith, R. C., and Baker, K.S., 1979, Penetration of UV-B and biologically effective dose-rates in natural waters, *Photochem. Photobiol.*, 32:367.

Stauffer, B., Hofer, H. Oeschger, H., Schwander, J., and Siegenthaler, U., 1984, Atmospheric CO_2 concentration during the last glaciation, *Annals of Glaciaology*, 5:160.

Sugimura, Y., and Suzuki, Y., 1988, A high-temperature catalytic oxidation method for the determination of non-volatile dissolved organic carbon in seawater by direct injection of a liquid sample, *Marine Chemistry*, 24:105.

Suzuki, Y., Sugimura, Y., and Itoh, T., 1985, A catalytic oxidation method for the determination of total nitrogen dissolved in seawater, *Marine Chemistry*, 16:83.

Tans, P. P., Fung, I. Y., and Takahashi, T., 1990, Observational constraints on the global atmospheric CO_2 budget, *Science*, 247:1431.

Toggweiler, J. R., 1989, Is the downward dissolved organic matter (DOM) flux important in carbon transport?, *In*: "Productivity of the ocean: Past and Present," W. H. Berger, V. Smetacek, and D. Wefer, eds., Dahlem Workshop Report, John Wiley and Sons, Chichester.

UNESCO, 1991, Report of the Second Session of the Joint JGOFS-CCCO Panel on Carbon Dioxide, April 1991, Paris.

Venrick, E. L., McGowan, J. A., Cayan, D. R., and Hayward, T. L., 1987, Climate and chlorophyll *a*: Long-term trends in the Central North Pacific Ocean, *Science*, 238:70.

Volk, T, and Hoffert, M. I., 1985, Ocean carbon pumps: Analysis of relative strengths and efficiencies in ocean-driven atmospheric CO_2 changes, *In*: "The Carbon Cycle and Atmospheric CO_2: Natural Variations Archean to Present," E. Sundquist and W. S. Broecker, eds., Geophysical Monograph 32, American Geophysical Union.

Volk, T., and Liu, Z., 1988, Controls of CO_2 sources and sinks in the earth scale surface ocean: Temperature and nutrients, *Global Biogeochem. Cycles*, 2:73.

Walsh, J. J., 1989, How much shelf production reaches the deep sea?, *In*: "Productivity of the Ocean: Present and Past," W. H. Berger, V. S. Smetacek, and G. Wefer, eds., John Wiley & Sons, Chichester.

Watson, A. J., Robinson, C., Robinson, J. E., Williams, P. J. leB., and Fasham, M. J. R., 1991, Spatial variability in the sink for atmospheric carbon dioxide in the North Atlantic, *Nature*, 350:50.

RESPIRATION: TAXATION WITHOUT REPRESENTATION?

Richard J. Geider

College of Marine Studies
University of Delaware
Lewes, DE 19718

INTRODUCTION

The role of phytoplankton respiration under conditions where mixed layer depth (z_{mix}) exceeds euphotic zone depth (z_{eu}) and phytoplankton growth is light-limited is well established. Sverdrup (1952) demonstrated the importance of respiration in determining the onset of the spring bloom and Wofsy (1983) considered the balance between respiration and photosynthesis in determining phytoplankton abundance in turbid waters. In stably stratified waters where $z_{mix} < z_{eu}$, algal respiration is often considered to be small relative to gross photosynthesis, and has been largely neglected as a potential sink for primary production. Microbial loop processes are thought to dominate remineralization of organic matter under conditions of stable phytoplankton abundance in a stably stratified euphotic zone (Frost, 1987; Fasham et al., 1990).

The general neglect of respiration as a component of phytoplankton energy budgets may arise from the common assumption that the rate of dark respiration equals only 10% of the rate of light-saturated photosynthesis (Parsons et al., 1984). Although this assumption appears to be valid for microalgae under light-limited and nutrient-saturated conditions, the ratio of dark respiration (r_d) to light-saturated photosynthesis (P_m) varies both with taxon under consideration and environmental conditions (Geider and Osborne, 1989). Under nutrient-saturated conditions, cyanobacteria are characterized by $r_d:P_m < 0.1$, while dinoflagellates are characterized by $r_d:P_m > 0.25$, with diatoms, prymensiophytes and chlorophytes having intermediate values (Geider and Osborne, 1989). Comparison of dark respiration with gross photosynthesis may be inappropriate when phytoplankton photosynthesis is light-limited and the realized rate of photosynthesis is much less than P_m. It is also inappropriate to assume $r_d < < P_m$ under nutrient-limited conditions. The ratio $r_d:P_m$ increases dramatically under nutrient-limitation or starvation to values of $r_d:P_m > 0.5$ (Ryther, 1954; 1956; Osborne and Geider, 1986). The importance of algal respiration under suboptimal conditions for phytoplankton growth in the sea is emphasized by Lancelot and Mathot's (1985) estimate that respiration of carbohydrate energy reserves accounted for 50% of gross CO_2 fixation in an experiment conducted in Belgian coastal waters.

Dark respiration is often thought of as an unavoidable debit in the energy balance for phytoplankton growth, with the implication that avoiding respiratory losses would be

Primary Productivity and Biogeochemical Cycles in the Sea
Edited by P.G. Falkowski and A.D. Woodhead, Plenum Press, New York, 1992

advantageous. However, energy provided by mitochondrial respiration allows nutrient assimilation and biosynthetic processes to be extended into darkness when light-generated reductant and ATP are not available. The observed coupling of protein synthesis with carbohydrate consumption in darkness provides evidence for a major role of dark respiratory processes in phytoplankton growth (Foy and Smith, 1980; Cuhel et al., 1984). In addition, glycolysis, the oxidative pentose phosphate pathway, and the tricarboxylic acid cycle link photosynthetic production of sugars with cell biosynthetic pathways (Raven, 1984). Finally, respiratory use of energy storage products may be important in the transient response of phytoplankton to changes of environmental conditions, such as decreases in irradiance and increases in nutrient availability.

THE BIOGEOCHEMICAL CONTEXT

Biogeochemical cycles are driven by the flux of organic matter which accompanies uncoupling of photosynthesis from respiration. The factors responsible for this uncoupling determine the development of the spring bloom, the depth of the euphotic zone, and the rates of accumulation and export of organic matter. In many oceanic regions, the rates of photosynthesis and respiration appear to be approximately balanced over time scales of days to weeks, but significant diel uncoupling can occur. In oligotrophic regions, the light-period increase of dissolved oxygen is often matched by an almost equal decrease at night (Tijssen, 1979; Williams et al., 1983; Grande et al., 1989; Oudot, 1989). Based on a 10-day time series, Oudot (1989) calculated that the net daily increase of O_2 equalled only 11% of the diel signal in the tropical North Atlantic. Even during spring blooms, when phytoplankton biomass is accumulating rapidly, the diel signal in O_2 concentration can be pronounced (Tijssen and Eijgenraam, 1982).

Diel changes of particulate organic carbon and beam attenuation within the euphotic zone are readily measurable. Eppley et al. (1988) observed a 15-25% increase of particulate organic carbon (POC) during a light period, with a corresponding decrease in the dark period at a depth of 30 m in the Eastern North Pacific subtropical gyre (28 °N, 155 °W). Changes of similar magnitude in beam attenuation (corrected for absorption by water) were reported at 28 °N, 142 °W (Siegel et al., 1989). Two-fold dawn-dusk variations of particulate matter were observed in the tropical Atlantic (POC measurements of Postma and Rommets, 1979) and equatorial Pacific (beam attenuation measurements of Cullen et al., 1991).

These diel changes of POC and bean attenuation have been interpreted in terms of phytoplankton production and microheterotroph consumption of organic matter (Siegel et al., 1989; Cullen et al., 1991). If considered at all, phytoplankton respiration was assumed to be small relative to gross photosynthesis. However, the growth and grazing rates required to describe the diel changes in the concentration of particulate matter are extremely, perhaps unrealistically, high (Cullen et al., 1991). For example, Smith et al. (1984) invoked a phytoplankton-specific photosynthesis rate of 0.2 h^{-1} (2.4 d^{-1} during the 12-hour light period) to obtain a two-fold change of particulate organic carbon over a light-dark cycle in a steady-state model ecosystem consisting solely of phytoplankton and protozoan grazers. Even higher specific production rates would need to be invoked if a significant fraction of the particulate carbon is detrital (Cullen et al., 1991).

A model of phytoplankton growth/microzooplankton grazing (Smith et al., 1984) may be able to account for more modest (15 to 25%) diel changes of particulate matter observed in the subtropical gyres (Eppley et al., 1988; Siegel et al., 1989) if moderate phytoplankton growth rates are assumed (about 1 d^{-1}). To obtain significant diel changes of the concentration of particulate organic carbon in a microbial loop requires that high, biomass-specific rates of phytoplankton photosynthesis are balanced by high rates of microheterotrophic activity. Conversely, processes in the microbial loop cannot account for large diel changes of POC changes if phytoplankton growth rates are low. Recent observations indicating that absolute growth rates ($\mu = 0.15$ to 0.45 d^{-1} in the Sargasso Sea near Bermuda) (Goericke, 1990) and relative growth rates of phytoplankton ($\mu/\mu < 0.2$ in the North Pacific near Hawaii) (Falkowski et al., 1991) can be low in the oligotrophic ocean may require a re-evaluation of the microbial loop in generating diel POC changes.

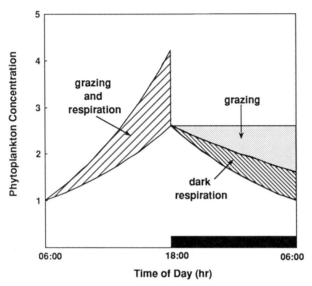

Fig. 1.   Diel change in phytoplankton carbon calculated from Eq. 1 with $p=0.16$ h^{-1} during 12 hours of daylight, $r=-0.04$ h^{-1} and $g=-0.04$ h^{-1}. Indicated in hatched areas are the amount of phytoplankton carbon lost to grazing and respiration during the day, the night loss of phytoplankton carbon to grazing alone and the additional night loss due to respiration.

Given the constraints on small particle interactions in a dilute milieu, it may be difficult for grazing by microzooplankton to control the abundance of rapidly growing phytoplankton. For example, one can calculate a maximum clearance rate of 0.49 d^{-1} for the oligotrophic central gyres, given a flagellate population of 700 cells ml^{-1} and a filtration rate of 0.0007 ml flagellate^{-1} (Landry et al., 1984). A clearance rate of 0.34 d^{-1} can be calculated for the equatorial Pacific, based on a mean microflagellate concentration of 180 flagellates ml^{-1} reported by Chavez et al. (1990) assuming the maximum filtration rate of 0.0019 ml flagellate^{-1} d^{-1} reported by Fenchel (1982). If phytoplankton growth rates are only 0.5 d^{-1} and phytoplankton account for 25% of the particulate matter, then a relative diel POC change of about 4% would be expected in the absence of significant phytoplankton respiration or other night-time losses. This value can be contrasted with the 15 to 25% change observed in the North Pacific subtropical gyre (Eppley et al., 1988) and the 100% change observed in the tropical Atlantic (Postma and Rommets, 1979).

As a first-order approximation, we can assume that the diel signal of POC can be attributed exclusively to changes in phytoplankton carbon, with microheterotroph and detrital pools independent of time. Then, the phytoplankton carbon pool size will be determined by:

$$dP/dt = (p-r-g)P \qquad (1)$$

where P is the phytoplankton carbon concentration, p is the biomass-specific photosynthesis rate (h^{-1}), r is the biomass-specific respiration rate (h^{-1}), and g is the biomass-specific grazing rate (h^{-1}). The growth rate of phytoplankton is $\mu = p - r$. In the steady state, phytoplankton carbon must increase sufficiently during 12 hours of daylight to support the subsequent respiration and grazing losses at night. Thus, in the steady-state, $p = 2(r + g)$.

How much of the diel POC signal can be attributed to phytoplankton respiration? Maximum rates of dark respiration up to 0.04 h^{-1} have been observed in phytoplankton cultures (Fig. 1). During 12 hours of darkness, a 0.04 h^{-1} dark respiration rate would require a 42% reduction in phytoplankton carbon. Although sufficient to account for a large fraction of the diel POC signal in oligotrophic gyres, phytoplankton respiration at rates of 0.04 h^{-1} could not account for the two-fold diel change of organic matter reported for equatorial waters.

The combination of phytoplankton respiration and microbial loop processes might be sufficient to account for diel signals in the concentration of particulate organic matter. Assuming a 12:12 L:D cycle, the magnitude of the diel change of phytoplankton carbon is given by:

$$P_{12}/P_0 = \exp(12[r+g]) \qquad (2)$$

For example, a photosynthetic rate of 0.12 h^{-1} (during 12 hours of daylight), combined with a respiration rate of 0.03 h^{-1} and grazing rate of 0.03 h^{-1}, would allow a doubling of phytoplankton carbon during the light period to be balanced by a similar decline at night. Then, the phytoplankton growth rate would equal 0.72 d^{-1}, with 50% of the night-time decrease of phytoplankton carbon due to phytoplankton respiration and the remainder to grazing. Partitioning the diel POC signal into contributions of phytoplankton respiration, grazing, and export will require measurements of phytoplankton growth rate, phytoplankton carbon content, and the diel changes in concentrations of dissolved gases, with appropriate controls to insure water mass homogeneity during the course of the observations.

CATEGORIES OF O_2 CONSUMING PROCESSES IN PHYTOPLANKTON

The rate of conversion of radiant energy into biochemical energy is called gross primary production, and can be considered as the fundamental energetic constraint on productivity of the biosphere. However, it is the rate of net primary production that determines the role of phytoplankton in global geochemical cycles, including the availability of phytoplankton biomass to higher trophic levels and for export from the photic zone. The loss processes which account for the difference between gross and net phytoplankton production include photorespiration, the Mehler reaction, and dark respiration (Badger, 1985).

Oxygen-consuming processes which are limited to illuminated conditions include photorespiration and the Mehler reaction (pseudocyclic electron flow). Photorespiration consists of the light-dependent uptake of O_2 by ribulose bisphosphate carboxylase-oxygenase with the subsequent exudation of glycolate or oxidation of glycolate with CO_2 release by the photorespiratory carbon oxidation cycle (Badger, 1985; Beardall, 1989). Since photorespiration involves the metabolism of Calvin cycle intermediates, it can be considered as a factor that limits the efficiency of gross photosynthesis. Reductions of both the initial slope and light-saturated rate of photosynthesis are expected under photorespiratory conditions. Unlike dark respiration, which typically has a role in conserving energy as ATP, photorespiration is primarily an energy-dissipating mechanism (Osmond, 1981). Photorespiration is suppressed in marine microalgae through operation of a CO_2-concentrating mechanism, (Beardall, 1989); however, pumping inorganic carbon into an algal cell requires energy which partially offsets the gain in efficiency provided by high internal CO_2 concentrations. Photorespiration may occur at up to 15% of the rate of gross oxygen evolution (Raven and Beardall, 1981). Photorespiration was recently reviewed in a special issue of Aquatic Botany (1989; Volume 34). Photorespiration should not be confused with the Mehler reaction (i.e., the light-dependent photoreduction of O_2) (Badger, 1985), which can be a major sink for O_2, occurring at rates that may equal the rate of gross oxygen evolution (Raven and Beardall, 1981). Much of the recent controversy concerning the measurement of net and gross photosynthesis of phytoplankton centers on the roles of photorespiration and the Mehler reaction (Bender et al., 1987; Grande et al., 1989). Although gross photosynthesis is strictly defined as the rate of O_2 evolution by photosystem 2, for most practical purposes the quantity of concern for biological oceanographers is the rate of oxygen evolution in the light after accounting for photorespiration and the Mehler reaction. Photorespiration and the Mehler reaction may play important roles in protecting the photosynthetic apparatus from damage (Osmond, 1981). Further consideration of these processes is beyond the scope of this review, which will be limited to the role of dark respiration in phytoplankton physiology and ecology.

The pathways of dark respiration include glycolysis, the pentose phosphate pathway, the tricarboxylic acid (TCA) cycle, and the mitochondrial electron transport chain. Dark respiration plays three major roles in phytoplankton metabolism: the generation of ATP, the generation of reductant (NAD(P)H), and the provision of essential carbon skeletons for lipid, nucleic acid, and protein synthesis (Raven, 1984). The TCA cycle supplies ATP via substrate-level phosphorylations and reductant (NADH). The mitochondrial electron transport system (ETS) consumes NADH to produce additional ATP. Although O_2 is the major electron acceptor in phytoplankton metabolism, the role of nitrate and nitrite reduction as sinks for NAD(P)H should not be ignored. Nitrate reduction to nitrite requires the transfer of two electrons and nitrite reduction to ammonium requires transfer of an additional six electrons. The reductant is provided by the oxidation of NAD(P)H which may be generated by photosynthesis in illuminated chloroplasts or by the pentose phosphate pathway in the cytoplasm. Thus, at times the generation of CO_2, and the production and consumption of reductant may be uncoupled from the consumption of O_2.

Since Burris (1980) reviewed respiration and photorespiration at the last Brookhaven Symposium on Primary Productivity in the Sea, a new category of respiration, termed chlororespiration, has been identified. Respiration and photosynthesis take place on the same membrane system in cyanobacteria, with plastoquinone serving as a common electron carrier (Hirano et al., 1980). Inhibition of respiration in cyanobacteria by red light (wavelength = 700 nm) has been attributed to a direct interaction with photosynthesis (Hoch et al., 1963; Jones and Myers, 1963). Thus, photosystem I competes with O_2 as an electron acceptor reducing O_2 consumption in the light. The same mechanism was proposed to account for the Kok effect in *Chlamydomonas reinhardtii* (Healey and Myers, 1971). Recently, dark oxygen consumption by chloroplasts involving oxidation of NAD(P)H at the expense of O_2 was identified in thylakoid membranes of *Chlorella pyrenoidosa* and *Chlamydomonas reinhardtii* (Bennoun, 1982; Peltier et al., 1987). Like cyanobacterial respiration, chlororespiration apparently shares plastoquinone (PQ) with the photosynthetic electron transfer chain. Peltier and Sarrey (1988) concluded that inhibition of chlororespiration by light was responsible for the Kok effect in *Chlamydomonas reinhardtii*. The terminal oxidase of the chlororespiratory electron transfer chain has not been identified, although it is apparently inhibited by potassium cyanide (KCN) but not by salicyl hydroxamic acid (SHAM) nor by antimycin A in *Chlamydomonas reinhardtii* (Bennoun, 1982; Peltier et al., 1987) and by SHAM but not KCN in *Chlorella pyrenoidosa* (Bennoun, 1982). Chlororespiration has been estimated to occur at a rate equal to 10-20% of the rate of dark respiration (Bennoun, 1982; Peltier et al., 1987).

THE RELATIVE RATES OF PHOTOSYNTHESIS, RESPIRATION, AND GROWTH

Only one study has been conducted which allows a direct determination of the importance of mitochondrial respiration in the light and darkness to the energetics of growth and photosynthesis of a microalga. The data of Weger et al. (1989) allow one to calculate that respiration consumed 60% of gross photosynthesis in the diatom *Thalassiosira weisflogii* grown in nutrient-sufficient semi-continuous culture on a 12:12 L:D cycle at an irradiance of 150 μmol photons m^{-2} s^{-1} at 18 °C. Under these light-limited conditions (μ=0.21 d^{-1}), the particulate organic carbon (POC) concentration increased by 40% from 56.6 to 79 μg C ml^{-1} during the light period, with a decline to 70 μg ml^{-1} over the subsequent dark period. Thus, dark respiration consumed 40% of the net photosynthate accumulated during the preceding light period. The biomass specific dark respiration rate was only 0.01 h^{-1}, well within the range encountered in microalgae (Fig. 2). Mitochondrial oxygen consumption in the light occurred at 33% of the rate of gross photosynthesis. Assuming that dissolved organic carbon was neither released nor consumed in this experiment and that O_2 fluxes are proportional to CO_2 fluxes, net and gross photosynthesis rates during the light period of 22.4 and 33.6 μg C ml^{-1} $(12h)^{-1}$ were calculated. The total respiration during the light and dark periods of 11.2+9.0=20.2 μg C ml^{-1} was 60% of the gross photosynthetic rate. One would expect respiration to account for an even larger proportion of gross photosynthesis in a nutrient-limited and light-saturated surface mixed layer.

The preceding discussion suggests that a substantial correction for phytoplankton respiration may be necessary to determine the rate of net phytoplankton photosynthesis if gross photosynthesis is going to be estimated from bio-optical models (Bidigare, this volume) or fluorescence techniques (Kiefer, this volume).

QUANTITATIVE SIGNIFICANCE OF DARK RESPIRATION TO PHYTOPLANKTON GROWTH

The Relationship Between Respiration and Growth

Significant progress in modeling phytoplankton growth was made in the late 1970s and 1980s (Bannister, 1979; Shuter, 1979; Kiefer and Mitchell, 1983; Laws et al., 1985; Sakshaug et al., 1989). The basis for these models is an energy budget in which the sum of growth plus respiration is set equal to the rate of gross photosynthesis (Geider et al.,

1986; Geider, 1990). Although much progress has been made in refining the parameterization of photosynthesis in terms of the optical properties of phytoplankton and the characterization of the underwater light field (Bidigare, this volume), one major problem with applying the models *in situ* is estimating the respiration rate. Parameterization of phytoplankton respiration is largely based on correlation as opposed to mechanistic insight or understanding. In fact, there has been little progress since Shuter (1979) introduced a linear relation between respiration and growth. Shuter (1979) assumed that respiration had two components, a constant maintenance metabolic rate, and a variable rate associated with cell synthetic activities:

$$r = \mu_o + \xi\mu \tag{3}$$

where r is the biomass-specific energy consumption rate (d^{-1}), μ_o is the maintenance metabolic rate (d^{-1}), ξ is the dimensionless cost of synthesis, and μ is the growth rate (d^{-1}). The concepts of maintenance metabolism and the cost of cell synthesis were inferred from the linear relation often observed between respiration and growth in studies of energetics of bacterial growth (Tempest and Neijssel, 1984).

One important implication of Eq. 3 is that high respiration rates are expected to be associated with high growth rates. Variations in the relation between respiration and growth rates may be attributed to (1) environmentally induced variations in the rate of maintenance metabolism, (2) changes in bulk biochemical composition and particularly in the fraction of cell mass accounted for by storage products, (3) differences in rate of functional biomass synthesis during light and dark periods, and (4) ability of microbes to dissociate catabolism from anabolism to a variable extent.

Maintenance Metabolic Rate

The maintenance metabolic rate is the minimum energy consumption rate required to maintain the viability of microorganisms. Maintenance metabolism includes the turnover of macromolecules, volume regulation, and maintenance of solute gradients between the cell and the suspending medium (Penning de Vries, 1975; Amthor, 1984). As formulated in Eq. 3, maintenance metabolism is assumed to be independent of growth rate. Empirical verification of this assumption is lacking, and some processes which have been included in maintenance metabolism, such as adaptation to a changing environment (Amthor, 1984), may be better characterized as a cost of biosynthesis.

Several methods have been employed to estimate the maintenance metabolic rates of microalgae: (1) extrapolating a plot of dark respiration rate (r_d) against growth rate (μ) to $\mu=0$; (2) measuring the asymptotic value of respiration after prolonged (20-30 hours) darkness; and (3) extrapolating a plot of growth rate versus irradiance (E_o) to $E_o=0$. The first method yields the highest estimates of maintenance metabolic rate of 0.01-0.4 d^{-1} (Geider and Osborne, 1989). Problems in using this approach to examine the energetics of bacterial growth have been discussed by Tempest and Neijssel (1984). In particular, the respiration rates of nutrient-limited bacteria growing under conditions of energy sufficiency are much greater than the rates of respiration in energy-limited bacteria growing under conditions of nutrient sufficiency (Tempest and Neijssel, 1984). Either the maintenance metabolic rate depends on the growth limiting factor and on growth rate as advocated by Pirt (1982), or some factors other than maintenance metabolism contribute to the respiration rate at $\mu=0$ (Tempest and Neijssel, 1984). Similar problems may face application of this approach to studies of algal growth energetics, and thus, independent estimates of maintenance metabolism are required.

The remaining two methods yield comparable estimates of maintenance metabolic rate. The asymptotic value of dark respiration in *Chlorella* after prolonged (~24h) darkness is about 0.025 d^{-1} (Myers, 1947; Groebellar and Soeder, 1985). The observed decline of particulate organic carbon in axenic cultures of *Skeletonema costatum* (Varum et al., 1986) leads to an estimate of maintenance metabolic rate of <0.01 d^{-1}. Survival of microalgae for weeks to months in darkness (Smayda and Mitchell-Innes, 1974; Hellebust and Lewin, 1977) provides additional support for low maintenance requirements. Extrapolating the light-limited growth rate to zero irradiance leads to estimates of a maintenance metabolic rate of <0.002 to 0.1 d^{-1} (Van Leire and Mur, 1979; Pirt, 1986; Geider and Osborne, 1989).

Direct and indirect estimates of respiration rates when synthetic requirements are minimal lead to maintenance metabolic requirements which are generally much less than 0.1 d^{-1}. Unfortunately, measuring rates of oxygen consumption or the decline of particulate organic matter provide no information on the processes which contribute to maintenance metabolism. To gain a better understanding of the role of respiration in the physiology of phytoplankton requires direct measurement of the processes which contribute to respiration.

Protein Turnover

Protein turnover is a costly biochemical transformation because of the large amounts of ATP consumed in forming peptide bonds (Penning de Vries, 1975). Measurement of protein turnover is not straightforward, and there are no observations of protein turnover for marine phytoplankton despite its potential importance as a sink for energy and the uncertainty over the extent to which ^{14}C labelling of phytoplankton protein represents net or gross protein synthesis (Cuhel et al., 1984). External trap methods (i.e., employing high concentrations of extracellular amino acids to prevent reincorporation of radioactively labelled amino acids in pulse-chase experiments) have been employed successfully with bacteria. Unfortunately, this approach is inappropriate for measuring protein turnover of phytoplankton because intracellular amino acid pools do not readily exchange with extracellular pools in these organisms (Richards and Thurston, 1980). An alternative approach labels amino acids with ^{3}H$_2$O in the α-carbon position by transamination reactions (Humphrey and Davies, 1975; 1976). Upon incorporation into proteins, the ^{3}H on the α-carbon is no longer exchangeable, but once protein is degraded into component amino acids exchange with ^{3}H$_2$O insures rapid loss of ^{3}H (Humphrey and Davies, 1975). Thus, protein degradation is measured by the decline of ^{3}H activity during the chase period. This approach has been used only once with a microalga, *Chlorella* sp. by Richards and Thurston (1980). They found little evidence for protein turnover in exponentially growing cultures. However, rapid turnover was observed in stationary phase cultures in which 85% of cell protein was degraded at a rate of 0.035 h^{-1} and 15% degraded at a rate of 0.011 h^{-1}. Thus, the overall protein turnover rate would be 0.03 h^{-1}.

The minimum energy required for protein turnover, based on cleavage and resynthesis of the peptide bond, is 0.25 g glucose per g protein (0.2 g C in glucose/g C in protein) (Penning de Vries, 1975). This amount represents 80% of the cost of protein synthesis from glucose and ammonium. The maintenance respiration associated with protein turnover can be estimated at 0.07 d^{-1}, based on the protein turnover rate of 0.72 d^{-1} determined by Richards and Thurston (1980) for nitrogen-starved *Chlorella sp.*, and assuming that protein accounts for 50% of cell carbon. This value may overestimate the cost of protein turnover if protein accounts for significantly less than 50% of cell carbon, as is the case in nitrogen- or phosphorus-limited phytoplankton, or if protein turnover in growing cells occurs at rates <0.03 h^{-1} measured under nitrogen-starved conditions.

Table 1. The Conversion of Glucose to Protein with Either Nitrate or Ammonium as the Inorganic Nitrogen Source

Nitrogen Source	$\dfrac{\text{mol glucose-C consumed}}{\text{mol protein-C produced}}$	Cost of Synthesis
NO_3^-	1.29[*]	0.29[**]
NH_4^+	1.98[*]	0.98

[*] Calculated from Table 4 of Penning de Vries et al. (1974) assuming that protein is 50% carbon by weight and glucose is 40% carbon by weight.

[**] The cost of synthesis (ξ) has units of mol CO_2 evolved per mol Protein-C synthesized from glucose and the indicated nitrogen source.

Cost of Biosynthesis

The concept of the cost of biosynthesis was derived from early observations of a constant relation between yield and substrate consumption in bacteria (Tempest and Neijssel, 1984). A theoretical approach for calculating the cost of synthesis from an examination of biochemical pathways, and ATP and NADH requirements for transport of nutrients and biosynthesis of macromolecules was provided by Penning de Vries et al. (1974). The cost of synthesis depends primarily on the biochemical composition of the biomass and the redox state of the nitrogen source (i.e., NH_4^+ versus NO_3^-) (Penning de Vries et al., 1974). The concept has been applied to the heterotrophic growth of microalgae (Raven, 1976), dark synthesis of protein from carbohydrate energy reserves (Post et al., 1985; Geider and Osborne, 1989), and physiological adaptation during algal growth (Shuter, 1979). Protein synthesis is a major component of the cost of biomass synthesis. Penning de Vries et al. (1974) calculated that 0.29 mol CO_2 is released per mol C incorporated into protein from glucose when NH_4^+ is the N-source (Table 1). This value increases to 0.98 mol CO_2 evolved per mol protein-C produced when NO_3^- is the N-source.

The relationship between mitochondrial respiration and rates of protein and lipid synthesis in the light has not been investigated. Raven (1976; 1984) calculates, however, that the maximum rate of mitochondrial respiration observed in microalgae is generally insufficient to support the energy demands for maximum growth. A direct supply of reductant and ATP from photosynthetic light reactions is, thus, required to make up any shortfall in energy requirements of a rapidly growing microalga in the light, and the potential exists for meeting all of the energy demands for growth from photosynthesis. However, respiratory pathways will still be required to provide carbon skeletons for lipid, protein, carbohydrate, and nucleic acid synthesis.

Respiration in the light may play an important role in the response of nutrient-limited cells to nutrient enrichment. Recent experiments with *Selenastrum minutum* indicate that mitochondrial respiration is required to support protein synthesis when ammonium is provided to N-limited cells, and NO_3^- and NO_2^- reduction as well as protein synthesis when NO_3^- or NO_2^- are provided (Weger et al., 1988; Weger and Turpin, 1989). Respiration rates have been found to be proportional to protein synthesis or ammonium

Table 2. Comparison of Theoretical and Observed Costs of Biomass Synthesis in Heterotrophically Metabolizing Microalgae with Glucose as both Carbon and Energy Source

| Category | N-source | Cost of Synthesis | |
		Theory	Observation
Protein Synthesis	NO_3^-	0.98	0.92-1.4
	NH_4^+	0.29	0.30-0.52
Growth	NO_3^-	0.67-0.76	0.75-1.4
	NH_4^+	0.32-0.46	0.40-1.2

Theoretical costs for protein synthesis were taken from Table 1, and theoretical costs for heterotrophic growth were taken from Shuter (1979) and Raven and Beardall (1981). Observations for dark synthesis of protein from carbohydrate energy reserves and heterotrophic growth of various microalgae are from the literature review summarized in Table 5 of Geider and Osborne (1989). The cost of protein and biosynthesis in these studies includes a maintenance component.

assimilation following the addition of ammonium to several nitrogen-limited chlorophytes (Syrett, 1953; Weger et al., 1988). Respiration (CO_2 evolution and O_2 consumption) is stimulated in both the light and darkness by addition of NH_4^+ to nitrogen-limited *S. minutum* (Guy et al., 1989). Enhanced activity of the TCA cycle activity after ammonium addition is associated with protein synthesis, with much of the carbon entering the TCA cycle as oxaloacetate and acetyl-CoA and exiting as α-ketoglutarate (Guy et al., 1989); also the rate of ß-carboxylation by phosphoenolpyruvate in nitrogen-limited *S. minutum* was found to be linearly related to the rate of ammonium assimilation (Vanlerberghe et al., 1990).

Significantly, mobilization of energy reserves to allow continued protein synthesis in darkness does not necessarily imply a high respiration rate. Table 1 indicates that remineralization to CO_2 only accounts for 22% of the dark glucose consumption when NH_4^+ is the N-source for protein synthesis, increasing to 50% when NO_3^- is the N-source.

The implication of Eq. 3, that the cost of biomass synthesis is met by respiration, may not be valid in the light. However, respiration associated with protein synthesis in darkness, and heterotrophic growth of microalgae should conform with Eq. 3. Limited data on protein synthesis from carbohydrate energy reserves in cells growing on a light:dark cycle, and of heterotrophic growth of microalgae, suggests that respiration occurs at rates equal to, or in excess of, the theoretical minima (Table 2). The cost of synthesis calculated in Table 2 may exceed the minimum theoretical values because the former have not been corrected for maintenance metabolic requirements, or because the respiratory pathways are operating at less than maximum efficiency (Raven, 1976; Raven, 1984; Geider and Osborne, 1989).

Application of the Concepts of Maintenance and Growth Respiration to a Marine Diatom

Observations of respiration in *Skeletonema costatum* (Underhill, 1981) can be adequately described in terms of Eq. 3 in cells cultured in NO_3^--limited chemostat cultures at 6 °C and 16 °C, but fail to conform to this model at 26 °C. Rates of respiration in a 12-hour dark period were linearly related to growth rate at 6 °C and 16 °C with a slope of 0.66 (Fig. 2), consistent with the cost of biomass synthesis with NO_3^- as the N-source (Shuter, 1979; Raven and Beardall, 1981). The intercept yielded a realistic value for the maintenance metabolic rate of 0.005 h^{-1} (0.12 d^{-1}). Significantly, Underhill (1981) reported that NO_3^- was undetectable throughout the light:dark cycle in cells cultured at low growth rates, but accumulated in darkness at high growth rates where the linear relationship between respiration and growth broke down. It is likely that the accumulation of carbohydrate energy reserves in the light was insufficient to support the high rates of NO_3^- assimilation throughout the dark period at high growth rates.

Respiration rates of *S. costatum* at 26 °C were significantly elevated above the apparent requirements for growth at low growth rates (Fig. 2). Cells grown at 26 °C were characterized by a low chlorophyll a:carbon ratio (Underhill, 1981), and thus, a high potential biomass-specific photosynthesis rate. The elevated respiration rates may have provided *S. costatum* with a means of disposing of excess energy while maintaining a high proportion of cell carbon devoted to the photosynthetic apparatus.

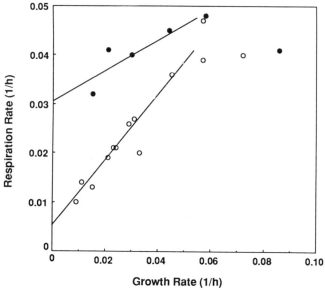

Fig. 2.   Relationship between respiration rate and growth rate of *Skeletonema costatum* in nitrate-limited chemostat cultures maintained on a 12:12 light:dark cycle with a light period irradiance of 315 μmol photons m^{-2} s^{-1} at temperatures of 6, 16 °C (open symbols) and 26 °C (closed symbols). Data of Underhill (1981) with respiration obtained from the dark loss of particulate carbon. Respiration appears to be tightly coupled to growth at 6 and 16 °C, with a slope of about 0.66 consistent with the cost of synthesis for cell growth with NO_3^- as a N-source. Significantly, Underhill (1981) reported that NO_3^- was undetectable throughout the L:D cycle in cells cultured at low growth rates, but that NO_3^- accumulated in darkness at high dilution rates. Enhanced respiration at low growth rates at 26 °C (closed symbols) suggests uncoupling of respiration from requirements for growth.

A similar uncoupling of respiration from growth to that observed in *S. costatum* at 26 °C has been reported in a bacterium by Tempest and Neijssel (1984). Elevated respiration rates were observed in nitrogen-, sulfate-, and phosphate-limited cultures of the bacterium *Klebsiella aerogenes*, but close coupling of respiration and growth were observed in energy-limited cultures. Tempest and Neijssel (1984) dismissed the explanation of an enhanced maintenance metabolic requirement associated with macromolecular turnover in nutrient-limited cultures, and proposed several alternative means of dissipating energy including (1) futile cycles, (2) modification of respiratory chain components or bypassing energy conserving steps on the respiratory chain, and (3) increased ion pumping. Some of these means of dissipating energy were adaptive in that they would allow an enhanced response to nutrient availability.

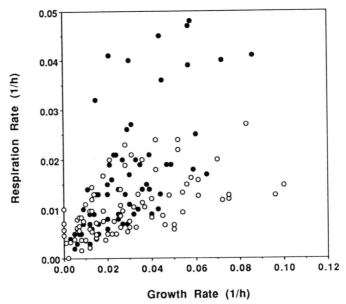

Fig. 3.   Relation between respiration rate and steady-state light- or nutrient-limited growth rate of marine and freshwater microalgae. Closed symbols indicate data obtained from the dark period loss of particulate carbon in cells grown on a light:dark cycle (Laws and Caperon, 1976; Laws and Wong, 1978; Laws and Bannister, 1980; Underhill, 1981). Open symbols indicate data obtained from dark O_2 consumption rate assuming a respiratory quotient of unity (Myers and Graham, 1971; Pickett, 1975; Falkowski and Owens, 1980; Geider et al., 1985; 1986; Osborne and Geider, 1986; Herzig and Falkowski, 1989).

In conclusion, although Eq. 3 may, at times, provide a reasonable description of dark respiration rates in microalgae, it is not universally applicable. It is not sufficient to fit a straight line to a plot of respiration rate against growth rate and assume that the slope and intercept are physiologically meaningful parameters (Tempest and Neijssel, 1984). In addition, a better knowledge of the conditions under which the model does not apply (i.e., when respiration is uncoupled from growth, as in *S. costatum* at 26 °C), and an explanation for uncoupling of respiration from growth are required.

REGULATION OF PHYTOPLANKTON RESPIRATION

The Biomass-specific Rate of Dark Respiration

To evaluate the role of dark respiration in algal physiology and biogeochemical cycles, it is useful to obtain absolute rates of dark respiration in units of grams CO_2 evolved per gram cell carbon per hour (i.e., g C g^{-1} C h^{-1}, or for convenience h^{-1}). Fig. 3 summarizes much of the data relating dark respiration rates and growth rates in steady-state nutrient-saturated and nutrient-limited phytoplankton cultures. The data was obtained from either (1) the rate of loss of particulate organic carbon during a prolonged (typically 12 hour) dark period (Laws and Caperon, 1976; Laws and Wong, 1978; Laws and Bannister, 1980; Underhill, 1981) or (2) the short term (<30 minute) rate of O_2 consumption assuming a respiratory quotient of unity (Myers and Graham, 1971; Pickett, 1975; Harris and Piccinin, 1983; Geider et al., 1985, 1986; Falkowski et al., 1985; Osborne and Geider, 1986). Biomass-specific respiration rates range from <0.01 h^{-1} to 0.05 h^{-1}. Although a linear relation between respiration and growth rates anticipated in light of Eq. 3 has often been observed in a single species, when all of the data from various investigations are considered together there appears to be no universal relationship between respiration and growth. The scatter in Fig. 3 may arise, in part, from an inappropriate comparison of mean growth rate determined over a 24 light:dark cycle with night loss of particulate carbon or instantaneous rate of dark O_2 consumption. Additional scatter may arise from the widely differing growth conditions employed by various investigators. Respiration rates in nutrient-sufficient but light-limited cultures often lie near the bottom of the range at all growth rates (Verity, 1982). Respiration rates are often directly correlated with growth rates in nutrient-limited but light-saturated phytoplankton at low growth rates, but independent or inversely related to growth at higher growth rates (Laws and Caperon, 1986; Laws and Wong, 1988; Underhill, 1981). In one study conducted under nitrate-limited conditions at light saturation and high temperature (26 °C), respiration rates were high (>0.03 h^{-1}) and largely independent of growth rate (Underhill, 1981).

Understanding the role of respiration in phytoplankton growth in the sea depends on our ability to model or predict respiration from environmental and physiological correlates, because phytoplankton respiration is difficult to measure in nature. Fig. 3 shows that the rate of respiration is a variable and potentially large fraction of phytoplankton energy balance. Therefore, simple assumptions that respiration is either (1) small compared to growth, or (2) linearly related to growth are not universally valid. Making sense out of the scatter in Fig. 3 requires an understanding of the regulation of dark respiration and its relation to growth. As a starting point, the following section summarizes current understanding of the regulation of algal respiration.

Possible points of regulatory control of dark respiration include the supply of substrate for glycolysis, glycolysis itself, transport of organic acids across the mitochondrial membrane, the TCA cycle, and mitochondrial electron transport (Lambers, 1985). Here I consider the possible regulation of phytoplankton respiration by (1) polysaccharide supply, (2) demand for energy, (3) demand for intermediates, and (4) the degree of engagement of various pathways of mitochondrial electron transport.

Substrate Supply

The principle substrates for phytoplankton respiration are sugars manufactured during photosynthesis and stored as polyglucans. Limitations on substrate supply are likely to regulate respiration in phytoplankton under nutrient-saturated but light-limited conditions. Increased rates of respiration following exposure to bright light (Brackett et al., 1953; Myers and Graham, 1971; Harris, 1978; Falkowski and Owens, 1978) is, at least in some cases, due to increased availability of substrate (Falkowski et al., 1985). Stimulation of respiration in some chlorophytes by the addition of exogenous glucose provides additional evidence for substrate limitation of respiration (Emerson, 1926; Syrett, 1951; Syrett and Fowden, 1952).

In cells regularly exposed to a light:dark cycle, energy storage products can accumulate in the light (Varum et al., 1986) allowing continued nutrient assimilation and protein synthesis in darkness (Foy and Smith, 1980; Cuhel et al., 1984). Polysaccharide accumulation is most pronounced in cells exposed to saturating irradiance. Lancelot et al. (1986) found that dark protein synthesis by *Phaeocystis pouchetii* was controlled, in part, by the previous light history. A linear relation was found between the rate of dark protein synthesis and the number of hours during which irradiance exceeded 8 J cm^{-2} h^{-1}. Similarly, Cuhel et al. (1984) concluded that significant protein synthesis at night by natural marine phytoplankton depended on prior exposure to irradiances which saturated photosynthesis. When the photoperiod is short relative to the dark period, carbohydrates accumulated in the light can be completely used in darkness, whereas when the photoperiod is long relative to the dark period, accumulation of energy reserves in the light may exceed the demand during darkness (Foy and Smith, 1980; Varum et al., 1986).

Skeletonema costatum cultured on 6:18 to 18:6 light:dark cycles accumulated ß1-3 glucans which accounted for about 50% of total cellular carbon at the end of the light period (Varum et al., 1986). From 42% to 89% of this carbohydrate storage product was metabolized during the subsequent 6-18 hours of darkness. Unfortunately, Varum et al. (1986) could not quantify the respiratory loss of organic carbon, and measurements of particulate organic carbon concentration at the start and end of the dark period suggest that essentially all of the glucan carbon was transformed into other intracellular polymers. In fact, the data indicated an increase of particulate organic carbon during darkness. This finding contrasts with results of Cuhel et al. (1984) who found an 18% loss of particulate organic carbon in an 8-hour dark period in *Dunaliella tertiolecta* growing on a 16:8 L:D cycle.

The diurnal pattern of carbohydrate synthesis and utilization observed in algal cultures also occurs in natural waters. Redistribution of photosynthetically fixed ^{14}C from polysaccharide to protein during darkness is often observed (Lancelot and Mathot, 1985). More specific information is available from GC/MS studies of rates of synthesis of individual monosaccharides and amino acids in ^{13}C productivity experiments (Hama, 1988). These experiments indicate a high rate of glucose synthesis during the day, with glucose probably accumulated as an energy reserve polymer. Ratios of glucose to total monosaccharide synthesis varied from 0.4-0.9, and ratios of carbohydrate to protein synthesis varied from 0.5-2.8, in an upwelling region off Japan (Hama, 1988). Carbon flow from glucose to other monosaccharides and amino acids was observed during the night, indicating carbon flow from energy reserves to proteins and structural polysaccharides. In addition, comparisons of carbon and nitrogen assimilation often yield ratios of C:N accumulation in the light which are greatly in excess of the Redfield ratio stoichiometry of 7:1. In Delaware Bay, for example, Pennock (1987) observed light-saturated photosynthesis to nitrogen assimilation ratios averaging 34:1. However, this imbalance was counteracted by low ratios of photosynthesis to nitrogen assimilation at low irradiances and by dark nitrogen assimilation. When integrated over the photic layer, C:N uptake ratios averaged 13:1 during the light period, decreasing to 8.5:1 over the entire day:night cycle. These observations indicate a short-term uncoupling of carbon and nitrogen assimilation, suggesting a role for protein synthesis associated with dark respiration in maintaining nearly balanced growth over longer periods (i.e., 1 day).

Demand for Energy and Intermediates

The relative concentrations of the adenylates ATP, ADP, and AMP are considered to play a major role in metabolic regulation (Atkinson, 1977), and adenylates are thought to play a role in controlling plant respiration (Wiskich and Dry, 1985). Control of respiration by ADP supply, or ATP demand is referred to as respiratory control (Wiskich and Dry, 1985). Limitations of the rate of mitochondrial respiration by ADP supply could, in turn, limit TCA cycle activity, although engagement of the cyanide-resistant, alternative electron transport pathway may allow continued operation of the TCA cycle in the absence of oxidative phosphorylation (Wiskich and Dry, 1985; Lance et al., 1985).

A major stimulator of energy demand in phytoplankton is the provision of inorganic nitrogen to nitrogen-limited populations. For example, addition of nitrate increased dark O_2 consumption by 50% in nitrate-limited cultures of the chlorophyte *Scenedesmus minutum*, and addition of ammonium more than doubled the dark respiration rate (Elrifi and Turpin, 1987). Syrett (1953) found that dark assimilation of ammonium was accompanied by enhanced respiration in nitrogen-limited *Chlorella vulgaris*, and that respiration and ammonium assimilation continued at high rates until either ammonium was completely removed from the medium or a carbon reserve within the cells was exhausted.

Glycolysis, the oxidative pentose phosphate pathway, and the TCA cycle are essential biosynthetic pathways leading to lipid, proteins, and nucleic acids. An estimated 0.1-0.15 carbon atoms are lost as CO_2 per unit carbon incorporated from glucose into cell material from essential biosynthetic reactions involving operation of the complete TCA cycle (Raven, 1984). Some of the quantitatively most significant of these biosynthetic pathways involve amino acid synthesis.

Under nitrogen-limitation, biosynthesis of proteins is likely to be limited by nitrogen supply rather than by the availability of carbon skeletons. In fact, one common bioassay for evaluating nitrogen limitation in phytoplankton relies on enhanced TCA cycle activity when inorganic nitrogen is added to a sample. This bioassay involves measuring the stimulation of dark $^{14}CO_2$ assimilation in the presence of added nitrogen (Elrifi and Turpin, 1987), where the enhanced CO_2 assimilation occurs primarily as a consequence of amino-acid synthesis using polysaccharide reserves as both energy and carbon sources (Syrett, 1956; Mortain-Bertrand et al., 1988). It is tempting to suggest that respiration rates during the dark period in nitrogen-limited chemostat cultures are limited by the supply of inorganic nitrogen in the dilution water, which in turn, controls the rate of protein synthesis. Thus, the linear relation between dark carbon loss and dilution rate often observed in nitrogen-limited chemostats (Laws and Caperon, 1976; Laws and Wong, 1978; Laws and Bannister, 1980) may reflect a limitation imposed by growth in a continuous culture. Observations of Turpin and coworkers indicate that respiration can support NO_3^-, NO_2^-, and NH_4^+ assimilation into proteins when these nutrients are supplied to N-limited cells (Weger et al., 1988; Weger and Turpin, 1989). These observations contrast with the situation in light-limited cultures in which the supply of carbohydrates may limit respiratory activity. If correct, pulses of new production in nitrogen-limited systems may be coupled to enhanced dark respiration. Specifically, respiration may be stimulated by the addition of a new source of nitrogen to populations from the upper region of the photic zone that have accumulated carbohydrate energy reserves.

ETS Activity and Engagement of the Cytochrome Pathway

Regulation of respiration rate by energy demand requires that the cytochrome path of mitochondrial electron transfer is not rate limiting. Rate limitation of the cytochrome path is assessed in terms of the degree of engagement of that pathway obtained from the ratio of rates of *in vivo* O_2 consumption to *in vitro* electron transport system activity as measured by the reduction of tetrazolium dyes (Packard, 1985):

$$E = R/V_m \qquad\qquad (4)$$

where R is the respiration rate, V_m is the maximum ETS activity as measured by tetrazolium reduction, and E is the degree of engagement of the electron transfer chain (i.e., equal to the decimal fraction of V_m attained *in vivo*). Application of this equation requires that O_2-consuming enzymes not associated with the energy-conserving, mitochondrial electron-transport system are minimal.

When R and V_m are expressed in comparable units, E is commonly assumed to equal 0.15 for phytoplankton with a range from 0.06-0.3 (Kenner and Ahmed, 1975). Observations that *in vivo* O_2 consumption rates of phytoplankton equal only 10-30% of the maximum *in vitro* respiratory activity of the electron transport system (ETS) during exponential growth, and can be as low as 6% under nutrient deficient conditions (Kenner and Ahmed, 1975) provide evidence for a large excess of respiratory ETS capacity over growth requirements. Much higher ratios of *in vivo* O_2 consumption to *in vitro* ETS activity in bacteria, protozoa, and zooplankton (Packard, 1985) indicate that the low ratios found in microalgae are due to actual limits on the degree of engagement of the ETS in the phytoplankton studied by Kenner and Ahmed (1975) rather than to possible methodological artifacts. However, Kenner and Ahmed's (1975) comparisons of *in vivo* respiration with *in vitro* ETS activity were conducted with cells acclimated to a limited range of environmental conditions and may not be representative of the full range of values which may be obtained in phytoplankton. A comparison of the observed ratios of dark respiration to growth of about 0.2 to 0.5 in resource-saturated phytoplankton with a mean ratio of *in vitro* ETS activity to growth of 0.67 obtained by Blasco et al. (1982) suggests that the commonly assumed value of E=0.15 may greatly underestimate the actual degree of engagement. More data on the relation between ETS activity and phytoplankton respiration rate is required because a value of E approaching unity would indicate that respiration rate could be limited by ETS activity.

CYANIDE-RESISTANT RESPIRATION

The energetic costs of maintenance and growth are often converted to a respiration rate by assuming the maximum permissible efficiency for mitochondrial electron transport of $ATP/O_2 = ATP/CO_2 = 6$. However, respiration can proceed with $ATP/O_2 < 6$ if energy conserving steps of the mitochondrial electron transport chain are bypassed. Operating in parallel with the energy-conserving pathways in plant mitochondria are non-energy conserving pathways which are insensitive to the normal respiratory inhibitors. These pathways include the cyanide (KCN)-resistant alternative pathway and rotenone-resistant NADPH reduction (Lance et al., 1985). If operating in series, the rotenone-and cyanide-resistant pathways would allow complete uncoupling of mitochondrial oxygen consumption from ATP production.

Cyanide-resistant respiration appears to be a common attribute of aquatic plants (Raven, 1984). There are two components of KCN resistant respiration: these are a salicylhydroxamic-acid-sensitive respiration which is assigned to the alternative oxidase, and residual respiration (often 10-50% of the KCN resistant respiration) which may be due to peroxisomal oxidation of fatty acids (Møller et al., 1988). Information on the extent of cyanide-resistant respiration in microalgae is fragmentary, and largely limited to colorless forms or heterotrophically cultured cells (Webster and Hackett, 1965, 1966; Grant and Hommersand, 1974). A significant fraction of respiration can be cyanide resistant in *Chlorella* (Emerson, 1926; Syrett, 1951; Sargent and Taylor, 1972) and *Chlamydomonas* (Peltier and Thibault, 1985; Goyal and Tolbert, 1989). There is less information on phytoplankton. Although respiration of the dinoflagellate *Gonyaulax polyhedra* was largely cyanide resistant (Hochachka and Teal, 1964), respiration in the diatom *Thalassiosira weisflogii* and the prymnesiophyte *Isochrysis galbana* was sensitive to cyanide but insensitive to the inhibitors of the alternative pathway (Falkowski et al., 1985).

Given the wide distribution of the cyanide-resistant pathway in microorganisms, and the potential of this pathway to short-circuit ATP production, more information is needed on the extent and degree of engagement of this pathway in phytoplankton. Recently, monoclonal antibodies have been raised to the alternative oxidase, allowing its identification in phylogenetically diverse organisms (Elthon et al., 1989; L. McIntosh, personal communication). However, the presence of the alternative oxidase, as indicated by KCN resistant but SHAM sensitive respiration, or cross-reactivity with monoclonal antibodies, gives no indication of whether the oxidase is engaged *in vivo*. In fact, it is generally assumed that the alternative pathway is not engaged unless the cyanide-sensitive pathway is saturated or restricted. The non-energy conserving pathways appear to play a regulatory role in higher plants by allowing the removal of reducing power, independently of control by energy charge or phosphate (Lance et al., 1985; Lambers, 1985). These pathways could potentially play a similar role in phytoplankton growing under high irradiance and nutrient limitation.

Until recently, all information on the extent of cyanide-resistant respiration was obtained through the use of inhibitors: KCN to block the energy-conserving cytochrome pathway, and salicylhydroxamic acid (SHAM) to block the cyanide resistant alternative pathway (Møller et al., 1988). However, use of SHAM and KCN to assess the relative magnitudes of electron flow through the cytochrome and alternative pathways is often suspect because of uncertainty about their specificity (Møller et al., 1988; Brouwer et al., 1986; Miller and Obendorf, 1981). Recently, Guy et al. (1989) showed that the discrimination against ^{18}O by alternative oxidase was substantially greater than discrimination by cytochrome oxidase, providing a new method for determining the rates of electron flow through the two mitochondrial electron-transport pathways. Discrimination ranged from 17.1-19.4 $°/_{oo}$ for plant cytochrome oxidase (i.e., for plant material to which SHAM was added), rising to 23.5-25.4 $°/_{oo}$ for the alternative oxidase (i.e., for plant material to which KCN was added) (Guy et al., 1989; Weger et al., 1990). Fractionation of oxygen isotopes during respiration of *Phaeodactylum tricornutum* (Guy et al., 1989) and nitrogen-limited *Chlamydomonas reinhardtii* (Weger et al., 1990) indicated little scope for engagement of the alternative oxidase. Thus, limited evidence suggests that the alternative oxidase is present, but not engaged in phytoplankton.

MITOCHONDRIAL RESPIRATION IN THE LIGHT

"...there are many potential ways in which photosynthesis and respiration can interact, and one can expect the mode and degree of interaction to vary with the organism and conditions used." (Healey and Myers, 1971)

Use of the stable oxygen isotope $^{18}O_2$ allows simultaneous determinations of rates of gross oxygen evolution and oxygen consumption in illuminated samples. However, determining the rate of mitochondrial respiration in illuminated cells has confounded plant physiologists because of the potential simultaneous occurrence of other oxygen and carbon exchange processes including photosynthesis, photorespiration, and the Mehler reaction (Graham, 1980; Badger, 1985). Some investigators report that the rate of oxygen consumption does not vary between light and darkness (Brown, 1953; Peltier and Thibault, 1985; Weger et al., 1989), while others have noted reductions in oxygen consumption, particularly at low irradiances (Brown and Webster, 1953; Hoch et al., 1963; Bate et al., 1988), and up to a 10-fold enhancement of oxygen consumption at high irradiances (Brown and Webster, 1953; Glidewell and Raven, 1975; Brechignac and Andre, 1984; Süeltemeyer et al., 1986; Brechignac and Furbank, 1987; Grande et al., 1989).

Based on inhibition of mitochondrial respiration by KCN and SHAM, the consensus of opinion appears to be that dark respiration proceeds at nearly the same rate in light and darkness (Glidewell and Raven, 1975; Peltier and Thibault, 1985; Bate et al., 1988). Enhanced O_2 consumption in illuminated conditions is often attributed to photorespiration at low TCO_2 concentration and/or the Mehler reaction at high TCO_2 (Brown and Webster, 1953; Glidewell and Raven, 1975). However, both inhibition (Bate et al., 1988) and stimulation (Weger et al., 1989) of mitochondrial respiration in the light has been reported. There is a need to differentiate between changes in mitochondrial respiration rates that occur within minutes following dark-to-light and light-to-dark transitions and changes that occur over longer periods. In the absence of photorespiration and the Mehler reaction, the rate of oxygen consumption does not change during the first 5 to 10 minutes following such transitions (Brown, 1953; Peltier and Thibault, 1985; Weger et al., 1989), but light enhancement of mitochondrial respiration has been inferred from increases in the rate of dark oxygen consumption following illumination with bright light (Falkowski et al., 1985). Direct confirmation of light-enhanced mitochondrial respiration has been obtained by Weger et al. (1989). Stimulation of mitochondrial respiration through an increase in substrate supply apparently accompanies photosynthesis in light-limited phytoplankton.

MEASURING PHYTOPLANKTON RESPIRATION IN NATURE

Estimating phytoplankton respiration in nature is complicated by the presence of heterotrophic microorganisms. In fact, microheterotrophs are often assumed to dominate dark bottle oxygen uptake. Consequently, investigators have modified the ^{14}C technique to indirectly estimate phytoplankton respiration. These charges usually involve labelling cells with ^{14}C in the light and examining the subsequent loss of particulate ^{14}C activity in darkness.

Recent observations employing this technique have been made in the North Pacific central gyre (28 °N 155 °W). Grande et al. (1989) found that dark loss of PO^{14}C (12-24 hours) accounted for 26-36% of ^{14}C assimilation into particulate matter during the preceding 12-hour incubation in the light. For stations occupied on different days, but on the same cruise, Laws et al. (1987) reported an average 24% loss of PO^{14}C during 12 hours of darkness. Interpreting the dark loss of ^{14}C in terms of phytoplankton respiration and microbial loop processes is not straight-forward, and is complicated by the potential for high rates of dark ^{14}C fixation (Li and Dickie, 1991; Taguchi et al., 1988).

Li and Harrison (1981) introduced a new twist into methods for estimating phytoplankton catabolism by examining time courses of protein, carbohydrate, and lipid synthesis during a 32-hour incubation with arctic phytoplankton in continuous light. ^{14}C incorporation into protein calculated from the sum of thirteen 2-3 hour incubations was the same as that measured in a continuous 32-hour incubation. In contrast, ^{14}C accumulation in lipids and polysaccharides calculated from the sum of short-term incubations exceeded that from the 32-hour incubation. This difference was attributed to respiration of ^{14}C labelled photosynthate. About 35% of total ^{14}C fixed into particulate organic matter, 40% of the ^{14}C fixed into carbohydrates, and 33% of the ^{14}C fixed into lipids was calculated to be respired during the 32-hour incubation.

Lancelot and Mathot (1985) conducted a similar experiment with a mixed diatom assemblage. They found no catabolism of protein, but significant respiration and mobilization of carbohydrates and lipids. They modelled their results with the following equation:

$$dR/dt = p_R \, a_{CO2} - k_r \, R \qquad (5)$$

where R is the activity of the reserve polymer fraction (either carbohydrate or lipid), p_R is the absolute rate of synthesis of R, a_{CO2} is the specific activity of the extracellular CO_2 pool which is assumed equal to the specific activity of the precursor pool for R, and k_R is the relative catabolic rate of the reserve polymer, R. Lancelot and Mathot (1985) showed that a poorly defined pool of intermediates and a well-defined pool of ß1-3 glucans become uniformly labelled during the light, supporting the assertion that the specific activity of intracellular pools approached that of the inorganic CO_2 pool. Using Eq. 4, they calculated catabolic rates for carbohydrates of 0.08 h^{-1} during darkness and 0.2 h^{-1} during the light, with lower catabolic rates for lipids of 0.01-0.025 h^{-1} in both light and darkness. From these observations, Lancelot and Mathot (1985) calculated that 65% of gross CO_2 fixation is catabolized with 42% during the light period and 23% during darkness. Accumulation of carbon skeletons in proteins accounts for 43% of the dark carbohydrate loss with the remainder presumably respired to provide energy for protein synthesis.

In experiments with an oligotrophic ocean phytoplankton community, Laws et al. (1987) found that ^{14}C in protein increased from 30% of PO^{14}C at end of light period to 45% by dawn. Given a 24% loss of PO^{14}C during darkness (Laws et al., 1987), and the reported ratios of ^{14}C-protein:PO^{14}C, one obtains a 14% net increase of ^{14}C labeled protein occurred during the dark period. Thus, all of the dark PO^{14}C loss was from non-protein fraction, as would be expected from respiration of carbohydrate and lipid reserves. A 60% reduction in non-protein ^{14}C would have occurred during darkness, 14% of which ended up in the protein fraction. The 24% loss of particulate ^{14}C in darkness cannot be attributed exclusively to phytoplankton, since microheterotrophs are also likely to preferentially metabolize carbohydrates and lipids.

RESPIRATION AND $^{14}CO_2$ ASSIMILATION

"If the rate of respiration is higher than the rate of photosynthesis, the intermixing of the two processes has thus a considerable influence on measurements of photosynthesis with the ^{14}C technique." Steemann Nielsen (1955)

The common assumption that assimilation of $^{14}CO_2$ into particulate matter during short-term experiments (< 30 minute) approximates gross photosynthesis may often be in error. When introducing the technique to oceanographic investigations, Steemann Nielsen warned of the possibility of significant underestimation of gross photosynthesis at low irradiances by the ^{14}C technique. Recent observations indicate that ^{14}C uptake can significantly underestimate net photosynthesis at higher irradiances as well. For example, the measured growth rate[*] of 0.14 h^{-1} in nutrient-saturated *Thalassiosira pseudonana* cultured at an irradiance of 2200 μmol photons m^{-2} s^{-1}, was about double the light-saturated $^{14}CO_2$ assimilation rate of 0.076 h^{-1} (Cullen and Lewis, 1988). In this experiment, the measured rate of photosynthesis inferred from ^{14}C assimilation was not sufficient to support the measured growth rate. This is clearly impossible. J. J. Cullen (personal communication) attributes the discrepancy between the rates of ^{14}C assimilation and net carbon fixation to a reduction in the specific activity of the intracellular pool of $^{14}CO_2$ by cold $^{12}CO_2$ arising from respiration, however, it is difficult to conceive a mass balance in which this is possible.

Similarly, Li and Goldman (1980) reported that $^{14}CO_2$ assimilation underestimated growth in two of five phytoplankton species examined. The rate of $^{14}CO_2$ assimilation underestimated steady-state, ammonium-limited growth rates by up to 40% in *Dunaliella tertiolecta* at a growth rate of 1 d^{-1}, and 100% in *Pavlova lutheri* at a growth rate of 1.4 d^{-1}. Li and Goldman (1980) attributed these differences to "physiological stress" that may have accompanied transfer from a chemostat to the batch mode used for measuring $^{14}CO_2$ assimilation. The discrepancy between net CO_2 fixation and $^{14}CO_2$ uptake is further aggravated by the requirement that a fraction (typically 5%) of $^{14}CO_2$ fixation occur by anaplerotic (i.e., non-photosynthetic) carboxylations associated with replenishment of TCA cycle intermediates (Raven, 1984; calculations based on Vanlerberghe et al., 1990). The results of Li and Goldman (1980) and Cullen and Lewis (1988) suggest the need for employing rigorous carbon balance experiments in re-evaluating the ^{14}C method, with particular emphasis on the role of respiration and the sources of respiratory CO_2.

Results from pulse-chase experiments support the hypothesis that there is efficient refixation of CO_2 evolved from respiratory processes in illuminated phytoplankton. In an early experiment intended to reconcile apparent differences between ^{14}C and O_2 estimates of primary productivity in the oligotrophic ocean, Ryther (1954) found considerable respiration of newly formed photosynthate, concluding that measuring photosynthesis by ^{14}C uptake could lead to a considerable underestimate of gross photosynthesis. A loss of up to 40% of the initial ^{14}C activity was observed within six hours following a transfer from light to darkness. Specific respiratory loss rates from the organic ^{14}C pool equalled -0.085 h^{-1} in *Chlamydonomas* sp., and -0.12 h^{-1} in *Nitzchia* sp. under nutrient-starved conditions.

In a subsequent experiment, Ryther (1956) measured the rate of loss of particulate ^{14}C from uniformly labelled *Dunaliella euchlora* in light and darkness; ^{14}C loss after 24 hours in darkness amounted to about 20% of the initial activity. In contrast, loss of particulate ^{14}C from the illuminated bottles was $< 1.5\%$. Similar results were obtained in cells from exponentially growing cultures in which the ratio of net O_2 evolution to dark O_2 consumption was 12:1 and stationary phase cultures in which this ratio was only 2:1.

---

[*] The growth rate will equal the biomass-specific net photosynthesis rate under conditions of balanced growth, if excretion of dissolved organic carbon is negligible.

These results were interpreted by Ryther (1956) to indicate that respiration proceeded at similar rates in illuminated and darkened cells, but that refixation of respiratory $^{14}CO_2$ occurred in the light. If, as Ryther (1956) maintained, respiration was occurring at 50% of the gross photosynthesis rate in nitrogen-limited stationary phase cultures, but all of the CO_2 evolved in respiration was refixed, then ^{14}C assimilation would significantly underestimate the gross photosynthesis rate in these experiments.

CONCLUSION

What We Know About Phytoplankton Respiration

It is appropriate to end with a summary of what we do and do not know about phytoplankton respiration. First, we know that respiration is often a large component of the energy budget of marine phytoplankton species in laboratory cultures. Biomass-specific respiration rates range from 0.02 to 1.2 d^{-1}. Second, there is no simple parameterization of respiration rate as a function of either the rates of growth or gross photosynthesis. Respiration accounts for from <10% to >60% of gross photosynthesis. Third, the rate of respiration depends on both the availability of substrate and the demand for intermediates. For example, respiration is stimulated by exposure to high irradiance and by enrichment of nutrient-limited phytoplankton. Fourth, the significance of phytoplankton respiration in nature has been underscored by investigations employing biochemical fractionation of organic ^{14}C in time series observations. However, the interpretation of ^{14}C results is sometimes ambiguous because of (1) significant rates of dark $^{14}CO_2$ assimilation, (2) heterotrophic remineralization of ^{14}C-labelled organic matter, and (3) uncertainty about the interaction of respiration and photosynthesis in determining $^{14}CO_2$ assimilation. Phytoplankton respiration rates in nature cannot be obtained routinely, but can only be inferred from carefully controlled experiments. Finally, we know something of the complexity of O_2 consuming and CO_2 evolving processes in microalgae. However, most of this information has been obtained for freshwater chlorophytes, and there have been few studies of the regulation of phytoplankton respiration.

What We Do Not Know About Phytoplankton Respiration

Many questions remain to be answered. How much of the diel variation in the concentration of particulate organic carbon in the ocean can be attributed to phytoplankton respiration? Under what conditions is phytoplankton respiration coupled to biomass synthesis, and under what conditions is respiration uncoupled from biosynthesis? How much of the variability in the relation between rates of respiration and growth is determined by the capacity for dark respiration, the ability to store and tap energy reserves, and/or the rate of nutrient assimilation? How common and how important is recycling of CO_2 between respiratory and photosynthetic pathways? What are the relative contributions of photorespiration, the Mehler reaction, and mitochondrial respiration to O_2 consumption in the light? Under what conditions, if any, is protein turnover a major contributor to phytoplankton respiration? What is the ecophysiological significance of differences in the respiratory physiology of cyanobacteria, diatoms and dinoflagellates?

At present, we do not know what proportion of gross photosynthetic carbon production is remineralized by phytoplankton. Phytoplankton respiration may account for remineralization of as little as 10% to greater than 50% of gross photosynthesis. More attention needs to be paid to phytoplankton respiration in our conceptual models of upper ocean carbon cycling. Field programs which include simultaneous measurements of diel signals in particulate organic carbon, dissolved oxygen (and potentially TCO_2), and phytoplankton growth rate should allow us to infer phytoplankton photosynthesis and respiration, micrzooplankton grazing and export production. Of the required information, measuring phytoplankton growth rate *in situ* is likely to be the most difficult.

When compared with the study of the photosynthetic physiology of marine phytoplankton, the examination of respiratory physiology has been largely neglected. If we are to fully understand the physiological bases of the ecology of the phytoplankton, it will be necessary to rectify this situation.

REFERENCES

Amthor, J. S., 1984, The role of maintenance respiration in plant growth, *Plant Cell Environ.*, 7:561.

Atkinson, D. E., 1977, Cellular energy metabolism and its regulation, Academic Press, New York.

Badger, M. R., 1985, Photosynthetic oxygen exchange, *Ann. Rev. Plant Physiol.*, 36:27.

Bannister, T. T., 1979, Quantitative description of steady-state, nutrient-saturated algal growth, including adaptation, *Limnol. Oceanogr.*, 24: 79.

Bannister, T. T., and Laws, E. A., 1980, Modeling phytoplankton carbon metabolism, *in*: "Primary Productivity in the Sea," P.G. Falkowski, ed., Plenum Press, New York.

Bate, C., Süeltemeyer, D. F., and Fock, H. P., 1988, $^{16}O_2/^{18}O_2$ analysis of oxygen exchange in *Dunaliella tertiolecta*. Evidence for the inhibition of mitochondrial respiration in the light, *Photosynthesis Research*, 16:219.

Beardall, J., 1989, Photosynthesis and photorespiration in marine phytoplankton, *Aquatic Botany*, 34:105.

Bender, M., Grande, K., Johnson, K., Marra, J., Williams, P.J. LeB., Sieburth, J., Pilson, M., Langdon, C., Hitchcock, G., Orchardo, J., Hunt, C., Donaghay, P., and Heinemann, K., 1987, A comparison of four methods for determining planktonic community production, *Limnol. Oceanogr.*, 32:1085.

Bennoun, P., 1982, Evidence for a respiratory chain in the chloroplast, *Proc. Natl. Acad. Sci. USA*, 79:4352.

Blasco, D., Packard, T. T., and Garfield, P. C., 1982, Size dependence of growth rate, respiratory electron transport system activity, and chemical composition in marine diatoms in the laboratory, *J. Phycol.*, 18:58.

Brackett, F. S., Olson, R. A., and Crickard, R. G., 1953, Respiration and intensity dependence of photosynthesis in *Chlorella*, *J. Gen. Physiol.*, 36:529.

Brechignac, F., and Andre, M., 1984, Oxygen uptake and photosynthesis of the red macroalga, *Chondrus crispus*, in seawater, *Plant. Physiol.*, 75:919.

Brechignac, F., and Furbank, R. T., 1987, On the nature of the oxygen uptake in the light by *Chondrus crispus*, Effects of inhibitors, temperature and light intensity, *Photosynthesis Research*, 11:45.

Brouwer, K. S., Van Valen, T., Day, D. A., and Lambers, H., 1986, Hydroxamate-stimulated O_2 uptake in roots of *Pisium sativum* and *Zea mays*, mediated by a peroxidase, *Plant Physiol.*, 82:236.

Brown, A. H., 1953, The effects of light on respiration using isotopically enriched oxygen, *Amer. J. Bot.*, 40:719.

Brown, A. H., and Webster, G. C., 1953, The influence of light on the rate of respiration of the blue-green alga Anabaena, *Amer. J. Bot.*, 40:753.

Burris, J. E., 1980, Respiration and photorespiration in marine algae, *in*: "Primary Productivity in the Sea," P.G. Falkowski, ed., Plenum Press, New York.

Chavez, F. P., Buck, K. R., and Barber, R. T., 1990, Phytoplankton taxa in relation to primary production in the equatorial Pacific, *Deep-Sea Res.*, 37:1733.

Cuhel, R. L., Ortner, P. B., and Lean, D. R. S., 1984, Night synthesis of protein by algae, *Limnol. Oceanogr.*, 29:731.

Cullen, J. J., and Lewis, M. R., 1988, The kinetics of algal photoadaptation in the context of vertical mixing, *J. Plankton Res.*, 10:1039.

Cullen, J. J., Lewis, M. R., Davis, C. O., and Barber, R. T., 1991, Photosynthetic characteristics and estimated growth rates indicate grazing is the proximate control of primary production in the equatorial Pacific, *J. Geophys. Res.* (in press).

Elrifi, I. R., and Turpin, D. H., 1987, Short-term physiological indicators of N deficiency in phytoplankton: A unifying model, *Mar. Biol.*, 96:425.

Elthon, T. E., Nickels, R. L., and McIntosh, L., 1989, Monoclonal antibodies to the alternative oxidase of higher plant mitochondria, *Plant Physiol.*, 89:1311.

Emerson, R., 1926, The effect of certain respiratory inhibitors on the respiration of *Chlorella*, *J. Gen. Physiol.*, 10:469.

Eppley, R. W., Swift, E., Redalje, D. G., Landry, M. R., and Haas, L. W., 1988, Subsurface chlorophyll maximum in August-September 1985 in the CLIMAX area of the North Pacific, *Mar. Ecol. Prog. Ser.*, 42:289.

Falkowski, P. G., Dubinsky, Z., and Santostefano, G., 1985, Light-enhanced dark respiration in phytoplankton, *Verh. Internat. Verein. Limnol.*, 22:2830.

Falkowski, P. G., Dubinsky, Z, and Wyman, K., 1985, Growth-irradiance relationships in phytoplankton, *Limnol. Oceanogr.*, 30:311.

Falkowski, P. G., and Owens, T. G., 1978, Effects of light intensity on photosynthesis and dark respiration in six species of marine phytoplankton, *Mar. Biol.*, 45:289.

Falkowski, P. G., Ziemann, D., Kolber, Z., and Bienfang, P. K., 1991, Role of eddy pumping in enhancing primary production in the ocean, *Nature*, 352:55.

Fasham, M. J. R., Ducklow, H. W., and McKelvie, S. M., 1990, A nitrogen-based model of phytoplankton dynamics in the oceanic mixed layer, *J. Mar. Res.*, 48:591.

Fenchel, T., 1982, Ecology of heterotrophic flagellates. II Bioenergetics and growth, *Mar. Ecol. Prog. Ser.*, 8:225.

Foy, R. H., and Smith, R. V., 1980, The role of carbohydrate accumulation in growth of planktonic Oscillatoria species, *British Phycol. J.*, 15:139.

Frost, B. W., 1987, Grazing control of phytoplankton stock in the open subarctic Pacific Ocean: A model assessing the role of mesozooplankton, particularly the large calanoid copepods Neocalanus spp, *Mar. Ecol. Prog. Ser.*, 39:49.

Geider, R. J., 1990, The relationship between steady-state phytoplankton growth and photosynthesis, *Limnol. Oceanogr.*, 35:971.

Geider, R. J., and Osborne, B. A., 1989, Respiration and microalgal growth: A review of the quantitative relationship between dark respiration and growth, *New Phytol.*, 112:327.

Geider, R. J., Osborne, B. A., and Raven, J. A., 1985, Light dependence of growth and photosynthesis in Phaeodactylum tricornutum (Bacillariophyceae), *J. Phycol.*, 21:609.

Geider, R. J., Osborne, B. A., and Raven, J. A., 1986, Growth, photosynthesis and maintenance metabolic cost in the diatom Phaeodactylum tricornutum at very low light levels, *J. Phycol.*, 22:39.

Geider, R. J., Platt, T., and Raven, J. A., 1986, Size dependence of growth and photosynthesis in diatoms: A synthesis, *Mar. Ecol. Prog. Ser.*, 30:93.

Glidewell, S. M., and Raven, J. A., 1975, Measurement of simultaneous oxygen evolution and uptake in *Hydrodictyon africanum*, *J. Exp. Bot.*, 26:479.

Goericke, 1990, Pigments as ecological tracers for the study of the abundance and growth of marine phytoplankton, PhD. Thesis, Deptartment of Organismic and Evolutionary Biology, Harvard University, Cambridge, Massachusettes.

Goyal, A., and Tolbert, N. E., 1989, Variations in the alternative oxidase in *Chlamydomonas* grown in air or high CO_2, *Plant Physiol.*, 89:958.

Graham, D., 1980, Effects of light on 'dark' respiration, *in*: "The Biochemistry of Plants," Volume 2, D.D. Davies, ed., Academic Press, New York.

Grande, K. D., Marra, J., Langdon, C., Heinemann, K., and Bender, M. L., 1989, Rates of respiration in the light measured in marine phytoplankton using an ^{18}O isotope-labelling technique, *J. Exp. Mar. Biol. Ecol.*, 129:95.

Grande, K. D., Williams, P. J. LeB., Marra, J., Purdie, D. A., Heinemann, K., Eppley, R. W., and Bender, M. L., 1989, Primary production in the North Pacific gyre: A comparison of rates determined by ^{14}C, O_2 concentration and ^{18}O methods, *Deep-Sea Res.*, 36:1621.

Grant, N. G., and Hommersand, M. H., 1974, The respiratory chain of Chlorella protothecoides. I. Inhibitor responses and cytochrome components of whole cells, *Plant Physiol.*, 54:50.

Groebellar, J. U., and Soeder, C. J., 1985, Respiration losses in planktonic green algae in raceway ponds, *J. Plankton Res.*, 7:497.

Guy, R. D., Berry, J. A., Fogel, M. L., and Hoering, T. C., 1989, Differential fractionation of oxygen isotopes by cyanide-resistant and cyanide-sensitive respiration in plants, *Planta*, 177:483.

Hama, T., 1988, ^{13}C-GC-MS analysis of photosynthetic products of the phytoplankton population in the regional upwelling area around the Izu Islands, Japan, *Deep-Sea Res.*, 35:91.

Harris, G.P., 1978, Photosynthesis, productivity and growth: The physiological ecology of phytoplankton, *Arch. Hydrobiol.*, 10: 1.

Harris, G. P., and Piccinin, B. B., 1983, Phosphorus limitation and carbon metabolism in a unicellular alga: Interactions between growth rate and the measurement of net and gross photosynthesis, *J. Phycol.*, 19:185.

Healey, F. P., and Myers, J., 1971, The Kok effect in *Chlamydomonas reinhardtii. Plant Physiol.*, 47:373.

Hellebust, J. A., and Lewin, J., 1977, Heterotrophic nutrition, *in*: "The Biology of Diatoms, Botanical Monographs Volume 13, D. Werner, ed., University of California Press.

Herzig, R., and Falkowski, P. G., 1989, Nitrogen limitation in Isochrysis galbana (Haptophyceae), I. Photosynthetic energy conversion and growth efficiencies, *J. Phycol.*, 25:462.

Hirano, M., Satoh, K., and Katoh, S., 1980, Plastoquinone as a common link between photosynthesis and respiration in a blue-green alga, *Photosynthesis Res.*, 1:149.

Hoch, G., Owens, O. V. H., and Kok, B., 1963, Photosynthesis and respiration, *Arch. Biochem. Biophys.*, 101:171.

Hochachka, P. W., and Teal, J. M., 1964, Respiratory metabolism in a marine dinoflagellate, *Biol. Bull.*, 126:274.

Humphrey, T. J., and Davies, D. D., 1975, A new method for the measurement of protein turnover, *Biochem. J.*, 148:274.

Humphrey, T. J., and Davies, D. D., 1976, A sensitive method for measuring protein turnover based on the measurement of 2-^3H labelled amino acids in protein, *Biochem. J.*, 156:561.

Jones, L. W., and Myers, J., 1963, A common link between photosynthesis and respiration in a blue-green alga, *Nature*, 199:670.

Kenner, R. A., and Ahmed, S.I., 1975, Correlation between oxygen utilization and electron transport activity in marine phytoplankton, *Mar. Biol.*, 33:129.

Kiefer, D. A., and Mitchell, B. G., 1983, A simple, steady-state description of phytoplankton growth based on absorption cross-section and quantum efficiency, *Limnol. Oceanogr.*, 28:770.

Lambers, H., 1985, Respiration in intact plants and tissues: Its regulation and dependence on environmental factors, metabolism and invaded organisms, *in*: "Higher plant cell respiration," R. Douce, and D.A. Day, eds., Springer-Verlag, Berlin.

Lance, C., Chauveau, M., and Dizengremel, P., 1985, The cyanide-resistant pathway of plant mitochondria, *in*: "Higher plant cell respiration," R. Douce, and D.A. Day, eds., Springer-Verlag, Berlin.

Lancelot, C., and Mathot, S., 1985, Biochemical fractionation of primary production by phytoplankton in Belgian coastal waters during short- and long-term incubations with ^{14}C-bicarbonate, *Mar. Biol.*, 86:219.

Lancelot, C., Mathot, S., and Owens, N. J. P., 1986, Modelling protein synthesis, a step to an accurate estimate of net primary production: *Phaeocystis pouchetii* colonies in Belgian coastal waters, *Mar. Ecol. Prog. Ser.*, 32:193.

Landry, M. R., Haas, L. W., and Fagerness, V. L., 1984, Dynamics of microbial plankton communities: Experiments in Kaneohe Bay, Hawaii, *Mar. Ecol. Prog. Ser.*, 16:127.

Laws, E. A., and Bannister, T. T., 1980, Nutrient- and light-limited growth of Thalassiosira fluviatilis in continuous culture, with implications for phytoplankton growth in the ocean, *Limnol. Oceanogr.*, 25:457.

Laws, E. A., and Caperon, J., 1976, Carbon and nitrogen metabolism by *Monochrysis lutheri*: Measurement of growth-rate-dependent respiration rates, *Mar. Biol.*, 36:85.

Laws, E. A., DiTullio, G. R., and Redalje, D. G., 1987, High phytoplankton growth and production rates in the North Pacific subtropical gyre, *Limnol. Oceanogr.*, 34:905.

Laws, E. A., Jones, D. R., Terry, K. L., and Hirata, J.A., 1985, Modifications in recent models of phytoplankton growth: Theoretical developments and experimental examination of predictions, *J. Theor. Biol.*, 114:323.

Laws, E. A., and Wong, D. C. L., 1978, Studies of carbon and nitrogen metabolism by three marine phytoplankton species in nitrate-limited continuous culture, *J. Phycol.*, 14:406.

Li, W. K. W., and Dickie, P. M., 1991, Light and dark ^{14}C uptake in dimly-lit oligotrophic waters: relation to bacterial activity, *J. Plankton Res.*, 13:29.

Li, W. K. W., and Goldman, J. C., 1980, Problems in estimating growth rates of phytoplankton from short term ^{14}C assays, *Microb. Ecol.*, 7:113.

Li, W. K. W., and Harrison, W. G., 1982, Carbon flow into the end-products of photosynthesis in short and long incubations of a natural phytoplankton population, *Mar. Biol.*, 72:175.

Miller, M. G., and Obendorf, R. L., 1981, Use of tetraethylthiuram disulfide to discriminate between alternative respiration and lipoxygenase, *Plant Physiol.*, 67:962.

Møller, I. A., Bérczi, A., van der Plas, L. H. W., and Lambers, H., 1988, Measurement of the activity and capacity of the alternative pathway in intact plant tissues: Identification of problems and possible solutions, *Physiologia Plantarum*, 72:642.

Mortain-Bertrand, A., Descolas-Gros, C., and Jupin, H., 1988, Pathway of dark inorganic carbon fixation in two species of diatoms: Influence of light regime and regulator factors on diel variations, *J. Plankton Res.*, 10: 199.

Myers, J., 1947, Oxidative assimilation in relation to photosynthesis in *Chlorella*, *J. Gen Physiol.*, 32:103.

Myers, J., and Graham, J. R., 1971, The photosynthetic unit in Chlorella measured by repetitive short flashes, *Plant Physiol.*, 28:282.

Osmond, C. B., 1981, Photorespiration and photoinhibition: Some implications for the energetics of photosynthesis, *Biochim. Biophys. Acta*, 639:77.

Osborne, B. A., and Geider, R. J., 1986, Effect of nitrate-nitrogen limitation on photosynthesis of the diatom *Phaeodactylum tricornutum* Bohlin (Bacillariophyceae), *Plant Cell Environ.*, 9:617.

Oudot, C., 1989, O_2 and CO_2 balances approach for estimating biological production in the mixed layer of the tropical Atlantic Ocean (Guinea Dome area), *J. Mar. Res.*, 47:385.

Packard, T. T., 1985, Measurement of electron transport activity of microplankton, *Adv. Aquat. Microbiol.*, 3:207.

Parsons, T. R., Takakashi, M., and Hargrave, B., 1984, Biological Oceanographic Processes, 3rd Edition, Pergamon Press, New York.

Peltier, G., Ravenel, J. and Verméglio, A., 1987, Inhibition of a respiratory activity by short saturating flashes in *Chlamydonomas*: Evidence for a chlororespiration, *Biochimica et Biophysica Acta*, 893:83.

Peltier, G., and Sarrey, F., 1988, The Kok effect and light-inhibition of chlororespiration in *Chlamydomonas reinhardtii*, *FEBS Lett.*, 228:259.

Peltier, G., and Thibault, P., 1985, O_2 uptake in the light in *Chlamydomonas*, Evidence for persistent mitochondrial respiration, *Plant. Physiol.*, 79:225.

Penning de Vries, F. W. T., 1975, The cost of maintenance processes in plant cells, *Ann. Bot.*, 39:77.

Penning de Vries, F. W. T., Brunsting, A. H. M., and van Laar, H. H., 1974, Products, requirements and efficiency of biosynthesis: A quantitative approach, *J. theor. Biol.*, 45:339.

Pennock, J. R., 1987, Temporal and spatial variability in phytoplankton ammonium and nitrate uptake in the Delaware estuary, *Est. Coastal Shelf Sci.*, 24:2.

Pickett, J. M., 1975, Growth of Chlorella in a nitrate-limited chemostat, *Plant Physiol.*, 55:223.

Pirt, S. J., 1982, Maintenance energy: A general model for energy-limited and energy sufficient growth, *Arch. Microbiol.*, 133:300.

Pirt, S. J., 1986, The thermodynamic efficiency (quantum demand) and dynamics of photosynthetic growth, *New Phytol.*, 102:3.

Post, A. F., Loogman, J. G., and Mur, L. R., 1985, Regulation of growth and photosynthesis by *Oscillatoria agardhii* grown with a light/dark cycle, *FEMS Microbial Ecology*, 31: 97.

Postma, H., and Rommets, J. W., 1979, Dissolved and particulate organic carbon in the North Equatorial Current of the Atlantic Ocean *Neth. J. Sea Res.*, 13:85.

Raven, J. A., 1976, Division of labour between chloroplast and cytoplasm, *in*: "The Intact Chloroplast," J. Barber, ed., Elsevier/North Holland Biomedical Press.

Raven, J. A., 1984, "Energetics and transport in aquatic plants," Alan R. Liss, Inc., New York.

Raven, J. A., and Beardall, J., 1981, Respiration and photorespiration, *Can. Bull. Fish. Aquat. Sci.*, 210:55.

Richards, L., and Thurston, C. F., 1980, Protein turnover in *Chlorella fusca* var. *vacuolata*: Measurement of the overall rate of intracellular protein degradation using isotopic exchange with water, *J. Gen. Microbiol.*, 121:49.

Ryther, J. H., 1954, The ratio of photosynthesis to respiration in marine plankton algae and its effect upon the measurement of productivity, *Deep Sea Res.*, 2:134.

Ryther, J. H., 1956, Interrelation between photosynthesis and respiration in the marine flagellate *Dunaliella euchlora*, *Nature*, 178:861.

Sakshaug, E., Andresen, K., and Kiefer, D. A., 1989, A steady state description of growth and light absorption in the marine planktonic diatom *Skeletonema costatum*. *Limnol. Oceanogr.*, 34:198.

Sargent, D. F., and Taylor, C. P. S., 1972, Terminal oxidases of Chlorella pyrenoidosa, *Plant Physiol.*, 49:775.

Shuter, B., 1979, A model of physiological adaptation in unicellular algae, *J. Theor. Biol.*, 78:519.

Siegel, D. A., Dickey, T. D., Washburn, L., Hamilton, M. K., and Mitchell, B. G., 1989, Optical determination of particulate abundance and production variations in the oligotrophic ocean, *Deep-Sea Res.*, 36:211.

Smayda, T. J., and Mitchell-Innes, B., 1974, Dark survival of autotrophic planktonic diatoms, *Mar. Biol.*, 25: 195.

Smith, R. E. H., Geider, R. J., and Platt, T., 1984, Microplankton productivity in the oligotrophic ocean, *Nature*, 311:252.

Steemann Nielsen, E., 1955, The interaction of photosynthesis and respiration and its importance for the determination of ^{14}C-discrimination in photosynthesis, *Physiol. Plant.*, 8:945.

Süeltemeyer, D. F., Klug, K., and Fock, H. P., 1986, Effect of photon fluence rate on oxygen evolution and uptake by *Chlamydomonas reinhardtii* suspensions grown in ambient and CO_2-enriched air, *Plant Physiol.*, 81:372.

Sverdrup, H. U., 1952, On conditions for the vernal blooming of phytoplankton, *J. Cons. Explor. Mer.*, 18:287.

Syrett, P.J., 1951, The effect of cyanide on respiration and the oxidative assimilation of glucose by C. vulgaris, *Ann. Bot.*, 15:473.

Syrett, P.J., 1953, The assimilation of ammonium by nitrogen-starved cells of *Chlorella vulgaris*, Part 1. The correlation of assimilation with respiration, *Ann. Bot.*, 17:1.

Syrett, P.J., 1956, The assimilation of ammonium by nitrogen-starved cells of Chlorella vulgaris 4, The dark fixation of carbon dioxide, *Physiol. Plant.*, 9:165.

Syrett, P.J. and Fowden, L., 1952, The assimilation of ammonia by nitrogen-starved cells of Chlorella vulgaris, Part 3. The effect of the addition of glucose on the products of assimilation, *Physiol. Plant.*, 5:558.

Taguchi, S., Ditullio, G. R., and Laws, E. A., 1988, Physiological characteristics and production of mixed layer and chlorophyll maximum phytoplankton populations in the Caribbean Sea and western Atlantic Ocean, *Deep-Sea Res.*, 35:1363.

Tempest, D. W., and Neijssel, O. M., 1984, The status of Y_{ATP} and maintenance energy as biologically interpretable phenomena, *Ann. Rev. Microbiol.*, 38:459.

Tijssen, S. B., 1979, Diurnal oxygen rhythm and primary production in the mixed layer of the Atlantic Ocean at 20 °N, *Neth. J. Sea. Res.*, 13:79.

Tijssen, S. B., and Eijgenraam, A., 1982, Primary and community production in the southern bight of the North Sea deduced from oxygen concentration variations in the spring, 1980, *Neth. J. Sea. Res.*, 16:247.

Underhill, P. A., 1981, Steady-state growth rate effects on the photosynthetic carbon budget and chemical composition of a marine diatom, Ph.D. Dissertation, University of Delaware.

Van Leire, L., and Mur, L. R., 1979, Growth kinetics of *Oscillatoria agardhii* Gomont in continuous culture, limited in its growth by the light energy supply, *J. Gen. Microbiol.*, 115:153.

Vanlerberghe, G. C., Schuller, K. A., Smith, R. G., Feil, R., Plaxton, W. C., and Turpin, D. H., 1990, Relationship between NH_4^+ assimilation rate and *in vivo* phosphoenolpyruvate carboxylase activity, *Plant Physiol.*, 94:284.

Varum, K. M., Østgaard, K., and Grimsrud, K., 1986, Diurnal rhythms in carbohydrate metabolism of the marine diatom *Skeletonema costatum* (Grev.) Cleve, *J. Exp. Mar. Biol. Ecol.*, 102:249.

Verity, P. G., 1982, Effects of temperature, irradiance and day length on the marine diatom *Skeletonema costatum* (Grev.) Cleve. IV. Growth, *J. Exp. Mar. Biol. Ecol.*, 60:209.

Webster, D. A., and Hackett, D. P., 1965, Respiratory chain of colorless algae. I. Chlorophyta and Euglenophyta, *Plant Physiol.*, 40:1091.

Webster, D. A., and Hackett, D. P., 1966, Respiratory chain of colorless algae, II. Cyanophyta, *Plant Physiol.*, 41:599.

Weger, H. G., Birch, D. G., Elrifi, I. R., and Turpin, D. H., 1988, Ammonium assimilation requires mitochondrial respiration in the light, *Plant Physiol.*, 86:688.

Weger, H. G., Guy, R. D., and Turpin, D. H., 1990, Cytochrome and alternative pathway respiration in green algae, *Plant Physiol.*, 93:356.

Weger, H. G., Herzig, R., Falkowski, P. G., and Turpin, D. H., 1989, Respiratory losses in the light in a marine diatom: Measurements by short- term mass spectrometry, *Limnol. Oceanogr.*, 34:1153.

Weger, H. G., and Turpin, D. H., 1989, Mitochondrial respiration can support NO_3^- and NO_2^- reduction during photosynthesis, *Plant Physiol.*, 89:409.

Williams, P. J. LeB., Heinemann, K. R., Marra, J., and Purdie, D. A., 1983, Comparison of ^{14}C and O_2 measurements of phytoplankton production in oligotrophic waters, *Nature*, 305:49.

Wiskich, J. T., and Dry, I. B., 1985, The tricarboxylic acid cycle in plant mitochondria: Its operation and regulation, *in*: "Higher Plant Cell Respiration," R. Dounce and D.A. Day, eds., Springer-Verlag, New York.

Wofsy, S. C., 1983, A simple model to predict extinction coefficients and phytoplankton biomass in eutrophic waters, *Limnol. Oceanogr.*, 28:1144.

BACTERIOPLANKTON ROLES IN CYCLING OF ORGANIC MATTER: THE MICROBIAL FOOD WEB

Jed Fuhrman

Department of Biological Sciences
University of Southern California
Los Angeles, CA 90089-0371

INTRODUCTION

More than a decade has passed since the realization that bacteria are quantitatively important consumers of organic carbon in marine food webs. The basic information on the significance of the microbial food web was put forth eloquently by Pomeroy (1974), who pieced together data from a variety of sources that all indicated a major role of small heterotrophs consuming dissolved and particulate material. However, these ideas did not gain wide recognition until the high abundance of marine bacteria was shown by epifluorescence microscopy (Ferguson and Rublee, 1976; Hobbie et al., 1977), and the bacterial heterotrophic production was shown to be large (i.e.,10-30%) compared to primary production (Hagström et al., 1979; Fuhrman and Azam, 1980; 1982). With reasonable estimates of bacterial growth efficiency (i.e., near 50%), it became clear that heterotrophic bacteria consume an amount of carbon equivalent to approximately 20-60% of total primary production. Williams (1981) reached this conclusion when he synthesized the extant results on bacterial biomass and production. He also showed that "normal" well-known processes and mechanisms could lead to as much as 60% of the primary production becoming dissolved organic carbon (DOC), and subsequently, being taken up by bacteria. Azam et al. (1983) formalized the concept of the microbial loop by which significant quantities of organic matter are produced or processed through prokaryotic and very small eukaryotic organisms, eventually feeding into the larger macrozooplankton.

As more data have accumulated, we have learned more about the distributions in space and time of bacterial biomass and production, but the new information has not substantially changed the conclusions and implications of Williams (1981) and Azam et al. (1983). Recent reviews have synthesized these data, such as the comprehensive overview of freshwater and marine biomass and production presented by Cole et al. (1988). From a variety of systems and methods, these authors found that bacterial heterotrophic production averaged about 20% of primary production by volume, and 30% on an areal basis, confirming that the earlier studies examined reasonably typical systems. These authors also showed that where data are sufficient to compare bacterial production to that of zooplankton (freshwater only), bacterial production is about twice as high as that of zooplankton. It should be noted that new methods to measure bacterial heterotrophic

Primary Productivity and Biogeochemical Cycles in the Sea
Edited by P.G. Falkowski and A.D. Woodhead, Plenum Press, New York, 1992

production, such as leucine incorporation (Kirchman et al., 1985; Chin Leo and Kirchman, 1988; Simon and Azam, 1989) have independently confirmed the validity of the previous techniques and results.

One substantial change is the increase in the estimates of bacterial biomass from cell counts and size measurements. Studies over the past few years have suggested that the conversion from apparent cell volume to cell carbon in marine bacteria is about three-fold higher than previously determined factors for relatively large bacteria grown in culture, with conversion factors typically 300-500 fg C μm^{-3} (Bratbak, 1985; Bjørnsen, 1986; Lee and Fuhrman, 1987; cf., Simon and Azam, 1989). This finding suggests that many of the previous estimates of biomass should be revised upward. Unexpectedly, Lee and Fuhrman (1987) found that cells with apparent volumes ranging from 0.036 to 0.073 um^3, typical in the sea, all had about 20 fg C and 5 fg N per cell, indicating a higher conversion factor for the smaller cells. These apparently constant cellular biomasses are independent of size measurements within the specified limits, so these figures can be used without applying particular conversion factors (Cho and Azam, 1990). I emphasize that the conversion factors are empirically derived and refer only to the apparent cell volume, as measured by epifluorescence microscopy of preserved cells. Changes in cell shape and size during preparation may be significant, so these factors are probably not appropriate for live cells (e.g., Lee and Fuhrman's high factor for the smallest bacteria seems far too high for living cells; Cho and Azam, 1990). Also, these conversion factors are highly dependent on the exact procedures used to measure cell size, because of problems such as "halos" around fluorescent cells; therefore, they should be applied with great care to reproduce the original conditions and criteria of measurement.

LIMITS OF SECONDARY PRODUCTION

The concepts of heterotrophic production and heterotrophic consumption of organic matter have sometimes been confusing. Consumption is the cumulative uptake of all organic compounds (often defined in units of carbon) by the organisms in question, and production is the synthesis of biomass of those organisms, which is the consumption minus losses from respiration and release of undigested material. It is important to define which organisms one is discussing. Commonly, the term "secondary production" is used interchangeably with heterotrophic production. The sorts of misconceptions that can arise are illustrated below.

When it became known that bacterial consumption is equivalent to about half of primary production, most workers seemed to take this information as meaning that bacteria get half of the production, and that the other half is available for other organisms. But, as bacterial heterotrophic production was measured in a variety of locations, some appeared to have higher heterotrophic consumption than primary production, at least for part of the year (e.g., Scavia and Laird, 1987). This finding was unexpected, because the primary producers are the source of organic matter, and it was thought that the heterotrophs could not consume more than the amount of organic carbon produced. If local primary production is lower than that of heterotrophs, import of organic matter is necessary. Therefore, researchers such as Scavia and Laird (1987) explained their unexpected results as being due to factors such as bacterial consumption of carbon fixed months earlier when primary production was in excess of bacterial consumption (the data already seemed to rule out lateral import of organic matter). The problem is not totally new. Sorokin (1971) tried to explain his similar results as being due to underestimates of primary production and also import of dissolved organic matter. Although his work was criticized for several reasons (Banse, 1974), it was taken for granted that bacterial heterotrophic consumption should be lower than primary production. However, this is not the case.

Strayer (1988) discussed this concept of the maximum limits of secondary production in response to Scavia and Laird (1987). He pointed out that the primary factor that should be considered a "limit" is that respiration cannot exceed primary production; the production of the consumers, per se, is not constrained to be less than 100% of primary production because carbon is recycled more than once. With the relatively high efficiencies of microbial systems, the results can be surprising. Imagine, for example, a system in which all the consumers have 60% efficiency and there is no import or export; furthermore, this system is in steady state, so all primary production eventually is respired within the system, and the biomass of all components remains constant over time. If this system has 100 units of primary production per day, what is the heterotrophic production? The primary consumers eat 100 units daily, respiring 40 and producing 60 units of biomass; the secondary consumers eat 60 units daily, respiring 24 and producing 36 units of biomass; the tertiary consumers eat 36, respiring 14.4 and producing 21.6 units daily. Already, heterotrophic production of the first three consumer trophic levels (60 + 36 + 21.6 units) exceeds primary production (100 units), and I have only accounted for about 78% of the respiration in this example. Calculations described by Strayer show that in a system such as this, the heterotrophic production (from all trophic levels) adds up to 150% of primary production, and the total uptake of organic matter by consumers ("demand") is 250% of primary production.

Such a calculation is very sensitive to the efficiency of the consumers. At 50% efficiency, heterotrophic production exactly equals the primary production, and consumer demand is 200% of primary production. At 20% efficiency, heterotrophic production is only 25% of primary production, and consumer demand is 125%. No matter what the efficiency, the consumer demand exceeds primary production, because there is always some recycling (i.e., the same carbon passes through more than one consumer). Real systems do not have simple monotonic dynamics, and there are import and export terms as well as variable efficiencies. However, this approach also can be illuminating in other ways; for example, Strayer (1988) calculated an average efficiency of 41% for all consumers (including metazoa and fish) in Mirror Lake, New Hampshire. This value is higher than many ecologists have thought.

We can rightfully ask how all this relates to bacterial production, as opposed to that of all heterotrophic trophic levels combined. Bacteria do not fit well into the concept of a trophic level; bacteria consume DOM released from all other organisms, including themselves (see the following section), so they occupy several trophic levels at once. Further, bacterial carbon oxidation cannot exceed primary production. It is possible to envision scenarios whereby the heterotrophic production of bacteria alone exceeds primary production. Therefore, this analysis shows that there is nothing wrong with data indicating bacterial consumption (or even production) in excess of primary production while some such situations may be better explained by processes such as import or temporal disequilibrium (Scavia, 1988), this recycling argument should be considered. Also, a high bacterial production rate does not suggest that there is nothing left for the other consumers, simply that a large fraction of the carbon flow is being funneled through the bacterial portion of the food web. It is noteworthy that high bacterial production does not detract from the significance of protozoa or macrozooplankton in food web or biogeochemical processes; this production simply provides an alternate pathway for carbon flow, and this pathway also connects to the zooplankton. This subject is covered very well in the flow analysis models of Scavia (1988), Vezina and Platt (1988), Hagström et al. (1988), Ducklow (1991), and McManus (1991).

The recycling hypothesis of Strayer (1988) points to an important conclusion about release rates of DOM in relation to primary production. It has been often stated that if bacterial consumption is equivalent to, say, 60% of primary production, then 60% of the primary production must somehow (directly or via food web processes) become DOM that is suitable for bacterial uptake, while the remaining 40% has other fates. However, this discussion shows that total heterotrophic consumption generally exceeds primary production, often by a large margin. Thus, the ratio of bacterial consumption to primary production is not the same as the percent of primary production that becomes DOM that can be used by bacteria. A carbon flow through the bacteria equivalent to 60% of primary production could represent a relatively small portion of the total carbon flow through consumers, which can exceed 200% of the primary production. This fact is implicit in the flow models cited in the previous paragraph. Still, it is useful to measure bacterial production as a relatively easy means of estimating bacterial DOM consumption and respiration.

Viruses and DOM Release Mechanisms

The mechanisms by which DOM is released from various plankton are controlling factors in bacterial productivity and related processes. At this point, it is useful to consider the definition of dissolved as opposed to particulate matter. The cutoff between DOM and POM is somewhat arbitrary, and functionally, is usually defined by the types of filters employed to separate these two components (Sharp, 1973). It is common to define all material passing 0.2 or 0.4 um pore size membrane filters as DOM, including tiny particles that may be considered "colloidal." POM is usually defined as material caught on a glass-fiber filter with a pore size no smaller than 0.7 μm (e.g., Whatman GF/F). It is noteworthy that the 0.2 - 0.7 μm "no-man's land" between "DOM" and "POM" includes many or most of the marine bacteria (Lee and Fuhrman, 1987).

Williams (1981) and Azam et al. (1983) summarized most of the mechanisms that are the most important in DOM release from organisms. First, there is release directly from healthy photosynthesizing phytoplankton, the subject of a great deal of literature (see excellent review by Williams, 1990). Second, there is the release from "unhealthy" phytoplankton, where unhealthy may mean organisms subjected to nutrient depletion, excessive or inadequate light, physical or chemical cell damage, or infection by bacteria or viruses; there are few data on DOM release that pinpoint these causes in natural systems, although crashing blooms may suffer several of these processes. A third source, DOM release mediated by zooplankton, has been studied in recent years, and there is clear evidence that processes such as "sloppy feeding" (or spillage of prey contents) and excretion or egestion are important routes in marine systems. The idea is that when many zooplankton feed, a major portion of the organic matter in the captured organisms does not become assimilated or respired, but instead is released as DOM or particulate organic matter (POM). Significant DOM or specific compound release associated with zooplankton feeding has been measured directly (Lampert, 1978; Copping and Lorenzen, 1980; Fuhrman, 1987; Roy et al., 1989). Similar results have been obtained from mesocosm experiments; mesocosms that include zooplankton support much more bacterial production than controls without zooplankton (Eppley et al., 1981; Roman et al., 1988). Additional experiments were reported by Jumars et al. (1989), who suggested that a great deal of the release is because of undigested material. Within the microbial food web, protozoa feeding upon bacteria and cyanobacteria release substantial amounts of DOM (Hagström et al., 1988), which can lead to immediate recycling via bacterial uptake. Many field measurements do not partition release from separate mechanisms, and simply follow the fate of added ^{14}C-bicarbonate into DOM or bacteria. Such studies may include discussions implying that they are measuring DOM released directly from healthy phytoplankton, but all mechanisms, those mediated by zooplankton (including protozoa), or other undefined processes are included.

Viruses and DOM Release

A newly recognized mechanism for DOM release is viral infection. Viruses are much more abundant in marine systems than was previously thought, commonly more abundant than bacteria (Proctor et al., 1988; Sieburth et al., 1988; Bergh et al., 1989; Proctor and Fuhrman, 1990; Børsheim et al., 1990; Bratbak et al., 1990; Suttle et al., 1991). Viruses are dynamic, changing rapidly in abundance in relation to other biological parameters (Bratbak et al., 1990; Børsheim et al., 1990). Proctor and Fuhrman (1990) estimated the bacterial and cyanobacterial mortality caused by viruses; they observed fully developed, mature viruses within marine microorganisms by transmission electron microscopy, and reported that about 30% of the total mortality of marine cyanobacteria and 60% of the mortality of free-living heterotrophic bacteria in coastal and open ocean environments may be due to viral infection. For bacteria living in sedimenting particulate material, the percentages were similar (Proctor and Fuhrman, 1991). These first mortality estimates are somewhat speculative, being based upon limited background data. Sieburth et al. (1988) showed that viral abundance was high and infection was common during an intense coastal algal bloom. Suttle et al. (1991) showed that viruses infecting a variety of phytoplankton could readily be isolated from seawater, and that adding a high molecular- weight concentrate, including viruses, significantly reduced growth of natural phytoplankton assemblages. Several potential consequences of viral infection on nutrient cycling were discussed by Proctor and Fuhrman (1990; 1991), Bratbak et al. (1990), and Suttle et al. (1991).

Lytic viruses act by infecting a host, multiplying within the host cells, and then bursting (lysing) the host cells to release the progeny viruses. One major consequence of this lysing is release of DOM (plus some POM). What had been a functioning cell ends up as a "soup" - this released material consists of cellular components, including proteins, nucleic acids, amino acids, carbohydrates, lipids, cofactors, cell wall components, and assorted complexes, in addition to viruses themselves. Viruses are generally small enough (20-200 nm in typical linear dimension; Ackerman and DuBow, 1987) also to be considered operationally dissolved. The extent to which the lysed cell becomes POM depends on the cell's size and structure; bacteria probably form primarily DOM (even large pieces of the cell wall are probably functionally dissolved), while large phytoplankton, such as diatoms, may yield significant particulate debris. For most cells, the cytoplasm probably becomes DOM. The amounts of DOM are probably a function of the type of virus; some may degrade certain constituents to "cannibalize" the cell, while others may allow the retention of most normal cell functions up until the time of lysis (Ackerman and Dubow, 1987). All viruses rely on cell machinery to function, so significant disruption of most essential functions before lysis is unlikely.

Thus, cells infected with viruses are converted into virus particles plus dissolved and particulate debris. If we assume that, over the long-term, the system is in steady state (i.e., no net increase or decrease in the abundance of cells or viruses), then, on average, each infection has a net yield of one virus that successfully infects another cell. This virus probably represents only about 1% or less of the original cell biomass. All progeny viruses, plus the remains of the cell (about 99% of the original cell), become dissolved and, particulate detritus, eventually are mineralized; a large portion of mineralization probably occurs via DOM and bacterial uptake. Thus, viral infection is an extremely efficient way to convert biomass into dissolved and particulate detritus thus available to bacteria.

Phytoplankton Infection

There are limited data on the viral infection of phytoplankton, so it is premature to make quantitative estimates. However, the data on cyanobacterial infection (Proctor and Fuhrman, 1990) and the effects described by Sieburth et al. (1988) and Suttle et al. (1991) suggest that viral infection of phytoplankton may be a significant mechanism in overall DOM release. It is not clear to what extent these virus-related processes were included in earlier measurements of release of labeled DOC from phytoplankton after incubation with ^{14}C. We do not know how late into infection the cells continue to fix CO_2; if the cells are not photosynthesizing during the incubation period, DOM release will not appear in the measurement. In any case, DOM release probably changes as a function of cell type, virus type, and environmental conditions.

Viral Infection of Bacteria

The infection of bacteria by viruses has a significant potential impact on the cycling of organic matter in marine systems. As described above, infection efficiently converts biomass into detritus, and bacterial detritus is most likely to be dissolved. Therefore, infection converts bacteria into DOM, the exact opposite of the process of bacterial growth. This DOM is available for bacterial uptake, so viral infection of bacteria should result in a cycling of organic matter between bacteria and DOM. However, because bacteria do not convert DOM into biomass with 100% efficiency (gross growth efficiencies are in the range of 10-70%, and may be near 30-50% in natural systems; Cole, 1982; Bjørnsen and Kuparinen, 1991), there is an oxidative loss each time the DOM is incorporated into biomass. Still, with widespread viral infection of bacteria, the same carbon may be taken up by bacteria several times before it is ultimately mineralized. Although the net effect is to convert DOM into CO_2 (and probably also other inorganic nutrients), this cycling increases the uptake rate of DOM by bacteria, compared to a system without viruses. Although this increase is balanced by release from other bacteria, this process can have an important impact on measurements and interpretation of microbial processes as well as the overall functioning of the system.

Figure 1 compares two models of hypothetical marine food webs that differ in the cause of mortality of bacteria: one has all mortality from nanozooplankton grazers, and the other has 50% of the mortality from viruses and 50% from nanozooplankton. The figure shows the biomasses of each compartment in steady state, and the export of macrozooplankton C production is balanced by an input of CO_2. The numbers next to the arrows indicate carbon flux rates in arbitrary units; the production rate of a compartment is equal to the sum of its outputs, excluding the output to CO_2. Steady state is achieved because the sum of fluxes into each compartment equals the sum of the fluxes out. These models include most of the newly recognized features of microbial food webs (except algal mixotrophy), including "sloppy feeding" by all grazers and recycling of carbon. Both food webs have the same phytoplankton grazing losses partitioned among nanozooplankton (i.e., protozoa smaller than 20 μm that consume prokaryotic and small eukaryotic organisms; Sherr and Sherr, 1991), microzooplankton (i.e., larger protozoa and small metazoa up to about 200 μm diameter that consume primarily nanoplankton), and macrozooplankton (metazoa that consume microplankton). The 30% direct release from phytoplankton may include leakiness, stressed cells, "deliberate" release, and viral lysis, and is well within the range cited in the literature, as are other rates and processes in the model. The growth efficiencies of some compartments vary slightly (<2%) between models to achieve balance. Biomasses are not necessary for such a model, but would be needed to determine turnover rates. The cells infected with viruses are assumed to become

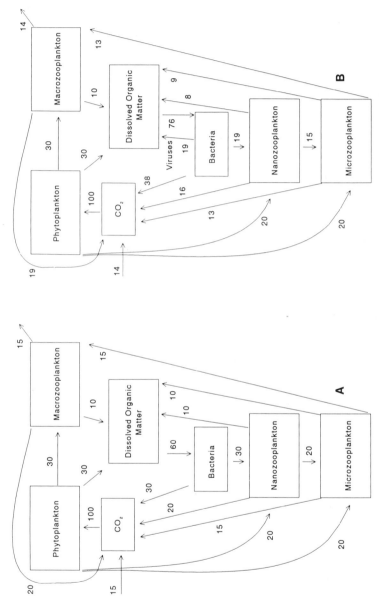

Fig. 1.   Two hypothetical carbon-flux models of marine food webs without (model A) or with (model B) a virus compartment.   At steady-state, model B leads to increased gross bacterial production but decreased availability of bacterial production to higher trophic levels.   Bacterial cells infected with viruses are assumed to degrade completely to DOC.

converted completely to DOC. Compared to the non-virus model, the presence of the viruses leads to 27% increases in both the secondary production and carbon mineralization rates of the bacteria, yet they cause a 37% decrease in the export of bacterial carbon to the nanozooplankton, and an ultimate loss of 7% in the macrozooplankton production. Therefore, significant bacterial losses to viruses has the somewhat unexpected result of leading to increased gross bacterial production, yet decreased availability of bacterial carbon to higher trophic levels.

Qualitatively, significant infection of bacteria and cyanobacteria may lead to very different composition of the DOM pool. Bacteria are rich in nitrogen and phosphorus compared to phytoplankton, probably due to their relatively high nucleic acid content (Fuhrman and Azam, 1982; Paul and Carlson, 1984; Simon and Azam, 1989). Thus, infection can lead to increased availability of readily utilizable DON and DOP, and can alter the uptake and regeneration of nutrients that can control primary productivity.

Viruses are generally thought to be very specific for their hosts, i.e., in most known cases, they infect only one species or strain of host; rarely do they infect different species, and even then, it is most often within a genus (Ackerman and Dubow, 1987). If this observation holds for natural marine populations, it probably has implications on the species compositions and dominance of natural microbial populations (Lenski, 1988). Specifically, viruses may help maintain the diversity of natural populations by preventing any one kind from becoming a dominant through competition, so viral processes may help explain the "paradox of the plankton" (Hutchinson, 1961). The reasoning is that virus infection is highly dependent on the host-cell abundance; infection spreads very rapidly in dense populations and much less so in sparse ones. If a significant portion of total mortality is due to viruses as the preliminary data suggest, then the rarer organisms benefit, because they are relatively dilute, and hence, encounter their viruses infrequently.

However, this specificity means that to understand viral processes in natural systems, we will need to know diversity and species compositions of natural populations, as well as the host specificities of the viruses. Until now, most marine microbiological studies have treated the bacteria like a uniform black box. This treatment has permitted us to deal with bulk properties such as biomass and total production with moderate success, but this is a huge oversimplification. With traditional identification techniques that require culturing the bacteria, learning the species compositions and their variability had been almost impossible because of problems with such cultures. Among these problems, the worst is that only about 1% or less of native populations grow readily in standard media (Jannasch and Jones, 1959; Ferguson et al., 1984). There also is the labor and expense in growing and identifying individual cultures, and we frequently encounter new organisms whose relationships to known organisms are unclear. However, recent molecular biological techniques based on similarities of nucleic acid sequence, that do not require cultures, are beginning to demonstrate the diversity and species compositions of marine planktonic bacteria (Pace et al., 1986; Giovannoni et al., 1990; Lee and Fuhrman, 1990; in press; Schmidt et al., 1991; DeLong et al., 1990; Britschgi and Giovannoni, 1991). These new studies show that marine planktonic systems are highly diverse and varied, with organisms frequently unlike any known previously, including a variety of Eubacteria, but also novel Archaebacteria (Fuhrman et al., unpublished). Lee and Fuhrman (1990; in press) show that different habitats (temperate coastal, open ocean, coral reef) have different bacterioplankton species compositions and stratified open ocean waters have stratified species compositions. It is likely that these organisms have a wide variety of metabolic processes, possibly including ones that are unknown. Much more work is needed before we can conceptually or mathematically model virus infections.

The Biogeochemical Roles of DOM

In the past few years there has been a renewed interest in the roles of DOM in marine food webs. Suzuki et al.(1985) and Sugimura and Suzuki (1988) reported that concentrations of these substances are much higher than was previously thought, with substantial and biogeochemically interesting variations with depth. The new data imply that these substances are highly active biologically, and also closely match the distribution of apparent oxygen utilization, suggesting that DOC and DON may have great significance in large-scale transport and cycling of organic matter. The new measurements are reported to be different from previous data, because these authors used high-temperature oxidation in the presence of a particular platinum catalyst. This is claimed to oxidize organic matter that is resistant to the previous standard techniques of UV-or persulfate-oxidation. Furthermore, these authors reported that a large portion of the new DOC has very high molecular weight, tens of thousands to millions of daltons. These data are under active debate within the marine chemistry community, nevertheless, they have stimulated some worthwhile thinking and discussion on the subject (Williams and Druffel, 1988; Jackson, 1988). It is beyond the scope of this review to discuss the merits of the new measurements, but I will describe some points relevant to DOM and bacterial roles in biogeochemical cycles.

Virus contribution to DOM. One question that comes immediately to mind involves the potential contribution of viruses to DOM measurements. Do the newly discovered viruses contribute significantly to the high molecular weight DOM? A simple calculation from virus abundance and biomass estimates suggests that virus particles themselves contribute a very small part to DOM. Abundance is about 10^8 to 10^{11} viruses per liter (Bergh et al., 1989; Proctor and Fuhrman, 1990; Børsheim et al., 1990), and even generous estimates put the carbon content at below 0.5 fg C per virus (a diameter of 100 nm and carbon density of 500 fg C per um^3 yields 0.26 fg C per virus): this figure yields a maximum carbon content of 0.05 to 50 ug C per liter. Comparing this value to the estimated 1 mg or more of high molecular weight DOC in seawater (Sugimura and Suzuki, 1988), it appears that the viruses contribute, at most, a small portion. However, processes relating to virus infection may still be significant in generating this high molecular weight DOM. As discussed previously, virus infection causes host cells to burst, releasing dissolved and small particulate cell debris. One could easily imagine that a large portion of the released material would be macromolecules, such as proteins and DNA, which make up the majority of the total cell mass (Simon and Azam, 1989). Also, we do not know to what extent degraded viruses contribute to high molecular weight DOM (or even if degraded viruses are present in any substantial amounts). Thus, while intact viruses contribute only a small amount to high molecular weight DOM, virus-related processes may contribute significantly.

DOM and new production. With a substantial portion of total primary production becoming DOM and with the observation of significant gradients of DOM concentrations, it is reasonable to ask how DOM fits into the picture of export and import of new production (*sensu* Dugdale and Goering, 1967; Eppley and Peterson, 1979). This picture includes both the vertical mixing of DOM to and from deeper waters, and its horizontal transport between adjacent ocean regions. Some recent global models of production included DOM import and export, and generally, inclusion of this feature provides a much better empirical fit to current observations than do models including only particulate matter (Toggweiler et al., 1988; Sarmiento et al., 1988; Bacastow and Maier-Reimer, 1991). This finding suggests that bacterial utilization of DOM imported from distant locations is an important factor in marine systems. A related but separate point involves the recent observation that photochemical degradation (near the sea surface) of old refractory DOM can generate small biologically labile molecules, such as formaldehyde and pyruvate (Kieber et al., 1989); therefore it is not only the new DOM that is potentially important in use of imported organic matter. This photochemical process, in addition to import of fresh DOM, may have particular significance in oligotrophic waters where photosynthetic production is low, and import of DOM from distant locations may fuel a non-trivial portion of the local primary production of particulate matter, in this case, by heterotrophic bacteria.

Even before the revival of interest caused by Suzuki et al.(1985) and Sugimura and Suzuki (1988), it was recognized that DON and DOP are potentially very important sources of N and P, which are the nutrients that are thought to limit primary productivity. Jackson and Williams (1985) reasoned that with concentrations of DON and DOP usually hundreds of times greater than the respective inorganic ions, the use of even a tiny portion of the organics could be a major input of nutrient. Either direct utilization of DON or DOP by phytoplankton, or first, uptake by heterotrophic bacteria, and then, production of inorganics by food web processes, could be a substantial fraction of nutrient input fueling new production in oligotrophic systems. The use of organic sources by phytoplankton is well-known (Flynn and Butler, 1986), although components that turn over rapidly, such as amino acids, may be regenerated rather than new production. This mechanism needs to be considered in measurements and models of marine production.

Solubilization of sinking POC. DOM plays a role in transport of organic matter both vertically and horizontally. A part of this transport involves the sinking flux of organic matter out of the euphotic zone. Until recently, most researchers considered the only mechanism of such flux to be sinking of particulate matter, and the decrease of sinking particulate flux being due to oxidation of the particles by large particle-feeding animals or attached microorganisms. However, recent evidence suggests that the decrease in flux of particulate matter may instead be caused largely by solubilization of the POM to DOM, and that bacteria subsequently use that DOM (Cho and Azam, 1988). This conclusion was based upon estimates of the distribution with depth of bacterial production; the high production rates of free-living suspended bacteria suggest that DOM is available to fuel this production, and the only obvious source is solubilization of POM as it sinks (however, an alternate possible source not considered by these authors is advection of DOC). Their interpretation was shown to be consistent with sediment trap results (Karl et al., 1988), as well as the distribution of certain radionuclides (Cho and Azam, 1988). It is not known how the POM becomes solubilized, but Cho and Azam suggested that the bacteria attached to the particles may be hypersolubilizing the particles by the action of exoenzymes. The term hypersolubilization indicates that the solubilization rate is much more rapid than the rate at which the resident bacteria could use the released substrate. Such a process may be viewed as a wasteful and inefficient use of scarce resources, which simply diffuse away. However, this process may make the carbon unavailable to larger consumers, yet still available to other bacteria, such as the descendants and relatives of the attached bacteria. Therefore, this suggestion is based upon a sociobiological interpretation. An alternative, but not mutually exclusive, explanation is that viral infection of microorganisms in the sinking particles may lead to solubilization due to cell lysis (Proctor and Fuhrman, 1991). It also is possible that the bacteria release exoenzymes not to solubilize the particle, but to inactivate viruses before they can infect a cell; in this case, the digestion of substrate is a side effect.

Bacterial Competition for Inorganic Nutrients

Until recently, it was generally assumed that the major role of bacteria, is in the process of regeneration, i.e., bacterial release of inorganic nutrients from organic ones.

Few researchers considered the possibility that bacteria may take up inorganic nutrients in marine systems, despite the fact that it has been known for some time that bacteria are responsible for a significant fraction of phosphate uptake in freshwater (Currie and Kalff, 1984) and marine systems (Harrison et al., 1977). However, recent measurements in natural systems, using the stable isotope ^{15}N and the short-lived radioisotope ^{13}N, showed that heterotrophic bacteria are responsible for perhaps 1/3 or even 1/2 of the total uptake of ammonium in marine systems (Wheeler and Kirchman, 1986; Fuhrman et al., 1988; Suttle et al., 1990). While such uptake does not necessarily occur all the time, its significance is undeniable. There also is evidence that bacteria can compete for nitrate as well (Horrigan et al., 1988). These results are likely to have significant impact on our thinking of marine systems, because nitrogen is thought to be the primary limiting nutrient in the sea.

The conditions under which bacteria are likely to take up inorganic nitrogen are well-described by Goldman et al. (1987). Thus, if the organic substrate being used by bacteria as a carbon and energy source has sufficient nitrogen for growth (after accounting for C losses by respiration), the bacteria will use that organic N alone; however, if there is insufficient nitrogen in this organic substrate, the bacteria will take up inorganic nitrogen to fulfill their needs. This hypothesis makes sense if the bacteria are limited by the available carbon and energy, as has been suggested by Kirchman (1990) and Kirchman et al. (1990). In natural systems, however, the types of substrates are undoubtedly much more varied and complex than in laboratory cultures. Therefore, while there is always apparently an adequate amount of DON (10s of micromolar in surface waters; Suzuki et al., 1985) for bacterial N needs, most of it may be in a chemical form that prevents easy use; thus, the bacteria may benefit from taking up ammonium, even if it is in extremely low concentration (sub-micromolar). It is noteworthy that marine bacteria are very rich in nitrogen, with C:N ratios of about 4 (Lee and Fuhrman, 1987); therefore, it is not surprising that they have high affinity uptake mechanisms for N. Kirchman et al. (1989) suggest that the supply of amino acids is an important determinant of whether or not the bacteria are net producers or consumers of ammonium.

A potentially confusing point is that several studies have shown that the bacterial size fraction contributes a significant amount to the total release of ammonium, as measured by isotope dilution (e.g., Glibert, 1982). Release seems to contradict the data showing bacterial uptake of these nutrients. However, such studies measure the gross release of ammonium, not net release, so even if bacteria were a net sink for ammonium, only the release would show. Possibly some bacteria may take up ammonium while others are releasing it, as has been seen in laboratory systems (Tupas and Koike, 1990). This is another example where the diversity of microbial populations needs to be considered in biogeochemical studies. As an extreme example, some marine bacteria may be photoheterotrophic, obtaining their carbon from organic sources, but their energy from sunlight (Britschgi and Giovannoni, 1991). Such bacteria have greatly reduced respiratory C needs and require either a very low C:N ratio of their organic substrate, or otherwise may need to take up substantial inorganic N to fulfill their growth demand. Such bacteria growing on the same substrate as chemoheterotrophs (i.e., conventional heterotrophs that use organic C for both energy and biomass) may need inorganic N, even if the chemoheterotrophs can meet their demand from the organic substrate alone. A similar situation may occur if certain bacteria can use particular DON sources, while others cannot.

There are several implications of bacterial uptake of inorganic N, and most of which were covered by Wheeler and Kirchman (1986), Fuhrman et al. (1988), and Horrigan et al. (1988). The first involves the measurement of f-ratios (*sensu* Eppley and

Peterson, 1979) from field measurements of nitrogen uptake. It is common for the uptake of nitrate and ammonium to be measured to estimate the new production and regenerated production, respectively. While it is often acknowledged that this ignores N_2 fixation (new production) and phytoplankton uptake of urea or other N compounds (regenerated production), few, if any researchers, refer to the possibility of uptake by bacteria. The data from natural systems suggest that bacteria take up ammonium far more than they do nitrate (Wheeler and Kirchman, 1986). This finding suggests that the estimates of regenerated production (from ammonium uptake) may be higher than the true amount because heterotrophic bacteria are responsible for some of the uptake, yet it is usually assumed that ammonium uptake reflects regenerated production by phytoplankton alone. Hence, new production may be a larger fraction of the total than one would calculate, based upon the measurements of uptake. The extent of the overestimation is not easy to calculate (without explicit measurements) because (a) the bacterial uptake is not a simple function of the total, and (b) only about half of the bacteria are collected on the glass-fiber filters typically used in ^{15}N uptake experiments (Lee and Fuhrman, 1987).

A new aspect of this subject is that the relative uptake of ammonium into bacteria vs. larger phytoplankton is highly dependent upon the concentration of ammonium (Suttle et al., 1990). With increasing concentrations, a larger fraction goes into larger phytoplankton. For example, in a sample from Long Island Sound, an experimental increase in the ammonium concentration from undetectable (< 80nm) to 0.5 uM changed the % uptake into the > 1 um fraction from 50% to 80%. Seasonal trends of concentration and size-fractionated uptake showed a similar pattern. Comparable results were obtained with phosphate uptake from the Sargasso Sea. The results are particularly applicable to understanding the fate of nutrient pulses, such as may occur as organisms pass through nutrient patches. A major implication is that addition of nutrients in pulses (as opposed to steady, low concentrations) allows a larger fraction of the nutrients to be used by large phytoplankton rather than by bacteria. A simple food web model showed that this, in turn, leads to a larger proportion of the primary productivity reaching higher trophic levels (Suttle et al., 1990). In general, this phenomenon may help explain the differences between the food webs of high nutrient vs. low nutrient environments, i.e., coastal vs. offshore or seasonal variations; in general, the low nutrient waters are dominated by prokaryotes.

Bacterial Dominance of Oligotrophic Waters

Until recently, most oceanographers thought that the biomass of heterotrophic bacteria is small relative to that of phytoplankton, conforming to the idea of a trophic pyramid, with highest biomass in the primary producers. This configuration appears to be true in eutrophic and mesotrophic waters (e.g., Williams et al., 1981; Azam et al., 1983). However, several observations have suggested that bacterial biomass is a much larger portion of the total in oligotrophic waters (e.g., Linley et al., 1983; Bird and Kalff, 1984; Laws et al., 1984). Comprehensive depth profiles showed that in oligotrophic waters, such as those with surface chlorophyll concentrations < 0.2 ug per liter, the bacterial biomass of the integrated euphotic zone exceeds that of phytoplankton, and often bacterial biomass is higher even at the depths where phytoplankton biomass is highest (Fuhrman et al., 1989; Dortch and Packard, 1989; Cho and Azam, 1990; Li et al., in press). This is part of a larger pattern whereby bacterial biomass changes relatively little with trophic status, while phytoplankton biomass changes much more (Bird and Kalff, 1984; Cole et al., 1988).

The dominance of biomass does not imply a dominance of production. In fact, estimates of bacterial growth and heterotrophic production in these oligotrophic waters are low, with population doubling times about a week or more (Fuhrman et al., 1989). Still, even with this slow growth, the bacteria consume an amount of organic carbon that is a large fraction of primary production. We expected that a relatively small phytoplankton biomass must turn over much faster than the bacteria in order to feed the larger bacterial biomass, which is apparently the case in oligotrophic waters (Fuhrman et al., 1989). The mechanisms by which this occurs are unknown, and it is difficult to reconcile the observations with traditional grazing processes alone; either the phytoplankton must release a very large fraction of their total production as DOM to support the observed bacterial production, or there is significant input of exogenous DOM into these oligotrophic waters (Fuhrman et al., 1989).

The broadest implications of this dominance involve size distribution, trophic structure, and their influence on the amounts and fates of primary production. Many of the relevant implications are discussed elsewhere in this review, such as the effective sequestration and suspension of nutrients in bacterial biomass. Other implications are practical ones, and need to be considered by those measuring the biological, physical, and chemical properties of oligotrophic waters. These implications include the observation that because only about half of the bacteria are caught on the glass-fiber filters commonly used to collect POC (Lee and Fuhrman, 1987), such measurements may miss a great deal of the biomass in oligotrophic waters. Altabet (1990) reported that particles from the open Altantic ocean passing a GF/F filter, but caught on a 0.2-um aluminum oxide filter, represented as much as 30-40% of the organic C and N, 60% of the ammonium uptake, and 80% of the nitrate uptake in some samples. Cho and Azam (1990) calculated that bacterial biomass in oligotrophic waters averages 40% of total POC. Also, bacteria in oligotrophic waters possess about 90% of the total biological surface area (Fuhrman et al., 1989); this result has significance for uptake studies and the fate of materials, such as radionuclides and iron, that adsorb to surfaces. Finally, Stramski and Kiefer (1990) demonstrated that optical properties, such as scattering and absorption, can have a very large bacterial component in oligotrophic waters; this finding has significance for interpreting data from optical remote sensing.

The reasons for the dominance of bacterial biomass in oligotrophic waters are not clear, but may partly be a result of superior competition for inorganic nutrients, as described above. On one hand, this dominance probably results from the balances between the growth and loss rates of the bacteria, and, on the other hand, phytoplankton. These processes lead to the relative constancy of bacterial biomass and the much greater fluctuation of phytoplankton biomass with changing nutrient status. Cho and Azam (1990) suggest that the primary factor in dominance is that bacteria have a lower limit of abundance in surface waters of about 3×10^5 cells per ml in surface waters. This is an important component, but understanding the causes requires knowing what determines the upper limits and the lower limits of the biomasses of both groups. I feel that loss rates are particularly important, because bacteria and phytoplankton have overlapping growth rates. Bacterial growth rates are more extreme, which might lead to more rather than less variability.

Loss processes for bacterial abundance are thought to be primarily protozoan grazing and viral infection (Azam et al., 1983; McManus and Fuhrman, 1988; Proctor and Fuhrman, 1990). Both processes are highly density-dependent, with more rapid mortality as biomass increases. This is due to a combination of functional responses plus the fact that both protozoa and viruses can increase in abundance as fast (or nearly so) as the bacteria, so preventing the bacteria from becoming very abundant. The lower limits for bacterial abundance probably represent the range where these loss processes become much less efficient, as is known for protozoa (Cho and Azam, 1990). This speculation implies that the general range of bacterial biomass is primarily set by loss processes, although the actual biomass within this range still reflects the balance between growth and loss. In addition, bacteria always have some DOM available to support growth, so slow removal processes, such as scavenging onto particles, can be balanced by some growth. In the deep sea, bacterial abundance drops to about 5×10^4 per ml (Fuhrman et al., 1989), therefore, there are loss processes that can reduce the abundance to that level. These losses probably include scavenging onto sinking particles, and possibly cell lysis by the induction of prophage (latent bacteriophage residing as DNA within the host genome; Ackerman and Dubow, 1987). Deep DOM is older and much less concentrated than in surface waters (Suzuki and Sugimura, 1988; Druffel et al., 1989), suggesting it supports less bacterial growth.

The upper limit of phytoplankton abundance is very high, being several orders of magnitude above the lower limit (Bird and Kalff, 1984). The upper limit does not appear to be set by grazers, probably because the population of larger phytoplankton can increase much faster than their metazoan grazers can keep up, and many phytoplankton are too large to be effectively grazed by protozoa (or they may have defenses against protozoa). Massive blooms of phytoplankton may be controlled by viruses (Sieburth et al., 1988), or growth may be stopped by nutrient limitation and become diluted by water mixing, or sink out. The lower limit of phytoplankton abundance may be a function of grazers and viruses no longer being able to reproduce efficiently on such sparse prey (as with bacteria), but the larger phytoplankton also have sinking losses independent of population size. Thus, the biomass of the larger phytoplankton can drop to extremely low levels. It is noteworthy that the small prokaryotic phytoplankton (cyanobacteria and prochlorophytes) are similar in size to bacteria, yet drop to much lower abundances (by 4 orders of magnitude) at some locations or some depths within the euphotic zone (Stockner and Antia, 1986; Olson et al., 1990); such low abundances are probably not caused by predation or viruses, and are a result of scavenging and mixing losses exceeding growth. I can only speculate as to why the other bacteria do not drop to such low levels: perhaps it is their ability to use DOM, or more generally, because these other bacteria are an extremely diverse group from which organisms can be selected to adapt to most conditions, while the cyanobacteria and prochlorophytes have very specific growth requirements of light and nutrients that are frequently difficult to meet.

Feedback Between Phytoplankton and Bacteria

Several processes described previously, such as bacterial competition with phytoplankton for inorganic nutrients, probably lead to feedback loops. One such loop that has received attention recently seems to be a paradox: when phytoplankton release organic C (particularly, when stressed by low nutrients), they are stimulating their bacterial competitors to take up inorganic nutrients and possibly outcompete them, inhibiting their growth (Currie and Kalff, 1984; Bratbak and Thingstad, 1985). It is not clear whether this release of organic matter is adaptive. Bjørnsen (1988) suggested that the release of organics can be passive, essentially due to leaky membranes, and no adaptation need be explained. However, there also are possible adaptive advantages. Under certain circumstances, the released material may serve an allelopathic (Sharp et al., 1979) or metal-chelating function (Murphy et al., 1976), although it is unlikely that these possibilities can explain most of the release.

Azam and Ammerman (1984 a,b) and Azam and Cho (1987) suggested that bacteria may hover in clusters near individual phytoplankton cells in a commensal relationship, whereby the release of DOC benefits the algae: the phytoplankton release C- and energy-rich compounds that are taken up by the bacteria, and the bacteria break down organic

N and P compounds into forms readily used by the phytoplankton. A major impetus for this suggestion came from the observation of multiphasic kinetics in bacterial populations, showing that bacteria possess functional (and possibly, constitutive) uptake mechanisms adapted not only for low nM concentrations of organic nutrients, but also for the uptake of micromolar or higher concentrations (Azam and Hodson, 1981). The authors suggested that this observation shows that the bacteria frequently encounter such concentrations, because it serves no purpose to have uptake mechanisms that are not used. Because these bacteria are free-living in an environment where the average concentrations of organic nutrients are nanomolar, the implication is that microzones of enhanced organic nutrients commonly exist near algae or decaying particles. Such clusters have yet to be observed, but attempts to do so have probably suffered from a sort of Heisenberg Uncertainty Principle: the act of trying to observe such clusters is likely to destroy them. Still, Mitchell and Fuhrman (1989) have demonstrated that significant microbial patchiness on cm scales is readily measured, and Alldredge and Cohen (1987) have measured persistent mm-scale chemical (depleted oxygen) patches near marine snow.

Mitchell et al. (1985) suggested that physical and biological processes acting to break up such clusters (or prevent them from forming in the first place) probably limit their occurrence to non-motile phytoplankton in very calm portions of the water column. These processes include the difficulty of bacteria in keeping up with fast-swimming (or sinking) algae, the difficulty of exuding enough material to maintain an enriched zone in the presence of diffusion, and the destruction of microzones by small (mm to cm), turbulent eddies, which are thought to sweep through and mix most ocean waters (except very stratified layers or calm regions) in seconds to minutes. Jackson (1989) modeled the interactions of bacteria with leaky algae and marine snow, and concluded that even transient interactions of chemotactic bacteria with enhanced microzones can be an advantage to the bacteria; however, he did not include turbulence in his model. Azam and Cho (1987) suggested a scenario in which these constraints may not fully apply: (1) algae exude long-chain polysaccharides that remain associated with the algal cells, (2) algae exude small molecules (including informational molecules like cyclic AMP) that diffuse out of this matrix and attract bacteria, (3) bacteria become loosely attached to the polysaccharides while they are digesting them, and (4) the polysaccharides help bind and localize any remineralized N and P produced by the bacteria. If this scenario holds true, then the diaphanous cloud of polysaccharides around the algal cells would dampen the effects of diffusive losses and turbulent disruption of the enhanced microzones.

Williams (1984) and Sherr et al. (1988) suggested that normal food-web processes may offset the apparent disadvantages of phytoplankton feeding the bacteria. They suggested that nutrients in the bacteria are regenerated rapidly by protozoan grazing, so no clustering of bacteria near phytoplankton needs to be invoked. But here again, processes similar to those described by Azam and Cho (1987) may be taking place. Some benefit may occur because the bacteria are making organic N and P, or other compounds, such as vitamins, available to the phytoplankton. Several researchers noted that some algae will grow only in the presence of bacteria, often due to the need for vitamins or other undefined factors (Wood and Van Valen, 1990).

Wood and Van Valen (1990) suggested that the release of organic matter may be part of an adaptive mechanism involving (1) overflow of organic matter production in nutrient-limited cells that are maintaining a high synthetic ability to be able to resume rapid growth when they encounter nutrients (i.e., in a patch), (2) protection of the photosynthetic apparatus under fluctuating (high) light conditions of cells lacking photorespiration, and (3) enhanced patch or community selection via beneficial stimulation of nearby heterotrophs. This latter point includes the suggestion that bacteria, which can enhance the local growth of algae, will benefit from increased algal growth and organic release, while bacteria that have a net negative effect (e.g., via nutrient competition) will be selected against.

Williams (1990) made an additional important point: the release of organic compounds by larger phytoplankton may keep nutrients suspended in the water column by maintaining nutrients as dissolved material or bacteria that act as non-sinking, yet rapidly cycling reservoirs of nutrients. In low-nutrient stratified waters, sinking of larger phytoplankton cells and zooplankton fecal pellets is the major mechanism of loss for nutrients, and mixing of new nutrients from deeper waters is highly restricted. Therefore, any means that increases the residence time of N and P in the euphotic zone is beneficial for the plankton, as long as N and P are biologically available. The microbial food web is such a mechanism, because the organisms within it are so small that they hardly sink at all, yet trophic interactions make the N and P from biomass available to other organisms. Therefore, it may be an advantage to feed the bacteria and have N and P cycle through the microbial food web rather than simply enter the larger phytoplankton and sink out of the euphotic zone.

This hypothesis of DOM release to assist nutrient suspension may not provide a full rationalization for release, because model data, such as those of Frost (1984), suggest that the benefit may be small (see discussion in Williams, 1990), but it certainly can help offset any disadvantages. This concept also may explain some of the differences between biomass distributions of oligotrophic and eutrophic systems (Fuhrman et al., 1989; Cho and Azam, 1990). It might have particular significance in waters where production is limited by iron (Martin and Fitzwater, 1988), although the evidence of such limitation still is under debate. Iron tends to bind to surfaces, and thus, is rapidly lost from surface waters on sinking particles (Bruland, 1983), even more so than N and P bound in biomass. Because bacteria have a large surface area, and thus, have the potential to bind a great deal of iron, it may be particularly advantageous for all plankton when bacteria are abundant, because they help keep iron suspended in the euphotic zone and available to primary producers.

Many of these possible advantages of DOM release need not be restricted to direct release by phytoplankton. In the cases of patch or community selection, as described by Wood and Van Valen (1990), or suspension of nutrients (Williams, 1990), the benefits to the whole plankton community of having an active microbial food web would operate, no matter what the source of DOM.

SUMMARY

1. The concept of the microbial food web is alive and well. Heterotrophic bacteria consume 50% or more of the primary production, but food-web processes and recycling allow much of this to pass to other organisms. While respiration by consumers is limited to 100% of the primary production (in closed systems), the total consumption and heterotrophic production by heterotrophs can greatly exceed 100%, due to recycling.

2. We still do not have exact information on the relative importance of different sources of DOM. There still are many questions about its release from phytoplankton. Zooplankton-mediated processes are important, and those related to viral infection are likely to be.

3. Viral infection of bacteria can increase gross bacterial production (via DOM release and recycling) yet decrease flux of this production to higher levels. The viruses also may have important roles in shaping the species composition and biogeochemical processes of microbial communities.

4. DOM must be included in conceptual and physical models of biogeochemical cycling, including the export and import of C, N, and P. The new DOM described by Suzuki et al. (1985) and Sugimura and Suzuki (1988) is probably particularly significant in this regard. Even old DOM can be used after photochemical transformation to labile compounds. The solubilization of sinking POC, and its degradation by free-living suspended bacteria, is a major unexplored route.

5. Bacteria compete with phytoplankton for the uptake of ammonium and phosphate. At higher concentrations of nutrients, such as during pulses, the larger organisms do proportionately better. This fact should affect our thinking about oligotrophic vs. eutrophic systems and, in general, nutrient cycling, because it adds to the potential for feedback loops between bacteria, phytoplankton, and grazers.

6. Bacterial biomass often exceeds that of phytoplankton in oligotrophic systems, which suggests phytoplankton have higher relative turnover rates. The bacteria are significant particulate reservoirs of N and P, effectively keeping these nutrients in suspension and available to other organisms via the food web. Measurements of particulate matter in oligotrophic waters need to include procedures to retain bacteria, which are missed with conventional glass fiber filters.

7. Feedback processes between bacteria and phytoplankton, occurring on both the microscale and macroscale, are very interesting, but largely speculative. It may be adaptive for phytoplankton to release high-energy organic compounds, even though bacteria might use these as sources of C and energy, and outcompete the phytoplankton for inorganic nutrients. Possible benefits include (a) bacteria breaking down DON and DOP into forms available to other plankton (directly or via food web processes), and (b) bacteria helping to suspend N, P, and maybe Fe in euphotic zone, so that less is lost via sinking.

CONCLUSIONS

Research into the microbial food web has advanced considerably in the past decade. We cannot hope to understand material and energy cycling in marine systems without intimate knowledge of the tiniest organisms and dissolved organic substances. The microbial processes are intertwined with those of the larger organisms as an essential part of an integrated whole. Progress in the future will go beyond the current mode of lumping all the diverse bacteria into a "black box," and will start to deal with individual components and processes.

ACKNOWLEDGMENTS

I wish to thank Farooq Azam, Tim Hollibaugh, Ake Hagström, Peter J. leB. Williams, John Hobbie, Dick Eppley, Doug Capone, Sarah Horrigan, George McManus, Curtis Suttle, Jim Mitchell, SangHoon Lee, and Lita Proctor, for stimulating and informative discussions. I thank David Kirchman, Farooq Azam, Michael Landry and an anonymous reviewer for reviewing the manuscript. This work was supported by NSF grants OCE8996136 and OCE8996117.

REFERENCES

Ackerman, H.-W., and Dubow, M. S., 1987, "Viruses of Prokaryotes. Vol. 1. General Properties of Bacteriophages," CRC Press, Boca Raton.

Alldredge, A. L., and Cohen, Y., 1987, Can microscale patches persist in the sea? Microelectrode study of marine snow, fecal pellets, *Science*, 235:689.

Altabet, M.A., 1990, Organic C, N, and stable isotopic composition of particulate matter collected on glass-fiber and aluminum oxide filters, *Limnol. Oceanogr.*, 35: 902.

Azam, F., Fenchel, T., Gray, J. G., Meyer-Reil, L. A., and Thingstad, T., 1983, The ecological role of water-column microbes in the sea, *Mar. Ecol. Prog. Ser.*, 10: 257.

Azam, F., and Ammerman, J. W., 1984a, Cycling of organic matter by bacterioplankton in pelagic marine ecosystems, *in:* "Microenvironmental Considerations, Flows of Energy and Materials in Marine Ecosystems," M. J. R. Fasham, ed., Plenum Publishing Company, New York.

Azam, F., and Ammerman, J. W., 1984b, Mechanisms of organic matter utilization by marine bacterioplankton, Lecture notes on coastal and estuarine studies, *in:* "Marine Phytoplankton and Productivity," O. Holm-Hansen, L. Bolis, and R. Gilles, eds., Springer-Verlag, Berlin.

Azam, F., and Cho, B. C., 1987, Bacterial utilization of organic matter in the sea, SGM 41, Ecology of Microbial Communities, Cambridge Univ. Press.

Azam, F., and Hodson, R. E., 1981, Multiphasic kinetics for D-glucose uptake by assemblages of natural marine bacteria, *Mar. Ecol. Prog. Ser.*, 6: 213.

Bacastow, R., and Maier-Reimer, E., 1991, Dissolved organic carbon in modeling oceanic new production, *Global Biogeochem. Cycles.*, 5: 71.

Banse, K., 1974, On the role of bacterioplankton in the tropical ocean, *Mar. Biol.*, 24:1.

Bergh, O., Børsheim, K. Y., Bratbak, G., and Heldal, M., 1989, High abundance of viruses found in aquatic environments, *Nature.*, 340: 467.

Bird, D. F., and Kalff, J., 1984, Empirical relationships between bacterial abundance and chlorophyll concentration in fresh and marine waters, *Can. J. Fish. Aquat. Sci.*, 41: 1015.

Bjørnsen, T. K., 1986, Automatic determination of bacterioplankton biomass by image analysis, *Appl. Environ. Microbiol.*, 51: 1199.

Bjørnsen, P. K., 1986, Bacterioplankton growth yield in continuous seawater cultures, *Mar. Ecol. Prog. Ser.*, 30: 191.

Bjørnsen, P. K., 1988, Phytoplankton exudation of organic matter: Why Do Healthy Cells Do It?, *Limnol. Oceanogr.*, 33: 151.

Bjørnsen, P. K., and Kuparinen, J., 1991, Determination of bacterioplankton biomass, net production and growth efficiency in the Southern Ocean, *Mar. Ecol. Prog. Ser.*, 71: 185.

Børsheim, K. Y., Bratbak, G., and Heldal, M., 1990, Enumeration and biomass estimation of planktonic bacteria and viruses by transmission electron microscopy, *Appl. Environ. Microbiol.*, 56: 352.

Bratbak, G., 1985, Bacterial biovolume and biomass estimations, *Appl. Environ, Microbiol.*, 49: 1488.

Bratbak, G., and Thingstad, T.F., 1985, Phytoplankton-bacteria interactions: An apparent paradox?, Analysis of a model ecosystem with both competition and commensalism, *Mar. Ecol. Prog. Ser.*, 25: 23.

Bratbak, G., Heldal, M., Norland, S., and Thingstad, T.F., 1990, Viruses as partners in spring bloom microbiol trophodynamics, *Appl. Environ. Microbiol.*, 56:1400.

Britschgi, T., and Giovannoni, S. J., 1991, Phylogenetic analysis of a natural marine bacterioplankton population by rRNA gene cloning and sequencing, *Appl. Environ. Microbiol.*, 57: 1707.

Bruland, K., 1983, Trace elements in seawater, *in*: "Chemical Oceanography," J. P. Riley, and R. Chester, eds., Academic Press, New York.

Chin-Leo, G., and Kirchman, D. L., 1988, Estimating bacterial production in marine waters from the simultaneous incorporation of thymidine and leucine, *Appl. Environ. Microbiol.*, 54: 1934.

Cho, B. C., and Azam, F., 1988, Major role of bacteria in biochemical fluxes in the ocean's interior, *Nature*, 332: 441.

Cho, B., and Azam. F., 1990, Biogeochemical significance of bacterial biomass in the ocean's euphotic zone, *Mar. Ecol. Prog. Ser.*, 63: 253.

Cole, J. J., 1982, Interactions between bacteria and algae in aquatic ecosystems, *Ann. Rev. Ecol. Syst.*, 13: 291.

Cole, J. J., Findlay, S., and Pace, M. L., 1988, Bacterial production in fresh and saltwater ecosystems: A cross-system overview, *Mar. Ecol. Prog. Ser.*, 43: 1.

Copping, A. E., and Lorenzen, C. J., 1980, Carbon budget of a marine phytoplankton-herbivore system with carbon-14 as tracer, *Limnol. Oceanogr.*, 25: 873.

Currie, D. J., and Kalff, J., 1984, The relative importance of bacterioplankton and phytoplankton in phosphorus uptake in freshwater, *Limnol. Oceanogr.*, 29: 311.

DeLong, E. F., Wickham, G. S., and Pace, N. R., 1990, Phylogenetic stains: Ribosomal RNA-based probes for the identification of single cells, *Science*, 243: 1360.

Dortch, Q., and Packard, T., 1989, Differences in biomass structure between oligotrophic and eutrophic marine ecosystems, *Deep Sea Res.*, 36: 223.

Druffel, E. R. M., Williams, P. M., and Suzuki, Y., 1989, Concentrations and radiocarbon signatures of dissolved organic matter in the Pacific Ocean, 16: 991.

Ducklow, H. W., 1991, The passage of carbon through microbial foodwebs: Results from flow network models, *Mar. Microb. Food Webs.*, 5: 129.

Dugdale, R. C., and Goering, J. J., 1967, Uptake of new and regenerated form of nitrogen in primary production, *Limnol. Oceanogr.*, 12: 196.

Eppley, R. W., and Peterson, B. J., 1979, Particulate organic matter flux and planktonic new production in the deep ocean, *Nature*, 282: 677.

Eppley, R. W., Horrigan, S. G., Fuhrman, J. A., Brooks, E. R., Price, C. C., and Sellner, K., 1981, Origins of dissolved organic matter in Southern California coastal waters: Experiments on the role of zooplankton, *Mar. Ecol. Prog. Ser.*, 6: 149.

Ferguson, R. L., and Rublee, P., 1976, Contribution of bacteria to standing crop of coastal plankton, *Limnol. Oceanogr.*, 21: 141.

Ferguson, R. L., Buckley, E. N., and Palumbo, A. V., 1984, Response of marine bacterioplankton to differential filtration and confinement, *Appl. Environ. Microbiol.*, 47: 49.

Flynn, K.J., and Butler, I., 1986, Nitrogen sources for the growth of marine microalgae: role of dissolved free amino acids, *Mar. Ecol. Prog. Ser.*, 34: 281.

Frost, B. W., 1984, Utilization of phytoplankton production in the surface layer, *in*: "Global Ocean Flux Study Workshop, US National Research Council."

Fuhrman, J. A., 1987, Close coupling between release and uptake of dissolved free amino acids in seawater studied by an isotope dilution approach, *Mar. Ecol. Prog. Ser.*, 37: 45.

Fuhrman, J. A., and Azam, F., 1980, Bacterioplankton secondary production estimates for coastal waters of British Columbia, Antarctica, and California, *Appl. Environ. Microbiol.*, 39: 1085.

Fuhrman, J. A., and Azam, F., 1982, Thymidine incorporation as a measure of heterotrophic bacterioplankton production in marine surface waters: Evaluation and field results, *Mar. Biol.*, 66: 109.

Fuhrman, J. A., Horrigan, S. G., and Capone, D. G., 1988, The use of ^{13}N as tracer for bacterial and algal uptake of ammonium from seawater, *Mar. Ecol. Prog. Ser.*, 45: 271.

Fuhrman, J. A., Sleeter, T. D., Carlson, C. A., and Proctor, L. M., 1989, Dominance of bacterial biomass in the Sargasso Sea and its ecological implications, *Mar. Ecol. Prog. Ser.*, 57: 207.

Giovannoni, S. J., Britschgi, T. B., Moyer, C. L., and Field, K. G., 1990, Genetic diversity in Sargasso Sea bacterioplankton, *Nature*, 345: 60.

Glibert, P. M., 1982, Regional studies of daily, seasonal, and size fractionation variability in ammonium regeneration, *Mar. Biol.* 70: 209.

Goldman, J. C., Caron. D. A., and Dennett, M. R., 1987, Regulation of gross growth efficiency and ammonium regeneration in bacteria by substrate C:N ratio, *Limnol. Oceanogr.*, 32: 1239.

Hagström, Å., Larsson, U., Horstedt, P., and Normark, S., 1979, Frequency of dividing cells, a new approach to the determination of bacterial growth rates in aquatic environments, *Appl. Environ. Microbiol.*, 37: 805.

Hagström, Å., Azam, F., Andersson, A., Wikner, J., and Rassoulzadegan, F., 1988, Microbial loop in an oligotrophic pelagic marine ecosystem: Possible roles of cyanobacteria and nanoflagellates in the organic fluxes, *Mar. Ecol. Prog. Ser.*, 49: 171.

Harrison, W. G., Azam, F., Renger, E. H., and Eppley, R. W., 1977, Some experiments on phosphate assimilation by coastal marine phytoplankton, *Mar. Biol.*, 40: 9.

Hobbie, J. E., Daley, R. J., and Jasper, S., 1977, Use of Nuclepore filters for counting bacteria by fluorescence microscopy, *Appl. Environ. Microbiol.*, 33: 1225.

Horrigan, S. G., Hagström, A., Koike, I., and Azam, F., 1988, Inorganic nitrogen utilization by assemblages of marine bacteria in seawater culture, *Mar. Ecol. Prog. Ser.*, 50: 147.

Hutchinson, G. E., 1961, The paradox of the plankton, *Amer. Nat.*, 45: 137.

Jackson, G., 1988, Implications of high dissolved organic matter concentrations for oceanic properties and processes, *Oceanography*, 1: 28.

Jackson, G. A., 1989, Simulation of bacterial attraction and adhesion to falling particles in an aquatic environment, *Limnol. Oceanogr.*, 34:514.

Jackson, G. A., and Williams, P. M., 1985, Importance of dissolved organic nitrogen and phosphorus to biological nutrient cycling, *Deep Sea Res.*, 32: 223.

Jannasch, H. W., and Jones, G. E., 1959, Bacterial populations in sea water as determined by different methods of enumeration, *Limnol. Oceanogr.*, 4: 128.

Jumars, P. A., Penry, D. L., Baross, J. A., Perry, M. J., and Frost, B. W., 1989, Closing the microbial loop: Dissolved carbon pathway to heterotrophic bacteria from incomplete ingestion, digestion, and absorption in animals, *Deep Sea Res.*, 36: 483.

Karl, D. M., Knauer, G. A., and Martin, J. H., 1988, Downward flux of particulate organic matter in the ocean: A particle decomposition paradox, *Nature*, 332: 438.

Kieber, D. J., McDaniel, J. A., and Mopper, K., 1989, Photochemical source of biological substrates in seawater: Implications for carbon cycling, *Nature*, 341: 637.

Kirchman, D. L., 1990, Limitation of bacterial growth by dissolved organic matter in the subarctic Pacific, *Mar. Ecol. Prog. Ser.*, 62: 47.

Kirchman, D. L., Keil, R. G., and Wheeler, P. A., 1989, The effect of amino acids on ammonium utilization and regeneration by heterotrophic bacteria in the subarctic Pacific, *Deep Sea Res.*, 36: 1763.

Kirchman, D. L., Keil, R. G., and Wheeler, P. A., 1990, Carbon limitation of ammonium uptake by heterotrophic bacteria in the subarctic Pacific, *Limnol. Oceanogr.* 35: 1258.

Kirchman, D. L., K'Nees, E., and Hodson, R. E., 1985, Leucine incorporation and its potential as a measure of protein synthesis by bacteria in natural aquatic systems, *Appl. Environ. Microbiol.*, 49: 599.

Lampert, W., 1978, Release of dissolved organic carbon by grazing zooplankton, *Limnol. Oceanogr.*, 23: 831.

Laws, E. A., Redalje, D. G., Haas, L. W., Bienfang, P. K., Eppley, R. W., Harrison, W. G., Karl, D. M., and Marra, J., 1984, High phytoplankton growth and production rates in oligotrophic Hawaiian coastal waters, *Limnol. Oceanogr.*, 29: 1161.

Lee, S., and Fuhrman, J. A., 1987, Relatioships between biovolume and biomass of naturally derived marine bacterioplankton, *Appl. Environ. Microbiol.*, 53: 1298.

Lee, S., and Fuhrman, J. A., 1990, DNA hybridization to compare species compositions of natural bacterioplankton assemblages, *Appl. Environ. Microbiol.*, 56: 739.

Lee, S., and Fuhrman, J. A., Spatial and temporal variation of natural bacterioplankton assemblages studied by total genomic DNA cross-hybridization, *Limnol. Oceanogr.*, in press.

Lenski, R. E., 1988, Dynamics of interactions between bacteria and virulent bacteriophage, *Adv. Microb. Ecol.*, 10:1.

Li, W. K. W., Dickie, P. M., Irwin, B. D., and Wood, A. M., Biomass of bacteria, cyanobacteria, prochlorophytes, and photosynthetic eukaryotes in the Sargasso Sea, *Deep Sea Res.*, in press.

Linley, E. A. S., Newell, R. C., and Lucas, M. I., 1983, Quantitative relationships between phytoplankton, bacteria, and heterotrophic microflagellates in shelf waters, *Mar. Ecol. Prog. Ser.*, 12:

Martin, J. H., and Fitzwater, S. E., 1988, Iron deficiency limits phytoplankton growth in the north-east Pacific subarctic, *Nature*, 331: 341.

McManus, G. B., 1991, Flow analysis of a planktonic microbial food web model, *Mar. Microb. Food Webs.*, 5: 145.

McManus, G. B., and Fuhrman, J. A., 1988, Control of marine bacterioplankton populations: Measurement and significance of grazing, *Hydrobiologia*, 159: 51.

Mitchell, J. G., and Fuhrman, J. A., 1989, Centimeter scale vertical heterogeneity in bacteria and chlorophyll *a*, *Mar. Ecol. Prog. Ser.*, 54: 141.

Mitchell, J. G., Okubo, A., and Fuhrman, J. A., 1985, Microzones form the basis for a stratified microbial ecosystem, *Nature*, (316): 58.

Murphy, T. P., Lean, D. R. S., and Nalewajko, C., 1976, Blue-green algae: their excretion of iron-selective chelators enables them to dominate other algae, *Science*, 192: 900.

Olson, R. J., Chisholm, S. W., Zettler, E. R., Altabet, M. A., and Dusenberry, J. A., 1990, Spatial and temporal distributions of prochlorophyte picoplankton in the North Atlantic Ocean, *Deep Sea Res.*, 37: 1033.

Pace, N. R., Stahl, D. A., Lane, D. L., and Olsen, G. J., 1986, The analysis of natural microbial populations by rRNA sequences, *Adv. Microbiol. Ecol.*, 9: 1.

Paul, J. H., and Carlson, D. J., 1984, Genetic material in the marine environment: implication for bacterial DNA, *Limnol. Oceangr.*, 29: 1091.

Pomeroy, L. R., 1974, The ocean's food web, a changing paradigm, *Bioscience*, 24: 499.

Proctor, L. M., and Fuhrman, J. A., 1990, Viral mortality of marine bacteria and cyanobacteria, *Nature*, 343: 60.

Proctor, L. M., and Fuhrman, J. A., 1991, Roles of viral infection in organic particle flux, *Mar. Ecol. Prog. Ser.*, 69: 133.

Proctor, L. M., Fuhrman, J. A., and Ledbetter, M. C., 1988, Marine bacteriophages and bacterial mortality, *EOS Trans. Am. Geophys. Union.*, 69: 1111.

Roman, M. R., Ducklow, H. W., Fuhrman, J. A., Garcide, C., Glibert, P. M., Malone, T. C., and McManus, G. B., 1988, Production, consumption and nutrient cycling in a laboratory mesocosm, *Mar. Ecol. Prog. Ser.*, 42: 39.

Roy, S., Harris, R. P., and Poulet, S. A., 1989, Inefficient feeding by *Calanus helgolandicus* and *Temora lingicornis* on *Coscinodiscus wailesii*: quantitative estimation using chlorophyll-type pigments and effects on dissolved free amino acids, *Mar. Ecol. Prog. Ser.*, 52:145.

Sarmiento, J. L., Toggweiler, J. R., and Najjar, R., 1988, Ocean carbon cycle dynamics and atmospheric pCO2, *Philos. Trans. R. Soc. London, Ser. A.*, 325: 3.

Scavia, D., 1988, On the role of bacteria in secondary production, *Limnol. Oceanogr.*, 33: 1220.'

Scavia, D., and Laird, G. A. 1987, Bacterioplankton in lake Michigan: Dynamics, controls, and significance to carbon flux, *Limnol. Oceanogr.*, 32: 1017.

Schmidt, T. M., DeLong, E. F., and Pace, N. R., 1991, Analysis of a marine picoplankton community by 16S rRNA gene cloning and sequencing, *J. Bacteriol.*, 173: 4371.

Sharp, J. H., 1973, Size classes of organic carbon in seawater, *Limnol. Oceanogr.*, 18: 441.

Sharp, J. H., Underhill, P. A., and Hughes, D. J., 1979, Interaction (allelopathy) between marine diatoms: *Thalassiosira pseudonana* and *Phaeodactylum tricornutum*, *J. Phycol.*, 15: 353.

Sherr, B. F., Sherr, E. B., and Hopkinson, C. S., 1988, Trophic interactions within pelagic microbial communities: Indications of feedback regulation of carbon flow, *Hydrobiologia*, 159: 19.

Sherr, E. B., and Sherr, B.F., 1991, Planktonic microbes: Tiny cells at the base of the ocean's food web, *Trends Ecol. Evol.*, 6: 50.

Sieburth, J. M., Johnson, P. W., and Hargraves, P. E., 1988, Ultrastructure and ecology of *Aureococcus anophagefferens* gen. et sp. nov. (Chrysophyseae): The dominant picoplankter during a bloom in Narragansett Bay, Rhode Island, Summer 1985, *J. Phycol.*, 24: 416.

Simon, M., and Azam, F., 1989, Protein content and protein synthesis rates of planktonic marine bacteria, *Mar. Ecol. Prog. Ser.*, 51: 201.

Sorokin, Y. I., 1971, Bacterial populations as components of oceanic ecosystems, *Mar. Biol.*, 11: 101.

Stockner, J. G., and Antia, N. J., 1986, Algal picoplankton from marine and freshwater ecosystems: A multidisciplinary perspective, *Can. J. Fish. Aquat. Sci.*, 43: 2472.

Stramski, D., and Kiefer, D. A., 1990, Optical properties of marine bacteria, Ocean Optics X. Conf. Proc. Int. Soc. for Optical Engineering, Orlando, FL. 1302: 250.

Strayer, D., 1988, On the limits to secondary production, *Limnol. Oceanogr.*, 33: 1217.

Sugimura, Y., and Suzuki, Y., 1988, A high-temperature catalytic oxidation method for the determination of non-volatile dissolved organic carbon in seawater by direct injection of liquid sample, *Mar. Chem.*, 24: 105.

Suttle, C. A., Fuhrman, J. A., and Capone, D. G., 1990, Rapid ammonium cycling and concentration-dependent partitioning of ammonium and phosphate: implications for carbon transfer in planktonic communities, *Limnol. Oceanogr.*, 36: 424.

Suttle, C. A., Fuhrman, J. A., and Capone, D. G. 1990, Rapid flux and concentration dependent partitioning of ammonium in marine plankton communities, *Limnol. Oceanogr.*, 35: 424.

Suttle, C. A., Chan, A. M., and Cottrell, M. T., 1991, Use of ultrafiltration to isolate viruses from seawater which are pathogens of marine phytoplankton, *Appl. Environ. Microbiol.*, 57: 721.

Suzuki, Y., Sugimura, Y., and Itoh, T., 1985, A catalytic oxidation method for the determination of total nitrogen dissolved in seawater, *Mar. Chem.*, 16: 83.

Toggweiler, J. R., 1988, Is the downward dissolved organic matter flux important in carbon transport? Productivity of the ocean: Present and past (Dahlem Conference), W. H. Berger, V. S. Smetacek, and G. Wefer, New York, John Wiley.

Tupas, L., and Koike, I., 1990, Amino acid and ammonium utilization by heterotrophic marine bacteria grown in enriched seawater, *Limnol. Oceanogr.*, 35:1145.

Vezina, A. F., and Platt, T., 1988, Food web dynamics in the ocean, Part 1. Best estimates of flow networks using inverse methods, *Mar. Ecol. Prog. Ser.*, 42: 269.

Wheeler, P. A., and Kirchman, D. L., 1986, Utilization of inorganic and organic nitrogen by bacteria in marine systems, *Limnol. Oceanogr.*, 31: 998.

Williams, P. J. l., 1981, Incorporation of microheterotrophic processes into the classical paradigm of the planktonic food web, Kieler Meeresforsch., Sonderh. 5: 1.

Williams, P. J. l., 1984, Bacterial production in marine food chains, Emperor's new suit of clothes? Flows in energy and materials in marine ecosystems: Theory and practice, M. J. R. Fasham, ed., Plenum, New York.

Williams, P. J. l., 1990, The importance of losses during microbial growth: Commentary on the physiology, measurement and ecology of the release of dissolved organic material, *Mar. Microb. Food Webs.*, 4: 175.

Williams, P. M., and Druffel, E. R. M., 1988, Dissolved organic matter in the oceans: Comments on a controversy, *Oceanography*, 1: 14.

Wood, A. M., and Van Valen, L. M., 1990, Paradox lost? On release of energy-rich compounds by phytoplankton, *Mar. Microb. Food Webs.*, 4: 103.

REGENERATION OF NUTRIENTS

William G. Harrison

Biological Oceanography Division
Bedford Institute of Oceanography
Box 1006, Dartmouth, Nova Scotia
CANADA B2Y 4A2

INTRODUCTION

In the ten years since this subject was last reviewed (Harrison, 1980), significant progress has been made in understanding remineralization processes in the sea, where they occur, which organisms are responsible, and what role regenerated nutrients play in primary productivity. A proliferation of research on oceanic nutrient cycles in the 80s produced numerous excellent books and review papers on the subject (Morris, 1980; Platt, 1981; Williams, 1981; Fogg, 1982; Azam et al., 1983; Carpenter and Capone, 1983; Ducklow, 1984; Fasham, 1984; Smetacek and Pollehne, 1986; Blackburn and Sørensen, 1988). Most of this literature has emphasized small scale to mesoscale processes within the oceanic euphotic zone and focussed on the role of microbial communities and their components in nutrient regeneration.

Growing interest in problems on the larger scale (regional to global, seasonal to decadal), however, has broadened the scope of research to consider nutrient regeneration in the context of the major biogeochemical cycles in the ocean, including waters and their dissolved and particulate constituents, below as well as within the euphotic zone (Jahnke, 1990). The conceptual model for this work remains the now classical papers by Dugdale and Goering (1967) and Eppley and Peterson (1979) on 'new' and 'regenerated' production and their relationship to export of biogenic matter from the oceanic euphotic zone.

At the conclusion of the 1980 Brookhaven symposium (Falkowski, 1980), several outstanding questions focussing principally on small-scale processes were identified for future study (Harrison, 1980). This review attempts to answer those questions in light of new discoveries over the last decade. With regard to the larger scale problems, it is clear that the exportable fraction of biological production is dependent on the efficiency of nutrient regeneration in the euphotic zone; what is presently known about regeneration efficiency and its variation in space and time are the subjects of the remainder of this paper. Discussion will be largely confined to aspects of the nitrogen cycle because of its central role in the biogeochemistry of the oceans and because much of what we have learned in the past decade about nutrient regeneration has come from nitrogen studies.

Primary Productivity and Biogeochemical Cycles in the Sea
Edited by P.G. Falkowski and A.D. Woodhead, Plenum Press, New York, 1992

The Organisms

In the early 80s, a 'new paradigm' for the planktonic foodweb was proposed, based on the view that microheterotrophs (bacteria and their consumers) accounted for considerably more of the energy transfer and nutrient cycling within the plankton community than the previous concept [phytoplankton-zooplankton-fish] held (Williams, 1981; Azam et al. 1983). This paradigm has been termed the 'microbial loop' and has had a major impact on the direction of research on nutrient regeneration. Interestingly, the earliest views held that nutrient regeneration was solely within the domain of bacteria (Sverdrup et al., 1942) but later shifted to the role of the larger grazers, i.e. metazoans (Harris, 1959). The focus has now effectively come full circle and microheterotrophs are again at the center of interest (Harrison, 1980; Bidigare, 1983; Billen, 1984). Theoretical analysis as well as experimental work over the last 10 years have clearly established the dominant role of the microheterotrophic community in nutrient regeneration. Research has largely concentrated on the heterotrophic bacteria and their predators.

Bacteria. The controversy over the role of bacteria as primary nutrient remineralizers or as algal competitors for inorganic nutrients heightened as more information became available on the physiology and chemical composition of bacteria and the chemical nature of their substrates. Both Williams (1981) and Azam et al. (1983) speculated that bacteria may be inefficient nutrient remineralizers because of their high growth efficiencies, high nutrient demand, and capabilities to effectively utilize nutrients present at low levels. Billen (1984), drawing from earlier studies, suggested that the question of whether bacteria were a 'source' or 'sink' for inorganic nutrients could be determined from simple mass-balance considerations. Nitrogen excretion (E), as an example, should occur if the C:N compositional ratio of the bacteria (C:N_b), corrected for C-losses from respiration (R), exceeds the C:N ratio of the utilized substrate (C:N_s); nitrogen uptake (-E) should occur if the C:N_s > C:N_b:

$$E = [R/(1-GGE)] \times [(1/C:N_s) - (GGE/C:N_b)] \qquad (1)$$

where GGE = gross growth efficiency as carbon (Goldman et al., 1987a; Caron and Goldman, 1988; Goldman and Dennett, 1991). Billen (1984) and Lancelot and Billen (1985) provided some limited supporting experimental evidence; however, Goldman et al. (1987a) and Goldman and Dennett (1991) thoroughly investigated this relationship with cultured, natural marine bacteria and concluded that actively growing bacteria with GGEs typically in the range of ~ 40-60% would be ineffective N-remineralizers when subsisting on substrates with C:N ratios >8-10, a value they considered reasonable for natural organic substrates dissolved in seawater. They determined that recycling efficiencies (NH_4-N reminerilized/substrate-N available) varied from zero when C:N_s was > 10 to over 80% when C:N_s was 1.5; Tezuka (1990) found similar relationships in studies of freshwater bacteria. It has proven difficult, however, to confirm this relationship in the field because of inabilities to reliably characterize the utilizable fraction of the organic matter dissolved in seawater. Another implication of Goldman's work has been that under conditions of elevated substrate C:N ratios, bacteria become effective competitors with phytoplankton for NH_4. The latter circumstance was deduced (Eppley et al., 1977; Laws et al., 1985) or directly demonstrated in the field (Wheeler and Kirchman, 1986; Fuhrman et al., 1988; Kirchman et al., 1989). Focusing on the interactions of amino acids and ammonium in bacterial N-nutrition, Kirchman et al. (1989), found that dissolved free amino-acids (DFAA) are preferentially used over NH_4, and that high DFAA concentrations may actually repress the enzymes mediating NH_4 utilization. They further suggested that only after biosynthetic needs are met does amino acid catabolism occur with concomitant production of NH_4; when DFAA concentrations are low, the NH_4 enzymes are derepressed and NH_4 utilization occurs. On the other hand, Goldman and Dennett (1991) argue that two separate metabolic processes may be operating simultaneously, one involving the uptake and catabolism of amino acids, and the other the uptake of NH_4 and a carbon substrate. Based on their findings, NH_4 uptake is tightly coupled with the availability of a suitable carbon substrate, even in the presence of amino acids.

The view that substrate C:N ratio largely determines whether bacteria excrete or utilize NH_4 has not been universally accepted (Tupas and Kioke, 1990, 1991). Using ^{15}N isotope dilution techniques, Tupas and Koike (1991) showed that natural assemblages of heterotrophic bacteria simultaneously assimilated and produced NH_4. They suggested that metabolic requirements, rather than substrate compositional ratios, govern the utilization of C and N-compounds. In their view, amino acids are assimilated primarily for energy and C-skeletons; nitrogen is largely excreted but subsequently reassimilated as NH_4. However, their experimental results do not appear to contradict the model proposed by Goldman and Dennett (1991) in which substrate compositional ratio plays an important role in the net uptake or release of NH_4.

Micro-grazers. The question of whether bacteria or their grazers are the major nutrient remineralizers in the ocean has been the subject of much debate over the last decade, dating back to the classic work of Johannes (1964, 1965) on protozoan nutrient excretion. Probably no other area of research regarding biological controls on nutrient regeneration has been more active. Much of this work has dealt with the distribution, abundance, growth, and excretion of the heterotrophic bactivorous and herbivorous protozoans and has been reviewed extensively (Stout, 1980; Taylor, 1982; Sherr and Sherr, 1984; Fenchel, 1988; Caron and Goldman, 1990; Caron, 1991; Berman, 1991). Early evidence suggesting that protozoans may be the dominant remineralizers was largely circumstantial and based on their high abundance in coastal and oceanic waters, the fact that they are major consumers of bacteria and autotrophic picoplankton, their high biomass-specific metabolism, and field observations that often associated high nutrient regeneration rates with the particle size class in which they fall (Caron and Goldman, 1990). More direct evidence has come from laboratory experiments using cultured microflagellates and ciliates (e.g., Sherr et al., 1983; Gast and Horstmann, 1983; Verity, 1985; Goldman and Caron, 1985; Goldman et al., 1985; Caron et al., 1985; Anderson et al., 1986, Berman et al., 1987; Goldman et al., 1987b; Caron et al., 1988). From these laboratory studies, it has become clear that protozoan biomass-specific nutrient excretion rates are higher, often by an order of magnitude or more, than rates for most metazoans. Goldman, Caron and co-workers have carried out a particularly thorough study of the physiology of nitrogen and phosphorus excretion by the omnivorous microflagellate, *Paraphysomonas imperforata* (Lucas). Basing their experimental approach on the model that nutrient excretion is determined by the nutritive quality of the prey and the nutrient demand/growth efficiency of the predator (see eq. 1 above), they showed that nutrient regeneration efficiencies varied in a predictable way from essentially zero to as high as 70%. Nitrogen regeneration efficiencies, for example, were highest when bacteria served as prey because of their lower C:N compositional ratio compared to that of the algae, i.e. more 'excess' nitrogen was available relative to the compositional requirements of the predator. Similarly, excretion was higher when the predator was fed on nutrient-replete algae than when fed nitrogen-limited prey; the predator clearly conserved the limiting nutrient. Regeneration efficiencies were found to vary not only with the chemical composition and physiological condition of the prey but also with the growth stage of the predator; efficiencies were lower during active growth (15-25%) than during the later stationary stage of growth (>50%). Manipulations of bacteria and microflagellates together in nitrogen-limited algae cultures (Caron et al., 1988) clearly demonstrated that bacteria effectively competed for inorganic nitrogen when their substrate C:N ratio was elevated (by the addition of glucose); however, the introduction of a bactivore resulted in lowered bacterial numbers as a result of grazing, NH_4 regeneration by the protozoan, and the subsequent elimination of nutrient-limitation of algal growth. In parallel experiments where the bacterial substrate C:N was lowered (by the addition of glycine), the bacteria became the major N-remineralizers.

Although these experiments represent very simplified predator-prey systems, their ecological implications are, nonetheless, significant. Firstly, its is clear from these studies that both bacteria and microheterotrophic grazers can be important nutrient remineralizers. Currently, however, mass balance arguments and our present limited knowledge of the chemical composition of predators, prey, and dissolved substrates suggest that micro-grazers are likely the more important remineralizers in nature. Secondly, despite the emphasis on micro-grazers as important nutrient remineralizers, they are still effective conservers of the growth-limiting nutrient, particularly during active growth; regeneration

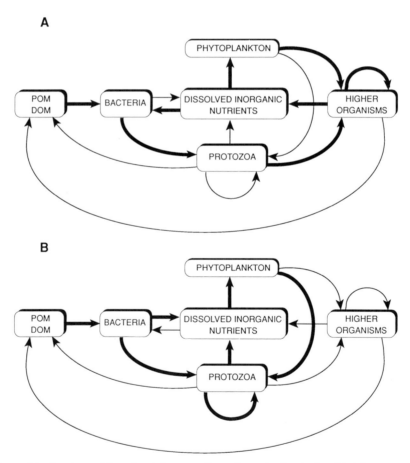

Fig. 1. Idealized models of nutrient cycling through the 'microbial loop'. Arrow thickness indicates relative importance of pathway for nutrient flow; A. Nutrient *link* - characterized by high efficiency of transfer of nutrient to higher organisms. B. Nutrient *sink* - characterized by high efficiency of nutrient remineralization (redrawn from Caron, 1991).

efficiencies rarely exceed 50%. Thirdly, with regeneration efficiencies no greater than 50%, a relatively complex, multi-trophic microbial food web is necessary to account for the >90% regeneration efficiencies thought to characterize the open ocean (see further discussion below). Fourthly, the debate about the microbial food web as a nutrient 'link' or 'sink' (Fig. 1) would appear oversimplified because it is clear that both processes are operating simultaneously; it is simply a matter of degree (Caron, 1991).

An interesting alternative cause of mortality of marine bacteria (and cyanobacteria) and potential mechanism for regenerating nutrients has been linked with the incidence of viral infection of bacteria (Proctor and Fuhrman, 1990). Observations in a variety of coastal and oceanic habitats showed a surprisingly high incidence of viral infections of natural heterotrophic bacteria. This opens the possibility of a significant nutrient source when the infected cells eventually lyse. Principal products of bacterial lysis (i.e. high molecular weight, low C:N ratio compounds such as nucleic acids and proteins), interestingly enough, seem to fit the chemical profile of the newly described class of dissolved organic compounds in surface seawater (Suzuki et al., 1985; Sugimura and Suzuki, 1985).

Metazoans. The general view that metazoan nutrient regeneration is less important than that of the microheterotrophs has not changed substantially in the last decade (Harrison, 1980; Bidigare, 1983; Longhurst and Harrison, 1989; Caron and Goldman, 1990). Recent experimental work, however, suggests that vertically migrating species, by virtue of their diel movements and metabolic patterns, may transport a significant fraction of their regenerated nitrogen (NH_4) below the euphotic zone and out of access of the primary producers (Longhurst and Harrison, 1988; Longhurst et al., 1989). Estimates from data in the literature on migrant populations from various ocean regions suggested that this 'active' vertical N-flux may be comparable to the 'passive' particulate-N flux measured by sediment traps in some situations (Longhurst and Harrison, 1988); passive particle fluxes are thought to represent 10-20% of the phytoplankton nitrogen productivity at the sea surface (Eppley and Peterson, 1979; Knauer et al., 1990; Lohrenz et al., 1991). The significance of sedimenting particles to nutrient regeneration will be discussed in more detail below.

Aggregates. The often observed high metabolic activity in oceanic microautrophic and heterotrophic communities, in what appears to be a nutritionally dilute environment, has long puzzled researchers. McCarthy and Goldman (1979) were the first to provide an explanation for the high phytoplankton growth rates in nutrient-impoverished oceanic waters by hypothesizing that the primary mode of nutrient supply to phytoplankton may be in the form of infrequent, short-lived but concentrated 'patches' of regenerated nutrients excreted by macro-grazers. Lehman and Scavia (1982) later provided empirical evidence in support of the nutrient patchiness idea; however, Jackson (1980) and Williams and Muir (1981) argued that such patches would dissipate too quickly by molecular diffusion and turbulence to be of benefit to phytoplankton. Problems of consumer-grazer densities and encounter probabilities also were recognized (Currie, 1984). Researchers have therefore sought a more plausible explanation for the existence of 'microzones' of materials and biological activity.

Flocculant macroscopic (0.05-10cm) particles, i.e. 'marine snow' (Alldredge and Silver, 1988) have been suggested as one likely source of these microzones of activity. With regard to their importance in nutrient regeneration, macroaggregates are known to be enriched, often by orders of magnitude above adjacent seawater levels, in organic matter (Riley, 1970), microheterotroph numbers, both bacteria and micrograzers (Caron et al., 1982), inorganic nutrients (Shanks and Trent, 1979) and metabolic activity

(Alldredge and Cohen, 1987); they are widely distributed in coastal and oceanic waters. In coastal waters, there is evidence that nutrient regeneration within mass flocculations of declining diatom blooms is important in sustaining productivity above levels in the adjacent waters (Gotschalk and Alldredge, 1989). The conclusion that regenerated nutrients fuelled the aggregate productivity was based on the co-occurence of maximum productivities and the appearance of regenerated nitrogen (NH_4) in the flocs and by its preferential utilization. These findings suggest an alternative to conventional adaptive strategies of algal aggregation (see, Smetacek, 1985), i.e. a possible survival mechanism for diatoms in the euphotic zone under post-bloom nutrient-limited conditions.

Goldman (1984a,b) considered another important but smaller size-class of aggregates, i.e. the 'microaggregates', in the micron size range, composed of similar but somewhat less complex interacting communities of autotrophs and heterotrophs (Fig. 2). He proposed that these floating 'oases' of interacting organisms in close physical proximity could help explain the high nutrient regeneration efficiencies attributed to a particle size class much smaller than represented by macroaggregates (see Field Studies below). Also significant was Goldman's observation that microaggregates would have a significantly longer residence time in the surface ocean by virtue of their size and density, and, therefore, have a potentially greater impact on nutrient regeneration and other biochemical

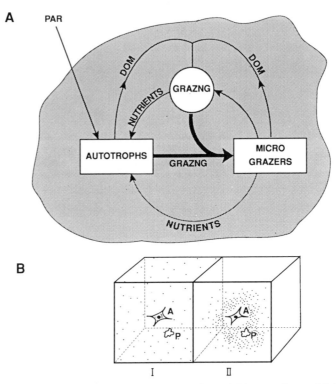

Fig. 2.   Two conceptual models of small-scale microbial interactions which are thought to facilitate nutrient remineralization. A. Micro-aggregates ($<5\mu m$ in size) of organic matter (redrawn from Goldman, 1984a). B. Bacterial 'clustering' around biogenic particles, I=random distribution, II=clustering of bacteria around algal cell (A) or other organic particle (P) (redrawn from Azam and Ammerman, 1984).

processes than the large aggregates. Presently, limitations in sampling and measurement methodologies preclude testing of the 'aggregate-spinning wheel' hypothesis although some progress is being made in small-scale analysis of the larger aggregates (Alldredge and Cohen, 1987).

An analogous but less complex association has been proposed for phytoplankton-bacteria interactions, the so-called 'cluster hypothesis' (Azam and Ammerman, 1984). Here, the bacterial microenvironment is structured with respect to its substrate source, in this case dissolved organic matter (DOM) released as algal exudates, in that the bacteria use motility and chemotaxis to optimize their position in the substrate field, resulting in a non-uniform 'clustered' distribution on the scale of a few to 100 μm around the source (Fig. 2). Any sustained DOM source can presumably serve as a focal point for clustering behavior; however, the phytoplankton-bacteria association was viewed by the authors as particularly important since it may represent a co-evolved commensal relationship, where the algae provide structural components and energy for the bacteria, and the bacteria, through mineralization of DOM, provide the algae with essential inorganic nutrients. Moreover, this hypothesis provides yet another plausible mechanism for the maintenance of microzones of high nutrient concentrations to support rapid phytoplankton growth, as suggested by McCarthy and Goldman (1979). Fuhrman (this volume), however, notes that algal exudates are generally deficient in essential nutrients (i.e. nitrogen, phosphorus) and may not be the principal nutrient source for algae that Azam and Ammerman envisaged. As in the case of microaggregates, bacterial clustering cannot be experimentally verified at present although elements of this hypothesis have been explored theoretically (Jackson, 1987).

Kepkay and Johnson (1989) described another form of aggregation with potential importance for nutrient regeneration in surface waters. These researchers have shown in a variety of ocean regions that coagulation of colloidal dissolved organic matter on bubble surfaces initiates a rapid elevation of microbial respiration, up to an order of magnitude above background levels, which may be an important mechanism for recycling a globally significant carbon and nutrient reservoir.

Field Studies

Quantitative field studies of the cycling of nitrogen gained considerable momentum after the introduction of stable isotope tracer methods in the early 60s (Harrison, 1983b). However, regenerative fluxes were inferred rather than measured (Dugdale and Goering, 1967) and not until the introduction of the 'isotope-dilution' method in the late 70s (Harrison, 1978; Caperon et al., 1979) and its later refinement (Glibert et al., 1982b) were direct measurements possible. Regeneration measurements have concentrated on the principal product of organic-N catabolism, NH_4, but methods for urea and amino acids have recently been described (Fuhrman, 1987; Hansell and Goering, 1989; Slawyk et al., 1990) as well as for other nutrients (Harrison, 1983a). One of the early conclusions from the NH_4 regeneration studies was that uptake and regeneration processes were in approximate balance much of the time (Harrison, 1978), a point which has been confirmed in numerous subsequent studies (Glibert, 1982; Harrison et al. 1983; Cochlan, 1986; Owens et al. 1986; Probyn, 1987; Hanson et al., 1990) and lends support to Dugdale and Goering's (1967) earlier conceptual model. There are situations, however, where the short-term assimilation and regeneration fluxes do not balance (e.g. Paasche, 1988). The application of 'regenerated production' measurements to infer NH_4 remineralization will be discussed further below.

The isotope-dilution method provided the first direct measurements of NH_4 regeneration by natural microplankton assemblages (Harrison, 1978). Apparent from this and later studies was that NH_4 regeneration by microplankton usually exceeded the contribution from other sources, e.g. metazoans and benthos (e.g. Harrison, 1978; Harrison et al., 1983; Hopkinson et al., 1987), supporting similar conclusions drawn from laboratory studies and allied field investigations as discussed previously. The introduction of particle size-fractionation techniques provided a means of better assessing the relative contribution of various functional groups within the microplankton community, e.g. for discriminating bacteria from the larger heterotrophs (Azam and Hodson, 1977). Published studies have employed a variety of screens sizes but results have been relatively consistent in showing the prevalence of the nanoplanktonic ($<20\mu m$) and smaller forms in NH_4 regeneration (Table 1). Contrary to the evidence already discussed that micro-grazers are the principal N-remineralizers, many of the ^{15}N field studies have shown significant NH_4 regeneration (as much as 50% or greater) associated with particles <1 μm in size, and presumably, heterotrophic bacteria. (e.g. Glibert, 1982; Harrison et al., 1983, Hopkinson et al., 1987; Probyn, 1987; Tupas and Kioke, 1991). Spatially, the importance of the bacterial size fraction in nitrogen regeneration has been shown to increase with depth in the euphotic zone (Probyn, 1987; Hanson et al., 1990) and temporally, to increase in association with the decline of blooms (Harrison, 1978; Glibert, 1982). As mentioned previously, field studies have also established that bacteria can be efficient NH_4 consumers as well as producers (Wheeler and Kirchman, 1986; Kirchman et al., 1989; Tupas and Koike, 1991; Chisholm, this volume).

With regard to organic aggregates, at least one study has directly measured NH_4 regeneration in marine snow, confirming earlier suggestions that they are sites of rapid nutrient recycling (Glibert et al. 1988). In four samples from the Gulf stream, Glibert and co-workers observed high but variable NH_4 concentrations (>1 $\mu molN$ l^{-1}) and regeneration rates (1 to 8 $\mu molN$ l^{-1} h^{-1}) orders of magnitude greater than rates measured in surrounding waters (<0.001 $\mu molN$ l^{-1} h^{-1}). Nitrate was absent and uptake rates correspondingly low, as has been shown for coastal phytodetrital aggregates (Gotschalk and Alldredge, 1989).

The above mentioned links between organisms, their interactions, and nutrient regeneration have helped to explain some of the larger scale patterns commonly observed in plankton communities, e.g. the well-defined vertical layering of primary producers and their grazers (Longhurst and Harrison, 1989). Zonal (confined largely to the upper euphotic zone) nutrient regeneration (e.g. King, 1984), for example, may contribute to the often observed vertical displacement of the primary productivity maximum above the algal biomass maximum (Harrison, 1990) and supports the idea of a distinctive 'two-layer' euphotic zone (Dugdale and Goering, 1967). Goldman (1988) offers an interesting contemporary perspective of the two-layer euphotic zone as it relates to nutrient cycling and microbial growth.

PROCESSES ON THE LARGE SCALE

Laboratory and field studies in the last 10 years have provided a wealth of new information on the magnitude, variation, and sources of nutrient regeneration from the small to the mesoscale (organismal to community level with characteristic times-scales of minutes to days). On the larger scale (regional to global, seasonal to interannual), however, progress has been much slower. Two basic approaches have been pursued: *in vitro* incubation experiments and bulk property analysis, each with characteristically different time-and space-scales (Eppley, 1989; Platt et al., 1989; this volume).

Table 1. Size fractionation studies of NH$_4$ regeneration. Numbers are percent (ranges in parentheses) of *total* microplankton regeneration associated with organisms smaller than the stated screen size.

Location	<1μm	<3μm	<5μm	<10μm	<15μm	<35μm	<45μm	<50μm	Reference
INSHORE									
Chesapeake Bay	20 (15-40)				58 (30-85)		90		Glibert, 1982
Olso Fjord							8 (0-37)		Paasche and Kristiansen, 1982
Kaneohe Bay						77			Caperon et al., 1979
COASTAL									
Mid Atlantic Bight	74 (60-95)								Harrison et al., 1983
South Atlantic Bight	28 (20-35)			32 (29-35)				40 (34-45)	Hanson and Robertson, 1988
S. California Bight	39					89			Harrison, 1978
Benguella Upwelling	42 (11-83)	58 (31-83)			94 (85-100)				Probyn, 1987
OCEANIC									
NW Atlantic			67 (50-100)	70 (10-97)		82 (10-100)			Glibert, 1982; Glibert et al., 1988
Great Barrier Reef					66 (40-88)				Hopkinson et al., 1987

Incubation Experiments. Large-scale studies of nutrient regeneration based on isotope tracer incubations have been limited to a few seasonal studies, largely confined to coastal waters (e.g. Glibert, 1982; Harrison, 1983a), and relatively broad but sparse geographical coverage encompassing coastal to oligotrophic oceanic waters (see references cited above). Despite the fact that the methodologies for direct measurement of nutrient regeneration have been available for more than a decade, the tedious nature of these measurements has meant that no more than a handful of scientists worldwide routinely make these measurements today. As a consequence, the global database has grown little since the methodology was introduced.

Much of what has been learned about the large scale features of nitrogen regeneration in the surface ocean has come indirectly from the more extensive tracer studies of NH_4 assimilation, based on the presumption that NH_4 assimilation reflects regenerative fluxes (Dugdale and Goering, 1967). Eppley and Peterson (1979) and Eppley (1981) extended Dugdale and Goering's concept of 'new' (NO_3-based) and 'regenerated' (NH_4-based) production to consider nitrogen uptake and recycling on a global basis. From an empirical relationship between primary productivity and new:regenerated production ratios, they suggested that recycling efficiencies (regenerated production/total production) in the euphotic zone can range from ~50% in coastal waters to >90% in the oligotrophic

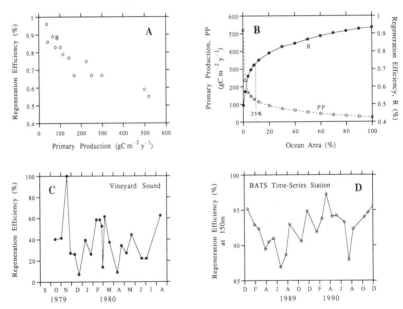

Fig. 3.   Variability of nutrient regeneration efficiency in the euphotic zone.   A. Relationship between regeneration efficiency (based on ^{15}N data) and primary production (Eppley, 1981).   B. Variation in primary production and nutrient regeneration efficiency (based on sediment trap data) by ocean surface area; vertical line shows that 25% of the global primary production occurs in areas representing only 10% of the ocean surface area, however, for 90% of the ocean area, >75% of the primary production is remineralized in the euphotic zone (Berger et al., 1989).   C. Seasonal variability in regeneration efficiency (based on ^{15}N data) in coastal waters (Glibert et al., 1982).   D. Seasonal variability in regeneration efficiency (based on sediment trap data) in oceanic waters, note expanded vertical scale (Lohrenz et al., 1991).

open ocean (Fig. 3) and are ~80% globally (Table 2). Subsequent tracer studies have generally confirmed these patterns although the database is still distressingly small, particularly for the open ocean.

Sediment Traps. Sediment traps deployed at the base of the euphotic zone have provided a much larger body of data from which nutrient regeneration in the upper ocean can be inferred. Under the assumption that sedimenting particulates represent the principal pathway of organic matter loss from the surface ocean (but, see discussion below), the export flux of particulates is a reflection of the inefficiency of nutrient remineralization in the euphotic zone and thus, can be used to scale regeneration efficiencies. The flux of particulate organic matter caught in traps is, indeed, an index of the coupling between the surface and the ocean interior (Platt et al., this volume). Trap data has provided a much clearer picture of spatial and temporal variations in nutrient regeneration (Fig. 3). Regional patterns and magnitudes of regeneration efficiency from traps have been similar to estimates from tracer incubations, i.e. efficiencies increase (the fractional sinking fluxes decrease) from inshore to offshore and with primary production level (e.g. Knauer et al., 1984, 1990). Wassmann (1990), however, has pointed out that trap fluxes may be higher (and regeneration efficiencies lower) at high latitudes than expected from the primary production level, due to the much shorter growth season compared with temperate/tropical climes (see also, Harrison and Cota, 1991). Global estimates of nutrient regeneration effiency (Berger et al., 1987; Martin et al., 1987) are consistent with Eppley and Peterson's tracer-based value of ~80% (Table 2).

Chemical Balance Models. Nutrient remineralization rates in the euphotic zone can also be derived from calculations of the annual oxygen balance and stoichiometric equivalencies (Musgrave et al. 1988; Spitzer and Jenkins, 1989). Efficiency estimates derived by this method have been significantly lower than tracer or trap estimates (Table 2). Similarly, lower efficiencies were obtained from calculations based on NO_3 flux (Jenkins, 1988a; but see Lewis et al., 1986).

Nutrient Recycling Below the Euphotic Zone

Measurements of oxygen consumption in the deep ocean have been useful in deducing regeneration efficiencies in the euphotic zone. Aphotic-zone estimates of nutrient regeneration in the deep ocean from oxygen consumption are consistent with chemical balance estimates from shallow waters and, again, yield regeneration efficiencies lower than those from incubations or traps (Table 2). For example, Packard et al. (1988) estimate from electron transport system (ETS) enzyme activities in the deep ocean that as little as 12-67% of the globally averaged primary production is remineralized in the euphotic zone; compared with the value of 80-85% based on sediment trap fluxes (Eppley and Peterson, 1979; Martin et al., 1987; Berger et al., 1987, 1989).

Various explanations have been offered for the apparent discrepancy between the (high) euphotic zone regeneration efficiencies derived from incubation and sediment trap fluxes and the (low) estimates based on chemical balances. Altabet (1989), for example, suggested that local downward mixing of suspended particulate matter, a source of particulate loss not previously considered, could significantly increase the organic flux and decrease regeneration efficiency in the euphotic zone (see Table 2). Downward flux of dissolved organic matter has also been considered an important but previously ignored local export term which would result in lowered euphotic zone regeneration efficiencies (Suzuki et al., 1985; Sugimura and Suzuki, 1988; Toggweiler, 1989). The implications of their DOM findings, however, are not without controversy (Jackson, 1988). Longhurst and Harrison (1988) and Longhurst et al. (1989) speculate that a small but significant

Table 2. Regional/global estimates of nutrient regeneration efficiencies (R) in the oceanic euphotic zone.

Method	Region	Depth (m)	Prim. Prod. (gC m^{-2} y^{-1})	R (%)	Reference
EUPHOTIC ZONE ESTIMATES					
[15]N:Prim.Prod.	Open Ocean	<200	25	94	Eppley and Peterson, 1979
	Transitional	<200	51	87	
	Divergence/Subpolar	<200	73	82	
	Inshore	<200	124	70	
	Neritic	<200	365	54	
	Global	<200	**19-24Gt**	**80-82**	Platt and Harrison, 1985
	Open Ocean (Sargasso Sea)	<100	82	69	Lewis et al., 1986
	Open Ocean (Sargasso Sea)	<100	125[1]	81	
Sediment Traps	Open Ocean (N. Central Pacific)	<150	120	83-94[2]	Knauer et al., 1990
	Open Ocean	100	130	86	Martin et al., 1987
	Coastal Zone	100	250	83	
	Upwelling	100	420	80	
	Global	100	**51Gt**[1]	**85**	Altabet, 1989
	Open Ocean (Sargasso Sea)	100	125[1]	85	
		100	125[1]	75[3]	
	Open Ocean (Sargasso Sea)	150	125	86-97[2]	Lohrenz et al., 1991
	Open Ocean (N. Central Pacific)	150	120	84-96[2]	Knauer et al., 1990
	Global	100	**27Gt**[1]	**80**	Berger et al., 1987,1989
O$_2$ Production Model	Open Ocean (Sargasso Sea)	<100	125[1]	69-77	Musgrave et al., 1988
	"	<100	125[1]	57-66	Spitzer and Jenkins, 1989
Turbulent NO$_3$ Flux	"	<100	125[1]	97	Lewis et al., 1986
NO$_3$ Flux Model	"	Mixed Layer	125[1]	64	Jenkins, 1988a
SUB-EUPHOTIC ZONE ESTIMATES					
Electron Transport System	**Global**	200-Bottom	**19-51Gt**[4]	**12-67**	Packard et al., 1988
[3]H/[3]He Inventory	Open Ocean (Sargasso Sea)	100-1000	125[1]	60	Jenkins, 1982
O$_2$ Consumption Model	"	100-400	125[1]	54-68	Jenkins and Goldman,1985
[3]H/[3]He Inventory	"	100-750	125[1]	73-81	Sarmiento et al., 1990
Radium-228 Inventory	"	100-750	125[1]	34	

[1]Annual primary production off Bermuda (BATS Station) from Lohrenz et al. (1991). [2]Range of values from seasonal study. [3]Includes downward mixing of suspended particulate organic matter. [4]Low value from Eppley and Peterson (1979), high value from Martin et al. (1987).

transport of nutrients out of the euphotic zone may be mediated by the migration of diel zooplankton, a flux not included in sediment trap estimates.

Martin et al. (1987) speculated that the 'missing' organic production in the subtropical Atlantic may be supplied from outside the region by transport along isopycnal surfaces; this would result in an apparent decrease in regeneration efficiency by elevating the local euphotic zone lost term. Sarmiento et al. (1990b) later estimated that lateral import of organic matter may account for almost half of the nitrogen remineralized in the thermocline of the region. Similarly, Jenkins (1988b) suggested that a significant fraction of the annual NO_3 flux supporting primary production is not provided locally by vertical mixing but is supplied by horizontal transport along isopycnal surfaces.

Others (Jenkins and Goldman, 1985; Goldman, 1988; Jenkins, 1988a) explain the discrepancy between incubation/trap-derived regeneration efficiencies and those based on chemical balance models by suggesting that much (maybe most) of the annual production in the euphotic zone is fuelled by episodic mixing events generally missed by conventional oceanographic sampling methods (i.e. poorly resolved in space and time). Questions remain, however, as to the exact mechanisms involved (Bigg et al., 1989) and why sediment traps measurements in particular should miss these transient events since they are by nature integrative (Altabet, 1989; Lohrenz et al., 1991). Evidence is accumulating, nonetheless, that these episodic events occur and influence primary production (Glover et al., 1988). However, fundamental differences in the time and space scale characteristics of the various approaches summarized in Table 2, may be the most important consideration in attempts to resolve these discrepancies (Platt et al., 1989; Platt et al., this volume). No single explanation has been adequate thus far.

From an ecological standpoint, high regeneration efficiencies within the euphotic zone, inferred from incubation and trap measurements, have important implications with regard to the community structure and complexity of the remineralizer populations. Based on experimental evidence, Goldman et al. (1985), and from a theoretical perspective, King (1987), regeneration efficiencies on the order of 80-90% are not possible unless extremely complex, multi-trophic level microbial communities are present. Although often considered characteristic of the open ocean (e.g. Sorokin, 1981), such a complex community structure is clearly counter to the concept of the 'microbial loop' as a significant link for nutrient transfer up the food chain (Williams, 1981; Azam et al., 1983; Caron, 1991). Goldman (1988) offered a relatively simple but provocative explanation for the maintenance of both high regeneration efficiencies and high organic export based on the concept of a two-layered euphotic zone (Dugdale and Goering, 1967). In his model, the upper euphotic zone is characterized by relatively constant and high regeneration efficiencies, approaching 100%, with 'regenerated' nutrients supporting high growth rates of small plankton. The lower euphotic zone, in contrast, is essentially isolated from the upper layer and characterized by infrequent, episodic injections of 'new' nutrients which support the growth of larger but more transient plankton populations. These 'ephemeral eutrophication zones' presumably account for the high export fluxes which elude conventional oceanographic sampling methods but are detected by the integrative bulk property measurements.

Below the euphotic zone, nutrient regeneration rates decrease with depth (Fig. 4). Martin et al. (1987) estimated from trap data that ~90% of the particulate matter leaving the euphotic zone is remineralized in the upper 1500m in the Pacific; Berger et al. (1987; 1989) came to similar conclusions from analysis of a much larger data set including other ocean regions. Moreover, particulate nitrogen is selectively lost (relative to carbon) as indicated by the progressively increasing regeneration C:N molar ratios (Fig 4). The decomposition processes which contribute to the observed decrease in particle flux with depth have been the subject of much study (Silver and Gowing, 1991). Recent evidence does not conform to the commonly held view of microbial colonization and direct remineralization of the large sinking particles but suggests that the particles are first fragmented and solubilized at depth (either abiotically or by microbes), leaving residual

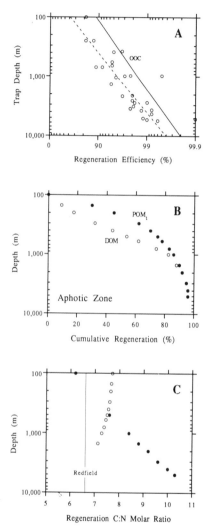

Fig. 4.   Nutrient regeneration in the deep ocean. A) Nutrient regeneration efficiency (based on sediment trap data) with depth as a percentage of surface primary production, open circles and dashed line are data and model fit compiled by Suess (1980), solid line represents model fit of composite data sets from the Pacific, open ocean composite (OOC), compiled by Martin et al. (1987). B) Cumulative nutrient regeneration (%) of sedimenting particulate matter (POM$_t$) and dissolved organic matter (DOM) below the euphotic zone (the aphotic zone). POM$_t$ regeneration based on OOC model (Martin et al., 1987), DOM regeneration approximated from depth dependent changes in DOM concentrations from Pacific data of Suzuki et al. (1985); data points represent best fit of all profile data combined. C) C:N molar ratios of organic matter remineralization; symbols and data sources as in (B).

pools of fine particulates and utilizable dissolved organic matter accessible to the primary remineralizers, thought to be mesopelagic free-living bacteria (Karl et al., 1988; Cho and Azam, 1988; but see Banse, 1990).

Remineralization of dissolved organic carbon and nitrogen also decreases with depth although the gradient below the euphotic zone (100-1000m) is not apparently as steep as that for particulates (Fig. 4). Because dissolved organic matter concentrations are much higher than particulate concentrations (Suzuki et al., 1985; Sugimura and Suzuki, 1988), however, dissolved organics may play a more important role in nutrient regeneration. Toggweiler (1989), for example, estimated that DON remineralization may account for as much as 80% of the NO_3 produced below the thermocline in the Pacific if the Suzuki and Sugimura data are correct. In contrast to the compositional changes in particulate matter, C:N regeneration ratios of DOM appear to increase slightly with depth. The changing chemical nature of the DOM with depth (Suzuki et al., 1985; Jackson, 1988; Toggweiler, 1989) and the degree to which it influences or is influenced by interactions with meso-pelagic biota (Cho and Azam, 1988; Banse, 1990) is an area requiring additional research.

Little of the surface primary production reaches the seafloor of the deep ocean; Berger et al. (1987) estimate between 1-10% depending on depth. Martin et al. (1987) found <4% of the particulate organic matter flux from the euphotic zone reaching a depth of 5000m (see Fig. 4). Walsh et al. (1988), concentrating on particle fluxes below the mesopelagic zone, estimated that 25-50% of the organic carbon flux at the often observed mid-water maximum (~2000m) was further remineralized before reaching the seafloor. Once on the seafloor, microbial colonization appears to be rapid and decomposition occurs at a rate as high as 1-2% day^{-1} (Lochte and Turley, 1988). The seafloor may in fact be the primary site of bathypelagic nutrient remineralization (Jahnke and Jackson, 1987). The dynamics of DOM in the bathypelagic zone are poorly understood although it is thought to be chemically modified compared to near surface DOM (Suzuki et al., 1985; Sugimura and Suzuki, 1988); its specific role in deep ocean nutrient remineralization, however, awaits further research.

Ecosystem Models

Modelling efforts aimed at describing the energetics and nutrient cycling of the microbial foodweb have advanced considerably over the past 10 years, from simplified bacteria-algae interactions (Bratbak and Thingstad, 1985) to consideration of complete microbial systems (e.g. Ducklow and Taylor, 1991; Moloney and Field, 1991). Models have the capability of 'bridging the gap' between the small scale and large scale processes and are an essential tool for understanding nutrient regeneration in the context of the major biogeochemical cycles (SCOR, 1990).

Few upper ocean ecosystem models in use before the 1980s incorporated terms explicitly for nutrient regeneration (Harrison, 1980); however, laboratory and field studies over the last decade have clarified many of the poorly understood processes linked with the regeneration of nutrients. To what extent therefore do contemporary modelling efforts reflect this new information?

Recently developed numerical and analytical simulation models have incorporated terms for, and have attempted to differentiate between the contributions of metazoans and microbes to the regeneration of nitrogen in studies of seasonal succession in pelagic plankton communities (e.g. Moloney et al., 1986; Frost, 1987; Newell et al., 1988). Other numerical simulation models (Vezina and Platt, 1987), and models based on network analysis (Ducklow et al., 1989), have moved a step further by incorporating terms for bacteria utilization as well as regeneration of inorganic nitrogen. Fasham et al. (1990) have developed perhaps the most complete nitrogen-based ecosystem model of plankton dynamics in the oceanic mixed layer to date. Efforts to incorporate this biological submodel into a coupled atmosphere-ocean general circulation model (GCM) are now underway (Sarmiento et al., 1990a).

The level of sophistication of these new generation GCMs will permit a clearer visualization of the impacts of non-steady state, episodic, 'new' nitrogen inputs on ecosystem structure and function, the significance of which is being actively debated (Jenkins and Goldman, 1985; Lewis et al., 1986; Goldman, 1988; Jenkins, 1988a; Bigg et al., 1989; Sarmiento et al., 1990b). The complementary non-steady state flux of regenerated nutrients arising from inhomogeneities in plankton population distributions (Goldman, 1984a,b) or in their metabolic activities (McCarthy and Goldman, 1979) at present are difficult to represent in mathematical terms. As a result, our capabilities to model nutrient regeneration at that level of complexity still seem far beyond the horizon. In that regard, we seem to have made very little progress (Harrison, 1980).

WHERE DO WE GO FROM HERE?

Important questions remain concerning nutrient regeneration on the small to mesoscale. Continued research is needed, in general, to more clearly establish the role of the 'microbial loop' in nutrient regeneration. Specifically, when and under what circumstances are bacteria principally producers or consumers of inorganic nutrients? To what extent can their role as nutrient 'source' or 'sink' be determined from the chemical nature of their substrates? More laboratory studies are needed on bacterial nutritional physiology; in the field, a better characterization of bacterial substrates is essential. More information is also required on bacterial (and algal) predators and their role in nutrient regeneration. Research efforts should be focussed on micro-grazer growth, food selectivity and excretion physiology, both in the laboratory and in the field. More controlled, multi-species (bacteria-bactivore-algae) studies will be required to elucidate the important interactions. As well, methods development is needed to better detect, characterize and evaluate the role of macro- and micro-aggregation of particles in nutrient regeneration. At present, information is largely lacking on the temporal and spatial distributions of all the above organisms and processes; more field work is clearly needed. Improvements in field methodologies need serious attention.

On the regional to global scale, large discrepancies exist among the various estimates of nutrient regeneration efficiencies of the oceanic euphotic zone. Reconciliation of these differences will require a better understanding of what each of the methods actually measures. To what extent can these discrepancies be attributed to fundamental differences in the time/space scales represented by the estimates? For example, more highly resolved (spatial and temporal) sampling will be necessary to bring the 'instantaneous' incubation measurements more in line with the 'integrative' chemical balance measurements. New data on dissolved organic matter and its distribution in the upper ocean challenge the view that sedimenting particles are the principle means by which organic matter is exported to the deep ocean. Nutrient regeneration processes in the meso- and bathypelagic zones are even less well defined. What are the relative roles of particles and dissolved organic matter in deep ocean nutrient remineralization? Which organisms are more important, bacteria or bactivores? How quickly and to what extent are organically-bound nutrients remineralized on the sea floor? Clearly, the next decade has many challenging questions to answer.

400

REFERENCES

Alldredge, A. L. and Cohen, Y., 1987, Can microscale chemical patches persist in the sea? Microelectrode study of marine snow, fecal pellets, *Science*, 235: 689.

Alldredge, A. L. and Silver, M. W., 1988, Characteristics, dynamics and significance of marine snow, *Prog. Oceanogr.*, 20: 41.

Altabet, M. A., 1989, Particulate new nitrogen fluxes in the Sargasso Sea, *J. Geophys. Res.*, 94: 12,771.

Anderson, O. K., Goldman, J. C., Caron, D. A., and Dennett, M. R., 1986, Nutrient cycling in a microflagellate food chain. III. Phosphorus dynamics, *Mar. Ecol. Prog. Ser.*, 31, 47.

Azam, F. and Ammerman, J. W., 1984, Cycling of organic matter by bacterioplankton in pelagic marine ecosystems: microenvironmental considerations, *in:* "Flows of Energy and Materials in Marine Ecosystems," M.J.R. Fasham, ed., Plenum Press, London.

Azam, F., Fenchel, T., Field, J. G., Gray, J. S., Meyer-Reil, L. A., and Thingstad, F., 1983, The ecological role of water-column microbes in the sea, *Mar. Ecol. Prog. Ser.*, 10: 257.

Azam, F. and Hodson, R. E., 1977, Size distribution and activity of marine microheterotrophs, *Limnol. Oceanogr.*, 22: 492-501.

Banse, K., 1990, New views on the degradation and disposition of organic particles as collected by sediment traps in the open sea, *Deep-Sea Res.*, 37: 1177.

Berger, W. H., 1989, Appendix. Global maps of Ocean Productivity, *in:* "Productivity of the Ocean: Present and Past," W.H. Berger, V.C. Smetacek and G. Wefer, eds., John Wiley & Sons, Chichester.

Berger, W. H., Fischer, K., Lai, C., and Wu, G., 1987, Oceanic productivity and organic carbon flux. Part 1. Overview and maps of primary production and export production. Univ. of California, San Diego, SIO Reference 87-30.

Berger, W. H., Smetacek, V. S., and Wefer, G., 1989, Ocean productivity and paleoproductivity-an overview, *in:* "Productivity of the Ocean: Present and Past," W.H. Berger, V.S. Smetacek and G. Wefer, eds., John Wiley & Sons, Chichester.

Berman, T., 1991, Protozoans as agents in planktonic nutrient cycling, *in:* "Protozoa and Their Role in Marine Processes," P.C. Reid, C.M. Turley and P.H. Burkill, eds., Springer-Verlag, Berlin.

Berman, T., Nawrocki, M., Taylor, G. T., and Karl, D. M., 1987, Nutrient flux between bacteria, bactivorous nanoplanktonic protists and algae, *Mar. Microbial Food Webs*, 2: 69.

Bidigare, R. R., 1983, Nitrogen excretion by marine zooplankton, *in:* "Nitrogen in the Marine Environment," E.J. Carpenter and D.C. Capone, eds., Academic Press, New York.

Bigg, G. R., Jickells, T. D., Knap, A. H., and Serriff-Dow, R., 1989, The significance of short term wind induced mixing events for "new" primary production in subtropical gyres, *Oceanol. Acta*, 12: 437.

Billen, G., 1984, Heterotrophic utilization and regeneration of nitrogen, *in:* "Heterotrophic Activity in the Sea," J.E. Hobbie and P.J. LeB. Williams, eds., Plenum Press, New York.

Blackburn, T. H. and Sørensen, J., 1988, "Nitrogen Cycling in Coastal Marine Environments," John Wiley & Sons, Chichester.

Bratbak, G. and Thingstad, T. F., 1985, Phytoplankton-bacteria interactions: an apparent paradox? Analysis of a model system with both competition and commensalism, *Mar. Ecol. Prog. Ser.*, 25: 23.

Caperon, J., Schell, D., Hirota, J., and Laws, E., 1979, Ammonium excretion rates in Kaneohe Bay, Hawaii, measured by a ^{15}N-isotope dilution technique, *Mar. Biol.*, 54: 33.

Caron, D. A., 1991, Evolving role of protozoa in aquatic nutrient cycles, *in*: "Protozoa and Their Role in Marine Processes," P.C. Reid, C.M. Turley and P.H. Burkill, eds., Springer-Verlag, Berlin.

Caron, D. A., Davis, P. G., Madin, L. P., and Sieburth, J. McN., 1982, Heterotrophic bacteria and bactivorous protozoa in oceanic macroaggregates, *Science*, 218, 795.

Caron, D. A. and Goldman, J. C., 1988, Dynamics of protistan carbon and nutrient cycling, *J. Protozool.*, 35: 247.

Caron, D. A. and Goldman, J. C., 1990, Protozoan nutrient regeneration, *in*: "Ecology of Marine Protozoa," G.M. Capriulo, ed., Oxford University Press, New York.

Caron, D. A., Goldman, J. C., Anderson, O. K., and Dennett, M. R., 1985, Nutrient cycling in a microflagellate food chain: II. Population dynamics and carbon cycling, *Mar. Ecol. Prog. Ser.*, 24: 243.

Caron, D. A., Goldman, J. C., and Dennett, M. R., 1988, Experimental demonstration of the role of bacteria and bactivorous protozoa in plankton nutrient cycles, *Hydrobiol.*, 159: 27.

Carpenter, E. J. and Capone, D. G., 1983, "Nitrogen in the Marine Environment," Academic Press, New York.

Cho, B. C. and Azam, F., 1988, Major role of bacteria in biogeochemical fluxes in the ocean's interior, *Nature*, 332: 441.

Cochlan, W. P., 1986, Seasonal study of uptake and regeneration of nitrogen on the Scotian Shelf, *Cont. Shelf. Res.*, 5: 555.

Currie, D. J., 1984, Microscale nutrient patches: Do they matter to the phytoplankton?, *Limnol. Oceanogr.*, 29: 211.

Ducklow, H. W., 1984, Geographical ecology of marine bacteria: physical and biological variability at the mesoscale, *in*: "Current Perspectives in Microbial Ecology," M.J. Klug and C.A. Reddy, eds., Am. Soc. Microbiol., Washington, D.C.

Ducklow, H. W., Fasham, M. J. R., and Vezina, A. F., 1989, Derivation and analysis of flow networks for open ocean plankton systems, *in*: "Network Analysis in Marine Ecology," F. Wulff, J.G. Field and K.H. Mann, eds., Springer-Verlag, Berlin.

Ducklow, H. W. and Taylor, A. H., 1991, Modelling - session summary, *in*: "Protozoa and Their Role in Marine Processes," P.C., Reid, C.M. Turley and P.H. Burkill, eds., Springer-Verlag, Berlin.

Dugdale, R. C. and Goering, J. J., 1967, Uptake of new and regenerated forms of nitrogen in primary productivity. *Limnol. Oceanogr.*, 12: 196.

Eppley, R. W., 1981, Autotrophic production of particulate matter, *in*: "Analysis of Marine Ecosystems," A.R. Longhurst, ed., Academic Press, London.

Eppley, R. W., 1989, New production: history, methods, problems, *in*: "Productivity of the Ocean: Present and Past," W.H. Berger, V.S. Smetacek and G. Wefer, eds., John Wiley & Sons, Chichester.

Eppley, R. W. and Peterson, B. J., 1979, Particulate organic matter flux and planktonic new production in the deep ocean, *Nature*, 282, 677.

Eppley, R. W., Sharp, J. H., Renger, E. H., Perry, M. J., and Harrison, W. G., 1977, Nitrogen assimilation by phytoplankton and other microorganisms in the surface waters of the Central North Pacific, *Mar. Biol.*, 39, 111.

Falkowski, P. G., 1980, "Primary Productivity in the Sea," Plenum Press, New York.

Fasham, M. J. R., 1984, "Flows of Energy and Materials in Marine Ecosystems," Plenum Press, London.

Fasham, M. J. R., Ducklow, H. W., and McKelvie, S. M., 1990, A nitrogen-based model of plankton dynamics in the oceanic mixed layer, *J. Mar. Res.*, 48: 591.

Fenchel, T., 1988, Microfauna in pelagic food chains, *in*: "Nitrogen Cycling in Coastal Marine Environments," T.H. Blackburn and J. Sorensen, eds., John Wiley & Sons, Chichester.

Fogg, G. E., 1982, Nitrogen cycling in sea waters, Phil. Trans. R. Soc. Lond. B., 296: 511.

Frost, B. W., 1987, Grazing control of phytoplankton stock in the open subarctic Pacific Ocean: a model assessing the role of mesozooplankton, particularly the large calanoid copepods *Neocalanus* spp., *Mar. Ecol. Prog. Ser.*, 39: 49.

Fuhrman, J., 1987, Close coupling between release and uptake of dissolved free amino acids in seawater studied by an isotope dilution approach, *Mar. Ecol. Prog. Ser.*, 37: 45.

Fuhrman, J. A., Horrigan, S. G., and Capone, D. G., 1988, Use of ^{13}N as a tracer for bacterial and algal uptake of ammonium from seawater, *Mar. Ecol. Prog. Ser.*, 45: 271.

Gast, V. and Horstmann, U., 1983, N-remineralization of phyto- and bacterioplankton by the marine ciliate *Euplotes vannus*, *Mar. Ecol. Prog. Ser.*, 13: 55.

Glibert, P. M., 1982, Regional studies of daily, seasonal and size fraction variability in ammonium remineralization, *Mar. Biol.*, 70: 209.

Glibert, P. M., Dennett, M. R., and Caron, D. A., 1988, Nitrogen uptake and NH_4 regeneration by pelagic microplankton and marine snow from the North Atlantic, *J. Mar. Res.*, 46: 837.

Glibert, P. M., Goldman, J. C., and Carpenter, E. J., 1982a, Seasonal variation in the utilization of ammonium and nitrate by phytoplankton in Vineyard Sound, Massachusetts, USA, *Mar. Biol.*, 70: 237.

Glibert, P. M., Lipschultz, F, McCarthy, J. J., and Altabet, M. A., 1982b, Isotope dilution models of uptake and remineralization of ammonium by marine plankton, *Limnol. Oceanogr.*, 27: 639.

Glover, H. E., Prezelin, B. B., Campbell, L., Wyman, M., and Garside, C., 1988, A nitrate-dependent *Synechococcus* bloom in surface Sargasso Sea water, *Nature*, 331: 161.

Goldman, J. C., 1984a, Oceanic nutrient cycles, *in*: "Flows of Energy and Materials in Marine Ecosystems: Theory and Practice," M.J. Fasham, ed., Plenum Press, New York.

Goldman, J. C., 1984b, Conceptual role for for microaggregates in pelagic waters, *Bull. Mar. Sci.*, 35: 462.

Goldman, J. C., 1988, Spatial and temporal discontinuities of biological processes in pelagic surface waters, *in*: "Toward a Theory on Biological-Physical Interactions in the World Ocean," B.J. Rothschild, ed., Kluwer Academic Publishers.

Goldman, J. C. and Caron, D. A, 1985, Experimental studies on an omnivorous microflagellate: implications for grazing and nutrient regeneration in the marine microbial food chain, *Deep-Sea Res.*, 32: 899.

Goldman, J. C., Caron, D. A., Anderson, O. K., and Dennett, M. R., 1985, Nutrient cycling in a microflagellate food chain: I. nitrogen dynamics, *Mar. Ecol. Prog. Ser.*, 24: 231.

Goldman, J. C., Caron, D. A., and Dennett, M. R., 1987a, Regulation of gross growth efficiency and ammonium regeneration in bacteria by substrate C:N ratio, *Limnol. Oceanogr.*, 32: 1239.

Goldman, J. C., Caron, D. A., and Dennett, M. R., 1987b, Nutrient cycling in a microflagellate food chain: IV. Phytoplankton-microflagellate interactions, *Mar. Ecol. Prog. Ser.*, 38: 75.

Goldman, J. C. and Dennett, M. R., 1991, Ammonium regeneration and carbon utilization by marine bacteria grown on mixed substrates, *Mar. Biol.*, In press.

Gotschalk, C. C. and Alldredge, A. L., 1989, Enhanced primary production and nutrient regeneration within aggregated marine diatoms, *Mar. Biol.*, 103: 119.

Hansell, D. A. and Goering, J. J., 1989, A method for estimating uptake and production rates for urea in seawater using [^{14}C] urea and [^{15}N] urea, *Can. J. Fish. Aq. Sci.*, 46: 198.

Hanson, R. B. and Robertson, C. Y., 1988, Spring recycling of ammonium in turbid continental shelf waters off the southeastern United States, *Cont. Shelf Res.*, 8: 49.

Hanson, R. B., Robertson, C. Y., Yoder, J. A., Verity, P. G., and Bishop, S. S., 1990, Nitrogen recycling in coastal waters of southeastern U.S. during summer 1986, *J. Mar. Res.*, 48: 641.

Harris, E., 1959, The nitrogen cycle in Long Island Sound, *Bull. Bingham Oceanogr. Collect.*, 17: 31.

Harrison, W. G., 1978, Experimental measurements of nitrogen remineralization in coastal waters, *Limnol. Oceanogr.*, 23: 684.

Harrison, W. G., 1980, Nutrient regeneration and primary production in the sea, in: "Primary Productivity in the Sea," P.G. Falkowski, ed., Plenum Press, New York.

Harrison, W. G., 1983a, Uptake and recycling of soluble reactive phosphorus by marine microplankton, *Mar. Ecol. Prog. Ser.*, 10: 127.

Harrison, W. G., 1983b, Use of isotopes, in: "Nitrogen in the Marine Environment," E.J. Carpenter and D.G. Capone, eds., Academic Press, New York.

Harrison, W. G., 1990, Nitrogen utilization in chlorophyll and primary productivity maximum layers: an analysis based on the f-ratio, *Mar. Ecol. Prog. Ser.*, 60: 85.

Harrison, W. G. and Cota, G. F., 1991, Primary production in polar waters: relation to nutrient availability, *Polar Res.*, In press.

Harrison, W. G., Douglas, D., Falkowski, P., Rowe, G., and Vidal, J., 1983, Summer nutrient dynamics of the Middle Atlantic Bight: nitrogen uptake and regeneration, *J. Plankt. Res.*, 5: 539.

Hopkinson, C. S., Jr., Sherr, B. F., and Ducklow, H. W., 1987, Microbial regeneration of ammonium in the water column of Davies Reef, Australia, *Mar. Ecol. Prog. Ser.*, 41: 147.

Jackson, G. A., 1980, Phytoplankton growth and zooplankton grazing in oligotrophic oceans, *Nature*, 284: 439.

Jackson, G. A., 1987, Simulating chemosensory responses of marine microorganisms, *Limnol. Oceanogr.*, 32: 1253.

Jackson, G. A., 1988, Implications of high dissolved organic matter concentrations for oceanic properties and processes, *Oceanogr.*, 1: 28.

Jahnke, R. A., 1990, Ocean flux studies: a status report, *Rev. Geophys.*, 28: 381.

Jahnke, R. A. and Jackson, G. A., 1987, Role of sea floor organisms in oxygen consumption in the deep North Pacific Ocean, *Nature*, 329: 621.

Jenkins, W. J., 1982, Oxygen utilization rates in North Atlantic subtropical gyre and primary production in oligotrophic systems, *Nature*, 300: 246.

Jenkins, W. J., 1988a, Nitrate flux into the euphotic zone near Bermuda, *Nature*, 331: 521.

Jenkins, W. J., 1988b, The use of anthropogenic tritium and helium-3 to study subtropical gyre ventilation and circulation, *Phil. Trans. R. Soc., London, Ser. A*, 325: 43.

Jenkins, W. J. and Goldman, J. C., 1985, Seasonal oxygen cycling and primary productivity in the Sargasso Sea, *J. Mar. Res.*, 43: 465.

Johannes, R. E., 1964, Phosphorus excretion as related to body size in marine animals: the significance of nannozooplankton in nutrient regeneration, *Science*, 146: 923.

Johannes, R. E., 1965, Influence of marine protozoa on nutrient regeneration, *Limnol. Oceanogr.*, 10: 434.

Karl, D. M., Knauer, G. A., and Martin, J. H., 1988, Downward flux of particulate organic matter in the ocean: a particle decomposition paradox, *Nature*, 332: 438.

Kepkay, P. E. and Johnson, B. D., 1989, Coagulation on bubbles allows microbial respiration of oceanic dissolved organic carbon, *Nature*, 338: 63.

King, F. D., 1984, Vertical distribution of zooplankton glutamate dehydrogenase in relation to chlorophyll in the vicinity of the Nantucket Shoals, *Mar. Biol.*, 79: 249.

King, F. D., 1987, Nitrogen recycling efficiency in steady state oceanic environments, *Deep-Sea Res.*, 34: 843.

Kirchman, D. L., Keil, R. G., and Wheeler, P. A., 1989, The effect of amino acids on ammonium utilization and regeneration by heterotrophic bacteria in the subarctic Pacific, *Deep-Sea Res.*, 36: 1763.

Knauer, G. A., Martin, J. H., and Karl, D. M., 1984, The flux of particulate matter out of the euphotic zone. *in*: "Global Ocean Flux Studies: Proceedings of a Workshop," National Academic Press, Woods Hole, MA.

Knauer, G. A., Redalje, D. G., Harrison, W. G., and Karl, D. M., 1990, New production at the VERTEX time-series site, *Deep-Sea Res.*, 37: 1121.

Lancelot, C. and Billen, G., 1985, Carbon-nitrogen relationships in nutrient metabolism of coastal marine ecosystems, *Adv. Aquat. Microbiol.*, 3:263.

Laws, E. A., Harrison, W. G., and Ditullio, G. R., 1985, A comparison of nitrogen assimilation rates based on ^{15}N uptake and autotrophic protein synthesis. *Deep-Sea Res.*, 32: 85.

Lehman, J. T. and Scavia, D., 1982, Microscale patchiness of nutrients in plankton communities, *Science*, 216: 729.

Lewis, M. R., Harrison, W. G., Oakey, N. S., Hebert, D., and Platt, T., 1986, Vertical nitrate fluxes in the oligotrophic ocean, *Science*, 234: 870.

Lochte, K. and Turley, C. M., 1988, Bacteria and cyanobacteria associated with phytodetritus in the deep sea, *Nature*, 333: 67.

Lohrenz, S. E., Knauer, G. A., Asper, V. L., Tuel, M., Michaels, A. F., and Knap, A. H., 1991, Seasonal variability in primary production and particle flux in the northwestern Sargasso Sea: U.S. JGOFS Bermuda Atlantic time-series study, *Deep-Sea Res.*, In press.

Longhurst, A. R., Bedo, A., Harrison, W. G., Head, E. J. H., Horne, E. P., Irwin, B., and Morales, C., 1989, NFLUX: a test of vertical nitrogen flux by diel migrant biota, *Deep-Sea Res.*, 36: 1705.

Longhurst, A. R. and Harrison, W. G., 1988, Vertical nitrogen flux from the oceanic photic zone by diel migrant zooplankton and nekton, *Deep-Sea Res.*, 35: 881.

Longhurst, A. R. and Harrison, W. G., 1989, The biological pump: Profiles of plankton production and consumption in the upper ocean, *Prog. Oceanogr.*, 22: 47.

Martin, J. H., Knauer, G. A., Karl, D. M., and Broenkow, W. W., 1987, VERTEX: carbon cycling in the northeast Pacific, *Deep-Sea Res.*, 34: 267.

McCarthy, J. J. and Goldman, J. C., 1979, Nitrogenous nutrition of marine phytoplankton in nutrient-depleted waters, *Science*, 203: 670.

Moloney, C. L., Bergh, M. O., Field, J. G., and Newell, R. C., 1986, The effect of sedimentation and microbial nitrogen regeneration in a plankton community: a simulation investigation, *J. Plankt. Res.*, 8: 427.

Moloney, C. L. and Field, J. G., 1991, Modelling carbon and nitrogen flows in a microbial plankton community, *in*: "Protozoa and Their Role in Marine Processes," P.C., Reid, C.M. Turley and P.H. Burkill, eds., Springer-Verlag, Berlin.

Morris, I., 1980, "The Physiological Ecology of Phytoplankton," Univ. of California Press, Berkeley.

Musgrave, D. L., Chou, J., and Jenkins, W. J., 1988, Application of a model of upper-ocean physics for studying seasonal cycles of oxygen, *J. Geophys. Res.*, 93: 15,679.

Newell, R. C., Moloney, C. L., Field, J. C., Lucas, M. I., and Probyn, T. A., 1988, Nitrogen models at the community level: plant-animal-microbe interactions, *in*: "Nitrogen Cycling in Coastal Marine Environments," T.H. Backburn and J. Sⲫrensen, eds., John Wiley & Sons, Chichester.

Owens, N. J. P., Mantoura, R. F. C., Burkill, P. H., Howland, R. J. M., Pomroy, A. J., and Woodward, E. M. S., 1986, Nutrient cycling studies in Carmarthen Bay: phytoplankton production, nitrogen assimilation and regeneration, *Mar. Biol.*, 93: 329.

Paasche, E., 1988, Pelagic primary production in nearshore waters, *in*: "Nitrogen Cycling in Coastal Marine Environments," T.H. Backburn and J. Sⲫrensen, eds., John Wiley & Sons, Chichester.

Paasche, E. and Kristiansen, S., 1982, Ammonium regeneration by microzooplankton in the Oslofjord, *Mar. Biol.*, 69: 55.

Packard, T. T., Denis, M., Rodier, M., and Garfield, P., 1988, Deep-ocean metabolic CO_2 production: calculations from ETS activity, *Deep-Sea Res.*, 35: 371.

Platt, T. 1981, "Physiological Bases of Phytoplankton Ecology,", Can. Bull. Fish. Aq. Sci., 210, Ottawa.

Platt, T. and Harrison, W. G., 1985, Biogenic fluxes of carbon and oxygen in the ocean, *Nature*, 318: 55.

Platt, T., Harrison, W. G., Lewis, M. R., Li, W. K. W., Sathyendranath, S., Smith, R.E.H., and Vezina, A., 1989, Biological production of the oceans: the case for a consensus. *Mar. Ecol. Prog. Ser.*, 52: 77.

Probyn, T. A., 1987, Ammonium regeneration by microplankton in an upwelling environment, *Mar. Ecol. Prog. Ser.*, 37: 53.

Proctor, L. M. and Fuhrman, J. A., 1990, Viral mortality of marine bacteria and cyanobacteria, *Nature*, 343: 60.

Riley, G. A., 1970, Particulate organic matter in seawater, *Adv. Mar. Biol.*, 8: 1.

Sarmiento, J. L., Fasham, M. J. R., Slater, R., Toggweiler, J. R., and Ducklow, H. W., 1990a, The role of biology in the chemistry of CO2 on the ocean, *in*: "Chemistry of the Greenhouse Effect," M. Farrell, ed., Lewis Publ. In press.

Sarmiento, J. L., Thiele, G., Key, R. M., and Moore, W. S., 1990b, Oxygen and nitrate new production and remineralization in the North Atlantic subtropical gyre, *J. Geophys. Res.*, 95: 18,303.

SCOR, 1990, Joint Global Ocean Flux Study: Science Plan, JGOFS Report No. 5.

Shanks, A. L. and Trent, J. D., 1979, Marine snow: microscale nutrient patches, *Limnol. Oceanogr.*, 24: 850.

Sherr, B. F. and Sherr, E. B., 1984, Role heterotrophic protozoa in carbon and energy flow in aquatic ecosystems, *in*: "Current Perspectives in Microbial Ecology," M.J. Klug and C.A. Reddy, eds., Am. Soc. Microbiol., Washington, D.C.

Sherr, B. F., Sherr, E. B., and Berman, T., 1983, Grazing, growth, and ammonium excretion rates of a heterotrophic microflagellate fed with four species of bacteria, *Appl. Environ. Microbiol.*, 45: 1196.

Silver, M. W. and Gowing, M. M., 1991, The "particle" flux: origins and biological components, *Prog. Oceanogr.*, 26: 75.

Slawyk, G., Raimbault, P., and L'Helguen, S., 1990, Recovery of urea nitrogen from seawater for measurement of ^{15}N abundance in urea regeneration studies using the isotope-dilution approach, *Mar. Chem.*, 30: 343.

Smetacek, V., 1985, Role of sinking in diatom life-history cycles: ecological, evolutionary and geological significance, *Mar. Biol.*, 84: 239-251.

Smetacek, V. and Pollehne, F., 1986, Nutrient cycling in pelagic systems: A reappraisal of the conceptual framework, *Ophellia*, 26: 401.

Sorokin, Y. I., 1981, Microheterotrophic organisms in marine ecosystems, *in*: "Analysis of Marine Ecosystems," A.R. Longhurst, ed., Academic Press, London.

Spitzer, W. S. and Jenkins, W. J., 1989, Rates of vertical mixing, gas exchange and new production: Estimates from seasonal gas cycles in the upper ocean near Bermuda, *J. Mar. Res.*, 47: 169.

Stout, J. D., 1980, The role of protozoa in nutrient cycling and energy flow, *Adv. Microbiol. Ecol.*, 4: 1.

Suess, E., 1980, Particulate organic carbon flux in the oceans-surface productivity and oxygen utilization, *Nature*, 28: 260.

Sugimura, I. and Suzuki, Y., 1988, A high-temperature catalytic oxidation method for the determination of non-volatile dissolved organic carbon in seawater by direct injection of a liquid sample, *Mar. Chem.*, 24: 105.

Suzuki, Y, Sugimura, Y., and Itoh, T., 1985, A catalytic oxidation method for the determination of total nitrogen dissolved in seawater, *Mar. Chem.*, 16: 83.

Sverdrup, H. U., Johnson, M. W., and Fleming, R. H., 1942, "The Oceans. Their Physics, Chemistry, and General Biology," Prentice-Hall, Englewood Cliffs, N.J.

Taylor, G. T., 1982, The role of pelagic heterotrophic protozoa in nutrient cycling: a review, *Ann. Inst. Oceanogr.*, 58(S): 227.

Tezuka, Y., 1990, Bacterial regeneration of ammonium and phosphate as affected by the carbon:nitrogen:phosphorus ratio of organic substrates, *Microb. Ecol.*, 19: 227.

Toggweiler, J. R., 1989, Is the downward dissolved organic matter (DOM) flux important in carbon transport?, *in*: "Productivity of the Ocean: Present and Past," W.H. Berger, V.S. Smetacek and G. Wefer, eds., John Wiley & Sons, Chichester.

Tupas, L. and Koike, I., 1990, Amino acid and ammonium utilization by heterotrophic marine bacteria grown in enriched seawater, *Limnol. Oceanogr.*, 31: 998.

Tupas, L. and Koike, I., 1991, Simultaneous uptake and regeneration of ammonium by mixed assemblages of heterotrophic marine bacteria, *Mar. Ecol. Prog. Ser.*, 70: 273.

Verity, P. G., 1985, Grazing, respiration, excretion and growth rates of tintinnids, *Limnol. Oceanogr.*, 30: 1268.

Vezina, A. F. and T. Platt, 1987, Small-scale variability of new production and particulate fluxes in the ocean, *Can. J. Fish. Aq. Sci.*, 44: 198.

Walsh, I., Dymond, J., and Collier, R., 1988, Rates of recycling of biogenic components of settling particles in the ocean derived from sediment trap experiments, *Deep-Sea Res.*, 35: 43.

Wassmann, P., 1990, Relationship between primary and export production in the boreal coastal zone of the North Atlantic, *Limnol. Oceanogr.*, 35: 464.

Wheeler, P. A. and Kirchman, D. L., 1986, Utilization of inorganic and organic nitrogen by bacteria in marine systems, *Limnol. Oceanogr.*, 31: 998.

Williams, P. J. LeB., 1981, Incorporation of microheterotrophic processes into the classical paradigm of the planktonic food web, *Kieler Meeresforsch. Sonderh.*, 5:1.

Williams, P. J. LeB., and Muir, L.,R., 1981, Diffusion as a constraint on the biological importance of microzones in the sea, *in*: "Ecohydrodynamics," J.C.J. Nihoul, ed., Elsevier, Amsterdam.

GRAZING, TEMPORAL CHANGES OF PHYTOPLANKTON CONCENTRATIONS, AND THE MICROBIAL LOOP IN THE OPEN SEA

Karl Banse

School of Oceanography, WB-10
University of Washington
Seattle, WA 98195

INTRODUCTION

With the advent of maps of chlorophyll distribution in the world ocean (Feldman et al., 1989; Lewis, this volume), parochial questions can be asked on a global basis: (1) where, (2) when, and (3) why does phytoplankton occur in the open sea, and (4) how much is found? The first and fourth questions will be addressed briefly at the outset. The paper will treat the second and third questions at length, emphasizing the annual and seasonal time scales. Grazing will be shown to be a key variable that largely has been underrated. In addition to affecting phytoplankton, grazing will be recognized as central to the supply of substrate for the microbial loop.

The global answer to the first question, about the location of high phytoplankton concentration in the open sea, can be inferred from correlating satellite maps of phytoplankton pigment with those of enhanced vertical water motions. The congruence essentially confirms the outlines of the issue by Hentschel and Wattenberg (1930) and Kesteven and Laevastu (1957-1958: Fig. 19; map reprinted in Hela and Laevastu, 1961: Fig. 37). The cause, now partially quantified, is upward transport of nutrients, especially NO_3^-, by physical forces. After uptake by phytoplankton, the production-controlling bound N is continually lost to depth by gravity acting on particles, against the concentration gradient of the dissolved inorganic moieties, as well as by eddy (turbulent) diffusion of dissolved organic N following the gradient. The correlation between upward nitrate flux and primary production was treated by, e.g., Eppley et al. (1979; see also Platt et al., 1989), after Dugdale and Goering (1967) introduced the concepts of "new" and "regenerated" production and popularized the way to distinguish them by routine measurement.

The general answer to the fourth question, about the maintenance of a particular concentration of phytoplankton, has to explain the balance between the rates of gain of phytoplankton to, and loss from, a volume of water or below a unit of sea surface. The balance comprises the interactions of cell physiology with the abiotic environment and those of phytoplankton population biology with the abiotic and biotic environment. Global solutions are not yet in hand, although mathematical models are showing the way (e.g., Riley et al., 1949; Riley, 1965; Kremer and Nixon, 1978; Frost, 1987; Fasham

Primary Productivity and Biogeochemical Cycles in the Sea
Edited by P.G. Falkowski and A.D. Woodhead, Plenum Press, New York, 1992

et al., 1990). Note that the concentration of phytoplankton times the average growth rate per unit mass yields the production rate of the water, which is of biological oceanographic and geochemical interest.

The answers about the when and why of temporal changes, the second and third questions, also are tied to the balance between rates of gain and loss. At a location in the open sea, the gain term for the concentration of phytoplankton in the euphotic zone, which adds cells per unit of sea surface, is first of all cell division, followed by the effect of water movement (advection or eddy diffusion from a rich area). Losses of cells are principally due to sinking, advection, and diffusion to an area poor in cells, including the depths below the euphotic zone, or grazing; death of phytoplankton from age, parasites, or viruses seems to be rare. This paper will (1) show the importance of grazing for understanding the presence or absence of change on temporal scales of longer than a few days and geographically on basin-wide scales; (2) mention the mechanisms of animal control of phytoplankton population change on a near-organismic scale; (3) outline the consequences of animal feeding for the generation of dissolved organic matter that fuels the so-called microbial loop; and (4) review the historical interactions among phytoplankton and zooplankton researchers that led to the present level of understanding. The treatment will be based almost entirely on carbon and chlorophyll, rather than on the biology of phytoplankton species.

THE ROLE OF GRAZING IN THE CONCENTRATION BALANCE

This section first will delineate the geographic domains of interest and juxtapose regions of slow and small seasonality with those of pronounced changes of chlorophyll, emphasizing the changes in the mixed layer (above the pycnocline), as they may be sensed by a satellite. Pigment concentration substitutes for phytoplankton biomass, which is the parameter of principal concern; changes of the carbon/chlorophyll ratio, as may be driven by temperature, nutrition, and community composition, are neglected. The section then estimates the rates of the terms that affect the temporal changes for periods longer than a few days and concludes with the relation of temporal to spatial changes on the meso-scale (several tens to hundreds of kilometers).

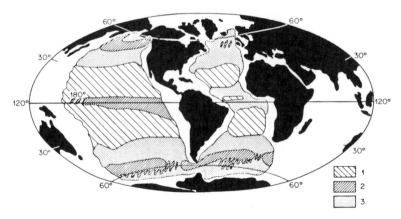

Fig. 1. Estimated extent of Domains 1-3 in the Atlantic and Pacific oceans. 1) Domain 1 with low seasonality of phytoplankton concentration, nutrient-depleted; 2) Domain 2 with low seasonality of phytoplankton concentration, nutrient-replete; 3) Domain 3 with high seasonality of phytoplankton concentration.

Figure 1 depicts my view of the major subdivisions of the Pacific and Atlantic oceans, i.e., Domains 1 and 2 having relatively little seasonal changes (the seasonal maximum of the monthly averages being two to three times the minimum), and Domain 3 having significant change, characterized in particular by a regular, markedly seasonal phytoplankton "bloom". Domains 1 and 2 are differentiated by the permanent depletion of the mixed layer of appreciable concentrations of nutrients in the former versus their presence in the latter. My view is based on general oceanographic knowledge sketched below; full documentation and the discussion of geographic particulars would require too much space, thereby detract from the main point of the paper, i.e., the mechanisms underlying the patterns.

In Fig. 1, Domain 1 is easily recognized as essentially comprising the oligotrophic subtropical gyres. The lack of marked seasonality has been documented best in the Pacific (e.g., Bienfang and Szyper, 1981; Dandonneau and Gohin, 1984; Hayward, 1987). On the equatorial side in the Pacific, the domain adjoins the low-latitude part of the nutrient-rich Domain 2 in the area of year-round divergence (cf. pigment distribution in Banse and Yong, 1990). The Atlantic tropical divergence associated with the equatorial current system is more seasonal, as signified by the weakening or disappearance of the counter current during winter (Knauss, 1963), and is not discussed.

Domain 3 adjoins Domain 1 on the poleward side. Here, winter cooling and partial overturn of the water column inject nutrients into the euphotic zone, thus temporarily relieving the inherent nutrient depletion of the sea caused by gravity. Depending on the seasonal change of incident irradiance, the transparency of the water, and the delay between overturn and re-stratification, the peak of phytoplankton concentration may occur in late winter (cf. Bermuda at 32°N, Menzel and Ryther, 1961; Platt and Harrison, 1985; former Ocean Weather Station *E* in the western North Atlantic at 35°N, Riley, 1957) or as late as May (central Norwegian Sea, former Ocean Weather Station *M*, Sverdrup, 1953).

In the higher-latitude parts of Domain 2, the spring bloom and a marked seasonality of phytoplankton are absent in spite of the pronounced seasonality of the abiotic environment; macro-nutrients (N, Si, P) are always replete, although the concentrations may be lower in summer than during winter. The concentrations of chlorophyll during winter are appreciable (relative to the summer values), due to the absence of a deep mixed layer. The underlying processes are understood best for the eastern subarctic Pacific (Frost 1987; see Anderson and Munson, 1972, for regional aspects). In Fig. 1, the subantarctic part of the Southern Ocean (between the subtropical and antarctic convergences [the latter often called the Polar Front]) also is part of Domain 2. Pigment observations from the southeastern Atlantic and the Indian sectors (Fukuchi, 1980) of this circumpolar belt are almost without exception low. Further, crossings of the Drake Passage (B.W. Frost, personal communication, from data of El-Sayed, 1967) suggest relatively constant pigment means year-round, which as in the subarctic Pacific might result from the relatively low latitude and the broad occurrence of shallow mixed layers around much of Antarctica (cf. Fig. 96b in Levitus, 1982, based on sigma-t criteria instead of temperature). As will be noted later, appreciable concentrations of phytoplankton in winter are one of the requirements for suppression of spring blooms. Smetacek et al. (1990) provide further references to the rarity of spring blooms in the Southern Ocean, but the definitive seasonal study of the subantarctic Domain 2 is wanting. Finally, new observations indicate that areas in the temperate North Atlantic do not become nutrient-(nitrate) depleted during summer (Fig. 1; personal communication of B.W. Frost).

Note that the biomass and the production rate of phytoplankton in Domains 1 and 2 are based largely on small cells (e.g., Frost, 1987; El-Sayed, 1988 [in part]; Peña et al., 1990; Smetacek et al., 1990). Microzooplankton (e.g., protozoans, copepod nauplii), the consumers, have high division and growth rates.

The Average Daily Rate of Seasonal Change of Phytoplankton

In addition to the geographic pattern and the extent of the seasonal range, the rate of seasonal change is important. For the latitudinal belts between 10° and 40° in the North Atlantic and Pacific oceans, J. Yoder (personal communication) summarized monthly pigment means from the CZCS Global Data Set, including the observations on the shelves. From these data, I find for the North Pacific, by division of the monthly maximum (December/January) by the monthly minimum (September) and the interval of three months (since December is nearly as high as January), a chlorophyll change of 1.7% d^{-1}. In contrast, the analogous change in the North Indian Ocean in the same latitudinal belt is 11% d^{-1} (June to August). In conclusion, the slight daily change that is to be expected in Domains 1 and 2 is small relative to the instantaneous growth rate of the phytoplankton k ($k/0.693$ = division rate) and will, therefore, be neglected.

The Effect of Horizontal Advection or Diffusion on Phytoplankton Concentrations

The principal argument about the role of grazing will be developed below by a model with only a vertical dimension. Therefore, a diversion about the small role of horizontal physical processes (advection from currents, eddy diffusion), relative to the algal division rates, is necessary. For the central Gulf of Alaska at the position of former Ocean Weather Station P (*Papa*), Frost (in press) calculated from the uneven distribution of surface phytoplankton on a grid of about 50 km x 50 km that the daily change due to advection would amount to about 10 to 20% of mean chlorophyll (Chl), while that due to turbulent diffusion would be a few percent. However, he noted that for large distances, the mean Chl does not change (cf. Anderson et al., 1977) so that over several days, the advective change must average out to near-zero. The same argument holds for Domain 2, the central subtropical gyres, in view of the small horizontal gradients over vast reaches (cf. Shulenberger, 1978 [Fig. 7]; Hayward, 1987). In the equatorial part of Domain 2, the east-west gradients relative to the mean concentrations (read from maps in Love, 1970; 1974) and the north-south gradients (Peña et al., 1991: Table 1) of chlorophyll over scales of several 100 km are smaller by an order of magnitude or more than those caused by unevenness in Frost's grid in the Gulf of Alaska. While acknowledging the presence of gradients for the equatorial part of Domain 2, as well as the higher horizontal advection (east-west) and the stronger horizontal shear (north-south), I estimate that the physical horizontal terms will change Chl by not more than a few percent per day. For the subantarctic part of Domain 2, I only can assume that the horizontal terms are about as small as in the subarctic parts, because maps of chlorophyll on a suitable scale do not exist. Therefore, the horizontal terms will be neglected.

The Temporal Balance of Phytoplankton and Zooplankton in Domains 1 and 2

The following formulates a balance sheet for four typical situations for Domains 1 and 2, with day as the unit of time. Consider a stratified ocean composed of a mixed layer above and the deep water below, which are separated by a pycnocline. The latter is placed at the bottom of the euphotic zone; the argument would not change greatly were the pycnocline, say, twice as deep. The integrated concentration of chlorophyll in the upper layer is Chl_u (mass m^{-2}) and its concentration is uniform with depth, but there is no pigment in the lower layer (Fig. 2A). The advection (a) of water from below, as in a divergence, dilutes Chl_u. For example, if the upward movement is 1 m d^{-1} and the mixed layer is 50 m deep, as may occur near the equator, the daily change of Chl_u will be 0.02 d^{-1}. The effect of vertical mixing (m) is treated in the same way. The loss of live cells (not pigment in fecal pellets) by sinking (s) as individual cells or included in marine snow (Silver et al., 1986) also is expressed as a fraction of Chl_u. The rates of algal instantaneous growth and grazing loss are k and g, respectively.

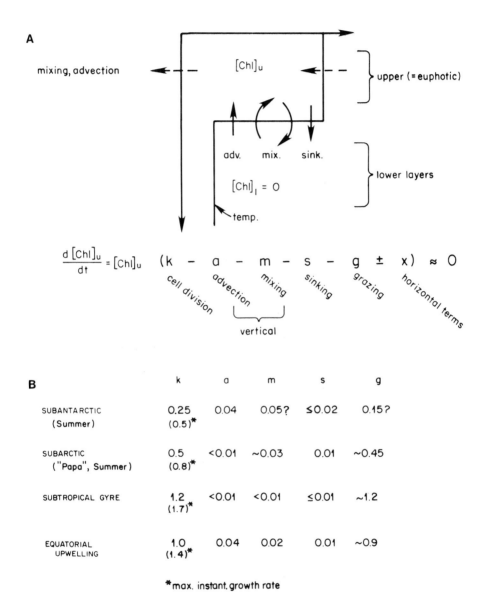

$$\frac{d[Chl]_u}{dt} = [Chl]_u \ (k - a - m - s - g \pm x) \approx 0$$

	k	a	m	s	g
SUBANTARCTIC (Summer)	0.25 (0.5)*	0.04	0.05?	≤0.02	0.15?
SUBARCTIC ("Papa", Summer)	0.5 (0.8)*	<0.01	~0.03	0.01	~0.45
SUBTROPICAL GYRE	1.2 (1.7)*	<0.01	<0.01	≤0.01	~1.2
EQUATORIAL UPWELLING	1.0 (1.4)*	0.04	0.02	0.01	~0.9

*max. instant. growth rate

Fig. 2. The role of grazing for the concentration balance. A) Model used in deriving the balance in Fig. 2B. B) Numerical values for the balance in four parts of Domains 1 and 2. Growth and grazing as instantaneous rates (d^{-1}); advection, mixing, and sinking as fractions of [Chl]$_u$ (mass m^{-2}) added or removed (d^{-1}). Rates in brackets in column under k calculated from Eppley (1972), taking temperature and day length into account.

Note that the few modern estimates of k, used below, are not based on averages through the entire euphotic zone, but extend roughly to the 10% light depth, so that they are not strictly comparable to the estimates of a, m, and s, which refer to the entire photic zone; calculating mean k would approximately halve the values of k in Fig. 2B and, hence, would not change any of the four principal points to be made below. Further, the term g as used here includes algal death from parasites, etc., but since the marine literature is almost devoid of observations to that effect, I assume that "natural death" (so to speak) is rare; qualifications will be reviewed under Domain 3. The equation in Fig. 2A defines the processes for the local time change of Chl_u; for the balance in Fig. 2B, recall that the horizontal terms are neglected, and the daily rate of seasonal change is set to zero.

The rates in Fig. 2B were chosen from the literature, but discussion of the choices for a and m are omitted because k dominates and essentially determines g, except in the Subantarctic. Domain 1, Pacific subtropical gyre: k, Laws et al. (1987, average for upper 30 m, i.e. down to approximately 23% of incident irradiance (I_o) (see also earlier data for oligotrophic regions in Walsh, 1976); a, estimated from Hayward (1987); m, Welschmeyer and Lorenzen (1985; for a similar m off Bermuda, see Fasham et al., 1990); s, Welschmeyer and Lorenzen (1985). - Domain 2, Pacific equatorial divergence: k, estimated from Cullen et al. (in press; 0.7 and 1.46 d^{-1} by two methods, respectively, at $\geq$ 10% I_o); a, estimated (see text above); m, 0.01 to 0.02 d^{-1} (B. W. Frost, personal communication); s, guessed. - Domain 2, Subarctic Pacific at station P, summer: k, control bottles in iron enrichment experiments by Martin et al. (1989; "... wrapped in 3 layers of plastic bags .... in lucite deck-top incubator ..." as re-evaluated in Banse (in press); a and m, Frost (1987); s, N.A. Welschmeyer (personal communication). - Domain 2, Subantarctic, summer: k, control bottles in iron enrichment experiments of de Baar et al. (1990) and Martin et al. (1990, ambient light in deck incubator; de Baar et al. used artificial light, apparently of about 1/10 of that irradiance), as re-evaluated in Banse (in press); a, B. W. Frost, personal communication; m, by extrapolation from station P, but considering that the vertical stratification tends to be less; regarding s, Huntley et al. (1991: footnote 8) cite other authors that "... sinking during blooms is generally < 5% of the net primary production ...", i.e. < 0.05 P; because k (0.25 d^{-1}, from Fig. 2B) also can be expressed as mg C (mg C x d)$^{-1}$, i.e., k = P/B, with B being the algal biomass, the daily loss as fraction of B is < 0.05 x 0.25. This is an overestimate as the model in Fig. 2A extends to the bottom of the euphotic zone where the algal division rate approaches zero. A similarly low estimate of s is implied by Smetacek et al. (1990) for areas outside krill occurrence. Finally, in the Barents Sea during its pronounced spring bloom, "massive sedimentation of *Phaeocystis*" meant that $\leq$ 0.02 Chl_u sedimented daily (Wassmann et al., 1990).

The principal result of the modeling exercise (Fig. 2B) is that the measured algal division rates and the relatively small physical terms can be reconciled only by very large death rates (g) of phytoplankton, when the seasonal change (as d^{-1}) is slow, as observed. Based on our present knowledge, this death must be from grazing. The only experimental test available for any of the sites is for station P where indeed, grazing rates are as high as claimed here; significantly, they are due to microzooplankton, not net-collected animals like juvenile or adult copepods (Strom and Welschmeyer, 1991). Indirect evidence to the same effect is provided by Cullen et al. (in press) for the equator.

Oversimplifying the equation in Fig. 2A for the sake of discussion to:

$$d[Chl]/dt = [Chl]\ (k - d)$$

shows that the change of algal concentrations in the two domains cannot be understood from the interaction of algal physiology with the abiotic environment alone. When investigating periods of a few days and longer, the equation with its two unknowns cannot be solved by studying k only. The two domains, to which the simplification of the equation applies, occupy roughly half of the ocean.

A closer look at Fig. 2B suggests a marked temperature dependence of k and g. Eppley (1972) quantified it for k (see brackets under Growth in Fig. 2B). As explained below, however, the temperature dependence of g has several causes. The following new points emerge from Fig. 2B:

(1) The realized growth rates of the phytoplankton in the allegedly oligotrophic central gyres is remarkably high in absolute terms and relative to the "Eppley value" (Eppley, 1972; supporting data in Banse, in press). The proximate explanation has to be an efficient recycling of nutrients in the euphotic zone due to animal grazing, directly or via the microbial loop. This is confirmed by the small supply of nutrients by physical mechanisms (vertical physical terms times nutrient gradients, Hayward, 1987), or the small ratio of uptake rates of nitrate versus those of ammonium (new versus regenerated production, or f-ratio). Thus, Fig. 2B is interpreted to mean that the animals themselves hold the key to the overall rate in warm, nutrient-depleted seas. In contrast, in warm nutrient-rich water like the equatorial divergence, limits on the the algal division rate other than the supply of nutrients like N, Si, or P seem proximately to control the overall rate of the system, e.g., average cell size and taxonomic differences. Even so, the absence or slowness of seasonal change can be achieved only by a large grazing term.

(2) In cold nutrient-rich water during summer, it is again the lid on algal growth rate that proximally drives the system. However, relative to the biological terms, including grazing and regeneration, the physical terms are more important than in warm water. This is particularly so near the freezing point (subantarctic area) but also is pronounced in parts of regions like the subarctic Pacific, where algal growth rate during summer is as low as in the controls of Martin et al. (1990; median, 0.15 d^{-1}, see Banse, in press), or is generally low as during winter.

(3) Because g equals the volume swept clear (F) by grazers (from $F = Vg/N$, with V [bottle size] and N [numbers of grazers in bottle] set to unity), the last column in Fig. 2B is numerically the same as the fraction of each liter that is cleared of particles every day, provided that cell division occurs at a uniform rate throughout 24 hr. Because this is unlikely, even more water needs to be cleared daily than indicated in Fig. 2B (McAllister, 1969; Lampert, 1987). Qualitatively, this can be seen by assuming less grazing during the day time, as it may be caused by lowering of g through the absence of diel migrators; then, some newly- formed algal cells can further photosynthesize before being eliminated, and the additional organic matter has to be removed at night. In any case, concentrations of live and inorganic particles, and thus the optical properties of the mixed layer (cf. Huntley et al., 1987), will be affected heavily by that amount of water passing daily through sieves or filters. This process is much more active in warm than in cold water (Fig. 2B).

(4) As reviewed below, the temperature-correlated grazing produces dissolved organic matter during feeding or from the subsequent growth processes

of the zooplankton, besides regenerating nutrients directly through animal metabolism. The result is a temperature correlation for the rate of cycling in the microbial loop that has nothing to do with temperature control of microbial metabolism.

The Temporal Balance of Phytoplankton and Zooplankton in Domain 3

Domain 3 comprises the areas of the open sea with pronounced seasonal blooms of phytoplankton. The blooms of the temperate and subpolar seas are usually thought to be tied to the seasonal cycle of incident light, as best typified by the Critical Depth concept of Sverdrup (1953). This concept combines water transparency with the seasonally changing mixed-layer depth and the phytoplankton physiology (24-hr compensation irradiance) to disallow or predict the rapid cell division needed for overcoming significantly the losses of cells and increasing the phytoplankton concentration. Smetacek and Passow (1990) noted that Sverdrup recognized grazing among the loss terms, but that subsequent users of the concept tended to overlook the animals. Their role will be stressed here. Besides the spring blooms in seas (and lakes), other blooms may result from removal of nutrient limitation of algal division rates, e.g., when increased vertical mixing injects nutrient-rich water into an impoverished euphotic zone, as in fall in the temperate oceans, or in spring in areas that adjoin equatorial Domain 1 on the poleward sides and possibly form largely circumglobal belts in each hemisphere. Again, the physics appear to set the stage. This subsection, however, shows that removal of growth limitation from light or nutrients is only a necessary condition for the onset of a bloom but may not be sufficient by itself.

Using an analytical model, Evans and Parslow (1985) investigated the minimum condition needed for simulating the difference between the annual cycle of phytoplankton in the north-hemispheric parts of Domains 2 and 3 (Fig. 1) and found that the depth of the mixed layer during the preceding winter is most critical. Neglecting the usually small sinking losses, they acknowledged that the fast increase of k due to the sudden increase of mean underwater light in the mixed layer, caused by stratification of the water column, sets the stage for a bloom. They showed, however, that the phytoplankton can escape grazing control and flourish only if there is little zooplankton around. That, in turn, is caused by a deep mixed layer in winter, which results in a very low mean underwater irradiance leading to very little algal growth. In consequence, animal production is smaller than the loss to predation, and zooplankton concentration declines. Because of the normally deep mixed layer of the North Atlantic, spring blooms are the rule.

Using a numerical model developed for 50°N in the central Gulf of Alaska (Frost, 1987), B.W Frost (personal communication) investigated the issue further. An essential feature of the model, which is based on field observations and shipboard experiments, is the large role of grazing of fast-growing small zooplankton on small phytoplankton cells that dominate the biomass and production in the subarctic Pacific. Both are preyed upon by copepod-like animals. The Pacific model was applied to the North Atlantic Weather Station *India* (59°N; 19°W), using the annual cycles of incident irradiance and temperature in the mixed layer of the site. The result was a low pigment content during winter, a pronounced spring bloom, and the variable chlorophyll level of the summer, all in principle as observed. Then, substitution of only the shallow mixed layer of the Pacific locality yielded the subarctic Pacific pattern for *India*, while substitution of only the Pacific irradiance still led to a spring bloom.

Unless being stirred by wind or surface cooling, deep "mixed layers", as defined by uniform profiles of temperature or density, are not turbulent. Therefore, under suitably quiet conditions, temporary blooms by small (slowly sinking) phytoplankton may develop in the North Atlantic in the middle of the winter in the euphotic part of the water column (e.g., Ryther and Hulburt, 1960). Similar observations were made by Riley (1957) in the northern Sargasso Sea. In both situations, quiet periods do not last for many days, but such "false starts" provide some food for overwintering zooplankton. Intermittent food-supply for the animals and resulting reproduction may be enough to cause, later in the year, the occasional absence of the normal spring bloom in the northern North Atlantic, as hypothesized by Steemann Nielsen (1962).

The role of grazing during the spring bloom itself is illustrated by another model (Fasham et al., 1983), developed for interpreting comprehensive observations in the Celtic Sea. The authors stated for calculating the temporal change of algal concentrations "it is necessary to postulate a high grazing rate from the early stages of the bloom", which, however, was not suffciently studied during the field work. The same importance of grazing can be seen from modeling an annual open-sea cycle of temperate phytoplankton by Jamart et al. (1977; 1979). Finally, phytoplankton spring blooms may be terminated by grazing instead of nutrient exhaustion followed by sinking of cells. The effect of grazing was first documented by Harvey et al. (1935) for the western English Channel, followed by the large study by Cushing (1963) for the western North Sea. During this period of termination, algal division rates may continue to rise (as in Cushing, 1963) or be steady, while the population growth rate is strongly negative (cf. Fig. 4, herein).

Spring blooms have attracted much attention so that there also are suggestions of material losses from other than grazing or settling of cells after nutrient exhaustion. Recent references to large relative declines of Chl_u or of cells from sinking live phytoplankton, without preceding nutrient exhaustion (from sticking-together of cells) are by, e.g., Alldredge and Gotschalk (1989); Jackson (1990); Kjørboe et al. (1990); and Passow (1991). Some earlier accounts of very high losses from cell mortality (as contrasted from grazing; e.g., Jassby and Goldman, 1974, in a lake) did not consider feeding by nanozooplankton and cannot be taken at face value. There are, however, observations of mass occurrence of empty, unbroken frustules of large pelagic diatoms (i.e., presumably not ingested; up to 80% of the suspended cells) from the North Sea (Cushing, 1955) and a suggestion of a daily loss of "moribund", but otherwise intact cells of at least 10-15% of the crop in the Gulf of Panama (Smayda, 1966; with other tropical observations). It remains to be seen whether parasitism or extra-cellular digestion by dinoflagellates or ciliates (see below) are the cause of such major occurrences.

In Domain 3, after macronutrients have become exhausted in the late spring or in summer, concentrations of phytoplankton may change little with time for several months in spite of appreciable cell division rates. These rates are primarily controlled by the rate of ammonium regeneration within the mixed layer. Cells sizes tend to be small so that sinking rates will be small, as are other physical losses (cf. Fig. 2B). For long periods during summer, the rate of cell division is principally balanced by the rate of grazing also over vast parts of Domain 3.

Fall blooms in Domain 3 depend on nutrient injection upon the diminishing of thermal stratification. This relaxation of nutrient control on algal k has to occur while incident irradiance, combined with water transparency and mixed-layer depth, still allow vigorous cell divisions. However, animal grazing also may hold a key to occurrence or failure: Riley (1967) calculated that the mean underwater irradiance for offshore Long Island Sound had to be appreciably higher in fall than in the late winter and spring for a fall bloom to occur. Apparently, the principal differences against the late-winter bloom were higher phytoplankton respiration and grazing losses in autumn, but not algal species composition or light adaptation.

In conclusion, even in Domain 3, seasonal change of phytoplankton concentration cannot be understood from studies of k alone or, by implication, of physics and physiology.

On Meso-Scale Spatial Relations between Phytoplankton and Zooplankton

While this paper focuses on time changes, temporal and spatial relations actually are intertwined in the free water because of eddy diffusion and advection. On a basin-wide scale, high concentrations of phytoplankton and zooplankton clearly go hand in hand (see Question 1, Introduction), but this is not always the case during the seasonal change at a single station or during a cruise on scales of 10^1 to 10^2 km. Negative correlation may be especially apparent when net samples of zooplankton, which miss the nanozooplankton and the juveniles of the larger species, are compared with chlorophyll concentrations. The seasonal alternation of phytoplankton and zooplankton, particularly during the period favorable for phytoplankton growth, is now understood as caused by grazing effects (Harvey et al., 1935). The spatial alternation, however, led to the theory of animal exclusion, i.e., avoidance of high concentrations of phytoplankton by zooplankton, principally through periodic vertical movement (diel migration) between water layers of different speed and direction, the stay in the upper layers to be inversely proportional to the amount of phytoplankton (Hardy, 1936). The theory engendered experimental work that did not support it, including the search for horizontal zooplankton movements (Bainbridge, 1953), and much discussion, which lead finally to rejection, as reviewed by Beklemishev (1957); it was abandoned by Hardy himself (1953). The theory, however, introduced animal behavior into the context of plant-animal relations among the plankton. For example, Riley (1976) and Evans et al. (1977) quantified interactions of animal behavior (periodic vertical migration), horizontal water movement, and grazing on phytoplankton, which generate horizontal patchiness.

Of late, the vertical relation within the euphotic zone of larger non-migratory animals, like some copepods, to phytoplankton is being discussed, the alternatives being accumulation in the maximum of biomass versus that of photosynthesis (e.g., Roman et al., 1986). The need for ample information about the vertical distribution of zooplankton was stressed by Longhurst and Williams (1979) in connection with modeling phytoplankton growth. Hamner (1985) further articulated the need to consider behavior in the context of the present paper, as well as in other aspects. Without doubt, more autecology (knowledge of species) than is available now will have to be used soon also in plankton production studies - whole animals do differ from their carbon and nitrogen content, as do phytoplankters from chlorophyll.

MECHANISMS OF PHYTOPLANKTON CONTROL BY GRAZING

This section uses examples to demonstrate the heavy influence of grazing on cell numbers and community composition but does not discuss the large amount of literature on mechanisms of feeding that has developed during the last two decades.

Phytoplankton Cell Numbers as Affected by Grazing

For a hypothetical example of grazing in an algal culture, Table 1A illustrates that animals need to ingest only a few cells to have a huge impact on the rate of increase of

Table 1. Numerical Examples Of The Effect Of Grazing On Phytoplankton Population Growth

A. Approximate cell numbers in dependence of grazing pressure after 1 and 5 divisions (N_1 and N_5) in a culture with 1000 cells initially, and number of cells ingested by the animals; k and g, instantaneous growth and grazing rates (unit: per generation time)

N_1	N_5	Cells eaten after 5 divisions	g	k-g
2,000	32,000	0	0	$\ln(2.0) = 0.693$
1,800	18,900	3,200	0.105	$\ln(1.8) = 0.588$
1,500	7,600	4,700	0.288	$\ln(1.5) = 0.405$
1,200	2,500	4,200	0.511	$\ln(1.2) = 0.182$

B. Approximate cell numbers of 5 algal species (a-e; increasing in size from a to e), each with 1000 cells initially, after 5 divisions under 3 grazing regimes. Non-selective feeding assumes $g = 0.10$ (see also A; adapted from B.W. Frost)

Alga	k	Grazing None N_5	Non-Selective ($g = 0.10$) N_5	Selective N_5	g
a	0.30	4,500	2,700	800	0.35
b	0.25	3,500	2,200	1,000	0.25
c	0.20	2,700	1,600	1,400´	0.13
d	0.15	2,100	1,300	1,900	0.02
e	0.10	1,600	1,000	1,600	0.01

their food. The reason is that the cells removed early on cannot contribute to the exponential population growth. Note that after the first division, as expected, the greatest numbers of cells is consumed when the rate of grazing equals the algal growth rate (Table 1a, last line). After the fifth division, however, an intermediate peak of ingestion is shown at $g = 0.288$ d^{-1}. This peak is caused by the interaction of the increasing clearing rate (g) with the decreasing algal mean concentration (not shown), such that, with an extended experiment at a constant k, the maximal ingestion (gain to the animals) is achieved at lower and lower clearing rates (K. Osgood, personal communication). A similar calculation to that in Table 1A was made by Braarud (1935) to show that, besides cell division, consumption of cells must be highly important for maintaining algal concentrations.

Division Rates of Grazers and Phytoplankton

To nip an incipient phytoplankton bloom in the bud, the individual grazers must be able to enhance their ingestion rate without quickly reaching satiation and/or to reproduce quickly and increase their numbers. I am unable to review functional feeding relations and reproduction (cf. Fenchel, 1982, for flagellates; Fenchel, 1980, for ciliates) but note that the small grazers that eat the generally small phytoplankton of Domains 1 and 2, can, at optimal (satiating) food levels, grow as fast or (greatly) faster than algae of the same size (flagellates: Fenchel 1982; review by Capriulo, 1990: Table 4.3; Ciliates, reviews by Banse 1982a; 1982b: Fig. 5). The maintenance of the balance under fluctuating weather has been modeled by Frost (in press) for station P; optimal feeding conditions do not seem to be required for the population of small grazers to keep up with the phytoplankton.

Phytoplankton Cell Sizes as Affected by Grazing

Grazing not only strongly influences cell concentrations, but also community composition. The range of phytoplankton cell sizes in most seas is large, as size easily extends from diameters of 1 μm to about 100 μm in samples of a few 100 ml. With identical cell shape, the equivalent mass range of $1:10^6$ corresponds among mammals to that between a shrew and the elephant. Single-celled species even may approach 1 mm in size (e.g., the diatom *Ethmodiscus rex*). Therefore, most grazers cannot cope with the entire range of prey but exert a size-selective grazing pressure. Moreover, at least among crustacean species, a given developmental (size) stage tends to select larger cells out of mixtures (review of experimental work, Mullin, 1963), so that for each phytoplankton species, a different balance between division and death rates may be struck. As one might glean from the size dependence of algal division rates, smaller species must be exposed to higher mortality. This is caused, in spite of the preference for larger cells within a grazer species, by the large size range within the grazer assemblages. In consequence, the population growth rate of a slowly dividing phytoplankton species may exceed that of a small, often-dividing alga (Table 1B). Some copepods may feed on the most abundant particles within the accessible range without much size selection (e.g., Turner and Tester, 1989).

The grazing (clearing) rate on large phytoplankton species is, in fact, much lower than indicated by the community grazing value of g in Fig. 2B: members of the dinoflagellate genus *Pyrocystis* of 0.25 to 0.35 mm diameter, in nutrient-depleted subtropical mixed layers near 26°C, divided approximately every 13 days (k approx. 0.05 d^{-1}, median for 6 populations; range over all populations, 4-22 d; Swift et al., 1976). The reason that large species with their necessarily low division rates persist must be because large size (perhaps helped by rarity) provides a refuge from predation. As implied by the equation in Fig. 2A, large species also must swim or possess flotation mechanisms (near-neutral buoyancy) to keep the sinking losses very small.

It was stated above that the size range of phytoplankton cells is too large to be handled by most grazers (e.g., ciliate, crustacean, and larvacean species; see, however, the non-selective pteropods, feeding by a parachute-like mucous sheet, and the salps). The implication is that the prey has to be appreciably smaller than most grazers. Recently, numerous heterotrophic dinoflagellates were shown to catch and digest extracellularly by a pseudopodial "pallium" so that cells or colonies like the diatom *Chaetoceros*, much larger than the dinoflagellate, can be used (Jacobson and Anderson, 1986; another mechanism in Hansen, 1991). Lessard (1991) suggests that such dinoflagellates may materially affect diatom blooms, and I add that the numerous unbroken, empty frustules mentioned earlier may be due to extracellular digestion. Furthermore, within the physically accessible size range of food for a species, considerable selectivity by flagellates, ciliates, and copepods in respect to the kind or quality of food has been experimentally demonstrated. While this may not be surprising for protozoa that phagocytize one food item at a time, "wholesale suspension feeders" like copepods may also fairly efficiently separate differently-sized food (Paffenhöfer 1988; Price, 1988), or even select between exponentially growing and stationary cells of the same diatom species (Cowles et al., 1988).

In conclusion to this section, similar to chlorophyll concentrations of the entire community (equation in Fig. 2A), the change of cell numbers with time cannot be understood from algal physiology alone. Also, small grazers can keep small phytoplankton at bay over the course of the year over vast stretches of the oceans. The mechanisms sketched above, however, neither explain the occasional increased concentrations of larger cells in, e.g., the open Gulf of Alaska (Domain 2; Clemons and Miller, 1984) nor the regular spring blooms of large-celled phytoplankton on continental shelves or in lakes.

GRAZING, DISSOLVED ORGANIC MATTER, AND THE MICROBIAL LOOP

The pelagic microbial loop comprises the trophic pathway(s) of dissolved organic matter (DOM) through (1) uptake by heterotrophic bacteria and subsequent degradation through metabolism by the bacteria themselves and (2) the food chain (web) of predators connected with them (Azam et al., 1983). The percentage of phytoplankton net production (dissolved and particulate) passing through this loop depends on the composition of the food web, including that of the phytoplankton. The rate at which the loop operates depends on the rate (less so on the quality) of the DOM supply, but not on temperature limitation of bacterial metabolism or growth: The *in situ* division rates of the free-living bacteria, one to at best few div. d^{-1}, are well below those that would be attained at optimal substrate levels. (In very cold water, though, Pomeroy and Wiebe, 1988, have noted temperature limitation of bacterial metabolism.) Nitrogen and carbon do not cycle identically, *inter alia* because of CO_2 release from respiration (e.g., Lancelot and Billen, 1985; Vézina and Platt, 1988; Ducklow, 1991); the issue is neglected below, in part because "sloppy feeding" by metazoans generates DOM mechanically and, presumably, without elemental fractionation.

This section shows that the main source of DOM in the pelagic domain may be the feeding of animals on phytoplankton and other organisms (Fig. 3); the hypothesis of large-scale, direct solubilization of organic detritus by bacteria (Azam et al., 1990) is not sufficiently quantified for estimating its overall role (cf. Banse, 1990, for the mesopelagial). Figure 3 develops the diagram by Ducklow (1983; vertically arranged boxes on the left side, with the same growth efficiencies), likewise depicting a generic foodweb. The principal differences from Ducklow's views are that (1) only < 10% of plant net production is exuded by the phytoplankton and (2) instead, much DOM is lost in front of the mouths of the animals ("sloppy feeding"). Low exudation, as percentage of the rate of primary production (not of biomass, as suggested by Bjørnsen, 1988) is supported by, e.g., Lancelot and Billen (1985), Fasham et al. (1990), Huntley et al. (1991: Footnote 9), and Sharp (1991). Methodological problems that can lead to false low values were discussed by Williams (1981), but I note that recent papers with high values (e.g., Wolter, 1982; Lancelot, 1983) by their design did not exclude DOM generation from grazing on ^{14}C-labeled phytoplankton. Some references to sloppy feeding can be found in Williams (1981). Additional papers on a carnivorous crustacean (Dagg, 1974) and a copepod fed with two diatoms (Copping and Lorenzen, 1980), which both found high losses, support my choice of large percentage values in Fig. 3 (additional references in Eppley et al., 1981; Riemann et al., 1986; Poulet et al., 1991). The major uncertainty for loss of DOM by animals during ingestion concerns the protozoans that take up whole food by phagocytosis and void the undigested fraction from vacuoles. On the basis of conservative bacterial biomass estimates and high growth efficiencies, Hagström et al. (1988) concluded for an enclosed oligotrophic community that zooflagellates and small ciliates released about half of the ingested organic C. The authors regarded this as the most likely explanation of the carbon budget in a carboy, which I presume contained only few copepod-sized animals. On the other hand, Caron et al. (1985) reported that only about 1/10 of the removed POC appeared as DOC in a flagellate culture. Similarly, low values were found in cultures of one flagellate (Andersson et al., 1985) and in each of two ciliates (Taylor et al., 1985; Verity, 1985; in the latter study, from consideration of assimilation efficiency). In any case, metazoans lose additional DOM as urea (excretion), leakage through gills, during molting, and from feces (Vézina and Platt, 1988; theoretical arguments for potentially high losses are given in Jumars et al., 1989). Finally, a high overall contribution by zooplankton (50-60% of primary production, minus release of DOC by lysis of algal cells and uptake of ambient [other] DOC by bacteria) was suggested with hesitation for an eutrophic lake (Riemann and Søndergaard, 1986); in their view, zooplankton would be especially important in generating DOC when being in control of algal population dynamics, as discussed herein.

Figure 3 also shows a short circuit from bacteria to the DOM pool but does not evaluate it because of lack of data. It may come about by exo-enzymes and the diffusive loss to the bacteria of DOM while degrading particulate substrate, or by loss of bacterial cell material upon lysis after virus infection (for the latter, see Proctor and Fuhrman, 1990; Heldal and Bratback, 1991; Fuhrman, this volume). The latter paper estimated that 2-24% of the bacterial population might be lysed per hour. Considering 24-hr averages, however, the normal value in Domains 1 and 2 must be $\leq$ 0.02 hr^{-1}, because 0.02 hr^{-1} corresponds to 2.3 div. d^{-1}, i.e. is at the high end of the usually assumed range of division rates of free-living bacteria. With the high bacterial biomass in oligotrophic waters, relative to the concentrations of phytoplankton (Fuhrman et al., 1989; Cho and Azam, 1990, and below), the bacterial division rates cannot be much larger, and hence, neither can their death rates (from lysis, as well as grazing!).

Figure 3 lets about 50% of phytoplankton net production pass through the bacteria. The fraction was 40% in Ducklow's (1983) food web, and > 50% in Joint and Morris (1982: Fig. 1b), who applied essentially the same literature as I did to a food web with three boxes containing living organisms (not counting the bacteria). Judging from actual measurements, the fraction seems to be independent of the oligotrophic-eutrophic gradient (Azam et al., 1990), i.e. it has to do with the composition of the food web. In Fig. 3, the high percentage of 50% is the consequence of using large terms for DOM losses from sloppy feeding (as in Joint and Morris, 1982), and may be on the high side of the possible range (but see Hagström et al., 1988; Small et al., 1989, among recent supportive papers based on measurements). Also in Fig. 3, production by net-collected zooplankton (upper boxes) is about as large as that by bacteria, which would be a slightly lower ratio than in Ducklow's (1983) web when a reasonable gross growth efficiency is applied to his "herbivores". Both estimates are contrary to the summary of freshwater observations by Cole et al. (1988), which suggests that pelagic bacterial production tends to be almost twice that of net-collected zooplankton. In any case, however, granted the uncertainty of the numbers below the three boxes in the upper part of Fig. 3, the relatively small role of exudation of DOM by the phytoplankton can be regarded as established, as it was tentatively suggested by Pomeroy (1984). Hence, the microbial loop depends first of all on animal feeding, and it is the animals that proximally connect the mostly nitrate-based new production with the regenerated production. Therefore, zooplankton has to be studied, not only for understanding the temporal changes of concentrations of phytoplankton and heterotrophic bacteria via the grazing effect, but also for assessing the supply of nutrients (minerals for phytoplankton, DOM for bacteria).

As an aside about sloppy feeding and using the model by Ducklow et al. (1989) as an example, I wish to point out the difficulties of operational definitions of budget terms in models versus problems of measurements, and the resulting problems of comparing data. Ducklow et al. defined zooplankton excretion as a fixed percentage of ammonium excretion (e.g., 10% for metazoans, Table 8A.1), and ammonium excretion as a fixed percentage of the input, i.e., of the measured ingestion; the two excretion terms clearly

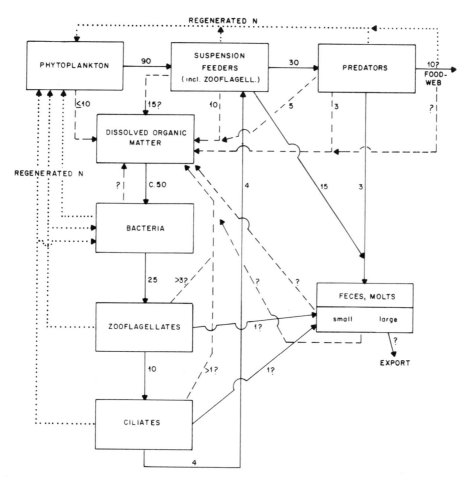

Fig. 3. Flow of material among major categories (as boxes) in the euphotic zone over deep water. Full lines, particulate organic material; broken lines, dissolved organic material; dotted lines, inorganic nutrients. Input of nutrients supporting "new production" from deep water, the atmosphere, and nitrogen (N_2) fixation, not shown. Numbers near arrows, percent of primary production.

referred to assimilated material. Indeed, ingestion for their observations was measured as assimilation of labeled food (Roman et al., 1986) and did not include losses in front of the mouth. Sloppy feeding, as a potentially important mortality term for the phytoplankton, as well as a DOM source, would have been incorporated had the grazing measurement been based on disappearance of whole cells; instead of the very large reported sinking losses of phytoplankton cells, the model would have provided for more food within the euphotic layer. With a grazing definition based on cell counts, however, ingestion would be incorrectly modeled, and true excretion and calculated rates of growth processes of the animals would be biased.

The correlation of animal action with temperature over much of the ocean (Fig. 2B) introduces a new type of temperature dependence into the operation of the microbial loop, which acts through direct release of nutrients by the animals, or the generation of DOM. I suggest that, at least over the vast expanse of Domain 1 and likely also during summer in Domain 3, the rate of primary production (principally regenerated production) is proximally due to the animals who can realize fast clearing rates at higher temperatures. Note that g, as in Fig. 2B, is the product of animal concentration and specific clearing rate, but I presume that the concentration of microzooplankton in the subtropical gyres is less than in low-latitude upwelling regions or the temperate zone; hence, the increase of g with temperature in Fig. 2B does reflect higher activity. In contrast to the animals, the bacteria in these warm oligotrophic waters must be on the average in poor shape, and their temperature dependence cannot control the rate of DOM break-down. Fuhrman et al. (1989) studied the ratio of bacterial biomass to small phytoplankton in the Sargasso Sea, and Cho and Azam (1990) compared the ratio in the North Pacific central gyre. The bacterial biomass (as C) in the two subtropical gyres is two to three times that of the phytoplankton. Both papers remark that a high steady state ratio of bacteria/phytoplankton biomass requires bacterial division rates to be several ties smaller than those of the primary producers. At particular sites in the oligotrophic seas, substrate limitation of bacterial growth may be less pronounced than at other places (Weisse and Scheffel-Möser, 1991). Note that my view about the key role of the animal release of DOM in Domain 1 differs from that by, e.g., Cushing (1980), who stresses the role of high temperature for regeneration of nutrients from DOM and, by implication, for bacterial metabolism.

In conclusion, the physically-driven rate of new production ultimately controls the geographic location and rate of total production (Eppley et al., 1979; Platt et al., 1989, about the f-ratio) because of the inherent tendency to nutrient depletion of the surface layers from gravity's pull. The rate of nitrate supply from below, as the principal determinant of the rate of new production, depends on physical oceanographic processes that are not temperature-dependent as are biological rates. Yet, the correlation with temperature that appears presumably rests in biological processes (Fig. 2B). Reflecting on future directions of research, I note the total production is usually several times the new production because of the regeneration of ammonium, especially in the vast Domain 1. Part of the ammonium is directly excreted by the animals feeding on phytoplankton but normally, more is contributed via bacteria and their consumption in the microbial loop. The bacteria require DOM as substrate, which is principally provided by animal action. Since bacterial metabolism in the euphotic zone of Domain 1, by and large, is substrate-limited (see the low bacterial division rates), the animals may be the key for appreciating in full the growth rates and population dynamics of the heterotrophic bacterioplankton and the phytoplankton, unless there is, in fact, large-scale solubilization of organic detritus by bacteria (Azam et al., 1990). I suggest that future system-directed studies of primary productivity and biochemical cycles in the sea should focus on the animals more than has been historically the case.

HISTORICAL INTERACTIONS IN MARINE PHYTOPLANKTON AND ZOOPLANKTON RESEARCH

This section sketches the development of marine phyto- and zooplankton research, as far as it pertains to the understanding of the processes that cause and maintain the observed temporal changes of phytoplankton, and to the role of animals in the microbial loop. The earlier years are stressed, since time has provided more perspective.

Phytoplankton Dynamics

The first datum regarding the quantitative study of relations between concentrations and processes among the plankton is Hensen's (1887) work that introduced the counting of specimens from catches obtained with reproducible methods toward the goal of understanding the "physiology of the sea". In his paper, most of the key ideas about the generic marine foodweb (minus the microbial loop) and the desirable measurements of concentrations and rates are discussed. Apart from Hensen's key issue, that fish production must bear some relation to phytoplankton production, he states that annual production is obtainable from summing all losses or measuring the rate of production. He presents grazing experiments (admitted to have been crude), which are interpreted from the copepods' point of view, speculation about superfluous feeding, a combination of data on life span and fecundity of copepods to estimate daily mortality for the steady-state case, and the extension to the entire plankton of the concept that the net increase or decrease is the balance between the daily population growth and loss. Few rates were derived, but the estimate of primary production of about 150 g $m^{-2}yr^{-1}$ ash-free dry weight for Kiel Bight was close to the truth as we know it now.

By about 1910, temperate seasonal cycles of phytoplankton, as well as zooplankton, were well-known; the issue of regional differences had been brought into sharp focus (Gran, 1912; Lohmann, 1912), after it was raised by the Plankton-Expedition of 1889, and the question about the causes posed (cf. Mills, 1989). Furthermore, division rates in the sea had been estimated for diatoms (cf. Lohmann, 1908) and were determined for a few dinoflagellates (e.g., Apstein, 1911; Gran, 1912). The respiration rates of copepods also had been measured and the rates of ingestion inferred (albeit incorrectly) during the first decade of the century (review by Pütter, 1925). Thus, rates of change entered the scene quickly. For decades, however, there was no follow-up to Hensen's (1887) prospectus of a *process study* of the food web, neither by himself and his school (see the epilogue to Hensen's school by Brandt, 1925), nor elsewhere.

The first large-scale attempt to understand phytoplankton-zooplankton interactions from correlation with hydrography failed. In the northwest European waters in the interval 1902-1908, the International Council for the Exploration of the Sea (ICES) studied hydrography and plankton quarterly for five years at about 330 stations with largely uniform methods; about 11,400 vertical, divided net hauls were inspected (Kyle, 1910). The decision not to follow Hensen and enumerate, but in spite of warnings, (Apstein, 1905), only to estimate abundance, was taken because the expected result did not seem to justify the enormous investment in counting. No understanding of the main problems of the day, i.e., the reasons for the seasonal changes and regional differences of the phytoplankton, was obtained. Clearly in hindsight, however, the status of the field was at fault, since the hauls of the German cruises were enumerated (see Brandt, 1925) without leading to startling results either. The result of long-range import, though, was that concentrations were not available even for the common net plankton of the ICES area in spite of the large expenditure of ship time.

The discovery of the nanoplankton by Lohmann (1903), and especially his comprehensive seasonal study published in 1908, led ICES, under the leadership of H.H. Gran, to collect water samples of phytoplankton during the quarterly cruises for the spring of 1912, to study the spring bloom. Point (bottle) samples were by then expected to show more clearly the relation between environment and phytoplankton abundance than the integrating net hauls ever could. The common species in about 1,200 samples were counted (Gran, 1915), but, even with this concerted effort, the mechanism behind the spring bloom could not be elucidated.

Gran's (1915) paper demonstrates well the attitude of the time toward the conditions of growth of phytoplankton, which was to hold sway for several decades. The algal division rate was driven by light, temperature, and nutrients, and the population change was affected by turbulence, horizontal advection, and sinking; four lines were given to the animals as unknowns. Nathanson's view (1908), that even more important than phytoplankton removal by vertical turbulence or death from mixing with low-salinity or cold water, was the destruction by animal grazing, largely was not accepted. For example, animals, as grazers, neither were mentioned in the entirely static monograph of the distribution of Atlantic nanoplankton by Lohmann (1920), nor in Bigelow's (1926) monograph of the Gulf of Maine, nor by Nathanson's former collaborator Gran (1932) in a review of the status of [phyto-] plankton research (see, however, Braarud, 1935, in Gran's institute). Nathanson (1908) further noted that the algae never would have the opportunity to exhaust the nutrients fully because of the zooplankton, a task made easier for the animals since the algal division rates would have to decline at very low nutrient values while, at the same time, grazing would release new nutrients. Also this insight was lost. Apparently, Nathanson's limited resonance was caused by other, unacceptable concepts of his plankton production, as well as by his low rank in the hierarchy (Mills, 1989).

The refinement of the concept of nutrient-limitation earlier in the century and the introduction of the colorimetric method for phosphate determination in the early 1920s (cf. Mills, 1989) apparently reinforced the restricted view about the processes driving phytoplankton dynamics. Examples are the review of the field by Gran (1932), the presentation of the results of the "Meteor" Expedition 1925-1927 (Hentschel and Wattenberg, 1930; Hentschel, 1933), and the discussion of conditions of plankton production in the Danish waters by Steemann Nielsen (1940). By 1940, rate measurements (O_2 method) had already been used by Steemann Nielsen and several other researchers.

The treatment of nutrient supply to the phytoplankton essentially was restricted, at least during about the first third of this century, to that of today's new production, since the discussed sources were those below the mixed layer or in rivers; very rarely was regeneration in the euphotic zone alluded to (e.g., Nathanson, 1908). Only in the second third of this century did the zooplankton become recognized broadly as another source of nutrients (e.g., Harvey et al., 1935; Harvey, 1955; Riley, 1956; new experimental work: Harris, 1959), instead of merely being mentioned. The development led to the vast, important literature about regenerated production, following Dugdale and Goering (1967), which also involves the microbial action. At the latest, by the time of the review by Harrison (1980), 4/5 or so of algal nutrient needs in the open sea commonly were considered to be based on regenerated production. Even so, temporal changes of phytoplankton continued to be treated without consideration of grazing, beyond mentioning it, even when zooplankton samples had been taken concomitantly by nets (Halldal, 1953; Holmes, 1956; Riley, 1957; all from the same type of collections on ocean weather ships).

The Nanozooplankton

In the context of this paper, the nanozooplankton was introduced by Lohmann (1903; 1908) who described numerous colorless flagellates, including dinoflagellates, down to about 3 μm size, and counted live specimens of about $\geq$ 10 μm size routinely in his about bi-weekly collections near Kiel (Lohmann, 1908). He enumerated colorless live flagellates of $\geq$ 5 μm size on his section through the North and South Atlantic (Lohmann, 1912; 1920). For the entire size range covered in the Kiel study (through copepods, from net hauls), calculation of the plasma volumes (i.e., cell volumes minus the vacuoles of diatoms and *Noctiluca*; for conversion to phytoplankton C, see Strathmann, 1967) permitted a meaningful comparison of mass between producers and consumers, resulting in the first estimates of the relative roles of protozoans and metazoans over the course of a year: The ratios of monthly means of plasma volumes (mm^3 [100 liter]$^{-1}$) of phytoplankton (diatoms) : protozoans : metazoans were 58.7(22.3) : 7.0 : 33.5 and 10.5(1.4) : 2.8 : 19.8 for April to October and November to February, respectively (calculated from Lohmann, 1908, but moving his colorless dinoflagellates from phytoplankton to protozoans). Next, Lohmann assumed "the small daily ration" for metazoans of 1/10 of their plasma volume, and for protozoans of 1/2, and calculated a conspicuous role of protozoans (roughly half of the consumption of phytoplankton), thus anticipating what we know now. However, by limiting the daily consumption to $\leq$ 1/3 of the algal plasma volume (this being the estimated average daily increase), he deprived himself of a full estimate of the phytoplankton dynamics under the influence of grazing. In addition, the use of 250 x or 500 x magnification in an ordinary light microscope prevented the recognition of the picoplankton, including the small flagellates that would turn out to be principal bacteriovores, and would have limited the accuracy of any budget, even if experimental determinations of rates had been available.

A similar effort was not to be undertaken for about half a century. Quite to the contrary, the efficiency of the work devoted to the nanozooplankton diminished because of inadequate methods. While Gran (1912) counted common, larger nanoplankton in live samples under medium magnification on shipboard, the previously mentioned ICES study of spring 1912 relied on Flemming's solution (chromic and osmic acids; Gran, 1915) that stains the cells so that autotrophic and heterotrophic species cannot be distinguished; the same holds for the often-used Lugol's KI - I$_2$ solution. Other researchers, e.g., Hentschel (1932) also enumerated live samples with 250 x magnification and, thus, was limited regarding small cells, but did distinguish colorless and chlorophyll-bearing flagellates. The treatment of his data, however, was static (Hentschel, 1933). An especially unfortunate episode was the sharp critique of the centrifuge, employed widely through the 1910s and 1920s for concentrating the nanoplankton, by Steemann Nielsen (1932). It led Hart (1942) in his Antarctic phytoplankton studies on the *Discovery* to switch to nets, after having investigated the nanoplankton with a centrifuge at 119 stations; his results, which for decades were the major source on Antarctic phytoplankton distribution, became strongly biased toward large or spiny species.

The introduction of the inverted microscope (Utermöhl, 1931), important as it was, continued to restrict routine counting to about 300-500 x magnification and, thus, tended to exclude the cells below 3-5 μm from consideration. Phytoplankton workers, though, mostly continued to enumerate small animals like tintinnid ciliates along with the plant cells (e.g., Holmes, 1956), which were missed by the zooplankton collections that normally relied on nets. During this period, however, these data normally were not utilized beyond making remarks about occurrence. In conclusion, a principal weakness in methodology between about 1910 and the recognition of the microbial loop in the last third of the century was that studies on phytoplankton in water samples and on zooplankton collected with vertically integrating nets were not joined; too often, even the sampling was not done at the same time and stations.

Essentially the full size range of planktonic heterotrophs, from bacteria to large zooplankters and at the same stations and times, first was covered by Soviet studies (cf. Vinogradov et al. 1972; 1977). In the English-speaking world, the microbial loop (named by Azam et al., 1983) was brought to common attention by Pomeroy (1974). Since then, the entire size range of phyto- and zooplankton, down to the small end of the spectrum, has become accessible to routine, accurate enumeration and microscopical mass estimates through the introduction of fine, physically flat filters, the epifluorescent microscope, and appropriate staining techniques. Using these methods, heterotrophic bacteria and eucaryotes are easily distinguishable from photoautotrophs (Hobbie et al., 1977; Porter and Feig, 1980). In addition, isotopic methods separate bacterial growth from that of phytoplankton (e.g., Staley and Konopka, 1985) and permit measurement of species-specific grazing rates (Lessard and Swift, 1985); other incubations can determine growth of, and grazing rates upon, bacteria, phytoplankton, and nano- (micro-) zooplankton in bulk (e.g., as chlorophyll) or by species (Landry and Hassett, 1982). Much of this knowledge has recently been reviewed by Fenchel (1988). In principle, we now seem to be more or less set!

The Inexplicable Delay of Synthesis

The red thread through this last section is the question why the effects of grazing on population dynamics of phytoplankton were recognized fully only halfway a century after Hensen (1887), and why phytoplankton dynamics during much of the following half century, more often than not, were still studied as if division rates (driven by the abiotic environment), mixing, and sinking were the only processes affecting the temporal changes.

As previously stated, the key ideas about the pelagic food web can be found in Hensen (1887), although the ranking of the processes was largely obscure. Yet, as recounted by Mills (1989), the principal conceptual advances concerning phytoplankton - zooplankton interactions were made much later, by Harvey et al. (1935; with the quantitative addendum by Fleming, 1939), followed by Riley (1946), and Cushing (1955; 1959a). The field collections for these treatments were guided by fairly firm hypotheses, i.e., models in a more or less rigorous form came first so that as many people as seemed to be needed could be put onto the job. Later models (e.g., Riley, 1965; Jamart et al., 1979) used existing data. Remarkably enough, the papers relied on net-collected zooplankton, with the exception of Cushing (1955), but why the nanoplankton was disregarded in the design of those studies is difficult to understand. Anyway, in hindsight, it seems to me that if not the paper by Harvey et al. (1935), then the 19-page treatment of the phytoplankton-zooplankton balance on Georges Bank (Riley, 1946) should have been the turning point in quantitatively treating and understanding the planktonic food web and should have had an immediate profound influence on the field. Figure 4, slightly modified from the original 1946 figures, shows clearly that the net production of phytoplankton does not match (= predict) the algal population changes (Fig. 4B), principally because of grazing, and underlines that even in Domain 3, phytoplankton dynamics cannot be understood only from cell division rates. It is beside the point that Riley's data base recently was critiqued (Davis, 1987). Further, I noted above that the small animals were not considered, and point out here that the scale of the ordinate of Fig. 4A is too high, as converting C to chlorophyll yields pigment concentrations that are several times too large (see also Davis, 1987).

The sociological puzzle about the delay in acceptance of the new paradigm is that Riley (1946) published in a widely distributed journal at a time when few people practiced oceanography and, therefore, still could read most of the few journals that existed in the field (cf. Wyrtki, 1990), as well as study the older literature. Yet for several decades to follow, in this country and elsewhere, phytoplankton and zooplankton research remained loosely coupled. This occurred although in the United States since about 1950, the number of oceanographers and fishery scientists increased tenfold, with the concomitant opportunity for the > 9/10 who entered the field since then to disregard intellectual ballast and start out afresh. It did not happen, although Riley taught, Harvey (e.g., 1942) elaborated on his earlier papers, Cushing (1955) showed by field and model studies that "the changes in zooplankton numbers lie at the very center of the dynamic cycle", and in a subsequent paper (1958) gave a clear exposition of the issues treated in the present paper at a pace-setting symposium in 1956 that was attended by most of the workers of the day concerned with natural phytoplankton, and Steele (1958), like Riley and Cushing, in his models considered algal division as largely being compensated by grazing loss (see also Steele, 1974). It did not happen, although more popular articles (so to speak) about the close balance between cell division and grazing were written by Steemann Nielsen (1958) and Cushing (1959b, based on Cushing, 1959a). Later studies, by Cushing and his colleagues on the phytoplankton in a *Calanus* patch in the North Sea (Cushing, 1963), and comprehensive Soviet plankton investigations of the 1960s and 1970s were already mentioned. When, however, by the 1970s integrated field approaches to the marine plankton began in North America (references in Walsh, 1977), the message of Fig. 4 had not been broadly accepted. While Mills (1989) discusses mostly personal (or sociological) reasons connected with Riley for the slow acceptance of the new view, the number of authors mentioned above suggests that the delay may have deeper roots.

The reasons for the lateness in acceptance are not obvious, so that it is difficult to point to a remedy. Perhaps, the prevailing use in most models of lowest, common denominators like carbon or chlorophyll often made the approach more attractive to earth scientists, including chemical oceanographers, than to the plankton workers who often were biologists rather than trained oceanographers. Certainly, early on, the biologists scoffed at the organismic over-simplification of the models and, therefore, were not apt to change their own ways, i.e., to study the trees instead of the forest in spite of the

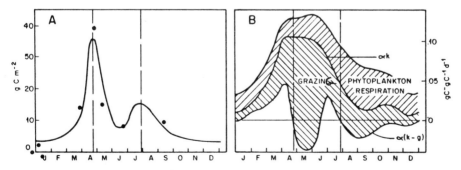

Fig. 4. Seasonal cycle of phytoplankton on Georges Bank. From Riley (1946), with vertical lines added. A) Observed and calculated cruise means. Vertical lines aligned with the two calculated peaks. B) Calculated specific rates of gross production, algal net production (two upper lines), and population change (lower line).

introduction of methods for assessing chlorophyll concentrations and later the ^{14}C-method for measuring photosynthesis; without doubt, many missed the oceanographic message over the mathematical approach. The weaknesses were not solely those of the marine phytoplankton workers. I noted (1982c) regarding experimental ecosystems (mesocosms) that in both the marine and fresh water realms during the 1960s and 1970s, the design of synecological studies (as carbon, chlorophyll, or other categories well above the species level) usually under-represented the zooplankton; I implied insufficiently broad reading of the literature as one reason (about, e.g., predation and community structure), but stating it again more broadly, the message had not been taken up.

Presently, the biologists often feel uneasy about the inability of directly testing model predictions when concentrations of abstract model entities like "herbivores" are not amenable to measurement (cf. Murdoch, 1966; Rigler, 1975). In contrast, earth scientists like biogeochemists do not generally appreciate the complications of the food web, which are rooted in life cycles and the size ranges of organisms. For example, how are we/they to reconcile that cell division rates of a chain-forming diatom depend on cell size, while the grazers and Stoke's law "see" the size of the entity? More basically, the essentially unicellular phytoplankton, entirely endowed with the same "enzyme" chlorophyll and analyzed accordingly, might be treated as one category, while nothing called "zooplankton" or "herbivores" should be in a rigorous model, especially since there is little temptation to lump all animals in the absence of routine measurement of rates of community ingestion or growth, as there is for phytoplankton. Note that the dilution method of measuring algal growth rate and grazing for the usually dominant small plankton (Landry and Hasset, 1982; results, e.g., in Strom and Welschmeyer, 1991), when based on pigments, comes close to the "lump-all" approach of measuring photosynthesis. Its results, by the way, should convince the empiricists among the phytoplankton workers that grazing is indeed most important for understanding phytoplankton dynamics.

CONCLUSION

What about the future? I mentioned above with regard to methodology for nanoplankton and smaller size classes that *"in principle*, we now seem to be more or less set." For the issues treated in this paper as a whole, however, I see obstacles that might slow or impede progress, i.e., (1) attitudes, (2) system-inherent problems, and (3) the lure of new methods currently being developed.

Addressing the first obstacle, I wonder whether the teachers of biological oceanography, including myself, during the last four or five decades (since the theory was developed by Riley, Cushing, and Steele) have to accept considerable blame for the neglect of grazing effects on temporal and spatial distributions of phytoplankton, and more generally, of feedback between zooplankton and phytoplankton. If so, change will be slow in coming, because even if all teachers changed their ways henceforth, the manpower pool in biological/chemical oceanography turns over slowly. Without new attitudes, however, new methods will lose much of their potential impact. Improvements of outlook would result from teaching biological oceanography in more process-oriented ways, including enhanced use of differential equations; from the publication of the near-perfect textbook (i.e. several of us should try it, in spite of the small market that will discourage publishers); and from embellishing the old art of teaching with the aid of new tools, e.g., software (distributed like books) for models of the subjects treated herein, which allow experimentation in classrooms and lab sections for the mutual education of teachers and students.

Considering the second obstacle, zooplankton research is more difficult than phytoplankton research because of the greater variety in sizes and processes among the animals, including long generation times and the concomitant larger spatial scales to be studied, and because of the obstacles to rate measurements already alluded to. Fortunately, the nano- and picoplankton fractions of the zooplankton are least affected by these problems and very often, are the important size fraction for many processes. However, their population dynamics cannot be fully understood without considering their predators, which is the same story as the phytoplankton-zooplankton interaction treated in the present paper, but on a larger spatial and temporal plane because of the longer life span of the predators. The overall goal has to be feedback models of phytoplankton-zooplankton interactions, where the size composition of the grazers (or predators) affects that of the phytoplankton (or "herbivores"), but where, conversely, the size composition and abundance of the food affects the individual growth rates of the animals and thus shifts their size composition. This partially physiological feedback has to be further converted into demographic predictions, i.e., a change in animal numbers. Ultimately, will comprehensive models of global change, which may calculate a sea surface temperature change of a few degrees that likely would shift the distributions of dominant zooplankton differentially, be able to predict the animal-mediated effects on phytoplankton and the resulting changes in nutrient and CO_2 fluxes, or are nutrient-phytoplankton-grazer-predator relations inherently of a chaotic nature even in a large aquarium (mesocosm)?

Regarding the third obstacle, new instrumentation, that has been or is being developed, is apt to assess routinely and quickly larger size ranges in any sample than the old approaches, to deliver more counts for any sample than is possible from manual enumeration, and to provide more collecting points. This will narrow confidence limits about single points, as well as establish the variance around the means for an aggregate of stations, both in space and in time. Until artificial intelligence can be used to count and identify specimens, however, electronic devices will most likely be based on registering common denominators, analogous to the venerable chlorophyll measurement, and perhaps acknowledge the biological (and functional) variety by differentiating and enumerating by, for example, volume or diameter. Since the data will be amenable to machine-processing and handling, the attraction and temptation to collect ever more will be enormous even when the novelty wears off. The biologists (who should and must learn to use these new tools), however, must keep in mind and tell their colleagues from other subdisciplines that, e.g., among common temperate suspension-feeding copepods ("herbivores"), an adult female of *Pseudocalanus* sp. has the same size as a juvenile (copepodite 1) *Calanus* sp. Yet, their function is different, one adding members to the population by converting food into eggs, the other just growing and entering the next size class. Moreover, both feed, respire, defecate, etc., but the rates differ by about a factor of two, in this case the bigger species (*Calanus*) being faster (Frost, 1980). Because exponential growth reigns for many individual invertebrates and among populations, such differences matter! Moreover, what if a co-occuring particle of the same size and body form were a predatory copepod?

DEDICATION

The reader may wonder whether this paper with its emphasis on zooplankton is not an epilogue to the Eppley Era of marine primary productivity studies. Only history can tell. For us to arrive at the present juncture, however, Dick Eppley's papers during the last 30 years often served as roadmarks or even turning points. Therefore, I dedicate this essay with genuine admiration, and also affection to Dick, as well as to Jean Eppley.

ACKNOWLEDGEMENTS

I wish first of all to thank P. Falkowski for arranging the symposium held in honor of R.W. Eppley and for inviting me to participate. I am grateful to NASA for their support (Grant No. NAGW-1007) and to B.W. Frost and E.J. Lessard for hints regarding the literature. Contribution no. 1905 of the School of Oceanography, University of Washington.

REFERENCES

Alldredge, A. L., and Gotschalk, C.C., 1989, Direct observations of the mass flocculation of diatom blooms: Characteristics, settling velocities and formation of diatom aggregates, *Deep-Sea Res.*, 36:159.

Anderson, G. C., Lam, R. K., Booth, B. C., and Glass, J. M., 1977, A description and numerical analysis of the factors affecting processes of production in the Gulf of Alaska, NOAA 03-5-022-67 Environmental Assessment Alaskan Continental Shelf, VII:477.

Anderson, G. C., and Munson, R. E., 1972, Primary productivity studies using merchant vessels in the North Pacific Ocean, *in*: "Biological Oceanography of the Northern North Pacific Ocean," A.Y. Takenouti, ed., Idemitsu Shoten, Tokyo.

Andersson, A., Lee, C., Azam, F., and Hagström, Å, 1985, Release of amino acids and inorganic nutrients by heterotrophic marine microflagellates, *Mar. Ecol. Prog. Ser.*, 23:99.

Apstein, C., 1905, Die Schätzungsmethode in der Planktonforschung, *Wiss. Meeresunters., Abt. Kiel,* N.F. 8:103.

Apstein, C., 1911, Biologische Studie über *Ceratium tripos* var. *subsalsa* Ostf, *Wiss. Meeresunters., Abt. Kiel,* N.F. 12:137.

Azam, F., Fenchel, T., Field, J. G., Gray, J. S., Meyer-Reil, L. A., and Thingstad, F., 1983, The ecological role of water-column microbes in the sea, *Mar. Ecol. Prog. Ser.*, 10:257.

Azam, F., Cho, B. C., Smith, D. C., and Simon, M., 1990, Bacterial cycling of matter in the pelagic zone of aquatic ecosystems, *in*: "Large Lakes," M.M. Tilzer, and C. Serruya, eds., Springer-Verlag, Berlin.

Bainbridge, R., 1953, Studies on the relationships of zooplankton and phytoplankton, *J. Mar. Biol. Ass. U.K.*, 32:385.

Banse, K., 1982a, Cell volumes, maximal growth rates of unicellular algae and ciliates, and the role of ciliates in the marine pelagial, *Limnol. Oceanogr.*, 27:1059.

Banse, K., 1982b, Mass-scaled rates of respiration and intrinsic growth in very small invertebrates, *Mar. Ecol. Prog. Ser.*, 9:281.

Banse, K., 1982c, Experimental marine ecosystem enclosures in a historial perspective, *in*: "Marine Mesocosms: Biological and Chemical Research in Experimental Ecosytems," G.D. Grice, and M.R. Reeve, eds., Springer-Verlag, New York.

Banse, K., 1990, New views on the degradation and disposition of organic particles as collected by sediment traps in the open sea, *Deep-Sea Res.*, 37:1177.

Banse, K., Rates of phytoplankton cell division in the field and in iron enrichment experiments, *Limnol. Oceanogr*, 36(8), in press.

Banse, K., and Yong, M., 1990, Sources of variability in satellite-derived estimates of phytoplankton production in the eastern tropical Pacific, *J. Geophys. Res.*, 95:7201.

Beklemishev, C. W., 1957, The spatial interrelationship of marine zoo- and phytoplankton (In Russian) *Tr. Inst. Okeanol., Akad. Nauk*SSSR, 20:253. Translation *in*: "Marine Biology," B.N. Nikitin, ed., Am. Inst. Biol. Sci., Washington, D.C. (1959).

Bienfang, P. K., and Szyper, J. P., 1981, Phytoplankton dynamics in the subtropical Pacific Ocean off Hawaii, *Deep-Sea Res.*, 28:981.

Bigelow, H.B., 1926, Plankton of the offshore waters of the Gulf of Maine, *Bull. U.S. Bureau Fish.*, 40:1.

Bjørnsen, P. K., 1988, Phytoplankton exudation of organic matter: Why do healthy cells do it?, *Limnol. Oceanogr.*, 33:151.

Brandt, K., 1925, Victor Hensen und die Meeresforschung, *Wiss. Meeresunters., Abt. Kiel*, N.F. 20:49.

Braarud, T., 1935, The "Øst" Expedition to the Denmark Strait 1929. II. The phytoplankton and its conditions of growth, *Norske Vidensk.-Akad. Oslo, Hvalrådets Skr.* 10.

Capriulo, G. M., 1990, Feeding-related ecology of marine protozoa, *in*: "Ecology of Marine Protozoa," G.M. Capriulo, ed., Oxford University Press, Oxford.

Caron, D. A., Goldman, J. C., Andersen, O. K., and Dennett, M. R., 1985, Nutrient cycling in a microflagellate food chain: II. Population dynamics and carbon cycling, *Mar. Ecol. Prog. Ser.*, 24:243.

Cho, B. C., and Azam, F., 1990, Biogeochemical significance of bacterial biomass in the ocean's euphotic zone, *Mar. Ecol. Prog. Ser.*, 63:253.

Clemons, M. J., and Miller, C. B., 1984, Blooms of large diatoms in the oceanic, subarctic Pacific, *Deep-Sea Res.*, 31:85.

Cohen, E. B., Grosslein, M. D., Sissenwine, M. P., Steimle, F., and Wright, W. R., 1982, Energy budget of Georges Bank, *Can. Spec. Publ. Fish. Aquat. Sci.*, 59:95.

Cole, J.,J., Findlay, S., and Pace, M. L., 1988, Bacterial production in fresh and saltwater ecosystems: A cross-system overview, *Mar. Ecol. Prog. Ser.*, 43:1.

Copping, A. E., and Lorenzen, C. J., 1980, Carbon budget of a marine phytoplankton-herbivore system with carbon-14 as a tracer, *Limnol. Oceanogr.*, 25:873.

Cowles, T. J., Olson, R. J., and Chisholm, S. W., 1988, Food selection by copepods: Discrimination on the basis of food quality, *Mar. Biol.*, 100:41.

Cullen, J. J., Lewis, M. R., Davis, C. O., and Barber, R. T., Photosynthetic characteristics and estimated growth rates indicate grazing is the proximate control of primary production in the equatorial Pacific, *J. Geophys. Res.*, in press.

Cushing, D. H., 1955, Production and a pelagic fishery, *Fish. Invest. Lond.*, Ser. II, 18(7):1.

Cushing, D. H., 1958, The effect of grazing in reducing the primary production: A review, *Rapp. Cons. Intern. Expl. Mer.*, 144:149.

Cushing, D. H., 1959a, On the nature of production in the sea, *Fish. Invest. London*, Ser. II, 22(6):1.

Cushing, D. H., 1959b, The seasonal variation in oceanic production as a proplem in population dynamics, *J. Cons. Int. Expl. Mer.*, 23:455.

Cushing, D. H., 1963, Studies on a *Calanus* patch. V. The production cruises in 1954: Summary and conclusions, *J. Mar. Biol. Ass. U.K.*, 43:387.

Cushing, D. H., 1980, Production in the central gyres of the Pacific, *Intergov. Oceanogr. Comm., Techn. Ser.*, 21:31.

Dagg, M. J., 1974, Loss of prey body contents during feeding by an aquatic predator, *Ecology*, 55:903.

Dandonneau, Y., and Gohin, F., 1984, Meridional and seasonal variations of the sea surface chlorophyll concentration in the southwestern tropical Pacific (14 to 32°S, 160 to 175°E), *Deep-Sea Res.*, 31:1377.

Davis, C. S., 1987, Components of the zooplankton production cycle in the temperate ocean, *J. Mar. Res.*, 45:947.

de Baar, H. J. W., Buma, A. G. J., Nolting, R. F., Cadée, G. C., Jacques, G., and Tréguer, P. J., 1990, On iron limitation of the Southern Ocean: experimental observations in the Weddell and Scotia Seas, *Mar. Ecol. Prog. Ser.*, 65:105.

Ducklow, H. W., 1983, Production and fate of bacteria in the oceans, *BioScience*, 33:494.

Ducklow, H.,W., 1991, The passage of carbon through microbial foodwebs: Results from flow networkmodels, *Mar. Microb. Food Webs*, 5:129.

Ducklow, H. W., Fasham, M. J. R., and Vézina, A. F., 1989, Derivation and analysis of flow networks for open ocean plankton systems, *in*: "Coastal and Estuarine Studies," F. Wulff, J.G. Field, and K.H. Mann, eds., Springer-Verlag, Berlin.

Dugdale, R. C., and Goering, J. J., 1967, Uptake of new and regenerated forms of nitrogen in primary productivity, *Limnol. Oceanogr.*, 12:196.

El-Sayed, S. Z., 1967, On the productivity of the southwest Atlantic Ocean and the waters west of the Antarctic Peninsula, *Antarct. Res. Ser.*, 11:15.

El-Sayed, S. Z., 1988, Productivity of the Southern Ocean: A closer look, *Comp. Biochem. Physiol.*, 90B:489.

Eppley, R. W., 1972, Temperature and phytoplankton growth in the sea, *U.S. Fish. Bull.*, 70:1063.

Eppley, R. W., Renger, E. H., and Harrison, W. G., 1979, Nitrate and phytoplankton production in southern California coastal waters, *Limnol. Oceanogr.*, 24:483.

Eppley, R. W., Horrigan, S. G., Fuhrman, J. A., Brooks, E. R., Price, C. C., and Sellner, K., 1981, Origins of dissolved organic matter in southern California coastal waters: Experiments on the role of zooplankton, *Mar. Ecol. Prog. Ser.*, 6:149.

Evans, G. T., and Parslow, J. S., 1985, A model of annual plankton cycles, *Biol. Oceanogr.*, 3:327.

Evans, G. T., Steele, J. H., and Kullenberg, G. E. B., 1977, A preliminary model of shear diffusion and plankton populations, *Scot. Fish. Res. Rept.*, 9.

Fasham, M. J. R., Ducklow, H. W., and McKelvie, S. M., 1990, A nitrogen-based model of plankton dynamics in the oceanic mixed layer, *J. Mar. Res.*, 48:591.

Fasham, M. J. R., Holligan, P. M., and Pugh, P. R., 1983, The spatial and temporal development of the spring phytoplankton bloom in the Celtic sea, April 1979, *Prog. Oceanogr.*, 12:87.

Feldman, G., and 12 co-authors, 1989, Ocean color, Availability of the Global Data Set, *EOS (Tr. Am. Geophys. U.)*, 70:634, 640.

Fenchel, T., 1980, Suspension feeding in ciliated protozoa: Feeding rates and their ecological significance, *Microb. Ecol.*, 6:13.

Fenchel, T., 1982, Ecology of heterotrophic microflagellates, II. Bioenergetics and growth, *Mar. Ecol. Prog. Ser.*, 8:225.

Fenchel, T., 1988, Marine plankton food chains, *Ann. Rev. Ecol. Syst.*, 19:19.

Fleming, R. H., 1939, The control of diatom populations by grazing, *J. Cons. Int. Expl. Mer*, 14:210.

Frost, B. W., 1980, The inadequacy of body size as an indicator of niches in the zooplankton, *in*: "Evolution and Ecology of Zooplankton Communities," W. C. Kerfoot, ed., University Press of New England, Hanover.

Frost, B. W., 1987, Grazing control of phytoplankton stock in the open subarctic Pacific Ocean: A model assessing the role of mesozooplankton, particularly the large calanoid copepods *Neocalanus* spp, *Mar. Ecol. Prog. Ser.*, 39:49.

Frost, B. W., The role of grazing in nutrient-rich areas of the open sea, *Limnol. Oceanogr.*, 36(8), in press.

Fuhrman, J. A., Sleeter, T. D., Carlson, C. A., and Proctor, L. M., 1989, Dominance of bacterial biomass in the Sargasso Sea and its ecological implications, *Mar. Ecol. Prog. Ser.*, 57:207.

Fukuchi, M., 1980, Phytoplankton chlorophyll stocks in the Antarctic Ocean, *J. Oceanogr. Soc. Japan*, 36:73.

Gran, H. H., 1912, Pelagic plant life, Ch. VI, *in*: "The Depths of the Ocean," J. Murray, and J. Hjort, eds., McMillan, London.

Gran, H. H., 1915, The plankton production in the North European waters in the spring of 1912, *Bull. Plankt. Cons. Int. Expl. Mer.*, 1912.:5.

Gran, H. H., 1932, On the conditions for the production of plankton in the sea, *Rapp. Cons. Int. Expl. Mer.*, 75: 37.

Hagström, Å., Azam, F., Andersson, A., Wikner, J., and Rassoulzadegan, F., 1988, Microbial loop in an oligotrophic pelagic marine ecosystem: Possible roles of cyanobacteria and nanoflagellates in the organic fluxes, *Mar. Ecol. Prog. Ser.*, 49:171.

Halldal, P., 1953, Phytoplankton investigations from Weather Ship M in the Norwegian Sea, 1948-49, *Norske Vidensk.-Akad. Oslo, Hvalrådets Skr.*, 38.

Hamner, W. M., 1985, The importance of ethology for investigations of marine zooplankton, *Bull. Mar. Sci.*, 37:414.

Hansen, P. J., 1991, *Dinophysis*: a planktonic dinoflagellate genus which can act both as a prey and a predator of a ciliate, *Mar. Ecol. Prog. Ser.*, 69:201.

Hardy, A. C., 1936, Plankton ecology and the theory of animal exclusion, *Proc. Linn. Soc.*, 148:64.

Hardy, A. C., 1953, Some problems of pelagic life, *in*: "Essays in Marine Biology (Richard Elmhirst Memorial Lectures)," S.M. Marshall and A.P. Orr, eds., Oliver and Boyd, Edinburgh.

Harris, E., 1959, The nitrogen cycle in Long Island Sound, *Bull. Bingham Oceanogr. Coll.*, 17:31.

Harrison, W. G., 1980, Nutrient regeneration and primary production in the sea, *in*: "Primary Productivity in the Sea," P.G. Falkowski, ed., Plenum Press, New York.

Hart, T. J., 1942, Phytoplankton periodicity in Antarctic surface waters, *Discovery Repts.*, 21:261.

Harvey, W. H., 1937, Note on selective feeding by Calanus, *J. Mar. Biol. Ass. U.K.*, 22:97.

Harvey, H. W., 1942, Production of life in the sea, *Biol. Rev.*, 17:221.

Harvey, H. W., 1955, "The Chemistry and Fertility of Sea Waters," Cambridge University Press, Cambridge.

Harvey, W. H., Cooper, L. H. N., Lebour, M. V., and Russell, F. S., 1935, Plankton production and its control, *J. Mar. Biol. Ass. U.K.*, 20:407.

Hayward, T. L., 1987, The nutrient distribution and primary production in the central North Pacific, *Deep-Sea Res.*, 34:1593.

Hela, I., and Laevastu, T., 1961, "Fisheries Hydrography," Fishing News, London.

Heldal, M., and Bratbak, G., 1991, Production and decay of viruses in aquatic environments, *Mar. Ecol. Prog. Ser.*, 72:205.

Hensen, V., 1887, Ueber die Bestimmung des Plankton's oder des im Meere treibenden Materials an Pflanzen und Tieren, *Fünfter Ber. Komm. Unters. dtsch. Meere Kiel (1882 bis 1886)*, XII-XVI:1.

Hentschel, E., 1932, Die biologischen Methoden und das biologische Beobachtungs-material der "Meteor" Expedition, *Wiss. Ergebn. Dtsch. Atl. Exp. "Meteor" 1925-1927*, 10:1.

Hentschel, E., 1933, Allgemeine Biologie des Südatlantischen Ozeans, Das Pelagial der obersten Wasserschicht, *Wiss. Ergebn. Dtsch. Atl. Exp. "Meteor" 1925-1927,* 11(1. Lief.):1.

Hentschel, E., and Wattenberg, H., 1930, Plankton and Phosphat in der Oberflächenschicht des Südatlantischen Ozeans, *Ann. Hydrogr.,* Berlin. 58:273.

Hobbie, J. E., Daley, R. J., and Jasper, S., 1977, Use of Nuclepore filters for counting bacteria by fluorescence microscopy, *Appl. Environ. Microbiol.,* 33:1225.

Holmes, R. W., 1956, The annual cycle of phytoplankton in the Labrador Sea, 1950-51, *Bull. Bingham Oceanogr. Coll.,* 16:1.

Huntley, M. E., Lopez, M. D. G., and Karl, D. M., 1991, Top predators in the Southern Ocean: A major leak in the biological carbon pump, *Science,* 253:64.

Huntley, M. E., Marin, V., and Escritor, F., 1987, Zooplankton grazers as transformers of ocean optics: a dynamic model, *J. Mar. Res.,* 45:911.

Jackson, G. A., 1990, A model of the formation of marine algal flocs by physical coagulation processes, *Deep-Sea Res.,* 37:1197.

Jacobson, D. M., and Anderson, D. M., 1986, Thecate heterotrophic dinoflagellates: feeding behaviour and mechanism, *J. Phycol.,* 22:249.

Jamart, B. M., Winter, D. F., Banse, K., Anderson, G.,C., and Lam, R. K., 1977, A theoretical study of phytoplankton growth and nutrient distribution in the Pacific Ocean off the Northwestern U.S. coast, *Deep-Sea Res.,* 24:753.

Jamart, B. M., Winter, D. F., and Banse, K., 1979, Sensitivity analysis of a mathematical model of phytoplankton growth and nutrient distribution in the Pacific Ocean off the northwestern U.S. coast, *J. Plankton Res.,* 1:267.

Jassby, A. D., and Goldman, C. R., 1974, Loss rates from a lake phytoplankton community, *Limnol. Oceanogr.,* 19:618.

Joint, I. R., and Morris, R. J., 1982, The role of bacteria in the turnover of organic matter in the sea, *Oceanogr. Mar. Biol. Ann. Rev.,* 20:65.

Jumars, P. A., Penry, D. L., Baross, J. A., Pery, M. J., and Frost, B. W., 1989, Closing the microbial loop: Dissolved carbon pathway to heterotrophic bacteria from incomplete ingestion, digestion and absorption in animals, *Deep-Sea Res.,* 36:483.

Kesteven, G. L, and Laevastu, T., 1957-1958, "The Oceanographic Conditions for Life and Abundance of Phytoplankton in Respect to Fisheries," Food and Agriculture Organization of the United Nations, Fisheries Division, Biology Branch, FAO/57/6/4144 and FAO/58/5/3749 (FB/58/T13).

Kjørboe, T., Andersen, K. P., and Dam, H. G., 1990, Coagulation efficiency and aggregate formation in marine phytoplankton, *Mar. Biol.,* 107:235.

Knauss, J. A., 1963, Equatorial current systems, *in:* "The Sea," Vol. 2, M.N. Hill, ed., Wiley, New York.

Kremer, J. N., and Nixon, S. W., 1978, "A Coastal Marine Ecosystem: Simulation and Analysis," Springer-Verlag, Berlin.

Kyle, H. M., 1910, Résumé des observations sur le plankton des mers explorées par le Conseil pendant les années 1902-1908, 1ere partie., *Bull. Trimestr. Cons. Int. Expl. Mer.*

Lampert, W., 1987, Vertical migration of freshwater zooplankton: Indirect effects of vertebrate predators on algal communities, *in:* "Predation: Direct and Indirect Impacts on Aquatic Communities," W.C. Kerfoot, and A. Sih, eds., University Press of New England.

Lancelot, C., 1983, Factors affecting phytoplankton extracellular release in the Southern Bight of the North Sea, *Mar. Ecol. Prog. Ser.,* 12:115-121.

Lancelot, C., and Billen, G. 1985. Carbon-nitrogen relationships in nutrient metabolism of coastal marine ecosystems, *in*: "Advances in Aquatic Microbiology," Vol. 3, H.W. Jannasch, and P.J.L. Williams, eds., Academic Press, London.

Landry, M. R., and Hassett, R. P., 1982, Estimating the grazing impact of marine microzooplankton, *Mar. Biol.*, 67:283.

Laws, E. A., DiTullio, G. R., and Redalje, D. G., 1987, High phytoplankton growth and production rates in the North Pacific subtropical gyre, *Limnol. Oceanogr.*, 32:905.

Lessard, E. J., 1991, The trophic role of heterotrophic dinoflagellates in diverse marine environments, *Mar. Microb. Food Webs*, 5:49.

Lessard, E. J., and Swift, E., 1985, Species-specific grazing rates of heterotrophic dinoflagellates in oceanic waters, measured with a dual-label radioisotope technique, *Mar. Biol.*, 87:289.

Levitus, S., 1982, "Climatological Atlas of the World Ocean," *U.S. Dept. of Commerce, Natl. Oceanic Atmosph. Admin., NOAA Prof. Pap.* 13, 173 pp.

Lewis, M., this volume.

Lohmann, H., 1903, Neue Untersuchungen über den Reichthum des Meeres an Plankton und über die Brauchbarkeit der verschiedenen Fangmethoden, *Wiss. Meeresunters., Abt. Kiel,* N.F. 7:1.

Lohmann, H., 1908, Untersuchungen zur Festellung des vollständigen Gehaltes des Meeres an Plankton, *Wiss. Meeresunters., Abt. Kiel,* N. F. 10:129.

Lohmann, H., 1912, Untersuchungen über das Pflanzen- und Tierleben der Hochsee, *Veröff. Inst. Meeresk. Univ. Berlin,* N.F. A.1: 1.

Lohmann, H., 1920, Die Bevölkerung des Ozeans mit Plankton nach den Ergebnissen der Zentrifugenfänge während der Ausreise der "Deutschland" 1911, *Arch. Biontol.* 4(3):1.

Longhurst, A. R., and Williams, R., 1979, Materials for plankton modelling: Vertical distribution of Atlantic zooplankton in summer, *J. Plankton Res.*, 1:1.

Love, C. M., 1970, "EASTROPAC Atlas", Vol. 4, *U.S. Fish Wildl. Serv. Circ.*, 330.

Love, C. M., 1974, "EASTROPAC Atlas", Vol. 8, *U.S. Fish Wildl. Serv. Circ.*, 330.

Martin, J. H., Gordon, R. M., Fitzwater, S. E., and Broenkow, W. W., 1989, VERTEX: phytoplankton/iron studies in the Gulf of Alaska, *Deep-Sea Res.*, 36: 649.

Martin, J. H., Fitzwater, S. E., and Gordon, R. M., 1990, Iron deficiency limits phytoplankton growth in Antarctic waters, *Global Biogeochem. Cycles*, 4:5.

McAllister, C. D., 1969, Aspects of estimating zooplankton production from phytoplankton production, *J. Fish. Res. Bd. Canada*, 26:199.

Menzel, D. W., and Ryther, J. H., 1961, Annual variations in primary production of the Sargasso sea off Bermuda, *Deep-Sea Res.*, 7:282.

Mills, E. L., 1989, "Biological Oceanography: An Early History, 1870-1960," Cornell University Press, Ithaca, NY.

Mullin, M. M., 1963, Some factors affecting the feeding of marine copepods of the genus *Calanus, Limnol. Oceanogr.*, 8:239.

Murdoch, W. M., 1966, "Community structure, population control, and competition"--a critique, *Amer. Nat.*, 100:219.

Nathanson, A., 1908, Beiträge zur Biologie des Plankton, I. Über die allgemeinen Produktionsbedingungen im Meere, *Intern. Rev. ges. Hydrobiol.*, 1:37.

Paffenhöfer, G. A., 1988, Feeding rates and behavior of zooplankton, *Bull Mar. Sci.*, 43:430.

Peña, M. A., Lewis, M. R., and Harrison, W. G., 1990, Primary productivity and size structure of phytoplankton biomass on a transect of the equator at 135°W in the Pacific Ocean, *Deep-Sea Res.*, 37:295.

Peña, M. A., Lewis, M. R., and Harrison, W. G., 1991, Particulate organic matter and chlorophyll in the surface layer of the equatorial Pacific Ocean along 135°W, *Mar. Ecol. Prog. Ser.*, 72:179.

Platt, T., and Harrison, W. G., 1985, Biogenic fluxes of carbon and oxygen in the ocean, *Nature*, 318:55.

Platt, T., Harrison, W. G., Lewis, M. R., Li, W. K. W., Sathyendranath, S., Smith, R. E., and Vézina, A. F., 1989, Biological production of the oceans: The case for a consensus, *Mar. Ecol. Prog. Ser.*, 52:77.

Pomeroy, L. R., 1974, The ocean's food web, a changing paradigm, *BioScience*, 24:499.

Pomeroy, L. R., 1984, Significance of microorganisms in carbon and energy flow in marine ecosystems, *in*: "Current Perspectives in Micobial Ecology," J.M. Klug and C.A. Reddy, eds., Am. Soc. Microbiol., Washington.

Pomeroy, L. R., and Wiebe, W. J., 1988, Energetics of microbial food webs, *Hydrobiologia,* 159:7.

Porter, K. G., and Feig, Y. S., 1980, The use of DAPI for identifying and counting aquatic microflora, *Limnol. Oceanogr.*, 25:943.

Poulet, S. A., Williams, R., Conway, D. V. P., and Videau, C., 1991, Co-occurrence of copepods and dissolved free amino acides in shelf sea waters, *Mar. Biol.*, 108:373.

Price, H. J., 1988, Feeding mechanisms in marine and freshwater plankton, *Bull. Mar. Sci.*, 43:327.

Proctor, L. A., and Fuhrman, J. A., 1990, Viral mortality of marine bacteria and cyanobacteria, *Nature*, 343:60.

Pütter, A., 1926, Die Ernährung der Copepoden, *Arch. Hydrobiol.*, 15:70.

Riebesell, U., 1991, Particle aggregation during a diatom bloom, II. Biological aspects, *Mar. Ecol. Prog. Ser.,* 69:281.

Riemann, B., Jørgensen, N. O. G., Lampert, W., and Fuhrman, J. A., 1986, Zooplankton induced changes in dissolved free amino acids and in production rates of freshwater bacteria, *Microb. Ecol.*, 12:247-258.

Riemann, B., and Søndergaard, M., 1986, Bacteria, *in*: "Carbon Dynamics in Eutrophic, Temperate Lakes," B. Riemann, and M. Søndergaard, eds., Elsevier, Amsterdam.

Rigler, F. H., 1975, The concept of energy flow and nutrient flow between trophic levels, *in*: "Unifying Concepts in Ecology," W.H. van Dobben, and R.H. Lowe-McConnell, eds., Junk, The Hague.

Riley, G. A., 1946, Factors controlling phytoplankton populations on Georges Bank, *J. Mar. Res.*, 6:54.

Riley, G. A., 1956, Oceanography of Long Island Sound, 1952-1954, IX. Production and utilization of organic matter, *Bull. Bingham Oceanogr. Coll.*, 15:324.

Riley, G. A., 1957, Phytoplankton of the North Central Sargasso Sea, 1950-52, *Limnol. Oceanogr.*, 2:252.

Riley, G. A., 1965, A mathematical model of regional variations in plankton, *Limnol. Oceanogr.*, 10 (Suppl.):202.

Riley, G. A., 1976, A model of plankton patchiness, *Limnol. Oceanogr.*, 21:873.

Riley, G. A., 1967, The plankton of estuaries, *in*: "Estuaries," G.H. Lauff, ed., *Amer. Assoc. Adv. Sci., Publ.*, 83.

Riley, G. A., Stommel, H., and Bumpus, D. F., 1949, Quantitative ecology of the plankton of the western North Atlantic, *Bull. Bingham Oceanogr. Coll.*, 12(3):1.

Roman, M. R., Yentsch, C. S., Gauzens, A. L., and Phinney, D. A., 1986, Grazer control of the fine-scale distribution of phytoplankton in warm-core Gulf Stream rings, *J. Mar. Res.*, 44:795.

Ryther, J. H., and Hulburt, E. M., 1960, On winter mixing and the vertical distribution of phytoplankton, *Limnol. Oceanogr.*, 5:337.

Sharp, J. H., 1991, Review of carbon, nitrogen, and phosphorus biogeochemistry, *Rev. Geophys.*, (Suppl.) Apr. 91:648.

Shulenberger, E., 1978, The deep chlorophyll maximum and mesoscale environmental heterogeneity in the western half of the North Pacific central gyre, *Deep-Sea Res.*, 25:1193.

Silver, M. W., Gowing, M. M., and Davoll, P. J., 1986, The association of photosynthetic picoplankton and ultraplankton with pelagic detritus through the water column (0-2000 m), *in*: "Photosynthetic Picoplankton," T. Platt, and W. K. W. Li, eds., *Can. Bull. Fish. Aquat. Sci.*, 214:311.

Small, L. F., Landry, M. R., Eppley, R. W., Azam, F., and Carlucci, A. F., 1989, Role of plankton in the carbon and nitrogen budgets of Santa Monica Basin, California, *Mar. Ecol. Prog. Ser.*, 56:57.

Smayda, T. J., 1966, A quantitative analysis of the phytoplankton of the Gulf of Panama. III. General ecological conditions, and the phytoplankton dynamics at 8°45'N, 79°23'W from November 1954 to May 1957, *Bull. Inter-Amer. Trop. Tuna Comm.*, 11:353.

Smetacek, V., and Passow, U., 1990, Spring bloom initiation and Sverdrup's critical-depth model, *Limnol. Oceanogr.*, 35:228.

Smetacek, V., Scharek, R., and Nöthig, E. M., 1990, Seasonal and regional variation in the pelagial and its relationship to the life history cycle of krill, *in*: "Antarctic Ecosystems, Ecological Change and Conservation," K.R. Kerry, and G. Hempel, eds., Springer-Verlag, Berlin.

Staley, J. T., and Konopka, A., 1985, Measurement of in situ activities of nonphotosynthetic microorganisms in aquatic and terrestrial habitats, *Ann. Rev. Microbiol.*, 39:321.

Steele, J. H., 1958, Plant production in the northern North Sea, *Scot. Home Dept., Mar. Res.*, 1958(7):1.

Steele, J. H., 1974, "The Structure of Marine Ecosystems," Harvard University Press, Cambridge, MA.

Steemann Nielsen, E., 1933, Über quantitative Untersuchung von marinem Plankton mit Utermöhls umgekehrten Mikroskop, *J. Cons. Int. Expl. Mer*, 8:201.

Steemann Nielsen, E., 1940, Die Produktionsbedingungen des Phytoplanktons im Übergangsgebiet zwischen der Nord- und Ostsee, *Medd. Komm. Danmarks Fisk.-Havunders., Ser. Plankton,* 3(4): 1.

Steemann Nielsen, E., 1958, The balance between phytoplankton and zooplankton in the sea, *J. Cons. Int. Expl. Mer.*, 23:178.

Steemann Nielsen, E., 1962, The relationship between phytoplankton and zooplankton in the sea, *Rapp. Cons. Int. Expl. Mer.*, 153:178.

Strathmann, R. R., 1967, Estimating the organic carbon content of phytoplankton from cell volume or plasma volume, *Limnol. Oceanogr.*, 12:411.

Strom, S. L., and Welschmeyer, N. A., 1991, Pigment-specific rates of phytoplankton growth and microzooplankton grazing in the open subarctic Pacific, *Limnol. Oceanogr.*, 36:50.

Sverdrup, H. U., 1953, On conditions for the vernal blooming of phytoplankton, *J. Cons. Int. Expl. Mer.*, 18:287.

Swift, E., Stuart, M., and Meunier, V., 1976, The *in situ* growth rates of some deep-living oceanic dinoflagellates: *Pyrocystis fusiformis* and *Pyrocystis noctiluca*, *Limnol. Oceanogr.*, 21:418.

Taylor, G. T., Iturriaga, R., and Sullivan, C. W., 1985, Interactions of bactivorous grazers and heterotrophic bacteria with dissolved organic matter, *Mar. Ecol. Prog. Ser.*, 23:129.

Turner, J. T., and Tester, P. A., 1989, Zooplankton feeding ecology: Nonselective grazing by the copepods *Acartia tonsa* Dana, *Centropages velificatus* De Oliveira, and *Eucalanus pileatus* Giesbrecht in the plume of the Mississippi River, *J. Exp. Mar. Biol. Ecol.*, 126:21.

Utermöhl, H., 1931, Neue Wege in der quantitativen Erfassung des Planktons (mit besonderer Berücksichtigung des Ultraplanktons), *Verh. int. Ver. theor. angew. Limnol.*, 5:567.

Verity, P. G., 1985, Grazing, respiration, excretion, and growth rates of tintinnids. *Limnol. Oceanogr.*, 30:1268.

Vézina, A. F., and Platt, T., 1988, Food web dynamics in the ocean, I. Best-estimates of flow networks using inverse methods, *Mar. Ecol. Prog. Ser.*, 42:269.

Vinogradov, M. E., Krapivin, V. F., Menshutkin, V. V., Fleyshman, B. S., and Shushkina, E. A., 1973, Mathematical model of the functions of the pelagial ecosystem in tropical regions (from the 50th voyage of the R/V VITYAZ), *Oceanology*, 13:704.

Vinogradov, M. E., and Menshutkin, V. V., 1977, The modeling of open-sea ecosystems, *in*: "The Sea," Vol. 6, E.D. Goldberg et al., eds., Wiley, New York.

Walsh, J. J., 1976, Herbivory as a factor in patterns of nutrient utilization in the sea, *Limnol. Oceanogr.*, 21:1.

Walsh, J. J., 1977, A biological sketchbook for an eastern boundary current, *in*: "The Sea," E.D. Goldberg et al., eds., Vol 6, Wiley, New York.

Wassmann, P., Vernet, M., Mitchell, B. G., and Rey, F., 1990, Mass sedimentation of *Phaeocystis pouchetii* in the Barents Sea, *Mar. Ecol. Prog. Ser.*, 66:183.

Weisse, T., and Scheffel-Möser, U., 1991, Uncoupling the microbial loop: Growth and grazing loss rates of bacteria and heterotrophic nanoflagellates in the North Atlantic, *Mar. Ecol. Prog. Ser.*, 71:195.

Welschmeyer, N. A., and Lorenzen, C. J., 1985, Chlorophyll budgets: Zooplankton grazing and phytoplankton growth in a temperate fjord and the central Pacific gyres, *Limnol. Oceanogr.*, 30:1.

Williams, P. J. leB., 1981, Incorporation of microheterotrophic processes into the classical paradigm of the planktonic foodweb, *Kieler Meeresforsch.*, Sonderh., 5:1.

Wolter, K., 1982, Bacterial incorporation of organic substances released by natural phytoplankton populations, *Mar. Ecol. Prog. Ser.*, 7:287.

Wyrtki, K., 1990, Becoming an oceanographer forty years ago, *Oceanography*, 3:39.

BIOSPHERE, ATMOSPHERE, OCEAN INTERACTIONS: A PLANT PHYSIOLOGIST'S PERSPECTIVE

Joseph A. Berry

Department of Plant Biology
Carnegie Institution of Washington
Stanford, CA   94305

INTRODUCTION

Recent evidence that the CO_2 concentration of the atmosphere is increasing (Keeling, 1973), coupled with simulation studies of possible climatic consequences of increasing CO_2 (Hansen et al., 1985) and evidence from ice cores that relates past changes in climate with changes in the concentration of CO_2 (Barnola et al., 1987), has stimulated a great deal of interest in the mechanisms that determine the CO_2 concentration of the atmosphere. Sampling programs reporting the composition of the atmosphere have been in place since the late 50s and the technologies for analysis of air recovered from ice cores provides a source of fossil air extending to approximately 160,000 years ago. These data provide convincing evidence that our present era is undergoing unprecedented change, and efforts to address this issue have made demands on many scientific disciplines not previously thought to have global aspirations. Biology, with its strong reductionist tradition has been slow to respond, and we find ourselves trying very hard to play catch-up with atmospheric scientists who have already made substantial progress towards including biological processes in their global studies. My goal here is to draw attention to some of the advantages of studying biological processes in closed systems, and to show that some properties of the global system can be understood by treating it as closed.

Many efforts to extend the study of traditional biological objects to larger scales flounder, because it becomes very difficult to define or measure the exchange of gases as the scale is expanded. The key issue here is the ability to enclose and manipulate the environment of the object. For example, we know a great deal about photosynthesis and respiration of single leaves of plants. It is more difficult, but possible, to enclose and control whole plants, trees, and parts of ecosystems (Field and Mooney, 1990). Aerodynamic techniques may be applied to observe net exchange of water vapor, CO_2, and heat in open systems approaching a km^2 (Baldocchi et al., 1988), and studies of the dynamics of the surface boundary layer using aircraft mounted sensors were used to estimate surface fluxes on a regional scale (Wofsy et al., 1988). However, as the scale of the measurement increases, complications due to heterogeneity, time resolution, and logistics take their toll. Despite heroic efforts, our knowledge of the gas exchange on geographically significant scales is very limited.

Primary Productivity and Biogeochemical Cycles in the Sea
Edited by P.G. Falkowski and A.D. Woodhead, Plenum Press, New York, 1992

However, it is very important to note that closure again occurs at the global scale. By invoking the assumption that the total volume of the atmosphere is constant, it is possible to simply and accurately measure net fluxes of gases between the atmosphere and other reservoirs, principally, the terrestrial biosphere and the upper oceans. It is also possible to take advantage of the mixing of the atmosphere to infer something about the spatial and temporal distribution of sources and sinks for CO_2 over the earth's surface (Tans et al., 1990). To put global biology on a firm footing, we also need to define suitable biological objects to associate with these sources and sinks. Satellite remote sensing offers clear promise in this area (Sellers et al., 1989).

Efforts to interpret data on atmospheric composition in terms of global biological processes hinge on the assumption that a sub-set of the earth system consisting of the atmosphere, biosphere, and upper ocean can be treated as a closed system. There are very discouraging statements on this issue. For example, Kasting et al. (1988) in discussing the constraints on CO_2 and O_2 concentration in the atmosphere acknowledges that "...living organisms play an important role in the exchange of CO_2 with the atmosphere...", but states that these activities will have little impact on the long-term steady-state established by the tectonic cycling of carbon. Broecker (1982), in discussing the basis for lower CO_2 concentrations during the glacial as opposed to interglacial periods (intervals of thousands of years), argues that "...the atmospheric CO_2 content on this time scale must be slave to the ocean's chemistry." These statements show that the problem is more complicated that it might at first appear. In fact, the global carbon cycle is comprised of a hierarchy of processes that operate at vastly different time-frames, from millions of years to less than one year. Different processes may be considered to dominate the "closed system", depending upon the time frame of interest (Sundquist, 1985). We are interested here in changes in the composition of the atmosphere that relax in the time-frame of decades or less, and processes that relax much more slowly are ignored here.

Most biologists would probably accept that, at least over the short-term, the steady-state concentrations of CO_2 and O_2 in the atmosphere is affected by the balance between photosynthesis and respiration. In this paper, I will review some physiological studies of closed systems, with special reference to their structure and kinetic properties. Then turning to the global system, I will consider how these physiological studies are relevant to the earth system, and review studies that relate atmospheric measurements to biospheric processes.

CLOSED PHYSIOLOGICAL SYSTEMS

If a photosynthetic organism is illuminated in a closed volume containing CO_2, but not O_2, it will produce O_2 as the CO_2 is consumed in reactions of net photosynthesis. Eventually, reservoirs of O_2 will reach a stable value depending on the initial charge of CO_2, the rate of CO_2 release (respiration), and the volume of the vessel. This is a highly simplified analog to planet earth, which presumably had a primeval atmosphere rich in CO_2 but lacking in O_2; now the Earth has an O_2-rich atmosphere at an apparent steady-state (Fig. 1).

Microcosms

Guy et al. (1987) recently examined biogenic atmospheres in so-called, "microcosm" experiments. In these experiments, mesophyll cells of the asparagus fern were suspended in He-sparged buffer containing approximately 250 μM HCO_3^- in a 400-ml vessel fitted with an O_2-electrode (Guy et al., 1990). The vessel was closed and illuminated, and the O_2 concentration within increases, eventually reached about air levels (Fig. 2), where there is a steady-state balance between uptake and release of CO_2 and O_2 (referred to as a compensation point in the physiological literature, see Woodrow and Berry, 1989). The steady-state balance, which is achieved at the CO_2 and O_2 compensation point in the microcosm, can be related to kinetic properties of the reactions that consume and produce CO_2 or O_2. As photosynthesis decreases the CO_2 concentration, the O_2 concentration simultaneously increases, according to:

$$CO_2 + H_2O \underset{k_r}{\overset{k_p}{\rightleftharpoons}} O_2 + [plant] \tag{1}$$

where k_p and k_r are rate constants for photosynthesis and respiration, respectively. If we treat each reaction as first order, (i.e. CO_2 uptake $= k_p[CO_2]$ or CO_2 release $= k_r[O_2]$) and impose the assumption that CO_2 uptake $= CO_2$ release, then:

$$\frac{[O_2]}{[CO_2]} = \frac{k_p}{k_r} \tag{2}$$

at the compensation point, Γ. Consistent with Eq. 2, the final O_2 concentration reached in the microcosm depends upon the initial CO_2 charge of the system, but the final $[O_2]/[CO_2]$ ratio is always constant at Γ (data not shown). In actual fact, the kinetic expressions are more complex, but the general concept still applies, and a rigorous kinetic treatment of the compensation point was developed (Laing et al., Woodrow and Berry, 1988).

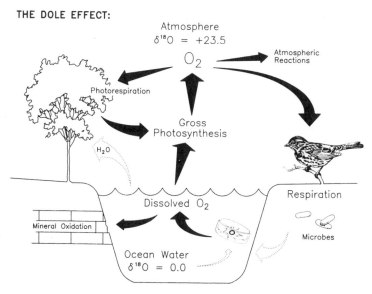

Fig. 1.   Scheme showing the geochemical cycling of oxygen. O_2 is produced from water during photosynthesis and is consumed in several biological and chemical reactions. Solid arrows represent transfer of the oxygen as O_2 and dotted arrows represent transfer as H_2O. The observation that the O_2 of the atmosphere is enriched in ^{18}O ($\delta^{18}O = 23.5$; $R_{O_2}/R_{H_2O} = 1.0235$) is referred to as the "Dole Effect" (Guy et al., 1986).

443

In the experiment shown in Fig. 2, great care was taken to prevent mixing of the O_2 within the vessel with that of the air, and the system was kept illuminated for different intervals of time before a sample of the medium was withdrawn for analysis of the isotope ratio ($^{18}O/^{16}O$) of oxygen of the O_2 and water in the vessel. As shown in Fig. 3, the isotope ratio (expressed as $\delta^{18}O$ relative to the water in the vessel) changes with time; initially it is close to that of the water in the vessel, and with time it becomes displaced by about 21.2 $^o/_{oo}$ (e.g. $R_{O_2} = 1.0212 \times R_{H_2O}$, where R is the $^{18}O/^{16}O$ ratio).

As shown in Fig. 1, the $^{18}O/^{16}O$ ratio of atmospheric O_2 is 1.0235 times that of standard mean ocean water (SMOW) ($\delta^{18}O = +23.5^o/_{oo}$). This has been named the "Dole effect" in honor of its discoverer, Malcolm Dole (Dole, 1935). Due to some complicating factors that I will consider later, the Dole effect on a global scale is about 2 $^o/_{oo}$ different from the microcosm scale. This a very small difference for such a vast jump of scale, and closer examination of the microcosm might give some insight into the macrocosm.

The phenomenon shown in Fig. 3 can be referred to as isotopic compensation. If reactions that produce or consume O_2 in the microcosm discriminate between stable isotopes, the isotopic ratio of the steady-state pool of O_2 will change to compensation for these isotopic differences such that the rates of production and consumption of the

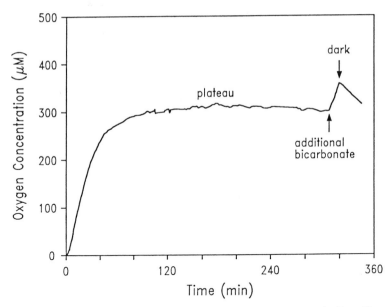

Fig. 2.   An oxygen electrode trace of an *Asparagus* mesophyll cell microcosm experiment. After degassing, 412 μM NaHCO$_3$ was added. This supported an increase in the O_2 concentration to 310 μM. Samples for $\delta^{18}O$ analysis were withdrawn at 32, 178, and 305 min. At 307 min, additional bicarbonate was injected to demonstrate that the cells were still photosynthetically active. At 320 min the light was turned off. The initial rate of O_2 uptake was 7 μmole O_2 (mg Chl)$^{-1}$h^{-1}, and the rate of dark respiration was 8.0 μmole O_2 (mg Chl)$^{-1}$h^{-1}. The temperature was 26°C and the chlorophyll concentration was 8.9 μg/ml (Guy et al., unpublished).

respective isotopic species are equal. The slow change in the $\delta^{18}O$ of O_2 in the microcosm eventually coming to a steady-state (Fig. 3) shows the approach to isotopic compensation. A quantitative description of this steady-state can be based on the isotope effects associated with the uptake and production of O_2. For example, if we define $\alpha_{PSII} = 16\,k/^{18}k$ (where the k's are separate first-order rate constants for the two isotopic species) for the reaction producing O_2 from water, and $\alpha_{O_2 asc}$ is the corresponding ratio for the oxidase reactions (respiratory enzymes) that reduce O_2 to water, then:

$$R_{O_2} \times \frac{1}{\alpha_{O_2 asc}} = R_U \quad \text{and} \quad R_{H_2O} \times \frac{1}{\alpha_{PSII}} = R_P$$

where R_{O_2} and R_{H_2O} are the $^{18}O/^{16}O$ ratios of O_2 and H_2O in the microcosm and R_U and R_P are the isotopic ratios of the O_2 that is taken up or produced, respectively. As discussed above, when the system is at steady-state the rates of uptake and production are equal and $R_U = R_P$. Thus:

$$\delta^{18}O = \frac{R_{O_2}}{R_{H_2O}} - 1 = \frac{\alpha_{O_2 asc}}{\alpha_{PSII}} - 1 \tag{3}$$

at the isotopic compensation point.

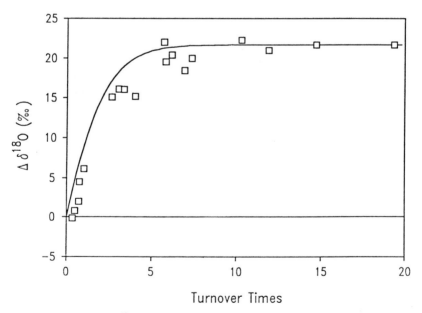

Fig. 3. Change in the $\delta^{18}O$ of dissolved O_2 in *Asparagus* cell microcosm experiments. Data from six separate experiments are combined. The time axis is normalized for each experiment according to the apparent turnover time (20-60 min. in these experiments) of the O_2 pool (the final O_2 concentration divided by the initial rate of O_2 production). The curve is a theoretical fit to the data (Guy et al., unpublished).

Table 1. Isotope Effect Associated with Processes Resulting in O_2 Production and O_2 Uptake by Plants. Discrimination in O_2 Uptake was Measured by the Substrate Analysis Method of Guy et al., 1990. Discrimination in O_2 Production was Determined by Direct Analysis of the Product O_2.

Enzyme System	Isotope Effect α
Rubisco oxygenase	1.0208
Glycolate oxidase	1.0222
Mitochondrial, O_2 uptake	1.0202
Mehler reaction, O_2 uptake	1.0151
Photosystem II, O_2 production	0.9997

Guy et al. (1987) explored the determinants of the Dole effect by assessing the discrimination associated with specific biochemical reactions that participate in the steady-state of the microcosm. Table 1 reports the isotope effect (α) in O_2 production or O_2 uptake by several enzymes and enzyme systems that are known to contribute to the oxygen exchange. Like the global system, all O_2 in the microcosm is produced by PSII. Respiratory O_2 uptake by plants includes both heterotrophic respiration and photorespiration. The latter is the dominant mechanism in this system, and this pathway involves O_2 uptake by mitochondria, and by the enzymes, Rubisco oxygenase, and glycolate oxidase; the Mehler reaction also is a possibility (Berry et al., 1978). To use these data to calculate the expected $\delta^{18}O$ at the isotopic compensation point according to Eq. 3, we need to come up with a weighted average estimate for $\alpha_{O_2,asc}$, which takes into account the participation of these different mechanisms of O_2 uptake that are occurring simultaneously. To do this, we need to know the stoichiometry. Berry et al. (1978) showed that for each CO_2 produced in photorespiration there are 3.5 O_2 taken up; 2 by Rubisco oxygenase, 1 by glycolate oxidase, and .5 by mitochondria (ignoring the Mehler reaction). From these results we estimate that $\alpha_{O_2,asc} = 1.0215$. Substituting this estimate and $\alpha_{PSII} = .9997$ directly in Eq. 3, we obtain the expected value of $R_{O_2} R_{H_2O} = 1.0212$ or $\delta^{18}O = 21.2 \, ^o/_{oo}$. The observed value was approximately 21.2 $^o/_{oo}$. Similar studies by Rooney (1988) found good agreement of the $\delta^{13}C$ reached at steady-state in a microcosm and that predicted from studies of carbon isotope discrimination in the relevant biochemical reactions.

Macrocosms

Increasing the scale to that of a terrarium, for example, introduces two important complications. First, there is an increased contribution by heterotrophic respiration (as opposed to photorespiration in the microcosm), which can be accommodated by adjusting the stoichiometric ratios used to calculate $\alpha_{O_2,asc}$. Assuming that photorespiration is about 30% of total O_2 uptake, we obtain $\alpha_{O_2,asc} = 1.0205$, not very different from 1.0212 used above. The other complication is that when we have leaves in air (as opposed to photosynthetic cells immersed in stirred water), we need to consider the effects of evaporation on the source water for PSII. Water containing ^{18}O has a lower vapor pressure than ^{16}O-H_2O. Thus, the water near the sites of evaporation in leaves may be enriched in ^{18}O relative to that supplied to the plant, and the apparent value of $\alpha_{PSII} \neq 1$. This effect depends on environmental conditions such as humidity, and may result in local gradients of as much as 20 $^o/_{oo}$.

Yakir et al. (in press) demonstrated that CO_2 may be used as a probe for the isotopic composition of chloroplast water. The oxygen atoms of CO_2 can exchange with those of water. Ordinarily, this is a slow process, but chloroplasts contain an enzyme, carbonic anhydrase, that dramatically accelerates this exchange reaction. Thus, CO_2 comes to isotopic equilibrium with the same pools of water that are the source of O_2, and measurements of the $\delta^{18}O$-CO_2 can be used to estimate the chloroplast water and correct for added isotope effects associated with hydrologic processes on the isotopic composition of photosynthetic O_2 (the apparent value of α_{PSII}).

Figure 4 shows an experiment in which sunflower leaves were enclosed in a cuvette that could be maintained at different relative humidities; the leaves were illuminated and permitted to evaporate (transpire) water at a given humidity for sufficient time to reach steady-state. The cuvette was then flushed with O_2-free helium containing CO_2, and the O_2 produced by PSII was trapped and analyzed for its $\delta^{18}O$ value (triangles, Fig. 4B). Because there is no discrimination in O_2 production, this gives the composition of chloroplast water. A sample of CO_2 that was permitted to exchange with the leaf was those of the O_2 produced in the same experiments. Figure 4A also shows the $\delta^{18}O$ value

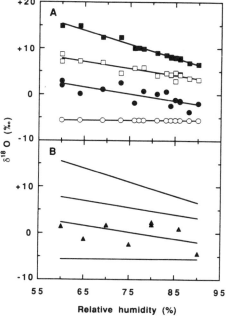

Fig. 4.   Water compartmentation in sunflower leaves. A) An attached sunflower leaf was enclosed in a gas exchange chamber at relative humidities shown and constant conditions were maintained until isotopic steady-state was attained. The chamber was then flushed with CO_2-free air, the respired CO_2 was collected, and its $\delta^{18}O$ value was determined (closed circles). Total leaf water (open squares) and stem water (open circles) were then extracted to determine their $\delta^{18}O$ values. The isotopic composition of the water at the evaporating surfaces (closed squares) was calculated using the Craig model. B) Leaves were treated as in A, but under a helium/CO_2 mixture (1500 μl l^{-1} CO_2). Photosynthetic O_2, evolved from water in the chloroplasts was trapped, concentrated, and fed directly into a mass spectrometer to determine $\delta^{18}O$. The results (triangles) are superimposed on the best fit lines reported in A (Yakir et al., in press).

analyzed for its $\delta^{18}O$ value, and this was used to calculate the apparent $\delta^{18}O$ value of the water it had exchanged with (taking into account the equilibrium isotope effect in the exchange of CO_2 and water). These $\delta^{18}O$ values (filled circles, Fig. 4A) agree well with water obtained by taking samples of the leaf (open squares), of water at the sites of evaporation (calculated from a model), and of water supplied to the leaf (open circles). The apparent value of α_{PSII} can be inferred from the $\delta^{18}O$ value of CO_2 in the system.

The Earth System

The average $\delta^{18}O$ of CO_2 of the earth's atmosphere is about 1 $\%_o$ heavier than would be expected if it were at equilibrium with seawater. This difference indicates that the chloroplast water of the terrestrial biosphere is probably enriched (relative to SMOW) and may account for part of the discrepancy noted earlier between the Dole effect in microcosms and that observed on the planetary scale. Furthermore, the isotopic composition reached in the global system is consistent with the isotope fractionating properties of the major enzyme system believed to be involved in the biospheric cycling of O_2. This is also true for CO_2 (Yakir et al., in press). The commonly held view is that systems should be described in terms of the properties of the next lower level of organization; however, this view does not apply here, because the state of the global environment is at least, in part, determined by biochemical properties of enzymes that can be measured in test tubes.

Recent studies by Michael Bender and colleagues provide a very interesting historical view of this closed system. They extracted and determined the $\delta^{18}O$ value of O_2 from ice cores (Bender et al., 1985; Sowers et al., 1989). These studies show that the $\delta^{18}O$ value of atmospheric O_2, which was fairly constant over the past few thousand years, increased by about 1.5 $\%_o$ during glacial advances. This increase coincides with a similar increase in the $\delta^{18}O$ of carbonates in benthic formanifera obtained from cores of ocean sediments. Both of these changes can be ascribed to changes in the isotopic composition of the liquid water in the Earth system. When isotopically light precipitation (snow) accumulated in the ice sheets, the abundance of ^{18}O in liquid water remaining in the oceans increased, and the steady-state isotopic composition of atmospheric O_2 changed by a corresponding amount. This study not only shows that the Earth system has behaved essentially as a closed system with respect to O_2, at least over thousands of years, but that these changes in the $\delta^{18}O$ value of O_2 provide a useful cross check on the chronologies assigned to ice cores and to those of ocean sediments. More subtle information on the properties of the steady-state may also be contained in this record.

Bender et al. (1985) also examined the time rate of change in the $\delta^{18}O-O_2$ following the last deglaciation. If we assume that there was a fairly abrupt change in the $\delta^{18}O$ of water in the active hydrological cycle at that time, there should be a corresponding change in the $\delta^{18}O$ value of O_2 entering the atmosphere, and the $\delta^{18}O$ value of atmospheric O_2 should adjust to a new steady-state value with a time constant that depends on O_2 content of the atmosphere and the rate of photosynthesis (Fig. 3). Based on current estimates of global photosynthesis, the O_2 of the atmosphere should have a turnover time of two to four thousand years. With sufficiently precise measurements, this approach might be used to determine the rate of O_2 production by photosynthesis (combined marine and terrestrial) in the Earth system.

O_2, CO_2, and the Oceans

Transformations of the O_2 cycle (Fig. 1) that involve redox changes are mostly linked to the carbon cycle. Redox changes in other elements (sulfur, nitrogen, and iron) also occur, but the fluxes are smaller than (and are generally linked to) carbon fixation or decomposition. Therefore, we should expect fairly strict stoichiometry between change in O_2 and CO_2. This clearly depends on the redox level of the organic compounds formed or consumed. For biomatter, the CO_2:O_2 ratio is about 1:1.1, while oil-based fuels are 1:1.3-1.5, and natural gas is 1:1.8-2.0. Because most of the turnover of carbon compounds in the Earth system is biological, changes in O_2 should be about equal and opposite to changes in CO_2.

However, because of differential solubility of CO_2 and O_2 in the ocean, the amount of CO_2 remaining in (or reaching) the atmosphere may not equal that of O_2 even though the chemical equations are balanced. Furthermore, the chemical kinetics and physical dynamics that govern CO_2 exchange with the ocean are complex (for reviews see Takahashi, 1989; Broecker and Peng, 1982). As summarized by Keeling (1988), the effective solubility of CO_2 in seawater of 24°C is 600-fold that of O_2, and the rate at which surface waters come into equilibrium with the atmosphere is about 20 times slower for CO_2 than for O_2. These differential interactions of CO_2 and O_2 with the oceans has the effect of making the "volume" of the earth system that is available to CO_2 much larger than that available to O_2. Further, the effective volume for CO_2 is divided into different compartments with different mixing times. Therefore, the coupling between CO_2 and O_2 in the atmosphere depends on where the reaction occurs (in the ocean or on land) and on the relaxation time of the process in question. The stoichiometry of changes in atmospheric CO_2 and O_2, therefore, is complex and highly dependent on the time-frame of the measurements.

However, precise measurements of O_2 mole fraction changes might provide new insight, and an instrument capable of measuring O_2 concentration to ppm resolution is described by Keeling (1988). For example, it might be possible to resolve whether CO_2 injected into the atmosphere by combustion of fossil fuel is taken up by ocean or biospheric sinks from studies of O_2:CO_2 stoichiometry. CO_2 taken up by the terrestrial biosphere would be replaced by O_2, but if it were taken up by ocean chemistry, there should be no corresponding change in O_2. With respect to the subject of these proceedings, one of the most interesting ideas explored by Keeling (1988) is that O_2 exchange associated with marine photosynthesis and respiration should be more closely coupled to the atmosphere than the corresponding CO_2 exchange, leading to larger seasonal changes in the O_2 than in the CO_2 content of marine air. Preliminary measurements (R. Keeling, personal communication) confirm this prediction, and show that high-resolution measurements of atmospheric O_2 may be a useful approach for studies of marine photosynthesis. Measurements of the O_2 content of air from ice cores also is in progress (Sowers et al., 1989).

SOURCES AND SINKS FOR CO_2

The slow mixing of the atmosphere and the fact that sources and sinks are neither uniformly distributed over the surface of the earth nor uniformly active in time, results in local departures of the CO_2 concentration from the mean global value. Networks of stations for sampling the composition of the atmosphere have been established by National Oceanic and Atmospheric Administration (Boulder, CO, USA), the Scripps Institution of Oceanography/UG (Scripps Inst., CA, USA) and the CSIRO/NERDDC (Australia). Each analyzes flask samples of air taken at intervals of 1 to 3 weeks from 20 to 30 stations distributed around the globe. Data from these networks establish latitudinal and seasonal variations in the CO_2 concentration that can be related to the location and seasonal activity sources and sinks. Other gases, including $^{13}CO_2$, $C^{18}O_2$, CO, O_2, and CH_4 also are being analyzed.

Analysis of these data depends on atmospheric transport models that can simulate the dispersion of tracer species in the atmosphere. It is noteworthy that these are also closed system models derived from atmospheric general circulation models. Generally, these are calibrated using species such as CFCs, ^{222}Rn, and ^{85}K (produced in reprocessing of nuclear fuel rods) that have well-defined sources. The models are tuned to obtain the observed latitudinal and vertical distribution of the tracer, and then are applied to analyze the distribution of CO_2. The most straightforward (but not the easiest) approach is to use the model in the inverse mode adjusting the location and strength of sinks to fit the observed global patterns of CO_2 concentration. There are a great many possible solutions, and this approach has only been attempted using 2-d models (Tans et al., 1989). As discussed by Tans (in press) one of the major limitations of this approach is the paucity of observational data to constrain the model. More intensive sampling of vertical profiles and the boundary layer over the continents is needed. Alternatively, the sources and sinks may be prescribed and the model output compared to the observed concentration fields. Some recent analyses use a combined approach constraining the model by prescribing some of the sources and sinks, and adjusting others to fit the data (e.g. Tans et al., 1990).

The Terrestrial Biosphere

Several approaches were used to specify CO_2 exchange of the terrestrial biosphere. Fung et al. (1983) provides a good discussion of the problem. They note that the total of photosynthetic CO_2 uptake and CO_2 release by plant respiration and microbial decomposition for the biosphere is approximately zero over the course of a year (i.e. the biosphere is at steady-state). However, in most locations, growth of vegetation and decomposition occur nonuniformly throughout the year. Therefore, the net CO_2 flux varies depending on the seasonal activity of these processes. The magnitude of this net flux and its variation with time impacts the local concentration of CO_2. However, there is very little direct information on the net ecosystem CO_2 exchange over appropriate time-frames, and it has been necessary to approach this problem by estimating the seasonal fluxes of CO_2 uptake and CO_2 release separately, and then taking the difference.

To understand the approaches to this problem, it is important to define some terms. Total photosynthetic CO_2 uptake (as measured in leaf gas exchange experiments) is referred to as gross primary productivity (GPP). Net primary productivity (NPP) is generally obtained from plant growth studies, and can be considered as GPP minus plant respiratory losses. The latter may vary a lot, but it is often assumed that GPP = 2 x NPP. Net ecosystem production (NEP) is NPP minus decomposition.

Ideally, one would like to have models or measurements for the gross flux of each of the three major processes (GPP, plant respiration, and decomposition) to estimate NEP over the year. Most productivity estimates, however, are in terms of NPP. One approach calculates the total area covered by vegetation types classified into different types (biomes) and multiplies that by an average productivity value for each, derived from point studies (Whittaker and Likens, 1975). Another approach uses climate-productivity algorithms such as those developed by Leith (1975) together with world maps of important climatic variables. Both estimates are essentially empirical and static (they give only the seasonal total NPP, not its time-dependence). Modelers then developed climatic algorithms to estimate the time-dependence of NPP and decomposition to obtain NEP. Fung et al. (1983) noted, "...the uncertainties in these seasonal uptake and release curves may be as large as or larger than the uncertainties in the NPP estimates...".

Despite these uncertainties, several early modeling studies (Fung et al., 1983; Gillette and Box, 1986; Pearman and Hyson, 1986) were able to obtain good agreement with the observed concentration patterns. However, the seasonal flux balance in these models is arbitrarily specified. Fung et al. (1987) introduced an important advance to further constrain the system. They used the seasonal change in a satellite measured greenness index to specify the seasonal pattern of NPP on a cell by cell basis. Heimann and Keeling (1989) went one step further by using an empirical relationship to calculate NPP from the photosynthetically active solar radiation absorbed by plants (computed from the greenness index and satellite measurements of cloud cover), rather than using estimates from the literature.

These innovations are substantial improvements in the models used for global CO_2 balance analysis because satellite measurements replace some of the parameters that had been available for arbitrary adjustment. However, it is important to note some shortcomings. First, the satellite index is most directly related to GPP not NPP, and second, these analyses do not take into account losses in GPP that may be attributed to stress, for example, unfavorable temperatures, drought stress, or nutrient limitations that do not affect the greenness index.

There is considerable scope to improve the empirical approaches used to estimate the seasonal NEP fluxes associated with the terrestrial biosphere. For example, models of surface energy balance and hydrology (Sellers et al., 1986) are being coupled with mechanistic models of photosynthesis and stomatal conductance (Collatz et al., 1990), and to models of respiration and decomposition. These approaches have the additional benefit that the models predict other parameters such as surface temperature and evapotranspiration that could be separately verified, thus, further constraining the system. Models that predict the exchange of isotopic species may also be useful to further constrain these systems.

The Oceans

I can add little to what other authors have already said on the exchange of CO_2 with the oceans. The most fruitful approach to date has been to construct a global map of the pCO_2 of the surface ocean, and use the wind fields and local atmospheric CO_2 concentrations generated by the mixing model to estimate net exchange with the surface ocean. The penultimate result in this area at this writing is the paper of Tans et al. (1990), which uses such a model to come to the controversial conclusion the the terrestrial biosphere is acting as a sink for industrial CO_2. There is hope that a resolution on this question can be obtained from analysis of isotopic species (Keeling et al., 1989). This analysis takes advantage of the fact that CO_2 exchange with the ocean gives a very different isotopic signature than exchange with the terrestrial biosphere. However, the resolution of this method depends critically on the $\delta^{13}C$ values assigned to CO_2 entering and leaving the biosphere. Because the $\delta^{13}C$ of the atmosphere has been becoming progressively more negative, there will be a difference between that entering and that leaving if there is a time lag.

The role of marine photosynthesis in these exchanges is unclear. On one hand, I could argue that the ocean is essentially a closed system with respect to nutrients, and, therefore, also to CO_2. On the other hand, we know that marine productivity acts as a so-called biological pump, with decomposition of sinking detritus adding to the CO_2 concentration of deep water and depleting that of surface water. This gradient would relax if marine productivity were to cease. Takahashi (1989) observed that the average CO_2

concentration of the whole ocean is considerably higher than that of the present atmosphere. So, presumably, the concentration of CO_2 would increase if we could disrupt marine photosynthesis.

CONCLUSIONS

To conclude this brief review, I believe that there is a need for a great deal more input from biologists on the analysis of global CO_2 fluxes. However, more important, the atmospheric scientists have provided a rigorous context for the biology. In this review, I explained the advantages of closed systems. In the comparatively simple system involving compensation between photosynthesis and photorespiration in the microcosm, there is remarkably good agreement of the concentrations and isotopic composition of CO_2 and O_2 reached at steady state with that predicted by independent kinetic measurements conducted *in vitro*. Also, as I explained, the global system conforms to similar kinetic relationships developed for the microcosm. The simplicity of the system is not the fundamental reason that concentrations and isotope ratios are so predictable; rather the predictability comes from the structure of the system. At steady-state, it is assumed that the fluxes producing and consuming the substance in question are equal and opposite. Thus, it is possible to write kinetic expressions for the concentration or isotope ratio that are independent of the flux and, to a large extent, independent of scale. From this perspective, we can see the specter of fundamental biochemical mechanisms on a global scale, which encourages those of us who use biochemical kinetics to explain global biosphere dynamics.

ACKNOWLEDGEMENTS

I would especially like to thank James Collatz, Chris Field, Rob Guy, Barry Osmond, and Dan Yakir who collaborated on the experiments described here and provided needed guidance. This is CIW-DPB publ. no 1120.

REFERENCES

Baldocchi, D. D., Hicks, B. B., and Meyers, T. P., 1988, Measuring biosphere-atmosphere exchanges of biologically related gas with micrometeorological methods, *Ecol.*, 69:1331.

Barnola, J. M., Raynaud, D., Korotkevich, Y. S., and Lorius, C., 1987, Vostok ice core provides 160,000-year record of atmospheric CO_2, *Nature*, 329:408.

Bender, M. L., Labeyrie, L., Raynaud, D., and Loris, C., 1985, Isotopic composition of atmospheric O_2 in ice linked with deglaciation and global primary productivity, *Nature*, 318:349.

Berry, J. A., Osmond, C. B., and Lorimer, G. H., 1978, Fixation of O_2 during photorespiration, *Plant Physiol.*, 62:954.

Broecker, W. S., 1982, Glacial to Interglacial changes in ocean chemistry, *Prog. Oceanog.* 11:151.

Broecker, W. S., and Peng, T-. H., 1982, Tracers in the sea, Lamont-Doherty Geological Observatory, Palisades, New York.

Collatz, G. J., Ball, J. T., Grivet, C., and Berry, J. A., 1990, Regulation of stomatal conductance and transpiration: A physiological model of canopy processes, *Ag. and Forest Meterol.*, 54:107.

Dole, M., 1935, The relative atomic weight of oxygen in water and air, *J. Am. Chem Soc.*, 57:2731.

Field, C. B., and, Mooney, H. A., 1990, Leaf chamber methods for measuring photosynthesis under field conditions, *Remote sensing Reviews*, 5:117.

Fung, I., Prentice, K., Matthews, E., Lerner, J., and Russel, G., 1983, Three-dimensional tracer model study of atmospheric CO_2: Response to seasonal exchanges with the terrestrial biosphere, *J. Geophys. Res.*, 88:1281.

Fung, I. Y., Tucker, C. J., and Prentice, K. C., 1987, Application of advanced very high resolution radiometer vegetation index to study atmosphere-biosphere exchange of CO_2, *J. Geophys. Res.*, 92:2999.

Gillette, D. A. and Box, E. A., 1986, Modeling seasonal changes of atmospheric carbon dioxide and carbon 13, *J. Geophys. Res.*, 91:5287.

Guy, R. D., Fogel, M. F., Berry, J. A., and Hoering, T. H., 1986, Isotope fractionation during oxygen production and consumption by plants, *in*: "Progress in Photosynthesis Research," J. Biggins, ed., Vol III, 597-600, Martinus Nijhoff, Dordrecht.

Guy, R. D., Berry, J. A., Fogel, M. L., and Hoering, T. C., 1989, Differential fractionation of oxygen isotopes by cyanide-resistant and cyanide-sensitive respiration in plants, *Planta*, 177:483.

Hansen, J., Russell, G., Lacis, A., Fung, I., and Rind, D., 1985, Climate response times: Dependence on climate sensitivity and Ocean mixing, *Science*, 229:857.

Heimann, M. and Keeling, C. D., 1989, A three-dimensional model of atmospheric CO_2 transport based on observed winds: 2. Model description and simulated tracer experiments, *in*: "Aspects of Climate Variability in the Pacific and Western Americas," D.H. Peterson, ed., *Geophys Monogr. Ser.*, 55:237.

Kastings, J. F., Toon, O. B., and Pollock, J. B., 1988, How climate evolved on the terrestrial planets, *Scientific American*, 258(2):90.

Keeling, C. D., 1973, Industrial production of carbon dioxide from fossil fuels and limestone, *Tellus*, 25:174.

Keeling, C. D., Bacastow, R., Carter, A. F., Piper, S. C., Whorf, T. P., Heimann, M., Mook, W. G., and Roeloffzen, H., 1989, A three-dimensional model of atmospheric CO_2 transport based on observed winds: 1. Analysis of observational data, *in*: "Aspects of Climate Variability in the Pacific and Western Americas, D.H. Peterson, ed., *Geophys Monogr. Ser.*, 55:165.

Keeling, R. D., 1988, Measuring correlations between atmospheric oxygen and carbon dioxide mole fraction: A preliminary study in urban air, *J. Atmos. Chem.*, 7:153.

Laing, W. A., Ogren, W. L., and Hageman, R. H., 1974, Regulation of soybean net photosynthetic CO_2 fixation by the interaction of CO_2, O_2, and ribulose-1,5-bisphosphate carboxylase, *Plant Physiol.*, 54:678.

Leith, H., 1975, Primary production of the major vegetation units of the world, *in*: "Primary Productivity of the Biosphere," Ecol. Stud. Vol. 15, H. Leith and R.H. Whittaker, eds., Springer-Verlag, New York.

Rooney, M. A., 1988, Short-term carbon isotope fractionation in plants, Ph.D. Thesis, University of Wisconsin-Madison.

Pearman, G. I. and Hyson, P., 1986, Global transport and inter-reservoir exchange of carbon dioxide with particular reference to stable isotopic distributions, *J. Atmos. Chem.*, 4:81.

Sellers, P. J., Mintz, Y., Sud, Y. C., and Dalcher, A., 1986, A simple biosphere model (SiB) for use within general circulation models, *J. Atmos. Sci.*, 42:505.

Sellers, P. J., Hall, F. G., Strebel, D. E., Asrar, G., and Murphy, R. E., 1989, Satellite remote sensing and field experiments, *in*: "Remote Sensing of Biospheric Function," Ecol. Stud. Vol. 79, R.J. Hobbs and H.A. Mooney, eds., Springer-Verlag, New York.

Sowers, T., Bender, M., and Raynaud, D., 1989, Elemental and isotopic compostion of occluded O_2 and N_2 in polar ice, *J. Geophys. Res.*, 94:5137.

Sundquist, E. T., 1985, Geological perspective on carbon dioxide and the carbon cycle, *in*: "The Carbon Cycle and Atmospheric CO_2: Natural Variation Archean to Present," E.T. Sundquist and W.S. Broecker, eds., *Geophys. Monogr. Ser. Vol. 32* AGU, Washington, D.C.

Takahashi, T., 1989, The carbon dioxide puzzle, *Oceanus*, 32:22.

Tans, P. P., Fung, I. Y., and Takahashi, T., 1990 Observational constraints on the global atmospheric CO_2 budget, *Science*, 247:1431.

Tans, P. P., An observational strategy for assessing the role of terrestrial ecosystems in the global carbon cycle: Scaling down to regional levels, *in*: "Scaling Problems in Climatic Change Research," C.R. Field and J.R. Ehleringer, eds., Academic Press, New York, in press.

Whittaker, R. H., and Likens, G. E., 1975, The biosphere and man, *in*: "Primary Productivity of the Biosphere," Ecol. Stud. Vol. 15, H. Leith and R.H. Whittaker, eds., Springer-Verlag, New York.

Wofsy, S. C., Harriss, R. C., and Kaplan, W. A., 1988, Carbon dioxide in the atmosphere over the Amazon Basin, *J. Geophys. Res.*, 93:1377.

Woodrow, I. E. and Berry, J. A., 1988, Enzymatic regulation of photosynthetic CO_2 fixation in C_3 plants, *Ann. Rev. Plant Physiol. Plant Mol. Biol.*, 39:533.

Yakir, D., Berry, J., Giles, L., Osmond, B., Thomas, R., Aplications of stable isotopes to scaling biospheric photosynthetic activites, *in*: "Scaling Problems in Climatic Change Research," C.R. Field and J.R. Ehleringer, eds., Academic Press, New York, in press.

READING THE SEDIMENTARY RECORD OF THE OCEAN'S PRODUCTIVITY

W.H. Berger and J.C. Herguera

Scripps Institution of Oceanography
University of California, San Diego
La Jolla, California 92093 MS 0215

INTRODUCTION

The basic controls on ocean productivity are poorly understood both biologically and geologically. In fact, we do not know the global patterns of productivity very well, either with regard to the rates of primary production (that is, the amount of carbon fixed in the photic zone each year), or with regard to the types of primary production (that is, the kinds of organisms involved).

Maps of primary productivity (Koblentz-Mishke et al., 1970; Berger et al., 1987a) have error margins of about 2 for large regions. Elementary questions about the controls on the rates of production are being hotly contested (Martin and Fitzwater, 1988; Martin et al., 1990; Banse, 1991), and the role of picoplankton (Platt and Li, 1986) is in dispute. Likewise, the long-term regulation of the fertility of the ocean is entirely open to discussion, including the identity of the "true" limiting nutrient (phosphate versus nitrate: Codispoti, 1989; Shaffer, 1989; iron: Martin, 1990), and also the importance of changing circulation and deep mixing (Keir, 1988; Boyle, 1990). Recently, the potential significance of dissolved organic matter within the carbon cycle of the productive zone has emerged as another topic of contention (Toggweiler, 1989; Bacastow and Maier-Reimer, 1991).

The evidence of long-term changes in ocean productivity from the sedimentary record has made important contributions to these various on-going discussions. Reconstruction of ocean productivity is increasingly used to test hypotheses about possible causes of atmospheric CO_2 fluctuations (Shackleton et al., 1983; Sarnthein et al., 1987, 1988; Mix, 1989a,b). Recently, for example, the idea that the Southern Ocean was much more productive during glacial times (Keir, 1990, and refs. therein; Martin, 1990), was confronted with unfavorable evidence from sediments (Mortlock et al., 1991). Such reconstructions are entirely based on the development of proxy indicators of productivity, and their application to the record. Thus, the uncertainties surrounding the use of proxies from sediments must be appreciated, if results of reconstructions are to be used properly.

Here we explain the basic approach used in the reconstruction of ocean productivity. We focus on two proxy indicators of production: organic matter content and foraminiferal shells. We shall mention others in passing. A more extensive list of proxies may be found in the reports by Bruland et al. (1989) and Herbert et al. (1989) (see also

Primary Productivity and Biogeochemical Cycles in the Sea
Edited by P.G. Falkowski and A.D. Woodhead, Plenum Press, New York, 1992

455

Elderfield, 1990). Our aim is to outline the methodology of productivity reconstruction, to provide a sense of the various assumptions involved. As examples, we use evidence from Quaternary sediments, especially on the glacial-interglacial contrast in productivity in low latitudes. We should have some understanding of the experiments performed by Nature in the late Quaternary, before we explain the more ancient record.

ORGANIC MATTER CONTENT AS PRODUCTIVITY INDEX

Primary Productivity and Organic Carbon in Sediments

Organic matter is generally a minor constituent of marine sediments, ranging from a fraction of a percent (by weight) in most deep-sea deposits to several percent in upwelling regions (Romankevich, 1984; Premuzic et al., 1982; Calvert, 1987; Emerson and Hedges, 1988). The observation that the pattern of organic content on the whole matches the pattern of productivity (compare Figs. 1 and 2) is the basis for using organic content as a proxy for productivity history. Applying this principle to the past is fraught with difficulties. Even just mapping the present distributions is not straightforward, because of the rapid change of organic carbon content in the uppermost few centimeters of sediment (e.g., Emerson et al., 1985). Difficulties increase as we travel back into the more distant periods (Pedersen and Calvert, 1990). Here we stay clear of the discussion on the importance of oxygen content of deep waters to the preservation of organic matter (see Emerson et al., 1985; Bralower and Thierstein, 1987; Emerson and Hedges, 1988; Reimers, 1989). Burial of organic carbon is largely independent of oxygenation, unless one deals with anaerobic conditions, combined with low sedimentation rates (Stein, 1991).

Approach of Müller and Suess (1979)

Müller and Suess (1979) compiled data for abundance and sedimentation rates of organic matter for the present seafloor, and calibrated these patterns against surface productivity of the ocean. Their empirical equation is as follows:

$$PaP = C * rho * (1\text{-}phi) / 0.0030 \ S^{0.3} \tag{1}$$

where PaP is the paleoproductivity in $gCm^{-2}y_{-1}$, C is the bulk organic carbon in percent of dry weight, rho is dry sediment density, phi is porosity (in percent of total), and S is the sedimentation rate. The numerator denotes the weight of the organic carbon. The denominator represents a moderate correction for sedimentation rate (which is itself correlated with organic content: Heath et al., 1977; Müller and Suess, 1979; Bralower and Thierstein, 1987).

Applying their equation to a core taken off NW Africa, Müller and Suess found that during interglacial stages productivity was about the same as today, but that it was substantially increased during glacial periods between 2 and 3 (Fig. 3). Increased coastal upwelling is shown, because of increased winds and currents, resulting from a stronger temperature gradient in the North Atlantic (cf. CLIMAP, 1976). Such increased glacial upwelling, which may be considered typical for continental margins in low latitudes, had a marked effect on the pCO_2 of the atmosphere (Sarnthein et al., 1987; 1988).

One weakness of the calibration used by Müller and Suess (1979) is that they used primary production rather than flux to the seafloor, that is, they did not correct for depth of water. Depth modifies the amount of carbon delivered from the photic zone, by the $1/z$ relationship (Suess, 1980). New calibrations, taking into account the $1/z$ effect, are

SYNTHETIC PRIMARY PRODUCTIVITY

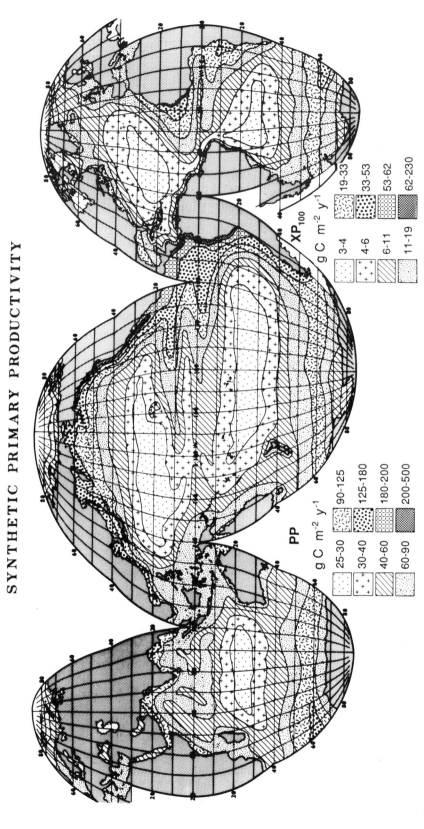

Fig. 1. Generalized productivity map, based on nutricline distribution (phosphate at 100 m), distance from shore (eddy effect), and latitude (sunlight). From Berger et al.(1987a). Export at 100 m (legend to right) calculated from Eq 2.

457

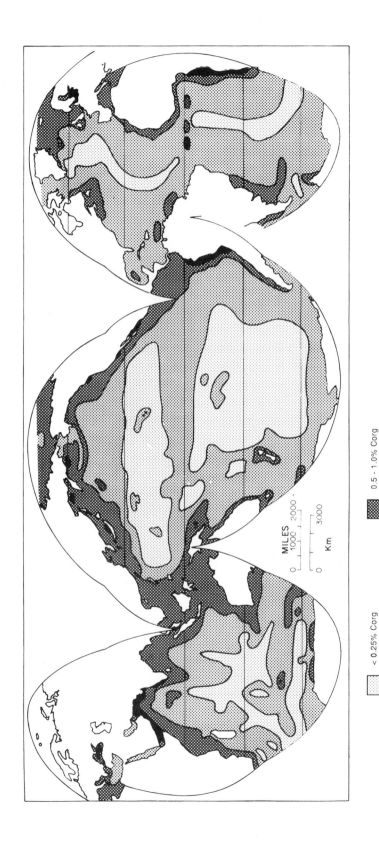

Fig. 2.   Generalized distribution of organic matter content in marine sediments. From Romankevich (1984), simplified and with minor additions.

MILES
0   1000   2000
0   3000
Km

□ < 0.25% Corg

▨ 0.25 - 0.5% Corg

▨ 0.5 - 1.0% Corg

■ >2% Corg

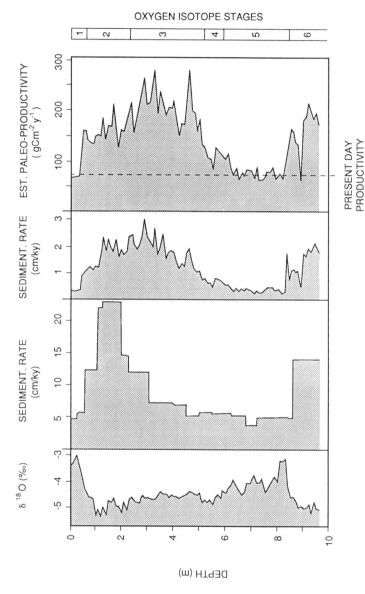

Fig. 3.   Paleoproductivity estimates for Meteor Core 12392-1 on the continental rise off the Spanish Sahara, NW Africa. Sedimentation rate and organic carbon content are the input variables, from which productivity is calculated according to Eq 1. The oxygen isotope stratigraphy is for scale (note Sub-Stage 5e just below 8 m depth in the core, with an age of 124,000 years). After Müller and Suess (1979) and Müller et al. (1983) (from Berger et al., 1989).

provided by Sarnthein et al. (1987; 1988) and by Stein (1991). In the Müller-Suess equation, the depth effect is hidden within the sedimentation-rate correction, because this rate is correlated with depth.

Constraints on the Conversion Factor

If the flux of organic matter to the seafloor is compared to the organic matter content within the sediment, a reasonably good correlation emerges, without any further corrections (Fig. 4). The flux to the seafloor is calculated as follows:

$$\text{Jsf} = 2\sqrt{PP*PP}/100 * (1/z+0.025) \tag{2}$$

where the first term denotes the export at 100 m depth as a function of primary production PP (Berger and Wefer, 1990), and the second term, in parentheses, contains the modifier for depth of deposition (Suess, 1980), with the constant portion representing that part of the export which is resistant to decay. The depth factor z is measured in units of 100 m. Note that Jsf and PP are related through an exponent of 1.5 on PP (which is between the one of 1.4 given by Betzer et al., 1984, and the one of 1.9 in Eppley and Peterson, 1979). The regression equation linking Jsf and the buried carbon Corg (in percent) is:

$$\log(C_{org}) = 0.7 \log(\text{Jsf})-0.24 \tag{3}$$

Thus, to calculate the change in Jsf from a change in $C_{org}/$, we have to take the factor for C_{org} to the power of 1.4 (that is, the inverse of 0.7).

When reconstructing ocean productivity, we are usually concerned with the time series in any one core. For this case, most corrections become unimportant as long as the sediment type in the section does not change drastically (if it does, quantitative reconstructions become questionable anyway). The one aspect that matters, then, is that

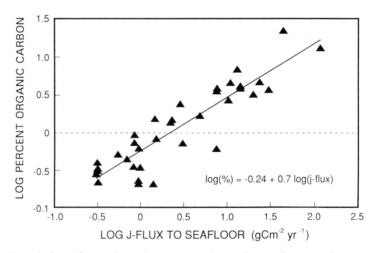

Fig. 4. Correlation of organic carbon content in marine sediments with organic matter flux to the seafloor. Data from Bralower and Thierstein (1990; Table 2); anaerobic conditions are omitted.

organic content reflects flux to the seafloor. Eq 2 shows that a change in primary production is related to a change in the flux of the seafloor through the exponent of 1.5 (that is, $\sqrt{PP}*PP$). This is close to the conversion exponent for C_{org} to Jsf (1/0.7; Eq 3), so that the two conversions nearly cancel, and the exponent is close to unity:

$$PaP = k * C_{org}^{0.9} \tag{4}$$

For the data set analyzed, the empirically derived constant k is somewhere near 200, if the exponent is taken as unity. However, k varies considerably depending on local circumstances, so that it should not be computed for large regions. Instead, it should be allowed to cancel by working with the ratios between PP and C_{org}, for each particular site.

Equation 4 states that a doubling of C_{org} in a given core indicates a change in productivity of 1.9, at the time. The equation of Müller and Suess (1979) may be reduced to an analogous relationship:

$$PaP = k * C_{org}^{0.5} \tag{5}$$

by making use of the correlation between sedimentation rate and carbon content, and assuming that bulk density and porosity do not vary much in the same core (Berger et al., 1989). Clearly, this square-root rule is more conservative than Eq 4 in deriving a change of productivity from a change in C_{org}. In this case, a doubling of C_{org} would show an increase of productivity of 1.4. This equation may be more appropriate for the coastal ocean, where highly pulsed productivity leads to amplified changes in Jsf (Berger and Wefer, 1990).

On average, it may be assumed that the truth lies somewhere in the middle between Eqs 4 and 5. This assumption is supported by the equations of Sarnthein et al. (1987), and of Stein (1991), which yield exponents of 0.6 and 0.8, respectively, using the simplifications suggested. Thus, a doubling of C_{org} means an increase in the productivity of between 1.4 and 1.9, with the most likely value being between 1.6 and 1.7.

Based on this analysis, we propose an exponent of 0.7 to the factor describing a change in C_{org} in any one core ("point-seven" rule-of-thumb), as the most likely exponent yielding an acceptable estimate of the change in paleoproductivity:

$$PaP = k * C_{org}^{0.7} \tag{6}$$

By this rule, we interpret a factor of 3 in the range of C_{org} (Fig. 3) as corresponding to a factor of 2.2 in the range of productivity.

There is great uncertainty on the amount of organic carbon within slope sediments that is brought in through redeposition, by bottom-near transport, rather than by delivery from above, from primary production. In places, much of the carbon may, in fact, be derived from productivity on the shelf, and from continental erosion (Prahl and Muehlhausen, 1989; Walsh, 1989; Stein, 1991). We prefer the point-seven rule (which makes no pretense as to precision) to an "exact" equation linking organic carbon of uncertain origin to productivity values based on calibration sets with large error margins. This rule should work best for sediment sections of moderate length, and with moderate changes in facies, in any one core.

The Problem of Decay within the Sediment

Estimates of paleoproductivity are affected by decay of organic matter within the sediment. Decay proceeds rapidly first, in sediments close to the seafloor, and then progressively more slowly, as the more accessible (and the more digestible) material is removed (Froelich et al., 1979). The apparent half-life of the remaining organic matter is typically about the same as the time scale considered, counting down from the present. For example, in box cores from the equatorial Pacific, the apparent half-life of organic matter is between 2,000 to 12,000 years (Berger et al., 1983; Price, 1988), with organic carbon values typically ranging from 0.1 to 0.4 percent. The underlying assumption in giving a half-life is that the trend produced by decay can be separated from changes in the rates of supply of organic matter, in the last 20,000 years (Fig. 5A).

The box core data show a drop in supply of organic matter at the end of the last glacial period, between 2 and 2.5 (log difference of 0.3 to 0.4; Fig. 5b). By the point-seven rule, a drop in productivity between 1.6 and 1.9 is shown. The change occurs over a rather short interval, so that the effects from decay can be neglected in this estimate. Note that the two cores shown at the same scale (ERDC112Bx and PLDS68Bx)

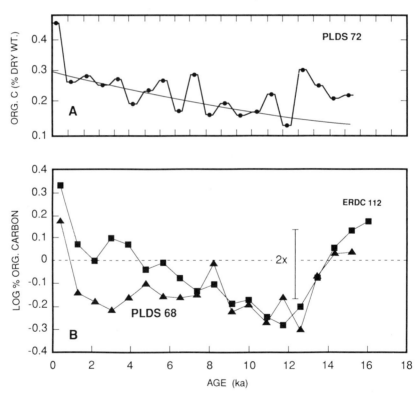

Fig. 5.   Organic matter content in three box cores from the equatorial Pacific. a) Core PLDS72Bx (1°N, 109°W, 3626m), from Berger et al. (1983); b) Cores ERDC112Bx (western equatorial Pacific, 1.38°S, 159.14°E, 2169m) and PLDS68Bx (eastern equatorial Pacific, 1.02°N, 105.30°W, 3650m), showing high values in pre-Holocene sediments (older than 10 ka) and in surface sediments (note logarithmic scale); data from Price (1988).

have absolute values that are surprisingly close to each other, considering the difference in the productivity of overlying waters (2). This reenforces our earlier statement that the calculation of absolute values of PP, from given values of C_{org}, is rather unreliable. However, these records agree on the magnitude of the change during deglaciation.

Though glacial values are reached at the bottom of the cores, they may not be the maximum glacial values. The results from longer cores (ERDC84P/PG, range 0.14% to 0.38%, uppermost two meters; PLDS76PG, range 0.1% to 0.25%, uppermost meter only; Price, 1988) show that our estimate (between 2 and 2.5) for the overall change in carbon content is conservative. Thus, 1.9 (by the point-seven rule) for the glacial-interglacial contrast in the productivity of the equatorial Pacific is a reasonable guess, based on these data. The results of Pedersen (1983) show C_{org} factors near 2.5 for a site west of the Galapagos Island, and a factor of 3 for the south-central Panama Basin, which translates into productivity contrasts of 1.9 to 2.2, between glacial and interglacial conditions. Prahl et al. (1989) also show a factor of 3 for C_{org} contrast, near 139°W, at the equator. A portion of this increase is because of greater eolian input of organic matter during glacial time. Again, a factor of 2 is indicated, for the change in productivity.

Surprisingly, the various estimates based on organic carbon content are reasonable, when compared with other methods (as we shall show). After all, only a small fraction of the carbon originally delivered gets buried within the sediment, and to begin with, the supply to the seafloor was a small fraction of the primary production. However, even in areas of extremely low sedimentation rates, with very little organic carbon remaining in sediments, correlations between cores can be very good, indicating that a primary signal is being preserved (Finney et al., 1988).

FORAMINIFERA AS PRODUCTIVITY INDICATORS

Ratio of Benthic to Planktonic Foraminifera

The abundance of benthic organisms is a function of the supply of organic matter to the seafloor (e.g., Rowe, 1983). Thus, it should be possible to read the history of supply of organic matter from the record of benthic foraminifera in the sediment. This concept was successfully applied off NW Africa, building on the earlier studies using organic carbon content (Lutze et al., 1986). Traditionally, geologists have considered the abundance of benthic foraminifera (or, more commonly, the benthic-to-planktonic ratio) to be an indicator of depth of deposition (Haake et al., 1982; van der Zwaan et al., 1990; and refs. therein). The reason that this is a viable proposition is that the supply of organic matter decreases in the deep ocean, because primary productivity is lower in the blue ocean than in the coastal ocean, and because the loss of organic carbon during settling is much greater in the deep ocean than over the margins.

Many of the data given by paleontologists studying deep-sea sediments contain information on the abundance of both planktonic and benthic foraminifera. The planktonic forms are less sensitive to changes in productivity than the benthic ones; thus, the ratio of benthics to planktonics (B/P) changes with productivity of overlying waters, at any given depth (Berger and Diester-Haass, 1988). A major problem in using B/P as a productivity index is the interference from differential dissolution of calcareous shells, which tends to enrich effected samples in benthic foraminifera (Parker and Berger, 1971). We analyzed data given in Phleger et al. (1953), for the central northern Atlantic, and found that the ratio between B/P and organic flux to the seafloor is by no means constant (as one might expect) but varies greatly from place to place, presumably because of winnowing and lateral input.

Let us assume that the local disturbing factors stay constant through time (which is admittedly risky). We should then be able to extract a valid signal related to regional (as opposed to local) conditions. We test this idea by analyzing some sites from above the lysocline (Fig. 6). To extract the presumed signal (and make it comparable between sites), we normalize the paleo-indices to their range of variation within each core. Productivity is represented as the normalized B/P, and temperature as the normalized weighted average of the latitude associated with each planktonic species, that represents its mid-range of biogeographic distribution (as given in Fig. 3 of Vincent and Berger, 1981). The species used for the temperature index are *Globigerina bulloides, G. eggeri, G. inflata, G. pachyderma, Globigerinoides rubra, G. sacculifera, G. menardii-tumida* group, and *G. truncatulinoides* ("*G. eggeri*" is synonymous with *Neogloboquadrina dutertrei*).

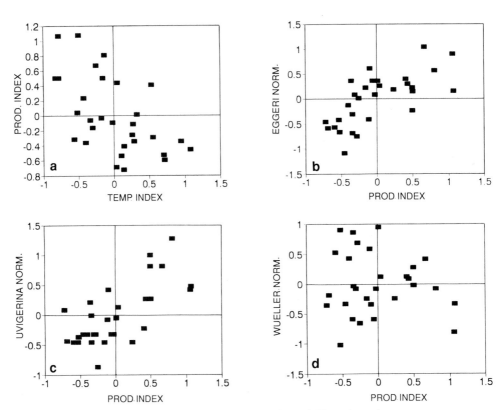

Fig. 6. The ratio of benthic to planktonic foraminifera in calcareous ooze as an indicator of productivity, as seen in three cores raised by the *Albatross*, in the equatorial Atlantic (Swedish Deep Sea Expedition). a) productivity (standardized B/P ratio) as a function of temperature (standardized weighted latitude of planktonics); b) percent *Globigerina eggeri* (normalized) versus productivity; c) relationship between *Uvigerina hollicki* and productivity; d) percent *Planulina wuellerstorfi* versus productivity. Data from Phleger et al. (1953), Cores 242, 243, and 246, first 10 samples each, downcore. (Sample no. 4 in Core 242 was omitted, because of an anomalously low foraminifera count.)

The results of the analysis show that there is a strong tendency for a negative correlation of benthic/planktonic ratios abundance with temperature in overlying waters (as recorded by planktonic foraminifera). More food was delivered to the seafloor when the waters above were colder. The factors of change in B/P are 8 for Core 242, 3.5 for Core 243, and 4.5 for Core 246 ([mean+sd]/[mean-sd]). Assuming (conservatively, for this case) that new production changes as the square of primary production (Eppley and Peterson, 1979), we would not expect factors greater than 4, in the range of Jsf, for a range of a factor of 2 in productivity. The ranges in B/P, when taken to reflect changes in food supply to the seafloor, therefore would seem to represent an exaggerated response to productivity changes in overlying waters (especially if planktonic shell also supply increases with productivity). In any case, it should prove difficult to calibrate the B/P index to productivity, although a strong positive correlation may be safely assumed, as we shall see.

The B/P index may be highly susceptible to pulsation of productivity; benthic foraminifera would benefit from pulsation (through more efficient delivery of food to the seafloor), while planktonic foraminifera would not (except species without symbionts, and only at the time of the bloom). A high sensitivity of B/P to seasonality is suggested by the fact that B/P is ten times higher in sites north of the Sargasso Sea, than in tropical sites with roughly the same productivity and at the same depth. In addition, the abovementioned redeposition effect, which presumably is increased during times of low sea level, should influence the magnitude of B/P ranges.

Changes in Species Composition as an Index

That planktonic foraminifera can yield interesting clues about changes in productivity has been appreciated for some time. For example, *G. bulloides* has long been known as an indicator of upwelling (Parker, 1973; Ganssen and Sarnthein, 1983; Prell and Curry, 1980). In the data at hand, the species *G. eggeri* (*N. dutertrei*) shows a positive relationship to productivity, as seen in the benthic/planktonic ratios (Fig. 6b). A ratio of this species to *Pulleniatina obliquiloculata* (which is less sensitive to productivity changes) has been used to argue for a drop in productivity from glacial to postglacial time, in the western equatorial Pacific (Berger, 1977). This drop has now been confirmed by studies on benthic foraminifera (see below).

We may assume that the composition of the benthic assemblage must change as the supply of food changes, because of differential growth and differential predation pressures. There is evidence for this type of response, especially in the continental margins (Altenbach and Sarnthein, 1989; Lutze et al., 1986). The genus Uvigerina, among the deep-sea benthics, has long been suspected of indicating conditions of high food supply (and possibly low oxygen values) (Douglas and Woodruff, 1981; Loubere, 1987; 1991). The correlation of the abundance of *Uvigerina hollicki*, with the B/P index, supports the concept (Fig. 6c). In contrast, there is little evidence that *Planulina wuellerstorfi* is a indicator species (Fig. 6d), although its relative abundance increases under conditions of low productivity, in the western equatorial Pacific (Burke et al., in press). (This does not preclude an increase in the absolute flux of shells, during times of higher productivity.)

Reconstructions of glacial productivity in the Atlantic recently were presented by Mix (1989a,b), based on CLIMAP-type transfer functions applied to the distribution of planktonic foraminifera. Similar reconstructions were attempted by Lapenis et al. (1990), who, in addition, used the patterns of glacial temperature anomalies (CLIMAP, 1976) to draw the paleoproductivity contours. Mix's reconstructions may be interpreted as showing a general increase of glacial productivity for the Atlantic Ocean, between 1.1 and 1.2

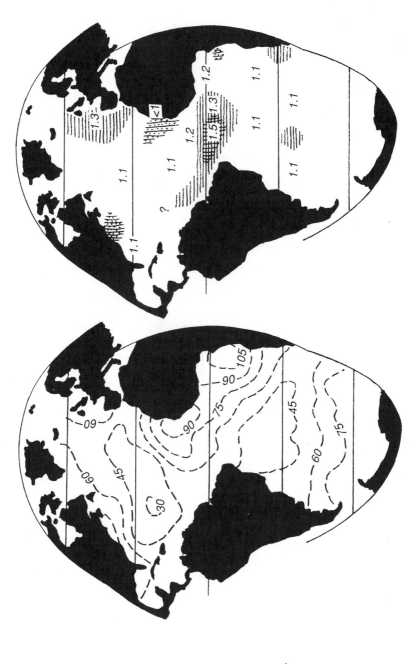

Fig. 7. Paleoproductivity reconstruction from planktonic foraminifera. Left: Productivity of present-day Atlantic, calculated from transfer functions based on calibration of the distribution of planktonic foraminifera on the seafloor with the overlying productivity (as given in Fig. 1). From Mix (1989a). Right: Factors of change, for the productivity values derived from 18 ka assemblages. From Mix (1989a).

(Fig. 7). For the central equatorial Atlantic, Mix obtained 1.5 for the glacial-postglacial contrast. Note that in Mix's map, there is a drop in productivity in glacial time, in a large area off NW Africa. This result does not agree with that from other evidence (e.g., organic matter content). A similar drop is shown off the East Coast, and in a sample off Angola. In these cases also, the results are unlikely to be correct, but rather point to some difficulty in the method of computation.

Two problems arise when using planktonic foraminifera for the reconstruction of ocean productivity: one is because of the effects of differential dissolution, the other is because of a strong present-day correlation between temperature and productivity patterns in the tropical and temperate realm. These problems are discussed by Mix (1989a) who believes they can be eliminated by judicious use of samples and statistics.

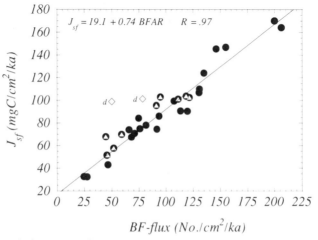

Fig. 8.   Accumulation rates of benthic foraminifera (BFAR) as a function of calculated flux of organic matter to the seafloor (Jsf). The flux calculation is based on Eq 2 (see text). The calibration equation shows linear regression. Open triangles mark depths between 4 and 4.5 km; diamonds denote >4.5 km (these values not used in the regression). Data from Herguera and Berger (in press).

While Mix's conclusions may be correct for the calibration of the data (which yields the transfer equations), it is hard to see how the interference can be eliminated for past conditions. If, in the past, the correlations between productivity and temperature (or rates of dissolution) were different, the statistics are then no longer appropriate in the same sense as for the calibration. For example, increased supply of organic matter can by itself increase the effects of differential dissolution of carbonate. Also, changing seasonality, or marked changes in nutrient contents of subsurface waters, would affect the present relationship between cold anomalies at the surface and elevated levels of productivity, leading into non-analog conditions. The question then is which factor determines the species composition among the planktonic foraminifera: if temperature dominates (CLIMAP, 1976), conclusions on productivity could acquire large error margins.

Accumulation Rates of Benthic Foraminifera

The best way to reconstruct the supply of organic matter to the seafloor is to measure the rate of accumulation of benthic foraminifera (Herguera and Berger, in press; Herguera, in press). This measure circumvents many of the problems arising when using proportions, and it also is less sensitive to differential dissolution than the B/P ratio. In box cores from the Ontong-Java Plateau, from the eastern equatorial Pacific, and the South Atlantic, the postglacial accumulation rate of benthic foraminifera closely tracks the calculated supply of organic matter to the seafloor (Fig. 8).

The regression equation relating the flux of organic matter to the bottom ("food") to the benthic foram accumulation rate (BFAR) is as follows:

$$Jsf = 19.1 + 0.74 * BFAR \tag{7}$$

where Jsf is the flux as calculated from Eq 2. A similar equation is given in Herguera and Berger (in press), with an intercept of 30.7 and a coefficient of 1.07. In that case, Jsf was calculated from a somewhat different equation, that is, one in which the flux depends on the square of PP, rather than on the power of 1.5. Thus, in the case at hand, conversion of BFAR to productivity follows the point-seven rule (inverse of 1.5), while for the equation of Herguera and Berger (in press), the conversion depends on a square-root rule (inverse of power of 2). In analogy to Eq 6, we propose to calculate paleoproductivity as follows:

$$PaP = k * BFAR^{0.7} \tag{8}$$

where k depends on the particular depth of the site, and is between 1 and 0.5 for depths between 2000 m and 4000 m; k is found by empirical calibration. As suggested for the C_{org} index, it is the record of ratios that matters, at any one site.

To ascertain the change in flux of benthic foraminifera, from glacial to postglacial time, we plot the product of abundance (BF/g) and sediment accumulation rate, against depth of water, for the last glacial and for the core-tops. A striking pattern emerges for the Ontong Java Plateau (Fig. 9). Both Recent and Glacial BFAR-values show a regular decrease with depth. If we form the product of BFAR and depth (in analogy to the BZ index of Berger and Diester-Haass, 1988), we obtain a value of about 200 throughout the depth investigated. This means that the benthic foram flux is roughly proportional to 1/depth, as it should be if it depends on the food supply, and if this supply follows the 1/z rule of Suess (1980).

The offset between Recent and Glacial BFAR-values shows a higher rate of flux of benthic shells to the sediment, during glacial time, by a factor of 2. The point-seven rule (Eq 8) translates this factor into a change of a factor of 1.6, between productivity in the last glacial maximum and the late Holocene. In other words, relying on the linear relationship between Jsf and BFAR (Fig. 8), and on Eq 2 showing an enhanced response of Jsf to primary productivity, we must assume that BFAR will show the same amplified response to productivity change.

The estimate of 1.6 for the glacial-postglacial contrast on Ontong Java Plateau is on the conservative side. When calculating BFAR, we assumed a steady sedimentation rate within each core. If productivity was higher during the last glacial, the sedimentation rate also should have been higher. A factor of 1.3 has been suggested for this effect (Schiffelbein and Dorman, 1984). If we assume this factor, the calculated accumulation rate of benthic foraminifera during glacial time increases similarly. The offset between Recent and Glacial BFAR-values then should be 2.6 (not 2, as shown in Fig. 9). Consequently, the factor for productivity change then is 2.0, rather than 1.6, as estimated for the steady sedimentation scenario.

In summary, from these data productivity in the western equatorial Pacific was higher during the last glacial maximum, between 1.6 and 2. This result agrees with our estimate based on organic matter content (1.6 to 1.9; with the higher number more probable). We propose a change of 1.8 as the most likely value for the glacial-postglacial productivity contrast, in the western equatorial Pacific.

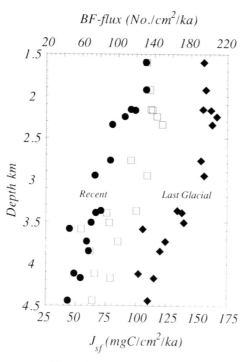

Fig. 9. Accumulation rate of benthic foraminifera as a function of depth, Ontong Java Plateau. Filled circles: core-top samples; open squares: mid-transition; filled diamonds: last glacial; bottom scale: estimated Jz-flux in mgC cm^{-2} ka^{-1}.

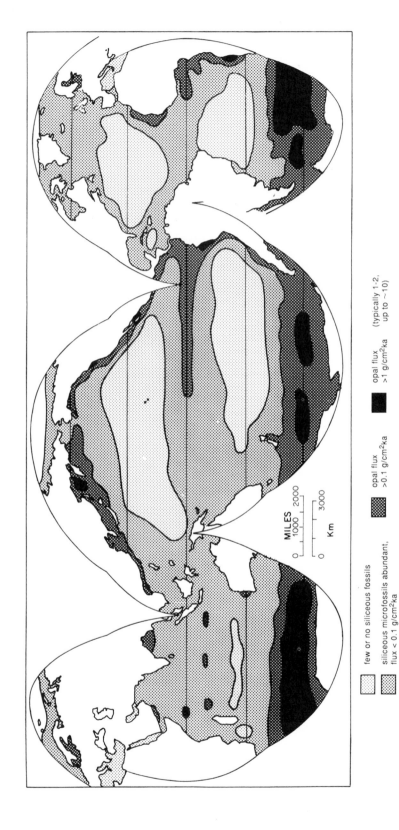

Fig. 10. Distribution of siliceous fossils on the seafloor. Light stipple: few or no siliceous fossils; medium stipple: abundant fossils (accumulation flux <0.1 g/cm²/ka); heavy stipple: opal flux 0.1 to 1 g/cm²/ka; black: flux >1g/cm²/ka. Data from various sources, mainly Lisitzin (1967), Leinen et al. (1986), Goll and Bjoerklund (1971; 1974), Riedel (1959), Hays (1967), DeMaster (1981), and Mortlock et al. (1991).

NOTES ON OTHER PROXY INDICATORS

Opaline Shells

The abundance of the opaline shells of diatoms and radiolarians on the seafloor correlates well with the overall patterns of productivity (Calvert, 1974; Heath, 1974; Berger, 1976; Fig. 10). The realization that these abundances could be used as productivity indicators took hold in the 1950s and 1960s (Pratje, 1951; Riedel and Funnell, 1964). Calvert (1966) showed that diatomaceous deposits in the Gulf of California are the result of upwelling alone (see also Byrne and Emery, 1960). Two decades earlier this relationship had not yet been recognized. In his classic paper on the Monterey Formation of California, Bramlette (1946) suggested that the most important factor in the accumulation of diatomaceous deposits was "...the drifting of this micro-plankton by currents from the open ocean into areas of deeper water and into cul-de-sacs along the coast..." Bramlette also suggested that the leaching of volcanic ash provided an important local source for much of the silica precipitated by the diatoms. In the global map of opal accumulation (Fig. 10), the towering dominance of the Antarctic system is conspicuous. At least one half of the ocean's opal is deposited here. This implies that preservation of siliceous fossils elsewhere will greatly depend on the efficiency with which the Antarctic

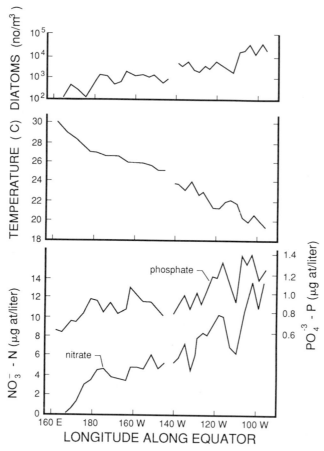

Fig. 11. East-west gradient in the concentration of large diatoms (retained in 80μ nets) in surface waters of the equatorial Pacific, as a function of the changing nutrient content. From Berger and Roth (1975), based on Desrosières (1969).

system extracts silica from the ocean. The state of the North Atlantic (which exports silica at present) also is important.

Along the equator in the Pacific the content of opal in sediments is considerably greater in the east than in the west, presumably because of the difference in supply of diatoms (Leinen, 1979; Leinen et al., 1986). A higher supply in the east may be inferred from the change in concentrations of diatoms in the water (Fig. 11). In turn, this gradient reflects the change in nutrient concentrations in surface waters, and the resulting gradient in productivity.

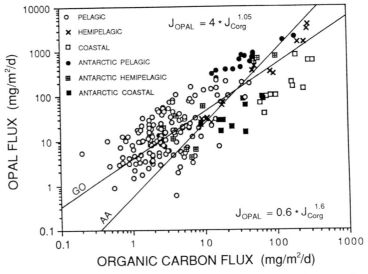

Fig. 12. Relationship between opal flux and organic carbon flux in traps from various regions of the ocean, fit by eye. GO: general ocean; flux of opal increases at a slightly greater factor than flux of organic carbon (exponent of 1.05). AA: Southern Ocean; flux of opal increases by a much greater factor than the flux of organic carbon (although scatter is large). Diagram from Milliman and Takahashi (in press); lines and equations were added.

Trapping data compiled by Milliman and Takahashi (in press) show that the flux of opal goes parallel with that of C_{org} over much of the ocean (Fig. 12). The relationship suggested by their diagram is an exponent of 1.05 on C_{org}, that is, there is a modest amplification of the factor of increase in C_{org} export, in the factor of increase of opal export. For the Southern Ocean this amplification is distinctly greater, and very significant (exponent of 1.6). Recalling that fluctuation in C_{org} export is itself amplified over productivity, we note that the amplification of productivity contrast by opal export contrast is a function of the setting, and varies somewhere between 1.5 and 2.4 in the present ocean (exponent on productivity). The range of the exponent is tied to the intensity of the seasonality (and the pulsation) of the productivity, and to the concentration of dissolved silicate.

Measurements of the concentration of biogenic opal in box cores from the equatorial Pacific (west: ERDC; east: PLDS) yield accumulation rates of 0.05 g/cm^2/ka for the west, and 0.34 g/cm^2/ka for the east (Herguera, in press). These rates are not very sensitive to the depth of deposition, unlike the rates for benthic foraminifera (which reflect organic matter supply). The lack of sensitivity to depth is because settling silica reaches the seafloor much more readily than settling organic matter, as is evident from the high ratios of opal to C_{org} in deep-sea traps (see Table 1 in Wefer, 1989).

The east-west contrast of a factor of six in opal flux, as measured in our box cores, is three times higher than the contrast in productivity. Thus, opal flux as a productivity index greatly amplifies the primary signal, along the equatorial Pacific. The cause of this amplification is that so much of the silica is redissolved, not as a proportion of what is coming down, but as a background loss, which is largely independent of the amount of material delivered. This process has the effect of greatly increasing the initial differences in supply. In the end, we compare residuals which are chiefly composed of shells and shell fragments resistant to dissolution. Such shells (that is, well-silicified large shells) are generated disproportionately in the more productive areas.

Other things being equal, we should expect evidence for a substantial increase in opaline flux during the last glacial period, in the western equatorial Pacific. If productivity increased by 1.8, as we propose, the opaline flux should have been greater by 5, if present geographic patterns are extrapolated back in time. In fact, no such change is seen (Herguera, in press). Instead, the opal flux stays about the same, or, more likely, is somewhat less during the glacial than now. This result, based on measurement of opal content, agrees well with observations on diatom abundance and preservation in cores from Ontong Java Plateau (Mikkelsen, 1979; Lange and Berger, 1991).

We conclude, from these observations, that preservation of diatoms deteriorated during glacial conditions, and that the flux to the sediment was decreased. Clues to possible causes may be found in the contrast of diatom sedimentation between the equatorial Pacific and equatorial Atlantic. At the same level of productivity, opaline sediments are notoriously less in the Atlantic than in the Pacific, because of a weaker silicification of tests in the Atlantic (as initially observed by Kolbe, 1955). We believe that the basic reason for this difference is the lower concentration of silicate, in subsurface waters of the Atlantic, compared with those in the Pacific (Berger, 1976). The glacial Pacific was more like the Atlantic today, that is, there was much less silicate dissolved in its intermediate waters. This proposition agrees well with the other observations suggesting more conditions like the Atlantic in the glacial Pacific, including the carbonate record (lowered carbonate compensation depth), and the $\delta^{13}C$ record (deep nutricline; Herguera et al., in press).

The main point of this discussion is that opal flux is unreliable as a quantitative index for changes in productivity, because it has a strong nonlinear relationship to supply, and because the supply itself is a function of the availability of dissolved silicate relative to phosphate, a ratio that changes regionally and through time. In addition, diatom shells are easily transported, and may be removed regionally from elevated areas and concentrated in depressions by bottom currents.

The species composition of diatom assemblages may contain useful information on paleoproductivity (Sancetta, 1979; Lange et al., 1990). A striking result of a recent study off Peru (Schrader and Sorknes, 1991) is that changes in productivity do not follow glacial-interglacial cycles, but change in an irregular fashion with respect to these cycles.

Carbonate Flux

The depression of the carbonate compensation depth (CCD) in the region of the equatorial Pacific, below the regional depth level, has long been considered prime evidence that the supply of carbonate to the seafloor increases with increasing productivity (Fig. 13). This pattern was first recognized by Arrhenius (1952), who used it to argue for substantially higher productivity during glacial time, when the CCD was depressed even further. Straightforward application of this argument for higher glacial productivity is complicated by interference from the increase in saturation of deep waters in the glacial Pacific, which decreases the rates of dissolution of carbonate (Berger, 1976; Hebbeln et al., 1990), simulating an increased supply.

The carbonate found within deep-sea sediments is almost all derived from the skeletal materials of coccolithophores and planktonic foraminifera. To some degree, the ratio of foraminifera to nannofossils is a measure of productivity, with sand content below the Sargasso Sea, for example, being considerably smaller than that on Ontong Java Plateau in the western equatorial Pacific. However, application of this relationship to the quantitative reconstruction of productivity is greatly hampered by the interfering effects of winnowing and dissolution on grain size (Johnson et al., 1977; Wu et al., 1990; Wu and Berger, 1991).

The effect of changing productivity on carbonate flux may be gauged by comparing the flux in a core from the Sargasso Sea (INMD50Bx, Somayajulu et al., 1984) with one from Ontong Java Plateau (ERDC125Bx, Berger et al. 1987b). Both the rates of accumulation of carbonate and of productivity in the two sites differ by about 1.8. Such a similarity in ratios is not seen when we consider the carbonate accumulation rate in the eastern equatorial Pacific. The highest rate (PLDS 72) exceeds that in the west (ERDC 125) by less than 1.2, even though the productivity differs by 2. Thus, the response of carbonate flux is strongly nonlinear with productivity (quickly flattening out at the high end).

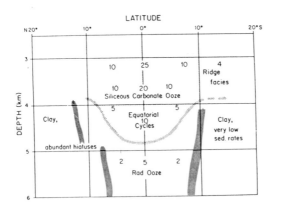
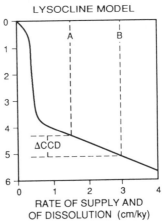

Fig. 13.   Depression of the carbonate compensation depth (CCD) below the regional level, under the high productivity region of the equatorial Pacific. Left: Post-Eocene generalized facies regime of the equatorial Pacific in a paleodepth-paleolatitude matrix. Numbers are sedimentation rates in mm/ka. Right: Model of CCD depression from increased productivity (Berger, 1978).

One reason for the postulated nonlinear relationship may be the effects of dissolution. Dissolution of carbonate increases with an increase in supply of organic matter (which lowers the pH of interstitial waters). We know from trapping results (Wefer, 1989) that carbonate fluxes differ greatly regionally and seasonally, in quality and in quantity, and that the ratio to organic matter flux is variable. Generally, the ratio of C_{org} to carbonate in the export changes in favor of organic matter, when productivity increases (Berger and Keir, 1984; Milliman and Takahashi, in press) (see Fig. 14).

From these data, we suspect that carbonate flux is unreliable as a quantitative indicator of productivity, except when the range of productivity is moderate, and when prevailing conditions do not change much. An example of variation that cannot be translated into productivity change is the transgression of a front over the site studied. An extreme case of such front migration would be the movement of the southern polar front, south of which little carbonate is deposited even though productivity is quite high.

Carbon Isotopes in Foraminifera

The export of organic carbon from the photic zone preferentially removes ^{12}C from the mixed layer, so that the ratio of ^{13}C to ^{12}C is higher here than in the deep waters below. Expressed in the δ-notation, the dissolved carbon in surface waters has $\delta^{13}C$ values

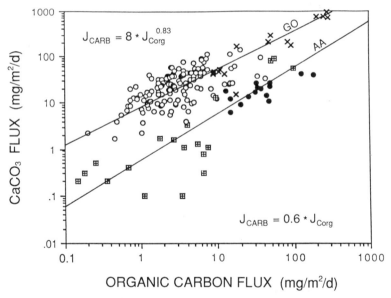

Fig. 14. Relationship between carbonate and organic carbon flux in traps from various regions of the ocean, fit by eye. GO, general ocean; flux of carbonate increases at a lesser factor than flux of organic carbon (exponent of 0.83). AA, Southern Ocean; flux of carbonate increases by the same factor as the flux of organic carbon (although scatter is large). Diagram from Milliman and Takahashi (in press); lines and equations were added.

that are typically 1 to 2 permil more positive than the $\delta^{13}C$ of deepwater carbon. Broecker (1973) first recognized that this difference is a measure of the intensity of biological pumping of carbon from surface waters to deep waters, and that a record of this process would be kept within the carbon isotopic composition of planktonic and benthic foraminifera. In his model of glacial CO_2 reduction, Broecker (1982) argued that the process was intensified during the last glacial (by the addition of phosphate to the ocean), and that the proof of this proposition would be in an increased difference in the $\delta^{13}C$ of planktonic and benthic foraminifera.

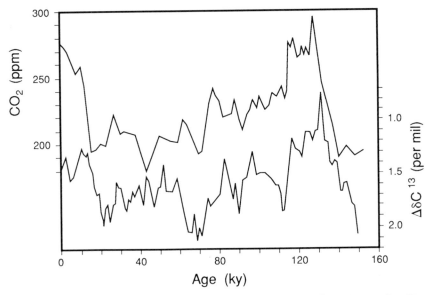

Fig. 15.   Carbon dioxide concentrations in the Vostok ice core from Antarctica (Barnola et al., 1987), compared with a productivity-related carbon isotope signal from the eastern tropical Pacific (difference between the $\delta^{13}C$ values of a planktonic foraminifer and a benthic one; Shackleton et al., 1983). From Berger et al. (1989).

Shortly after, Shackleton et al. (1983) applied the model to a record of a core raised in the eastern tropical Pacific. They plotted the difference in $\delta^{13}C$ between a planktonic foraminifer (*Neogloboquadrina dutertrei*) and a benthic foraminifer (*Uvigerina*), and interpreted the result in terms of strength of biological pumping, and resulting modification of atmospheric pCO_2. Their postdiction of pCO_2 levels was subsequently strikingly supported by results from ice-coring (Fig. 15).

The success of this exercise is the more surprising as neither *Neogloboquadrina* nor *Uvigerina* would seem to be ideal as monitors of surface waters and deep waters, respectively. The former is typical for regions (or seasons) of elevated productivity, the latter lives within the sediment and has a $\delta^{13}C$ signal that is strongly influenced by the supply of organic matter (according to Zahn et al., 1986). In addition, it is difficult to see why fluctuations in the particular region studied should reflect biological pumping for the entire ocean. Nevertheless, these results are strong evidence that the pCO_2 of the atmosphere and the $\delta^{13}C$ pattern in the ocean (that is, its productivity and its circulation) are closely associated.

We may ask, given the evidence for the glacial-postglacial productivity change in the western equatorial Pacific presented above, whether the $\delta^{13}C$ record in the area shows the expected change; it does indeed (Fig. 16). The plot compares the average $\delta^{13}C$ signal for three benthic records (*Planulina wuellerstorfi*, ERDC92Bx+ERDC88Bx+ERDC113P) with the equivalent signal from planktonic foraminifera (*P. obliquiloculata*, stacked record from seven cores above the lysocline).

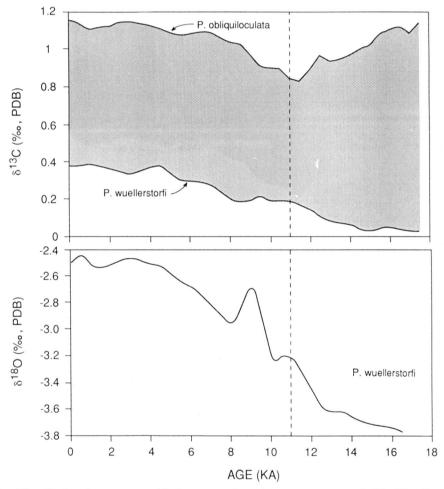

Fig. 16.   Carbon isotope record in box cores from the western equatorial Pacific. Oxygen isotope record was used for guidance, and the time scale was based on setting the midpoint of deglaciation equal to 11,000 ka. Planktonic record (*P. obliquiloculata*): stack of seven cores above the lysocline (data from Berger et al., 1987b). Benthic record (*P. wuellerstorfi*): stack of 3 cores near 2000 m depth (data from Herguera et al., in press). Lower panel: $\delta^{18}O$ record of *P. wuellerstorfi*, for stratigraphic control.

First of all, the deep water composition is not constant, but shows a change toward more positive values from glacial to postglacial time. This may be interpreted as extraction of organic carbon from the ocean, by growth of the biosphere on land (Shackleton, 1977). If so, the change shown (roughly 0.4o/oo, valid for depths below 1500 m) would indicate an extraction of roughly 1.2 percent of the ocean's carbon, corresponding to a little more than one-half of the present biosphere mass, or $500*10^9tC$.

The planktonic foraminifera also contain the overall change in $\delta^{13}C$, but they do not show more negative values in the glacial. Thus, the mixed layer was enriched in ^{13}C, to a greater degree than today. The $\delta^{13}C$ values of the planktonics do change toward more negative values with the onset of deglaciation, indicating the drop in productivity and in bio-pumping. They subsequently rise again, riding the long wave shown in the deepwater signal. This rise, then, is a reflection of the extraction of carbon, to the terrestrial biosphere.

In summary, the difference between benthic and planktonic $\delta^{13}C$ values images the overall change from high to low productivity, from glacial to postglacial time, in the western equatorial Pacific. However, we cannot readily translate this record into a quantitative estimate of productivity change, because it represents a balance between upwelling of dissolved inorganic carbon and downward transport of organic carbon. What it does tell us is that there must have been an increase in nutrients within the upwelled waters, without a corresponding increase in dissolved CO_2 (which would cancel the effect on $\delta^{13}C$). We suggest that nitrate was more abundant in these waters, because of an overall reduction in denitrification within the glacial ocean (Berger and Keir, 1984). The remainder of the increase in productivity (not attributable to change in nutrient concentration and therefore not reflected in the $\delta^{13}C$ signal) would have to be assigned to increased rates of upwelling and mixing.

CONCLUSIONS

A good proxy for productivity is a sediment parameter (e.g., organic matter content, or benthic foram flux), which can be calibrated against present-day primary production in a wide variety of oceanographic and sedimentologic settings. This robustness is some guarantee that the calibration will not change readily through time, as conditions change. The limited survey of proxy indicators offered here should make it clear that considerable uncertainties remain even for well-chosen proxies. Thus, statements about changes in productivity, and especially quantitative changes, are necessarily tentative. It is advisable to use several proxies simultaneously.

We have presented compelling evidence for productivity changes in the western equatorial Pacific (and also elsewhere along the equator), between glacial and postglacial time, by a factor of 2 or slightly less. Increased rates of upwelling and increased concentrations of nutrients within underutilized waters during glacial time, must both be considered as possible causes. The first factor is generally accepted, as a result of increased zonal winds (Rea et al., 1986; Arrhenius, 1988; Lyle, 1988; Lyle et al., 1988; Pisias and Rea, 1988; Molfino and McIntyre, 1991). The second also is a strong candidate, especially with respect to nitrate, since shelves play an important role in facilitating denitrification (Christensen et al., 1987). Shelf areas were greatly reduced during glacial times, so that denitrification was diminished, and export production was increased (Berger and Keir, 1984).

Also, we believe that Martin's (1990) iron mechanism, for stimulating glacial-age productivity, has merit (even though it may not work in the Southern Ocean; Berger and Wefer, in press). Well over 90% of the iron available for photosynthesis is introduced by the wind (Zhuang et al., 1990). On the whole, during glacial times winds were more active, and areas of wind-erosion more extensive (Sarnthein et al., 1982; Petit et al., 1990) (Although in detail, eolian response to climatic variation is complicated; Kutzbach, 1989). Thus, nutrients in the open ocean should have been used more efficiently, in those regions where they are underutilized today because of iron-limitation. The quality and mode of export could have also changed in those areas where nutrients are used to virtual depletion (due to changes in the trophic structure). In particular, phytoplankton blooms, and hence, pulsed export, could have been stimulated (cf. Banse, 1990). The effects of biological pumping on atmospheric pCO_2 would have been correspondingly increased, because recycling would have been less efficient.

Significantly, rates of opal accumulation are not parallel to the changes in productivity deduced from the other proxies (benthic foraminifera and C_{org}), for the equatorial Pacific. Thus, the glacial silica system was entirely different from today's, presumably because of a major drawing down of concentrations from increased deposition in the coastal ocean. If so, the attempt to test Martin's (1990) Antarctic hypothesis using changes in the rate of opal deposition (Mortlock et al., 1991) must be viewed with caution.

ACKNOWLEDGMENTS

W. H. Berger is indebted to many colleagues for discussions on the productivity of the ocean, at the Dahlem meeting in Berlin, 1988, and at the ASLO meeting on the same topic, San Marcos, 1991. Discussions with Carina Lange, on glacial-age diatom sedimentation in the equatorial Pacific, are gratefully acknowledged. Our work is supported by the Petroleum Fund of the American Chemical Society, and by the National Science Foundation (NSF OCE 90-17717).

REFERENCES

Altenbach, A. V., and Sarnthein, M., 1989, Productivity record in benthic foraminifera, *in*: "Productivity of the Ocean: Present and Past," W.H. Berger, V.S. Smetacek, and G. Wefer, eds., John Wiley and Sons, Chichester.

Arrhenius, G. O. S., 1952, Sediment cores from the east Pacific, *Rep. Swed. Deep Sea Exped. 1947-1948*, 5:1.

Arrhenius, G. O. S., 1988, Rate of production, dissolution and accumulation of biogenic solids in the ocean, *Palaeogeography, Palaeoclimatology, Palaeoecology*, 67:119.

Bacastow, R., and Maier-Reimer, E., 1991, Dissolved organic carbon in modeling oceanic new production, *Global Biogeochem. Cycles*, 5: 71.

Banse, K., 1990, Does iron really limit phytoplankton production in the offshore subarctic Pacific?, *Limnol. Oceanogr.*, 35:772.

Banse, K., 1991, Iron availability, nitrate uptake, and exportable new production in the subarctic Pacific, *J. Geophys. Res.*, 96 (C1):741.

Barnola, J. M., Raynaud, D., Korotkevich, Y. S., Lorius, C., 1987, Vostok ice core provides 160,000 year record of atmospheric CO_2, *Nature*, 329:408.

Berger, W. H., 1976, Biogenous deep-sea sediments: production, preservation and interpretation, *in*: "Treatise on Chemical Oceanography," Vol.5, J.P. Riley and R. Chester, eds., Academic Press, London.

Berger, W. H., 1977, Carbon dioxide excursions and the deep sea record: aspects of the problem, *in*: "The Fate of Fossil Fuel CO_2 in the Oceans," N.R. Andersen and A. Malahoff, eds., Plenum Press, New York.

Berger, W. H., 1978, Sedimentation of deep-sea carbonate: maps and models of variations and fluctuations, *J. Foram. Res.*, 8:286.

Berger, W. H., and Diester-Haass, L., 1988, Paleoproductivity: the benthic/planktonic ratio in foraminifera as a productivity index, *Marine Geol.*, 81:15.

Berger, W. H., Finkel, R. C., Killingley, J. S., and Marchig, V., 1983, Glacial-Holocene transition in deep-sea sediments: manganese spike in the east-equatorial Pacific, *Nature*, 303:231.

Berger, W. H., Fischer, K., Lai, C., and Wu. G., 1987a, Ocean productivity and organic carbon flux. Part I. Overview and maps of primary production and export production, Scripps Institution of Oceanography Ref. 87-30:1.

Berger, W. H., and Keir, R. S., 1984, Glacial-Holocene changes in atmospheric CO_2 and the deep-sea record, *in*: "Climate Processes and Climate Sensitivity," J.E. Hansen and T. Takahashi, eds., Geophys. Monogr. 29, American Geophys. Union, Washington, D.C..

Berger, W. H., Killingley, J.S., and Vincent, E., 1987b, Time scale of Wisconsin/ Holocene transition: oxygen isotope record in the western equatorial Pacific, *Quaternary Res.*, 28:295.

Berger, W. H., and Roth, P. H., 1975, Oceanic micropaleontology: progress and prospects, *Rev. Geophys. Space Phys.*, 13:561.

Berger, W. H., Smetacek, V. S., and Wefer, G., 1989, Ocean productivity and paleoproductivity - an overview, *in*: "Productivity of the Ocean: Present and Past," W.H. Berger, V.S. Smetacek, and G. Wefer, eds., John Wiley, Chichester.

Berger, W. H., and Wefer, G., 1990, Export production: seasonality and intermittency, and paleoceanographic implications, *Global and Planetary Change*, 3:245.

Berger, W. H., and Wefer, G., On the productivity of the glacial ocean: discussion of the iron hypothesis, *Limnology and Oceanography*, in press.

Betzer, P. R., Showers, W. J., Laws, E. A., Winn, C. D., DiTullio, G. R., and Kroopnick, P. M., 1984, Primary productivity and particle fluxes on a transect of the equator at 153°W in the Pacific Ocean, *Deep-Sea Res.*, 31:1.

Boyle, E. A., 1990, Quaternary deepwater paleoceanography, *Science*, 249:863.

Bralower, T. J., and Thierstein, H. R., 1987, Organic carbon and metal accumulation rates in Holocene and mid-Cretaceous sediments: palaeoceanographic significance, *in*: "Marine Petroleum Source Rocks," J. Brooks and A.J. Fleet, eds., *Geol. Soc. Spec. Publ.*, 26:345.

Bramlette, M. N., 1946, The Monterey Formation of California and the origin of its siliceous rocks, *U.S. Geol. Survey Prof. Paper*, 212:1.

Broecker, W. S., 1973, Factors controlling CO_2 content in the oceans and atmosphere, *in*: "Carbon and the Biosphere," G.M. Woodwell and E.V. Pecan, eds., AEC Symposium, 30:32.

Broecker, W. S., 1982, Ocean chemistry during glacial time, *Geochim. Cosmochim. Acta*, 46:1689.

Bruland, K. W., Bienfang, P. K., Bishop, J. K. B., Eglinton, G., Ittekkot, V. A. W., Lampitt, R., Sarnthein, M., Thiede, J., Walsh, J. J., and Wefer, G., 1989, Flux to the seafloor, *in*: "Productivity of the Ocean: Present and Past," W.H. Berger, V.S. Smetacek, and G. Wefer, eds., John Wiley, Chichester.

Burke, S. K., Berger, W. H., Coulbourn, W. T., and Vincent, E., Benthic foraminifera in box core ERDC 112, Ontong Java Plateau, *J. Foram. Res.*, in press.

Byrne, J. V., and Emery, K. O., 1960, Sediments of the Gulf of California, *Geol. Soc. America Bulletin*, 71:983.

Calvert, S. E., 1966, Accumulation of diatomaceous silica in the sediments of the Gulf of California, *Geol. Soc. America Bull.*, 77:569.

Calvert, S. E., 1974, Deposition and diagenesis of silica in marine sediments, *in*: "Pelagic Sediments on Land and Under the Sea," K.J. Hsü and H. Jenkyns, eds., Spec. Publ. Internat. Assoc. Sedimentologists, 1:273.

Calvert, S. E., 1987, Oceanographic controls on the accumulation of organic matter in marine sediments, *in*: "Marine Petroleum Source Rocks," J. Brooks and A.J. Fleet, eds., Geol. Soc. Spec. Publ., 26:137.

Christensen, J. P., Murray, J. W., Devol, A.H., and Codispoti, L. A., 1987, Denitrification in continental shelf sediments has major impact on the oceanic nitrogen budget, *Global Biogeochem. Cycles*, 1:97.

CLIMAP Project Members, 1976, The surface of the ice-age earth, *Science*, 191:1131.

Codispoti, L. A., 1989, Phosphorus vs. nitrogen limitation of new and export production, *in*: "Productivity of the Ocean: Present and Past," W.H. Berger, V.S. Smetacek, and G. Wefer, eds., Wiley-Interscience, Chichester.

DeMaster, D.J., 1981, The supply and accumulation of silica in the marine environment, *Geochim. Cosmochim. Acta*, 45:1715.

Desrosières, R., 1969, Surface macroplankton of the Pacific Ocean along the equator, *Limnol. Oceanogr.*, 14:626.

Douglas, R. G., and Woodruff, F., 1981, Deep sea benthic foraminifera, *in*: "The Sea, vol 7, the Oceanic Lithosphere," C. Emiliani, ed., Wiley-Interscience, New York.

Elderfield, H., 1990, Tracers of ocean paleoproductivity and paleochemistry: an introduction, *Paleoceanogr.*, 5:711.

Emerson, S., Fischer, K., Reimers, C., and Heggie, D., 1985, Organic carbon dynamics and preservation in deep-sea sediments, *Deep-Sea Res.*, 32:1.

Emerson, S., and Hedges, J. I., 1988, Processes controlling the organic carbon content of open ocean sediments, *Palaeogeogr., Palaeoclimat., Palaeoecol.*, 3:621.

Eppley, R. W., and Peterson, B. J., 1979, Particulate organic matter flux and planktonic new production in the deep ocean, *Nature*, 282:677.

Finney, B. P., Lyle, M. W., and Heath, G. R., 1988, Sedimentation at MANOP Site H (eastern equatorial Pacific) over the past 400,000 years: climatically induced redox variations and their effects on transition metal cycling, *Paleoceanogr.*, 3:169.

Froelich, P. N., Klinkhammer, G. P., Bender, M. L., Luedtke, N. A., Heath, G. R., Cullen, D., Dauphin, P., Hammond, D., Hartman, B., and Maynard, V., 1979, Early oxidation of organic matter in pelagic sediments of the eastern equatorial Atlantic: suboxic diagenesis, *Geochim. Cosmochim. Acta.*, 43:1075.

Ganssen, G., and Sarnthein, M., 1983, Stable-isotope composition of foraminifers: the surface and bottom water record of coastal upwelling, *in*: "Coastal Upwelling, its Sediment Record. Part A: Responses of the Sedimentary Regime to Present Coastal Upwelling," E. Suess and J. Thiede, eds., Plenum Press, New York.

Goll, R. M., and Bjoerklund, K. R., 1971, Radiolaria in surface sediments of the North Atlantic Ocean, *Micropaleontology*, 17:434.

Goll, R. M., and Bjoerklund, K. R., 1974, Radiolaria in surface sediments of the South Atlantic, *Micropaleontology*, 20:38.

Haake, F. -W., Coulbourn, W. T., and Berger, W. H., 1982, Benthic foraminifera: depth distribution and redeposition, *in*: "Geology of the Northwest African Continental Margin," U. von Rad, K. Hinz, M. Sarnthein, and E. Seibold, eds., Springer Verlag, Heidelberg.

Hays, J.D., 1967, Quaternary sediments of the Antarctic Ocean, *Progress in Oceanography*, 4:117.

Heath, G. R., 1974, Dissolved silica and deep-sea sediments, *in*: "Studies in Paleo-Oceanography," W.W. Hay, ed., Soc. Econ. Paleont. and Mineral., Spec. Pub., 20: 77.

Heath, G. R., Moore, T. C., and Dauphin, J. P., 1977, Organic carbon in deep-sea sediments, *in*: "The Fate of Fossil Fuel CO_2 in the Oceans," N.R. Andersen and A. Malahoff, eds., Plenum, New York.

Hebbeln, D., Wefer, G., and Berger, W.H., 1990, Pleistocene dissolution fluctuations from apparent depth of deposition in Core ERDC127P, west-equatorial Pacific, *Marine Geology*, 92:165.

Herbert, T. S., Curry, W. B., Barron, J. A., Codispoti, L. A., Gersonde, R., Keir, R. S., Mix, A. C., Mycke, B., Schrader, H., Stein, R., and Thierstein, H. R., 1989, Geological Reconstructions of Marine Productivity, *in*: "Productivity of the Ocean: Present and Past," W.H. Berger, V.S. Smetacek, and G. Wefer, eds., John Wiley, Chichester.

Herguera, J. C., Deep-sea benthic foraminifera and biogenic opal: glacial to postglacial productivity changes in the west-equatorial Pacific, *Marine Micropal.*, in press.

Herguera, J. C., and Berger, W. H., Paleoproductivity: glacial to postglacial change in the western equatorial Pacific, from benthic foraminifera, *Geology*, in press.

Herguera, J. C., Stott, L., and Berger, W. H., Glacial deep-water properties in the west-equatorial Pacific: bathyal thermocline near 2000 m depth, *Marine Geol.*, in press.

Johnson, T. C., Hamilton, E. L., and Berger, W. H., 1977, Physical properties of calcareous ooze: control by dissolution at depth, *Marine Geol.*, 24:259.

Keir, R. S., 1988, On the late Pleistocene ocean geochemistry and circulation, *Paleoceanogr.*, 3:413.

Keir, R. S., 1990, Reconstructing the ocean carbon system variation during the last 150,000 years according to the Antarctic nutrient hypothesis, *Paleoceanogr.*, 5:253.

Kolbe, R. W., 1955, Diatoms from equatorial Atlantic cores, *Repts. Swed. Deep-Sea Exped.*, 7 (3):149.

Koblentz-Mishke, O. I., Volkovinsky, V. V., and Kabanova, J. G., 1970, Plankton primary production of the world ocean, *in*: "Scientific Exploration of the South Pacific," W. Wooster, ed., National Academy of Sciences, Washington, D.C.

Kutzbach, J. E., 1989, Possible effects of orbital variations on past sources and transports of eolian material: estimates from general circulation model experiments, *in*: "Paleoclimatology and Paleometeorology: Modern and Past Patterns of Global Atmospheric Transport," M. Leinen and M. Sarnthein, eds., Kluwer Academic, Dordrecht.

Lange, C. B., and Berger, W. H., 1991, Diatoms fail to indicate the increased glacial productivity in the western equatorial, Pacific. Geol. Soc. America, Fall Meeting 1991, Abstracts.

Lange, C. B., Burke, S. K., and Berger, W. H., 1990, Biological production off Southern California is linked to climatic change, *Climatic Change*, 16:319.

Lapenis, A. G., Os'kina, N. S., Barash, M. S., Biyum, N. S., and Vasileva, Ye. V., 1990, Late Quaternary variations in the productivity of the ocean's biota, *Okeanologiya*, 30: 93. [In Russian].

Leinen, M., 1979, Biogenic silica accumulation in the central equatorial Pacific and its implications for Cenozoic paleoceanography: summary, *Geol. Soc. America Bull.*, Part I, 90:801.

Leinen, M., Cwienk, D., Heath, G. R., Biscaye, P., Kolla, V., Thiede, J., and Dauphin, J. P., 1986, Distribution of biogenic silica and quartz in recent deep-sea sediments, *Geology*, 14:199.

Lisitzin, A. P., 1967, Basic relationships in distribution of modern siliceous sediments and their connection with climatic zonation, *Internat. Geol. Rev.*, 9:631 (transl. fr. Russian).

Lisitzin, A. P., 1972. Sedimentation in the world ocean, *Soc. Econ. Paleont. Mineral. Spec. Publ.*, 17:1.

Loubere, P., 1987, Late Pliocene variations in the carbon isotope values of north Atlantic benthic foraminifera: biotic control of the isotope record, *Mar. Geol.*, 76:45.

Loubere, P., 1991, Deep-sea benthic foraminiferal assemblage response to surface ocean productivity gradient: a test, *Paleoceanogr.*, 6:193.

Lutze, G. F., Pflaumann, U., and Weinholz, P., 1986, Jungquartäre Fluktuationen der benthischen Foraminiferenfaunen in Tiefsee-Sedimenten vor NW-Afrika - eine Reaktion auf Produktivitätsänderungen im Oberflächenwasser, *Meteor Forschungs-Ergebnisse, Reihe C*, 40:163.

Lyle, M., 1988, Climatically forced organic carbon burial in equatorial Atlantic and Pacific oceans, *Nature*, 335:529.

Lyle, M., Murray, D. W., Finney, B. P., Dymond, J., Robbins, J. M., and Brooksforce, K., 1988, The record of late Pleistocene biogenic sedimentation in the eastern tropical Pacific Ocean, *Paleoceanogr.*, 3:39.

Martin, J. H., 1990, Glacial-interglacial CO_2 change: the iron hypothesis, *Paleoceanogr.*, 5:1.

Martin, J. H. and Fitzwater, S. E., 1988, Iron deficiency limits phytoplankton growth in north-east Pacific subarctic, *Nature*, 331:341.

Martin, J. H., Gordon, R. M., and Fitzwater, S. E., 1990, Iron in Antarctic waters, *Nature*, 345:156.

Mikkelsen, N., 1979, Diatoms in equatorial deep-sea sediments: sedimentation and dissolution over the last 20,000 years, *Nova Hedwigia*, 64:489.

Milliman, J. D., and Takahashi, K., Carbonate and opal production and accumulation in the ocean, *in*: "Global Surficial Geofluxes: Modern to Glacial," W. Hay, M. Meybeck, and T. Usselman, eds., National Research Council, Washington, D.C., in press.

Mix, A. C., 1989a, Pleistocene paleoproductivity: evidence from organic carbon and foraminiferal species, *in*: "Productivity of the Ocean: Present and Past," W.H. Berger, V.S., Smetacek, and G. Wefer, eds., John Wiley, Chichester.

Mix, A. C., 1989b, Influence of productivity variations on long term atmospheric CO_2, *Nature*, 337:541.

Molfino, B., and McIntyre, A., 1991, Precessional forcing of nutricline dynamics in the equatorial Atlantic, *Science*, 249:766.

Mortlock, R. A., Charles, C. D., Froelich, P. N., Zibello, M. A., Saltzman, J., Hays, J. D., and Burckle, L. H., 1991, Evidence for lower productivity in the Antarctic Ocean during the last glaciation, *Nature*, 351:220.

Müller, P. J., Erlenkeuser, H., and von Grafenstein, R., 1983, Glacial-interglacial cycles in oceanic productivity inferred from organic carbon contents in eastern North Atlantic sediment cores, *in*: "Coastal Upwelling, its Sedimentary Record. Part B: Sedimentary Records of Ancient Coastal Upwellings," J. Thiede and E. Suess, eds., Plenum Press, New York.

Müller, P. J., and Suess, E., 1979, Productivity, sedimentation rate, and sedimentary organic matter in the oceans - I. Organic carbon preservation, *Deep-Sea Res.*, 26A:1347.

Parker, F. L., 1973, Living planktonic foraminifera from the Gulf of California, *J. Foram. Res.*, 3:70.

Parker, F. L., and Berger, W. H., 1971, Faunal and solution patterns of planktonic foraminifera in surface sediments of the South Pacific, *Deep-Sea Res.*, 18:73.

Pedersen, T. F., 1983, Increased productivity in the eastern equatorial Pacific during the last glacial maximum (19,000 to 14,000 yr B.P.), *Geology*, 11:16.

Pedersen, T. F., and Calvert, S. E., 1990, Anoxia vs. productivity: what controls the formation of organic-carbon-rich sediments and sedimentary rocks?, *Amer. Assoc. Petrol. Geol. Bull.*, 74:454.

Petit, J. R., Mounier, L., Jouzel, J., Korotkevich, Y. S., Kotlyakov, V. I., and Lorius, C., 1990, Palaeoclimatological and chronological implications of the Vostok core dust record, *Nature*, 343:56.

Phleger, F. B, Parker, F. L., and Peirson, J. F., 1953, North Atlantic Foraminifera, Sediment cores from the North Atlantic Ocean, *Swedish Deep-Sea Exped. Repts.*, 7:1.

Pisias, N. G., and Rea, D. K., 1988, Late Pleistocene paleoclimatology of the central equatorial Pacific: sea surface response to the Southeast Trade Winds, *Paleoceanogr.*, 3:21.

Platt, T., and Li, W. K. W., eds., 1986, "Photosynthetic Picoplankton," Canadian Bulletin of Fisheries and Aquatic Sciences, 214:1.

Prahl, F. G., and Muelhausen, L. A., 1989, Lipid biomarkers as geochemical tools for paleoceanographic study, *in*: "Productivity of the Ocean: Present and Past," W.H. Berger, V.S. Smetacek, and G. Wefer, eds., John Wiley, Chichester.

Prahl, F. G., Muelhausen, L. A., and Lyle, M., 1989, An organic geochemical assessment of oceanographic conditions at MANOP Site C over the past 26,000 years, *Paleoceanogr.*, 4:495.

Pratje, O., 1951, Die Kieselsäureorganismen des Südatlantischen Ozeans als Leitformen in den Bodenablagerungen, *Deut. Hydr. Zeitschr.*, 4:1.

Prell, W. L., and Curry, W. B., 1980, Faunal and isotopic indices of monsoonal upwelling: Western Arabian Sea, *Oceanol. Acta.*, 4: 91.

Premuzic, E. T., Benkovitz, C. M., Gaffney, J. S., and Walsh, J. J., 1982, The nature and distribution of organic matter in the surface sediments of world oceans and seas, *Org. Geochem.*, 4:63.

Price, B. A., 1988, "Equatorial Pacific Sediments: Studies on Amino Acid, Organic Matter, and Manganese Deposition," Ph.D. thesis, University of California, San Diego.

Rea, D. K., Chambers, L. W., Chuey, J. M., Janecek, T. R., Leinen, M., and Pisias, N. G., 1986, A 420,000-year record of cyclicity in oceanic and atmospheric processes from the eastern equatorial Pacific, *Paleoceanogr.*, 1:577.

Reimers, C. E., 1989, Control of benthic fluxes by particulate supply, *in*: "Productivity of the Ocean: Present and Past," W. H. Berger, V. S. Smetacek, and G. Wefer, eds., John Wiley, Chichester.

Riedel, W. R., 1959, Siliceous organic remains in pelagic sediments, *Soc. Econ. Paleont. Mineral. Spec. Pub.*, 7:80.

Riedel, W. R., and Funnell, B. M., 1964, Tertiary sediment cores and microfossils from the Pacific Ocean floor, *Quat. J. Geol. Soc. London*, 120:305.

Romankevich, E. A., 1984, "Geochemistry of Organic Matter in the Ocean," Springer Verlag, Heidelberg.

Rowe, G. T., 1983, Biomass and production of the deep-sea macrobenthos, *in*: "The Sea, Vol. 8, Deep-Sea Biology," G. T. Rowe, ed., Wiley Interscience, New York.

Sancetta, C., 1979, Oceanography of the North Pacific during the last 18,000 years: evidence from fossil diatoms, *Marine Micropal.*, 4:103.

Sarnthein, M., Thiede, J., Pflaumann, U., Erlenkeuser, H., Fütterer, D., Koopmann, B., Lange, H. and Seibold, E., 1982, Atmospheric and oceanic circulation patterns off Northwest Africa during the past 25 million years, *in*: "Geology of the Northwest African Continental Margin," U. von Rad, K. Hinz, M. Sarnthein, and E. Seibold, eds., Springer Verlag, Heidelberg.

Sarnthein, M., Winn, K., and Zahn, R., 1987, Paleoproductivity of oceanic upwelling and the effect on atmospheric CO_2 and climatic change during deglaciation times, *in*: "Abrupt Climatic Change - Evidence and Implications," W.H. Berger and L.D. Labeyrie, eds., Reidel, Dordrecht.

Sarnthein, M., Winn, K., Duplessy, J.-C., and Fontugne, M. R., 1988, Global variations of surface ocean productivity in low and mid latitudes: influence on CO_2 reservoirs of the deep ocean and atmosphere during the last 21,000 years, *Paleoceanogr.*, 3:361.

Schiffelbein, P., and Dorman, L., 1986, Spectral effects of time-depth nonlinearities in deep sea sediment records: a deconvolution technique for realigning time and depth scales, *J. Geophys. Res.*, 91 (B3):3821.

Schrader, H., and Sorknes, R., 1991, Peruvian coastal upwelling: Late Quaternary productivity changes revealed by diatoms, *Mar. Geol.*, 97:233.

Shackleton, N. J., 1977, Carbon-13 in Uvigerina: tropical rainforest history and the equatorial Pacific carbonate dissolution cycles, *in*: "The Fate of Fossil Fuel CO_2 in the Oceans," N.R. Andersen, and A. Malahoff, eds., Plenum Press, New York.

Shackleton, N. J., Hall, M. A., Line, J., and Shuxi, C., 1983, Carbon isotope data in Core V19-30 confirm reduced carbon dioxide concentration in the ice age atmosphere, *Nature*, 306:319.

Shaffer, G., 1989, A model of biogeochemical cycling of phosphorus, nitrogen, oxygen, and sulphur in the ocean: one step toward a global climate model, *J. Geophys. Res.*, 94 (C2):1979.

Somayajulu, B. L. K., Sharma, P., and Berger, W. H., 1984. ^{10}Be, ^{14}C, U-Th decay series nuclides and ^{18}O in a box core from the central North Atlantic, *Marine Geol.*, 54:169.

Stein, R., 1991, "Accumulation of Organic Carbon in Marine Sediments," Springer Verlag, Berlin.

Suess, E., 1980, Particulate organic carbon flux in the oceans - surface productivity and oxygen utilization, *Nature*, 288:260.

Toggweiler, J. R., 1989, Is the downward dissolved organic matter (DOM) flux important in carbon transport?, *in*: "Productivity of the Ocean: Present and Past," W.H. Berger, V.S. Smetacek, and G. Wefer, eds., John Wiley, Chichester.

van der Zwaan, G. J., Jorissen, F. J., and de Stigter, H. C., 1990, The depth dependency of planktonic/benthic foraminiferal ratios: constraints and applications, *Marine Geol.*, 95:1.

Vincent, E., and Berger, W. H., 1981, Planktonic foraminifera and their use in paleoceanography, *in*: "The Sea, Vol. 7, the Oceanic Lithosphere," C. Emiliani, ed., Wiley-Interscience, New York.

Walsh, J. J., 1989, How much shelf production reaches the deep sea?, *in*: "Productivity of the Ocean: Present and Past," W.H. Berger, V.S. Smetacek, and G. Wefer, eds., John Wiley, Chichester.

Wefer, G., 1989, Particle flux in the ocean: effects of episodic production, *in*: "Productivity of the Ocean: Present and Past," W.H. Berger, V.S. Smetacek, and G. Wefer, eds., John Wiley, Chichester.

Wu, G., and Berger, W. H., 1991, Pleistocene $\delta^{18}O$ records from Ontong-Java Plateau: effects of winnowing and dissolution, *Marine Geol.*, 96:193.

Wu, G., Herguera, J. C., and Berger, W. H., 1990, Differential dissolution: modification of late Pleistocene oxygen isotope records in the western equatorial Pacific. *Paleoceanogr.*, 5:581.

Zahn, R., Winn, K., and Sarnthein, M., 1986, Benthic foraminiferal δ^{13}C and accumulation rates of organic carbon: Uvigerina peregrina group and Cibicidoides wuellerstorfi, *Paleoceanogr.*, 1:27.

Zhuang, G., Duce, R. A., and Kester, D. R., 1990, The dissolution of atmospheric iron in surface water of the open ocean, *J. Geophys. Res.*, 95 (C9):16,207.

DO MARINE PHYTOPLANKTON INFLUENCE GLOBAL CLIMATE?

P.M. Holligan

Plymouth Marine Laboratory
Plymouth PL1 3DH, U.K.

INTRODUCTION

The oceans have a major influence on the climate of the earth through effects on the global solar radiation budget, on meridional heat transport, and on the trace gas composition of the atmosphere. The climate of the ocean-atmosphere system is sensitive to variations of the solar constant and the orbital characteristics of the earth. However, the properties of surface ocean waters and of the marine atmosphere are modified also by the optical and biochemical properties of marine organisms, in particular, the phytoplankton. It is generally recognized that the global climate would have been quite different through geological time in the absence of life in the sea.

In attempting to explain past variations in climate or predict future trends, the question arises, therefore, as to whether or not marine phytoplankton have the ability to influence the climate system over shorter time scales. At one level, many of the basic climatic features and differences in state of the oceans can be explained in purely physical and physico-chemical terms (Broecker and Denton, 1989; Rind and Chandler, 1991), but at another level the dynamic behavior of the climate system and responses to perturbations may be sensitive to biological processes which affect atmosphere-ocean interactions.

Advances over the last two decades in understanding how the global climate has changed during the glacial-interglacial oscillations of the Pleistocene period (the last 1.6 million years) has provided a much stronger framework for examining hypotheses about possible relationships between ocean biology and climate. For this reason, I discuss this topic here primarily in the context of recent glacial cycles.

The importance of carbon dioxide as a greenhouse gas was recognized at the end of the last century and, in considering the Tertiary environment, Chamberlin (1898) stated that "...it would not seem to be beyond the competency of a great limestone-making epoch to enrich the atmosphere in carbonic acid sufficiently ... to thermally blanket the earth effectively to render all latitudes available to vegetal and animal life...". It was to be more than 60 years before marine scientists paid serious attention to this idea. Studies of North Atlantic sediments in the 1960s and 1970s showed how large-scale changes in ocean conditions, in particular latitudinal displacements of the polar front, occurred at the start and end of each glaciation and were accompanied by shifts in the distributions of

Primary Productivity and Biogeochemical Cycles in the Sea
Edited by P.G. Falkowski and A.D. Woodhead, Plenum Press, New York, 1992

calcareous organisms (McIntyre et al., 1972; Ruddiman and McIntyre, 1981). Subsequent work in other ocean areas, using opal and organic carbon as additional indicators, has demonstrated comparable variations in the abundance and productivity of diatoms and marine plankton in general (e.g. Lyle et al., 1988; Mortlock et al., 1991).

The next big advance in understanding Pleistocene changes in the earth's environment came with the reconstructions based on ice-core analyses of global temperature, as well as various properties of the atmosphere including levels of CO_2 (Barnola et al., 1987, 1991) and other trace gases, and transport of aeolian dust (Petit et al., 1990). Measurements of the relative abundance of carbon isotopes in biogenic matter in marine sediments have enabled direct comparisons to be drawn between changes in climate, in ocean circulation, and in ocean productivity (Mix, 1989; Raymo et al., 1990). The correlation between the temperature and $\delta^{13}C$ records provides strong evidence for a direct relationship between changes in climate and ocean biology (Fig. 1), although interpretation of the biological signal is a complex issue.

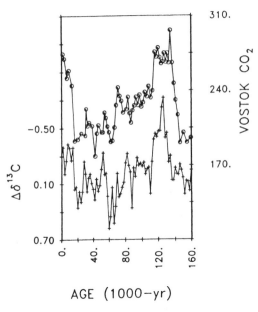

Fig. 1. Comparison for the last 160k years of the $\Delta\delta^{1.3}C$ ($^o/oo$) record from an ocean sediment core (+) and the atmospheric CO_2 (ppm) record from Antarctic ice cores (o). From Mix (1989), by permission.

As these new discoveries were being made, several biogeochemical models of the ocean carbon cycle were constructed to explain the variations in atmospheric CO_2 levels that are closely related to changes in global temperature (Berger and Keir, 1984; Broecker and Peng, 1989; Keir, 1988). The importance of high-latitude ocean areas for the drawdown of carbon dioxide from the atmosphere (e.g. Sarmiento et al., 1988), and of changes in ocean circulation in relation to CO_2 storage and vertical nutrient distributions

(Boyle, 1988) has been clearly established. Recently, the iron hypothesis of Martin (1990; this volume), which explains lower atmospheric CO_2 concentrations during glacial periods in terms of higher biological productivity due to aeolian dust (iron) fertilization of nitrate-rich regions of the oceans, has attracted much attention. However, models based on glacial-to-interglacial changes in the biological pump for CO_2 and in ocean circulation still do not fully meet certain restraints imposed by the ice core and marine-sediment records, particularly concerning the relative timing of global temperature and atmospheric CO_2 changes (Barnola et al., 1991) and the reconstructions of deep ocean chemistry. Therefore, alternative explanations not based on non-biological processes, such as the polar alkalinity hypothesis of Broecker and Peng (1989), continue to be given serious consideration. One weakness of the ocean biogeochemical models for carbon is that they fail to address adequately conditions at the ocean margins where there are significant inputs of terrestrial organic matter and where a large proportion of marine carbon burial occurs (Berner, 1989; Smith and Mackenzie, 1987).

Another major development in thinking about how the ocean biota might affect the global climate has been the theory of Charlson et al. (1987) that dimethylsulphide (DMS) from phytoplankton is the main source of cloud condensation nuclei in the marine atmosphere. Although there are many uncertainties about the quantitative effects of DMS on climate, and even about the directional effects in terms of the relative abundance of DMS-producing algae under glacial and interglacial conditions, the theory continues to attract much attention due to the importance of clouds for the radiation budget of the earth (Slingo, 1990).

Against this background of new knowledge as well as new uncertainties, this paper reviews briefly the properties and distributions of marine phytoplankton relevant to the global climate change. Some preliminary remarks about how the oceans affect global climate provide a general perspective for assessing the potential significance of marine algae, and in conclusion, some priorities for future research are identified that could improve the information on ocean biogeochemistry for predictive climate models.

ROLE OF OCEANS IN CLIMATE CHANGE

The present understanding of the causes of the ice ages was recently reviewed by Broecker and Denton (1989). They conclude that changes in climate are related to two or more different modes of operation of the ocean-atmosphere system, which, in turn, are driven by insolation fluctuations determined by the earth's orbit. The key features are water transport in the atmosphere and salt transport in the oceans, which together determine ocean density and circulation patterns. One of the most sensitive ocean regions is the North Atlantic where the active formation of deep water during interglacial periods alters the flushing times of the deep ocean basins and promotes the transport of heat poleward in the northern hemisphere. Changes in ocean circulation affect the exchange and partitioning of trace gases, such as CO_2, between the ocean and atmosphere, as well as the distribution of nutrients in the ocean, which, in turn, influence biological productivity. The potential for feedback effects on the composition of the atmosphere exists through both physical (e.g. the influence of water temperature on gas solubility) and biological (the biological carbon pump) mechanisms.

Broecker and Denton (1989) attributed changes is mean global temperature to several factors working in harmony, including changes in atmospheric levels of greenhouse gases, and in the albedos of the atmosphere and of the land/sea surface. The best estimates indicate that increases in greenhouse gas concentrations account for only about

2°C warming during the interglacials, which is probably less than half of the mean global value and much less than half of that at high latitudes. Albedo and cloud effects are much harder to quantify, although Harvey (1988) has estimated a 2-3°C change due to atmospheric aerosols. By comparison, the recent modelling studies of Rind and Chandler (1991) suggest that changes in the transport of heat by the oceans, coupled with variations in albedo due to ice cover, might account for a large part of the glacial-interglacial shifts in temperature. The involvement of non-linear processes and feedback effects in the control of global temperature means that such uncertainties and discrepancies are unlikely to be resolved soon.

Another major problem is the interpretation of relatively rapid (decadal) changes in global environment that are recorded in ice cores (Dansgaard et al., 1989), and of differences in the synchronicity of the temperature and CO_2 variations (Barnola et al., 1991). The first issue is indicative of the importance of sudden, but perhaps small, alterations in ocean circulation. The second, which concerns a definite lag in CO_2 changes with respect to temperature at the onset of glaciation but not at the onset of deglaciation, suggests that the oceans lose CO_2 rather more easily than they gain it, possibly due to the nature of the biological pump.

Before examining how marine phytoplankton might influence the levels of trace gases in the atmosphere, the albedo of the atmosphere and surface ocean waters, and even the heat budget of the ocean, it is worthwhile to consider how their distributions might change. Through evolution and, in very recent times, through manipulation by humans, the types and distributions of organisms can vary independently of climate. In this sense, phytoplankton represent a potential forcing mechanism for climate change, and indeed, their impact over geological time on the composition of the atmosphere is interpreted in this way (Lovelock, 1986). But, climate change itself also affects the surface ocean environment and the abundance and type of marine phytoplankton, giving rise to direct and indirect feedback effects on the climate system. In climate models, such non-linear and interacting biological processes are generally neither well defined nor explicitly treated. The ability to recognize and quantitatively describe variations in the global distributions of marine phytoplankton is fundamental to any attempt to understand how these organisms might affect climate.

PROPERTIES OF PHYTOPLANKTON THAT MAY AFFECT CLIMATE

Marine phytoplankton strongly influence the air-sea exchanges of carbon dioxide and sulfur compounds, and the albedo of surface ocean water. All three properties potentially affect the climate system but in rather different ways:

Carbon Fluxes

Next to water vapor, carbon dioxide is the major greenhouse gas, the main reservoir for which is the ocean. CO_2 is incorporated into organic matter and calcium carbonate by phytoplankton. Much of the organic matter is rapidly re-oxidized within the euphotic zone, but a small proportion ($\sim 10\%$ of net primary production) is transferred to deep water and the sediments, so maintaining an atmosphere-to-deep water gradient in CO_2 concentration, which represents the ocean carbon pump. The overall efficiency for the burial of organic carbon in ocean sediments is very low ($<0.1\%$, Smith and Mackenzie, 1987; Berner, 1989) with the result that most of the particulate organic carbon in deep water is eventually returned to the surface in upwelled water as CO_2 or dissolved organic carbon. By contrast, calcium carbonate is stable in surface water, with dissolution only occurring below the lysocline (the depth below which water is undersaturated with respect to $CaCO_3$) at considerable depth. About half of the area of the ocean bottom lies above the lysocline depth, and the global burial rate for calcite carbon, although uncertain (Milliman and Takahashi, 1991), is more than one magnitude greater than for organic carbon.

The major taxonomic and successional groups of phytoplankton (Margalef, 1978) are distinct in terms of carbon fluxes. A major part of the organic matter that sinks to deep water is made up of diatoms, which grow and sink relatively rapidly (Lampitt, 1985), and of colonial algae such as *Phaeocystis*, which are avoided by herbivores and often replace diatoms as nutrients become depleted. In many ocean areas, these forms are succeeded by coccolithophores, the dominant calcifying group of phytoplankton, which are a major constituent of chalks and limestones. Flagellates and picoplankton are typical of stratified, oligotrophic waters characterized by efficient recycling of both nutrients and carbon, and appear to contribute little to the vertical export of particulate organic or inorganic carbon.

The two forms of carbon burial in the oceans have different effects on the air-sea exchange of CO_2 and are coupled in various ways. The flux of organic matter to deep water and the sediments depletes surface CO_2 levels, leading to uptake from the atmosphere at least at the time scales of ocean ventilation (Sarmiento et al., 1988). By contrast, the formation of $CaCO_3$ reduces alkalinity so that, for a given rate of organic productivity, the capacity for surface water uptake of CO_2 is reduced (Tsunogai and Noriki, 1991), as demonstrated by the models of Berger and Keir (1984), Dymond and Lyle (1985), and Taylor et al. (1991). The overall capacity for carbonate burial in the oceans without a net change in alkalinity is determined by the rates of weathering of calcareous rocks. Internal compensation also occurs due to the dissolution of $CaCO_3$ in deep water associated with the oxidation of organic matter (Broecker and Peng, 1987). This process plays a major role in the adjustment of the oceanic CO_2 system at times of change in ocean circulation and/or productivity.

The burial efficiency for organic matter is correlated with the rate of sediment accumulation (Blackburn, 1991), because of, in part, the adsorption of organic matter onto mineral surfaces (Gordon and Millero, 1985) including biogenic calcite particles (Suess, 1973). This effect may be a more important factor than the rate of biological productivity in determining marine organic carbon burial. Nearshore estuarine, deltaic, and continental slope environments are characterized by rapid deposition of terrigenous sediments and are the major sinks for organic carbon in the oceans (Berner, 1989; Walsh, 1989; 1991). In the open oceans, the main sources of mineral particles are calcifying and silicifying planktonic organisms, and here, the processes controlling removal of organic and inorganic carbon from surface waters, and therefore, net exchange with the atmosphere, are intimately linked.

Models of the ocean CO_2 pump have been concerned mainly with organic carbon fluxes (e.g. Sarmiento et al., 1988). To improve models of the ocean carbon cycle that aim to explain the glacial-interglacial changes in atmospheric CO_2, a more comprehensive analysis is required both of the spatial and temporal variations in inorganic carbon fluxes (Fig. 2), including shallow water reef and sand accumulations under high sea level conditions (Milliman and Takahashi, 1991), and of the factors that determine the rate of organic carbon burial in marine sediments. One hypothesis that still needs to be critically evaluated is that global changes in the ratio of organic to inorganic productivity have a significant effect on the partitioning of CO_2 between the surface ocean and atmosphere (Berger and Keir, 1984) in a manner complementary to the physico-chemical polar alkalinity hypothesis of Broecker and Peng (1989). Any reduction in ocean alkalinity associated with increased calcification under conditions of higher global mean temperature and higher sea levels will tend to cause a net shift of CO_2 from the oceans to the atmosphere. Also, terrestrial inputs of organic matter (which presently exceed ocean burial, thereby making the ocean heterotrophic; see Smith and Mackenzie, 1987) and of suspended matter appear to play a vital role in ocean carbon budgets; the global significance of changes in organic carbon fluxes at the ocean margins in response to ice age cycles and to human activities is poorly understood.

Sulfur Emissions

Charlson et al. (1987) presented the hypothesis that the atmospheric oxidation products of dimethy lsulfide (DMS) derived from oceanic phytoplankton are the main source of cloud condensation nuclei (CCN) in the marine atmosphere, and therefore, of

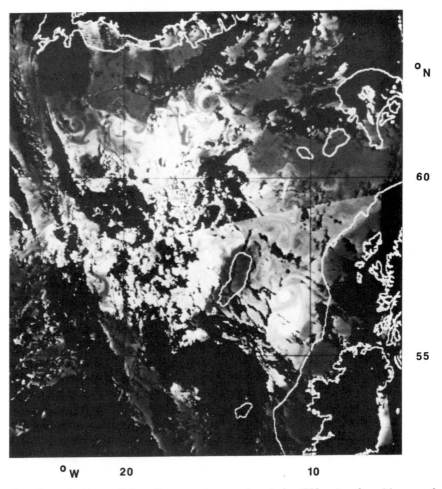

Fig. 2. Coastal Zone Color Scanner image (band 3, 550nm) of a bloom of the coccolithophore, *Emiliania huxleyi*, in the NE Atlantic Ocean between Iceland and Ireland in 1980. The image is a composite of atmospherically-corrected scenes for July 27 and August 1. Clouds and land have been masked black, and the 200m isobath is shown. The extensive bright area shows strong backscattering of light by the coccoliths.

global climatic significance in terms of cloud albedo. It is generally thought that any increase in low stratiform clouds over the oceans would lead to cooling (Slingo, 1989). Dimethylsulfoniopropionate (DMSP) is a major osmosolute in certain groups of phytoplankton including coccolithophores (Fig. 2), other prymnesiophytes and some flagellates (e.g. Turner et al., 1988; Keller et al., 1989), and is the precursor of DMS. Both DMSP and DMS are subject to photochemical and microbial degradation in the sea (e.g. Kiene and Bates, 1990). These processes affect the relationship between phytoplankton productivity and DMS emissions, but have not been investigated in detail.

A clear relationship between atmospheric DMS levels and CCN abundance was demonstrated for unpolluted marine environments (Ayers et al., 1991; Hegg et al., 1991). A positive correlation, based on satellite data for the NW Atlantic Ocean, between surface chlorophyll (phytoplankton biomass) and low cloud reflectance also was found (Falkowski et al., 1991). However, the nature and global significance of the postulated feedback loop between phytoplankton production of DMS and cloud albedo remains uncertain. The cloud-dependent changes in surface temperature and irradiance are likely to have only relatively weak effects on phytoplankton sulfur emissions so that variations in marine sulfur fluxes to the atmosphere may not be sufficiently large to have a significant impact on climate (Foley et al., 1991).

Ice-core measurements of methanesulfonic acid (MSA), the main oxidation product of DMS in the atmosphere, suggest that oceanic emissions of DMS were much higher during the last glacial period (Legrand et al., 1991). This observation raises two problems; first, any increase in DMS production during the ice ages suggests a positive feedback (destabilizing effect) on global temperature, and second, there is no strong ecological evidence for a greater abundance of DMS-producing algae at this time. Although primary production on the oceans is thought to have been somewhat higher during the ice age (Mix, 1989), the increase is attributed largely to more abundant diatoms, a group that produces little DMS (Keller et al., 1989). Furthermore, DMS and MSA have relatively short lifetimes in the atmosphere (1-2 days) so that, with more extensive sea ice cover during the glacial period, less rather than more MSA deposition on the ice cap would be expected for a given emission of DMS associated, for example, with phytoplankton blooms along the ice edge. Possibly, the higher wind strength and greater atmospheric dust loading significantly altered the processes affecting the oxidation and transport rates of sulfur compounds in the atmosphere, in which case the relationship between the ice-core record for MSA and phytoplankton DMS emissions is not straightforward.

In summary, there is strong evidence that algal DMS emissions and cloud albedo are related although no experimental test of the hypothesis has yet been made. The feedback link between cloud albedo and DMS production appears weak although other independently-varying environmental factors (e.g. nutrient availability, zooplankton grazing) may strongly influence the cycling of sulfur.

Optical Properties

Variations in the inherent optical properties of water related to the light absorption and scattering by phytoplankton cells alter the efficiency and vertical distribution of solar heating in the water column (Kirk, 1988). A part of the light energy absorbed by algal pigments is released as heat to the surrounding water, and an increase in the scattering path length within a given layer causes more heat to be absorbed directly by water. Some light also will be lost back to the atmosphere (i.e. greater water reflectance). Model results indicate that the seasonal cycle of phytoplankton in the ocean significantly impacts sea surface temperature, and, therefore, heat exchange with the atmosphere (Simonot et al., 1988).

A correlation between surface chlorophyll distributions and the rate of surface heating was demonstrated by Sathyendrenath et al. (1991). Such an effect is likely initially to enhance the growth rate of a surface phytoplankton population because of a reduction in the depth of the surface mixed layer, but will have the opposite effect once the nutrients have been depleted by inhibiting upward mixing across the seasonal thermocline. However, heating within sub-surface phytoplankton or particle-rich layers enhances upward convection (Lewis et al., 1983), and so tends to deepen the surface mixed layer.

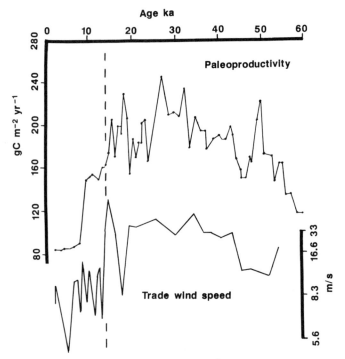

Fig. 3.   Estimated ocean palaeoproductivity (g C m⁻²) and average trade wind speed (m s⁻¹) for the northwest African upwelling region based on studies of marine sediment cores.   From Sarnthein et al. (1987), by permission.

Light scattering by coccoliths represents a special case of biological effects on surface ocean optics, with values for sub-surface reflectance exceeding 30% (Balch et al., 1991) compared to 3-5% in the absence of coccoliths. The implications in terms of efficiency of surface layer heating and of vertical temperature (density) profiles have not been analyzed quantitatively, but with the density of coccoliths exceeding 3.10^5 ml⁻¹ within blooms of *E. huxleyi* (Fig. 2), extreme conditions for biological warming and shallowing of the mixed layer (Kirk, 1988) are predicted. Shading will tend to preclude the growth of coccolithophores or other types of phytoplankton in the thermocline so that the secondary effects of sub-surface heating (Lewis et al, 1983) will be inhibited.

Any overall assessment of the influence of phytoplankton on climate must be based on a combination of quantitative analysis and modelling of the effects described above at a global scale. Much progress will be possible as new data sets from satellites on the global distributions of chlorophyll, of surface optical properties (diffuse attenuation coefficients and sub-surface reflectances), and of particular taxonomic groups (e.g. coccolithophores which contribute significantly to both carbon and sulfur fluxes) become available for the validation of models.

GLOBAL CHANGES IN PHYTOPLANKTON DISTRIBUTIONS

Studies of sediment cores from a wide range of oceanic environments have provided a wealth of information about local and regional changes in phytoplankton productivity in the past and, together with the results of geochemical models (Keir, 1988; Boyle, 1988), represent a crucial contribution to understanding how the global carbon cycle works. However, a comprehensive global picture of marine productivity during the last ice age still has not emerged, partly from the lack of a global framework for interpreting local or regional variations in the relative fluxes of opal, organic carbon, and carbonate, and in the abundance of individual species. The application of transfer functions based on the distributions of foraminifera and of organic carbon for estimating the productivity of the Atlantic Ocean during the last glaciation are discussed by Mix (1989); both methods indicate higher productivity ($\sim 20\%$) during the ice age but lack agreement in spatial and quantitative detail. Carbonate records are generally not a good measure of productivity, except in regions where the bottom depth lies well above that of excursions in the lysocline depth, due to the difficulty of distinguishing between variations in rates of carbonate sedimentation and of carbonate dissolution (Keir and Berger, 1985).

Low latitude cores usually suggest higher rates of productivity during glacial periods (e.g. Sarnthein et al., 1987), especially for upwelling regions where greater wind stress is thought to have enhanced inputs of nutrients to surface waters (Fig. 3). Studies of sediments from the tropical Pacific have shown that increases in content of organic carbon were accompanied by higher accumulation rates of either carbonate or opal (Lyle et al., 1988). The wider ecological and biogeochemical implications of shifts between calcifying and silicifying plankton communities at times of enhanced productivity remain uncertain.

Sediment records from higher latitudes show great changes in phytoplankton distributions with movements of the polar fronts. The result of the CLIMAP program (McIntyre et al., 1972; Ruddiman and McIntyre, 1981) in the North Atlantic demonstrated variations in the abundance of coccolithophores, as well as warm water foraminifera, which reflect poleward migration of the polar front as the climate warmed after the last glaciation (Fig. 4). Comparable latitudinal shifts of the region of high diatom productivity have been described by Mortlock et al. (1991) for the Southern Ocean.

The alternating advance and retreat of the sea across the continental shelves with the rise and fall of sea level represents a third major type of change in marine productivity. These shallow waters are generally considered to contribute 15 to 20% of present marine phytoplankton productivity. Under glacial conditions, the ocean margins are likely to have been biologically active (Fig. 3) but, on account of a much reduced spatial extent, their total productivity was probably less modern shelf systems because of a much reduced spatial extent.

How did the phytoplankton distributions in the oceans during the last glaciation differ from those of today? The evidence from sediment cores suggests that total productivity was higher (e.g. Mix, 1989), but such estimates do not usually take account of the reduced shelf area or the greater extent of sea ice at higher latitudes. Coccolithophores appear generally to have been less abundant under colder conditions (Fig. 4), and conversely, diatoms were probably more abundant, as would be expected with more wind-driven upwelling and surface mixing. Also, the greater input of aeolian dust to the oceans, as well as the ice caps (Petit et al., 1990), may have stimulated diatom growth as suggested by Martin (1990), although direct palaeoceanographic evidence for this hypothesis is still lacking.

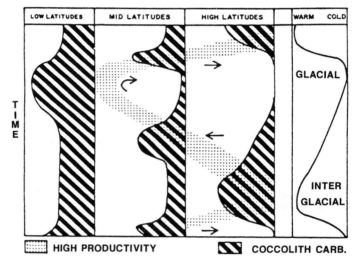

Fig. 4. Reconstructions based on CLIMAP data of latitudinal displacements of regions of high coccolithophore productivity in the North Atlantic during glacial-interglacial cycles. From McIntyre et al. (1972), by permission. The coccolithophore bloom shown in Fig. 2 corresponds to the interglacial high latitude maximum.

One possible interpretation is that global productivity has not changed significantly, but that, with deglaciation, the regions of oligotrophic conditions centered around the oceanic sub-tropical gyres increased in size as the polar fronts retreated polewards, as the shelves became flooded, and as the extent of margin and equatorial upwelling regions decreased with weaker winds. Under such a scenario, relative enhancement of pelagic calcite productivity is likely, thus increasing the ratio of inorganic to organic carbon in exported biogenic particulate matter, as well as the albedo of surface ocean waters. Similarly, higher emissions of DMS would be expected. Changes in organic carbon burial might be determined by sedimentation rates; the glacial ocean received more terrigenous material due to enhanced ice and atmospheric transport, as well as riverine erosion of the exposed continental shelves, and, therefore, was probably a more active sink for organic carbon of both terrestrial and marine origin. It is only through examining a combination of factors such as these through new observational and modelling studies that the overall impact of marine phytoplankton on climate can be fully assessed.

Finally, consideration must be given to changes in species distributions. Probably the most spectacular recent example is the spread of the coccolithophore, *Emiliania huxleyi*, which is now found abundantly from the arctic to the tropics, and from inshore waters and inland seas to the open ocean, and is possibly the most abundant calcifying organism on earth today. *E. huxleyi* is first described from ocean sediments about 270,000 years old (Fig. 5) and is thought to have appeared first in the tropics and subsequently spread to higher latitudes (Thierstein et al., 1977). It has the unique property of producing abundant calcite through the shedding of coccoliths, which affects ocean reflectance (Fig. 2), and is an active producer of DMS (Turner et al., 1988). Thus, within the time scale of about two Pleistocene ice age cycles, a new species has arisen which may be significantly influencing the climate system.

CONCLUSIONS

In considering the earth system as a whole, few scientists would doubt that the evolution of the global climate and of the biogeochemical properties of the biota are in some way inextricably linked (Lovelock, 1986; 1991). New geochemical models are

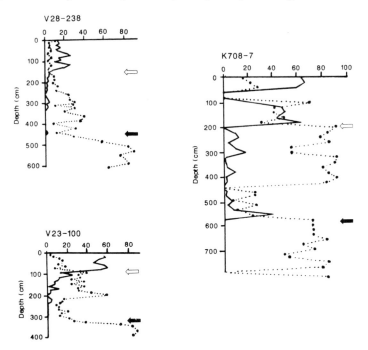

Fig. 5.   Low (1°N, V28-238), mid- (21°N, V23-100), and high (54°N, K708-7) latitude ocean sediment records for the relative abundance (%) of the coccolithophores, *Emiliania huxleyi* (solid lines) and *Gephyrocapsa caribbeanica* (dotted lines). The arrows mark the first appearance of *E. huxleyi* (268k years ago), and the reversal of dominance between *E. huxleyi* and *G. caribbeanica* (85-73k years ago). Adapted from Thierstein et al. (1977), by permission.

beginning to provide some clues about how particular components of the biota, such as marine phytoplankton, might act as a forcing variable on the climate system over geological time scales (Volk, 1989).

However, over shorter time scales that characterize, for example, the Pleistocene glacial cycles, it has proved difficult to establish tenable hypotheses that relate climate variations to changes in biogeochemical cycles (see Broecker and Peng, 1989) or other biological properties of the oceans. This issue is of particular scientific importance in the context of attempts to predict future climate changes under conditions of significant perturbation of the earth's environment by humans.

Various mechanisms by which marine phytoplankton influence global climate are recognized, such as effects on atmospheric CO_2 levels, cloud albedo, and ocean albedo. In each case, too little is known about the types of feedback mechanisms that link climate change (driven by external factors or by humans) to the functional responses of phytoplankton to determine quantitatively their climatic significance at a global scale. New research is needed to define and quantify such feedback processes (e.g. is marine organic carbon burial dependent on fluxes of suspended sediment from land? Is the ratio of organic to inorganic carbon in marine biogenic particulate material related to nutrient availability in surface waters? What are the environmental factors that cause DMS emissions to vary?), and to assess from palaeoecological evidence how they might have operated in the past.

Furthermore, just as the physical state or mode of the oceans is known to vary under different climatic conditions (Broecker and Denton, 1989), so will the biological mode. Global models for the biogeochemical cycling of several elements are beginning to be developed (Shaffer, 1989). In the context of climate studies, these models should be extended to include relevant optical parameterization for both the surface ocean and marine atmosphere, so that the overall impact of marine phytoplankton on the global radiation budget can be assessed. A change in the sum of phytoplankton properties, rather than in the separate components, is the proper basis for determining the degree to which phytoplankton influence the global climate.

REFERENCES

Ayers, G. P., Ivey, J. P., and Gillett, R. W., 1991, Coherence between seasonal cycles of dimethyl sulphide, methanesulphonate and sulphate in marine air, *Nature*, 349:404.

Balch, W. M., Holligan, P. M., Ackelson, S. G., and Voss, K. J., 1991, Biological and optical properties of mesoscale coccolithophore blooms in the Gulf of Maine, *Limnol. Oceanogr.*, 36:629.

Barnola, J. -M., Pimenta, P., Raynaud, D., and Korotkevich, Y. S., 1991, CO_2 - climate relationship as deduced from the Vostok ice core: a re-examination based on new measurements and on a re-valuation of the air dating, *Tellus*, 43B:83.

Barnola, J.- M., Raynaud, D., Korotkevich, Y. S., and Lorius, C., 1987, Vostok ice core provides 160,000-year record of atmospheric CO_2, *Nature*, 329:408.

Berger, W. H., and Keir, R. S., 1984, Glacial-Holocene changes in atmospheric CO_2 and the deep-sea record, *American Geophys. Union, Geophys. Mono. Series*, 29:337.

Berner, R. A., 1989, Atmospheric carbon dioxide levels over Phanerozoic time, *Science*, 249:1382.

Blackburn, T. H., 1991, Accumulation and regeneration: Processes at the benthic boundry layer, *In*: "Ocean Margin Processes in Global Change," R.F.C. Mantoura, J.-M. Martin, and R. Wollast, eds., John Wiley & Sons, Chichester.

Boyle, E. A., 1988, The role of vertical chemical fractionation in controlling late Quaternary atmospheric carbon dioxide, *J. Geophys. Res.*, 93:15701.

Broecker, W. S., and Denton, G. H., 1989, The role of ocean-atmosphere reorganisation in glacial cycles, *Geochim. Cosmochim. Acta*, 53:2465.

Broecker, W. S., and Peng, T. -H., 1987, The role of $CaCO_3$ compensation in the glacial to interglacial atmospheric CO_2 change, *Global Biogeochem. Cycles*, 1:15.

Broecker, W. S., and Peng, T. -H., 1989, The cause of the glacial to interglacial atmospheric CO_2 change: A polar alkalinity hypothesis, *Global Biogeochem. Cycles*, 3:215.

Chamberlin, T. C., 1898, The influence of great epochs of limestone formation upon the constitution of the atmosphere, *J. Geol.*, 6:609.

Charlson, R. J. , Lovelock, J. E., Andreae, M. O., and Warren, S. G., 1987, Oceanic phytoplankton, atmospheric sulphur, cloud albedo and climate, *Nature*, 326:655.

Dansgaard, W., White, J. W. C., and Johnsen, S. J., 1989, The abrupt termination of the Younger Dryas climate event, *Nature*, 339:532.

Dymond, J., and Lyle, M., 1985, Flux comparisons between sediments and sediment traps in the eastern tropical Pacific: Implications for atmospheric CO_2 variations during the Pleistocene, *Limnol. Oceanogr.*, 30:699.

Falkowski, P. G., Kim, Y., Kolber, Z., Wilson, C., Wirick, C., and Cess, R., 1991, Distinguishing between anthropogenic and natural factors affecting low-level cloud albedo over the North Atlantic Ocean, *Science*, submitted.

Foley, J. A., Taylor, K. E., and Ghan, S. J., 1991, Planktonic dimethylsuphide and cloud albedo: An estimate of the feedback response, *Climatic Change*, 18:1.

Gordon, A. S., and Millero, F. J., 1985, Adsorption mediated decrease in the biodegradation rate of organic compounds, *Microb. Ecol.*, 11:289.

Harvey, D. L. D., 1988, Climatic impact of ice-age aerosols, *Nature*, 334:333.

Hegg, D. A., Ferek, R. J., Hobbs, P. V., and Radke, L. F., 1991, Dimethylsulfide and cloud condensation nucleus correlations in the northeast Pacific Ocean, *J. Geophys. Res.*, 96:13189.

Herbert, T. D., Curry, W. B., Barron, J. A., Codispoti, L. A., Gersonde, R., Keir, R. S., Mix, A. C., Mycke, B., Schrader, H., Stein, R., Thierstein, H. R., 1989, Geological reconstructions of marine productivity, *In*: "Productivity of the Ocean: Present and Past," W. H. Berger, V. S. Smetacek, and G. Wefer, eds., John Wiley & Sons, Chichester.

Keir, R. S., 1988, On the late Pleistocene ocean geochemistry and circulation, *Paleoceanography*, 3:413.

Keir, R. S., and Berger, W. H., 1985, Late Holocene carbonate dissolution in the equatorial Pacific: Reef growth or neoglaciation? *In*: "Natural Variations in Carbon Dioxide and the Carbon Cycle, Archean to Present," E.T. Sundquist and W.S. Broecker, eds., *Geophys. Monogr. Ser.*, 32:208, AGU, Washington D.C.

Keller, M. D., Bellows, W. K., and Guillard, R. R. L., 1989, Dimethylsulphide production in marine phytoplankton, *In*: "Biogenic Sulphur in the Marine Environment," E.S. Saltzman and W.J. Cooper, eds., *Am. Chem. Soc. Symp. Ser.*, 393:167, ACS, Washington.

Kiene, R. P., and Bates, T. S., 1990, Biological removal of dimethylsulphide from sea water, *Nature*, 345:702.

Kirk, J. T. O., 1988, Solar heating of water bodies as influenced by their inherent optical properties, *J. Geophys. Res.*, 93:10897.

Lampitt, R. S., 1985, Evidence for the seasonal distribution of detritus to the deep-sea floor and its subsequent resuspension, *Deep-Sea Res.*, 32:885.

Legrand, M., Feniet-Saigne, C., Saltzman, E. S., Germain, C. Barkov, N. I., and Petrov, V. N., 1991, Ice-core record of oceanic emissions of dimethylsulphide during the last climate cycle, *Nature*, 350:544.

Lewis, M. R., Cullen, J. J., and Platt, T., 1983, Phytoplankton and thermal structure in the upper ocean; consequences of nonuniformity in chlorophyll profile, *J. Geophys. Res.*, 88:2565.

Lovelock, J. E., 1986, Geophysiology: A new look at earth science, *Bull. Amer. Meteorol. Soc.*, 67:392.

Lovelock, J. E., 1991, Geophysiology of the oceans, *In*: "Ocean Margin Processes in Global Change," R.F.C. Mantoura, J.-M. Martin, and R. Wollast, eds., John Wiley & Sons, Chichester.

Lyle, M., Murray, D. W., Finney, B. P., Dymond, J., Robbins, J. M., and Brooksforce, K., 1988, The record of late Pleistocene biogenic sedimentation in the eastern tropical Pacific Ocean, *Paleoceanography*, 3:39.

Margalef, R., 1978, Life-forms of phytoplankton as survival alternatives in an unstable environment, *Oceanol. Acta.*, 1:493.

Martin, J. H., 1990, Glacial-Interglacial CO_2 change: The iron hypothesis, *Paleooceanography*, 5:1.

Martin, J. H., this volume.

McIntyre, A., Ruddiman, W. F., and Jantzen, R., 1972, Southward penetrations of the North Atlantic polar front: Faunal and floral evidence of large-scale surface water mass movements over the last 225,000 years, *Deep-Sea Res.*, 19:61.

Milliman, J. D., and Takahashi, K., 1991, Carbonate and opal production and accumulation in the ocean, *In*: "Global Surficial Geofluxes: Modern to Glacial," T.M. Usselman, W. Hay, and M. Meybeck, eds., in press.

Mix, A. C., 1989, Pleistocene paleoproductivity: Evidence from organic carbon and foraminiferal species, *In*: "Productivity of the Ocean: Present and Past," W.H. Berger, V.S. Smetacek, and G. Wefer, eds., John Wiley & Sons, Chichester.

Mortlock, R. A., Charles, C. D., Froelich, P. N., Zibello, M. A., Saltzman, J., Hays, J. D., and Burckle, L. H., 1991, Evidence for lower productivity in the Antarctic Ocean during the last glaciation, *Nature*, 351:220.

Petit, J. R., Mounier, L., Jouzel, J., Korotkevich, Y. S., Kotlyakov, V. I, and Lorius, C., 1990, Palaeoclimatological and chronological implications of the Vostok core dust record, *Nature*, 343:56.

Raymo, M. E., Ruddiman, W. F., Shackleton, N. J., and Oppo, D. W., 1990, Evolution of Atlantic-Pacific ^{13}C gradients over the last 2.5 m.y., *Earth Planet. Sci. Letters*, 97:353.

Rind, D., and Chandler, M., 1991, Increased ocean heat transports and warmer climate, *J. Geophys. Res.*, 96:7437.

Ruddiman, W. F., and McIntyre, A., 1981, The north Atlantic Ocean during the last deglaciation, *Palaeogeo. Palaeoclim. Palaeoeco.*, 35:145.

Sarmiento, J. L., Toggweiler, J. R., and Najjar, R., 1988, Ocean carbon-cycle dynamics and atmospheric pCO_2, *Phil. Trans. R. Soc. Lond. A*, 325:3.

Sarnthein, M., Winn, K., and Zahn, R., 1987, Paleoproductivity of oceanic upwelling and the effect on atmospheric CO_2 and climatic change during deglaciation times, *In*: "Abrupt Climatic Change," W.H. Berger and L.D. Labeyrie, eds., D. Reidel Publ. Co.

Sathyendrenath, S., Gouveia, A. D., Shetya, S. R., Ravindran, P., and Platt, T., 1991, Biological control of surface temperature in the Arabian Sea, *Nature*, 349:54.

Shaffer, G., 1989, A model of biogeochemical cycling of phosphorus, nitrogen, oxygen, and sulphur in the ocean: One step toward a global climate model, *J. Geophys. Res.*, 94:1979.

Simonot, J. -Y., Dollinger, E., and Le Treut, H., 1988, Thermodynamic-biological-optical coupling in the oceanic mixed layer, *J. Geophys. Res.*, 93:8193.

Slingo, A., 1989, Sensitivity of the earth's radiation budget to changes in low clouds, *Nature*, 343:49.

Smith, S. V., and Mackenzie, F. T., 1987, The ocean as a net heterotrophic system: Implications from the carbon biogeochemical cycle, *Global Biogeochem. Cycles*, 1:187.

Suess, E., 1973, Interaction of organic compounds with calcium carbonate - II, Organo-carbonate association in recent sediments, *Geochim. Cosmochim. Acta*, 37:2435.

Taylor, A. H., Watson, A. J., Ainsworth, M., Robertson, J. E., and Turner, D. R., 1990, A modelling investigation of the role of phytoplankton in the balance of carbon at the surface of the North Atlantic, *Global Biogeochem. Cycles*, 5:1.

Thierstein, H. R., Geitsenauer, K. R., and Molfino, B., 1977, Global synchroneity of late Quaternary coccolith datum levels: Validation by oxygen isotopes, *Geology*, 5: 400.

Tsunogai, S., and Noriki, S., 1991, Particulate fluxes of carbonate and organic carbon in the ocean, Is the marine biological activity working as a sink of the atmospheric carbon?, *Tellus*, 43B:256.

Turner, S. M., Malin, G., Liss P. S., Holligan P. M., and Harbour, D. S., 1988, The seasonal variation of dimethylsulfide and dimethylsulfoniopropionate concentrations in nearshore waters, *Limnol Oceanogr.*, 33:364.

Volk, T., 1989, Sensitivity of climate and atmospheric CO_2 to deep-ocean and shallow ocean carbonate burial, *Nature*, 337:637.

Walsh, J. J., 1989, How much shelf production reaches the deep sea?, *In*: "Productivity of the Ocean: Present and Past," W.H. Berger, V.S. Smetacek, and G. Wefer, eds., John Wiley & Sons, Chichester.

Walsh, J. J., 1991, Importance of the continental margins in the marine biogeochemical cycling of carbon and nitrogen, *Nature*, 350:53.

ABSTRACTS

PALEOPRODUCTIVITY RECORDS FROM THE NORTHEAST ATLANTIC MARGIN: DIATOM EVIDENCE

F. Abrantes
Serviços Geológicos de Portugal
INGMAR
Rua Academia das Ciencias, 19 2°
1200 Lisboa, Portugal

The Iberian margin is an area of coastal upwelling created by summer northerly trade winds. The spatial patterns and intensity of this upwelling system are accurately recorded by the diatom abundances and taxonomic composition in the surface sediments.

Studies of opaline microfossil accumulation rates and diatom assemblage in latest Quaternary (~ 130 Ka to P.D.) sediment sequences collected off Portugal and northwest Africa indicate that coastal upwelling increased during glacial episodes (Isotopic Stages 2, 4, and 6) relative to interglacial intervals (Isotopic Stages 1, 3, and 5). During the late Holocene, primary productivity levels are similar to those observed for Stage 5, but in the early Holocene upwelling intensities have been weaker than they are today.

The paleoproductivity reconstruction based on the diatom record is supported by those revealed by two independent approaches: a carbon derived "new paleoproductivity" and the bottom-pore water difference in $\delta^{13}C$.

PHYTOPLANKTON ECOLOGY OF THE WESTERN ARABIAN GULF (PERSIAN GULF)

K.A. Al-Abdulkader and S.Z. El-Sayed
Texas A & M
College Station, TX 7784

An ecological study of the phytoplankton of the western Arabian Gulf was conducted between May 1985 and April 1988. Fourteen shallow-water stations were occupied, with depths not exceeding 25 m. Temperature varies between 16.6°C during February and 35.5°C during August. Two salinity regimes were observed in the western Gulf. The northern and middle regions have salinity values around $43^o/_{oo}$, while the southern region, primarily in the semi-enclosed Bay of Salwah, is characterized by higher salinity (around $55^o/_{oo}$).

Nutrient salts were low during all seasons, with slightly higher values recorded in the winter at most stations. The average nutrient values are as follows: $PO_4 = 0.027$ μg-at/L; $NO_3 = 0.150$ μg-at/L; $NH_4 = 0.460$ μg-at/L; and $SiO_4 = 2.950$ μg-at/L.

The phytoplankton standing crop, expressed in terms of chlorophyll a, was relatively high; biomass ranged from 0.16 to 5.2 mg/m^3, with an average of 0.90 mg/m^3. Northern and middle stations had higher chlorophyll a values in the fall and winter, and lower values during spring and summer. Southern stations generally had high chlorophyll a values at all seasons.

Phytoplankton primary production was high during the summer and fall of 1986, with an average of 5.12 and 4.71 mg C/m^3/hour, respectively. Low values were recorded in spring, with an average of 2.1 mg C/m^3/hour.

Seventy-seven species of diatom and dinoflagellates were identified. Diatom abundance ranged from 59% to 100% of the total phytoplankton count. Species variation depended on the location of the stations and the mesh-size of the nets used.

Annual primary production of the Arabian Gulf is estimated to be 136×10^6 tons of carbon. Employing theoretical considerations and different levels of trophic efficiency, we calculated the potential yields in terms of biomass of herbivores, primary, and secondary carnivores. From these calculations, the Arabian Gulf seems capable of sustaining severalfold the present commercial catch of finfish and shellfish.

THE BIOLOGICAL PUMP IN ISLAND-INDUCED EDDIES

J. Arístegui, S. Hernández-León, M.F. Montero,
S. Ballesteros, and G. Basterretxea
Facultad de Ciencias del Mar. Box 550
Las Palmas de Gran Canaria, Canary Islands, Spain

Eddies are important not only because they may store and transport large amounts of energy or alter the mesoscale circulation of an area, but because they may accelerate the vertical transport of carbon through the water column.

Downstream of the Canary Islands, cyclonic and anticyclonic eddies are generated. In cyclonic eddies, nutrient-rich water upwells into the euphotic layer, enhancing the uptake of inorganic carbon by autotrophs. These cold-core eddies are characterized by high productivity though low respiration rates, chlorophyll being concentrated mainly in the margins of the ring. By contrast, in the anticyclonic warm-core eddies, respiration rates are high because of the concentration of organisms which are consumed or sink down to deeper layers. When cyclonic and anticyclonic eddies occur together, the biological pump, i.e., the production and downward flux of organic carbon in the ocean due to the biological activity, may work faster than in the surrounding waters.

In the Canary Islands, these processes are especially important for several reasons: (1) cyclonic and anticyclonic eddies seem to be a common feature of the mesoscale circulation of the area, (2) new production generated in the surface waters of the cold-core eddies may be advected to the island shelf, and (3) phytoplankton-rich water transported offshore from the northwest Africa upwelling system can be trapped and sunk by anticyclonic eddies.

SIZE FRACTIONATED DON RELEASE IN A SUBESTUARY OF CHESAPEAKE BAY

D.A. Bronk and P.M. Gilbert
Horn Point Environmental Laboratory
P.O. Box 775
Cambridge, MD 21613

Total and low molecular weight (LMW, $< 10,000D$) DON release was studied in two size-fractions ($< 1.2\ \mu$ and $< 202\mu$) in the Choptank River during August, 1990. Nutrient and biomass concentrations, short-term NH_4^+ uptake and regeneration, and total and LMW DON release were measured every 3 h over 36 h in 20L carboys. DON release was measured by passing water that was incubated with $^{15}NH_4^+$ through a resin which retains inorganic ions, and then measuring the ^{15}N content of the isolated DON.

NH_4^+ concentrations decreased from 7.6 to 0.85 μM in the $< 1.2\mu$ carboy and from 7.64 to 0 μM in the $< 202\mu$ carboy over the 36 hours. No decrease in NO_3^- concentrations was seen in either carboy. Chl a concentrations increased from 0.35 to 1.39 μg l^{-1} in the $< 1.2\mu$ carboy, and from 1.25 to 1.74 μg l^{-1} in the $< 202\mu$ carboy. PN:chl a ratios decreased from 15.8 to 3.6 in the $< 1.2\mu$ carboy, reflecting an increase in autotrophic organisms. Total DON concentrations ranged from 23.43 to 26.27 μM in the $< 1.2\mu$ carboy and from 25.31 to 32.32 μM in the $< 202\mu$ carboy. LMW DON ranged from 9.6 to 45.3% of total DON in the $<1.2\mu$ carboy and from 16.7 to 42.3% in the $<202\mu$ carboy.

Preliminary data indicates that total DON release rates varied from 0.001 to 0.105 μmole l^{-1} h^{-1} in the $<1.2\mu$ carboy and from 0.09 to 0.645 μmole l^{-1} h^{-1} in the $<202\mu$ carboy. LMW DON release ranged from 0.003 to 0.08 μmole l^{-1} h^{-1} in the $<1.2\mu$ carboy and from 0.01 to 0.181 μmole l^{-1} h^{-1} in the $<202\mu$ carboy. Total DON release increased throughout the day in the $<202\mu$ carboy, reaching a peak at 1323 on the first day. Total DON release in the $<1.2\mu$ carboy showed considerably less variation. Expressed as h^{-1}, total DON release as a percentage of NH_4^+ uptake (V, h^{-1}) ranged from 0.5 to 21.4% in the $<1.2\mu$ carboy. However, in the $<202\mu$ carboy, total DON release occasionally exceeded the uptake of NH_4^+.

SOUTH BRAZILIAN CONTINENTAL SHELF: CHLOROPHYLL-*A*, PRIMARY PRODUCTION, AND THE RELATIONSHIP BETWEEN ABIOTIC AND BIOTIC PARAMETERS

A.M. Ciotti and C. Odebrecht
Phytoplankton Laboratory, Depto. Oceanografia
Cxp 474 Rio Grande RS 96200, Brazil

O. Moller, Jr.
Physical Oceanography Laboratory, Depto. Oceanografia
Cxp 474 Rio Grande RS 96200, Brazil

The south Brazilian continental shelf presents a complex hydrographic regime due to land drainage influence (Rio de la Plata and Lagoa dos Patos) and the proximity of the Subtropical Convergence Zone. Our aim is to determine the effect of the interacting water masses on biological production. During September 1988, we sampled 53 oceanographic stations between 31°30'S and 34°30'S along the continental shelf. The water masses were identified by TS diagrams and classified according to the Mamaiev mixture triangle method, adapted for continental shelf analysis. Chlorophyll-*a* and phaeopigments were analyzed by fluorometric methods, and we performed simulated *in situ* C14 primary production experiments on 13 stations. We also measured chlorophyll-*a* in three size classes (>20 μm, 1 to 20 μm, and <1 μm), dissolved inorganic nutrients (nitrate + nitrite, phosphate, ammonia, and silicate), light intensity, and phytoplankton and protozooplankton abundance and composition.

The spatial pattern of chlorophyll-*a* concentration was related to the distribution of the water masses. Higher values were observed in the Subantarctic Water (SAW) over the central areas of the continental shelf and nearshore in the Coastal Water (CW). The lowest values were observed in the Tropical Water (TW) near the continental slope. The mid-continental shelf showed the highest productive potential with integrated chlorophyll-*a* concentration around 50 mg/m^2 and a mean primary production rate of 105 mgC/m^2/h. Microplankton (>20 μm) represented the major component of phytoplankton chlorophyll-*a*, being mostly composed of *Rhizosolenia setigera*, *F. pungens*, *Chaetoceros* spp. The principal component analysis discriminated the central areas of the continental shelf due to the higher nitrate concentration, which suggests that the input of nitrate-nitrogen by SAW influenced primary production and biomass. Further north, the chlorophyll-*a* concentration decreased and the importance of *Protoperidinium* spp. and aloricate ciliates increased, suggesting a maturing process of this water mass. In the coastal zone, high values of both chlorophyll-*a* concentration and primary production rates occurred only near the mouth of Lagoa dos Patos and north, where the SAW appeared at shallower depths.

Lower chlorophyll-*a* concentration and primary production rates were found towards the continental slope. The results show the importance of SAW for biological primary production at the south Brazilian continental shelf, and a local influence of Lagoa dos Patos.

VARIABILITY IN PHYTOPLANKTON ABSORPTION SPECTRA: PHOTOADAPTATION AND MIXING IN VESTFJORD, NORWAY

Joan S. Cleveland
Naval Oceanographic and Atmospheric Research Laboratory
Code 331
Stennis Space Center, MS 39529-5004

Absorption spectra were measured for particles collected on glass filters in Vestfjord, Norway during September, 1990. Particles from this dinoflagellate-dominated phytoplankton community exhibited strong absorption peaks and shoulders in the absorption spectra at 415 nm. In the euphotic zone, the absorption signal was dominated by phytoplankton, with detrital or other nonphytoplanktonic particles dominating at depth.

Initially, the relative height and sharpness of the chlorophyll *a* peak (located between 464 and 470 nm) increased with depth within the euphotic zone, implying a photoadaptive increase in accessory pigment absorption as irradiance decreased. This change was also suggested by an increase in the ratio of chlorophyll *c* to chlorophyll *a* absorption. At the same time, the specific absorption coefficient at 435 nm decreased with depth; this decrease may have resulted from pigment-packaging effects if cellular pigment concentrations increased as part of the photoadaptive response. Following wind-induced mixing of the fjord during a storm, absorption spectra from the upper part of the water column were identical in shape and magnitude, indicating that there was physical mixing of the previously distinct photoadapted communities. These trends will be quantified through derivative analysis of the absorption spectra.

DIMETHYLSULFIDE PRODUCTION FROM A *PHAEOCYSTIS* BLOOM IN THE BELLINGSHAUSEN SEA, ANTARCTICA

K. Crocker, M. Ondrescek, and R. Petty
Department of Biological Sciences
University of California
Santa Barbara, CA 93106

Dimethylsulfide (DMS) and pigment samples were taken over two months (October-November 1990) in the Bellingshausen Sea, Antarctica, as part of the Icecolors' 90 ozone-hole cruise. During this period, a large ice-edge bloom of *Phaeocystis* developed over an extensive area. To estimate DMS production over the entire bloom, CTD casts provided samples over depth, while transects were taken following the ice edge, and from heavy ice to the open sea. Together, these data provide a three-dimensional overview of DMS production. DMS measurements correlate well with *in vivo* measurements, and with the presence of *Phaeocystis*. An estimate was made of DMS production over the duration of an ice-edge bloom.

NORTH PACIFIC PHYTOPLANKTON STOCKS IN GYRE, TRANSITION, AND SUBARCTIC WATERS ARE SIMILAR WHEN PICOPLANKTON ARE INCLUDED

M.-L. Dickson and P.A. Wheeler
College of Oceanography
Oregon State University
Corvallis, OR 97331-5503

Large gradients have been reported in nutrients, biomass, and rates of production between subtropical and subarctic oceanic regions. We measured chlorophyll concentrations as a function of depth from 20°N to 48°N along 152°W during March, 1991. Surface chlorophyll concentrations (for GF/F filters, nominal pore size 0.7 μm) ranged from 0.021 to 0.404 μg/L, with the lowest values in subtropical waters and the highest values in subarctic waters. Deep subsurface chlorophyll maxima were located at approximately 100 meters in subtropical and transition waters, but were absent in subarctic water. Relative picoplankton abundance at the chlorophyll maximum was highest in subtropical water (85% of total) and lowest in subarctic water (35-50% of total).

Chlorophyll standing stocks integrated from 0-200 meters showed a threefold increase between 28°N and 48°N for GF/F filters (9-26 mg/m^2). However, a significant fraction of chlorophyll in subtropical and transition waters passed through GF/F filters but was trapped on 0.2 μm filters. Comparison of integrated chlorophyll for phytoplankton trapped on the 0.2 μm filters showed a nearly constant standing stock between 28°N and 48°N (25-31 mg/m^2).

THE RELATION BETWEEN NEW PRODUCTION, AS ESTIMATED BY EXPORTED NITROGEN AND TOTAL PRIMARY PRODUCTION

J.N. Downs and J.W. Murray
Monterey Bay Aquarium Research Institute
160 Central Ave.
Pacific Grove, CA 93950

Under steady-state conditions, the f-ratio (new:total primary production), should be equivalent whether calculated with new production measured from allochthonous nitrogen uptake, or from particulate nitrogen export from the euphotic zone. In comparing several marine sites; Monterey Bay, the eastern equatorial Pacific, and the central north Pacific gyre, the f-ratios based on PON fluxes and Redfield ratio-converted ^{14}C-uptake primary production exhibit a direct relation to primary production; in general, higher f-ratios are found with the higher primary production and vice versa. However, some data sets show significant variation. At times, (early upwelling in Monterey Bay, likely a non steady-state) this relation does not hold, and over much of the year, primary production and the f-ratio are inversely related.

New production, estimated by exported nitrogen (PON flux), is not consistently correlated to the total production. It is more probably linked to the zooplankton grazer population, based on evidence from pigment analysis of sinking particulates that, in these areas, the principal source of organic flux is from grazed material (phaeopigment-rich) not from sinking phytoplankton (chlorophyll-rich). Particulate fluxes of organic carbon and nitrogen in Monterey Bay show much less seasonal variation than primary production. Similarly, settling volumes of zooplankton biomass show small seasonal variability. Therefore, the apparent inverse relation of the PON flux-derived f-ratio with primary production over time for Monterey Bay is largely due to a substantial variation in the primary production (days) and a slower, smaller response of the flux as mediated by the zooplankton population (months). The seasonal high production will fuel an accumulation of zooplankton biomass in the euphotic zone over the growing season. The relative amount of carbon to phaeopigment (an index of herbivory) in the flux changes with the production cycle. Data from four stations in the eastern equatorial Pacific show that increases in the zooplankton standing stock and trophic pathway can substantially influence particulate fluxes. Despite equivalent primary production at all stations, the particulate flux of PON, POC, and phaeopigment at one station was 2-3 times higher than at the others. This station also had a 3-fold higher larvacean abundance.

Systems where seasonal variation in phytoplankton and zooplankton production is profound must be viewed as being composed of several separate systems, with caution given to the presence or absence of steady state during any measurement. Seasonal growth cycles of phytoplankton and zooplankton in a temperate system assure that steady state is never attained. Generalizations to an "annual average" for one system must incorporate seasonal time-averaging.

DETECTION OF "SHINGLES" IN CZCS AND AVHRR IMAGES EAST OF CAPE HATTERAS

G. Garcia-Moliner, P.M. Stegmann, and J.A. Yoder
Graduate School of Oceanography
University of Rhode Island
Narragansett, RI 02882

"Shingles" (filaments) are ejections of Gulf Stream water which occur at the crest of meanders, and are opposite in direction to the flow of the stream. Evidence from AVHRR images of sea-surface temperature off the US eastern coast clearly shows the occurrence of numerous "shingles" flowing shoreward within any single image. Ocean color images (CZCS) show pockets of high pigment content located between the "shingles" and the north wall of the Gulf Stream. Coastal Zone Color Scanner images are available from 1978-1986, albeit on an irregular basis. AVHRR images for sea-surface temperature are available twice daily since 1982; before this, only radiance values were obtained. Nonetheless, the features of interest, i.e. "shingles", can still be detected from these data. Phytoplankton growth along the north wall of the Gulf Stream is sustained by the recycling of essential nutrients, as well as by nutrient transport resulting from upwelling. CZCS and AVHRR images off the northeast US coast were examined to determine the spatial/temporal occurrence of "shingles" in relation to upwelling mechanisms along the shelf break.

EVOLUTION OF SMALL PARTICLE DISTRIBUTION DURING THE NORTH ATLANTIC BLOOM

Wilford Gardner, Ian Walsh, and Mary Jo Richardson
Department of Oceanography
Texas A & M University
College Station, TX 77843

During the April/May leg of the North Atlantic Bloom Experiment, CTD/Transmissometer profiles were taken at 47°N, 20°W every morning and evening. Water samples were filtered to obtain particulate matter samples, to determine particle mass concentrations, and to calibrate the transmissometer. Particle concentrations in the surface mixed layer, calculated from beam attenuation coefficients, increased three-fold (from 0.17 mg/1 to 0.52 mg/1) over a two-week period, indicative of a phytoplankton bloom. The overall trend in the surface-layer beam attenuation during the second cruise (May/June) was flat, with the exception of a spike (concentrations > 1 mg/1), related to a storm and/or the edge of a mesoscale eddy. Superimposed on the April/May increase were diel variations (0.06 mg/1 to 0.12 mg/1), with evening highs and morning lows. Increases in particle mass easily account for the uptake of CO_2 (LDGO CO_2 group), suggesting that CO_2 is rapidly converted to the particulate phase, after which it may be converted to DOC directly, or incorporated into large aggregates, which could release DOC as they settle and are remineralized. Decreases in particle concentration are, presumably, associated with biological consumption, large-particle production, and subsequent particle settling (fecal pellets, aggregates), or consumption and bioactive transport by organisms migrating to deeper depths.

Changes in particle concentrations with time suggests a three-layer partitioning of the water column. Particle concentrations steadily increased in the surface 40 m, decreased between 40 and 400 m, and increased between 400 and 2000 m. We assume that these changes are related to particle production, settling, and transfer between large- and small-particle pools in the water column.

PHYSICAL/BIO-OPTICAL INTERACTION AND PRODUCTIVITY ESTIMATES FROM OCEANIC TIME SERIES

T.C. Granata, J. Wiggert, M. Hamilton, and T. Dickey
Ocean Physics Group
Department of Geological Sciences
University of Southern California
Los Angeles, CA 90089

J. Marra and C. Langdon
Lamont-Doherty Geological Observatory
Columbia University
Palisades, NY 10964

In situ beam transmissometer and dissolved oxygen data from the Sargasso Sea (20 m) were plotted as a function of PAR (photosynthetically available radiation), for a 14-day period in spring during the BIOWATT Experiment (1987). These pseudo P versus I curves reflect complicated diurnal trends in *in situ* beam attenuation and in dissolved oxygen which are produced by either advection, cloud cover, or both. Generally, particle concentrations are coherent with dissolved oxygen concentrations, except at the onset of advective periods. Peaks in particle concentration and dissolved oxygen lagged peaks in PAR by 4-6 h for all 14 days.

UNCOUPLING OF CELL DIVISION AND PIGMENT SYNTHESIS DURING MANIPULATIONS OF A NATURAL DIATOM POPULATION

Lisa M. Graziano
College of Marine Studies
University of Delaware
Lewes, DE 19958

The relationship between population growth and pigment synthesis was examined in Rehoboth Bay, Delaware, in August 1990 during a bloom dominated by small diatoms

(5 μm). Changes in cell number and the concentration of chlorophylls and carotenoids were examined in diluted samples (9:1 mixture of <0.45 μm filtered:unfiltered seawater) and in undiluted samples (screened for >211 μm organisms). This manipulation allowed estimation of the grazing impact of microzooplankton on the growth rate of phytoplankton. However, at most, microzooplankton grazing could account for a 12% reduction in phytoplankton growth rate, from 3.00 to $\approx$ 2.65 d^{-1}. Rather, varying rates of cell division and pigment synthesis were observed, indicating that there was nutrient limitation in undiluted samples. Cell division rate was only slightly affected by dilution, but rates of net pigment synthesis differed widely from estimates of cell-specific growth and loss rate. Chlorophyll a degradation was apparent in the undiluted light treatment, while in diluted samples, chlorophyll a accumulated at a rate nearly equal to cell division rate. Nutrient limitation affected the rates of chlorophyll a and c synthesis but not the synthesis rates of diatom carotenoids, fucoxanthin, and diadinoxanthin. Thus, a de-coupling in time of chlorophyll a synthesis from carotenoid synthesis and cell division complicated the use of taxon-specific pigments in assessing the effect of dilution on algal population dynamics, but provided insight into the nutrient status of the phytoplankton assemblage.

REGULATIONS OF PHOTOSYNTHESIS UNDER HIGH-FREQUENCY LIGHT FLUCTUATIONS

Richard Greene
Oceanographic and Atmospheric Sciences Division
Brookhaven National Laboratory
Upton, NY 11973

The light-saturated rate of photosynthesis under constant light can be increased by incubations under high-frequency (1 Hz) fluctuating light and dark cycles. Experimental results for the red alga *Chondrus crispus* suggest that this effect of fluctuating light can be attributed to increased rates of electron transport during the light interval and to post-illumination carbon fixation during the dark. Mathematical model simulations, which incorporate the key biochemical reactions in photosynthesis, also predicted enhanced photosynthetic rates under fluctuating light. It is suggested that fluctuating light effects, by acting at the level of the photosynthetic apparatus, may be a general phenomenon that stimulates photosynthesis by alleviating the rate limitation imposed by carbon metabolism under constant saturating light.

SOLUBILIZATION OF ORGANIC PARTICLES COLLECTED IN SEDIMENT TRAPS: EFFECT ON EXPORT PRODUCTION ESTIMATES

D.A. Hansell
Institute of Marine Sciences
University of California
Santa Cruz, CA 95064

J.N. Downs
Monterey Bay Aquarium Research Institute
160 Central Avenue
Pacific Grove, CA 93950

Sinking organic particles may be a source of DOM to subeuphotic zone waters through particle solubilization or via particle grazing. If solubilization is significant, then particulate organic carbon and nitrogen flux into sediment traps could be underestimated, resulting in underestimates of export production. VERTEX-style sediment traps were deployed at 5 depths, ranging from 60 to 210 m, in Monterey Bay, CA, and DOC and DON accumulation in the supernatant of the traps was determined by high-temperature catalytic oxidation techniques. Our results show that by considering accumulated particles only, the export flux of carbon and nitrogen may be significantly underestimated at the base of the euphotic zone. The increase in the DOC concentration of the trap supernatant was equivalent to a 57% increase in the organic carbon sedimentation flux at 60 m, while nitrogen export increased 94% when DON and NH_4^+ were considered. The contribution of dissolved components to total flux decreased with depth such that the particle flux was a much better indicator of total flux at 210 m.

The experiments did not resolve the contribution of swimmers to the release of DOC or DON in the trap supernatant, a problem that may be especially important near the euphotic zone where the concentrations of swimmers in the traps are maximal. A swimmer-free trap designed specifically to monitor the release of DOM from accumulated particles was tested in Dabob Bay, WA, and Monterey Bay, CA. Swimmers were segregated from the sinking particles with an efficiency of 85%. More field evaluations of the trap will be necessary to resolve the contribution of swimmers to DOM production in the supernatant.

THE ROLE OF CORAL REEFS IN GLOBAL OCEAN PRODUCTION

B.G. Hatcher
Department of Oceanography
Dalhousie University
Halifax, Nova Scotia, B3H 4J1 Canada

S.V. Smith
Department of Oceanography
University of Hawaii
Honolulu, HI 96822

C.J. Crossland
CSIRO Institute of Natural Resources and Environment
P.O. Box 225
Dickson, ACT 2602, Australia

Coral reefs cover some 600 x 10^3 km^{-2} of the earth's surface (0.17% of the ocean surface). First-order estimates show coral reefs to contribute about 0.85% of the estimated net CO_2 fixation rate of the global oceans. Gross CO_2 fixation is relatively high (about 700 x 10^{12} gC yr^{-1}), but most of this material is recycled within the reefs. Excess (net) production of organic material (E) is much smaller, about 20 x 10^{12} gC yr^{-1}. We estimate that 3 x 10^{12} gC yr^{-1} (15% of E) is buried in reef structure; 2 x 10^{12} gC yr^{-1} (10% of E) for sustained human harvest, and the remaining 75% of E for export from coral reefs to adjacent areas. Comparison of estimates for net production by reefs and their surrounding oceans indicates that the excess production by coral reefs is similar to new production in the photic zone of the oligotrophic oceans. Consequently, estimates for global ocean production should include as a first approximation areas of reefs with the surrounding ocean when assigning average net production rates. While there are significant uncertainties in these numbers, we conclude that organic production by reefs plays a relatively minor role in the global scale fluxes and storage of elements. In comparison, the companion process of biologically-mediated inorganic carbon precipitation represents a major role for reefs. While reef production does respond on local scales to variation in ocean climate, neither the absolute rates nor the amount accumulated into organic pools are either sensitive indicators or accurate recorders of climatic change in most reef systems. Similarly, the productivity of most reefs should be little affected by the environmental changes currently predicted from the Greenhouse Effect.

EFFECTS OF ULTRAVIOLET RADIATION ON MARINE PHYTOPLANKTON

E.W. Helbling, V. Villafañe, M. Ferrario, and O. Holm-Hansen
Polar Research Program A-002-P
Scripps Institute of Oceanography
University of California, San Diego
La Jolla, CA 92093

Our studies in both the Antarctic and tropical waters, addressed three major questions: (1) To what extent does ultraviolet radiation (UVR) reduce integrated rates of primary production, and what is the action spectrum for such inhibition of photosynthesis?; (2) Is there differential sensitivity to UVR in Antarctic species of phytoplankton, especially in regard to the microplankton, which are the food items preferred by krill?; (3) Are phytoplankton from Antarctic water more sensitive to UVR than phytoplankton from tropical waters, which seasonally receive much more UVR that Antarctic waters?

Our data from Antarctic waters show that solar UVR can reduce photosynthetic rates by up to 50% in the upper few meters of the water column, but there is no detectable effect below about 10 m. Approximately half of this inhibition is caused by UV-B (290-320 nm), and the rest by UV-A (320-400 nm). Tropical phytoplankton sampled from within the upper mixed layer show relatively little inhibition of photosynthesis due to UVR, in spite of much higher fluences of solar UV-B. Preliminary floristic analyses of samples from Antarctic waters suggest that the larger diatom species are particularly sensitive to UV-B.

ESTIMATES OF EXPORT PRODUCTION IN THE COASTAL NORTHERN BALTIC SEA

A.-S. Helskanen
Tvärmlnne Zoological Station
SF-90100 Hanko
Finland

Production and sedimentation of organic matter were studied during the spring and summer in 1983 and 1988, in the open sea zone off the SW coast of Finland. Net primary productivity was measured in situ (^{14}C incorporation), and sedimentation rates of particulate organic carbon (POC) and nitrogen (PON) were measured using cylindrical sediment traps moored below the euphotic layer in a 50-meter deep basin. Export production (P_e) was measured directly as sedimentation of POC and PON. Measurements were compared to the P_e values calculated by the power model presented by Wassmann (1990) for the boreal coastal zone of the North Atlantic. Potential new production during the spring was assumed to be equivalent to the depletion of nitrate from the whole water column. The estimated f-ratio (P_e/P_{total}) based on sedimentation measurements and NO_3-N depletion (in spring) varied between 0.43 to 0.85 during the spring and about 0.37 to 0.43 for the whole study periods. P_e calculated by the power model (Wassmann, 1990) gave considerably lower values. The time scale of the study (5-6 months) might not have been sufficient for accomplishing steady-state of the pelagic system, hence explaining a part of the discrepancy. Moreover, the validity of the power model for the coastal area of Northern Baltic Sea may have been hampered by shallowness and topography of the coastal area. The weak density stratification of the water column and movements of water masses caused elevated resuspension of organic matter from the benthos. Also, methodological differences both in primary production and sedimentation measurements probably influenced the estimates of f-ratio. In conclusion, caution is advised when sedimentation measurements are used to estimate export production in the shallow coastal areas, which are often subjected to intense mixing and nutrient loading from terrestrial, anthropogenic sources.

PHOSPHATE LIMITATION IN *ISOCHRYSIS GALBANA*

Ronny Herzig
Haifa University - Oranim
Tivon 36910, Israel

The effect of steady-state phosphorus limitation on photosynthetic characteristics was examined in the marine haptophyte *Isochrysis galbana* Green. Phosphate-limited chemostates were maintained at four dilution rates, ranging from 0.13 - 0.78 d^{-1}, under continuous irradiance levels of 175 μmole quanta m^{-2} s^{-1}, which saturated photosynthesis at all growth rates. Phosphate limitation led to an overall reduction in pigmentation and a decrease in the cellular concentration of reaction centers; however, the optical absorption cross-section, normalized to chlorophyll *a*, increased. Light-saturated photosynthetic rates normalized to chlorophyll *a* remained relatively constant. The PSII/PSI ratio increased from 2.18 to 2.66 over the dilution rates, and PSI decreased much faster than PSII. The average turnover time for light-saturated photosynthetic electron transport from water to carbon dioxide increased from 6.76 to 8.51 ms.

The dark respiration decreased twofold over this range of dilution rates. We also measured the cellular composition in the different growth rates and found a decrease in P and N content, but an increase in C content. Rubisco per cell and per chlorophyll remained constant in all the dilution rates examined.

THE RELATIONSHIP BETWEEN PHYTOPLANKTON ANNUAL CARBON PRODUCTION AND NEW NITROGEN PRODUCTION

R. L. Iverson
Department of Oceanography
Florida State University
Tallahassee, FL 32306

Phytoplankton annual carbon production is a linear function of new production, estimated from data obtained with nitrogen methods, and of export production, estimated from measurements of sediment-trap carbon flux. A plot of the f ratio as a function of total phytoplankton production, obtained using the linear new production equation, increased nonlinearly to an asymptotic value. The plot of percent new production obtained by Eppley and Peterson (1979) from six data values is a good approximation to the curve derived from the new production equation obtained with 41 values. Data on ^{14}C production in the Sargasso Sea and new, nitrogen production data pairs were distributed along the linear regression line, with average new production estimated from oxygen data in agreement with average new production estimated from nitrogen measurements.

AN EULERIAN MODEL OF PHYTOPLANKTON PHOTOSYNTHETIC RESPONSE IN THE UPPER MIXED LAYER

G. S. Janowitz and D. Kamykowski
Department of Marine, Earth and Atmospheric Sciences
North Carolina State University
Raleigh, NC 27695-8208

We combined an Eulerian model of an upper mixed layer forced by dynamic and convective processes with a responsive photoadaptation model of photosynthesis. Incident radiation was divided into photosynthetically active and inactive components. The model was initialized to simulate physical conditions in Lake Titicaca on August 12, 1982, but it applies equally well to oceanic conditions. The simulation provides the time course of water column temperature, phytoplankton photoinhibition, and relative carbon fixation of phytoplankton as depth contours for comparison with the field data. The predicted and observed temperature profiles generally agree; diurnal changes in specific humidity are required to more accurately specify heat loss. The predicted photoinhibition resembles the observed photosynthetic state of the natural community based on fluorescence, but the modeled recovery from photoinhibition is too rapid. This discrepancy also affects the comparison between predicted and observed bottle determination of carbon fixation. The biological formulations in the model provide approximate fits to field observations, but improved realism requires a biological initialization that is more specific to the natural phytoplankton community.

THE ROLE OF PLANKTON COMMUNITY METABOLISM IN REGULATING SEDIMENTATION OF ORGANIC MATTER IN A COASTAL ENVIRONMENT

W. Michael Kemp, E. M. Smith, and W. R. Boynton
Center for Environmental and Estuarine Studies
University of Maryland
Cambridge, MD 21613

We compared temporal variations in the production and respiration of coastal plankton community in the mesohaline region of Chesapeake Bay at time-scales of weeks with contemporaneous measurements of plankton trophic structure and deposition of particulate organic matter (POM) for the spring to summer. Seasonal cycles were punctuated by two major events of POM deposition, which were observed repeatedly over six years. The first event occurred in mid-April in association with the sinking of intact diatoms from the spring bloom, while the second occurred in mid-August at the time of peak phytoplankton production. Cool springtime temperatures retarded grazing and heterotrophic metabolism during March to April (despite a relatively high abundance of copepods), allowing the sedimentation of a large fraction of phytoplankton production. In late spring and early summer (May-July), the phytoplankton community was dominated by small flagellates, and most of the primary production was consumed and respired by rapidly growing communities of bacteria and protozoa, with little POM sinking to bottom waters. A comparison of rates of POM collection in traps deployed near the water surface, with those at the pycnocline, suggested that there is tightly coupled nutrient cycling in the upper euphotic zone during this period, with little new production. However, by mid-summer, heterotrophic activity was reduced; these relatively low rates of community respiration rates, coupled with high rates of primary production, result in increased rates of POM deposition. Net diel metabolism of the surface plankton community (NCM) was estimated as the sum of net daytime production (O_2) minus nighttime respiration (O_2). POM (total chl) sedimentation was highly correlated with NCM ($r^2 = 0.87$), but unrelated to gross primary production ($r^2 = 0.20$). Thus, we conclude that the timing and magnitude of organic matter losses from the euphotic zone are controlled as much by heterotrophic activity as by new rates of primary production.

FLUXES OF MATERIAL AND ENERGY BETWEEN CORAL REEFS AND THE OPEN SEA

T. Korpal and B. Lazar
H. Steinitz Marine Laboratory
The Interuniversity Institute
The Hebrew University of Jerusalem
Elat 88103, Israel

J. Erez
Institute of Earth Sciences
The Hebrew University of Jerusalem
Givat Ram, Jerusalem, Israel

The REEFLUX project is a multidisciplinary oceanographic research effort aimed at understanding the high productivity of coral reefs in oligotrophic water. The working hypothesis states that the main source of nutrients (nitrogen and phosphorous) for reef productivity is zooplankton transport from the open sea. Zooplankton is filtered from seawater by the reef community (corals, filter feeders, planktonivorous fish, and others). The digestion products in the form of ammonia nitrate and phosphate are used by the primary producers (symbionts, free living algae, and microalgae) to maintain the very high photosynthesis rates observed. Excess nutrients may "leak" from the reef towards the open sea to support its primary production.

The coral reef in the Nature Reserve of Elat, Red Sea, like other reef ecosystems, maintains very high gross primary productivity, which demands a constant supply of nutrients in the dissolved or particulate phase. During the stratified season (April-January), the concentration of dissolved nutrients in surface waters of the adjacent sea is extremely low. In such conditions zooplankton or particulate organic matter are probably the major sources of nutrients that support reef productivity, rather than the generally accepted dogma of complete nutrient recycling within the reef. Furthermore, the presence of a positive nutrient gradient from the reef to the open sea shows that the reef releases dissolved nutrients to the surrounding oligotrophic waters.

Preliminary results of the diurnal sampling in the reef water show that there is a sharp increase in nutrients for about 3 h at the beginning of the night. Bell-jar experiments indicate that a single coral colony with its associated fauna shows a similar diurnal pattern. The most intense nutrient release occurred just after sunset when respiration was highest. Before dawn, nutrient release and respiration are both close to zero.

This study shows that recycling of nutrients within the reef is not perfect and that significant quantities of dissolved nutrients leak from the reef water to the open sea. To compensate for this loss, we must evoke input of nutrients in the non-dissolved phase, most likely particulate organic matter in the form of oceanic zooplankton. The abundant filter feeders in the reef community extract plankton from the water and release dissolved nutrients. These findings support the working hypothesis of project REEFLUX.

THE INFLUENCE OF FLUCTUATING LIGHT ON THE GROWTH AND PHOTOSYNTHESIS OF *SKELETONEMA COSTATUM*

J. Kromkamp and M. Limbeek
Delta Institute for Hydrobiological Research
Vierstraat 28
3301 EA Yerseke, The Netherlands

The marine diatom *Skeletonema costatum* was grown in two light-limited continuous cultures with a 8h light:16h dark period. Although the total amount of light energy received by both cultures was the same, the way in which they received it was different. Culture 1 received one sinusoidal cycle (with the maximum 4h after the onset of light). In culture two, a pattern of vertical mixing was simulated: the light intensity changed in such a way that it simulated three times a mixing from the surface to the bottom of the euphotic zone and back. The simulation was performed by a computer-controlled stepper motor connected to a Venetian Blind system.

The results show that although both cultures were grown at identical, steady-state growth rates at equal total daily light dose, the photosynthetic physiology was quite different. The culture simulating vertical mixing had a higher absorption cross-section and a high light harvesting capacity (α per unit chlorophyll). Also, the photosynthetic unit size (PSU) was smaller (approximately 1000 molecules of chlorophyll/O_2 as opposed to 1800 in culture 1), but the turnover time (τ) was higher (35 ms as opposed to 22 ms in culture 1).

Some results of transient states (stepwise changes in irradiance) also are shown.

LIMITATION OF MARINE PHYTOPLANKTON BIOMASS YIELDS BY IRON

W. Lee
Department of Oceanography
Kunsan National University
Kunsan, 573-360 Republic of Korea

L.E. Brand
Division of Marine Biology and Fisheries
Rosentiel School of Marine and Atmospheric Science
University of Miami
4600 Rickenbaker Causeway
Miami, FL 33149

We measured biomass yields of 17 clones of marine phytoplankton in media with different concentrations of iron. The biomass yields of all the clones except two were limited by iron. All cyanobacteria tested were dead at iron concentration of 10^{-9} M, still limited severely at up to 10^{-7} M. Coastal dinoflagellates showed evident limitation of biomass yield by iron. Different levels of biomass yield limitations were found among the seven coccolithophorid clones. Among other clones, *Micromonas* (A2167) was only slightly limited by low concentrations of iron, while *Heterosigma akashiwo* (OLISTH) showed no growth below 10^{-10} M. In two oceanic clones, *Mantoniella* sp. (MC-1) and *Umbilicosphaera* (A2390), no significant limitation of biomass yields were measured at all the different concentrations of iron.

The various responses of the above 17 clones to the iron concentration gradients in the media imply the ecological importances of iron in seawater for the control of phytoplankton species composition and biomass distribution.

LIMITS TO MICROALGAL GROWTH

M. Levasseur
Institut Maurice Lamontagne
Ministere des Peches et des Oceans
C.P. 1000
Mont-Joli, Quebec G5H 3Z4 Canada

P.A. Thompson and P.J. Harrison
Department of Oceanography
University of British Columbia
Vancouver, V6T 1Z4 Canada

Our objectives were to determine: 1) the influence of different N forms (NH_4^+, NO_3, and urea) on the growth rate and on the nitrogen and carbon content of four microalgae, and 2) to relate the N form-induced physiological and biochemical differences to a shortage in energy (presumably reductant for NO_3^- reduction), or a rate-limiting step in the N uptake and/or assimilation. The use of NO_3^- instead of NH_4^+ resulted in a decrease in growth rate in *Thalassiosira pseudonana* and in significantly smaller N quotas and higher C:N ratios in *T. pseudonana* and *Chaetoceros gracilis*. Growth on urea instead of NH_4^+ resulted in a lower growth rate in *C. gracilis* and *Gymnodinium sanguineum* and in a lower N quota and a higher C:N ratio in most species tested. Growth on NO_3^- rather than on NH_4^+ often resulted in a smaller (less chlorophyll *a* cell^{-1}) and less efficient (lower quantum yield for oxygen production, higher chlorophyll *a* fluorescence yield) photosynthetic system, suggesting that NO_3^--grown cells were not energy-limited. On the other hand, since urea-grown cells have a reductant requirement similar to that of NH_4^+-grown cells, the urea-induced decrease in growth rate and/or N quotas cannot be attributed to reductant limitation. The rate-limiting step for growth is more likely the activity of nitrate reductase (NR), nitrite reductase (NiR), or urease. Our results also show that symptoms of N limitation in algae (low N quotas, high C:N ratios) may be introduced by the N form, even at saturating N concentrations.

BIOMASS OF BACTERIA, CYANOBACTERIA, PROCHLOROPHYTES, AND PHOTOSYNTHETIC EUCARYOTES IN THE SARGASSO SEA

W.K.W. Li, P.M. Dickie, and B.D. Irwin
Bedford Institute
P.O. Box 1006
Dartmouth, Nova Scotia, Canada B2Y 4A2

A.M. Wood
University of Oregon
Eugene, OR 97403

Bacteria, cyanobacteria, prochlorophytes, and photosynthetic eukaryotes were enumerated in depth profiles at a station in the northern Sargasso Sea occupied for 9 days during September, 1988. Carbon biomass of each group was inferred from cell abundance using conversion factors taken from the literature. Over the upper 200 m in the water column, carbon biomass occurred in the approximate proportion of 1:2:4:8 for cyanobacteria: prochlorophytes: photosynthetic eukaryotes: bacteria. Taken together, the three phytoplankton groups represented about the same amount of carbon biomass as the bacteria. This conclusion was validated by the independent measure of bulk chlorophyll *a* if the carbon:chl ratio was assumed to be about 44 in the nitrate-depleted layer and about 15 in the nitrate-rich layer. In reporting the biomass co-dominance of bacteria and phytoplankton, we do not deny that bacteria may dominate phytoplankton at other times and in other places in the oligotrophic ocean. Biomass co-dominance between these two trophic groups admits the possibility that oligotrophic bacterial assemblages may have high growth rates.

THE EFFECT OF RESUSPENSION ON PRIMARY PRODUCTION

Hugh L. MacIntyre
College of Marine Studies
University of Delaware
Lewes, DE 19958

In a physically-forced environment, the rate of net areal production may be driven by resuspension of benthic algae and sediment. Short-term (hourly) shifts in the number of photosynthetic organisms and in the attenuation of light may be greater than variation between days. The effect of resuspension on net areal production can be described by a simple model derived from that developed by Talling (New Phytologist 56:133-149, 1957). The model is based on four dimensionless parameters: relative changes in suspended chlorophyll (Chl') and the diffuse attenuation coefficient (k'); the ratio of total incident light to that saturating photosynthesis (I_o/I_k); and the relative photosynthetic capacities of benthic and suspended algae ($^BP/^SP$). Chl' and k' are not dependent; their relationship depends on the relative suspensibility of benthic algae and sediment. Where benthic algae are more easily suspended than the sediment, resuspension enhances net areal production. Otherwise, the decrease in light penetration as sediment is suspended outweighs the effect of the increase in the number of potential primary producers and production drops. The sensitivity of net areal production to resuspension is highest when I_o/I_k and $^BP/^SP$ are high and is lowest when I_o/I_k and $^BP^SP$ are low. The relationships described by the model are compared to experimental data of fluxes of chlorophyll, attenuation, incident light, and the photosynthesis-irradiance parameters of suspended and benthic algae.

THE EFFECT OF IRRADIANCE ON THE OPTICAL PROPERTIES OF TWO DIATOMS

M. McCarthy and B. Osborne
University College
Dublin, Ireland

Pigment in a cell suspension absorbs less light than the same quantity of pigment in solution. The difference is most obvious at the peaks of maximum absorption in the blue and red regions of the spectrum at $\lambda = 436nm$ and $\lambda = 675nm$. Although theoretical predictions showed that this Package effect is dependent on cell size and intracellular pigment concentrations, there are no detailed studies of the effects of ultrastructural changes on light absorption.

Variations in the magnitude of the Package effect in two diatoms *Thalassiosira pseudonana* and *Attheya decora* acclimated to different irradiances were assessed, using measurements of whole and disrupted cell-absorption spectra. Concomitant ultrastructural changes also were investigated using Transmission electron microscopy.

Cells grown at the lowest irradiance (10 μmolm^{-2}s^{-1}) showed the greatest Package effect, although the magnitude of the effect was less in the smaller celled *Thalassiosira pseudonana*. At the highest irradiances (70, 150, and 250 μmolm^{-2}s^{-1}), the ratios of the absorbance of whole cells: absorbance of sonicated cells (λ = 436nm or λ = 675nm) were similar. The ratios also were lower in the larger species *Attheya decora*, in agreement with predictions based on the Package effect.

Ultrastructural studies showed that the ratios of chloroplast volume: cell volume in the cells grown at the highest irradiance of 250 μmolm^{-2}s^{-1} were considerably lower than those grown under the lowest irradiance of 10 μmolm^{-2}s^{-1}. The relationship between ultrastructural changes and the extent of the Package effect was shown.

THE BIOGEOCHEMISTRY OF DIMETHYLSULFIDE IN ANTARCTIC COASTAL SEAWATER

Andrew McTaggart
School of Environmental Sciences
University of East Anglia
Norwich NR4 7TJ, U.K.

The opportunity to observe large-chemical changes in the ocean's vertical structure due to a biological process often means being at the right place at the right time.

Throughout 1988, various marine chemical constituents associated with algal growth were measured at a coastal site near Davis Station, East Antarctica. A variety of chromatographic techniques were developed and applied to determine amino-acids, nutrients, micronutrients (arsenic and iodine speciation), and several specific phytoplanktonic pigments and metabolites (organic sulphur compounds, biomethylated inorganic compounds, and fatty acids).

In the austral summer of 1988 and 1989, the colonial alga *Phaeocystis pouchetii* dominated the water column for two months. The highest surface dimethylsulfide concentration (410 nM) coincided with the growth maximum of *Phaeocystis pouchetii*. Nitrates and phosphates decreased by approximately 50% during the peak of the bloom. Organo-iodine and organo-arsenics were detected in the water column.

CAN WE PREDICT THE LOCAL/GLOBAL DISTRIBUTION OF DIMETHYLSULFIDE?

Patricia A. Matrai
RSMAS/MAC
University of Miami
4600 Rickenbacker Causeway
Miami, FL 33149

Dimethylsulfide (DMS) appears to be relatively widespread in oceanic waters. Concentrations of DMS in seawater vary significantly in space within specific areas. Detailed analysis failed to provide any simple correlation of DMS in seawater with chlorophyll *a* or primary production. Shipboard measurements of DMS in air and water, as well as surface chlorophyll, were taken in the North Atlantic in the summer of 1989 during NASA CITE-3's northern hemisphere missions. Large-scale changes associated with features such as the Gulf Stream can be seen; however, no striking correlations were observed among these parameters over time once the ship remained on a single station. ODAS, a spectro-radiometer mounted and operated from an aircraft, provided relative chlorophyll values derived from the radiances collected. Tropospheric DMS values were simultaneously collected in-flight by E. Saltzman and D. Cooper (RSMAS, Miami). Preliminary results from three missions show a decrease in relative chlorophyll numbers with increasing concentrations of atmospheric DMS. A bloom of the cyanobacterium *Oscillatoria* seen on the very surface at one station was not accompanied by an equivalent DMS signal in water nor in air. We analyzed the relationships between plant biomass, CITE-3 DMS, and shipboard DMS data for the northern hemisphere missions. Whether the concentrations of in-flight atmospheric DMS observed were the result of the algal communities surveyed by the spectro-radiometer or the meteorological conditions before, and during, each mission will be discussed.

MODELLING THE EFFECTS OF EDDY-INDUCED NUTRIENT PULSES ON PRIMARY PRODUCTION AND ALGAL TAXONOMIC COMPOSITION

Mark B. Meyers, John J. Walsh, and Frank E. Muller-Karger
Department of Marine Science
University of South Florida
St. Petersburg, FL 33701

Ship-based observations of Loop Current frontal eddies and rings showed that their interaction with the eastern and western Gulf of Mexico margins leads to shoaling of nitrate isopleths of ca. 150 m or more. This engenders elevated levels of primary production and chlorophyll biomass. It also should alter algal taxonomic composition, favoring larger-celled diatoms, with increases in the vertical flux of organic matter on a time scale of several days. CZCS images show tongues of elevated pigment biomass around the periphery of the Loop Current and its separated anticyclonic rings when these features abut the margins. This may reflect offshore advection of continental shelf waters as well as local elevated production.

It was hypothesized that sporadic pulses of nitrate into the euphotic zone will result in increased diatom populations with concomitant increases in new (and export) production. A one-dimensional model of new primary production is used to assess the impact of event-scale nitrate pulses, such as those found in cyclonic eddy features, on primary production. This model contains multiple algal groups, differing in their intrinsic rates of growth and respiration, photosynthesis-irradiance responses, and sinking characteristics.

A VERTICAL, TIME-DEPENDENT BIO-OPTICAL MODEL

John R. Moisan
Department of Oceanography
Old Dominion University
Norfolk, VA 23529-0276

A vertical, time-dependent bio-optical model was developed to investigate the depth-dependent processes associated with the plankton distributions in the California Coastal Transition Zone (CTZ). The biological components included in the model are silicate, nitrate, ammonia two phytoplankton size fractions, copepods, doliolids, euphausiids, and a detritus pool. The wavelength-dependent attenuation of the subsurface irradiance field caused by seawater, phytoplankton, and dissolved organic matter, is incorporated into this model as a depth-dependent energy flux which balances the phytoplankton energy uptake and the kinetic energy flux (ΔT) into the water. This allows the phytoplankton to modify the vertical temperature structure which, in turn, affects the rate of the biological processes. Vertical velocities (w) along the track of simulated Lagrangian drifters, which were derived from a 3-D primitive equation circulation model simulating the flow observed within the CTZ, are used to parameterize upwelling and downwelling. Parameterization of biological processes, such as light versus photosynthesis and zooplankton growth and fecundity rates, are based upon data obtained during the CTZ field experiments. Comparison of simulations, which were initialized with measured distributions of phytoplankton, zooplankton, and nutrients obtained from various regions (inshore, offshore, filament, and non-filament) in the CTZ, provide insight into the role of the biological and physical processes in determining the development of the subsurface chlorophyll maximum and other related features (i.e., regions of high zooplankton fecundity). Model verification is obtained from comparing the simulations to actual observations obtained by following a Lagrangian drifter in the CTZ during field experiments in 1988. Preliminary results suggest that the bio-optical model adequately simulates the photosynthetic processes which play a role in developing the subsurface chlorophyll maximum in the CTZ.

LIGHT HARVESTING IN UNIALGAL BLOOMS

N.B. Nelson and B.B. Prézelin
Department of Biological Sciences and Marine Science Institute
University of California
Santa Barbara, CA 93106

We developed a model for spectral light absorption coefficients of dinoflagellate cell suspensions based on mean cell dimensions, cellular pigment concentrations, and known pigment-protein absorption spectra. *In situ* algal absorption coefficients of unialgal dinoflagellate blooms can be discovered by determining pigment concentrations, cell densities, and cell dimensions from water samples. The model can be used to estimate photosynthetic light harvesting and algal contribution to light extinction in the water column. This method is sensitive to photoadaptive responses, and also allows us to assess the relative influences of light quality, accessory pigments, and packaging effects on algal light absorption. Results of the model are presented in the context of a hypothetical dino-flagellate bloom. The model may be extended to estimate primary productivity by adding a photosynthetic efficiency parameter, or including other algal taxa.

PHYTOPLANKTON LOSSES AND GAINS DURING COASTAL UPWELLING

S. Neuer, M.-L. Dickson and L. Fessenden
College of Oceanography
Oregon State University
Corvallis, OR 97331

We investigated phytoplankton growth and microzooplankton grazing in coastal water off Oregon during the upwelling period. Phytoplankton growth and microzooplankton grazing rates were determined by applying the dilution method for water collected from the mixed layer. Distinct patterns in relative nutrient and phytoplankton biomass levels (measured as chlorophyll) allow characterization of prebloom, bloom, and postbloom conditions. Both prebloom and bloom periods yielded high phytoplankton growth rates (0.6-1.5 d^{-1}, respectively) and grazing rates (around 0.7 d^{-1}), but prebloom periods were characterized by a smaller phytoplankton biomass than bloom periods. The nitrate-to-chlorophyll ratio of prebloom and bloom periods were much greater than 1 μmol/μg Chl. Postbloom situations also were characterized by high phytoplankton biomass, but had a nitrate-to-chlorophyll ratio smaller than 1 μmol/μg Chl. Phytoplankton growth rates were low (around 0.1-0.2 d^{-1}) or negative, and grazing rates were low or not measurable.

Our results show that microzooplankton grazing constitutes a significant loss factor during all stages of the upwelling. During prebloom and bloom periods, both phytoplankton growth and grazing rates are high, but during the bloom, phytoplankton growth exceeds grazing. During postbloom conditions, phytoplankton growth and microzooplankton grazing are reduced, but at times, grazing might exceed phytoplankton growth.

THE SCATTERING CHARACTERISTICS OF MICROALGAE

B. Osborne and P.F. Nolan
University College
Dublin, Ireland

Few detailed measurements of the angular scattering characteristics of phytoplankton have been made because of the practical difficulties in measuring scattering at very narrow angles, within a few degrees of the forward direction, and because of the prohibitive cost of off-the-shelf instrumentation. However, such measurements are required for many remote sensing applications and for resolving discrepancies between predicted and measured absorption characteristics. We describe here the construction and performance of a volume scattering device for measuring the volume-scattering characteristics of microalgal cells between $0°$ and $180°$.

The scattering device consists of a fixed circular table on which is mounted a cuvette containing the algal cells. A miniature silicon photodiode, operating in the photovoltaic mode, revolves in the horizontal plane of the laboratory with the cuvette at the center of the circular orbit of the photodiode. A single feedback resistor, connected between the output and inverting input of the operational amplifier (355), determines the gain of the amplifier and is of the order of 1 Megaohm for a photodiode laser with an output power of about 1mW.

A gear wheel, 6 inches in diameter, which is attached to the supporting arm of the photodiode, causes the spindle of a ten-turn 10K potentiometer to turn; this generates a voltage at the wiper of the potentiometer which is directly proportioned to the angle through which the photodiode has moved. A small motor drives the main 6-inch gear wheel, and data scattering intensity is angle of scattering, and is transferred to an XY chart recorder.

Two photodiode lasers are used as light sources with peak wavelengths at 690mm and 750mm. The beam diameter is about 4mm. The lasers are mounted on a small (3"x3") lab jack.

The positioning sensor and photodiode amplifier both generate voltage signals which can be readily processed by a computer, having a multichannel analogue to digital converter. The data then can be analyzed easily and transmitted to a mainframe computer for plotting.

TEMPORAL ALTERATION BETWEEN LIGHT- AND NUTRIENT-LIMITATION OF PHYTOPLANKTON PRODUCTION IN A COASTAL PLAIN ESTUARY

J.R. Pennock
University of Alabama
Marine Environmental Sciences Consortium
Dauphin Island, AL 36528

J.H. Sharp
University of Delaware
College of Marine Studies
Lewes, DE 19958

We examined the potential for light- and nutrient-limitation of phytoplankton production in the Delaware Estuary using a hierarchy of experimental approaches including small scale bioassay experiments, ecosystem analysis of nutrient concentration and stoichiometric ratios, light-limitation modeling and determination of planktonic elemental ratios.

Light was found to be the predominant regulator of phytoplankton growth throughout the estuary during the winter as a result of high turbidity and a well-mixed water column. However, during spring, phosphorus (P) was found to limit growth. This observation was confirmed at each of the experimental levels, and was related to several factors, including elevated $DIN:PO_4$ input ratios ($\geq 30:1$) in river waters, low rates of P regeneration, and the accumulation of P into phytoplankton. During summer, P no longer limited production. At this time, bioassay experiments and $DIN:PO_4$ ratios revealed the potential for nitrogen (N) limitation, particularly in the lower estuary, while particulate composition ratios and ecosystem nutrient flux estimates gave contradictory evidence. From these data, it appears that N was potentially limiting to phytoplankton biomass but that the constant flux of N from upstream and rapid N regeneration, maintained non-nutrient-limited, steady-state growth.

These data display a pattern of recurring system-wide variation in the factors that limit phytoplankton production over several annual sequences. Such temporal and spatial variations are related to both light availability--as regulated by incident light, suspended sediment concentration, and depth of the surface mixed-layer--and nutrient availability--as determined by riverine inputs in *in situ* biogeochemical processes.

PHYTOPLANKTON PHOTOSYNTHESIS-LIGHT RELATIONSHIP IN THE PAS ESTUARY, CANTABRIA, SPAIN

Luisa Perez
College of Marine Studies
University of Delaware
Lewes, DE 11958

J. C. Canteras
Laboratory of Ecology
University of Cantabria
39005-Santander, Spain

We investigated the relationship between the photosynthesis-irradiance (P-I) curves of natural phytoplankton assemblages and environmental conditions in a partially mixed estuary, the Pas. During an annual cycle, primary production was measured both by *in situ* and laboratory (light-saturation curve) incubation methods.

The inital slope (α) and the specific productivity at saturating light (P^B_{max}) of P-I curves changed significantly over the year. The parameter α ranged from 0.003 to 0.04 (mgC (mgChl)$^{-1}$ h^{-1} μE m^{-2}s^{-1}), and P^B_{max} ranged from 0.5 to 10 mgC (mgChl)$^{-1}$h^{-1}. Maxima for both α and P^B_{max} occurred when the community of phytoplankton was dominated by small cells (small diatoms, *Cryptomonas* sp.), and the minima were obtained when the turbidity was highest. In field (*in situ*) incubations, we found the highest photosynthetic efficiency when river flow was low and tidal level was high. The production normalized to chlorophyll *a* (P^B) correlated positively with salinity and temperature, and negatively with the nutrient concentrations.

PHYTOPLANKTON DYNAMICS AND THE VERTICAL FLUX OF ORGANIC CARBON IN THE MISSISSIPPI RIVER PLUME AND INNER GULF OF MEXICO SHELF REGION

D.G. Redalje and S.E. Lohrenz
University of Southern Mississippi Center for Marine Science
Stennis Space Center, MS 39529

G.L. Fahnenstiel
NOAA Great Lakes Environmental Research Laboratory
Ann Arbor, MI 48105

In this ongoing study, we are examining the degree to which elevated levels of nutrients carried to the coastal Gulf of Mexico waters in the Mississippi River Plume affect carbon cycling, phytoplankton growth and production, and photosynthesis-irradiance (P-I) properties. The river outflow enters the coastal Gulf of Mexico through several channels in the delta and gives rise to strong environmental gradients in salinity, nutrients, and light in the study region.

Initial results obtained during a July-August research cruise indicate that surface primary production in waters adjacent to the major channels in the river delta (2.4-3.1 gC m^{-3} d^{-1}) was greater than in locations between the channels (0.05-0.24 gC m^{-3} d^{-1}) or the inner Gulf shelf (0.5-0.7 gC m^{-3} d^{-1}). Phytoplankton C specific growth rates in the shallow euphotic zone were 0.8-2.3 d^{-1} for the river plume and 0.45-0.66 d^{-1} for the inner shelf. Photic zone C biomass was 0.08-0.1 gC m^{-3} in the plume and 0.8-1.5 gC m^{-3} in the shelf region. The vertical flux of organic C collected in free-floating sediment traps deployed beneath the photic zone (15m) was 0.27-0.33 gC m^{-2} d^{-1} in the plume region and 0.16-0.20 gC m^{-2} d^{-1} for the shelf region.

We found that P-I parameters were not correlated with horizontal gradients in light or nutrients, which led us to conclude that dynamic conditions in the study region moderated photosynthetic adaptive responses. Variations in irradiance and phytoplankton biomass were found to be more important than variations in P-I parameters in explaining observed patterns in integrated primary production estimated from a P-I model.

MODELING *IN SITU* PHYTOPLANKTON ABSORPTION SPECTRA FROM SPECTRAL REFLECTANCE: EFFECTS OF SPECTRAL BACKSCATTER

C.S. Roesler and M.J. Perry
School of Oceanography
University of Washington
Seattle, WA 98195

Mechanistic models of marine primary production require a spectrum of phytoplankton absorption coefficients as a primary forcing function, but direct *in situ* measurement of these coefficients is difficult. A two-component model was developed to determine spectral phytoplankton absorption coefficients from *in situ* measurements of spectral irradiance.

In the first module, total seawater absorption was calculated, using the radiative transfer equation (RTE) for the upper ocean from *in situ* measurements of spectral irradiance reflectance and modeled backscatter coefficients. Using measurements of *in situ* irradiance reflectance and absorption (measured on discrete water samples), the inverted RTE was used to calculate a set of spectral backscatter coefficients from an optically broad range of marine environments. Analysis of these backscatter spectra showed that the magnitude and spectral shape are relatively invariant between sites in the 350-600 nm range, and thus, a mean backscatter spectrum can be applied. In the spectral range 600-750 nm, the spectral shape of backscatter, although relatively constant among samples collected over a series of days at one site, varies between sites. Thus, a site-specific shape must be used in this spectral range. Application of a mean backscatter spectrum to independent reflectance data showed that variations in the spectral backscatter contribute negligibly to the calculation of total absorption within a cruise. The next approach for this module is the prediction of the spectral shape of backscatter in the 600-750 nm range from accessory measurements such as transmission, concentration of chlorophyll, and particle size frequency distributions.

In the second module (Roesler et al., 1989) total seawater absorption spectra were deconvolved into three components: phytoplankton, total organic detritus (dissolved and particulate), and water itself. With the complete model, phytoplankton spectral absorption coefficients can be estimated with less than 50% error.

OCEANIC PHYTOPLANKTON NEED LESS IRON

D.G. Swift
Graduate School of Oceanography
University of Rhode Island
Narragansett, RI 02882

W.G. Sunda and S.A. Huntsman
National Marine Fisheries Service
Southeast Fisheries Center
Beauford, NC 28516

Growth studies were undertaken with two diatoms, the oceanic diatom *Thalassiosira oceanica* and the estuarine species *Thalassiosira pseudonana*, using radiolabelled ^{55}Fe in Gulf Stream seawater having an EDTA-trace metal buffer. Intracellular Fe in exponentially growing cells was determined following treatment with titanium(III) EDTA-citrate reducing solution to dissolve adsorbed and noncellular iron. The two species showed different growth responses. *T. oceanica* grows at its maximum rate over a wide range of Fe concentration. Growth rate is reduced slightly (less than 20%) at the lowest Fe levels. Such low levels result in the growth rate of *T. pseudonana* dropping to almost zero. *T. pseudonana* shows a graded response to increased Fe, and a maximum growth rate that is 50% greater than that of *T. oceanica*. Fast growth by *T. oceanica* at lowest Fe levels is explained almost entirely by a much lower cell Fe requirement, or quota, rather than by more effective transport at low Fe conditions. *T. oceanica* cells at near-maximum specific growth rate of 1 day^{-1} have a cellular iron:carbon ratio of about 2×10^{-6}. *T. pseudonana* at the same growth rate requires several times this amount, and even more at its highest growth rate. Cellular Fe required for rapid growth of the oceanic diatom is substantially lower than values for coastal or estuarine species in this and other studies. Computations concerning the phytoplankton carbon that can result from use of a particular amount of Fe give different results; therefore, the Fe:C ratio for a genetically adapted low-iron oceanic phytoplankton species is used, rather than values for coastal species as were applied in the past.

VERTICAL FLUXES OF BIOGENIC PARTICLES AND ASSOCIATED BIOTA IN THE EASTERN NORTH PACIFIC: IMPLICATIONS FOR BIOGEOCHEMICAL CYCLING AND PRODUCTIVITY

G. T. Taylor
Marine Sciences Research Center
SUNY
Stony Brook, NY 11794-5000

D. M. Karl
School of Ocean and Earth Science and Technology
Department of Oceanography
University of Hawaii at Manoa
Honolulu, HI 96822

Previously published data on vertical fluxes of particulate carbon (PC), nitrogen (PN), organisms (MICRO), and extracted adenosine triphosphate (ATP) into screened sediment traps (335 μm) from the VERTEX 5 and ADIOS I programs are re-examined as they relate to biogeochemical cycling and oceanic productivity. Stations represent an oligoto mesotrophic gradient in total primary production (P_t), ranging from 245 to 1141 mg C m^{-2} d^{-1} and a gradient in PC flux from the euphotic zone, ranging from 12 to 164 mg C m^{-2} d^{-1} for particles <335 μm in diameter.

Attrition of vertical fluxes of ATP with depth was significantly greater than those of PC, PN, and MICRO. The flux of intact biota decreased rapidly in the upper 200 m, contributing as much as 52.4% at the most productive station and as little as 1.6% to the flux of PC at oligotrophic stations. Multiple regression analyses, expressing fluxes as functions of depth and P_T or new production, P_N, demonstrated that MICRO and ATP fluxes were more dependent on P_T, P_N, and depth than bulk PC or PN fluxes.

The present analysis illustrates that while sinking particulate organic matter (POM) undergoes rapid attrition in the upper water column, the fluxes of sedimenting biota decrease at even higher rates. Findings support the hypothesis that in oceanic waters, POM sinking from the euphotic zone rapidly becomes a poor habitat for associated microbes, and mechanisms other than remineralization by attached microbes must be invoked to explain observed fluxes and attrition rates. This study also supports the hypothesis that the vertical flux of intact organisms is a more sensitive and less ambiguous record of upper ocean processes than bulk flux measurements of total mass, PC, or PN.

LARGE-SCALE CROSS-SHELF VARIABILITY IN PIGMENT CONCENTRATIONS ALONG THE NORTH AMERICAN WEST COAST

A. Thomas
Atlantic Centre for Remote Sensing of the Oceans
Halifax, Nova Scotia

P.T. Strub
College of Oceanography
Oregon State University
Corvallis, OR 97331-5503

The seasonal and interannual variability of cross-shelf phytoplankton pigment concentrations along the west coast of North America was analyzed using six years (1979-83, 1986) of CZCS data from the West Coast Time Series. These data were first averaged into 10-daytime periods and then divided into six large-scale (approximately 400 km of coastline) regions within which the cross-shelf pigment concentrations were meridionally averaged. Regional differences in the temporal variability of cross-shelf pigment structure in the California Current were illustrated by comparing these regions. Seasonality was most strongly developed in regions north of Point Conception and was dominated by pigment concentrations exceeding 1.5 mg m^{-3} extending over 400 km offshore during April-May. Strong interannual variability was evident along the entire study area, from southern Baja to northern Washington. The dominant feature of this variability is the restriction of pigment concentrations greater than 1.5 mg m^{-3} to within 100 km of the coast over most of the North American west coast during 1983, an El Nino year. Correlations of the cross-shelf pigment structure to along-shore wind stress and U^3 in each of the regions was low, suggesting a weak relationship between wind forcing and pigment patterns on the time- and space-scales analyzed.

SHORT-TERM VARIATIONS OF GEOCHEMICAL PROFILES IN PORE WATERS OF LOWER ST. LAWRENCE ESTUARY SEDIMENTS

Gilles-H. Tremblay and Norman Silverberg
Institut Maurice-Lamontagne
Ministere des Peches et des Oceans
C.P. 1000, Mont-Joli, Quebec, Canada

Daily collections of box cores were made at a site in the middle of the Laurentian Trough, Lower St. Lawrence Estuary, to study the chemical responses of the sediments to short-term changes in the flux of fresh organic matter. Each of the 18 cores was finely subsampled under N_2 atmosphere and the pore waters separated and analyzed for nitrate, ammonium, silicate, phosphate, iron, and manganese.

Concentration profiles showed some variability, but preliminary results indicated some correlation between the depths of nitrate reduction and manganese oxide dissolution, and the period of highest carbon content in the sedimentation flux.

In addition, from companion cores, which were very finely subdivided, dissolved O_2, carbon and Mn also were determined.

LARGE AGGREGATE DISTRIBUTION DURING THE JGOFS NORTH ATLANTIC BLOOM EXPERIMENT

Ian Walsh and Wilford Gardner
Department of Oceanography
Texas A & M University
College Station, TX 77843

Profiles made during the North Atlantic Bloom Experiment with a Large Aggregate Profiling System (LAPS) recorded the concentration and distribution of particles in the marine snow-size range (>0.5 mm diameter) while simultaneously recording salinity, temperature, depth, and beam attenuation. Five profiles were made at the 47°N, 20°W station. To measure diel variations, four profiles were made within a 24-hour period (April 19-20, 1989); a fifth profile was made two days later.

We calculated aggregate abundance and size distribution from the film negatives. Aggregates were binned into six size ranges on the basis of their equivalent circular diameters at 0.5 mm intervals from 0.5 mm to >3.0 mm. Particle volume was calculated assuming sphericity and diameters equal to the mean of the size range, and, as a conservative estimate, the minimum size (3 mm) of the unbounded largest size bin. Mass concentration of large particles was calculated by using the particle diameter/density relationship of Alldredge and Gotschald (1988). Aggregate fluxes were calculated using the Stokes settling relationship (Gardner and Walsh, 1990).

Aggregate abundance, mass concentration, and flux profiles all showed similar distributions with high values in the surface mixed layer and below to 100-200 m, with a sharp decline at that level to values only 15 to 25% of the surface values. Aggregate abundance ranged from 5 to 20/1 near the surface, and 1 to 5/1 at depth. Aggregate mass concentration ranged from 5 to 15 μg/l near the surface and 1 to 5 μg/l at depth. The decrease in aggregate abundance correlated with the decrease in beam attenuation. However, the ratio of the aggregate mass concentration (AMC) to small particle concentration (SPC) (as determined by beam attenuation) changed with depth. In the surface layer through 300 m, the AMC was approximately 10% or less of the SPC, whereas below 300 m, the AMC was 50% of the SPC.

The fluxes calculated by LAPS agree reasonably well with the data from the moored (Honjo) and floating sediment trap (Martin) collected during the LAPS sampling. Transfer of particles from the small to large pools was suggested by the diel variation in aggregate abundance observed in the surface layer.

COSMOGENIC ^{32}P AND ^{33}P IN RAINWATER AND PLANKTON OFF BERMUDA AND IMPLICATIONS FOR PHOSPHORUS CYCLING

N.A. Waser and M.P. Bacon
Woods Hole Oceanographic Institution
Woods Hole, MA 02543

A.F. Michaels
Bermuda Biological Station for Research
Bermuda GE 01

Measurements of ^{32}P (half-life: 14.28 days) and ^{33}P (half-life: 25.3 days), two cosmic ray-produced phosphorus radioisotopes, were made at the Bermuda site in rainwater and zooplankton. The activity ratio ^{33}P/^{32}P was close to 1 in rainwater collected from precipitation which occurred during the spring of 1991. In contrast, the plankton collected offshore of the island with a 125-μm net had ratios ranging from 1.4 to 3. Assuming that isotopic fractionation is small and that precipitation is the only supply of the cosmogenic isotopes to the ocean, the increase in the $^{33}P/^{32}P$ ratio in plankton relative to rain can be interpreted as aging of the phosphorus in that pool, or in the pool from which it is derived. This is because of the differential radioactive decay rates of ^{32}P and ^{33}P. From this preliminary data set, we suggest that $^{33}P/^{32}P$ could be used as an *in situ* tracer to estimate rates of phosphorus export and the cycling among the various phosphorus pools in the upper ocean.

PHYSICAL/BIO-OPTICAL INTERACTIONS AND PRODUCTIVITY ESTIMATES FROM OCEANIC TIME SERIES

J. Wiggert, T. Dickey, and M. Hamilton
USC Ocean Physics Group
Los Angeles, CA 90089-0740

T. Granata
Department of Biology
Southeastern Massachusetts University
North Dartmouth, MA 02747

J. Marra and C. Langdon
Lamont-Doherty Geological Observatory
Columbia University
Palisades, NY 10964

D. Siegel and L. Washburn
Department of Geography
University of California at Santa Barbara
Santa Barbara, CA 93106

We collected *in situ* bio-optical and physical oceanographic data using the MVMS (Multi-Variable Moored System) 34N 70W in the Sargasso Sea for nine months in 1987 in the upper 160 meters. The sampled oceanographic parameters include temperature, currents, photosynthetically available radiation, beam attenuation coefficient, stimulated fluorescence, and dissolved oxygen concentration. Meteorological measurements were made, and sea surface temperature maps and Geosat altimetry data were used to investigate the mesoscale field associated with the dynamics of the Gulf Stream. Time series indicate seasonal cycling of heat and momentum fields, the diel cycle in bio-optical variables, episodic events related to advected water masses and local meteorological forcing, a phytoplankton bloom concentrated at 20 meters during the spring, and the development of sub-subsurface chlorophyll and particle maxima during the summer. Spectral analysis reveals a correspondence between internal wave and bio-optical variability. High-resolution time series of gross primary production were estimated with the Kiefer-Mitchell model using the time series of stimulated fluorescence and PAR as input. Particle production rate also was calculated using the time series of beam attenuation coefficient. By combining the two estimates, time series of Chl:C ratio were created. Our results show that high-frequency sampling (minimizing aliasing) is essential in revealing the dynamical changes occurring in upper ocean primary productivity and carbon flux.

DO PHYTOPLANKTON LIKE THE WEATHER?

E. A. Woods
Unit for Coastal and Estuarine Studies
Marine Science Laboratories
University College of North Wales
Menai Bridge, Gwynedd, U.K.

A 2-layer vertical process model of the seasonal cycle of phytoplankton and nitrogen, driven by a model of the seasonal thermocline, with a one-day time-step, was applied to the Northwest European Shelf.   The thermocline model is driven by meteorology and M2 tides, and provides the biological model with daily thermocline depths and incident solar irradiance.

The biological simulation includes equations for nutrient uptake and nutrient-controlled growth, net photosynthesis, and the recycling of nutrients by grazers and by benthic remineralization.   The simulation uses the cell-quota threshold-limitation theory for nutrient- and light- limited algal growth.

A series of weather simulations were run with meteorology from 1984, 1985, and 1986 to compare with simulations using the 25-year average meteorology or climate. Expectations that the simulations with the weather would show higher primary productivity than those with climate, because of increased new nutrient in the upper layer, were unfounded. The simulations with the weather often had lower annual primary production and standing stocks of chlorophyll than those simulations with climate.   At summer-stratified stations, this was due to light limitation occurring when storms pushed the thermocline deeper, despite increased nutrient exchanges across the thermocline.   The autumn bloom seen in the simulations with climate was not often seen in those with weather because the thermocline could be driven deeper very quickly, preventing the effective use of the new nitrogen in light-limited conditions.

The timing of the spring bloom in the weather simulations varied by up to 30 days, and the maximum concentrations of chlorophyll by up to 40%.

^{234}TH AS A TRACER FOR EXPORT NEW PRODUCTION IN MONTEREY BAY

Jennifer Young[1], James W. Murray[1], Barbara Paul[1],
Thomas Chapin[1], and Jan N. Downs[2]
[1]School of Oceanography
University of Washington
Seattle, WA 98195

[2]Monterey Bay Aquarium Research Institute
160 Central Avenue
Pacific Grove, CA 93950

Eppley urged that the naturally produced ratio-isotope ^{234}Th be carefully evaluated as a tracer for export new production. ^{234}Th might be ideal for this purpose because it is produced at a known rate in the water column, it has a strong affinity for particles, its half-life (24d) is appropriate for upper ocean carbon cycling, and because it can be analyzed at sea.   We studied this tracer in Monterey Bay which is an ideal location because there is a large seasonal signal in primary and new productivity.

We conducted seven cruises from June 1990 to April 1991.   The total program included primary (^{14}C) and new productivity (^{15}N) and the distribution and sediment trap fluxes of POC, PON, POP, Chl, Phaeo as well as hydrography and nutrients. We summarized dissolved, particulate and sediment trap [234]Th analyses.

As expected for a biologically productive location, there was a large deficiency in dissolved Th in the euphotic zone. Particulate Th was uniformly constant. The trap fluxes varied seasonally and were about twice as high in June than in November to February.

Our goal is to predict the Th flux using a simple scavenging model and to compare the predicted fluxes with the measured fluxes. Our approach is to calculate the [234]Th concentrations in inventory units (dpm m^{-2}) for a sequence of boxes that correspond with the trap depths. We determined that steady-state is a valid assumption for most of the time series. The predicted fluxes were systematically lower than the measured fluxes. Possible explanations are that there are hydrodynamic effects that cause overtrapping, or that the simple model that we have used does not allow for small particle repackaging into large particles.

The residence times of the C, N, total suspended matter, and [234]Th in the euphotic zone were compared.

PARTICIPANTS

Fatima Abrantes
Servicos Geologicos de Portugal
Rua Academia das Ciencias, 19 2°
1200 Lisbon, Portugal

Khaled Al-Abdulkader
Texas A & M
P.O. Box 8153
College Station, TX   77844

Jim Aiken
Plymouth Marine Laboratory
Prospect Place
Plymouth PL1 3DH
United Kingdom

Saul Alvarez-Borrego
CICESE
Apdo. Postal 2732 Ensenada
Baja California, Mexico

Thomas R. Anderson
Deacon Laboratory
Institute of Oceanographic Sciences
Brook Road, Wormley, Godalming
Surrey GU8 5UB, United Kingdom

Javier Aristegui
Facultad de Ciencias del Mar
Las Palmas G. C.
Canary Islands, Spain

Marcel Babin
Department of Biology
Laval University
Quebec (QC), Canada G1K 7P4

William Balch
Marine Biology and Fisheries
Rosensteil School

University of Miami
Miami, FL 33149-1098

Salome Ballesteros
Facultad Ciencias del Mar
Las Palmas G. C.
Canary Islands, Spain

Karl Banse
Department of Oceanography
University of Washington
Seattle, WA 98195

Richard Barber
Duke University Marine Laboratory
Beaufort, NC 28516

Gotzon Basterretxea
Facultad de Ciencias del Mar
Las Palmas G. C.
Canary Islands, Spain

Michael Behrenfeld
ASCI/EPA
2111 S.E. Marine Science Street
Newport, OR 97365-5260

Michael L. Bender
Graduate School of Oceanography
University of Rhode Island
South Ferry Road
Kingston, RI 02881

Wolfgang Berger
Scripps Institute of Oceanography
University of California, San Diego
La Jolla, CA 92093

Tom Berman
Kinneret Limnological Laboratory

(continued)
P.O. Box 345
Tiberias 14102
Israel

Joseph Berry
Carnegie Institute
290 Panama Street
Stanford, CA 94305

Robert Bidigare
Department of Oceanography
University of Hawaii
Honolulu, HI 96822

Paul Bienfang
Oceanic Institute
Makapuu Point
Waimanalo, HI 96795

Carl M. Boyd
Department of Oceanography
Dalhousie University
Halifax, Nova Scotia
Canada B3H 411

Walter R. Boynton
Chesapeake Biological Laboratory
University of Maryland System
Solomons, MD 20688-0038

Deborah A. Bronk
Horn Point Environmental Laboratory
University of Maryland System
Cambridge, MD 21613

Jeng Chang
Marine Sciences Research Center
State University of New York
Stony Brook, NY 11794

S. W. Chisholm
Ralph Parsons Laboratory
Massachusetts Institute of Technology
Cambridge, MA 02139

Forrest G. Chumley
DuPont Company
Experimental Station
Building 402, Room 4232
P.O. Box 80402
Wilmington, DE 19880-0402

Aurea Maria Ciotti
Laboratorio de Fitoplancton
Departo. Oceanografia
CX Posto 474
Rio Grande RS 96200 Brazil

Ramon Cipriano
5594 Paul Road
East Bethany, NY 14054

Joan S. Cleveland
Naval Oceanographic and
Atmospheric Research Laboratory
Stennis Space Center, MS 39529

William P. Cochlan
Marine Biology Research Division
Scripps Institute of Oceanography
University of California, San Diego
La Jolla, CA 92093

Lytha Conquest
The Oceanic Institute
P.O. Box 25280
Honolulu, HI 96825

Elizabeth M. Cosper
Marine Sciences Research Center
State University of New York
Stony Brook, NY 11794

Kenneth M. Crocker
Department of Biological Sciences
University of California
Santa Barbara, CA 93106

John Cullen
Bigelow Laboratory for Ocean Sciences
McKown Point
W. Boothbay Harbor, ME 04575

Michele DuRand
Biology Department
Woods Hole Oceanographic Institution
Woods Hole, MA 02543

Curtiss O. Davis
Jet Propulsion Laboratory
California Institute of Technology
Pasadena, CA 91109

Tom Dickey
Oceanic Physics Group
University of Southern California
Los Angeles, CA 90089-0740

Mary-Lynn Dickson
College of Oceanography
Oregon State University
Corvallis, OR 97331-5503

Michael Doall
Marine Sciences Research Center
State University of New York
Stony Brook, NY 11794

Jan Newton Downs
Monterey Bay Aquarium
Research Institute
Pacific Grove, CA 93950

Zvy Dubinsky
Department of Life Sciences
Bar Ilan University
Ramat Gan, Israel

Richard Dugdale
Department of Biological Sciences
University of Southern California
Los Angeles, CA 90089-0371

William M. Dunstan
Old Dominion University
Norfolk, VA 23508

Jeffrey A. Dusenberry
Ralph M. Parsons Laboratory
Department of Civil Engineering
Massachusetts Institute of Technology
Cambridge, MA 02139

Richard Eppley
1969 Loring Street
San Diego, CA 92109

David L. Eslinger
NASA/Goddard Space Flight Center
Greenbelt, MD 20771

Jonathan Erez
Department of Geology

Institute of Earth Sciences
The Hebrew University of Jerusalem
Jerusalem, Israel 91904

Paul Falkowski
Oceanographic and Atmospheric
Sciences Division
Brookhaven National Laboratory
Upton, NY 11973

Nicholas Fisher
Marine Sciences Research Center
State University of New York
Stony Brook, NY 11794

James A. Fee
Biochemistry Section and Stable
Isotope Resource
Los Alamos National Laboratory
Los Alamos, NM 87545

Lynne M. Fessenden
College of Oceanography
Oregon State University
Corvallis, OR 97331

Santiago Fraga Rivas
Instituto Espanol de Oceanografia
Apdo. 1552
36280 Vigo, Spain

Peter J.S. Franks
College of Oceanography
Oregon State University
Corvallis, OR 97331

Jed Fuhrman
Department of Biological Sciences
University of Southern California
Los Angeles, CA 90089

Graciela Garcia-Moliner
Graduate School of Oceanography
University of Rhode Island
Narragansett, RI 02882

David Galas
Office of Health and Environmental
Research
ER-70, GTN
U.S. Department of Energy
Washington, DC 20545

Wilford D. Gardner
Department of Oceanography
Texas A & M University
College Station, TX 77843

Ronald T. Garry
Marine Sciences Research Center
State University of New York
Stony Brook, NY 11794

Gilberto Gaxiola-Castro
CICESE
Apdo. Postal 2732, Ensenada
Baja California, Mexico

Richard Geider
College of Marine Studies
University of Deleware
Newark, DE 19718

Val Gerard
Marine Sciences Research Center
State University of New York
Stony Brook, NY 11794

Fernando Gilbes-Santaella
University of Puerto Rico
P.O. Box 5528 College Station
Mayagüez, Puerto Rico 00709

Ralf Goericke
Department of Chemistry
Woods Hole Oceanographic Institution
Woods Hole, MA 02543

Joel C. Goldman
Woods Hole Oceanographic Institution
Woods Hole, MA 02543

Timothy C. Granata
Department of Geological Sciences
Ocean Physics Group
University of Southern California
Los Angeles, CA 90089-0740

Lisa M. Graziano
College of Marine Studies
University of Delaware
911 Bauman Drive
Lewes, DE 19958

Monika Greene
Marine Sciences Research Center
State University of New York
Stony Brook, NY 11794

Richard Greene
Oceanographic and Atmospheric
Sciences Division
Brookhaven National Laboratory
Upton, NY 11973

Eleanor Grist
Biology Department
Brookhaven National Laboratory
Upton, NY 11973

Dennis A. Hansell
Institute of Marine Sciences
University of California
Santa Cruz, CA 95064

W. G. Harrison
Bedford Institute
Dartmouth, Nova Scotia
Canada B2Y 4A2

Bruce G. Hatcher
Department of Oceanography
Dalhousie University
Halifax, Nova Scotia
Canada B3H 4J1

John M. Hayes
Biogeochemical Laboratories
Indiana University
Bloomington, IN 47405-5101

Christina Hazzard
Department of Zoology
University of Hawaii
Honolulu, HI 96822

Anna-Stiina Heiskanen
Tvarminne Zoological Station
SF-10900 Hanko, Finland

Walter Helbling
Scripps Institute of Oceanography
Polar Research Program
University of California - San Diego
La Jolla, CA 92093-0202

Diane Henry
Howard Hughes Medical Institute
722 West 168th Street
New York, NY 10032

Ronny Herzig
University of Haifa, Oranim
Tivon 36910 Israel

Patrick Holligan
Plymouth Marine Laboratory
Prospect Place
Plymouth PL1 3DH
United Kingdom

David A. Hutchins
Institute of Marine Sciences/
Biology
University of California
Santa Cruz, CA 95064

Joji Ishizaka
National Research Institute for
Pollution and Resources
16-3 Onogawa, Tsukuba,
Ibaraki, 305 Japan

Richard L. Iverson
Department of Oceanography, B-169
Florida State University
Tallahassee, FL 32306-3048

Sigrun Jonasdottir
Marine Sciences Research Center
State University of New York
Stony Brook, NY 11794-5000

Burton H. Jones
Department of Biological Sciences
University of Southern California
Los Angeles, CA 90089-0371

Daniel Kamykowski
Marine Earth and Atmospheric Sciences
North Carolina State University
Raleigh, NC 27645

W. Michael Kemp
Horn Point Environmental Laboratory
University of Maryland
Cambridge, MD 21613

Mahlon Kennicutt II
GERG
Texas A & M University
College Station, TX 77845

Dale Kiefer
Department of Biological Sciences
University of Southern California
Los Angeles, CA 90089-0371

John T. O. Kirk
Division of Plant Industry
Commonwealth Scientific and Industrial
Research Organization
Canberra, Australia

Gary J. Kirkpatrick
Mote Marine Laboratory
Sarasota, FL 34236

Jacco Kromkamp
Delta Institute for
Hydrobiological Research
Vierstraat 28, 4401 EA Yerseke
NL-4401 EA The Netherlands

Robert Kunzig
251 Seaman Avenue - Apt. 4F
New York, NY 10034

J. Ruben Lara Lara
CICESE
Espinoza 843
Ensenada, B.C. Mexico

Julie LaRoche
Oceanographic and Atmospheric
Sciences Division
Brookhaven National Laboratory
Upton, NY 11973

Cindy Lee
Marine Sciences Research Center
State University of New York
Stony Brook, NY 11794-5000

SangHoon Lee
Oceanographic and Atmospheric
Sciences Division
Brookhaven National Laboratory
Upton, NY 11973

Wonho Lee
Department of Oceanography
Kunsan National University
Kunsan 573-360
Republic of Korea

Maurice E. Levasseur
Institut Maurice Lamontagne
Ministere des Pecheo et des Oceans
Mont-Joli, Quebec
Canada G5H 3Z4

Marlon Lewis
Department of Oceanography
Dalhousie University
Halifax, Nova Scotia
Canada B3H 4J1

Steven E. Lohrenz
University of Southern Mississippi
Stennis Space Center, MS 39529

Daray Lonsdale
Marine Sciences Research Center
State University of New York
Stony Brook, NY 11794

Jose M. Lopez
Department of Marine Sciences
University of Puerto Rico
Mayaguez Campus Box 5000
Mayaguez, Puerto Rico 00709-5000

Hugh L. MacIntyre
College of Marine Studies
University of Delaware
Lewes, DE 19958

Thomas C. Malone
Horn Point Environmental Laboratory
University of Maryland System
Cambridge, MD 21613

Walter F. Mangel
Biology Department
Brookhaven National Laboratory
Upton, NY 11973

John Martin
Moss Landing Marine Laboratory
Moss Landing, CA 95039

Helmut Maske
CICESE
Apdo. Postal 2732, Ensenada
Baja California, Mexico

Patricia A. Matrai
Marine and Atmospheric Chemistry
University of Miami
Miami, FL 33149

Mary McCarthy
Botany Department
University of Dublin
Dublin 4
Ireland

Andrew McTaggart
University of NSW
Department of Analytical Chemistry
University of East Anglia
Norwich, United Kingdom

Mark Meyers
Department of Marine Science
University of South Florida
St. Petersburg, FL 33701

Robert Millan Nunez
Facultad de Ciencias Marinas
Apdo. Postal 453, Ensenada
Baja California, Mexico

Richard L. Miller
National Aeronautics and
Space Administration
Stennis Space Center, MS 39529

Allen J. Milligan
Marine Sciences Research Center
State University of New York
Stony Brook, NY 11794

Greg Mitchell
NASA Headquarters
Washington, DC 20546

John R. Moisan
Center for Coastal Physical
Oceanography
Old Dominion University
Norfolk, VA 23529

Maria F. Montero
Facultad de Ciencias del Mar
Les Palmas C. G.
Canary Islands, Spain

Anne Mooney
Marine Sciences Research Center
State University of New York
Stony Brook, NY 11794

Andre Morel
University of Paris
Villefranche sur Mer, France

Francois Morel
Massachusetts Institute of Technology
15 Vassar Street
Cambridge, MA 02139

Richard J. Murnane
Atmospheric & Oceanic Sciences Prog.
Sayre Hall, Princeton University
Princeton, NJ 08544

James W. Murray
School of Oceanography
University of Washington
Seattle, WA 98195

Leonard Muscatine
Department of Biology
University of California
Los Angeles, CA 90024

David M. Nelson
College of Oceanography
Oregon State University
Corvallis, OR 97331

Norman B. Nelson
Department of Biological Sciences
University of California
Santa Barbara, CA 93106

Susanne Neuer
College of Oceanography
Oregon State University
Corvallis, OR 97331

Barbara Nieke
GIROQ

Laval University
Canada, G1K 7P4

Peter Nolan
Botany Department
University of Dublin
Dublin 4
Ireland

Rod L. Oliver
Murray-Darling Freshwater
Research Centre
Albury, NSW, Australia 2640

Robert J. Olson
Biology Department
Woods Hole Oceanographic Institution
Woods Hole, MA 02543

Thomas Osborn
IOC/UNESCO
7 Place de Fontenoy
75700 Paris, France

Vita Pariente
Department of Oceanography
Texas A & M University
College Station, TX 77842

Maria-Angelica Peña
Department of Oceanography
Dalhousie University
Halifax, Nova Scotia
Canada B3H 4J1

Luisa Perez
College of Marine Studies
University of Delaware
Lewes, DE 19958

Jonathan Pennock
Marine Environmental Sciences
Consortium
University of Alabama
Dauphin Island, AL 36528

Trevor Platt
Bedford Institute of Oceanography
Dartmouth, Nova Scotia
Canada B2Y 4A2

Donald B. Porcella
Electric Power Research Institute
Palo Alto, CA 94303

Joseph Ramus
Duke University Marine Laboratory
Beaufort, NC 28516

Greg Rau
MS 239-4, NASA-Ames
University of California
Santa Cruz, CA 94035

Donald Redalje
Center for Marine Science
University of Southern Mississippi
Stennis Space Center, MS 39529

John Reinfelder
Marine Sciences Research Center
State University of New York
Stony Brook, NY 11794

Edward H. Renger
Scripps Institute of Oceanography
University of California - San Diego
La Jolla, CA 92093

Rick A. Reynolds
Department of Biological Sciences
University of Southern California
Los Angeles, CA 90089-0371

Tammi L. Richardson
Department of Oceanography
Dalhousie University
Halifax, Nova Scotia
Canada B3H 4J1

Roel Riegman
Netherlands Institute for
Sea Research
Postbus 59, 1790 AB Den Burg
Texel, The Netherlands

Collin Roesler
School of Oceanography
University of Washington
Seattle, WA 98195

Toshiro Saino
Ocean Research Institute
University of Tokyo
1-15-1, Minami-dai, Nakano-ku
Tokyo 164 Japan

Ray Sambrotto
Lamont-Doherty Geological
Observatory
Palisades, NY 10964

Lionel Sanchez
Department of Oceanography
Texas A & M University
College Station, TX 77843

M.C. Eduardo Santamaria del Angel
Universidad Autonoma de Baj
California
Facultad de Ciencias Marinas
Apdo. Postal 453, Ensenada
Baja California, Mexico 22800

Jorge Sarmiento
Geophysics Dynamics
Princeton University
Princeton, NJ 08540

George W. Saunders
Environmental Sciences Division
ER-74 GTN
U. S. Department of Energy
Washington, DC 20585

Richard Setlow
Biology Department
Brookhaven National Laboratory
Upton, NY 11973

Erik M. Smith
Horn Point Environmental Laboratories
The University of Maryland System
P.O. Box 775
Cambridge, MD 21613

Sharon Smith
Oceanographic and Atmospheric
Sciences Division
Brookhaven National Laboratory
Upton, NY 11973

Heidi M. Sosik
Scripps Institute of Oceanography
University of California - San Diego
La Jolla, CA 92093

Petra M. Stegmann
Graduate School of Oceanography

(continued)
University of Rhode Island
Narragansett, RI 02882

F. William Studier
Biology Department
Brookhaven National Laboratory
Upton, NY 11973

Dariusz Stramski
Department of Biological Sciences
University of Southern California
Los Angeles, CA 90089-0371

Ajit Subramaniam
Marine Sciences Research Center
State University of New York
Stony Brook, NY 11794

Dorothy G. Swift
Graduate School of Oceanography
University of Rhode Island
Narragansett, RI 02882

Isabelle Taupier-Letage
Department of Geological Sciences
University of Southern California
Los Angeles, CA 90089-0740

Peter Thompson
Department of Oceanography
The University of British Columbia
Vancouver, B. C.
Canada V6T 1W5

Dennis Taylor
Bigelow Laboratory for
Ocean Sciences
West Boothbay Harbor, ME 04575

Gordon T. Taylor
Marine Sciences Research Center
State University of New York
Stony Brook, NY 11794

Andrew Thomas
Atlantic Centre for Remote Sensing
of the Oceans
Halifax, Nova Scotia
Canada B3J 2R7

Gilles-H Tremblay
Department of Fisheries and Oceans

Maurice-Lamontagne Institute
Mont-Joli, Quebec
Canada G5H 324

Ena Urbach
Massachusetts Institute of
Technology
Cambridge, MA 02139

Tracy Villareal
Woods Hole Oceanographic Institution
Woods Hole, MA 02543

Anya Waite
Department of Oceanography
University of British Columbia
Vancouver, B. C.
Canada V6T 1W5

Henry Walker
Environmental Protection Agency
27 Tarzwell Drive
Narragansett, RI 02882

Douglas Wallace
Oceanographic and Atmospheric
Sciences Division
Brookhaven National Laboratory
Upton, NY 11973

Ian Walsh
College of Geosciences
Texas A & M University
College Station, TX   77843

Nathalie A. Waser
Woods Hole Oceanographic Institution
Woods Hole, MA 02543

Paul Wassmann
Norwegia College of Fishery Science
University of Tromso
N-9001 Tromso, Norway

Dianne E. Wickland
NASA Headquaters
Washington, DC 20546

Jerry Wiggert
Ocean Physics Group
Department of Geological Sciences
University of Southern California
Los Angeles, CA 90089-0740

Cara Wilson
Department of Applied Sciences
Brookhaven National Laboratory
Upton, NY 11973

Michelle Wood
University of Oregon
Eugene, OR 97403

Avril Woodhead
Biology Department
Brookhaven National Laboratory
Upton, NY 11973

E. Woods
Unit for Coastal and
Estuarine Studies
University College of North Wales
Menai Bridge, Anglesey, Gwynedd
UK LL59 5EY

Jinping Wu
Department of Oceanography
University of British Columbia
6270 University Boulevard
Vancouver, B. C.
Canada V6T 1Z4

Sung R. Yang
Department of Biological Sciences
University of Southern California
Los Angeles, CA 90089-0371

Jeanette Yeh
Marine Sciences Research Center
State University of New York
Stony Brook, NY 11794

Stephan Zeeman
University of New England
Biddeford, ME 04005

John Zehr
Marine Sciences Research Center
State University of New York
Stony Brook, NY 11794

Erik Zettler
Woods Hole Oceanographic Institution
Woods Hole, MA 02543

David Ziemann
The Oceanic Institute
Honolulu, HI 96825